ORGANIC CHEMISTRY

AN INTRODUCTION EMPHASIZING BIOLOGICAL CONNECTIONS

ORGANIC CHEMISTRY

AN INTRODUCTION EMPHASIZING BIOLOGICAL CONNECTIONS

PRELIMINARY EDITION

I. DAVID REINGOLD
JUNIATA COLLEGE

HOUGHTON MIFFLIN COMPANY   BOSTON   NEW YORK

Senior Sponsoring Editor: Richard Stratton
Editorial Assistant: Marisa R. Papile
Senior Project Editor: Nancy Blodget
Associate Production/Design Coordinator: Lisa Jelly Smith
Senior Manufacturing Coordinator: Priscilla Bailey
Executive Marketing Manager: Andy Fisher

Printed in the U.S.A

ISBN: 0-618-072136

123456789-CRS-05 04 03 02 01

To Kay

whose willingness to be a homemaker has made possible
whatever professional success I may have had,
both in research and in teaching,
and who makes coming home fun;

to Colin and Ali

who have refused to live up to all the
stereotypes of what teenagers should be,
and of whom I am immensely proud;

to Evan

who was a wonderful baby and
would have been a wonderful kid;

and

to all my students over the years

who endured my idiosyncrasies
and helped me to become a better teacher.

Table of Contents

Chapter 10: Alkenes I

Chapter 11: Alkenes II

Chapter 12: Alkynes

Chapter 13: Substitution Reactions

Chapter 14: Structure Determination

Acknowledgments

This book was read by an army of reviewers during its birth. No book is perfect for anyone but the author (if that), and all readers have suggestions and objections. My readers have been no exception, and I am grateful for their many thoughtful suggestions. The readers are

Carol Angstadt, Hahnemann Medical School; Howard Angstadt, Sun Oil (ret); Nicole Ballew, Dartmouth College; Howard Black, Eastern Illinois University; Fran Blase, Haverford College; Mark Brodl, Knox College; Dee Casteel, Bucknell University; Julio DePaula, Haverford College; Jim Duncan, Lewis and Clark College; Molly Ettenger, Physician; Jean Fuller-Stanley, Wellesley College; Scott Hartsell, University of Wisconsin, Eau Claire; Steve Holmgren, Montana State University; Diane Husic, East Stroudsburg University; Janet Kirkley, Knox College; Mike Levett, Juniata College; Bill Loffredo, East Stroudsburg University; Mike Long, Huntingdon High School; Mitch Malachowski, University of San Diego; Jim McElroy, Chestnut Ridge High School; Nancy Mills, Trinity University; Tetsuo Otsuki, Occidental College; Greg Petsko, Brandeis University; Ruth Reed, Juniata College; Ray Reeder, Elizabethtown College; Dagmar Ringe, Brandeis University; Bill Russey, Juniata College; Rich Scamehorn, Ripon College; Paul Scudder, New College; Ernie Trujillo, Albright College; Scott Yeager, University of Vermont Medical School

Among these I must single out Howard Angstadt and Ernie Trujillo for timely and thorough reading and comments, and especially my Juniata colleague Bill Russey, who read every word and changed many of them, always for the better. To the extent that there is any elegance in my prose, it is likely attributable to him! Bill also wrote quite a few of the problems.

My student Amber Jade Helsel, Juniata '04, has been extremely diligent about finding grammatical errors and helping me polish my writing, and I thank her. I also thank Professor Dietmar Kuck of the University of Bielefeld for helping make my discussion of mass spectrometry more accurate.

Marisa Papile at Houghton Mifflin has been very helpful and patient in nursing this new author through the book publishing process.

I thank the National Science Foundation's CCLI Program, which provided the money that allowed me to take a year off to write this. Without them, this book would still be a figment of my imagination. Special thanks to Program Officer Susan Hixson, who looked past reviewers' ratings into their comments and was able to factor out irrelevant issues.

Finally, I want to thank Seyhan Ege, who generously shared my name with her editor while he was urging *her* to write this book. That trip turned into an excellent relationship with Richard Stratton of Houghton-Mifflin, who did not mince his criticisms but who was willing to listen to the other side and be persuaded that this approach is worth trying.

David Reingold
Huntingdon, PA

FOREWORD—TO THE INSTRUCTOR

This book differs from standard organic texts in three major ways. It spends more time on basic ("general") chemical concepts; it covers less of the traditional material of sophomore organic chemistry, focusing only on that material deemed necessary to understand biological processes; and coverage of the applications of organic chemistry to life is woven throughout the book rather than relegated to a few chapters near the end. The book is intended to be covered over the course of a full year. The early chapters represent review for many students, even though the material is not presented that way, and it is expected that one could cover many of the first eight chapters at a fairly rapid pace. Chapters 7 and 9 represent the real beginnings of new material, and that is where most students will need a slower pace.

Rationale

Why have I constructed such a book? As I see it, most students of chemistry fall into one of two categories. There are those who just need a little bit of chemistry—say, one year or less—and those who need two years or more. The one-year students, nurses, for example, frequently find a course on General, Organic and Biochemistry. There are already a great many books available for such a course. These books tend to offer fairly concise and elementary coverage of organic chemistry and will not adequately serve students requiring a deeper understanding of the subject.

It is for that other group of students that this book is intended. However, I believe that we in the chemical community should acknowledge that the majority of these students really have no intrinsic interest in chemistry. They want, by and large, to be life scientists. They are taking chemistry because they have to, and any concern they have in chemistry is in how it applies to biological systems. Obviously this does not describe *all* the students we find in our introductory classes, but it is probably fair to say that it describes 70–80% of them. What in fact do they find? At most schools, they find an introductory course that is relatively physical and mathematical in its approach; a course that covers much of the same material they saw in high school (admittedly at a higher level, but material that some members of the class already understand fairly well while others are overwhelmed by it); and a course that appears to jump, with little connection, from one topic to another. Then, if they survive, they move on to sophomore chemistry, where they are offered a course in Organic Chemistry that (at most schools) confronts them with much more than they would ever want or need to know about organic molecules, including exotic reactions (Wittig, Diels-Alder) that have no obvious applicability to biological systems, lengthy discussions on how to synthesize things in a laboratory, and, in most cases, relatively little about the organic chemistry of biological systems. This is, of course, not because their instructors have no interest in the application of organic chemistry to life, but because they usually run out of time before getting to those topics.

This book represents a different approach to the teaching of chemistry. It is based on two premises: 1) that the topic of organic chemistry, being relatively narrow and self-contained and by nature continually building an understanding from what has come previously, represents a more natural way to introduce the basic concepts of chemistry than the broad overview of the traditional freshman course; and 2) that the traditional sophomore organic course covers too much organic chemistry and not enough biochemistry for the majority of students taking the course. This book grew out of my experience at Juniata College, where we teach General Bioorganic Chemistry in the first year and Inorganic and Analytical Chemistry in the second year. Most of the students enrolled (except the chemistry majors) will take no more chemistry than that.* Thus all the *necessary* parts of organic chemistry must be covered in that first year. At the same time, because it comes first, I cannot assume the General Chemistry background commonly taken for granted by most organic chemistry texts. I have written this book first and foremost with that curriculum in mind, consciously trying to design a book that could be used on its own in the first-year course.

Other Possible Uses

Although I wrote the book for our "Organic First" curriculum, several other possible uses have become apparent as the book came together. For example, this book provides an ideal transition for underprepared students moving from the traditional first-year general chemistry course into a standard sophomore organic course. Although I do not go into all the organic details they might be expected to master, the description of how to solve basic problems is laid out more fully in this book than in any other I have seen. The current text could be used either as a supplement or as the main text. This text could also fill the needs of a short organic course meant for science-related majors other than chemistry and biology students, majors from such areas as allied health, agricultural, or engineering students—those who need some organic but perhaps not everything covered in a standard sophomore course. Finally, advanced high school students who have mastered Chemistry 1 and 2 and are looking for a further challenge in a new aspect of chemistry can use this book as an introduction to organic chemistry, pretty much on their own. Several students at our local high school have done this successfully.

Coverage

In examining this book, you may discover that I have left out your favorite topic. You're right, I have. I left out a lot of topics that I love as well. But I have tried very hard to apply to every topic the criterion: "Does a prospective life scientist *really* need to know this?" Obviously some doctors and some biologists, such as those engaged in molecular biology, do need to know much more organic chemistry than is presented here. Those students will certainly take more courses in organic chemistry and learn that extra material. But is it appropriate to teach that material to *all* introductory organic students? I believe not, and have chosen instead to focus on the basic material needed by all. Further, I believe that it is better for students to learn well a smaller amount of material than to encounter everything they could ever possibly use with most of them taking away very little (less is more). This philosophy has led to extensive application of a scalpel to the traditional organic material. I have deleted many topics entirely, and cut back on details of others. What survives, I believe, is the organic chemistry most practicing life scientists need to know, at the level they need to know it. Others could obviously have made different choices at certain points, but I hope there is general approval on the overall selection of topics.

Level

A major benefit of writing a textbook such as this—aimed at beginning students and presenting an overview of organic chemistry that will be supplemented later for those who need detailed organic knowledge—is that I need not shy away from a simplified presentation of most topics. I believe that it does no good, at this level and for this audience, to explain that the Grignard Reaction actually goes by an electron transfer mechanism, or that S_N1 reactions are complicated by ion pairs of various types. I am presenting organic chemistry that people can use in a way that makes it simple for them to use.

Style

The book is written in a colloquial, conversational style and speaks directly to the student. I recognize the danger that a combination of folksy style and simplification of topics may lead some faculty to believe this text is not rigorous enough for their students. For some this is undoubtedly true, but I truly believe that for the vast majority it is not, and I ask only that you carefully examine selected sections and judge for yourself. I believe I have presented organic chemistry with sufficient detail and rigor for the core audience of science majors. Within a limited set of reactions, I am still providing and expecting a thorough mechanistic understanding of the essence of organic chemistry.

Help!

Finally, I would like to say that I view this very much as a work in progress. I have made choices of what to cover and what to omit. For any choice I have made, there might have been a better one. I rely on you, the users, to help me make it better. Every misplaced comma or misspelled word makes the book harder to read; more importantly, concepts that are not explained as well as they should be need to be pointed out to me, and, if possible, reworded so as to explain it better. I am eager to receive all suggestions of any kind. I will not promise to accept them, but I do promise to take them seriously. Please email me at reingold@juniata.edu.

* Chemistry majors, of course, would continue with Physical Chemistry and various electives. In addition, there would need to be a course covering those aspects of organic chemistry left out of this book. At Juniata this course is called Organic Reactions and is required at the junior level.

FOREWORD—TO THE STUDENT

This book is an experiment. Virtually all of you readers are aspiring scientists of some type or another, so you understand what experiments are all about: you try something, wait around a bit to see what happens, and if the result is not what you wanted, you adjust things a little bit and try again. Since this is an educational experiment, the result is you, and since I am not there, it is up to you to communicate that result to me if I am to make this better for the next round. I have tried as hard as I can to make this the most understandable book there is on Organic Chemistry. Nevertheless there may be some places where I have fallen short of that goal. I need you to tell me where those places are and how they could be improved. It is fine to tell me that you did not understand the section on such-and-such. What is infinitely more useful to me is that you did not understand the section on such-and-such, but after working some problems and talking to the professor you finally figured it out, and if I had said "yadda yadda yadda" in the first place, none of that would have been necessary, and you would have understood it the first time around. I need to know what "yadda yadda yadda" is; you need to tell me, in as much detail as possible. My email is Reingold@juniata.edu and I am eager to hear from you.

At the end of each chapter is a selection of problems. The problems are very important, as this is your chance to see whether you can apply the new knowledge you have gained in the chapter. But some of them have no answers! How will you know if you have done them right? I know from years of experience that students desperately want to be reassured that they are doing things right, and therefore they want answers. There is nothing wrong with this, provided the answers are used exclusively for that purpose, but inevitably students look at the answers sooner than they ought to. The proper use of answers is to confirm that the answer you have *written down* is correct. You should *never* look at an answer until you have your own, and further, until you are confident it is right. The purpose of answers is *not* to show you how to do something that you have tried and failed to do. Organic chemistry is like playing the piano: you will not get better by listening to someone else play; you will get only slightly better by trying yourself, making lots of mistakes, and *then* listening to someone else play. The only way to get better is to keep practicing, over and over, until you can do it yourself without mistakes. If you have trouble answering a question, you should *not* look at the answer. Have someone give you a hint, nudge you in the right direction, but force yourself to get an answer on your own. If you are not sure it is right, find someone else who is having the same problem and argue about it until one of you convinces the other. Only when you are pretty sure you are right should you look at the printed answer.

When you get stuck and you do need a hint, you should not use the answer to provide it. Once you have seen the answer you cannot undo that look, and you have denied yourself the opportunity to arrive at that answer, by whatever tortured route, on your own.

Even under these constraints, students tend to overuse answers. Whether you have successfully argued with someone else or not, there is a (natural) tendency to be hesitant about your answer, and to defer to the voice of God (as personified by the answer book). The answer is not right, you think, until some authority figure has certified it as such. So when faced with conflicting answers, instead of arguing the point, you just look it up to find out who is right. This approach cheats you of a very valuable learning experience, that of defending your position in an argument where the actual answer is not known. For this reason I have deliberately left some of the questions with no answers. Please, please, do not ask your professor for the right answer! Take advantage of the opportunity to study with your friends and convince each other that your answer is the correct one.

There is another reason why I have left off some answers. As we get further and further along in the book, we will more and more often encounter questions which do not have one single right answer. Many questions have several equally good answers. If I write down a correct one, students may come to the conclusion that their answer is wrong if it is different, when in fact it might be just as good (or sometimes even better) than mine. The point is that any answer you find in an answer book is *a* right answer, but not necessarily *the* right answer; students are too quick to assume that if their answer does not correspond to the written one, it is wrong.

In preparing this book, I have debated with myself and with others about providing answers. Should I include no answers in the book; just a few answers to the toughest questions; about half of them; or all of them? I have finally decided on the third option—about half—because I have realized, again from years of experience, that no matter what I said in the preceding page, you will ignore me. The only way to prevent you from misusing answers is to leave them off. Furthermore, I am hoping that your professor will assign some of these questions for class discussion, and there will be no discussion if the answer is in the book.

Introduction

This is a course in the chemistry of carbon compounds, Organic Chemistry. YIKES! Isn't that the killer course that sophomores around the world dread? Why are they teaching it to us, students taking our first chemistry course? How will we survive?

Don't panic. Yes, organic is the course that sophomores around the world dread, although if you press further you will discover that once they got into it, a surprising number of them liked it. No, it is not a killer course. It used to be, when it was taught as an exercise in memorization, but these days most courses are taught with an emphasis on understanding what is going on, and if you approach it in that way, you will discover that everything fits together so neatly there is little need for memorization. Molecules usually do what makes sense, and if you learn to think the way they do, their choices will make sense to you too. You are getting this as a first course because your teachers have come to the conclusion that organic is a more sensible way to begin your college-level chemistry than the traditional rehash and amplification of high school chemistry. I hope that by the time you finish this course, you will agree.

Few of you taking this course intend to be chemistry majors. Most, I suspect, are taking it because you have to, because it is required for medical or some other professional school, or simply as a prerequisite for some course you want to take in another department, perhaps biology. Organic chemistry is the branch of chemistry most applicable to biology and life processes. Further, it involves only a small subset of the available elements. And understanding organic chemistry does not require a great deal of math. For all these reasons, organic makes a logical starting point on your chemical journey. Further, most of you have not seen much organic chemistry before, so the problem of half the class already "knowing" the material disappears.

In order to present organic chemistry at the introductory level I have had to remove some of the material traditionally covered in a sophomore-level course. I have also added some material at the front end to make sure you have the background necessary to understand the organic parts. And I have slanted the treatment very deliberately toward students interested in the life sciences. In other words, I am dealing mostly with the organic chemistry relevant to life processes. Thus, in a real sense, this is a treatment of BIOorganic chemistry.

For those of you who are interested in chemistry but not biology, do not despair. This course is a good introduction for you also. You will learn many of the ways that chemistry applies to biology, applications that your peers at other schools may not be getting, and frankly, a great deal of modern chemistry does interface with biology, so this will be good for you. But at the same time you will learn appropriate basic chemistry that can be applied to many other areas of science.

Chapter 1
Basic Concepts

Organic chemistry deals almost entirely with molecules. Before we start discussing it, then, we must first establish what we mean by a molecule. And before that, we must discuss atoms. In this chapter we will

- Learn about elements, atoms, and subatomic particles

- Encounter isotopes and see how they contribute to average atomic weights

- Discover the mass spectrometer and see how it helps us understand these concepts.

ELEMENTS

There are millions of different kinds of substances. Since antiquity, people have cherished the notion that all these myriad substances are ultimately composed of a much smaller number of basic building blocks, put together in various combinations. The ancient Greek philosophers speculated that there were only four such building blocks: earth, air, fire, and water. Unfortunately most Greek science was based on thought, not experiment, so it was many centuries before people realized that the story was more complicated. Much useful experimentation was done by medieval alchemists: while trying to turn other materials into gold, they discovered a number of substances that appeared to be elementary building blocks. We now call these substances elements. Examples known to the alchemists included gold, silver, copper, iron, tin, lead, and mercury.

How do we know whether something is an element or not? This was a very difficult question for early scientists. Aristotle defined an element this way: "An element is that into which other bodies can be resolved, but which cannot itself be resolved into anything simpler, or different in kind." Further, "Everything is either an element or composed of elements." That sounds nice, but when one kind of stuff changes into another kind of stuff, how do you know whether you have made it simpler or more complex? How do you know when you get to the stuff that cannot be further resolved? Consider a substance that is a combination of two elements: say, iron and lead. If you do some process that separates the iron and lead, and weigh the lead, it will weigh less than the combination you started with. As a general rule, if you do some process that turns A into B, and the B you get weighs less than A did, A cannot be an element. B is a component of A. B *might* be an element, but then again, it might not, depending on whether B itself can be further simplified.

Robert Boyle, a prominent scientist of the 1600s, said that an element will always gain weight when undergoing a chemical change. One must be careful in applying this rule. After all, iron turns into rust, which appears to be lighter than iron. But if you keep careful track of all the rust that comes from a piece of iron, you will discover that the rust actually weighs more than the iron it came from. True, rust is less *dense* than iron (a cubic centimeter of it weighs less than a cubic centimeter of iron), but we are talking about total quantity. Thus rust cannot be an element; iron might be.

By careful measurement and extensive study, it is not too hard to prove that something is not an element. However, it is almost impossible by this method to prove that something *is*: perhaps we have just not tried the right reaction yet. As Justus Liebig said in 1857, "Elements are considered to be simple not because we know they are, but because we do not know they are not." As we will see, this observation applies to most of science: you can prove something to be false, but you can almost never prove something to be true. The best you can do is prove that you have not yet been able to prove it false, so we will continue to believe it until someone does prove it false.

With much experience, we now have a list of over 110 elements, of which 90 or so are naturally occurring. You have most likely seen the chemist's version of a list of these elements, called the periodic table. We will presently learn why chemists tabulate the elements in this way.

Matter is composed of zillions of tiny particles called atoms. Sometimes these atoms congregate together; a chunk of material that contains only one kind of atom is called an element. For the most part during this course, we will consider only one other possible arrangement, one in which atoms of various types get glued together into clusters called molecules. A chunk of material that contains molecules that are all alike is called a *compound*. Thus, atom is to element as molecule is to compound; that is, **atoms are the microscopic units of elements, and molecules are the microscopic units of compounds**. Since molecules are atoms glued together, if we hope to understand molecules we ought to take a look first at atoms.

ATOMS

Just what are these atoms? How tiny are they? What do they look like? What do they do? Why do they do the things they do? These questions, especially the last two, form the basis of the science called chemistry. We will creep up on the latter questions, but the first ones can be answered now.

How Small?

Atoms are matter. They take up space, and they have weight. Not very much space, and not very much weight, but some. Every chemist has a favorite example for impressing upon students how small an atom is. Mine is that if you imagine taking a single drop of water and allowing it to spread out and out and out until the molecules in it are evenly distributed over the entire surface of the earth, there would still be more than 150,000 of them on this page. Later on we will calculate this; for now, you need simply to accept the fact that a molecule or atom is unimaginably small; the number of them in any sample you can see is unimaginably large. Fortunately, we never really need to deal with the actual size or weight of individual molecules. For all our purposes, all we need to know is *relative* size and weight. In other words, we will *compare* atoms and molecules to each other.

What Do They Look Like?

The original notion of atoms was that they were essentially marbles packed together—tiny spheres. Then it was discovered that there were positive and negative electrical charges associated somehow with these things. In particular, the electron was discovered, and it was found to be the home of the negative charge. The next step was thinking of the atom as being like tapioca pudding (actually, since the scientists were British, they used raisin pudding in their model), in which the overall blob was the atom, which carried a positive charge, and the little nubs (or raisins) were the electrons embedded in it. Then Ernest Rutherford did a key experiment. Rutherford had discovered a source of very small, positively charged particles, and he shot them at high speed at a target consisting of a thin foil of metal. If atoms are blobs, then a thin foil is a bunch of blobs pasted together; tiny, speeding, positively charged particles should go through the foil like a bullet through a Kleenex. Imagine Rutherford's astonishment when he discovered that a few of his "bullets" actually bounced off the Kleenex! To account for this, Rutherford postulated that an atom was not a blob, but that all the positive material in an atom, and effectively all of its mass, was concentrated in a tiny speck at the center of the atom (the *nucleus*) with the electrons whizzing around it at a distance. Thus, most of an atom is empty space, but there are occasional tough spots. Bullets that happen to hit a nucleus bounce back, while most of them go through without hindrance.

This picture of the atom, with some modification, is still accepted today. The nucleus contains essentially all the mass of an atom, but occupies only a tiny fraction of the atom's "space." A common analogy is a marble suspended in the middle of the Astrodome. Electrons "occupy" (or travel through) the rest of the space, but have virtually no mass.

What Are They Made Of?

A nucleus is actually composed of two different kinds of particles (if we ignore recent advances in subatomic physics): the proton and the neutron. Both have effectively the same mass. Again, this mass is so small that if we express it in terms of anything you know, we have to write a ridiculously small number. So, for convenience, we invent a new unit for the mass of a proton or neutron. We call it an *atomic mass unit*, 1 amu. It doesn't matter what it is in real terms, because we will be always comparing atoms to one another. These "amu" units are as good as anything. Protons and neutrons have essentially the same mass, 1 amu, but they differ in an important way. The difference is that a neutron has no charge, while a proton always has a positive charge. How big a positive charge is it? It is the smallest amount of plus charge there can be, so call it +1. As with mass, if the actual amount were expressed in terms of something you could easily identify with, the number would be so small as to be meaningless. So we'll stick with +1.

The third atomic particle is the electron. As mentioned before, electrons flit around in space outside the nucleus. An electron has a minus charge, equal in magnitude to the plus charge of the proton, so call it –1. Its mass is about 2000 times less than the mass of the proton or neutron. This is small enough to ignore for most purposes. So for us, the mass of an atom is simply the mass of the protons added to the mass of the neutrons. Since each of these has a mass of 1 amu, the mass of the atom (measured in amu) is thus the *sum* of the number of protons and neutrons. This is referred to as the *mass number* of the atom.

How many of each kind of particle should one expect to find in a nucleus? This depends on the kind of atom you are looking at. Indeed, in one sense, it *determines* what kind of atom you are looking at: **the identity of any atom is defined by the number of protons in its nucleus**. If there is only one proton, it is always hydrogen. Conversely, if it is hydrogen, there is exactly one proton. This is one of the few scientific statements to which there are no exceptions. The number of protons is a fundamental property of an atom. The number of protons is called the *atomic number* of an element. All nitrogen atoms have seven protons, or an atomic number of 7. If, by magic, one of these protons were to disappear and suddenly there were only 6 protons (this magic does occur occasionally in the upper atmosphere), it would no longer be a nitrogen atom, but now an atom of carbon, atomic number 6. The atomic number of an atom is sometimes written in the lower left corner of the symbol for the element: $_6$C. However, this does not convey any new information, and is done only for the convenience of the reader, sparing you the necessity of looking up the number of protons in a carbon atom (or the identity of the element with 6 protons). You normally see this symbolism only in the context of nuclear reactions, which we will not discuss. There are lists of elements in which one can look up the atomic number of any element. Conversely, one could list the elements in order of atomic number. The periodic table is such a list, although it is arranged in a peculiar way. Later we will discover the reason for its funny arrangement.

How many electrons do you find in an atom? Since atoms have no charge, the number of electrons must be identical to the number of protons. All hydrogen atoms have 1 proton, therefore 1 electron. What, then, do you call an atom with 1 proton and no electrons? Since it has one proton, it is by definition hydrogen, but with no electron to balance the proton, it has a + charge. An atom with a charge is called an "ion." This is H$^+$, a hydrogen *ion*. In general, an ion with a plus charge is called a cation (pronounced **cat**-eye-un, not cayshun).

What do you call the substance with 8 protons and 10 electrons? With 8 protons it must be oxygen, but with 10 electrons it has 2 minus charges. This is O^{2-}. It is also an ion. A negative ion is called an anion (**an**-eye-un, not anyun).

Test Yourself 1

How many electrons are in $_{17}Cl^-$?

The other particle in an atom is the neutron. Here the story gets messy. The number of protons is defined by (and defines) the identity of the element; the number of electrons is the same (if there is no charge). But the number of neutrons can vary. You can have an atom with one proton and no neutrons. This is a hydrogen atom. You can have an atom with one proton and one neutron. This is still a hydrogen atom. And an atom with one proton and two neutrons is *still* a hydrogen atom. The difference is in the *mass number*. This number is typically reported in the upper left corner of the element's symbol, and unlike the atomic number, it is not redundant. The three atoms mentioned above would be written as ^1H, ^2H, and ^3H. These are three different kinds of atoms: all hydrogens, and identical in most of their properties, but not completely identical. They are *isotopes* of each other. **Atoms of the same element (i.e., with the same atomic number, or the same number of protons) but different weights (i.e., different numbers of neutrons) are isotopes.** If you intend to convey to a reader a specific isotope of an atom, you have to specify it by writing the mass number on the atom, as in ^{238}U. You *may* include the atomic number, but this is not required:

$$_{92}^{238}U$$

By the way, you should be aware that ^2H is called "deuterium" and is often written as "D," and ^3H is called "tritium" and is given the symbol "T." These are the only isotopes that have their own symbols.

Test Yourself 2

OK, let's see how you are doing: How many protons, neutrons, and electrons are there in an atom of $^{238}U^{2+}$? Write down answers!

Relevance of Isotopes

A reasonable question to ask at this point (indeed, at every point) is, "Who cares?" What difference does it make whether there are isotopes or not? This is not a course in nuclear chemistry, but suffice it to say that certain isotopes, because of their neutron/proton ratio, spontaneously fall apart, releasing energy of various types. This property has led to, among other benefits, nuclear energy, carbon dating, and much of our knowledge about metabolism, as well as such problems as nuclear waste dumps, Three Mile Island, Chernobyl, and many cases of cancer. Nuclear bombs, of course, can be viewed as benefits or problems, depending on whether you are the donor or the recipient of one. We will see some uses of isotopes in labeling molecules later in this book. For now, let us concentrate on one rather basic consequence of isotopes.

Look on your periodic table (we'll return later to why the table is set up the way it is) and find chlorine. The atomic weight (more accurately, the atomic mass) of chlorine is listed as 35.453. I just finished telling you that each atom is composed of protons, neutrons, and electrons, and that each proton and neutron weighs 1 amu, and an electron weighs nothing. How can you create an atom weighing 35.453 amu from such pieces? *You can't!!* So what gives? Is the periodic table wrong? No. Did I lie? Yes, but only a little, not enough to account for this. The key to this apparent discrepancy is that there is a difference between the *mass number* of an atom and its *atomic mass*. The mass number refers to a *single* atom, and it will always be (to a first approximation) a whole number, because each atom is composed of protons and neutrons of mass 1, and electrons of mass 0. But the "atomic mass," the number reported on the periodic table, is not the mass of a single atom, but the average mass of a collection of atoms.

Imagine your next chemistry test has ten questions, each worth 10 points, and there is no partial credit. Each student in the class gets a 100, or 90, or 80, 70, 60, etc., but never anything in between. Nevertheless, the *average* grade on the exam could easily be 75, even though not one student made that score. The same is true with atoms: the *average* mass of a chlorine atom can be 35.453, even though no single chlorine atom has that mass, *provided there are different kinds of chlorine atoms present.* As it turns out, there are.

If you pick up a random collection of one million chlorine atoms, you will discover that about ¾ of them weigh 35, and the other ¼ weigh 37. (By the way, *none* of them weighs 36. Why not? No one knows. It's just one of the rules of the universe.) How many neutrons are present in each of these isotopes? What is the *average* weight of a chlorine atom? Like any other property, the average is the sum of all the weights divided by the number of atoms. In this case, (¾ x 1,000,000) x 35 + (¼ x 1,000,000) x 37 = 35,500,000 is the total weight of all 1 million atoms; 35,500,000/1,000,000 = 35.5 is their average weight.

Rearrange this math a bit:

¾ x 1,000,000 x 35 + ¼ x 1,000,000 x 37 =
(¾ x 35) x 1,000,000 + (¼ x 37) x 1,000,000 =
[(¾ x 35) + (¼ x 37)] x 1,000,000

This was how we got the total weight. Then we divided this number by 1,000,000 in order to get the average weight.

$$\frac{[(3/4 \times 35) + (1/4 \times 37)] \times 1,000,000}{1,000,000} = (3/4 \times 35) + (1/4 \times 37) = 35.5$$

In other words, it is not necessary to imagine any particular *number* of atoms in order to do this calculation: the average weight is simply the fraction that weigh 35 multiplied by 35, plus the fraction that weigh 37 multiplied by 37. This logic applies even if there are more than two types (isotopes) of atoms. Just be sure that you account for all of them: that is, the fractions, all put together, should add to one.

Now we have a problem: we have calculated that the average weight of a chlorine atom should be 35.5, but on the table it says 35.453. Why? Because I lied again: not *exactly* ¾ of the chlorine atoms weigh 35. We can figure out exactly what fraction does, knowing that there are two kinds, 35 and 37, and that the average weight is 35.453. However many weigh

35, the rest weigh 37. Thus we can call X the fraction that weighs 35 and Y the fraction that weighs 37, and use the same approach we used above:

$(X \times 35) + [Y \times 37] = 35.453$, where $X + Y = 1$ because there are only two kinds of chlorine.

Solve for X, and you will discover that $X = .7735$, meaning that 77.35%, or slightly more than ¾ of the chlorine atoms, weigh 35.

Note that for most situations, such as weighing something, when you are using a large number of atoms it is most appropriate to use the average weight of such atoms for your calculations. To all intents and purposes, the same average appears to apply everywhere on earth.

WHO CARES? YOU DO!!

Many elements occur as more than one isotope—indeed, many occur as quite a few isotopes. Uranium, for example, has 14 different known isotopes. Of these three occur naturally: ^{234}U has a natural abundance of 0.005%, effectively nothing; ^{235}U is 0.72% of natural uranium; and ^{238}U is the rest, 99.275% of it. The kind that is used in nuclear reactors is ^{235}U. Clearly one has to mine a lot of uranium to get a useful amount of the correct isotope. Plus there is the terrible problem of separating one from the other! It is not necessary to get pure ^{235}U to run a reactor, but it has to be a lot better than 0.7%. Uranium is not being formed at any significant rate, so every bit we mine and use up is not there for the next generation to use. This is called a non-renewable resource—just like coal and oil. Once it is gone, it is gone. Much effort these days is going into figuring out how to produce energy from renewable sources, such as sunlight, wind, tides, etc. So far these account for only a tiny fraction of the total energy used in the world, but there is hope that more research will lead to more efficient ways of harnessing renewable energy sources.

Evidence for Isotopes

There are a few methods of measurement, however, in which atoms do not behave as collections. One of these involves the technique called mass spectrometry. Consider a bowling alley with a powerful fan blowing across it about halfway down. Roll down the alley a bowling ball, a golf ball, and a ping-pong ball. It does not take any major insight to realize that the bowling ball will be essentially unaffected, the golf ball might curve a bit, and the ping-pong ball will be blown way off course. The message is that lighter moving particles will respond to an outside force more than heavier ones will. To apply this to molecules, we have to get them moving and apply some force. Without going into detail, simply accept that machines exist that are able to create moving molecules with positive charges on them ("ions"), and that these ions can be passed through a magnetic field which behaves like a blowing fan and tries to push them off course. The ions move in curved paths, with the lighter ones curving more. By lots of calculations and proper calibration, it is possible to figure out from the curvature of the path how heavy each particle must be. Further, in this machine (called a mass spectrometer), the ions have enough internal energy to occasionally break the glue holding individual molecules together.

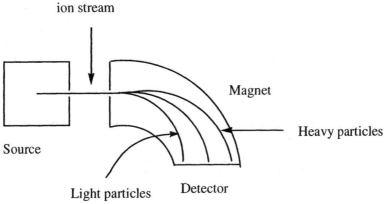

ion stream

Source

Magnet

Heavy particles

Light particles   Detector

Mass Spectrometry

Mass spectrometry requires generating molecules with positive charges ("ions") that still have the same weight as the original molecule. We can imagine how this might be done. Without going into detail, imagine that we could knock an electron out of some molecule. It would then be one electron short, so it would have a positive charge. But what would it weigh (technically, what would its mass be)? Just about the same as the original, since electrons have effectively no weight. The resulting ions are then sent through a magnetic field, and their path bends depending on their weight. We can figure out the weight of the particles as they emerge from the other end of the instrument by seeing how much the path curved. The resulting pattern is called a "mass spectrum." In a mass spectrum, the position of any line on the x axis is the mass of some molecule (technically, the mass-to-charge ratio, m/z, but since z is almost always 1 we can consider this the mass); the height represents the number of molecules that have that mass, which is directly proportional to the probability of finding that mass.

Imagine doing this with the molecule CH_3Cl. If we assume that all C atoms weigh 12 amu and all H atoms weigh 1 amu (not true, but pretty close), then the CH_3 group will always weigh 15 amu. The Cl atom, however, will weigh 35 amu in some of the molecules (how many?) and 37 amu in the rest. Thus there will be two different kinds of molecules, those that weigh 50 amu and those that weigh 52 amu. The mass spectrometer should be able to tell them apart, and also tell us how many of each there are, based on the strengths of the various signals. Shown below is the mass spectrum of CH_3Cl.

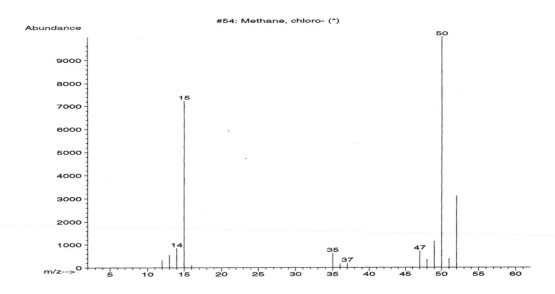

You can see that there is a peak (signal) at mass 50 of a certain height (intensity), and a peak at mass 52 that is about 1/3 the size of the peak at mass 50. A 3:1 ratio is just about what we would predict on the basis of the calculations we did using molecular weights.

If you study the mass spectrum more closely you notice a number of other features: There is also a small peak at 51. What causes this? Remember that this machine creates ions with enough internal energy to occasionally break them apart. The peak at 51 is caused by a few of the molecules that lose a hydrogen atom after they have lost their electron. Thus there are peaks at 51, 50, and 49 for loss of 1, 2, and 3 H's from the CH_3Cl that weighs 52. Of course the peak at 50 is covered up by the large peak from the CH_3Cl that weighs 50. It also loses H's, so there are peaks at 49, 48, and 47. The 49 peak is a combination of the loss of 1 H from 50 and the loss of 3 H's from 52. In addition to these pieces ("fragments"), there are also some molecules that break apart in the middle, into a CH_3 piece and a Cl piece. You can see the Cl fragments at 35 and 37, in a 3:1 ratio. You can also see the CH_3 fragment at 15, and some of the latter lose their H's to make pieces weighing 14, 13, and 12. If you were very clever you could reconstruct the molecule just knowing what fragments are produced. Chemists called mass spectroscopists have made careers of doing just this. The overall pattern produced by a molecule is called its "fragmentation pattern" and tends to be unique for a particular type of molecule. Substances are now routinely identified by checking the fragmentation pattern of an unknown against a huge library of known substances on a computer. For the time being, we will not worry about fragmentation of molecules, and our discussion of mass spectra will only concern the whole charged molecule (called the "parent" or "molecular ion").

Any random sample of Cl atoms will be about ¾ ^{35}Cl and about ¼ ^{37}Cl. Knowing this, can we predict what *pairs* of Cl atoms will look like? You probably already know that chlorine occurs naturally as Cl_2. What would the mass spectrum of Cl_2 look like? Consider how many peaks you would see, what weights they represent, and how big they would be. You have probably never done this, so even if you had a strong background in chemistry and you think you have learned nothing new in your reading so far, take a few minutes to commit to an answer by writing it down before you turn the page: draw a predicted mass spectrum for Cl_2. Don't worry about fragmentation, just concentrate on the pattern of the peaks produced by whole molecules.

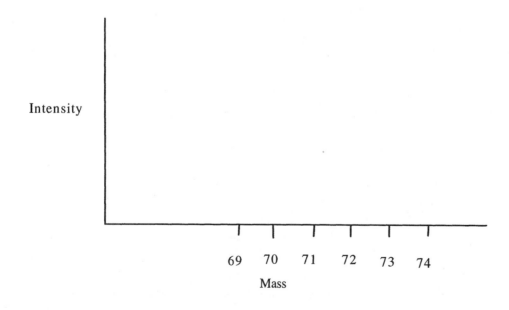

You turned the page anyway, didn't you? Now, listen carefully, folks: you cannot learn organic chemistry by seeing it done *for* you. You must work at doing things yourself. When I give you a problem to solve, you have to work at it, and if you can't get it, talk it over with some other students. Come up with an answer and write it down. It doesn't have to be right, but it should be your best effort. All this should occur *before* you look at the answer I have provided. That way, when you see the answer, you will see how it relates to what you were trying to do, and you are more likely to learn how to do the problem by understanding it that by simply memorizing what someone tells you to do. Now go back and try this problem for yourself.

The real mass spectrum of Cl_2 follows. If you did not predict this pattern correctly, try to follow this logic: If you choose a random handful of Cl atoms, ¾ or 75% of them will be ^{35}Cl. If you consider only those, and choose random partners for them, 75% of the time you will get a second ^{35}Cl. So what are the chances of getting two ^{35}Cl's? You multiply the probabilities: 75% x 75% = .75 x .75 = .56 or 56% of all the molecules should weigh 70 amu. Likewise, the chances of getting two ^{37}Cl's is .25 x .25 = 6%, having a weight of 74. The rest of the molecules must have one atom of each type and weigh 72. You could calculate the size of this peak by knowing that it must represent all the rest, or you could recognize that it consists of those 35's that added a 37 (75% x 25%) plus those 37's that picked up a 35 (25% x 75%) = 19% + 19% = 38%. There are no other possible combinations, so we will see peaks only at 70, 72, and 74.

MASS SPECTRUM OF Cl_2

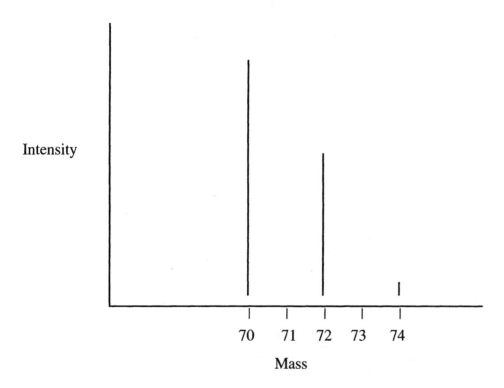

If you still have trouble seeing this, check out the addendum at the end of the chapter.

Test Yourself 3

Now for a real challenge. If you have followed all these arguments and *really understood them*, you should be able to do this problem. If you have trouble, then you are probably falling back into "high school memorize-stuff" mode. Now is as good a time as any to dump that approach, because it will not work in college.

Below is a mass spectrum of bromine molecules, Br_2. Without looking in a book, i.e., using only the information in the mass spectrum, figure out what are the naturally occurring isotopes of bromine and in what proportions they occur. (This is like looking at the Cl_2 spectrum and deducing that Cl comes as 75% ^{35}Cl and 25% ^{37}Cl.) Then using your answer and what we learned in this chapter, predict the appearance of molecules of BrCl in a mass spectrum. These are not trivial problems but *you can do them!*

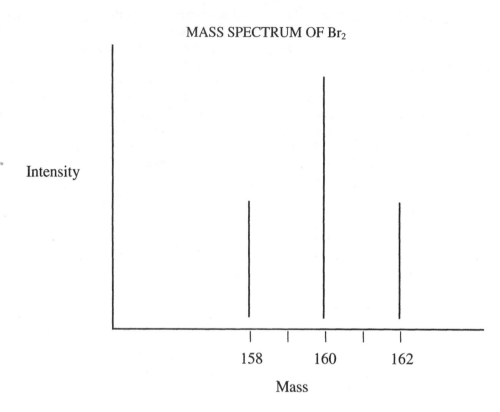

MASS SPECTRUM OF Br_2

Intensity

158   160   162

Mass

Here are some questions to help you through this problem. *Don't look at them until you have thoroughly given up on solving it on your own!*

1. You are looking at the pattern for bromine *doubles*, Br_2. What must be the heaviest kind of Br singles?
2. What is the lightest kind of Br singles?
3. Is there a third kind of Br singles?
   a. If there were Br singles weighing 80, wouldn't you expect some Br doubles weighing 159 and 161? Are there any? Is there any way to get Br doubles weighing 160 *without* imagining Br singles weighing 80? If you got them this other way, would you expect any kind of Br doubles other than what you see?

You should now know how many kinds of Br singles there are, and how much each weighs. Now for the tougher question: how many of each type are there?

1. If there were more 79's than 81's, which peak would be bigger, 158 or 162?
2. If there were more 81's than 79's, which peak would be bigger, 158 or 162?
3. If there were the same number of 79's and 81's how big would each peak be on the Br_2 chart?

12   Chapter 1   Basic Concepts

Now that you know the kinds of Br's there are, and (roughly) the fraction of each kind, and you already know the kinds of Cl's there are, and the fraction of each kind, figure out what a BrCl mass spectrum would look like!

Addendum

A visual approach to determining isotopic distribution in Cl_2.

Grab a chlorine atom. It will be either ^{35}Cl or ^{37}Cl, right? If it was a ^{35}Cl, we can grab a partner for it, and the partner will be either ^{35}Cl or ^{37}Cl. If the first one was a ^{37}Cl, the same is true. Thus there are four and only four ways you can grab a pair of Cl atoms:

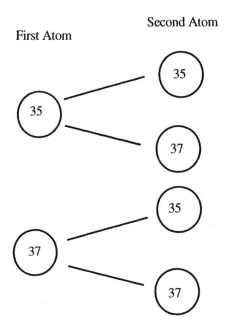

First Atom

Second Atom

What are the chances of each of these happening? Each time you grab an atom, the probability is 0.75 that it is ^{35}Cl and 0.25 that it is ^{37}Cl. So the chances of each of these occurrences are shown below.

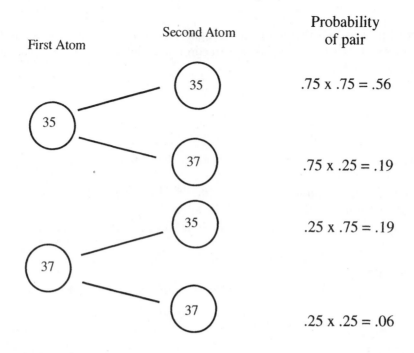

First Atom   Second Atom   Probability
of pair

35 — 35   .75 x .75 = .56

35 — 37   .75 x .25 = .19

37 — 35   .25 x .75 = .19

37 — 37   .25 x .25 = .06

Now, the first pair weighs 70, the second pair weighs 72, the third pair also weighs 72, and the bottom pair weighs 74. Thus you can easily calculate the probability of any weight:

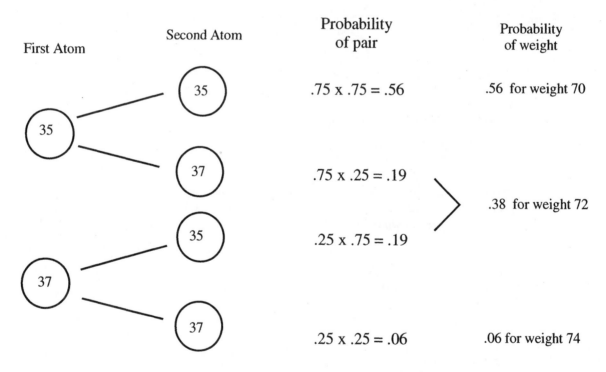

First Atom   Second Atom   Probability
of pair   Probability
of weight

35 — 35   .75 x .75 = .56   .56 for weight 70

35 — 37   .75 x .25 = .19

.38 for weight 72

37 — 35   .25 x .75 = .19

37 — 37   .25 x .25 = .06   .06 for weight 74

In the mass spectrum, the position of any line on the *x* axis corresponds to its weight (mass); the height is the probability of finding that mass. Thus we should expect to see a line at weight 70 that is some height, another line at weight 72 that is a bit shorter (2/3 as high, actually), and a very small line at weight 74. That's what we see.

Problems

(Answers are provided at the end of the chapter for italicized problems.)

1. Fill in the following table: (You will need to use a periodic table for this.)

atom or ion	atomic mass	protons	neutrons	electrons
Fe	54			
S^{2-}			18	
	44	20		18
	7		4	2
		7	8	7
Mg^{2+}	26			

2. a. The three predominant isotopes of silicon are of mass 28, 29, and 30, respectively. What is the difference between ^{28}Si and ^{30}Si that leads to a difference of 2 mass units?

   b. How many electrons does a single bromide <u>ion</u> contain? (A bromide ion is a bromine atom that has a single negative charge.)

   c. Predict the mass spectrum for methyl bromide, CH_3Br. Recall that naturally occurring Br is split about equally between isotopes of mass 79 and 81. Assume that C and H do not have isotopes other than 12 and 1 (they actually do, but in very minor amounts). Consider only the peaks for whole molecules, do not worry about fragments.

   d. Predict the mass spectrum for dibromomethane, CH_2Br_2 (no fragments).

3. Naturally occurring strontium exists as several isotopes:

mass (amu)	abundance (%)
84	0.5
86	9.9
87	7.0
88	82.6

   Estimate (by simply looking at the data) the mass of naturally occurring strontium.

   Calculate the mass of naturally occurring strontium to three significant figures.

4. Potassium (K) has a molecular weight of 39.10 and exists as three isotopes: 39, 40, and 41. Isotope 40 is present in only trace amounts (0.01%).

   a. Estimate (by simply looking at the data) the percent abundance of each of the other two isotopes.

b. *Calculate the percent abundance of each isotope.*

5. *Magnesium comes in three isotopes, 24, 25, and 26. The atomic weight of Mg is 24.3. If ^{26}Mg is present at 11%, what are the natural abundances of the other two isotopes?*

6. Predict the appearance of the mass spectrum of tribromoethane, $C_2H_3Br_3$, in the mass spectrometer (no fragments).

7. *Predict the appearance of chloroform, $CHCl_3$, in the mass spectrometer.*

8. Zinc has four naturally occurring isotopes, ^{64}Zn, ^{66}Zn, ^{67}Zn (4.1%), and ^{68}Zn (18.8%). The average atomic mass of Zn is 65.38. Predict the appearance of the mass spectrum of diethyl zinc, $C_4H_{10}Zn$. Draw a picture and label the locations and heights of the peaks.

9. *Given the following mass spectrum for the compound $C_4H_{12}Si$, what is the % distribution of the isotopes of silicon?*

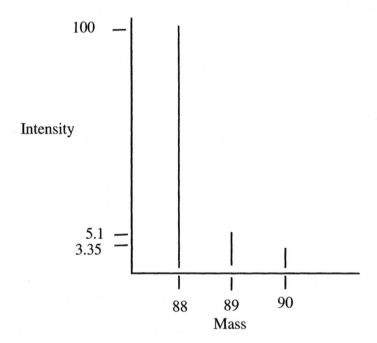

Mass Spectrum of $C_4H_{12}Si$

10. Based on the data in problem 9, calculate the average atomic mass of silicon.

11. Shown below is the mass spectrum of potassium ethoxide, KOC_2H_5 (only the whole molecule—fragments are not shown).

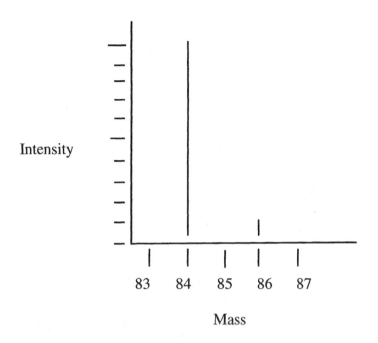

Intensity

83   84   85   86   87

Mass

a. Assuming that C, H and O have no isotopes other than 12, 1, and 16, what are the isotopes of K and what is their relative abundance?

b. Based on the information in part a, what is the average atomic weight of potassium?

12. a. Calculate the approximate distribution for the various isotopic forms of BrCl. Assume for now that the ratio $^{79}Br{:}^{81}Br$ is 1, and that that for $^{35}Cl{:}^{37}Cl$ is 3:1.

b. Based on this model, what should be the molecular weight of BrCl?

c. Using atomic weight data from the periodic table (where the atomic weight of Cl is 35.453 and Br is 79.904), what should be the molecular weight of BrCl? Is the result the same as what you got in part b? If not, why not?

d. Given the reported atomic weight for Br is 79.904, calculate more precisely the natural isotopic distribution for bromine.

e. Using the result from part d and the fact that the abundance of ^{35}Cl is actually 77.35%, derive a more accurate molecular weight for BrCl using isotopic distribution data. How does the result compare with your answer in part b?

Selected Answers

Internals

Test Yourself 1: There are 18 electrons in $_{17}Cl^-$

Test Yourself 2: You should have gotten 92, 146, and 90. If you didn't, reread the previous sections.

Mass Spectrum Problem: Br is 50% ^{79}Br and 50% ^{81}Br. For BrCl, see answer to Problem 12 a.

End of Chapter Problems

4. b.

95% 39 and 5% 41

5.

$24 X + 25 Y + 26 (0.11) = 24.3$
$X + Y + 0.11 = 1$

81% ^{24}Mg and 8% ^{25}Mg

7.

42% 118, 42% 120, 14% 122, 2% 124

9.

92.2% 28, 4.7% 29, 3.1% 30
Remember, the vertical axis is a count of how many molecules have a particular weight. % is amount of this one over the total amount. In this case the total amount is 108.45.

Chapter 2
Electronic Structure

Except for mass considerations, the nucleus really has relatively little effect on the properties of atoms. After all, it is buried deep in the middle of the atom, and by and large neighboring atoms see nothing but electrons. It is the electrons that determine the majority of an atom's properties, at least with respect to organic chemistry, and we must spend some time learning how the electrons are arranged around an atom. In this chapter we will do three things:

- Since most of our understanding of electronic structure comes from experiments involving light, we will take some time out first to make sure we understand the properties of light.

- Then we will show how the interaction of light with atoms leads us to suggest that energy levels exist in atoms.

- Finally we will use energy levels to explain the organization of the periodic table.

LIGHT AND ENERGY

Ordinary white light, such as that bouncing around wherever you are right now, is composed of many different kinds of light. Passing white light through a prism or a raindrop spreads it out into its component colors, called a spectrum, as shown below.

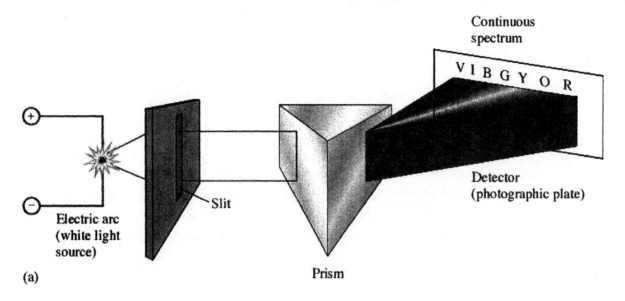

(a)

What you see in a spectrum is different colors of light, which are actually different *kinds* of light. Each kind (color) is characterized by a frequency and a wavelength, but to describe it, it is necessary to specify only one of these, since they are related by the equation

$$\lambda \nu = c$$

where λ (lambda) is the wavelength, ν (nu) is the frequency, and c is the speed of light (3 x 10^8 m/sec), which never changes. Thus if you know either the frequency or the wavelength, you automatically know the other. Further, as the frequency gets larger (higher), the wavelength necessarily gets smaller (shorter), and *vice versa*. (This is of course a necessary conclusion for waves traveling at a constant speed: the closer together the wave crests, clearly, the more often one will go past you!) Violet light has a shorter wavelength, and therefore a higher frequency, than red light.

It should be clear from the above equation that frequency must have the weird unit 1/sec, which is frequently written sec^{-1}. This might be pronounced "per second," but that's pretty awkward; more often the dummy word "cycles" is inserted, so it is pronounced "cycles per second" or "cps," where the word "cycles" disappears for mathematical calculations. Because of this awkwardness, the unit sec^{-1} has recently been replaced with the unit Hz, pronounced "hertz." So a frequency might be reported as 350 MHz ("megahertz") or 350 megacycles per second, but in a calculation you would use 350 x 10^6 sec^{-1}.

For many purposes it is sometimes useful to think of light as a stream of tiny particles called "photons." Photons of different colors of light are different from each other. The key difference is in the amount of energy carried by each photon. This amount of energy (E) is called a *quantum* (plural, *quanta*), and is given by the equation

$$E = h\nu$$

where h is Planck's constant and v is again the frequency of the particular kind of light. (Planck's constant = 6.63 x 10^{-34} J sec, but there is no need for you to remember that.) You should have no trouble showing from the previous two equations that another version of Planck's equation is

$$E = hc/\lambda$$

(*prove it*!). It should also be clear now that violet light carries more energy (packs a stronger punch) than red light.

The kind of light we can see amounts to only a very small portion of the many kinds of light there are. Light that is more energetic than visible violet light carries the name *ultraviolet* light, the same stuff only invisible to us and somewhat more energetic. Energetic enough to do significant damage to life [which is why we are lucky to have (for now) an ozone layer keeping most of it from getting here] but not nearly as energetic as the next type, X-rays, or the next, γ-rays (gamma-rays). These are all exactly the same stuff as visible light, just shorter in wavelength, higher in frequency, and more energetic. Going in the other direction, beyond red light we find *infrared* light, again the same stuff but *less* energetic; then microwaves, the same energy that heats your food, and radio waves. Yes, radio waves and γ-rays are both versions of the same phenomenon, simply carrying different punches. Light in its more general sense is usually designated by the term *electromagnetic radiation*. Visible light is *one* kind of electromagnetic radiation. All of this is illustrated below.

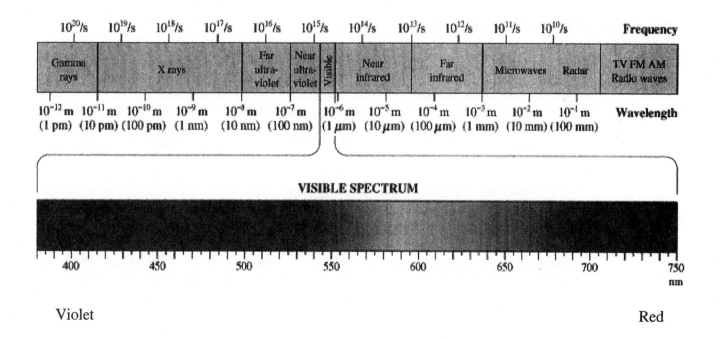

Violet

Red

ATOMS AND LIGHT

Visible light from the sun or a normal light bulb is composed of all the colors we can see, in a "continuous spectrum." But the light that comes from a neon lamp, or any other hot, gaseous element, is composed of discrete lines, as shown below. The light emitted consists of only a few, very specific wavelengths. Why? Why should there be any light at all, and why is it only of certain kinds?

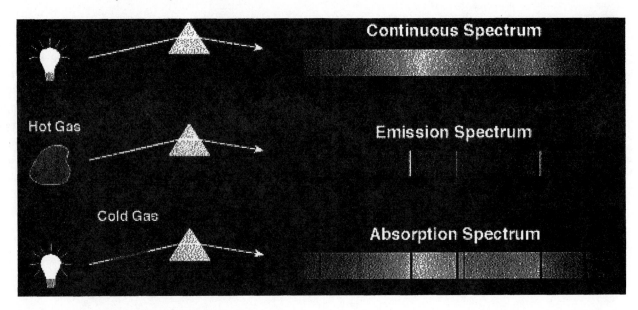

Atomic Absorption and Emission

The first question is fairly easy, once you recognize that light is energy, as we established in the previous section. The atoms in the lamp are giving off light because for some reason they have extra energy, and they want to get rid of it. How they acquired this extra energy is irrelevant—it might have been by heating, by zapping with electricity, by absorbing some other light, or some other way. The point is, if an atom has extra energy, and that amount of energy corresponds to some wavelength of light, the atom can get rid of it by emitting a photon of light.

Where does this energy in the atom come from? The energy comes from electrons. An atom's nucleus sits in the middle of the atom, but the electrons are whizzing around the outside. Whizzing takes energy: electrons have energy. Some have more than others. Electrons can gain and lose energy by absorbing or emitting light. If they absorb light, they end up with more energy than they had before, by the amount of energy in the photon they absorbed. Likewise, if they emit light, they end up with less energy than before, again by the amount of energy in the photon emitted.

Energy Levels

So why does neon, or hydrogen, emit only certain wavelengths of light and not others? Apparently because they are *able* to emit only certain amounts of energy and not others. But this does not make sense! It is like saying that you could walk 10 yards or 20 yards but you do not have the option of walking 15 yards. On the other hand, there are some areas of your experience where such restricted options do exist. For example, a bugle can emit sound of certain frequencies and not others (i.e., it can produce only certain notes). And when you are walking up or down steps, you have a limited set of options as to how far up or down you can go: if the steps are 6 inches high, you can go up 6, or 12, or 18 inches, but you cannot go up 8 or 9 or 10 inches. The reason you cannot go up 10 inches on such a staircase is that there is no place you can land 10 inches from where you start. If the distance you can travel is limited, it must mean that there are only certain stopping places (levels) where you can be.

By analogy, if an atom can emit only certain energies, it follows that there must be only certain "energy stopping places" in an atom. We'll call these "energy levels." Although this was a radical suggestion when Niels Bohr first made it in 1913, it is now widely accepted as fact, and Bohr later won the Nobel Prize for this and many other contributions to science. Bohr stated his theory this way: "An atomic system can, and can only, exist permanently in a certain series of states corresponding to a discontinuous series of values for its energy, and hence any change in the energy of the system, including emission and absorption of electromagnetic radiation, must take place by a complete transition between two such states." In other words, atoms have certain permitted energy levels, and jump back and forth between them by absorbing and emitting the amount of energy that corresponds to a difference in energy between two levels. Systems that are allowed to exist only in certain energy states are called "quantized." An atom wanting to increase its energy cannot absorb any old amount of energy, but only an amount that will allow it to land at an allowed level. It has to jump from one level to another, in a "quantum jump."

Thus, according to Bohr, an atom is like a staircase, with only certain allowed energy levels for its electrons. But this is no ordinary staircase. An ordinary staircase, with 6-inch steps, would allow you to rise by 6, 12, 18, 24, ... inches, in other words, 1, 2, 3, 4, ... times the height of a step. Further, if you started on the second step, the same options would exist as on the first step. Not so in an atom, says Bohr. The atomic staircase, which he described specifically for hydrogen, would have a second step (the bottom is considered the first step) that carried you 3/4 of the way to the top! But the next step is much smaller, and with it you would land 8/9 of the way to the top. The fourth step is 15/16 of the way to the top, and the next is 24/25 of the way, and there are an infinite number of additional steps, each smaller than the last! All this can be expressed quite simply by imagining yourself at the top of this staircase looking down to the bottom, a distance (actually, in this case an *energy*) B: the various steps n (where n is some integer) are B/n^2 away from you. Thus the first step is $B/1^2$ away, at the bottom. The second step is $B/2^2$, or $B/4$ (that is, 1/4 of B) from the top, thus 3/4 of the way from the bottom. The third step is $B/9$ from the top, the fourth is $B/16$, and so on. This can be expressed most easily by arbitrarily assigning the top step the energy 0, so each step will thus sit at an energy expressed by $-B/n^2$. The first several steps of a staircase like Bohr's are shown in the left margin. B is the energy difference, ΔE, between the bottom and the top, that is, it is the maximum energy that an electron can possibly have in this atom (more accurately, it is the energy required to yank the electron out completely).

Now, scientists have no way of going into a hydrogen atom and directly measuring its energy levels, so we have no way of knowing what the actual values themselves are. However, we *do* have a way of measuring differences *between* the energy levels, because when a hydrogen atom absorbs energy, an electron must be jumping from one level up to another. And when it emits energy, the reverse is true: it is jumping down instead of up, but by the same amount. And we can measure the amount of energy absorbed or emitted by looking at the wavelength of light involved.

So to determine the amount of energy emitted when an electron jumps from the 3 level to the 2 level, we need to calculate the difference in energy between these two levels. This is easy enough: The 3 level has $-B/9$ energy, and the 2 level has $-B/4$ energy. So the difference in energy, ΔE, is $-B/9 - (-B/4)$.

$$\Delta E = -B/9 - (-B/4) = [-4B - (-9B)]/36 = 5/36\ B$$

where the positive result can be interpreted to mean that energy is coming out. Actually this is unnecessary: since the electron is going from a high level to a lower one, it is obvious that energy is being emitted. All we really need from this calculation is the *amount* of energy, not its sign, so $5/36\ B$ (= 0.139 B) is the amount of energy separating the steps. If we were going from the 2 to the 3 level, the same amount of energy, $5/36\ B$, would be involved, but the calculation would give a negative answer. The sign is irrelevant as long as we recognize what is going on.

Now $5/36\ B$ is a meaningless number, since we don't know what B is: it could represent any amount of energy. Fortunately, using a bunch of complicated math we do not want to bother with, Bohr was able to calculate how much energy should be required to strip a hydrogen atom of its electron (in other words, B), and using this he determined that the 3 —>2 transition for hydrogen should emit light of wavelength 656.280 nm. Further, the 4 —>2 transition should emit light of wavelength 486.133 nm. The table below shows Bohr's calculated values next to the wavelengths of light that are actually observed for hydrogen.

Transition	Bohr's Calculation	Observed Wavelength
3—>2	656.280 nm	656.279 nm
4—>2	486.133	486.133
5—>2	434.048	434.047
6—>2	410.175	410.174
7—>2	397.009	397.007
8—>2	388.907	388.906

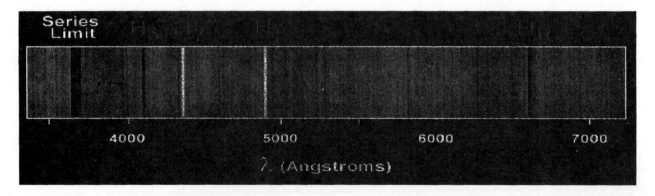

These are the lines found in the visible part of the spectrum. There is another series of lines (n—>1) that appears at higher energy, in the ultraviolet part of the spectrum, and more lines at lower energy in the infrared part of the spectrum. All of these match the predictions just as well. Agreement like this, to 5 significant figures, is very rare in science, and it persuaded other scientists that even though Bohr never explained *why* electrons should be restricted to specific energy levels, it must be so.

Test Yourself 1

To test your understanding of what we have discussed so far, take some time out now and figure out the numerical value of B. Then figure out the wavelength of light absorbed when an electron jumps from the 4—>9 level. Don't continue until you get an answer that makes sense. (How can you know whether it makes sense? You should be able to tell, without doing any calculations, whether your final answer should be smaller or larger than the numbers in the preceding table.) Remember that B is an energy and should therefore have energy units. The scientific unit for energy is Joules, J.

The answer is at the end of the chapter. Do not look at it until you have tried, really tried, to answer the question above. But also, do not continue until you have either solved this on your own or, at a minimum, understood the answer given. If you don't understand it, get help right away.

Bohr's calculations applied only to the hydrogen atom. For reasons that are too complicated to mention here, energy levels in other atoms are spaced differently. This has some immediate consequences: if the energy levels are spaced differently, the wavelengths of light absorbed and emitted by other atoms should be different from those of the hydrogen atom, and also from each other. Each element should have its own pattern of lines, and we should actually be able to identify an atom by its line pattern.

This is true, and there are instruments called *atomic absorption* and *atomic emission spectrometers* that are designed to determine both qualitatively and quantitatively what elements are present in a sample. An *inductively coupled plasma* instrument is a high-tech version of an atomic emission spectrometer. The same phenomenon allows us to determine what stars are composed of. Even the sun does not put out every wavelength of light. If you look closely at the light that comes from the sun (below), you will discover that there are gaps (called Frauenhofer lines) exactly at the wavelengths that hydrogen absorbs. This is caused by hydrogen in the outer part of the sun absorbing these wavelengths from the light as it goes by. It is obvious, then, that there must be a lot of hydrogen up there.

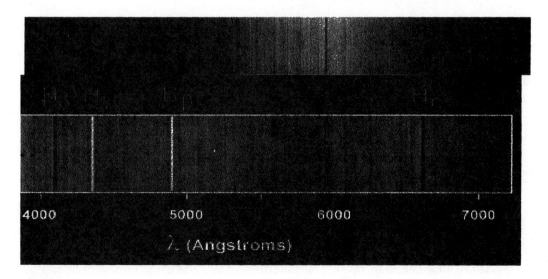

Orbitals

Bohr conceived of his energy levels as equivalent to orbits, like those of the planets around the sun. However, despite pictures in many books that suggest electrons whizzing in circles around a nucleus, this is no longer an accepted model. The word we use now for the energy levels is *orbitals*, a modification of the concept of orbits. Instead of conceiving the electron to be moving in a circle of defined radius around a nucleus, it is now understood that it can be located anywhere within some particular area. The area is called an orbital, and is often pictured as a fuzzy cloud around a nucleus. The reason for the change in the model is related to the fact that, as with light, it is sometimes necessary to view an electron as a wave and other times as a particle. We don't need to get into the details of this duality here. Still, we do need to add a little complexity to the concept of energy levels that we established in the previous section.

What we concluded before is that atoms have energy levels for their electrons. An electron can be in energy level 1,2,3,4, or any other, call it n. These numbers are called the "principal quantum numbers," a term you may hear later in your career. Here is the added complexity:

Rule 1: Energy levels for electrons are not all alike. **Each level n is actually composed of n^2 orbitals.** That is, there is 1 orbital in level 1, 4 orbitals make up level 2, 9 constitute level 3, and so on.

Rule 2: Even orbitals are not all alike. **Each energy level has orbitals just like the previous energy level, plus some extras of a different type, of slightly higher energy**, so the "levels" really are "bands" of energy. Each new type of orbital is assigned a letter of the alphabet. For historical reasons, we start with the letters s, p, d, and f, then continue up the alphabet. As it turns out, we currently have no need for letters higher than f, but higher orbitals still exist in theory.

What does this mean? Level 1 has one orbital. It must be of the first type, s.

Level 1:   ___ s

Level 2 has 4 orbitals. Of these 4, one is like the orbital in level 1 (thus, s); the rest are of a new type, p, with slightly higher energy. Thus level 2 has one s orbital and three p orbitals.

Level 2:   ___ ___ ___ p
             ___ s

Level 3 has 9 orbitals. Four of these are like the ones in level 2: one s and three p's. The rest are a new type, d. Thus level 3 has one s, three p, and five d orbitals.

Level 3:   ___ ___ ___ ___ ___ d
             ___ ___ ___ p
             ___ s

Level 4 has all of these plus seven f orbitals. Level 5 has all of those plus nine g orbitals. And so on. Note that wherever there is an s orbital, there is one of them; p orbitals always come in three's, d's come in five's, f's come in sevens, and so on.

Rule 3: Each orbital can hold up to two electrons. Also, electrons normally choose to have the lowest possible energy. These two statements actually have names. The first is called the Pauli exclusion principle, the second is called the Aufbau principle.

Using these rules, we can now specify where in an atom all the electrons are. For example, in a hydrogen atom, with one electron, it will naturally be in the lowest possible place, a 1s orbital. We express this as $1s^1$. Helium, with two electrons, has both of them in the 1s orbital, shown as $1s^2$. Lithium has three electrons. It cannot put the third electron in the 1s orbital, because there is only room for two electrons there, so the third electron must go in the next most favorable slot, the 2s orbital: Li has $1s^2 2s^1$. Nitrogen, with 7 electrons, has $1s^2, 2s^2, 2p^3$. Chlorine, with 17 electrons, has $1s^2 2s^2 2p^6 3s^2 3p^5$. This kind of list is called the "electron configuration" of an atom.

Rules 1–3 work well up to atom number 20, calcium, but then a new complexity arises.

Rule 4: There is some overlap between levels. This is actually predictable, and does not violate any of the other rules.

Recall that there is a big gap between energy levels 1 and 2 (step 2 is ¾ of the way to the top, remember?). The gap between 2 and 3 is much smaller, and the gap between 3 and 4 is smaller yet. So as you go up in energy levels, the next one above you is getting closer and closer. At the same time, however, each level is accumulating more and more kinds of orbitals, each at slightly higher energy than the last. It should not surprise us that at some point, the upper orbitals of some level will actually be higher in energy than the lower orbitals of the next level. The only question is, where does this start? The answer is, element 21. This overlap is shown in graphical form in the picture below.

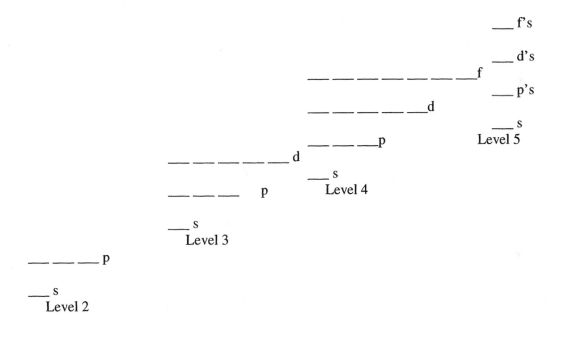

Relative Energies of Orbitals in the First Five Levels (Not to Scale)

As you can see, the 4s orbitals are actually lower in energy than the 3d orbitals. Thus electrons occupy the 4s orbitals before they are completely finished with the 3 level. Likewise, level 5 starts before level 4 is done. Indeed, level 6 actually starts before level 4 is done: the 6s level falls between the 5p and the 4f levels.

How can you remember all this? It is actually pretty easy, because there is a simple mnemonic device to help you. This is shown on the next page. Start at the lower left, follow the arrows and you will end up listing all the orbitals in the right order from an energy standpoint.

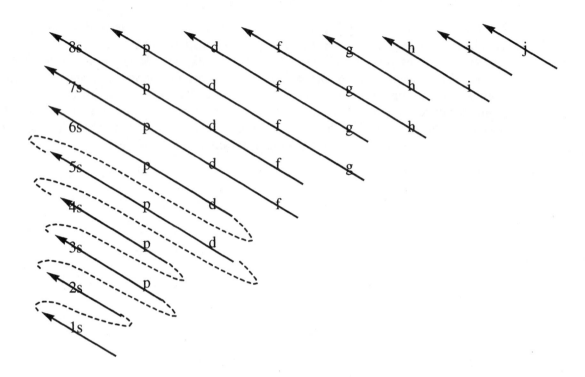

Now you should be able to assign levels to all the electrons in any element in the periodic table. Pick one, say lead, with 82 electrons. If you do not end with the last level being $6p^2$ you are doing something wrong.

THE PERIODIC TABLE

There is one more rule we need to complete our understanding of a large part of chemistry.

Rule 5: There is special stability associated with full levels. (Levels are sometimes called "shells.") Furthermore, an atom, being pretty gullible, thinks a level is full as soon as the next level above it begins to be occupied. If you look closely at the above chart, you will see that this occurs reliably whenever a p level fills up. *Thus there is special stability associated with a full set of p orbitals.*

Now let's do a simple exercise. We'll list the elements in order of atomic number, by name, horizontally, with two guidelines: we'll start a new row every time we begin a new level (shell); and we'll arrange the elements so that we are filling orbitals of the same kind in any given vertical column. Start:

H He

Woops! Time for a new row, because we are entering level number 2.

H   He
Li Be B   C   N   O   F   Ne

Time for another new row, because we are now starting level 3.

H   He
Li Be B   C   N   O   F   Ne
Na Mg Al Si P   S   Cl Ar

Another new row.

H   He
Li Be B   C   N   O   F   Ne
Na Mg Al Si P   S   Cl Ar
K   Ca

And now we run into trouble. The next element is Sc, but where should we put it? It belongs next to Ca if we are filling in order. On the other hand, it does not belong under Al, because Al has its last electron in a p orbital, and Sc has *its* last electron in a d orbital. The clever solution is to grab the whole group of p elements and drag them to the right to make room for the next 10 elements we are about to name:

H   He   H   He
Li Be ⎡B   C   N   O   F   Ne⎤   Li Be   B   C   N   O   F   Ne
Na Mg ⎢Al Si P   S   Cl Ar⎥   ⟶   Na Mg   Al Si P   S   Cl Ar
K   Ca ⎣   ⎦   K   Ca

Now we can continue:

H   He
Li Be   B   C   N   O   F   Ne
Na Mg   Al Si P   S   Cl Ar
K   Ca Sc Ti V   Cr Mn Fe Co Ni Cu Zn

Now we can add, still in the same row, the next group of p elements in their correct columns:

H   He
Li Be   B   C   N   O   F   Ne
Na Mg   Al Si P   S   Cl Ar
K   Ca Sc Ti V   Cr Mn Fe Co Ni Cu Zn Ga Ge As Se Br Kr

Finally we get to start a new row, as we add the 5s, 4d, and 5p electrons:

```
H  He
Li Be                                                    B  C  N  O  F  Ne
Na Mg                                                   Al Si P  S  Cl Ar
K  Ca Sc Ti V  Cr Mn Fe Co Ni Cu Zn Ga Ge As Se Br Kr
Rb Sr Y  Zr Nb Mo Tc  Ru Rh Pd Ag Cd In Sn Sb Te I  Xe
```

And again:

```
H  He
Li Be                                                    B  C  N  O  F  Ne
Na Mg                                                   Al Si P  S  Cl Ar
K  Ca Sc Ti V  Cr Mn Fe Co Ni Cu Zn Ga Ge As Se Br Kr
Rb Sr Y  Zr Nb Mo Tc  Ru Rh Pd Ag Cd In Sn Sb Te I  Xe
Cs Ba
```

Once again, trouble. The next element, La, should be next to Ba, but not under Y, because this one is now adding f electrons. So we'll play the same trick again:

```
H  He
Li Be                                                   B  C  N  O  F  Ne
Na Mg                                                  Al Si P  S  Cl Ar
K  Ca Sc Ti V  Cr Mn Fe Co Ni Cu Zn Ga Ge As Se Br Kr  ⟶
Rb Sr Y  Zr Nb Mo Tc  Ru Rh Pd Ag Cd In Sn Sb Te I  Xe
Cs Ba
```

```
H  He
Li Be                                                                        B  C  N  O  F  Ne
Na Mg                                                                       Al Si P  S  Cl Ar
K  Ca                   Sc Ti V  Cr Mn Fe Co Ni Cu Zn Ga Ge As Se Br Kr
Rb Sr                   Y  Zr Nb Mo Tc  Ru Rh Pd Ag Cd In Sn Sb Te I  Xe
Cs Ba
```

This should make room for the next 14 elements:

```
H  He
Li Be                                                                        B  C  N  O  F  Ne
Na Mg                                                                       Al Si P  S  Cl Ar
K  Ca                   Sc Ti V  Cr Mn Fe Co Ni Cu Zn Ga Ge As Se Br Kr
Rb Sr                   Y  Zr Nb Mo Tc  Ru Rh Pd Ag Cd In Sn Sb Te I  Xe
Cs Ba La Ce Pr Nd Pm Sm Eu Gd Tb Dy Ho Er Tm Yb
```

Hmm. This is what a periodic table ought to look like, but it is too long and skinny to be useful. It won't fit on a normal page with reasonable type size. So let's make a cosmetic change: let's take this last group that we put in and pull it out from the body of the chart, *recognizing that it really belongs here where we put it!!* To remind us where it belongs, we'll leave a gap in the chart, and put in asterisks. Like this:

```
H  He
Li Be                                                                        B  C  N  O  F  Ne
Na Mg                                                                       Al Si P  S  Cl Ar
K  Ca                   Sc Ti V  Cr Mn Fe Co Ni Cu Zn Ga Ge As Se Br Kr
Rb Sr                   Y  Zr Nb Mo Tc  Ru Rh Pd Ag Cd In Sn Sb Te I  Xe
Cs Ba La Ce Pr Nd Pm Sm Eu Gd Tb ....
```

```
H  He
Li Be
Na Mg                    ←          Sc Ti V Cr Mn Fe Co Ni Cu Zn  Ga Ge As Se Br Kr
K  Ca                                Y  Zr Nb Mo Tc Ru Rh Pd Ag Cd  In Sn Sb Te I  Xe
Rb Sr
Cs Ba
```

```
                                                                    B  C  N  O  F  Ne
                                                                    Al Si P  S  Cl Ar
```

La Ce Pr Nd Pm Sm Eu Gd Tb Dy Ho Er Tm Yb

```
H  He
Li Be                                          B  C  N  O  F  Ne
Na Mg                                          Al Si P  S  Cl Ar
K  Ca    Sc Ti V Cr Mn Fe Co Ni Cu Zn  Ga Ge As Se Br Kr
Rb Sr    Y  Zr Nb Mo Tc Ru Rh Pd Ag Cd  In Sn Sb Te I  Xe
Cs Ba *
```

      * La Ce Pr Nd Pm Sm Eu Gd Tb Dy Ho Er Tm Yb

Now we can add the next 10 elements, filling the 5d orbitals:

```
H  He
Li Be                                          B  C  N  O  F  Ne
Na Mg                                          Al Si P  S  Cl Ar
K  Ca    Sc Ti V Cr Mn Fe Co Ni Cu Zn  Ga Ge As Se Br Kr
Rb Sr    Y  Zr Nb Mo Tc Ru Rh Pd Ag Cd  In Sn Sb Te I  Xe
Cs Ba *  Lu Hf Ta W Re Os Ir Pt Au Hg
```

        * La Ce Pr Nd Pm Sm Eu Gd Tb Dy Ho Er Tm Yb

And then the next 6:

```
H  He
Li Be                                          B  C  N  O  F  Ne
Na Mg                                          Al Si P  S  Cl Ar
K  Ca    Sc Ti V Cr Mn Fe Co Ni Cu Zn  Ga Ge As Se Br Kr
Rb Sr    Y  Zr Nb Mo Tc Ru Rh Pd Ag Cd  In Sn Sb Te I  Xe
Cs Ba *  Lu Hf Ta W Re Os Ir Pt Au Hg  Tl Pb Bi Po At Rn
```

        * La Ce Pr Nd Pm Sm Eu Gd Tb Dy Ho Er Tm Yb

Now (at last!) on to the next level:

```
H  He
Li Be                                          B  C  N  O  F  Ne
Na Mg                                          Al Si P  S  Cl Ar
K  Ca    Sc Ti V Cr Mn Fe Co Ni Cu Zn  Ga Ge As Se Br Kr
Rb Sr    Y  Zr Nb Mo Tc Ru Rh Pd Ag Cd  In Sn Sb Te I  Xe
Cs Ba *  Lu Hf Ta W Re Os Ir Pt Au Hg  Tl Pb Bi Po At Rn
Fr Ra
```

        * La Ce Pr Nd Pm Sm Eu Gd Tb Dy Ho Er Tm Yb

Where does the next one go? It should be to the right of Ra. Remember, the one to the right of Ba is not Lu, but really La, in the group that we split out to the bottom. The next element, Ac, belongs right under La:

```
H   He
Li  Be                                              B  C  N  O  F  Ne
Na  Mg                                             Al Si  P  S  Cl Ar
K   Ca    Sc Ti  V  Cr Mn Fe Co Ni Cu Zn Ga Ge As Se Br Kr
Rb  Sr    Y  Zr Nb Mo Tc  Ru Rh Pd Ag Cd In Sn Sb Te  I  Xe
Cs  Ba *  Lu Hf Ta  W  Re Os Ir  Pt Au Hg Tl Pb Bi Po At Rn
Fr  Ra **
```

```
 * La Ce Pr Nd Pm Sm Eu Gd Tb Dy Ho Er Tm Yb
** Ac Th Pa  U  Np Pu Am Cm Bk Cf Es  Fm Md No
```

The next element is the first one filling the 6d orbitals, so it belongs with the other elements filling d orbitals.

```
H   He
Li  Be                                              B  C  N  O  F  Ne
Na  Mg                                             Al Si  P  S  Cl Ar
K   Ca    Sc Ti  V  Cr Mn Fe Co Ni Cu Zn Ga Ge As Se Br Kr
Rb  Sr    Y  Zr Nb Mo Tc  Ru Rh Pd Ag Cd In Sn Sb Te  I  Xe
Cs  Ba *  Lu Hf Ta  W  Re Os Ir  Pt Au Hg Tl Pb Bi Po At Rn
Fr  Ra ** Lr Rf Db Sg Bh Hs Mt
```

```
 * La Ce Pr Nd Pm Sm Eu Gd Tb Dy Ho Er Tm Yb
** Ac Th Pa  U  Np Pu Am Cm Bk Cf Es Fm Md No
```

And we have now run out of elements. So here is a complete list of chemical elements arranged in an entirely logical array. By looking at an element's place in the periodic table, you can tell what its electron arrangement is. (Advanced students may know that a few of the elements do not quite follow the pattern I have laid out, but we need not worry about that in this course.) There is, however, one change we still need to make before this table looks like the one in most books: the second element, helium, is in its proper place, in the same column as all the other elements with 2s electrons in their outer shell. But unlike all the others in this column, helium also has a *full* shell, making it extra stable (remember our last rule?) and much more like the far right-hand column in its chemical properties. Thus, most tables place helium in the top right corner (a few put it in both places!). So a more typical periodic table looks like the one that follows, and that is the one we will use.

H																	He
Li	Be											B	C	N	O	F	Ne
Na	Mg											Al	Si	P	S	Cl	Ar
K	Ca		Sc	Ti	V	Cr	Mn	Fe	Co	Ni	Cu	Zn	Ga	Ge	As	Se	Br	Kr
Rb	Sr		Y	Zr	Nb	Mo	Tc	Ru	Rh	Pd	Ag	Cd	In	Sn	Sb	Te	I	Xe
Cs	Ba	*	Lu	Hf	Ta	W	Re	Os	Ir	Pt	Au	Hg	Tl	Pb	Bi	Po	At	Rn
Fr	Ra	**	Lr	Rf	Db	Sg	Bh	Hs	Mt								

```
 * La Ce Pr Nd Pm Sm Eu Gd Tb Dy Ho Er Tm Yb
** Ac Th Pa  U Np Pu Am Cm Bk Cf Es Fm Md No
```

The same reasoning we just applied suggests that hydrogen is also special. True, like the others in its column it has one s electron in its outer shell, but unlike the others in its column, it is one short of a full shell. Thus you may see occasional periodic tables with hydrogen placed somewhere over the middle to remind you of its ambivalent nature. I will hope that you'll remember that without such an explicit reminder.

Problems

(Answers are provided at the end of the chapter for italicized problems.)

1. Which element has the same number of 4d electrons as 2p electrons?

2. Which element has the same number of electrons in three successive orbitals?

3. Which element has the same number of d electrons as s electrons?

4. Which element has twice as many p electrons as d electrons?

5. Which elements have the same number of p electrons as d electrons?

6. How does the model help to explain why the elements of a particular <u>column</u> might have some properties in common?

7. Give an example of a property that elements of the far right-hand column should have in common.

8. a. How many protons, neutrons, and electrons are there in $^{211}_{83}\text{Bi}^{+2}$?

   b. Show the electron occupancy ($1s^2$, etc.) of the species in part a.

   c. Name two elements from among the first 18 elements that have the same electron occupancy in their outer shell as the above species.

9. a. Write out the complete electron configuration for element 82.

   b. What second-row element does element 82 resemble in terms of its outermost electron shell?

10. *What number element will be the first to have a g electron?*

11. *How many total electrons will level number 8 hold?*

12. What is the evidence for building an atomic model where the energy needed to "boost" an electron from one energy level to the next higher energy level isn't the same from level to level (i.e., the energy difference isn't constant)?

13. What is the evidence that electrons cannot be "kicked" by energy of any possible value (i.e., the absorption of energy is quantized)?

14. How much energy (in Joules, J) is there in a photon of radiation from a radio station broadcasting at 92.3 megacycles/sec? Planck's constant is 6.63 x 10^{-34} J sec.

15. How much energy is there in a photon of radiation from a tanning bed, at a wavelength of 350 nm?

16. *It takes 4.184 Joules to raise the temperature of 1 mL of water by 1 °C (thus, 4.184 J/mL°C). A typical source of irradiation arrives at a target at the rate of about 10^{20} photons/sec. How long would you have to irradiate 10 mL of water with the sources in problems 14 and 15 in order to raise the temperature by 10 °C? Assume perfect efficiency and no dissipation of heat.*

17. What is the wavelength in *yards* of electromagnetic radiation with an energy equivalent of 7.29×10^{-27} J?

18. In a hydrogen atom, in which energy levels are described by the formula $E_n = -B/n^2$, which transition involves a longer wavelength of light: 8----->2 or 3----->1? Explain. You should be able to do this both with and without calculations.

19. Given that the 3--->2 transition for a hydrogen atom is calculated to occur at 656.280 nm, calculate the wavelength for the 5--->3 transition. Planck's constant (h) is 6.63 x 10^{-34} J sec.

20. *In order to generate light of wavelength 1094 nm, what transition must occur in a hydrogen atom? This will likely require some trial and error.*

21. a. Determine a value for B starting from the fact that the $4 \rightarrow 2$ energy transition for hydrogen emits light with a wavelength of 486.1 nm.

    b. At what frequency and wavelength would emission from the corresponding $12 \rightarrow 11$ transition occur?

Selected Answers

Internal Problems

Test Yourself 1:

Experience tells me that a lot of students have a great deal of trouble with this question, even though, when you get down to it, it involves no more than simple algebra. The problem is that it is a two-part, back-and-forth problem, and it involves some weird concepts.

The first thing we have to recognize is that the energy *levels*, $-B/n^2$, do not themselves represent *amounts* of energy. They are merely labels for points on an energy scale, much as 2 o'clock is a label for a point on the time scale. So when we say that the second step has energy $-B/4$, we cannot relate this to a quantity of energy any more than 2:00 represents an amount of time. You do not have an *amount* of energy until you consider what it takes to get from one level to another. Then you subtract the energy label for one state from the energy label for the other state to get an amount of energy that measures the difference between the two states. Take an example: You drive 60 miles between 3:00 and 5:00. What is your average speed? Speed is distance over time, right? So do you take 60/3:00 and subtract 60/5:00 to get a speed? Of course not, because even though 3:00 is a time, it is not an *amount* of time, right? So first you subtract 3:00 from 5:00 to get an *amount* of time, 2 hours, and divide your distance by this amount of time to get an average speed, 30 mph. In the same sense, you do not have an *amount* of energy until you subtract one energy level from another.

The other problem is that we have two equations involved:

$$\Delta E = -B/n_1^2 - (-B/n_2^2), \text{ and}$$

$$E = hc/\lambda$$

where the E in the second equation *is* an amount of energy and has the same value as the ΔE in the first equation. (Notice that to follow this, you have to really understand what these equations are telling you and not just manipulate letters as you may have done in the past.) So let's pick a line, say the 3→2 line at 656 nm, and work with it.

First, how much energy are we talking about? The light has a wavelength of 656 nm, so we can use $E = hc/\lambda$ to convert this into an energy. I gave you λ, and h and c are constants (values on pages 20 and 21), so there is only one variable in this equation. If you keep your units straight (10^9 nm = 1 m), you should get 3.03×10^{-19} J. Where did this energy come from? An electron dropped from the third to the second level of a hydrogen atom. It has less energy than it had before, and the extra was released as this photon of light. How much energy did it lose? The energy difference between the third and second energy levels!

$$\Delta E = -B/3^2 - (-B/2^2) = -B/9 - (-B/4) = 0.139B$$

as we calculated before. (Note that if you did the levels in reverse you would get the same number but negative. Since we are interested in the *amount* of the energy difference, we can take the absolute value and come up with the same number in either case. The sign can be interpreted as indicating whether the light is going in or coming out, but we already know *that* by whether we are going from 3→2 or from 2→3, so the sign has no use for us.) Now this amount of energy, $0.139B$, corresponds to the energy in the photon of light involved, 3.03×10^{-19} J. We can set these equal and solve for B and get its value: 2.18×10^{-18} J. This is the energy difference between the lowest energy level, level 1, and the highest, which in practical terms means the electron is gone; in other words, it's the amount of energy it would take to completely remove an electron from level 1.

Now what about the 4→9 line? This energy difference is

$$-B/9^2 - (-B/4^2) = 0.05B$$

This time we know what B is, so we convert this into an actual amount of energy, and then use $E = hc/\lambda$, where λ is our only variable this time, to convert that to a new wavelength. Do it now. You should get 1825 nm.

Those of you who are good at math will realize that we made a lot of unnecessary calculations in the above approach. I do it this way because it demands that you think about the various energy differences and amounts of energy. If you want to save some time, and do not mind mathematical manipulation, check this out:

$$\Delta E = -B/n_1^2 - (-B/n_2^2), \text{ and}$$

$$E = hc/\lambda, \text{ therefore}$$

$$hc/\lambda = B/n_2^2 - B/n_1^2$$

$$\text{or } hc/B = \lambda\,(1/n_2^2 - 1/n_1^2)$$

where n_1 and n_2 are the two energy levels involved. Since h, c, and B do not change, $\lambda\,(1/n_2^2 - 1/n_1^2)$ is a constant. Thus you can insert a λ and its corresponding n's and set this equal to a new λ and *its* corresponding n's. A much simpler math problem, but one that obscures the relationship of the energies developed above. [Note that if you put the n's in reverse order, your answer would show a negative wavelength, which of course is impossible. Just remove the minus sign because you recognize that you could legitimately have reversed the n's to make it come out positive.] Try this, and make sure you get the same number for the wavelength of the 4→9 jump.

End of Chapter Problems

10. 121

11. 128

16. First calculate how many Joules it will take:

$$4.184 \;\frac{J}{mL\,{}^{\circ}C} \;\times\; 10 \text{ mL} \;\times\; 10\,{}^{\circ}C \;=\; 418.4 \text{ J}$$

Then how many photons it would take to deliver that many Joules. For the radio waves,

$$418.4 \text{ J} \;\frac{photon}{6.11 \times 10^{-26} \text{ J}} \;=\; 6.85 \times 10^{27} \text{ photons}$$

Then turn this into time: about 2.17 years for the radio waves; about 7.4 seconds for the tanning bed

20. 6→3

Chapter 3
Bonding

Now that we understand how the periodic table is put together, let's take a look at some of the conclusions we can draw from it. In this chapter we will

- Establish the concept of bonding as a means for atoms to achieve stability

- Talk about the various ways organic molecules can be put together

- Discuss charges within molecules

- Learn the various ways chemists draw pictures of molecules

- Learn the functional groups.

IONIC COMPOUNDS

Start with the far right column of the periodic table. All these elements have full electron shells, as we have defined them (full s and p levels), meaning that they are especially stable. They are "happy," they do not "want" to do anything, they should be inert. And so they are. It is extremely difficult to get these elements to react with anything; indeed, to a first approximation, we can declare that they do not. So they are *essentially* inert. This column of the table is often referred to as the "inert gases." They are more often known as the "noble gases," probably meaning that they are aloof and unwilling to interact with others. This may be an unfair slam on nobility, but it is nevertheless a common name for the inert gases. Having a full shell, being like an inert gas, is the ultimate aspiration of all elements. All the other elements want to be like these (maybe *that's* why they are called noble!).

Now look at another column, the second-to-the-last column on the right. These elements are called the "halogens," and like the inert gases they also share a common characteristic: they are all one electron shy of having a full shell. Because having a full shell is the ultimate goal of all elements, it should come as no surprise that the halogens are all especially eager to acquire one electron. It may not be obvious yet how they might go about this, but it should not surprise you that they all behave in a similar way: whatever way one thinks of to gain an electron, the others will do it too.

> Of course, it was the similarity of properties of elements in the same "family" or column that led to construction of the periodic table in the first place, and only later did chemists come up with the theory of electron levels and orbitals. Mendeleev constructed the periodic table with no clue as to why properties should repeat the way they did, but it was clear to him that chlorine and bromine shared so many properties that there had to be some relationship between them.

Consider the column on the far left, the "alkali metals." These all have one more electron than would be desirable—if they had one less, they would have a full outer shell (that of the previous level). So all these elements are eager to lose an electron. Hmmm…the halogens are eager to gain an electron. What would happen if you mixed a sample of, say, chlorine (Cl) with, say, sodium (Na)? It does not take a rocket scientist, or even a chemist, to imagine that an electron might jump from the sodium, which wants to get rid of it, to the chlorine, which wants an extra one. What would we have then? The sodium atom would still have its 11 protons but now only 10 electrons. This leaves the atom with a positive charge, so that it technically should no longer be called a sodium *atom*, but rather a sodium *ion*, Na^+. Meanwhile the chlorine, still with 17 protons, now has 18 electrons, giving it a minus charge. This is called chlor**ide** ion (not chlor**ine** ion), Cl^-. These two ions are now attracted to each other by electrical charge, forming the substance *sodium chloride*, more commonly known as table salt. The attraction between oppositely charged ions is sometimes referred to as an *ionic bond*, but this is really a bad term that should be avoided, because a "bond" is usually expected to imply some specificity, as in "this atom is attached to that atom." An ionic interaction is not an attraction to a specific other ion but rather to all oppositely charged ions in the vicinity, so the term bond is a poor choice of words.

What else can we say about these ions? The sodium ion now has 11 protons attracting 10 electrons. The protons should be able to do a better job of attracting 10 electrons than 11: in other words, the electrons in the sodium *ion* should be held closer to the nucleus than was the case for the sodium *atom* with its 11 electrons. A sodium ion should be smaller than a sodium atom. This is indeed the case, although I will not try to explain how we prove it.

Similarly, the chloride *ion*, with only 17 protons to attract 18 electrons, should be *larger* than a chlorine atom, and this too is true.

Let's make some more predictions. Consider what might happen when you mix lithium with sulfur. Looking at the periodic table, you can see that Li (like Na) is eager to get rid of one electron. S, on the other hand, is *two* electrons shy of having a full shell. Again, it does not take a great mind to suggest that perhaps two lithium atoms could give up one electron each, while one sulfur atom accepts both, leading to a substance with the formula Li_2S.

Test Yourself 1: Make your own prediction for the combination of calcium (Ca) and fluorine (F).

Valence Electrons

It should be clear from these few examples that what matters is not how many electrons an atom has, but how many of them are in the outer shell. This is where behavior is determined. If the atom has just one or two electrons past a full shell (i.e., one or two in its new outer shell), it will try to get rid of them. If it is very close to filling a shell, it will try to gain some. **The outer shell, where all the business of chemistry happens, is called the *valence shell*.** You should also note the pattern that, in general, **atoms are trying to <u>fill</u> their valence shell, which in most cases means 8 electrons, thus the so-called *octet rule*.** (Hydrogen is an exception to this, of course, since its valence shell can hold only two electrons.) Finally, you should recognize that d and f orbitals are always a level or two below the valence shell, so (at least for our purposes, this will change when you get to inorganic chemistry) they do not have much effect on what goes on.

COVALENT BONDS

Not all partnerships are as easy to predict as the ones we have just been discussing. Sure, if one atom wants more electrons and another wants less, it is easy to predict that electrons will move. But what if both atoms want the same thing? Consider the case of two fluorine atoms encountering each other. Each wants to gain another electron, but neither wants to give one up. Is this a stalemate? Not at all. Atoms, like people, have learned how to share. The two fluorine atoms agree to a deal: each has one electron that would like a partner. If they each leave their lone electron in the vicinity of the other guy's lone electron, then it will seem as if each atom has access to both. This can be symbolized by the following picture:

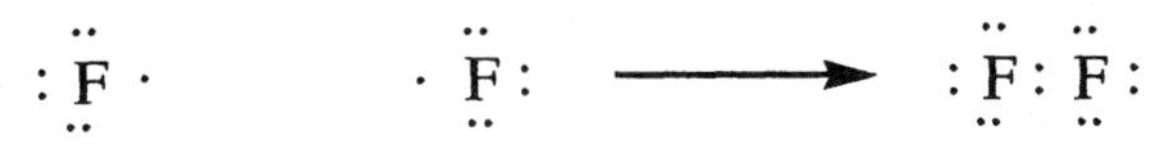

$$: \overset{\cdot\cdot}{\underset{\cdot\cdot}{F}} \cdot \qquad \cdot \overset{\cdot\cdot}{\underset{\cdot\cdot}{F}} : \qquad \longrightarrow \qquad : \overset{\cdot\cdot}{\underset{\cdot\cdot}{F}} : \overset{\cdot\cdot}{\underset{\cdot\cdot}{F}} :$$

Lewis Structures

The resultant partnership, a *molecule*, is two atoms of fluorine held together by a *bond*, in this case a *covalent bond*, meaning a sharing of electrons. It can be depicted in any of several ways:

$$: \overset{\cdot\cdot}{\underset{\cdot\cdot}{F}} : \overset{\cdot\cdot}{\underset{\cdot\cdot}{F}} : \qquad\qquad F : F \qquad\qquad F\!-\!F \qquad\qquad : \overset{\cdot\cdot}{\underset{\cdot\cdot}{F}}\!-\!\overset{\cdot\cdot}{\underset{\cdot\cdot}{F}} :$$

The first picture accounts for all the valence electrons. The second picture leaves off (the reader assumes them) all the valence electrons that are not shared ("unshared pairs", or "lone pairs"). The third picture replaces the two shared electrons with a line, a convention we will use throughout this book. The fourth is a combination of the first and the third. All of these are called *Lewis structures*, after Gilbert Lewis, who is responsible for many of our current notions of bonding. Although experienced chemists usually do not represent the lone pairs unless they are using them for some purpose, at this stage in your career it is probably a good idea to show them explicitly, at least for a while.

Notice how each fluorine atom can look around and see eight electrons. The fluorine atoms have seemed to gain an electron by sharing a pair with a neighbor. Every instance of sharing brings the appearance of having an extra electron. With that in mind, what would happen if you mixed some chlorine and some oxygen? Let's see: chlorine wants to gain one electron, and oxygen wants to gain two. Clearly no electron transfer will help the situation any: this calls for sharing. If each instance of sharing results in the impression of one extra electron, then chlorine wants to share once, and oxygen wants to share twice. This is easily accomplished by having the oxygen atom pair up with two different chlorine partners, sharing one electron with each:

$$:\overset{\cdot\cdot}{\underset{\cdot\cdot}{Cl}}\cdot \qquad \cdot \overset{\cdot\cdot}{\underset{\cdot\cdot}{O}}\cdot \qquad \cdot\overset{\cdot\cdot}{\underset{\cdot\cdot}{Cl}}: \qquad \longrightarrow \qquad :\overset{\cdot\cdot}{\underset{\cdot\cdot}{Cl}}\!-\!\overset{\cdot\cdot}{\underset{\cdot\cdot}{O}}\!-\!\overset{\cdot\cdot}{\underset{\cdot\cdot}{Cl}}:$$

Multiple Bonds

OK, but what if there were no chlorine around? What would oxygen do, all by itself? Well, it wants to share twice, so one possibility is to share twice with the same partner:

$$\overset{\cdot\cdot}{\underset{\cdot\cdot}{O}}: \qquad :\overset{\cdot\cdot}{O} \qquad \longrightarrow \qquad \overset{\cdot\cdot}{\underset{\cdot\cdot}{O}}::\overset{\cdot\cdot}{O} \quad or \quad \overset{\cdot\cdot}{\underset{\cdot\cdot}{O}}\!=\!\overset{\cdot\cdot}{\underset{\cdot\cdot}{O}}$$

This is what we call a double bond. Notice how each oxygen atom still seems to be surrounded by eight electrons. (Some of you may know that oxygen does not really look like this, but we will ignore that for the purposes of this course.) It is a small jump from here to suggest that nitrogen atoms can join together by a triple bond. Again, each atom can "see" an octet:

$$:N\!\equiv\!N:$$

Unfortunately (or, actually, fortunately, for otherwise the world would be far less rich in diversity), carbon cannot pair up with another carbon to make a quadruple bond. This will perhaps become clear later, but in any case carbon, in need of four electrons, needs to share all four of its valence electrons with partners, and this almost always involves more than one partner. It can make single, double, and triple bonds, and with always at least one site left over. Carbon thus finds itself involved in a huge variety of molecules, bound up with many different kinds of partners. This is one of the reasons why organic chemistry, the chemistry of carbon compounds, is so fascinating.

Electronegativity

We have seen situations in which an electron completely jumped from one atom to another, and other situations in which both atoms retained some ownership of the shared electrons. How does one know which will occur? The bottom line is that you really don't always know for sure, but you can get close by looking at a property called "electronegativity." This is a fairly loose term that refers to an atom's "desire" for electrons. Atoms with a strong desire for electrons have a high electronegativity, and are called electronegative. Atoms with a weak desire for electrons have a low electronegativity and are called electropositive. A scale has been developed that runs from about 0.7 to 4. This is shown below. *Do not memorize the numbers!* All you need to do is look at the trends, and it should be pretty obvious that the low numbers (electropositive elements) tend to be on the left of the periodic table and the high numbers (electronegative elements) tend to be on the right. Further, the numbers increase as you go up the table. So the lowest numbers are at the bottom left, and the highest are at the top right. Notice that the inert gases are missing from this chart, and that hydrogen does not fit the pattern of the elements in its column. (For this reason some people place hydrogen near the middle of the table, closer to other elements with similar electronegativities.)

So what? First of all, the general pattern should not surprise us, since we already know that the atoms on the right are trying to get more electrons and the ones on the left are trying to get rid of them. All this scale allows us to do is to rate elements roughly according to how *strongly* they want to do these things. And, we can look at a particular pair of elements and predict whether they will react with each other. As a rule of thumb, electropositive elements tend to react with electronegative elements but not with each other. Electronegative elements can react with electropositive elements *and also* with each other. Further, when electronegative elements react with electropositive elements, there is usually a transfer of electrons to create an ionic substance, also known as a *salt*. This usually occurs when elements in the first two columns (except hydrogen), that is, columns IA and IIA, react with elements from columns VIA and VIIA. In most other situations, any reaction that occurs involves a sharing of electrons, forming what we call *covalent bonds*. Organic chemistry concerns itself, for the most part, with carbon, hydrogen, nitrogen, oxygen, the halogens, and a smattering of others. Thus, nearly all of organic chemistry involves covalent bonds.

													H 2.1				
IA	**IIA**											**IIIA**	**IVA**	**VA**	**VIA**	**VIIA**	
Li 1.0	Be 1.5											B 2.0	C 2.5	N 3.0	O 3.5	F 4.0	
Na 0.9	Mg 1.2						VIIIB					Al 1.5	Si 1.8	P 2.1	S 2.5	Cl 3.0	
		IIIB	**IVB**	**VB**	**VIB**	**VIIB**				**IB**	**IIB**						
K 0.8	Ca 1.0	Sc 1.3	Ti 1.5	V 1.6	Cr 1.6	Mn 1.5	Fe 1.8	Co 1.8	Ni 1.8	Cu 1.9	Zn 1.6	Ga 1.6	Ge 1.8	As 2.0	Se 2.4	Br 2.8	
Rb 0.8	Sr 1.0	Y 1.2	Zr 1.4	Nb 1.6	Mo 1.8	Tc 1.9	Ru 2.2	Rh 2.2	Pd 2.2	Ag 1.9	Cd 1.7	In 1.7	Sn 1.8	Sb 1.9	Te 2.1	I 2.5	
Cs 0.7	Ba 0.9	La–Lu 1.1–1.2	Hf 1.3	Ta 1.5	W 1.7	Re 1.9	Os 2.2	Ir 2.2	Pt 2.2	Au 2.4	Hg 1.9	Tl 1.8	Pb 1.8	Bi 1.9	Po 2.0	At 2.2	
Fr 0.7	Ra 0.9	Ac–No 1.1–1.7															

Organic Structures

Since covalent bonds are partnerships between atoms, one can legitimately draw pictures of molecules identifying which atoms are attached to which other atoms. Organic molecules are covalent, and you need to get used to drawing pictures that show in various ways what is attached to what. You also need to get used to the kinds of shorthand that chemists use. In drawing organic structures, it is important to remember a few rules of thumb: under normal circumstances, carbon will form 4 bonds, nitrogen 3, oxygen 2, and the halogens and hydrogen 1. This is not to say that situations cannot exist where carbon has three bonds, or oxygen has one bond or three bonds, but when these situations arise it is a red flag that something unusual is up.

After a while it won't be necessary, but while you are starting out it will be helpful when creating structures to count electrons. Make sure your structure accounts for *all* the valence electrons the various partners bring to the molecule, plus or minus as many as are necessary to account for any charge present.

Try some examples. Start with a structure for the formula CH_4.

First we have to understand what the formula CH_4 means. It does NOT mean that we should write C–H–H–H–H. It is simply an expression of what is present in the molecule. It is our job to put the atoms together in a sensible way. The formula means we have one carbon atom and four hydrogen atoms to work with. That is what is written. But there is more. It also means that there are eight electrons to work with. How do I know that? Because carbon has four (valence) electrons and each hydrogen has one. So now our job is to put together one carbon, four hydrogens, and eight electrons in a sensible way. It is usually a good idea to start in the middle and build out; it should also be obvious that since hydrogen can have only two electrons and therefore one bond, it can never be in the middle. Every time you write down a hydrogen you have ended the molecule in that particular direction. So let's start with the carbon in the middle, and give it four bonds to hydrogens:

$$
\begin{array}{c}
\text{H} \\
| \\
\text{H} - \text{C} - \text{H} \\
| \\
\text{H}
\end{array}
$$

Lo and behold, this requires eight electrons, two for each bond. We are done. Each atom has what it wants, a full shell. This means an octet for most atoms, and a "duet" for each hydrogen. What is more, there is no other sensible way to do it.

Try H_2O. Again, eight electrons, six from the oxygen and one from each hydrogen. Again, put the O in the middle and attach the H's:

$$
\begin{array}{c}
\text{H} - \text{O} \\
| \\
\text{H}
\end{array}
$$

There are two points to make here. First, I'll bet you didn't do it that way. (Actually, I fear you didn't do it at all. Remember, you don't learn to play the piano by watching a pianist. YOU have to practice! You must always be writing these things down, doing them for yourselves, preferably before I do them for you.) Anyway, if you had written this molecule, you probably would have written it this way:

$$
\text{H} - \text{O} - \text{H}
$$

Does it matter? Not yet. Later (next chapter) we will begin to consider the actual shapes of molecules. For now, let us focus only on *connectivity*: what is attached to what. Both pictures show an oxygen attached to two hydrogens. So let's agree that the position on the page is irrelevant for now; what matters is the lines drawn between the atoms.

The second point is that we are not done yet. We had 8 electrons to account for, and there are only 4 indicated in the picture we drew. As we progress through this book, we will become less rigorous about this, but for now it is important that you keep track of *all* the valence electrons. We do this with dots:

$$H-\overset{\bullet\bullet}{\underset{\bullet\bullet}{O}}-H$$

Note that some books do the same thing with lines instead of dots, so you might see this picture:

$$H-\overline{\underline{O}}-H$$, where each line implies *two* dots.

The intended meaning is the same. I personally don't like this convention, because it is sometimes hard to distinguish a line for electrons from a minus-charge sign, but you will probably see it in some books, and you ought to be aware of what it means.

As always, when we finish we should check our work. Does each hydrogen have two and only two electrons? Does each larger atom have eight? OK, now we are done.

Isomers

Try another: C_2H_6O. Now we have more atoms and more electrons to work with. How many electrons? Count them! Then try to put the molecule together in a sensible way.

It turns out there are two ways you could do this. There are many ways you could have written these two molecules, but if you look carefully, they should all be equivalent to one of these two:

What does this tell us? There are two perfectly acceptable molecules with the same formula. **Molecules with the same atoms but connected differently are called isomers.** The kind of isomers we are discussing here, where atoms have different partners, that is, they differ in their connectivity, are called *constitutional* (or *structural*) isomers. There are also other kinds of isomers, where the partners are the same but the geometry is different—so-called geometric or stereoisomers—but that will come later. There are two constitutional isomers with the formula C_2H_6O. They are different substances, and a few months from now you will be able to rattle off a bunch of ways in which they are different. For now, you should recognize that a molecular formula of the type C_2H_6O is relatively useless to a chemist, because it does not sufficiently specify which molecule one is referring to. Of course, it should not surprise you to learn that as we get to bigger and bigger molecules, the ambiguity gets worse. For example, the structure $C_{10}H_{22}$ has 75 different isomers; with 25 carbons there are over 35 million!

Try C_5H_{12}. See how many different ways you can put this together. Do it now, before you continue. It turns out there are only three. I'll bet you came up with more. How come? It is difficult for beginners to see that apparently different pictures are really the same. For example, here are two structures you might have written.

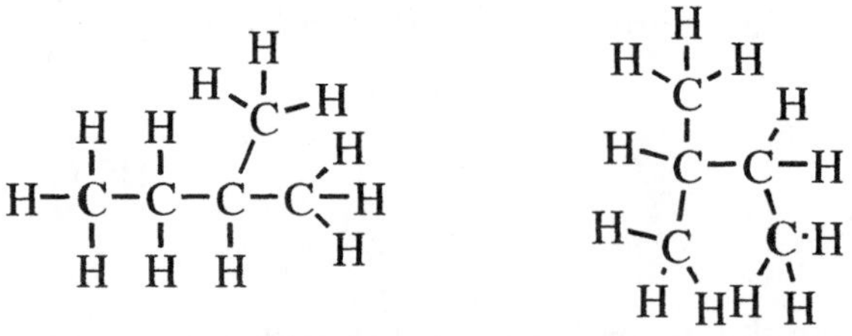

Can you tell that these are the same thing? It's not easy, I know. Imagine that the lines (bonds) are strings, and imagine picking each of these up by the bold carbon and letting the rest of the molecule dangle. Aren't you holding the same thing in each hand? Alternatively, talk your way through the molecule. Start with the bold carbon: "Carbon attached to three H's and one other carbon. Move to that carbon. In addition to the one we just came from, it is attached to two H's and one other C. Move to that one. It is attached to one H and two other C's, each of which has three H's." Can you say anything at all about one of these pictures that is not also true of the other?

Try another, NH₃O. Do it!

Formal Charges

Look more closely at that last example, NH_3O. You probably put it together with two hydrogens on the nitrogen and one on the oxygen. Everyone has an octet, everyone is happy. But there is another way we could have put this together, using the same number of electrons:

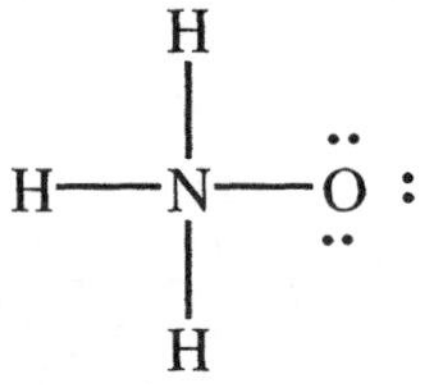

To make this structure, we used 14 electrons: one from each hydrogen, 5 from the nitrogen, and 6 from the oxygen, just as in the previous structure. Everyone has an octet, so everyone should be happy. But wait: there is a red flag lurking here. Nitrogen normally takes three bonds, and here it has four. Oxygen usually takes two bonds, and here it has one. What are the consequences of this? Up to now we have been counting all the electrons surrounding an atom to discover whether it "sees" eight of them, an octet. This is important and legitimate. But there is another form of bookkeeping that is also important, and that is to discover how many, of the electrons that an atom "sees," it also "owns." To determine *ownership*, we arbitrarily decide that all electrons shared are split between the partners; all electrons not shared are completely owned by the atom where they reside. So let's look at the structure again in this new way.

Nitrogen sees eight electrons, but how many does it own? One of the north ones, one of the south ones, one of the east ones, and one of the west ones. Thus four. But ordinary nitrogen has five valence electrons. Thus from the point of view of ownership, nitrogen is missing one, and has a positive charge. We call this a "formal charge" and indicate it with a + sign, often in a circle, near the nitrogen. Meanwhile, the oxygen owns one of the east electrons and all six of the others around it, thus seven. Ordinarily oxygen has six valence electrons. Thus this oxygen formally has a negative charge, which we indicate with a – sign in a circle. So a more accurate picture of this molecule would look like this:

$$
\begin{array}{c}
H \\
| \\
H-\overset{\oplus}{N}-\overset{..}{\underset{..}{O}}:\hspace{-2pt}\ominus \\
| \\
H
\end{array}
$$

Keeping track of charge is going to become extremely important for us. Knowing which atoms have charges, and what they can do to lose those charges, will help us make sense of the vast majority of the reactions we will encounter this year. Don't blow this off as a curiosity: you need to become very comfortable with assigning charges.

What effect do these charges have on a molecule? Does the presence of charges mean the molecule is charged? Does it mean the molecule can't exist? Does it mean it would rather be put together another way? The answers are yes and no. A molecule's charge is the sum of all formal charges. The molecule above, as a whole, is not charged, even though we can see *places* in it that are. Some molecules are charged (we call them ions then), and formal charges let us designate where in the molecule the charge is located. Molecules with charges, and neutral molecules with formal charges, DO exist. They are sometimes less stable than isomers without the charges, but not always, and even if they are, they can't always figure out how to change from one into the other. As a general rule, you should try to avoid distributing electrons in such a way as to create charges; if the molecule can be connected in the same way without charges, it is usually better. But if the connections of the molecule demand a charge, all you can do is to note that it is there.

COMMON MISTAKES: DON'T LET THIS HAPPEN TO YOU!

Many students will write structures like the following for C_2H_6O, for example:

$$
\begin{array}{ccc}
H & H & H \\
| & | & | \\
H-\underset{..}{\overset{}{C}}-\underset{|}{C}-\underset{..}{\overset{}{O}}-H \\
& H &
\end{array}
$$

Although it is true that the right number of atoms and electrons has been used, and each atom has its octet, there are also some problems. Remember, under ordinary circumstances, carbon has 4 bonds, nitrogen has 3, oxygen has 2, and the halogens and hydrogen 1. Here, there is a carbon atom with only three bonds, and an oxygen atom with three. This is a clear indication that something is amiss. The carbon on the left owns five electrons (one from each bond and both of the indicated electrons), but is supposed to own only four: it has a negative charge. The oxygen owns 5 electrons but is supposed to own 6, so it has a positive charge. At a minimum, these charges should be indicated: a structure with charges is incorrect if the charges are not shown. Even with the charges shown, this would be a bad structure. Although the structure violates no serious rules, the fact is that carbon atoms with less than 4 bonds are very unhappy campers, and molecules with atoms like that (although they *do* exist) have extremely short lifetimes, and should not be written as stable structures. As we will discover later, most of the reactions of organic chemistry can be understood by the desire of carbon atoms like this to relieve their pain. **Make sure your carbons always have four bonds in stable molecules!**

You are probably already getting tired of writing all those C's and H's. Join the club! Chemists have devised a couple of ways to save writing time, and you need to be able to read and write in these new styles. Of course, there is a cost to the time-savings: the new styles remove information, but it is information you can put back. The problem is that many students never learn to see information that is not explicit in the pictures. Don't let this happen to you!

Condensed Structures

The first approach to writing economy is called "condensed structures," in which we write groups of atoms as groups rather than specifying every bond. For example,

$$
\begin{array}{c}
\text{H} \\
| \\
-\text{C}-\text{H} \\
| \\
\text{H}
\end{array}
$$

would be rendered as —CH_3, and everyone would know what it means. Likewise —CH_2— should be obvious. The problem comes in grouping the groups. $CH_3CH_2CH_2CH_2CH_3$ probably gives you no trouble. Although I have seen this rendered as

$$
\begin{array}{ccccc}
\text{H} & \text{H} & \text{H} & \text{H} & \text{H} \\
| & | & | & | & | \\
\text{C}-\text{H}-\text{C}-\text{C}-\text{C}-\text{C}-\text{H} \\
| & | & | & | & | \\
\text{H} & \text{H} & \text{H} & \text{H}
\end{array}
$$

I am sure *you* wouldn't do that! You recognize it instead as

$$
\begin{array}{ccccc}
\text{H} & \text{H} & \text{H} & \text{H} & \text{H} \\
| & | & | & | & | \\
\text{H}-\text{C}-\text{C}-\text{C}-\text{C}-\text{C}-\text{H} \\
| & | & | & | & | \\
\text{H} & \text{H} & \text{H} & \text{H} & \text{H}
\end{array}
$$

Would you still recognize it if it were written $CH_3(CH_2)_3CH_3$? It sometimes is. But that was an easy one. Try this on for size:

$$(CH_3)_2CHCH(CH_3)CH_2CH_3.$$

Let me take you through this, but first, try to figure it out on your own. Remember, it is real: don't write a nonsensical molecule.

COMMON MISTAKES: DON'T LET THIS HAPPEN TO YOU!

Don't look at a structure like this and immediately start to write it down from left to right: CH_3-CH_3-CH-CH-CH_3-CH_2-CH_3. This will get you into all kinds of trouble, not the least of which is the bane of all organic chemistry profs, the dreaded five-bonded carbon. Structures like this have to be thought about, not just diagrammed as written.

Start on the left. It says $(CH_3)_2$. Can this mean CH_3–CH_3? No way, because once you have written that, there is no room for any more bonds, and the rest of the atoms must go somewhere! (Notice that a CH_3 group ALWAYS attaches to one and only one thing.) So the two CH_3 groups cannot be attached to each other, and therefore both must be attached to something else. Look further: the next thing mentioned is CH. Since C always has 4 bonds, and there is only one mentioned here, there must be three others. Two of them must be the two CH_3 groups we have been struggling with, and the third must be the rest of the molecule. So far we have

$$
\begin{array}{c}
CH_3 \\
| \\
H_3C-C- \\
| \\
H
\end{array}
$$

That's a good start. What's next? $CH(CH_3)CH_2CH_3$. Apparently the CH is attached to the stuff over on the left. But what do we do with that new CH_3? We can't place it in the middle of this chain, because CH_3's always have one and only one bond to them, and furthermore, if we put it to the right of the new CH and attempt to put the rest of the molecule to the right of that, we would be leaving the poor CH with too *few* bonds. Apparently this CH_3 is attached to the CH we just put in, but off to the side, while the rest of the chain continues on. That's why it was put in parentheses. Then comes the rest of the molecule, CH_2CH_3, which should be pretty obvious. Altogether, here is what we have:

$$
\begin{array}{ccccc}
H_3C & H & H & H \\
| & | & | & | \\
H_3C-C-&C-&C-&C-H \\
| & | & | & | \\
H & CH_3 & H & H
\end{array}
$$

Did you get that? If so, you're doing well!

Test Yourself 2: Now practice on this one: CH_3COCH_3.

Stick Figures

The second approach to shorthand, which we will use much more often, is stick figures. If I show you a picture, like this, you recognize it to be a picture of a person. You also know that at the end of the lines at the top there are hands, and on the hands there are fingers, even though I haven't drawn them, because, well, because you just know that they have to be there. They always are, unless you are dealing with an unusual specimen of a person who is missing a hand, in which case the picture would make some indication that the hand that your mind fills in is actually missing. Chemists do the same thing with chemical pictures, leaving off the parts that are not necessary

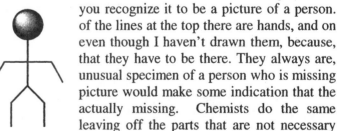

for specifying the object being drawn, recognizing that the reader will fill them in, and also recognizing that if any of these assumed parts are missing, there would have to be a special indication to that effect. The parts we need to specify are the big atoms: the carbons, oxygens, nitrogens, and so on. The parts we can assume are the hydrogens. Then we draw stick figures. The ground rules are that every corner and/or end of a line represents a carbon unless otherwise specified, and that all carbons have four bonds, nitrogens three, oxygens two, etc., unless otherwise noted. It then falls to the reader to fill in the hydrogens necessary for this to be true. For example, the following three pictures are identical. They contain exactly the same information. It's just that each successive rendering requires you to supply a little bit more information on your own.

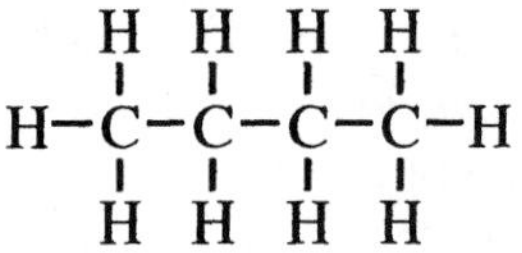

$$H-\overset{\overset{\displaystyle H}{|}}{\underset{\underset{\displaystyle H}{|}}{C}}-\overset{\overset{\displaystyle H}{|}}{\underset{\underset{\displaystyle H}{|}}{C}}-\overset{\overset{\displaystyle H}{|}}{\underset{\underset{\displaystyle H}{|}}{C}}-\overset{\overset{\displaystyle H}{|}}{\underset{\underset{\displaystyle H}{|}}{C}}-H \qquad CH_3CH_2CH_2CH_3 \qquad CH_3(CH_2)_2CH_3 \qquad$$

Below are some more comparisons. Study them to make sure you see what is happening. It will take you a while to get used to this, so for the time being, do not be embarrassed about filling in the blanks. If it helps you see what is going on, feel free to take the picture on the right and convert it into the one on the left, so that you can see everything you are supposed to see. You will gradually need to do this less and less.

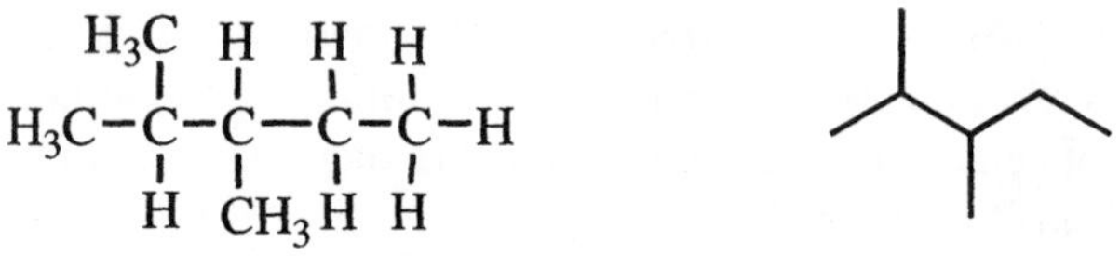

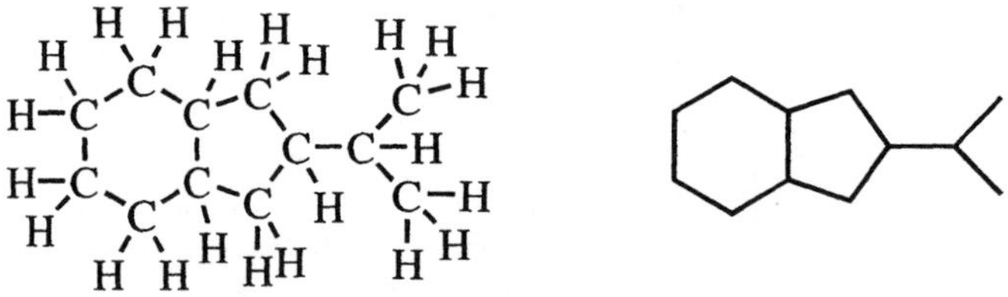

There is one rule of thumb for these pictures. This is not a big deal, but it is tradition: if, for whatever reason, you actually write the symbol of the atom at any place in one of these pictures, you should also write in the hydrogens associated with it. For example, it is not *wrong* to write pictures like the following ones, on the left, but chemists will recognize it as the work of an amateur. A professional would write it the way it is on the right. In other words, when you write the symbol of any atom, you should also fill in all the hydrogens attached to it.

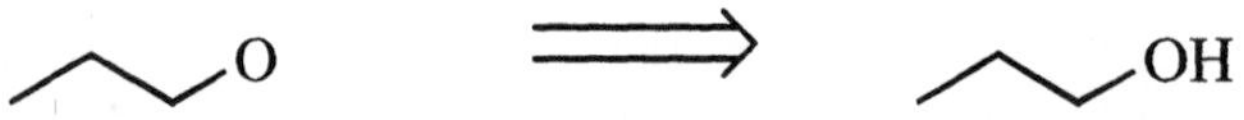

C−C−C−C ⟹

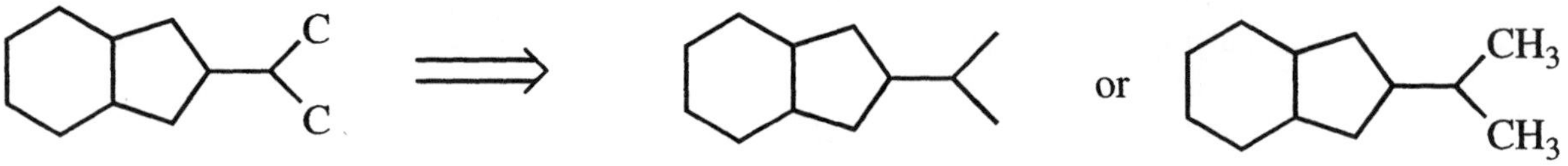

⟹ or

Let's continue building structures. See what you can make from the formula C_5H_{10}. Do it now! There are actually 13 possibilities, but several of them differ in subtle ways we have not yet discussed. You ought to be able to think of 10. Bet you can't!

The first thing we do is write all the carbons in a line, and fill in the hydrogens:

$$\begin{array}{ccccc} H & H & H & H & H \\ | & | & | & | & | \\ H-C & -C & -C & -C & -C-H \\ | & | & | & | & | \\ H & H & H & H & H \end{array}$$

You immediately discover that this requires 12 H's, and you only have 10, so this structure won't work. You are deficient in hydrogens, compared to a molecule that has as many hydrogens as are possible. You are missing two hydrogens, which is referred to as one "unit of hydrogen deficiency." (Many books refer to this as a "unit of unsaturation," but later on we will be using the term unsaturation in another context, so it is less confusing to use the term hydrogen deficiency. This has not yet caught on, however, so be prepared to see the term "units of unsaturation" in other books.) Every two hydrogens you are lacking constitutes one unit of hydrogen deficiency; this molecule, C_5H_{10}, has one. There are two ways to deal with the absence of hydrogens: you can make double bonds, somewhere, or rings. Both approaches allow you to create a molecule with two fewer hydrogens than the one shown. For example, if you remove the bold hydrogens below and connect the vacancies created, you get the molecule shown on the top right. If you remove the H's in italics you get the molecule on the bottom right. Both are perfectly legitimate structures for the formula C_5H_{10}. You can see that each double bond (or ring) accounts for an absence of two hydrogens—that is why it is called *one* unit (rather than two) of hydrogen deficiency.

See if you can find the other eight structures! *Hint*: When doing problems like these, it is best to be systematic. Draw all 5 carbons in a row, and see how many different places the double bond can be. Then draw 4 carbons in a row, with one coming off, in as many ways you can, and put the double bond in as many different places as you can. Do the same with 3 carbons in a row. And so on. After you have exhausted all the possibilities with chains, go for rings. Start with a 5-membered ring, then a 4-membered ring with one carbon attached, then a 3-membered ring, etc.

For a real challenge, try $C_5H_{10}O$. This also has one unit of hydrogen deficiency, but there are over 90 of these, of which you should be able to come up with over 70 at this stage!

In drawing molecules that fit a particular formula, you can see that it is helpful to know up front how many units of hydrogen deficiency there are. For example, if you were trying to draw a structure for $C_{17}H_{23}Cl_3O_7N_2$, how many rings and/or double bonds would be needed? Many books give you formulas for figuring this out, and many students memorize them and get right answers, but of course in a few weeks they have no recollection of how to do it. I prefer a logical approach to such questions. We can figure out formulas on the spot, never have to memorize anything, and always be able to reproduce them in the future. Here's how.

We start with the number of carbons, and figure out how many H's they would need if they were completely full of hydrogens. This is easy. Look at the molecule we studied earlier:

$$
\begin{array}{ccccc}
\text{H} & \text{H} & \text{H} & \text{H} & \text{H} \\
| & | & | & | & | \\
\text{H}-\text{C}-\text{C}-\text{C}-\text{C}-\text{C}-\text{H} \\
| & | & | & | & | \\
\text{H} & \text{H} & \text{H} & \text{H} & \text{H}
\end{array}
$$

This is full of H's right? How many are there? There are two for every carbon (the vertical ones), plus one more for each end, a total of 12. Thus any full molecule, with no hydrogen deficiency, will always have two H's for each carbon, plus two more: for n C's, there will be $(2n + 2)$ H's. This one is C_5H_{12}. That wasn't hard, was it? No memorization there, just a few seconds of thought. Good.

Will this work if the molecule is not in a straight line? Let's find out:

$$
\begin{array}{ccccc}
\text{H} & \text{H} & \text{H} & \text{H} & \text{H} \\
| & | & | & | & | \\
\text{H}-\text{C}-\text{C}-\text{C}-\text{C}-\text{C}-\text{H} \\
| & | & | & | & | \\
\text{H} & \text{H} & | & \text{H} & \text{H} \\
& & \text{H}-\text{C}-\text{H} \\
& & | \\
& & \text{H}
\end{array}
$$

I took the previous molecule and replaced one of the central hydrogens with a CH_3 group. What effect did it have? We lost a hydrogen, but we got it back, plus a CH_2 group. So we still have two hydrogens for every carbon, plus two extras. The formula still works.

Thus, any molecule containing only hydrogen and carbon can be evaluated to see how many hydrogens short it is: it should have $2n + 2$ hydrogens for n carbons, and every two hydrogens it is short represents one unit of hydrogen deficiency. Now all we have to do with a complicated formula is to figure out how the extra atoms affect the number of hydrogens necessary to fill up all the available spaces. We want to take the complicated formula I gave you above, $C_{17}H_{23}Cl_3O_7N_2$, and figure out how many hydrogens there would need to be to take care of 17 carbons, 3 chlorines, 7 oxygens, and 2 nitrogens; then we compare that to 23 and we will know right away how many units of hydrogen deficiency there are.

So now that we know what works if there are only C's and H's in your molecule, how do we deal with other atoms? What if there is a chlorine? Look at one.

$$\begin{array}{ccccccccccc} & H & & H & & H & & H & & H & \\ & | & & | & & | & & | & & | & \\ H- & C & - & C & - & C & - & C & - & C & -Cl \\ & | & & | & & | & & | & & | & \\ & H & & H & & H & & H & & H & \end{array}$$

This molecule has the formula $C_5H_{11}Cl$. We can tell by looking that it is "full" (there are no double bonds or rings), but we'd like to be able to do this without writing it out. This is easy: the molecule has a chlorine where a hydrogen used to be. The fact that the chlorine is there means that we need one less hydrogen than we thought to fill up the molecule. So the number of hydrogens needed is $(2n + 2)$ – (# of chlorines). The same should be true of the other halogens, F, Br, and I. Every halogen reduces by one the number of hydrogens needed. So actually the number of hydrogens needed is $(2n + 2)$ – (# of halogens). Again, no memorization, just a little thought.

Next, oxygen: how do we deal with that? Again, let's look at a "full" molecule that contains oxygen.

$$\begin{array}{ccccccccccc} & H & & H & & H & & H & & H & \\ & | & & | & & | & & | & & | & \\ H- & C & - & C & - & C & - & C & - & C & -OH \\ & | & & | & & | & & | & & | & \\ & H & & H & & H & & H & & H & \end{array}$$

We can see that the oxygen took the place where a hydrogen used to be, but at the same time it created a new place for a hydrogen. The number of slots for hydrogens has not changed by having the oxygen here. Alternatively, we can look at the formula of this molecule, $C_5H_{12}O$, and see that a full molecule that has oxygen requires the same number of hydrogens as a corresponding molecule with no oxygen (C_5H_{12}). So either by logic or by comparing formulas, but without memorization, you can conclude that oxygen can be ignored in calculations of hydrogen deficiency.

Last one: what do we do with nitrogen? Again, let's construct an example:

$$\begin{array}{ccccccccccc} & H & & H & & H & & H & & H & \\ & | & & | & & | & & | & & | & \\ H- & C & - & C & - & C & - & C & - & C & -NH_2 \\ & | & & | & & | & & | & & | & \\ & H & & H & & H & & H & & H & \end{array}$$

The nitrogen took the place of one hydrogen, but in doing so generated slots for *two* more. The presence of the nitrogen requires one more hydrogen than we thought to make the molecule "full." We can also see this by looking at its formula: $C_5H_{13}N$. Compared to C_5H_{12}, we have had to add an extra hydrogen because the nitrogen is there. So in general, every nitrogen adds an extra hydrogen to what is needed to fill a molecule. The number of hydrogens needed, then, is $(2n + 2)$ – (# of halogens) + (# of nitrogens).

So let's get back to the original challenge: if you were trying to draw a structure for $C_{17}H_{23}Cl_3O_7N_2$, how many rings and/or double bonds would you need? To be "full," this molecule with 17 carbons would require $(2 \times 17) + 2$ or 36 hydrogens, –3 for the chlorines, + 2 for the nitrogens, or 35 hydrogens total. It has 23. This leaves us 12 H's short of "full," or 6 units of hydrogen deficiency. We ought to be able to write structures for this molecule containing 6 double bonds, 6 rings, 3 double bonds and 3 rings, one triple bond (counts as two double bonds, of course) and 4 rings, or any other equivalent combination. Further, any structure we write that does *not* have 6 units of hydrogen deficiency will be wrong. Try it!

Functional Groups

It turns out that organic chemistry is a lot simpler than it looks, because reactions tend to occur only in specific places in a molecule, and in many cases, a great deal of the structure is irrelevant decoration with no chemical function. The places where the action happens are called "functional groups." Each functional group has a name and a specific set of reactions associated with it. You can understand the chemistry of many complicated molecules by recognizing the functional groups present, and if you understand the general properties of those functional groups, you will know most of what you need to know about the molecule. In the process of randomly generating structures with various formulas, you have already written down a number of functional groups. Now is as good a time as any to learn what they are. We will consider each of these in turn as we proceed through the book—for now, simply try to identify these functional groups in molecules you encounter. You ought to memorize this list. [Of course, I refer only to the first two columns of the list. As you learn how to name compounds, you will also learn how to produce the third column. The fourth column is merely for your amusement.]

The letter R in this table means either carbon or hydrogen. Some people think of it as "rest of the molecule."

Functional Group	Defining Feature	Simple Example	Complicated Example
alkene	C=C	2-butene	cholesterol
alkyne	C≡C	acetylene (ethyne)	3-methyl-4-nonyne
halide	C—X X = F, Cl, Br, I	methyl iodide	dieldrin

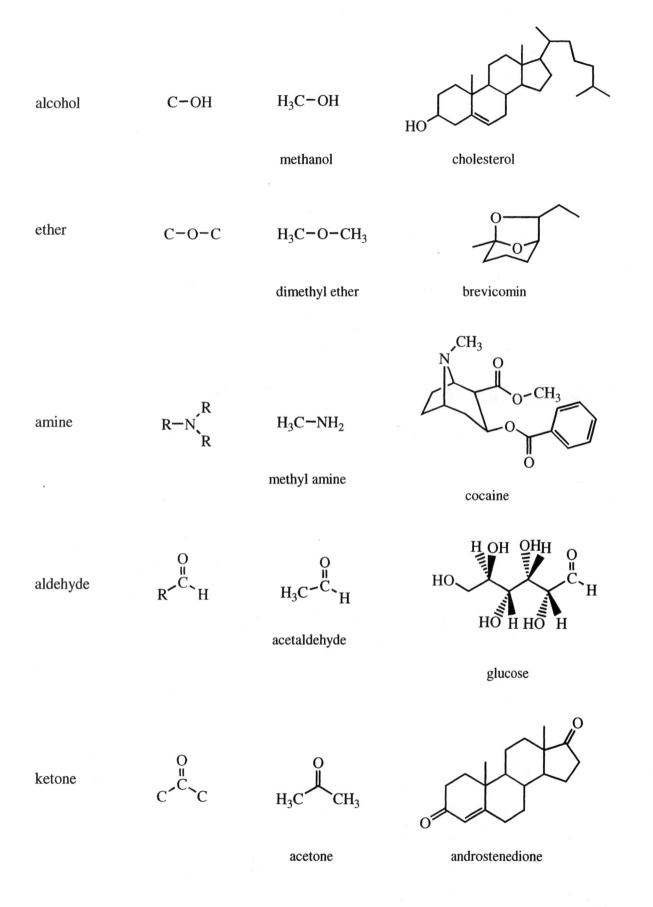

alcohol   C−OH H_3C−OH

methanol   cholesterol

ether   C−O−C H_3C−O−CH_3

dimethyl ether   brevicomin

amine H_3C−NH_2

methyl amine   cocaine

aldehyde

acetaldehyde   glucose

ketone

acetone   androstenedione

carboxylic acid

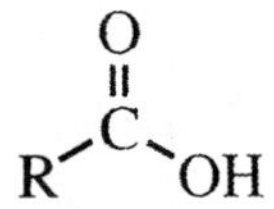

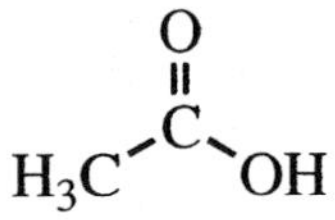

O
‖
R—C—OH

O
‖
H₃C—C—OH

acetic acid

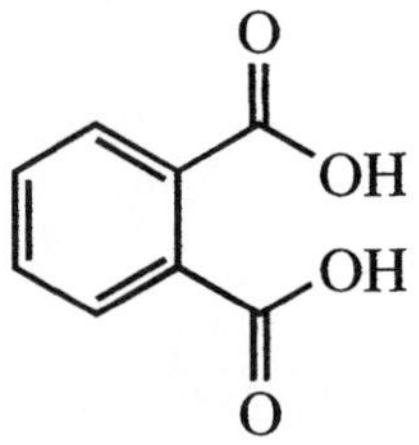

phthalic acid

acid chloride

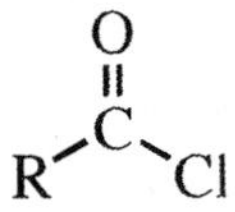

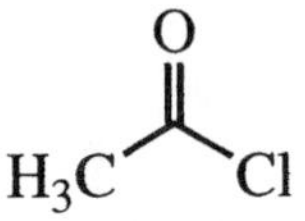

O
‖
R—C—Cl

O
‖
H₃C—C—Cl

acetyl chloride

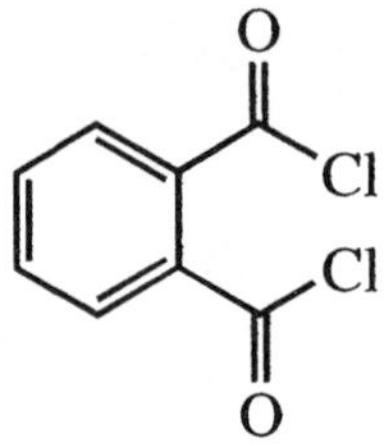

phthaloyl chloride

acid anhydride

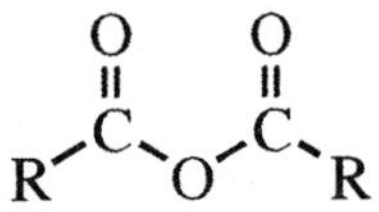

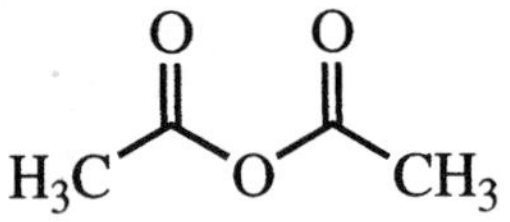

O   O
‖   ‖
R—C—O—C—R

O   O
‖   ‖
H₃C—C—O—C—CH₃

acetic anhydride

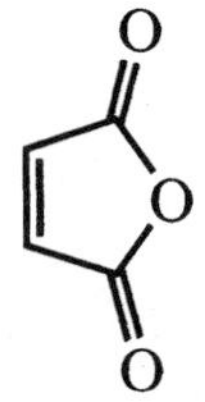

maleic anhydride

ester

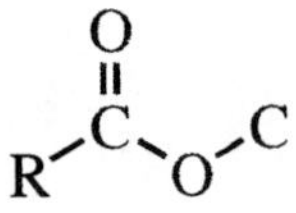

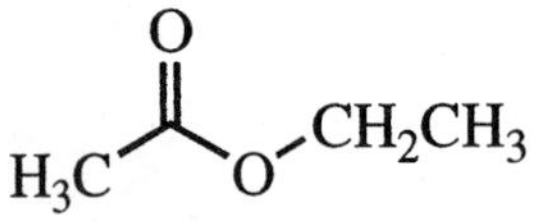

O
‖
R—C—O—C

O
‖
H₃C—C—O—CH₂CH₃

ethyl acetate

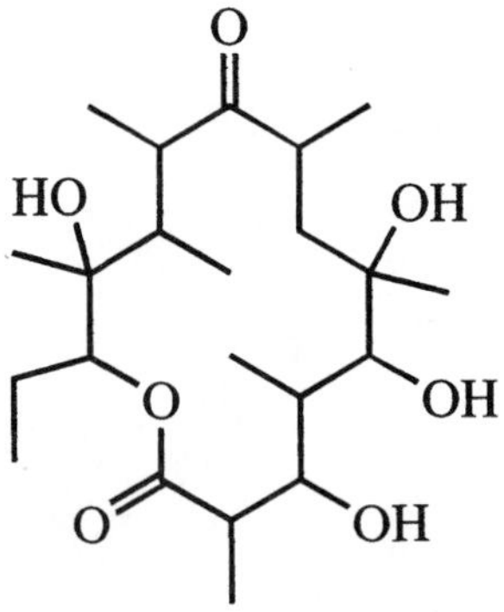

erythromycin (aglycone)

amide

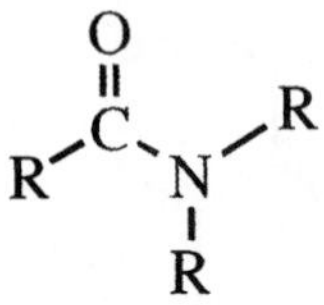

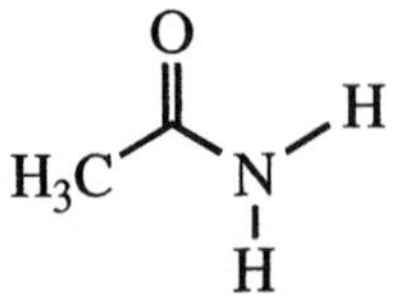

O   R
‖   /
R—C—N
     \
      R

O
‖   H
H₃C—C—N
        \
         H

acetamide

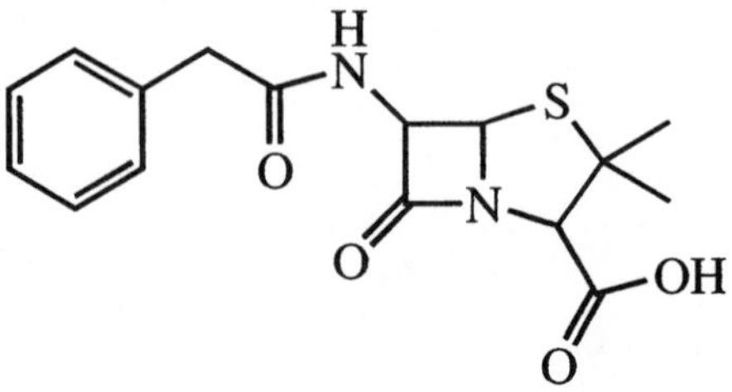

penicillin G

You should notice, of course, that many compounds contain more than one functional group. Cholesterol was listed as both an alkene and an alcohol: it *is* both. Dieldrin was shown as a halide, but it is also an alkene and an ether. See what other functional groups you can find in the above examples.

Problems

(Answers are provided at the end of the chapter for italicized problems)

1. Why is it not surprising that the alkali metals do not exist in diatomic form? Illustrate with electron dot structures.

2. Write chemical equations for the following reactions, showing electron dot structures for starting and ending materials:

   a. *silvery magnesium metal + purple iodine vapor reacts to give a white solid*

   b. silvery calcium metal + colorless oxygen gas reacts to give a white solid

   c. silvery sodium metal + yellow sulfur powder reacts to give a white solid

3. Draw possible Lewis structures of the following compounds:

   CH_4O H_2CO_3 NH_4^+ C_4H_6 H_2N_2O

4. *Draw Lewis structures for the following compounds. Include any formal charges that are necessary.*

   NCl_3 N_2O_3 HNO_3
   $C_2H_4O_2$ NH_4Cl $NaBF_4$

5. Write Lewis structures for the following: (Be sure to distinguish ionic interactions from covalent bonding.)

   CO, CO_2, CO_3^{2-}, CH_4, CH_4O (CH_3OH, methanol), CH_2O (formaldehyde), $HCOOH$ (formic acid, both oxygen atoms and one of the hydrogen atoms bonded to the carbon), C_2H_6, C_2H_4, CH_2Cl_2, NO_2^-, NO_3^-, SO_2, SO_3, SO_3^{2-}, SO_4^{2-}, $NaCl$

6. Write Lewis dot structures that correspond to each of the following formulas:

   a. CH_3NH_2 b. $ClCH_2COOH$ c. H_3BO_3

                                                    (B surrounded by the 3 oxygens)

7. *Write as many isomers as possible for the formula C_5H_{10}.*

8. Write as many isomers as possible for the formula $C_2H_4O_2$.

9. How many units of hydrogen deficiency are in each of the following? Draw several isomers of each.

a. $C_6H_{12}N_2I_4$ b. $C_5H_{11}NO$

10. How many units of hydrogen deficiency are in the following molecule? Draw one possible structure for this formula.

$$C_{10}H_{10}BrClN_4O_2$$

11. a. How many units of hydrogen deficiency are implied by the formula C_2H_2NOCl?

b. Write as many plausible structures (isomers) as you can for a compound with this formula. (I found 22 without rings and 28 with rings.)

12. Draw a single molecule that contains the following five functional groups. Label which group is which.

       alkyne; carboxylic acid; alcohol; ether; aldehyde

13. Identify by name each functional group present in the following molecules:

a.

b. $CH_3C{\equiv}CCH_2OCH_3$

c. $CH_3COCH_2CONHCH_3$

Selected Answers

Internal Problems

Test Yourself 1: CaF_2

Test Yourself 2:

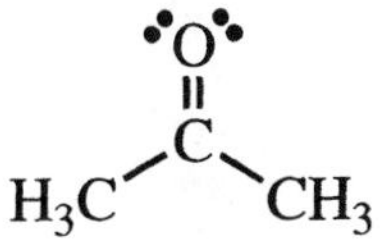

End of Chapter Problems

2. a. $Mg + I_2 \rightarrow MgI_2$

I–I   I⁻ Mg²⁺ I⁻

4.

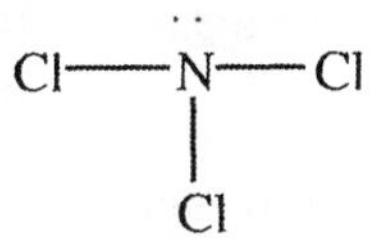

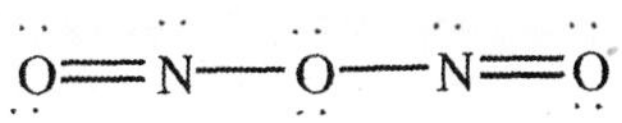

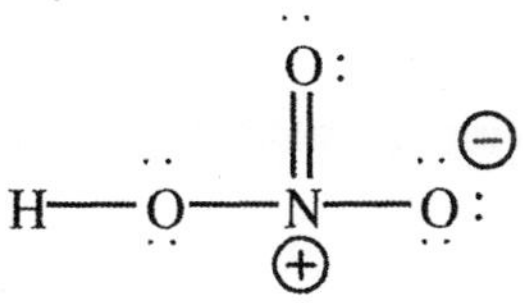

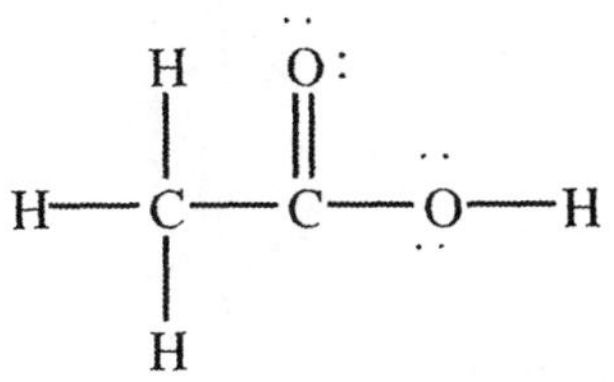

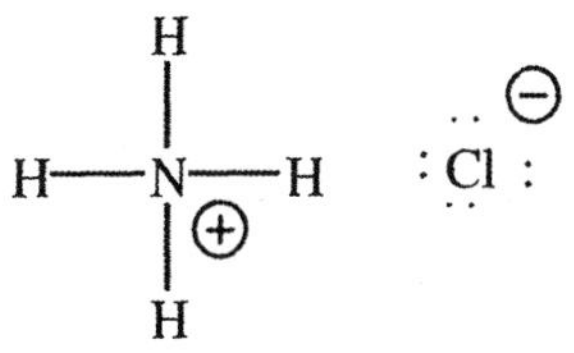

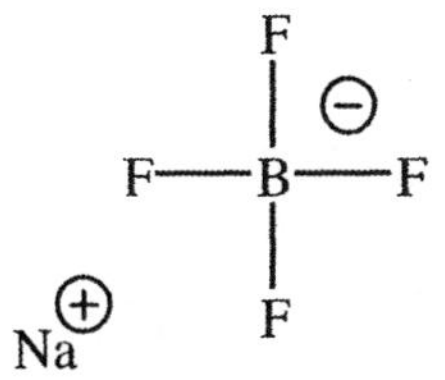

7.

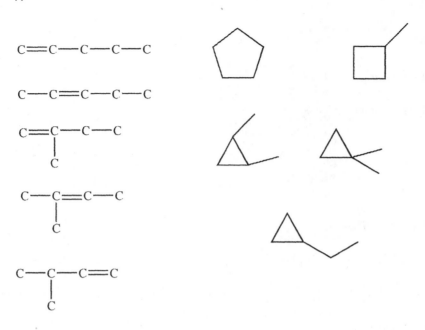

C=C—C—C—C

C—C=C—C—C

C=C—C—C
        |
        C

C—C=C—C
    |
    C

C—C—C=C
    |
    C

9. b.

One unit of HD. Three of what are probably more than 100 possible answers:

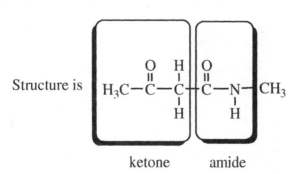

13. c.

Structure is

H₃C—C—C—C—N—CH₃

ketone   amide

Chapter 4
Molecular Shapes:
Hybridization

At one level a covalent bond is merely the sharing of electrons. But there are many other levels on which we can discuss bonding. In this chapter we will

- Learn some of the more sophisticated approaches to describing bonding, concentrating especially on the notion of "hybridization"

- Discuss the geometry of organic compounds

- Learn how to draw pictures that convey the information we want.

MOLECULAR ORBITALS

We have determined so far that a covalent bond is a sharing of electrons. In this section we want to consider this proposition in more detail. It will require us to consider more deeply the properties of orbitals.

An orbital is a home for electrons. It has a size and a shape. Like your home, it is a collection of places where you might be: you could be in the living room, the kitchen, the bedroom, etc. Saying that you are in your house does not specify exactly where you are in it, simply that you are somewhere inside. An orbital is similar, except that its borders are fuzzy. It is not obvious where an orbital ends. In fact, when you do the appropriate math, it turns out that it never quite ends. But to make the pictures useful, we usually outline the area where the electron has a fairly high probability of being. That is why different pictures of orbitals are not exactly the same: different authors may make arbitrarily different decisions of how much of the orbital they want to show. But all the pictures share the following properties: s orbitals are spherical, and p orbitals are dumbbell-shaped.

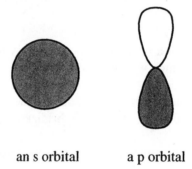

an s orbital   a p orbital

One of the beauties of using organic chemistry as the starting point in your study of chemistry is that, for most purposes, these are the only orbitals you need to deal with. The d orbitals are more complicated, but only crop up with element 21 and later, while the preponderance of organic chemistry deals with earlier elements.

You will notice that an s orbital is the same no matter how you look at it, but a p orbital has a specific "direction." There are three of them in each energy level (except the first), and the difference between the three is that they are pointed in different directions, all perpendicular to each other. These are usually referred to as p_x, p_y, and p_z (referring to a coordinate axis system). I will not specify which is which, because it does not matter. It is completely arbitrary which direction you select for x, y, and z, as long as they are perpendicular to each other.

Sigma Bonds

OK, with this as background, what is a bond? There are actually two types we need to deal with, σ and π. A σ (sigma) bond is simply an overlap of two orbitals along the line connecting the corresponding nuclei. Following are some pictures of what a σ bond might look like.

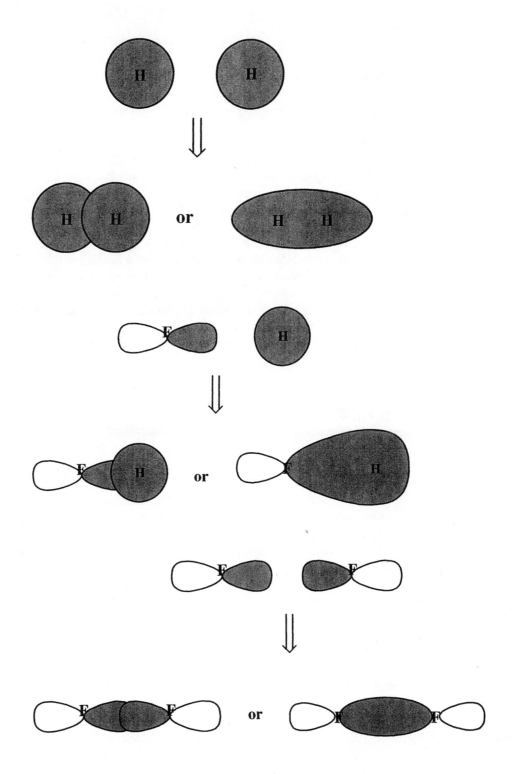

In each case we start with two atoms, each with its own orbital (an atomic orbital, s or p), and end up with an orbital covering both atoms, the molecule. This is called a molecular orbital. In this case, it is a σ molecular orbital. All share the property that if you look along the bond, from one atom to the next, and rotate the molecule, you can't see any change. Likewise if you hold one atom and rotate the other along the bond that connects them, nothing changes. This is called "cylindrical symmetry," and it is a property of all σ bonds.

Setting aside for the moment the discussion of π bonds, let's consider in detail the bonding for a carbon atom with only single bonds. If a bond is a sharing involving one electron in an atomic orbital on one atom with one electron in an atomic orbital on another carbon, then what bonding possibilities are available for carbon? According to our previous bookkeeping, carbon has the electron configuration $1s^2 2s^2 2p^2$. This would seem to make the single electrons in two p orbitals available for forming bonds, but no others: carbon should be able to form only two bonds. But we already know that, in almost all cases, carbon forms four bonds. How can we account for this? Two ways come to mind: we could throw out the model entirely (remember, everything in this book is nothing but a "best guess" based on our collective imaginations and experience, and it all might be wrong), or we could try to modify the model a bit. Normally scientists try the latter, especially when they are content with the theory they have, and want to retain as much of it as possible. When all else fails, they may throw a theory out, but not without first trying to fit new observations to the old model. So what shall we do? Recall that 2p orbitals are a bit higher in energy than 2s orbitals, but only a little. For the price of a little energy, carbon might send one of its 2s electrons up into the one 2p orbital that is still empty, thus creating a situation where there would be 4 electrons, each in its own orbital, available for bonding. *Provided* that the benefit of making extra bonds is greater than the cost of raising the energy of this one electron (which it is), the "expense" would be worth it to the carbon atom.

| 2p | * | * | ___ | $\Longrightarrow$ | 2p | * | * | * |
| 2s | * * | | | | 2s | * | | |

So now we are picturing a carbon atom with the following configuration in its outer shell: $2s^1$, $2p_x^1$, $2p_y^1$, $2p_z^1$. It is capable of forming single bonds with four other partners, just as we know carbon does. But there is still something unsatisfying about the model: it implies that one of the resulting bonds should be different from all the others, but in every measurement people have ever made on CH_4, for example, all four bonds seem to be identical. Further, it suggests that the three bonds formed with the p orbitals should be at 90° angles to each other, since p orbitals are all mutually perpendicular, and the bond to the s orbital could be almost anywhere (because the s orbital is a sphere), but when we actually look at CH_4 (methane)— don't worry about how we do this, but we can—we find that all the bonds are in fact about 109° from each other.

VSEPR Theory

We should pause for a moment here to try to understand why the bonds might be spread at angles of 109°. What an unusual number! Actually, it is the most natural thing in the world. If you were to take any four bulky objects and try to bring them together around one spot, as we are trying to do with four bonds fastened around the center of a carbon atom, they would naturally try to get as far away from each other as possible. Try this with balloons and see for yourself. Four mutually repelling objects automatically take up positions at roughly 109° angles from each other. On the other hand, three such objects tend to be separated by 120° angles, and two (obviously) go off in opposite directions, creating a 180° angle. When applied to atoms and angles, this notion is referred to as the Valence Shell Electron Pair Repulsion (VSEPR) model.

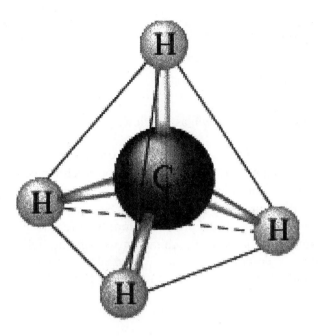

But the model we've so far developed based on electrons in orbitals does not predict 109° angles for four bonds, as noted above. This apparently demands another tweak in the theory. The new tweak is called *hybridization.*

sp³ Hybrids

What we are going to do is to "melt together" or "blend" the s and p orbitals and then make some new, different orbitals out of the mixture. Why can we do this? I want to remind you that there is nothing sacred about s and p orbitals. No one has ever seen one. *Scientists invented them!* They are a concept created to explain certain observations, but they are really only a model, a guess, as to what may be going on deep down in an atom. This particular guess is not sacred, and if a modified guess explains some things that the first one cannot, it is better than the first. Scientists always keep modifying guesses until they find a model that works with everything they want to explain.

So here's what we do. We currently have four orbitals, three of one kind (p) and one of another (s). We want four orbitals that are all alike, so we can make four bonds that are all alike. Therefore we take the four we have, toss them in a food processor, and grind them up. (In real life this corresponds to finding new solutions to a set of simultaneous equations.) What we pull out is four new orbitals, each of which must be a combination of the ones we put in; that is, one quarter s and three quarters p.

$$2p \quad \underset{*}{\underline{}} \quad \underset{*}{\underline{}} \quad \underset{*}{\underline{}} \qquad\qquad \Longrightarrow \qquad \underset{*}{\underline{}} \quad \underset{*}{\underline{}} \quad \underset{*}{\underline{}} \quad \underset{*}{\underline{}} \quad sp^3$$

$$2s \quad \underset{*}{\underline{}}$$

We call these "sp³" orbitals, after their parents, one s and three p orbitals. They look like distorted p orbitals, with one end bigger than it used to be and the other end smaller. In fact, the small end is so small that it is frequently omitted in pictures, although it is really there, and for some aspects of chemistry it turns out to be important. We will rediscover it in Chapters 12 and 15.

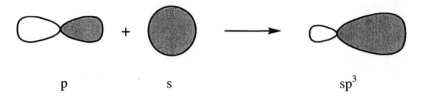

$$p \qquad\qquad s \qquad\qquad sp^3$$

A fancier picture of this is shown as A below.

We end up with four of these "hybrids" (four orbitals into the blender, four out!) and it turns out they point automatically at 109° angles from each other. Each of these new orbitals is now capable of making a σ bond with some other atom, using the big end of its orbital. What's more, the four bonds that result will be directed 109° from each other, exactly the angle actually observed. Thus, a carbon atom with four single bonds is described as being "sp^3 hybridized," and it uses sp^3 orbitals in its bonding, resulting in bond angles of 109°. This geometry is described as being "tetrahedral," because the ends of the orbitals, if connected, would describe a four-sided triangular pyramid, called a tetrahedron.

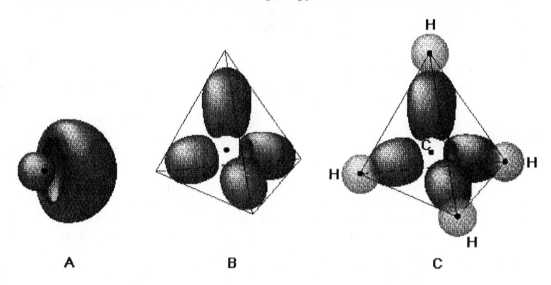

A   B   C

The same will be true of any other atom with electrons in four locations around it. This has to be phrased rather carefully, because a "location for electrons" is not necessarily the same as a bond. For example, a double bond is two bonds, but both occupy the *same* location in terms of direction from a particular atom (namely, pointed toward the partner), so it counts as only one location. On the other hand, a lone pair is not a bond at all, but it *is* a location for electrons. So ammonia, :NH₃, has only three bonds, but again four locations (the three bonds plus the lone pair). The nitrogen atom, like the carbon atom we just described, has to find four different places to put its four pairs of electrons, and it solves the problem in the same way carbon does: by becoming sp^3 hybridized. Thus the four locations for ammonia are also 109° apart.

Actually this is not completely true—the angle from H to N to H is really slightly less than 109°. This has been explained by suggesting that the lone pair, having no partner on the other end to squeeze it down, is a bit fatter than the other three pairs of electrons (those running from N to H), and therefore it pushes the other three a bit closer together to give itself more room. So the angle from H to N to lone pair is a bit *more* than 109°, and the HNH angle is a bit less. The measured HNH angle is 107°. The "squeezing effect" is even larger in water, where there are *two* lone pairs pushing the two OH bonds together, and the resulting HOH angle is only 105°. Still, both ammonia and water are most conveniently thought of as tetrahedral—just a little distorted.

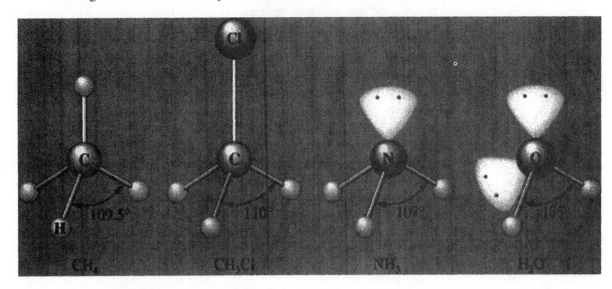

A linguistic problem arises with this designation, since the word "tetrahedral" officially refers to the geometry of the bonded partners, not the central atom. The methods we have of "looking" at molecules are capable of seeing nuclei (atoms), but not electrons, so we cannot actually see any lone pairs. Thus when we "look" at water, all we see are the H, the O, and the other H, with an angle of 105°, so any sensible person trying to name this arrangement would use the word "bent" rather than "tetrahedral." We know better, though: it is bent precisely because of the overall tetrahedral geometry with respect to its electrons, and if it weren't for those other electrons there would not be a "bend." Nevertheless, the official term for the geometry of water is "bent." Likewise, the official term for the geometry of ammonia is "pyramidal," rather then tetrahedral. In this book, however, we care very much about the invisible electrons, so we will often use the term tetrahedral in referring to the geometry of oxygen and nitrogen compounds.

In summary, any atom surrounded by four different regions of electrons will be sp^3 hybridized and have a tetrahedral geometry.

Pi Bonds

sp^2 Hybrids

So what happens if there are only *three* regions of electrons, as for example in $H_2C=CH_2$? VSEPR theory assures us that in these molecules, the three partners will be oriented 120° from each other, an arrangement referred to as "trigonal" (pronounced TRIG-gun-null, not try-GO-null). How do we account for this in terms of s and p orbitals? As before, we imagine that the carbon has promoted an electron from an s into the vacant p orbital, so we again have $2s^1 2p_x^1 2p_y^1 2p_z^1$. But this time, instead of putting all four of these into the food processor, we put in only three: the s and two of the p's. These we grind up to create three new orbitals, "sp^2" hybrids.

Conveniently, when you go through the appropriate math associated with this, the three new orbitals come out pointed at 120° angles from each other. Now, recall that there was a fourth orbital that we did not toss into the mix. It is still there, intact. The carbon atom still has four orbitals. (The number of orbitals before and after hybridization never changes. Four in, four out. Three in with one left over, three out with one left over.) The one we did not put in was a p orbital, so there is still a p orbital left on the carbon, and it still is the home of one of the electrons.

So how do we describe the bonding for an sp^2 atom? Each of the sp^2 orbitals makes a σ bond with some partner at the other end, either an H or the other C. This is shown in the picture labeled A below. Each carbon now has six electrons around it in these σ bonds, plus *one more electron,* alone in a p orbital. The two electrons in the p orbitals, one on each atom, then form another bond. This is the *second* bond of the double bond. You should notice immediately that this bond is different from all the other bonds you have seen so far: all other bonds have formed along the line defined by the two atoms being joined, and cylindrically symmetric. This one is neither. It forms simultaneously above and below the plane of the rest of the molecule, and rotating the molecule has a very visible effect. This kind of bond is given a new name, a π (pi) bond. It is illustrated as the bottom part of picture B.

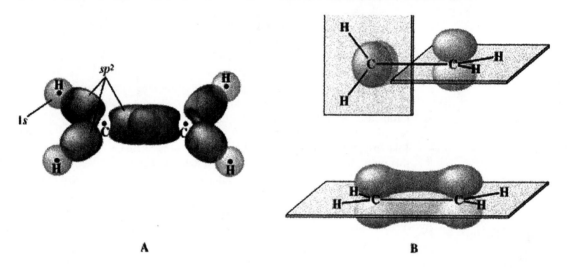

A   B

There is one consequence of π bonds that should be clear from the above description: if the π bond consists of two p orbitals lined up parallel to each other, overlapping side to side, then rotation of the individual atoms around a π bond should be very difficult. Such rotation would destroy the bond, as illustrated in the top part of picture B. This leads to the further consequence that many double-bonded structures exist in more than one form. For example, 2-butene, the molecule shown below, can have both CH_3 groups on the same side of the plane of the π bond, forming a "U" shaped molecule as below left, called "*cis,*" or on opposite sides, forming a zig-zag shaped molecule called "*trans.*" These are different molecules!

cis-2-butene   *trans*-2-butene

Any atom with electrons in three locations will be sp^2-hybridized. For example, BH_3 has three bonds and no lone pair, so only three locations for its electrons. It is sp^2-hybridized and trigonal. The three σ bonds are between sp^2 orbitals on boron and s orbitals on hydrogen. In this case there is an *empty* p orbital on the boron, pining for electrons but unable to do anything about it. Of course, if a friendly water molecule passed by and offered to share some of the electrons it wasn't using, the boron would be pleased to accept them. This would create a new bond between oxygen and boron. The boron would now have four locations for electrons around it, and would change shape to accommodate them. It would become sp^3- hybridized and tetrahedral. The new molecule would also have formal charges, as described earlier.

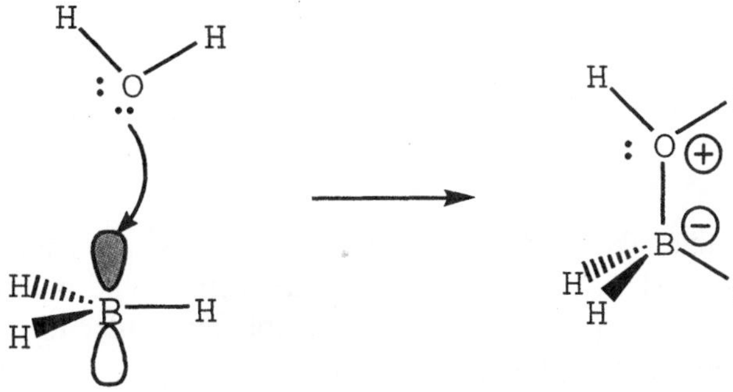

Molecules like BH_3, lacking octets but with no charge, are somewhat rare. They are called Lewis acids, and we will work with them later. For now, most of the molecules we will see with sp^2 hybridization will contain double bonds.

sp Hybrids

There is still another type of hybridization to consider. How do we describe an atom with only *two* locations for electrons around it, such as in HC≡CH? If you have followed the argument up to now, it should not surprise you that we will accomplish this by taking only two of the available orbitals, one s and one p, and grinding them up. This will create two "sp" hybrid orbitals, directed in a straight line, pointing in opposite directions. The angle between them is 180°.

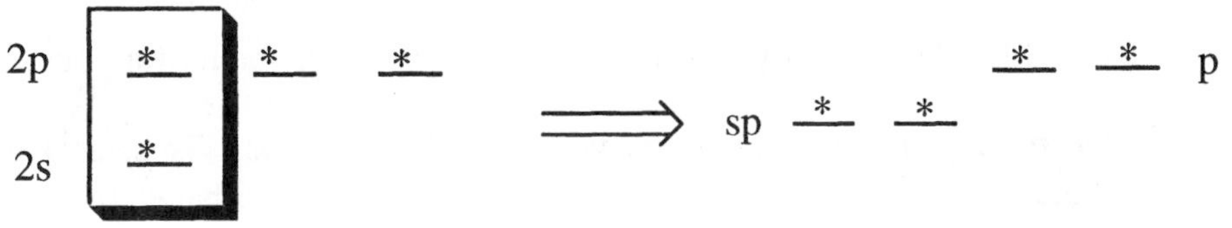

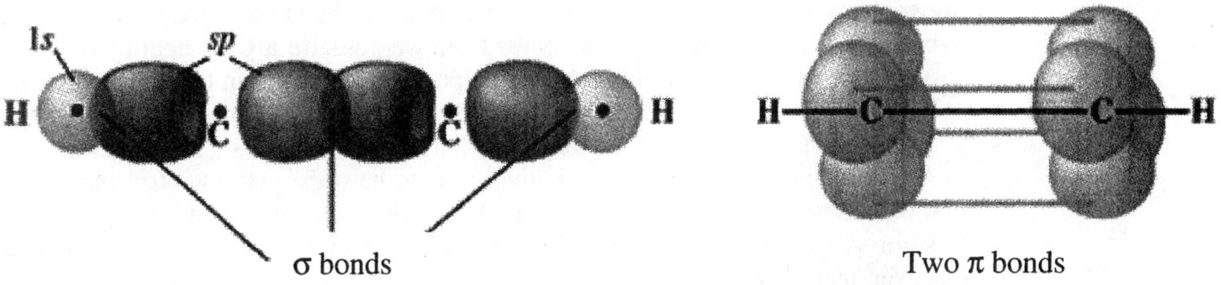

σ bonds

A

Two π bonds

B

This same type of hybridization (and therefore geometry) will be found whenever an atom has only two locations for electrons. For example, the carbon of CO_2 is also sp hybridized, because electrons are found in only two regions around it.

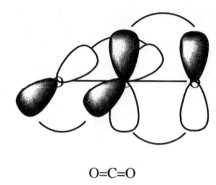

O=C=O

Notice that the carbon (in the middle) is participating in two different π bonds simultaneously, one to the left and one to the right, and that the p orbital that talks to the oxygen on the right is different from (perpendicular to) the p orbital that is talking to the oxygen on the left. Each oxygen is participating in one π bond, using a single p orbital, but the p orbital on the right oxygen happens to be perpendicular to the one on the left oxygen. The carbon is still sp hybridized.

Drawing Pictures

Now you can look at fairly complicated molecules and draw pretty decent pictures of them that show something about their true shapes. For example, consider the following molecule:

$$CH_3CH_2-C{\equiv}C-CH_2-\overset{\overset{\displaystyle O}{\|}}{C}-NH_2$$

Label the hybridization at each atom, then draw a new picture to show more accurately the shape of this thing, labeling appropriate angles. Go ahead! You can do this!

OK, I'll help, but really, you shouldn't read this until you try for yourself.

On the left, we have a carbon with 4 different bonds. This must be sp^3-hybridized, and therefore it has 109° angles. (Or, if you prefer, it must have 109° angles, so therefore it is sp^3-hybridized!) The next carbon also has four different partners, so it too is sp^3. The next one to the right has only two partners. Electrons will be in only two locations, so this is sp-hybridized. Likewise for the next one. Now we come to a CH_2, again with four different partners, so sp^3. The next carbon has three partners, electrons in three locations, so it is sp^2-hybridized. Then we come to the NH_2. What do *you* think? Many of you said sp^2, because you see three partners. What you forgot is that nitrogen also has a lone pair that is not drawn in on the above structure, but you know it has to be there. (Remember, earlier I suggested that for now you should always show the lone pairs. This is one of the reasons why.) Thus this nitrogen has electrons in *four* locations and should be labeled sp^3. [Much later we will discover that this is actually wrong, but for now there is no reason for you to think otherwise.] What about the oxygen? We know that it is engaged with the carbon in a double bond, which consists of a σ bond and a π bond, so at the very least it has one p orbital. It also has two lone pairs (again not written in on the structure, but you know they are there).

The natural thing to think is that it is sp^2-hybridized, and that is what I think too, but you know what? Nobody knows! Why not? We can tell the hybridization around an atom by measuring the angles it makes with its partners. This is done using a technique called X-ray crystallography, which locates atoms (more specifically, the nuclei of atoms). But this particular oxygen has only one partner, so we cannot tell for sure in what direction the other electrons might be headed. There is no angle to measure! Without knowing the geometry, we cannot specify the hybridization, so you need not label it. It will not hurt you to imagine it being sp^2, however.

So now we return to the structure and label the hybridizations:

$$CH_3CH_2-C\equiv C-CH_2-\overset{\overset{\displaystyle O}{\|}}{C}-NH_2$$

$$sp^3 \quad sp^3 \quad sp \quad sp \quad sp^3 \quad sp^2 \quad sp^3$$

Drawing Molecular Geometry

We know that this is not a very good representation of the molecule, however, because almost all the atoms are shown in a straight line from left to right, even though in most cases the angles are *not* 180°. Let's redraw this molecule in a more realistic fashion. A piece of advice if you are ever asked to do this: Try to keep the major atoms in the plane of the paper; the whole thing will be much easier to draw if you do.

It doesn't matter where you start. It's probably easiest to start in the middle somewhere—I would probably start with the triple bond—but let's imagine that you started on the left end. The carbon is tetrahedral, with 109° bond angles. Right away you have a problem. Right now, get out your model set and select or build a carbon with tetrahedral geometry. This is not a flat object. It is impossible to accurately represent this on a two-dimensional surface like this page, yet that is precisely what you have to do. So look at this thing. Select the carbon in the middle and any other two atoms attached. These three atoms are in a plane (three points *define* a plane, unless they are colinear, which these are not). Assume that this plane is the plane of the paper. You can draw these three atoms on your paper, like so:

Notice that the angle I have drawn is slightly larger than 90°. It is intended to be 109°. Whether it really is does not matter so much, but it should at a minimum be larger than 90°. Now, examine your model: while holding the first three atoms as drawn, where are the others? They are not in the plane of the paper. One is closer to you, the other is farther away. But there is more: look closely. They are both on the outside of the angle you already drew. Therefore it is nonsense to draw a picture in which there is anything drawn on the *inside* of this angle. In real life there is nothing there, and a picture that shows something there cannot be representing anything real.

So we have to represent our two other atoms on the outside of this angle, one up and one down, and somehow convey this on a piece of paper. The chemist's tradition is to do this with wedges and dashed wedges. The atom coming toward you, call it A, is shown at the end of a solid wedge, intended to evoke perspective that makes you think it is coming toward you. The other atom, B, is placed at the end of a dashed wedge, intended to make you think of it fading off into the distance. In keeping with thoughts of perspective, it would make sense to draw this as shown on the left, with the wedge getting bigger as it comes toward you, and a few books do so. But frankly, most chemists do not. If you look in any chemistry journal and find pictures like this, they are generally represented as shown on the right. It is the dashed line, and not the perspective of the wedge, that indicates that the bond is going back. It is unfortunate that chemists have settled on the picture on the right over the one on the left, but they have, and there is nothing I can do about it, so you might as well get used to it too.

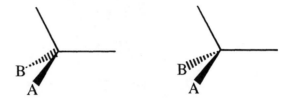

Now we have to represent the left-hand carbon in the molecule we are working on in the style shown above. This is a CH_3 group, so three of these lines end in hydrogens, and the fourth is the next carbon in the chain. In order to make our lives easier, let's make sure that the next carbon is one of the atoms in the plane of the paper, and since we are following the original molecule from left to right, we might as well put the second carbon to the right of the first one. The other three atoms are hydrogens:

Now we have to do the same thing with the second carbon, also sp^3-hybridized and tetrahedral. Again, we will draw two lines in the plane, and then one solid wedge and one dotted wedge. Where to start? Well, we already have one line in the plane, the one headed back to the first carbon. So all we need to do is draw a second line from this one, at an angle of 109°. It doesn't matter whether you do this up or down. Imagine a 90° angle and expand it a little, then add your solid and dotted wedges. Again, for simplicity, we will continue the chain in the plane of the paper and reserve the wedges for the hydrogens if at all possible.

Continuing to the right, the carbon we drew is sp-hybridized and linear. That does not mean that you should draw a horizontal line. It means that the next line coming out of it should be a continuation of the line that is already there. We can draw this as a triple bond. I'm going to shrink the picture now, to save space.

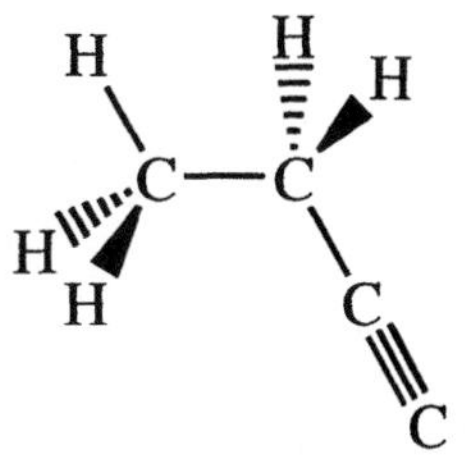

Again, this new carbon is sp-hybridized and linear, so yet another line is added in the same direction:

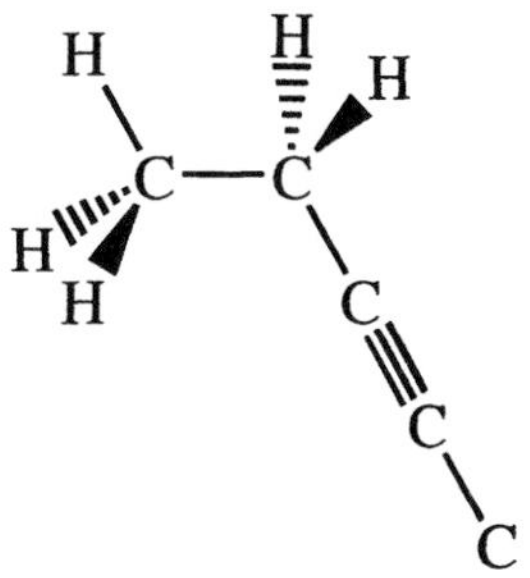

This latest carbon is sp³-hybridized and tetrahedral, so again we have to draw a 109° angle in the plane, then wedges on the outside of that angle. Again, it does not matter in which direction the new bond points, but the angle should be a bit larger than 90°:

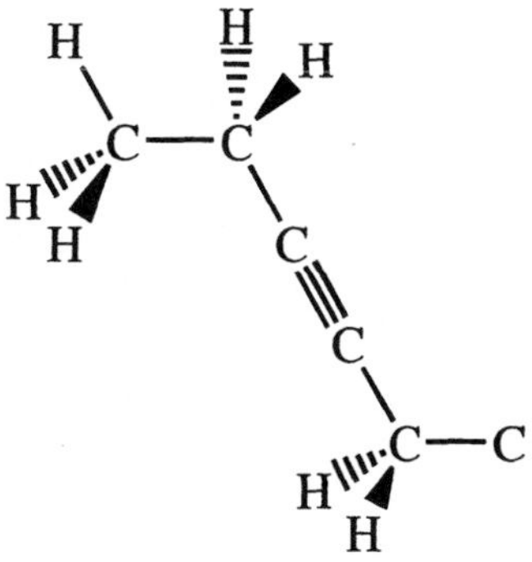

This carbon is participating in a double bond. It is sp²-hybridized and trigonal, requiring 120° angles. To be quite honest, most people can't distinguish between 109° and 120° in a picture. Once again, just be sure it is a bit larger than 90°. Drawing 120° angles is actually not that tough, because you are usually drawing three of them, and they should all be about equal. Remember, trigonal atoms are planar, so we can put all of these atoms in the plane. This would be very difficult to draw if you had not kept your chain in the plane of the paper in the first place.

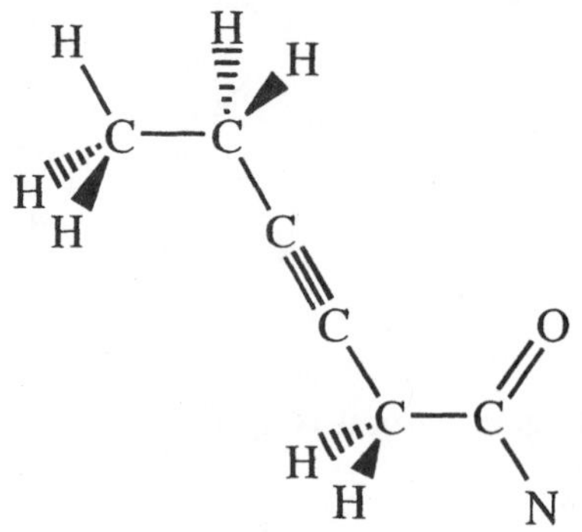

Only one atom left to deal with (remember, there is no geometry associated with the oxygen, so we do not have to show any!). The nitrogen is sp^3 and tetrahedral, so we will use our familiar technique of drawing one line at about 109° in the plane, then two wedges:

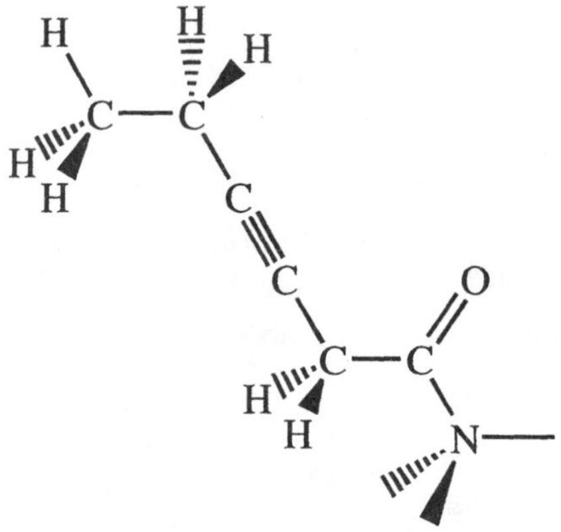

What do we put on the ends of these lines? We have two hydrogens and a lone pair of electrons. We can put the two hydrogens on any two places. The pair of electrons can be shown if you want, in the last position, or they can be ignored, since in point of fact they cannot be seen. And so you are done:

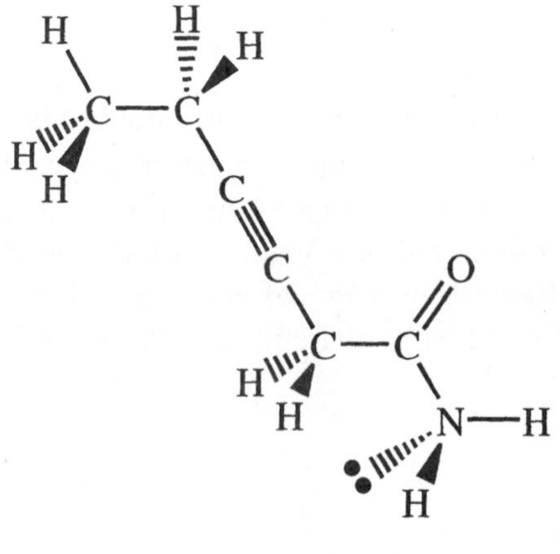

If I had been doing this, I probably would have started with the triple bond and made it horizontal. My picture would be the same as yours, but it would take up less space:

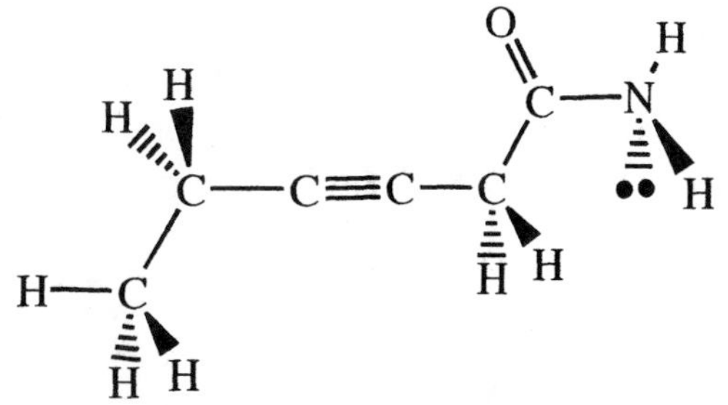

For completeness, it would do no harm to label a few angles in here:

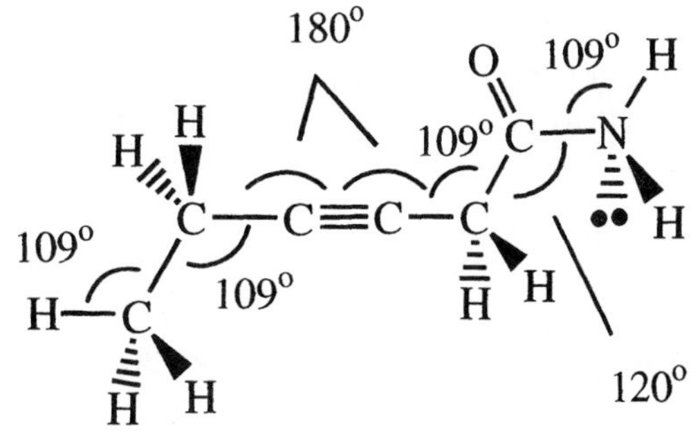

COMMON MISTAKES: DON'T LET THIS HAPPEN TO YOU!

Many students have a lot of trouble with the concept of a 180° angle. A 180° angle is a straight line, right? No it isn't! A 180° angle, like every other angle, is a relationship between *two* lines. It takes two lines, or three points, to make an angle. Just as it is nonsense to ask about a distance associated with a single point (the concept of distance requires two points, the distance from A to B), it is equally nonsense to discuss an angle associated with a single line or two points (an angle requires A to B to C). Nevertheless, many students, including some in your class, will write the following:

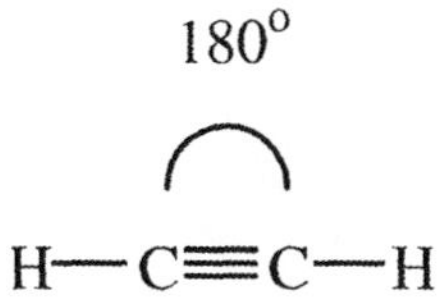

This does not mean anything, because the two carbons represent only *two* points, and the connection between them is a line. Even though it is a straight line, it makes no sense to describe a 180° angle between the two carbons, because there is no *other* line to relate to it. Only when you have two lines do you have an angle. There are *two* 180° angles in the above molecule, but not the one shown above:

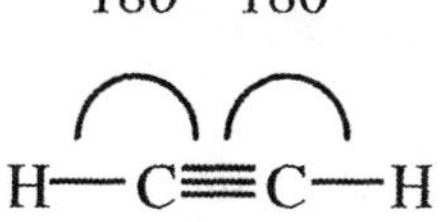

These are 180° angles because they describe the relationship between one line (connecting the C and the H) and another line (connecting the C and the C). **Don't show angles between two points!**

Another problem students have with 180° angles is drawing them. Now this you have trouble believing—everyone can draw a straight line, right? Not so! You would not believe how often I have seen the following:

180°

It is perfectly obvious that this is *not* 180°, yet many students have no qualms about writing such things. (Other students, past students, certainly none of you!) There is much about organic chemistry that requires drawing skill, and we each try to get by with what we have, but even the most limited artist among us is capable of drawing a line between two points and then continuing that line in the same direction without a bend. Similarly, it should be obvious to everyone that the angle below is less than 90°, yet many students will draw a picture like this and label it 109°.

109°

Your artistry does not have to be perfect—very few people can accurately draw a 109° angle—but *everyone* can draw a straight line, and everyone can draw an angle a little larger than 90°. **Make your pictures bear some relationship to reality!**

Bond Blow-Ups

After drawing a picture like the one you just drew, with all the right geometry, it would not be unreasonable if someone were to ask you to focus in on a particular group of atoms and describe the bonding involved. Let's focus in on the triply bonded part of the above molecule and show a blow-up of what is going on there.

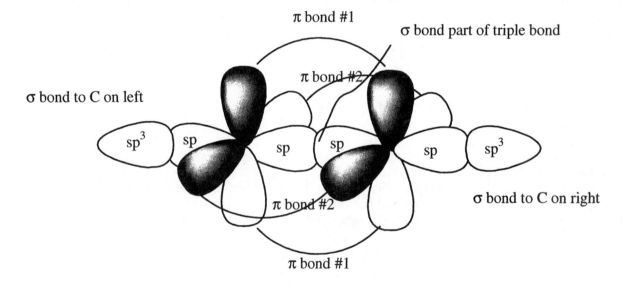

COMMON MISTAKES: DON'T LET THIS HAPPEN TO YOU!

Many students try to label bonds as sp³, sp², etc. This is wrong. Bonds are either σ or π. π bonds are composed of p orbitals lined up side by side. σ bonds are composed of orbitals pointing at each other. Often, as in the above example, the two orbitals are not of the same type. The σ bond on the left in the previous picture, for example, is composed of an sp orbital from one partner and an sp³ orbital from another partner. It makes no sense to label the bond as having a hybridization. **Hybridization refers to atoms and orbitals, not bonds.**

The concepts we have learned in this chapter will have important consequences as we explore the properties of organic molecules in later chapters.

Problems

(Answers are provided at the end of the chapter for italicized problems.)

1. Based on VSEPR theory, what shapes would you predict for the following species?

   a. PCl_3

   b. BCl_3

   c. CCl_3^+

2. Now that "molecular shape" has become another criterion to consider, how many isomers can you write for a molecule with the formula C_5H_{10}?

3. *Produce a geometrically accurate drawing of $(CH_3)_2C=CHCH_2C\equiv CCONHCH_3$. Indicate the expected bond angles with respect to each bond in the "main chain" and label each of these main-chain atoms with respect to its expected hybridization.*

4. For the following compound, label the hybridization at each non-hydrogen atom and draw a more realistic picture of it showing bond angles.

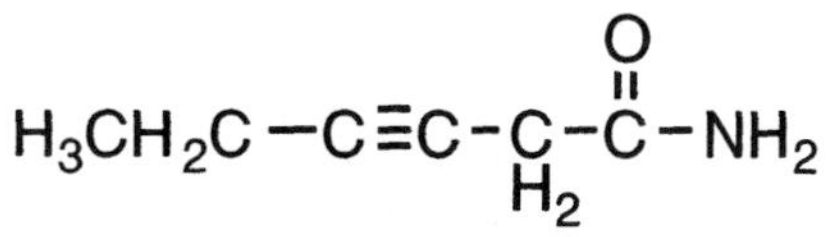

$$H_3CH_2C-C\equiv C-\underset{H_2}{C}-\overset{\overset{O}{\parallel}}{C}-NH_2$$

5. For the following molecule,

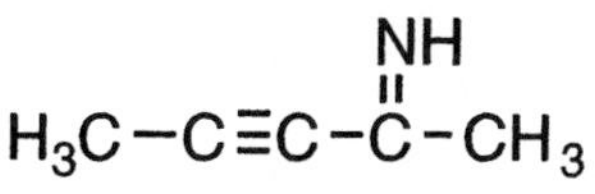

$$H_3C-C\equiv C-\overset{\overset{NH}{\parallel}}{C}-CH_3$$

   a. Identify the hybridization of each atom (except H's).

b. Redraw the molecule to show a more accurate representation of the geometry. Label one angle at each atom where an angle can be measured.

c. *Draw an orbital diagram (show and label the balloon-like orbitals) of the bonding between the carbon and the nitrogen.*

6. a. Show the hybridization of each non-hydrogen atom in the following molecule.

$$H_3C-\overset{H_2}{C}-\overset{H}{N}-C\equiv C-O-\overset{\overset{O}{\parallel}}{C}-H$$

b. Show a blow up of the bonding, using "balloon" orbitals, at the triple bond.

c. Redraw the molecule showing a more accurate representation of its geometry. Label one angle at each relevant atom.

7. *Show the bonding in allene, below. Pay careful attention to the geometry of the resulting structure.*

$$H_2C=C=CH_2$$

Selected Answers

3.

5. c.

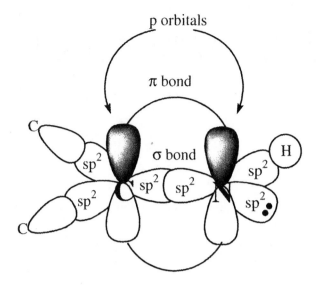

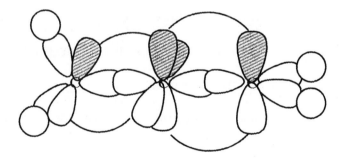

7.

Chapter 5
Polarity and Intermolecular Forces

Many of the properties of molecules (organic and otherwise) are determined by the forces that operate between them, so-called "intermolecular forces." In this chapter we will

- Identify the various types of intermolecular forces

- See how these forces lead to observable properties such as boiling points and solubilities

- Discuss a special force called the "hydrogen bond"

- Explain how all these forces lead to important biochemical consequences such as protein folding.

In Chapter 3 we looked briefly at the concept of electronegativity. Frankly, it wasn't of much use then: we used it simply to guess whether a bond would be ionic or covalent. It turns out that the concept of electronegativity is considerably more useful than that, and we will encounter some of its important uses now.

POLAR BONDS

Consider the hydrogen–chlorine bond, H—Cl. Hydrogen has an electronegativity of 2.1, while chlorine has an electronegativity of 3.0. This is not enough difference to cause us to suspect an ionic bond: we will not have H^+ and Cl^-, we will have a covalent bond, a sharing of electrons. On the other hand, the electronegativities are not the same, so the electrons will not be shared *equally*. The chlorine wants the electrons more than the hydrogen does, but not enough to steal them completely away. They will still share, but not equally. The chlorine takes a greater share of the electrons than the hydrogen does; the electrons are displaced somewhat in the direction of the chlorine. This leaves the chlorine with somewhat of a negative charge: not a whole charge, but a *partial* charge. How much? We don't know. But surely the electrons prefer the chlorine to the hydrogen, and the sharing is unequal. We call this a *polar* bond. There are two common ways to illustrate this. One is to indicate the partial charges that are on each atom, like this, where the Greek letter δ (delta) means "partial."

$$\delta^+ \quad \delta^-$$
$$H—Cl$$

The other is to show an arrow over the bond, where the arrow points to the negative end of the bond. To remind the reader that this is a polarity arrow, there is a cross at the tail (positive) end of it:

$$\longmapsto$$
$$H—Cl$$

It should be obvious that not only is the bond polar, but the HCl molecule, too, is polar; i.e., there is an end that is somewhat negative and another end that is somewhat positive. A molecule, or a piece of a molecule, that is polar is referred to as a "dipole" (two poles, get it?). The HCl bond is a dipole; the HCl molecule is a dipole.

Well, this is fine for a diatomic molecule (one with two atoms), but what do you do if there are more? Each bond is independent of all others. Each can be analyzed separately, to determine whether the bond in question is polar. Then all the bonds must be looked at as a package, to see what effect they have on the molecule. Take, for example, CO_2. Each bond is polar, because the oxygen is more electronegative than the carbon. The left bond is polar, with the negative end toward the O. The right bond is the same.

$$\longleftarrow\!\!\!\!\dashv \vdash\!\!\!\!\longrightarrow$$
$$O{=\!=\!=}C{=\!=\!=}O$$

POLAR MOLECULES

The CO_2 molecule has a concentration of charge on the sides (as drawn) and a shortage in the middle. But is the *molecule* polar? Does any end of the molecule have more charge than the opposite end? The right end is somewhat negative, to be sure, but so is the left, and by the same amount. Anywhere you look in this molecule, if you go through the center and out the other side by a similar amount, you come to a place that looks the same in terms of charge. The molecule is symmetric! There is no positive or negative end to the molecule, it is "non-polar." The two bond dipoles have canceled each other, because they are pointing in opposite directions. (Those of you who remember physics from earlier courses may realize that we are doing vector addition here, but it is not necessary that you treat it that way unless you want to get really quantitative about it.) Of course, if one of the oxygen atoms had been sulfur (the molecule O=C=S), then the dipoles would *not* cancel each other out: they are indeed pointed in opposite directions, but one is stronger than the other, and the molecule would end up with more negative charge on the oxygen end than the sulfur end.

Now consider water, H_2O. Again each bond is polar, this time with the negative end in the middle and the positive end on the outside. And again, it appears that the two bond dipoles cancel, leading to a non-polar molecule:

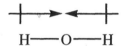

But wait! In the last chapter, we determined that water does not look like this! Remember, there are two lone pairs of electrons on the oxygen, and the four pairs of electrons repel each other, making the shape more or less tetrahedral, with an angle of approximately 109° between bonds (actually a bit less). Thus, the picture above is wrong: it should be written the following way.

Now it should be clear that the top of the molecule (as written, or, in general, the oxygen end) has more electron density than the bottom. The molecule as a whole has a dipole, not aligned with any of the bonds but running between them (again, vector addition would get you here). The vertical arrow below represents the molecular dipole, which is the sum of the bond dipoles. The bottom line is that water is polar, and the reason it is polar is that it is bent. If it were straight it would not be polar.

The message here is that you cannot tell about molecular polarity unless you know molecular geometry. That is why we had to wait until this chapter to discuss it. For a molecule to be polar, two things must be the case: first, at least one of the bonds must be polar; and second, the polar bonds must not cancel each other out. I should mention here that although the electronegativities of carbon and hydrogen are not the same, they are so close that carbon–

hydrogen bonds are generally considered to be non-polar. You should also notice that lone pairs do not have polarity in themselves: their effect is on the shape of the molecule.

INTERMOLECULAR FORCES

Now that we understand polarity, we can discuss the forces that hold molecules to each other. We have to be very careful about this, though: we are not discussing bonds, which are forces *within* a molecule that hold one atom to another and make the whole thing a discreet unit. Typical bonds have strengths somewhere between 60 and 100 kcal/mol (the significance of these units will become clear later). In this section I am instead discussing those forces that hold one molecule to another, *inter*molecular forces, where "inter" means "between." These are much weaker forces, something like 1–5 kcal/mol. We know there are such forces, because without them everything would be a gas, with all molecules flying around independently. Liquids and solids exist as such because something makes the molecules want to hang around together. There is some attractive force *between* molecules, and we want to figure out what this might be. (Note that these same forces can operate within the same molecule, from one chunk of it to another distant chunk, and when they do they are called *intra*molecular attractions.)

Intermolecular forces in general are called "van der Waals forces." There are actually several different kinds. They are all related, and all electrical in nature, but some are more obvious than others.

Dipole–Dipole

Let's start with the easy kind. We established in the previous section that certain molecules are polar. That is, the molecule as a whole, not just the bonds in it, can be treated as a blob with one end more positive than the other, which is more negative: a so-called "dipole."

It is not hard to imagine that when several of these get together they tend to line up so that the positive end of one is near the negative end of another, maximizing attractive interactions.

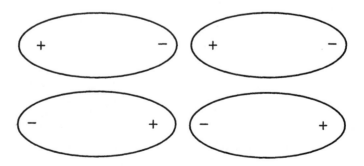

This is what we call dipole–dipole attraction. I hope it is pretty obvious. Below is an example of what it might look like in a real case, where the dotted lines represent the intermolecular (dipole–dipole) attractions.

$$H \quad\quad H$$
$$\overset{H}{\underset{|}{C}}=O \ -- \ \overset{H}{\underset{|}{C}}=O$$

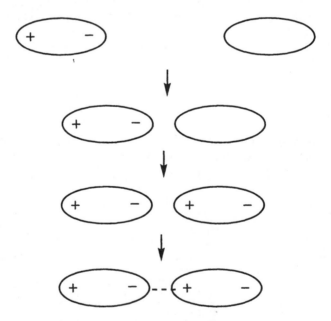

Dipole–Induced Dipole

The next level of intermolecular forces is what occurs when a polar molecule, a dipole, finds itself next to another molecule that is not polar. You would naturally think that there would be no particular attraction between them, but you would be wrong. How can there be attraction between a blob with + and − ends and another blob with no such poles? The answer lies in the fluid nature of electrons. A molecule (blob) is not like a piece of wood, just sitting there. It is a fluid blob, with jiggling nuclei sitting near their prescribed places and a sea of electrons surrounding them. The electrons are like a cloud of gnats around your head, always there, getting in your eyes and ears and nose, on the average symmetrically distributed, with no more gnats near your left ear than your right, but they are not locked in place. If you bring some insect repellent near your left ear, they will back off, creating a higher concentration of gnats near your right ear. The same is true with electrons in a molecule: if you take a non-polar molecule and bring up the positive end of a polar molecule, the electrons in the non-polar molecule will readjust in response so as to turn the non-polar molecule temporarily into a dipole, staying that way for as long as the polar molecule is around. The dipole (the polar molecule) *induces* the non-polar molecule to become a dipole, and while it is a dipole there can be attractions: Below is a cartoon representation of this.

This kind of interaction is called a "dipole–induced dipole" interaction.

This exhausts the kinds of interactions possible for polar molecules: the partner can be either polar or non-polar, and the force will be either dipole–dipole or dipole–induced dipole. But what possible attraction can there be between two non-polar molecules? Again, we know there is some, because many non-polar substances, like gasoline, are liquids, and others, like wax, are solids. What can possibly hold these molecules to each other?

London Forces

The answer lies in a trick phrase I inserted into the gnat analogy above: "The electrons are like a cloud of gnats around your head, always there, getting in your eyes and ears and nose, *on the average* symmetrically distributed, with no more gnats near your left ear than your right, but they are not locked in place." We saw that the gnat population could redistribute itself in response to some external stimulus. Those of you with experience with gnats know that an external stimulus is not really necessary. Although *on the average* there may be no more gnats around one ear than the other, at any random time there might be. This is entirely unpredictable, and if at some random time there are more gnats around your left ear than your right, at some other unpredictable time the reverse will be true. Nevertheless, the point is that even non-polar molecules can be polar, all by themselves, for brief instants. We call this polarity a temporary dipole. Once a non-polar molecule has become like this, even though the situation is brief and temporary, it behaves just like a real dipole in being able to persuade a neighbor molecule to respond; that is, it can induce a dipole in a neighbor. Of course, this also lasts only as long as the original molecule itself remains a dipole. What we end up with is an attraction called a "temporary dipole–induced dipole" interaction. Although these are brief and fleeting, and also weak compared to dipole–dipole attractions, there can be many of them along the surface of a non-polar molecule, and they are actually pretty effective at holding molecules to each other. Below is a cartoon representation of this force. It is also called a "London dispersion force," or just "London force." All molecules experience these forces, but they are secondary to dipole–dipole interactions in polar molecules. However, they do account for the fact that, in a rough way, the bigger the molecule, the more tightly it is able to stick to its neighbor. The key here is that this kind of brief attraction can happen in small areas on the surface of a molecule, and there can be many such occurrences at once wherever the molecules are touching.

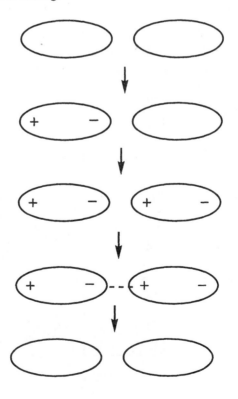

Comparisons

Well, that is a lot of speculation—is any of this real? Let me try to persuade you with a few comparisons. You should have a gut feeling that dipole–dipole interactions ought to be stronger than any of the other forces, since these are permanent electric charges rather than fleeting ones. Therefore, other things being equal, polar molecules ought to stick to each other better than non-polar ones. Recall, molecules that stick tightly to their neighbors are solids, those that stick loosely are liquids, and those that do not stick at all are gases. How can we tell how sticky molecules are? One way is to measure how hot we have to get them before they unstick. In other words, high melting and boiling points ought to be associated with molecules that stick to each other tightly; the lower the melting or boiling point, the more the molecules tend to ignore each other. Thus, polar molecules should have higher melting or boiling points than non-polar ones of the same general size and shape. (There are some other factors that go into melting points, but we need not worry about them here.) Indeed, if you look at the first two entries in the table below, you will find this to be the case. Br_2 is clearly non-polar, since neither end is more electronegative than the other, but ICl is polar, so ICl has a higher boiling point.

With respect to London forces, we can make two predictions. One is that larger molecules, with more surface area on which to interact and more electrons to play around with, should stick together better than smaller ones. The next four entries should persuade you that this is indeed the case. You should also predict that there could be a shape effect. Consider three everyday objects that occupy similar volumes; say, a basketball, a football, and a bologna (before slicing). One is a sphere, one a bit oblong, the third very elongated. If you put two (or more) basketballs side by side, they don't touch in many places. Footballs will pack somewhat better, while bolognas (think of cigars in a box) pack very nicely, touching each other all along the length of the things. The same logic applies to molecules. SF_6 (take my word for it) is a sphere; Br_2 is oblong, and $n\text{-}C_{10}H_{22}$ is a straight chain of carbons, very cigar-like in shape. As you can see from the table, there is a substantial shape effect, even though these three molecules weigh approximately the same, and thus have about the same volume.

Molecule	Molecular weight	mp	bp
Br–Br	160		59°C
I–Cl	162		97
F–F	38	−223	
Cl–Cl	71	−102	
Br–Br	160	−7	
I–I	254	113	
SF_6	146		−64
Br_2	160		59
$n\text{-}C_{10}H_{22}$	142		174

Solubility

Why are we wasting our time with this? What do intermolecular forces have to do with anything useful? As you have seen, intermolecular forces from one molecule to another of the same type determine whether a substance is a solid, liquid, or gas. But perhaps more importantly, intermolecular forces from one molecule to another of a *different* type determine solubility. What kinds of solutes (the substance being dissolved) dissolve in what kinds of solvents (the substance doing the dissolving, usually in significant excess), and why?

Start by examining what happens when a salt, like sodium chloride (ordinary table salt), dissolves in water. A salt is made up of ions with full, heavy-duty charges on particles. Some particles are positively charged; others are negative. These particles ("ions") attract each other strongly and arrange themselves in one of various types of arrays like the following, where the dark balls might be positive ions and the light ones negative. You should imagine this array continuing in the third dimension also, toward you and away from you. You can see that each dark ball will be surrounded by light ones of opposite charge, and vice versa. This should be an ideal situation, and it is hard to imagine anything better. The forces involved are very strong, and, indeed, it is extremely hard to break up such a *crystal* by heating it (i.e., salts have very high melting points).

Old vs. New Forces

What happens when one of these crystals dissolves in water? Each ion breaks out of the lattice it is in, and instead of being surrounded by ions of the opposite charge, it becomes surrounded by water molecules oriented so that the proper ends are pointing toward the ion. Each ion has lost the wonderful interactions it had with its neighbors, but these are replaced by new interactions with water. Now, interactions between ions and water are pretty good, but they are not as good as the old ones, since each ion used to have whole charges to surround it, and now it only has partial charges. On the other hand, look at this from the water's point of view: whereas before each molecule had other dipoles to interact with nearby, many of them now have something even better, a whole charge. So the ions are worse off than before, but the water is better off.

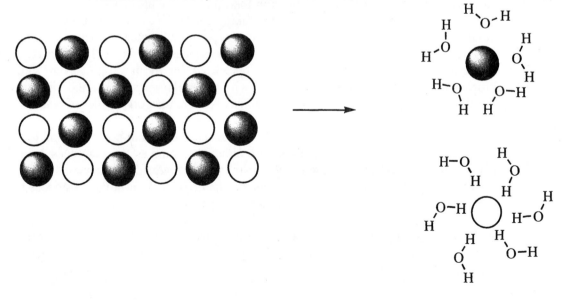

One might imagine that this leads to a stalemate, and it could, except for one factor we have not yet discussed. The ions in the crystal are locked into place, whereas after dissolution they are free to move around wherever they want. *Freedom of motion* is an important concept in science, and it has a special name: **entropy**. Entropy is a measure of the randomness (freedom) in a system, and randomness is good. Because this system has more entropy after dissolution than before, the salt dissolves.

The process described above works only because the ion–water interactions are strong enough to come close to compensating for the loss of the ion–ion interactions. Consider what would happen if you tried to dissolve salt in hexane (a component of gasoline). You would lose the ion–ion interactions of the salt, and replace them with new ones, but this time with ion–hexane interactions, which are pretty weak. It is also true that you would have entropy working in your favor. However, the new interactions are not similar enough to the old ones to allow this process to occur. As a general rule, you can expect salts to dissolve in water but not in non-polar substances. (Interaction forces are usually more powerful than entropy effects.)

Next, let's try a polar substance in another polar substance, let's say ethanol in water. You would lose ethanol–ethanol interactions, which are dipole–dipole interactions and not too bad, and you would also lose some water–water interactions, which are the same, but you would gain new water–ethanol interactions, which are also dipole–dipole interactions and just about the same as what you lost. What is more, you would gain entropy by allowing more randomness for the system. So, in general, you should expect a polar substance to dissolve in a polar solvent. You knew this, because you know that the ethanol in beer and wine does not separate from the water it is in.

What happens with a non-polar substance and a polar solvent, or vice versa? You lose the interactions in the non-polar substance, which are pretty weak London forces, and replace them with dipole–induced dipole interactions, which are a bit better. But you also lose the dipole–dipole interactions in the polar substance, which are pretty good, and replace them with dipole–induced dipole interactions in the mix. These are worse than what you started with. Again this appears to be a wash, and taking entropy into account, it turns out that there are in fact many instances of polar and non-polar substances mixing to some extent. However (and this is important), water is a special case, as we will see shortly, and the interactions it loses usually cannot be compensated for by new ones, even with entropy thrown in. In general, water does not mix with non-polar substances, and vice versa. You already knew this, because you have all seen gasoline floating on puddles of water.

Now consider a non-polar substance like cholesterol in a non-polar solvent like hexane. Again, the forces lost by both partners are about the same as the new ones gained in the mixture, but the mixture has more entropy than the separated substances. So you should expect dissolving (mixing) to occur. (Actually the drive for non-polar substances to be near each other is pretty strong, and has its own term: *hydrophobic interactions*.)

In summary, polar substances dissolve in polar solvents but not in non-polar solvents; non-polar substances dissolve in non-polar solvents but not in polar ones. Put another way, "like dissolves like."

WHO CARES? YOU DO!!

Consider what would happen to a poor molecule with two parts, one polar and the other not. An example might be a long chain of carbons and hydrogens (non-polar) attached to a piece with a charge, as shown below.

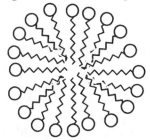

The part inside the brackets is non-polar and wants to hang around with other non-polar things. The part outside the brackets is charged, an ion, and wants to hang around water. Imagine you took this thing and stuck it in water. What would it do? (What would you do?) The clever thing for such molecules to do is to group together in such a way that all the non-polar parts are near each other while the polar parts are near water. There are two common ways for molecules to pull this off. The first is for them to form spheres with the polar "heads" at the surface and the non-polar "tails" hanging into the middle, as shown below. This way the tails all think they are in a non-polar environment, and the heads all think they are in water. What an ingenious solution! Spheres like this are called micelles.

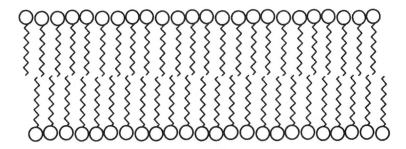

Can you think of a use for them? Consider washing your hands. You run them under water—anything that dissolves in water will dissolve and go away, but there is still stuff left. This must be somewhat non-polar stuff (e.g., oily) because it does not dissolve in water. How can you get rid of it? Perhaps if you had some micelles around, you could push the non-polar stuff into the middle of a micelle, where it would think it is in a non-polar solvent and dissolve, yet the whole sphere could still float happily in water and could be washed away! Micelles like the above are what make soap work.

The second way molecules find to accommodate the desire to be in both polar and non-polar environments is to form back-to-back rows, as shown below.

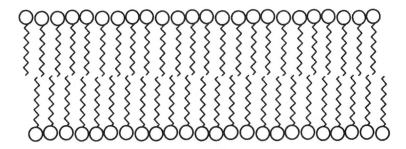

As in a micelle, the non-polar ends look around and see nothing but other non-polar things, and the polar heads look around and see nothing but water. If you can imagine a large array like this continuing in three dimensions and bending around to form the wall of a capsule, then you should see that you have created a whole system with water on the inside,

the coating, and then more water on the outside. In other words, you have made a capsule of water environment that is separated from the rest of the water environment by a non-polar barrier. This is a cell, like the biological kind, and real cell membranes are in fact composed of "bilayers" like the one shown above. The actual molecules involved are phospholipids, which look something like this:

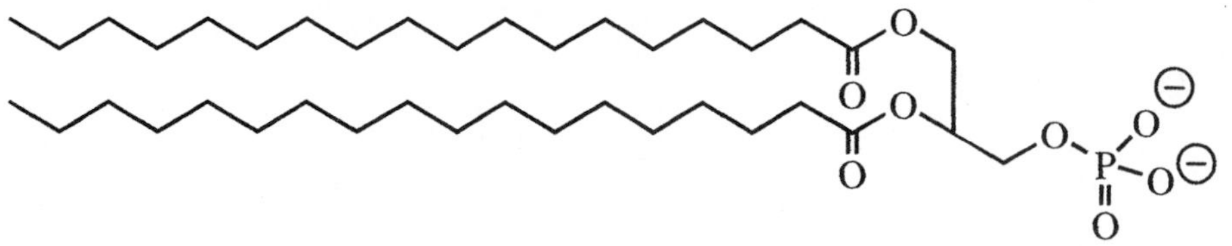

These molecules are not exactly like soap molecules, but they share with soap the characteristic of having polar heads and non-polar tails.

The desire of non-polar substances, or parts of substances, to stay together and avoid water is often referred to as a "hydrophobic" force. ("Hydrophobic" comes from the Greek, meaning "water-fearing.") The area inside a micelle or a lipid bilayer is called a hydrophobic region, and pieces of molecules that are non-polar are sometimes called hydrophobic groups. Polar molecules, or parts of molecules, are sometimes called "hydrophilic" ("water-loving").

A Special Force: The Hydrogen Bond

I mentioned above that water is special: the forces holding water molecules to each other are stronger than those holding most other polar molecules together. What is this all about?

There is a special intermolecular force called a "hydrogen bond" that operates only in very limited circumstances. It is unfortunate that the force is called a hydrogen bond because there is no "bond" in the normal sense of the word (although some debate exists about this), a bond being the glue that holds adjacent atoms together within the same molecule. A hydrogen bond is only about 1/10 to 1/20 as strong as a normal bond, but still considerably stronger than other intermolecular forces, on the order of 5 kcal/mol rather than 1. The hydrogen bond is a subset of dipole–dipole interactions and is a particular, identifiable interaction that operates from a lone pair on one <u>atom</u> to a hydrogen atom bonded (*really* bonded, covalently) to a different <u>atom</u>. The key feature of a hydrogen bond, and what makes the circumstances very limited and very special, is that *both* of the underlined atoms in the above sentence have to be fluorine, oxygen, or nitrogen. In a picture, a hydrogen bond is usually shown as a dotted line, ⋯⋯, as shown below.

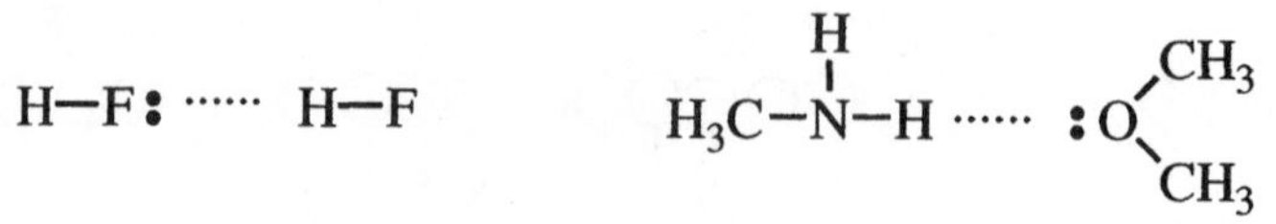

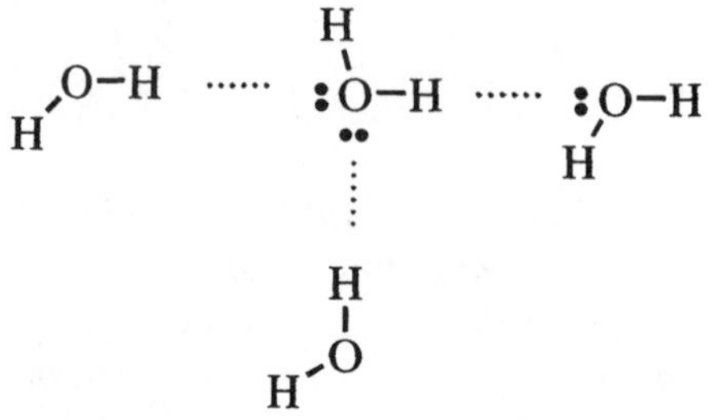

Notice that in the preceding pictures I am not showing all the lone pairs, only the ones I want to emphasize because they are involved in the hydrogens bonds I have illustrated. Hydrogen bonds are stronger than any other kind of intermolecular force, so molecules subject to this kind of interaction tend to stick to each other quite tightly, making melting and boiling points higher. For example, you have already seen a size trend in the halogens, where the smallest one, F_2, has the lowest boiling point. Based on this, you would expect other molecules that are similar to each other to follow a similar pattern (the smaller ones should have lower boiling points). As you can see from the chart and graph below, in the carbon series this is true, but in the other three series, the smallest member has a much higher boiling point than expected, because its molecules can hydrogen bond to each other.

Compound	bp	Compound	bp	Compound	bp	Compound	bp
CH_4	−164	NH_3	−33	H_2O	+100	HF	+33
SiH_4	−112	PH_3	−87	H_2S	−61	HCl	−87
GeH_4	−89	AsH_3	−55	H_2Ge	−42	HBr	−67
SnH_4	−52	SbH_3	−17	H_2Se	−2	HI	−35

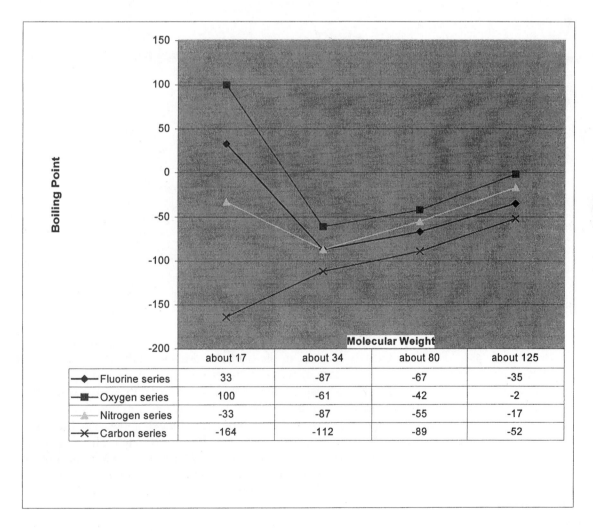

	about 17	about 34	about 80	about 125
◆ Fluorine series	33	-87	-67	-35
■ Oxygen series	100	-61	-42	-2
▲ Nitrogen series	-33	-87	-55	-17
✕ Carbon series	-164	-112	-89	-52

Another example is the comparison of CH_3OCH_3 with CH_3CH_2OH. These are isomers, compounds with exactly the same atoms, just arranged differently. The first, dimethyl ether, has lone pairs on an oxygen but no hydrogens bonded to F, O, or N. Thus, it could H-bond to other molecules that *do* have such hydrogens, but it cannot H-bond to itself. The second molecule, ethanol, also has lone pairs on an oxygen, and furthermore there is a hydrogen atom bonded to an oxygen. This molecule *can* form H bonds between molecules of the same substance. The boiling points bear out our prediction: dimethyl ether has a boiling point of –25 °C, while ethanol has a bp of +79°.

Even within the category of molecules that can hydrogen bond to each other, water is special, because it is the only substance that has both two lone pairs (on F, O or N) and also two H's (on F, O, or N). Thus, water has the ability to form four H bonds per molecule: two involving its own lone pairs, with hydrogens from other molecules, and two involving its hydrogens, with lone pairs from other molecules. Every hydrogen and lone pair can get into this act. The result is an extensive network of hydrogen bonds that can lock all the atoms in a rigid array. When solid (ice), this is how water is arranged. Below is a picture that shows this beautiful three-dimensional array. Notice the six-fold symmetry in the molecules, a symmetry that is reflected in the geometry of snowflakes.

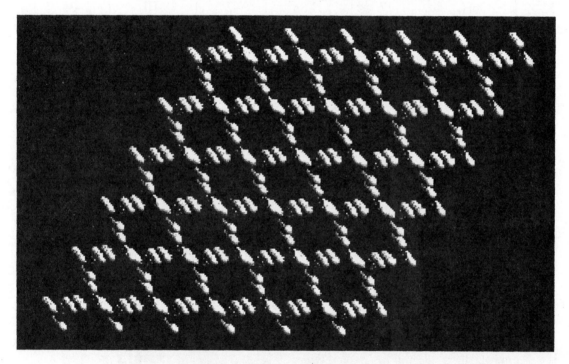

Another interesting feature of solid water is that there is a lot of empty space in its structure, as you can see. This is extremely unusual in the natural world—usually solids are packed more tightly than liquids, but in the case of water the reverse is true. Water expands when it freezes, because it settles into this extensively hydrogen-bonded structure with lots of space in it. Many of you have probably broken bottles of soda by putting them into the freezer to chill and forgetting about them. You may not have thought too much about it, but in a very real sense this particular feature of hydrogen bonds is responsible for much of life on earth: the fact that ice is less dense than water means that ice floats on water. This in turn means that when a lake or river freezes, it does not freeze all the way to the bottom. There is still liquid water at the bottom, allowing life processes to go on through the winter.

DNA Structure

WHO CARES? YOU DO!!

Hydrogen bonds are extremely important biologically. In many ways, they run all the important machinery of cells. You probably know that DNA is the genetic material that allows organisms to pass information down from one generation to the next. The way this happens is that DNA consists of a matched pair of molecules, rather like the two halves of a zipper, and most of the time it is zipped up into its double form (the double helix). Periodically (when a cell divides), the DNA unzips, and then each half independently creates a new second half to mate with, thus creating two zippers where there used to be one. How is this molecular zipper held together? Hydrogen bonds! Each strand of DNA is a stacked set of molecules with hydrogens and lone pairs of electrons pointing out. A corresponding molecule with lone pairs where the first one had hydrogens, and hydrogens where the first one had lone pairs, lines up perfectly. Below is a picture of one such "base-pair interaction." In this picture, as is traditional, nitrogens are shown in blue, oxygens red, and hydrogens white. The yellow structures are more normal pictures of the same things to help you get oriented. If the following picture is in black and white, there should be a color version of it (and the next several pictures) in the color pages of this book.

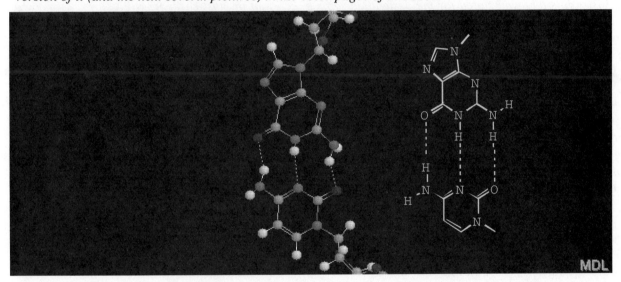

A series of these then line up as shown in the next picture:

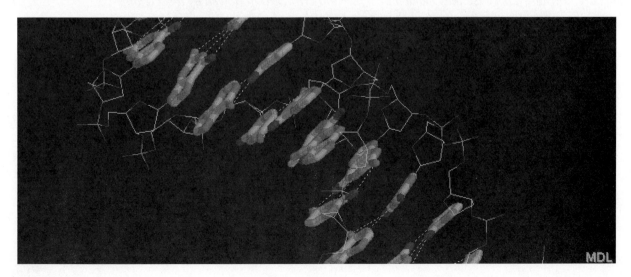

For a beautiful interactive look at a DNA strand, go to the following web site, http://www.umass.edu/microbio/chime/dna, and play around. You will need a plug-in called Chime, but it is free and well worth putting on your computer. The pictures above were stolen (legally) from this web site.

All the examples of hydrogen bonds I have shown so far are intermolecular; that is, they operate from one molecule to a different one. Hydrogen bonds are also possible within the same molecule, from one part of it to another ("intramolecular hydrogen bonds"). The ability to form a hydrogen bond with another part of a molecule would encourage those two parts to hang out near each other. In other words, hydrogen bonds can significantly affect the shapes of molecules.

Protein Structure

Why do shapes matter? Let me tell you about enzymes. Enzymes are the assembly-line workers of a cell. When it is time to make something you need (or take apart something you've eaten), the assembly line goes to work, passing molecules along from worker to worker. Each worker is an expert at what he does (tightening this bolt, snipping this bond, etc.) and is exquisitely well suited for the designated job, so well that in most cases he will refuse to do the job if the material handed to him is not exactly right. Enzymes have developed this expertise by acquiring the right shape, which allows for two things: first, recognition—the molecule being operated on has to fit just right into nooks and crannies in the surface of the enzyme (often compared to a lock and key)—and second, once inside, the enzyme has just the right scalpels, scissors, and staples stored at just the right places to do the operations called for. But how can this be, since an enzyme is nothing but a long piece of spaghetti? How does it acquire and hold a particular shape, such that every copy of the same enzyme has this particular shape and does the same job? Hydrogen bonds! The piece of spaghetti folds up in such a way as to maximize attractive interactions between various parts of the molecule, "intramolecular hydrogen bonds," and in so doing acquires a unique shape and function. There is, of course, more to it than that, but hydrogen bonds are major factors in the shapes enzymes take.

One example of this is shown on the next page. A common motif in enzyme structure is what is called the "alpha-helix," a spiral-shaped piece of protein that has hydrogen bonds running along its axis. I show on the next page a picture of an alpha helix, with the H-bonds in white (again in this picture, nitrogens are blue and oxygens are red, but this time hydrogens are not shown because they make the picture too crowded), followed by a skeleton drawing of hemoglobin, in which you ought to be able to see lots of alpha-helices (colored in red). Both of these pictures come from another nice web site you should explore, http://www.umass.edu/microbio/chime/hemoglob

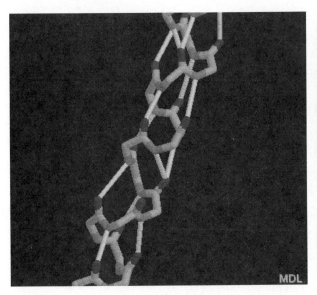
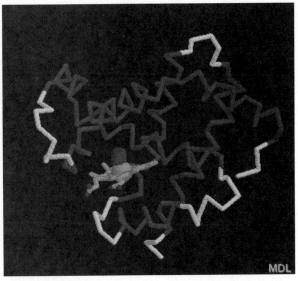

Another neat aspect of helices like these is that if you look down one you see a tube, with groups sticking out around the sides. Frequently enzymes are constructed so that in a helix the groups sticking out on one side are relatively polar, while the groups on the other side are hydrophobic. These tubes can therefore pack together in such a way as to leave the hydrophilic parts of the molecule soaking in water on the outside of the enzyme, while the hydrophobic parts are sequestered from the water and hidden on the inside of the enzyme. This is precisely what happens. The following image from the same web site shows an end-on view of one helix. The helix itself has been darkened so that only the side chains are shown. The gray part is hydrophobic, while the red and blue (oxygen and nitrogen) represents polar parts.

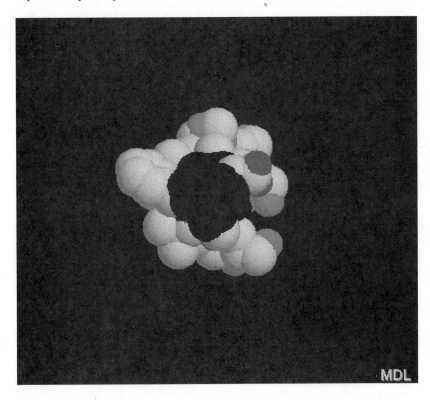

Problems

(Answers are provided at the end of the chapter for italicized problems)

1. a. *Show the geometry of each of the following molecules. Determine whether each is polar.*

   SF_2 CCl_2F_2 CH_3OCH_3 NH_3 $(CH_3)_3N$ CO_2 BF_3

   b. *Which pairs of the above molecules can form hydrogen bonds to each other?*

2. a. Determine whether each of the following molecules is polar.

   BF_3 NF_3 CCl_4 $CHCl_3$ $AlCl_3$ CH_3OCH_3 CH_3CH_2OH
   $CH_3CH_2CH_2CH_3$

   b. Name the chief intermolecular force responsible for holding groups of each of the above molecules together.

3. Which of the following pairs are capable of hydrogen bonding with each other? Illustrate the hydrogen bond, if there is one.

   NH_3 and CH_3OCH_3

   NF_3 and CH_3OCH_3

   CH_3CH_2OH and CH_3OCH_3

   CH_3CH_2OH and $CHCl_3$

   CH_3CH_2OH and NF_3

4. Consider the following compounds (**A**, **B**, and **C**) , all with similar molecular weights:

   A: $CH_3CH_2CH_2CH_3$ **B:** $CH_3CH_2CH_2OH$ **C:** CH_3COCH_3

   a. Which of the three should have the *lowest* boiling point? **A B C**

   Which should have the *highest* boiling point: **A B C**

   b. For each compound, identify the strongest force of attraction among the molecules (which tends to prevent boiling).

   A: Strongest force –

   B: Strongest force –

   C: Strongest force –

   c. Circle the compound or compounds that is/are capable of forming hydrogen bonds with water: **A B C**

Show how hydrogen bonding would occur in each case you identified. If *more than one type* of hydrogen bonding with water is possible in a given case, show *all* the possibilities.

5. *For the following species, identify (x or check mark or any other symbol) those that are polar. Identify those that can form hydrogen bonds with another identical molecule. Identify those that can form hydrogen bonds with ammonia (NH_3). Identify those that can form hydrogen bonds with dimethyl ether (CH_3OCH_3). For one species in this last category, draw a picture that shows the hydrogen bond.*

H_2O $AlCl_3$ NH_4^+ HBr

Is it polar?					
H-bond with self?					
H-bond with ammonia?					
H-bond with dimethyl ether?					

6. The acidic spray (largely composed of acetic acid, CH_3COOH) from a scorpion would not harm an attacking insect unless it contained 5% of caprylic acid, $CH_3CH_2CH_2CH_2CH_2CH_2CH_2COOH$. The function of this acid is to get past the waxy coating on the cuticle of the attacking insect, allowing the acetic acid to get inside and harm the attacker. The waxy coating is made of long chain alkanes containing 20–30 carbons. Explain why caprylic acid is able to penetrate the waxy coating while acetic acid is not. What kind of intermolecular forces are involved in the process that allows this penetration?

Selected Answers

1. a.

bent	tetrahedral	tet-bent-tet	tet	tet-tet	linear	trigonal
yes	yes	yes	yes	yes	no	no

b.

NH_3 can H-bond with each of the others, and each of the others can H-bond with NH_3 but no one else.

5.

H_2O $AlCl_3$ NH_4^+ HBr

	H_2O	$AlCl_3$		NH_4^+	HBr
Is it polar?	Yes	No	Yes	No	Yes
H-bond with self?	Yes	No	No	No	No
H-bond with ammonia?	Yes	No	Yes	Yes	No
H-bond with dimethyl ether?	Yes	No	No	Yes	No

Chapter 6
Quantities in Chemistry

So far we have been able to get through all our discussions without worrying about how much of anything we are dealing with. Of course, in real life we have to deal with quantities all the time. In this chapter we will

- Use an extended analogy to establish the concept of the mole

- Explain and practice how chemists use this method of counting molecules.

The main (almost sole) purpose of this chapter is to establish that the chemist's unit of counting is the mole. Yes, I know, most of you already know that. Almost all of you have solved problems using moles ("stoichiometry," pronounced "stoy-key-OMM-it-tree"); a fair number still could. Many of you thought you understood moles at the time; some of you still do. But in my experience, a great many students, while being able to manipulate numbers properly, have little clue as to what the concept of a mole is all about. This includes a fair number of those who think they understand it. Moles, for many beginning students, are a recipe to follow in order to get a right answer, rather than a concept that is logical and an obvious way to deal with huge numbers.

My approach to solving this problem is to avoid chemistry entirely and establish the concept in the context of real, tangible objects so that your brains will engage and think about it rather than turn into plug-and-chug mode. I urge you to take the following extended analogy seriously, to do the math and think hard about what I say—*even if you think you already understand moles*. The purpose of the analogy is to show you *why* you do the math that you do, to solidify the concept of what a mole really is. Please bear with me, we really are going somewhere with this. Don't blow it off: work through this section with me. When we get done, you may find that you have a better understanding of what this is all about. If not, you haven't lost anything!

I am assuming at this point that you are familiar with the number-handling material in Appendix 1.

THE PROBLEM WITH NUMBERS

We have previously established that atoms are unimaginably small. Recall, I mentioned that there are so many molecules in a single drop of water that we could spread them evenly all over the earth and still find nearly 200,000 of them on this page. It follows that the number of atoms or molecules in any visible sample of any material is humungous. If we are going to count them, we'd better establish some collective unit, like "dozen" (only bigger), that will allow us to discuss numbers of atoms or molecules without having to write out huge numbers. I have a proposal for one: let's invent a huge number and call it a gajillion.

What is a gajillion? I don't know and I don't care, and neither should you. It's a whopping big number, that's what, and that's all we need to know. But we can still use it. How?

Well, let's say you had a gajillion cars. How many engines do you have? I'll bet no one had any difficulty deciding that there would be a gajillion engines. Why? Because each individual car has one engine. A million cars have a million engines, and a gajillion cars have a gajillion engines, regardless of the actual value of the number "gajillion." This is not rocket science, or even chemistry; it is simple logic.

On the same gajillion cars, how many wheels do you have? Again, no one has trouble deciding there are 4 gajillion wheels. You could set this up mathematically as follows:

$$1 \text{ gajillion } \cancel{\text{cars}} \quad \frac{4 \text{ wheels}}{\cancel{\text{car}}} = 4 \text{ gajillion wheels}$$

Hmmm, but what if we started with 7.8 gajillion cars? Or 0.062 gajillion cars? No problem. Just do the same math. What if I told you that you had 15.73 gajillion wheels? How many cars do you have? Again, no problem, is it?

$$15.73 \text{ gajillion } \cancel{\text{wheels}} \quad \frac{1 \text{ car}}{4 \cancel{\text{wheels}}} = 3.93 \text{ gajillion cars}$$

We can do math involving gajillions, as long as we stay in gajillions, without having a care in the world what a gajillion is.

Counting by Weight

We can take this a step further: Let's say that a gajillion is the number of grains of salt in a ton of salt. Now the word "gajillion" is not just an abstract concept; it is one particular number; still big, still unknown, but with a specific value. Note that this does not make the previous example invalid: just because we define this number in terms of grains of salt, it is still possible to contemplate a gajillion cars, even if they would not fit on the continent. Now, we still do not know what a gajillion is, nor do we know the weight of a grain of salt. But suppose I told you that a grain of pepper, on average, weighs twice as much as a grain of salt. How much do a gajillion grains of pepper weigh? Well, a gajillion grains of salt weighs one ton, and each grain of pepper weighs twice as much as a grain of salt, so a gajillion of them must weigh twice as much as a gajillion grains of salt, right? So two tons. How about grains of X? Assume each of these weighs 7 times as much as a grain of salt. So a gajillion of them must weigh 7 tons, right? This is not hard stuff.

We can now count grains by weighing. How many grains of salt are there in 17.4 tons? Easy:

$$17.4 \; \cancel{\text{tons of salt}} \; \text{grains} \quad \frac{1 \; \text{gajillion salt grains}}{1 \; \cancel{\text{ton of salt}} \; \text{grains}} \quad = 17.4 \; \text{gajillion salt grains}$$

How many grains of X are there in 17.4 tons? Just as easy:

$$17.4 \; \cancel{\text{tons of X}} \; \text{grains} \quad \frac{1 \; \text{gajillion X grains}}{7 \; \cancel{\text{tons of X}} \; \text{grains}} \quad = 2.49 \; \text{gajillion X grains}$$

Now let's streamline our language a little. Instead of saying that a grain of X weighs 7 times as much as a grain of salt, let's define the weight of a grain of salt (which we don't know, remember?) as 1 salt grain unit (sgu). Now we can say that a grain of X weighs 7 sgu. And we already know that a gajillion of them weighs 7 tons. Hmm, this is pretty convenient. Because of the way we defined things, we don't even need to do any math: whatever something weighs in sgu's, a gajillion of them weighs that same number of tons. So grains of rice, which weigh 53 sgu's each, must also weigh 53 tons/gajillion.

Using Recipes

Now let's put you in charge of ordering supplies for the army. The standard recipe for making army rice, in traditional sadistic army fashion, is written in grains rather than in weight. So be it. Here it is:

50 grains rice + 10 grains salt + 1 grain pepper makes delicious army rice.

Will you order 50 tons of rice, 10 tons of salt, and 1 ton of pepper? I hope not, or you'll be busted to private! The recipe is not written in weight. Grains of rice, salt, and pepper do not weigh the same. You have to deal with this. But that's OK, because you already know how!

First of all, if the recipe says 50 grains of rice, 10 grains of salt, and 1 grain of pepper, it should not be difficult to see that it will work just as well with 50 gajillion grains of rice, 10 gajillion grains of salt, and 1 gajillion grains of pepper. Further, you know that salt weighs 1 sgu, or 1 ton/gajillion. Pepper weighs 2 sgu, or 2 tons/gajillion. Rice weighs 53 sgu, or 53 tons/gajillion. So 50 gajillion grains of rice weighs 2650 tons; 10 gajillion grains of salt weighs 10 tons; and 1 gajillion grains of pepper weighs 2 tons. So you should order these weights: 2650 tons of rice, 10 tons of salt, and 2 tons of pepper.

Perhaps there is a shortage of rice, and there is only 1000 tons available. How much salt and pepper do you need to go with it? Easy. First, we count how many grains of rice we have:

$$1000 \; \cancel{\text{tons}} \; \text{of rice} \quad \frac{1 \; \text{gajillion}}{53 \; \cancel{\text{tons}}} \quad = 18.87 \; \text{gajillion rice grains}$$

Notice that we do not use the recipe to make this count—it is true of 1000 tons of rice no matter what we plan to do with it.

Now we use the recipe to figure out how many grains of salt and pepper we need to go with this much rice:

$$18.87 \text{ gajillion rice grains} \quad \frac{10 \text{ salt}}{50 \text{ rice}} \quad = 3.77 \text{ gajillion salt grains}$$

$$18.87 \text{ gajillion rice grains} \quad \frac{1 \text{ pepper}}{50 \text{ rice}} \quad = 0.377 \text{ gajillion pepper grains}$$

Now we can calculate the weight of that much salt and pepper, using the conversions we already established before.

What if there were a shortage of all three ingredients? You can get only 1 ton of pepper, 566 tons of rice, and 3.74 tons of salt. Assuming that you cannot modify the recipe, what should you mix up?

Again, the key is to count what you DO have. Do this yourself, and confirm that you have 0.5 gajillion grains of pepper, 10.68 gajillion grains of rice, and 3.74 gajillion grains of salt. Now we have to figure out what will run out first when we start to make rice. One way to do this is to figure out how much of each ingredient you would need to completely use up one of them, say, the pepper. If you have *more* than you need of everything else, then pepper will run out first. If you have *less* than you need of some other ingredient, then pepper will not run out first; the other ingredient *might* be the one that will. Repeat with the new possibility to find out whether *it* will run out first. Use the recipe, like this:

$$0.5 \text{ gajillion pepper} \quad \frac{50 \text{ rice}}{1 \text{ pepper}} \quad = \quad 25 \text{ gajillion rice}$$

We would need 25 gajillion grains of rice to go with the pepper we have, but we have only 10.68 gajillion grains of rice. Clearly, we have enough pepper: rice will run out long before pepper. But we have not yet tested the salt—maybe it will run out before the rice. So let's test it. No point in testing against the pepper, since we already know we have more than enough of that. Test the salt against the rice:

$$10.68 \text{ gajillion rice grains} \quad \frac{10 \text{ salt grains}}{50 \text{ rice grains}} \quad = \quad 2.14 \text{ gajillion salt grains}$$

We would need 2.14 gajillion grains of salt to go with the rice we have; we have 3.74 gajillion grains, more than enough. So we will run out of rice before anything else, and we must base all future calculations on that.

Test Yourself 1

What weight of salt and pepper will be left over after you make all the rice you can?

DOING IT WITH CHEMICALS: MOLES AND STOICHIOMETRY

There is little left to say about how to do this with molecules, except to review what to most of you has probably become obvious: the "gajillion" we have been using above, defined as the number of salt grains in a ton, gets replaced by the "mole," defined as the number of hydrogen atoms in a gram. A mole is nothing more than a gajillion: a name for a humungous number. The abbreviation for mole is mol, which seems silly, since it only saves you one letter, but that adds up, and it also saves you from deciding whether to write "mole" or "moles."

Let's return to an earlier example: the following is a verbatim repeat from the beginning of the chapter, replacing "gajillion" with "mole."

Let's say you had a mole of cars. How many engines do you have? I'll bet no one had any difficulty deciding that there would be a mole of engines. Why? Because each individual car has one engine. A million cars have a million engines, and a mole of cars has a mole of engines, regardless of the actual value of the number "mole." This is not rocket science, or even chemistry; it is simple logic.

On the same mole of cars, how many wheels do you have? Again, no one has trouble deciding there are 4 moles of wheels. You could set this up mathematically as follows:

$$1 \text{ mol } \cancel{\text{cars}} \quad \frac{4 \text{ wheels}}{\cancel{\text{car}}} = 4 \text{ mol wheels}$$

Can we transfer this to chemistry? Sure! Here is the same section again, in chemical terms:

Let's say you had a mole of methane, CH_4. How many carbon atoms do you have? You should have no difficulty deciding that there would be a mole of carbon atoms. Why? Because each individual CH_4 molecule has one C atom. A million CH_4's have a million C's, and a mole of CH_4's have a mole of C's, regardless of the actual value of the number "mole." This is not rocket science, or even chemistry; it is simple logic.

On the same mole of CH_4's, how many H's do you have? Again, you should have no trouble deciding there are 4 moles of H's. You could set this up mathematically as follows:

$$1 \text{ mol } \cancel{CH_4} \quad \frac{4 \text{ H}}{1 \, \cancel{CH_4}} = 4 \text{ mol H}$$

OK? Likewise, the "sgu" that we defined in the first section gets replaced by "amu." We set the value of a mole to be the number of hydrogen atoms in 1 g of hydrogen. Since a hydrogen atom weighs 1 amu, and a mole of them weighs 1 g, the units "amu" and "g/mol" are interchangeable. (Recall an amu was defined as the weight of a proton or neutron. Since hydrogen has one proton and no neutrons, a hydrogen atom weighs 1 amu. Other atoms are heavier, and their weight in amu is the number of protons and neutrons added together: it is the weight of the atom relative to hydrogen.) If you know what any substance weighs in amu, you also know what a mole of it weighs in grams.

We never did figure out the number of salt grains in a ton, and by the same token we will hardly ever need to use the number of atoms in a mole. Of course, you already know what this number is: it is "Avogadro's number," 602,300,000,000,000,000,000,000 (6.023×10^{23}). But really, the only time you will ever need that (in this course) is if you want to play the game of counting molecules in a sample. Remember I told you earlier about how many molecules of water will be on your page? "My favorite [example] is that if you imagine taking a single drop of water and allowing it to spread out and out and out until the molecules in it are evenly distributed over the entire surface of the earth, there would still be nearly 200,000 of them on this page. Later on we will calculate this; for now, you need simply to accept the fact that the size of a molecule or atom is unimaginably small; the number of them in any sample you can see is unimaginably large." We could now do this calculation, it's easy.

First, we need to know that there are about 20 drops of water in a mL. And of course water, H_2O, weighs 18 amu or 18 g/mol. So

1 drop x 1 mL/20 drops x 1 g/mL x 1 mol/18 g x 6.023 x 10^{23} molecules/mol = 1.67 x 10^{21} molecules in a drop of water

Next we calculate the surface area of the earth. We need to know that the diameter of the earth is about 8000 miles, which means that the radius is about 4000 miles, and the surface area of a sphere is $4\pi r^2$. So

4000 mi x 5280 ft/mi x 12 in/ft = 2.53 x 10^8 in. This is the radius of the earth.

surface area = 4 x π x (2.53 x 10^8)^2 = 8.1 x 10^{17} in^2.

So we have 1.67 x 10^{21} molecules spread over 8.1 x 10^{17} in^2, or 2069 molecules/ in^2. This page is about 75 in^2, so there would be a bit more than 150,000 molecules on it.

So let's get back to stoichiometry and do some exercises. Here is one of the first reactions you may have seen in the past:

$$HCl + NaOH \rightarrow H_2O + NaCl$$

This is just like the recipe for rice we used previously. The groups of letters are molecules, and can be thought of as the "grains of something" that were in the recipe. "NaOH" in the recipe is a grain (molecule) of sodium hydroxide, and by adding the atomic weights of Na, O, and H we can determine that this grain weighs 23 + 16 + 1 = 40 amu, or 40 times as much as a hydrogen atom, or 40 g/mol.

First, we have to check the recipe (chemists call this an equation, which is unfortunate, since the two sides are not really equal) to ensure that we have accounted for all the ingredients. There must be the same number of H atoms, O atoms, Cl atoms, etc., on both sides of the equation. We call this "balancing the equation," and in most cases there is no way around trial and error. In this case the reaction is already balanced (2 H's, 1 O, 1 Na, 1 Cl). If there is no number in front of a chemical, it implies a "one."

How much HCl will it take to go with 60 g of NaOH? Here are the steps, and they will work for almost any stoichiometry problem: (1) Count your ingredients (in moles); if they are not presented to you in moles, convert them into moles. (2) Relate one chemical to the other, in moles, using the equation. (3) Change your new count into a weight, or whatever was requested. To count 60 g of NaOH you need a grams-to-mole conversion factor, which we just calculated to be 40 g/mol. So you should end up calculating that you have 1.5 mol of NaOH. You need 1.5 mol of HCl to react with this, and that weighs 54.75 g. Here's what this looks like, written out in gory detail:

$$60 \text{ g NaOH} \quad \frac{\text{mol}}{40 \text{ g}} \quad = \quad 1.5 \text{ mol NaOH} \qquad \text{This is what we start with}$$

$$1.5 \text{ mol NaOH} \quad \frac{1 \text{ mol HCl}}{1 \text{ mol NaOH}} \quad = \quad 1.5 \text{ mol HCl} \qquad \begin{array}{l}\text{This is what we will} \\ \text{need to go with the NaOH}\end{array}$$

$$1.5 \text{ mol HCl} \quad \frac{36.5 \text{ g}}{\text{mol}} \quad = \quad 54.75 \text{ g HCl} \qquad \text{This is how much that weighs}$$

Of course, one could string these calculations together in one line, but I like to stop after each to recognize what it is I have calculated.

What if the above equation had HI in it instead of HCl (and, of course, NaI as a product instead of NaCl)? You can go through the same process and get an answer. Go ahead, do it. The answer is the square root of 36864. Did you go through all the steps? You didn't have to! Of course, it doesn't hurt, but since we are encouraging thinking about what you are doing instead of rote data-processing, you should think about this. In the first problem we already determined that it would take 1.5 moles of HCl to react with 60 g of NaOH. It should be obvious that it will take the same amount, in number, of HI's as it did HCl's. So we can start the second problem two-thirds of the way through, and the only step left is to convert 1.5 moles of HI into grams.

What if the HI in our new equation was presented to us as a solution with a concentration of 3.7 moles/liter (mol/L, or M, pronounced "molar")? Easy enough: we have already established that we need 1.5 moles of HI (again starting 2/3 of the way through, because we have already done it). So we need to convert 1.5 moles into liters, and we have a handy conversion factor of 3.7 mol/L. So

$$1.5 \text{ mol x } 1\text{L}/3.7 \text{ mol} = 0.405 \text{ L} \text{ x } 1000 \text{ mL/L} = 405 \text{ mL}$$

Note: if you have not had concentrations before, or if you have struggled with them in the past, do not despair. A concentration is nothing more than a conversion factor between moles and volume. Just use it that way and you will be all right.

Test Yourself 2

How many g of NaCl are in 257 mL of a 0.783 M solution? Do it now!

If you have 25.0 mL of an unknown solution of HCl, and you determine that it takes 37.4 mL of a 0.686 M solution of NaOH to just react with it (no extra), what is the concentration of the original unknown solution? Do it now!

Hint: Calculate the number of moles in the NaOH solution. Since this reacts with HCl in a 1:1 ratio (we've just been using the equation), there must be exactly that many moles of HCl in the HCl solution. Divide moles by liters to get molarity.

If you have trouble doing these problems, seek help immediately!

Now, what if we used H_2SO_4 instead of HCl or HI? Our new equation looks like this:

$$H_2SO_4 + NaOH \rightarrow H_2O + Na_2SO_4$$

How much sulfuric acid will it take now to react with 60 g of NaOH? Careful! The equation is not yet balanced. It will not hurt to calculate that we have 1.5 moles of NaOH (which we already knew anyway), but we cannot relate this to the amount of H_2SO_4 needed until the equation is balanced. A few tries should result in the following balanced equation:

$$H_2SO_4 + 2\,NaOH \rightarrow 2\,H_2O + Na_2SO_4$$

which means that one molecule (or mole) of sulfuric acid will react with 2 molecules (or moles) of sodium hydroxide to make two waters and one sodium sulfate.

Now we can relate our quantities of chemicals:

1.5 mol NaOH x 1 H_2SO_4 / 2 NaOH = 0.75 mol H_2SO_4

and then convert that into grams:

0.75 mol H_2SO_4 x 98 g/mol = 73.5 g H_2SO_4.

Now we can get a little fancier. Given 100 mL of 3.0 M NaOH and 170 mL of 2.0 M H_2SO_4, how much Na_2SO_4 will be produced? This is no different from the rice problem: you just count what you have and make the right relationships. We have 100 mL of a NaOH solution. Can we count in moles? Yes, because we have been told the proper conversion factor, 3.0 mol/L. So:

100 mL x 1 L / 1000 mL x 3.0 mol / L = 0.30 mol of NaOH

Similarly,

170 mL x 1 L / 1000 mL x 2.0 mol / L = 0.34 mol of H_2SO_4

Now we proceed just like any other problem. First we have to figure out which will run out first:

0.30 mol NaOH x 1 H_2SO_4 / 2 NaOH = 0.15 mol H_2SO_4 needed, but we have more than that. So NaOH will run out first. We call this the "limiting reagent," and we need to base all future calculations on that.

0.30 mol NaOH x 1 Na_2SO_4 / 2 NaOH = 0.15 mol Na_2SO_4 expected x 142 g / mol = 21.3 g.

This is called the "theoretical yield," the amount you *ought* to get from this reaction. But most reactions do not actually work at maximum efficiency. So if you do this reaction and obtain only 18.0 g, how well did you do? "Percent yield" is defined as what you really got compared to what you should have gotten: actual over theoretical = 18.0 g / 21.3 g = 84.5% yield.

Some problems will require you to use a density, usually given in g/mL. This is no different from any other conversion factor, except that it varies from substance to substance; but if you have the density for a particular substance, you can easily convert weight into volume or *vice versa*.

There is a tremendous variety of ways people can construct stoichiometry problems, and you will be treated to many of them, but if you understand that all equations are written in moles and all counting needs to be done in moles, you will be able to do them all. Practice, practice, practice!

The bottom line, take-home message of this chapter is the following: chemists count in moles. Although many quantities are reported in weight, volume, or one of many other units, the most important information is the number of moles, and for most calculations whatever you are given will have to be converted into moles. Chemical equations relate chemicals in terms of moles. **To use an equation, you have to change whatever you are given into moles, compare the result to other chemicals in the equation (in moles), and then convert from moles into whatever final quantity is desired.** There are shortcuts around all this, but if you do it the way I've described you will always succeed.

WHO CARES? YOU DO!!

I recently overheard a student saying, "Why do I have to learn all this crap? All I'm going to do is swab arms and stick a needle in! I don't need to know all this!" So why am I making you learn all this?

First of all, say a prayer for the patients of the student I overheard. Nurses are not robots, and anyone who thinks good care means mechanically following protocols will be a lousy nurse. Even nurses occasionally receive inaccurate instructions, and if they have no feel for what they are supposed to be doing, what the chemicals are, and how much they are administering, there is bound to be a mistake somewhere along the line. Everyone in the health care industry needs to know how to weigh and measure things, and what these weights and measures mean. No matter what you end up doing in science, you will at some point have to make a solution of something. It is important that you know what you are doing, and that you understand something about chemical counting.

Combustion Analysis Problems

There is another context in which stoichiometry is important. It is not really any different from what you have seen already, but the point of view is inverted. An old-fashioned (but still-used) method of analysis of an unknown compound is to burn it and see what elements are in it, and in what quantities. For simplicity, let's consider a case where the compound in question contains only carbon and hydrogen. If you burn this thoroughly in oxygen, *all* of the carbon in the original sample will be converted into CO_2, and *all* of the hydrogen will become H_2O. Further (and this is very important), there can be no carbons or hydrogens (and therefore no CO_2 or H_2O) from any other source. Thus, by counting the carbons in the CO_2 product, and the hydrogens in the H_2O product, you have a direct count of the carbons and hydrogens in your unknown! The concept is simple; the difficult thing, for many students, is that you have to do the calculations *without* a balanced equation. Here's how:

Let's take 10.4 mg of our unknown. Burn the heck out of it and collect the CO_2 and H_2O generated. Imagine we find 31.2 mg of CO_2 and 17.0 mg of H_2O. What was the formula of the compound we started with?

How on earth do you do that? Just like any other stoichiometry problem: you count. Let's start by counting carbon atoms. We got 31.2 mg of CO_2, which calculates out (do it!) to 7.091×10^{-4} moles of CO_2 molecules. How many carbon atoms are there? Each CO_2 molecule has one carbon atom, so we must have 7.091×10^{-4} mol of carbon atoms. Every one of these came from the unknown, so now we know how many carbon atoms were in our sample. We can set this up in logical, linear fashion as follows:

$$31.2 \text{ mg } CO_2 \quad \frac{1 \text{ g}}{1000 \text{ mg}} \quad \frac{1 \text{ mol } CO_2}{44 \text{ g } CO_2} \quad \frac{1 \text{ C atom}}{1 \text{ } CO_2 \text{ molecule}} = 7.091 \times 10^{-4} \text{ mol C atoms}$$

Now we can do the same for the hydrogen:

$$17.0 \text{ mg } H_2O \quad \frac{1 \text{ g}}{1000 \text{ mg}} \quad \frac{1 \text{ mol } H_2O}{18 \text{ g } H_2O} \quad \frac{2 \text{ H atoms}}{1 \text{ } H_2O \text{ molecule}} = 1.889 \times 10^{-3} \text{ mol H atoms}$$

So now we know exactly how many carbon and hydrogen atoms were in our unknown. But that was in 10.4 mg of our unknown, which contained many zillions of molecules. What we really want to know is, how many carbons and hydrogens are in *each* molecule of our unknown? We have no way of knowing that, but we do have a way of making a guess. We know that in each molecule, the number of carbon and hydrogen atoms will be integers—there is no such thing as a fractional atom—so we can figure out the ratio of carbon to hydrogen in this substance and see what whole numbers would give the same ratio. You should recognize that the ratio of the numbers we have calculated will not change if we multiply or divide both by the same new number, so we can manipulate them at will, as long as we do the same thing to both. The best approach is usually to make one of these numbers be "one." The easiest way to do this, mathematically, is to divide both (or all, if there are more than two) by the smallest one.

If we divide both of our numbers by 7.091×10^{-4}, we get 1 mol of carbon atoms and 2.66 mol of hydrogen atoms. So the atoms in our unknown are in the ratio of 2.66 : 1, or, if you like, the formula appears to be $C_1H_{2.66}$. Now you all recognize this as nonsense, but you also recognize that once again, the ratio will not change if we multiply both of these numbers by the same new number. What will it take to make these whole numbers? Here is where art and mathematical common sense come into play: there is no rule that will always get you the "right" answer. Nevertheless, you should all recognize 2.66 as being essentially 2 and 2/3, and therefore, that multiplying it by 3 will get you to a whole number. If we multiply both numbers by 3, we get C_3H_8 as our formula.

Is this the right answer? I don't know. No one knows. It is a *possible* answer; more than that, it is the *simplest* possible answer. But if the molecule were in fact C_6H_{16}, we would have obtained exactly the same results. How can we know which it is? From this data alone, we can't. But we can report C_3H_8 as the simplest possible correct answer. This is called the *empirical formula* of the unknown, and it is the formula we will use until we determine that it is not correct. But we also know that the real answer (the *molecular formula*) has to be some multiple of the empirical formula, or the ratio would not be right.

Is there any way of finding out what the molecular formula is? Certainly: all you need to know is how much the molecule weighs, and you already know one method for determining that: put it in a mass spectrometer. If this molecule weighs 44 amu (g/mol), then C_3H_8 must be correct. If the molecular weight turns out to be 88, it must be C_6H_{16}. If it turns out to weigh 308 g/mol, what must the formula be?

We did this whole exercise on the assumption that the unknown contained only carbon and hydrogen atoms. What if it contained oxygen atoms in addition? Consider this problem:

An unknown contains carbon, hydrogen, and oxygen. A sample weighing 14.2 mg is burned completely in oxygen and the products are collected and weighed. The CO_2 collected weighs 30.04 mg and the H_2O weighs 14.75 mg. What is the empirical formula of the unknown?

Now we have a problem. We can easily calculate the number of moles of carbon atoms and hydrogen atoms in the unknown, but what do we do about the oxygen? Many students add up all the oxygen in the products and claim that it came from the unknown, but we know that cannot be true: after all, there was oxygen in the first problem, and *none* of it came from the unknown. We are adding oxygen for the burning process, so clearly, while *some* of the oxygen in the products originated in the unknown, we certainly cannot claim that all of it did. So knowing the amount of oxygen in the CO_2 and H_2O will not help us.

Now what? Elementary, my dear student. Put this problem on the shelf for a moment and consider the following story. Then we'll come back to this.

Another Digression: Sherlock Holmes

Sherlock Holmes and his friend Dr. John Watson arrived at Scotland Yard headquarters.

"I'm so glad to see you, Mr. Holmes," said Inspector Lestrade. "I sent for you because I have finally found a case even you can't solve."

"We'll see about that," said Holmes. "Tell me the facts."

"You recall the robbery at the jewelry factory last year? Of course you do. A crate containing some gold, some diamonds, some rubies, and some emeralds was stolen from the grounds of the ABC Jewelry Co. The company has not yet been able to collect on its insurance."

"Why not?" asked Holmes.

"You see, the crate had just arrived at the company. They had weighed it and found that it weighed 50 pounds (net), but they had not yet opened it, so they did not know how much of that weight was gold, how much was diamonds, rubies, or emeralds. Since each of these items is worth quite a different amount per pound, nobody knows how much the crate was worth. The insurance company won't pay a large amount, arguing that the crate may have contained mostly the least valuable material, but the company won't accept a small payment, arguing that the crate may have contained mostly the most valuable material.

"Through brilliant detective work, our men have been able to trace the crate to the XYZ Jewelry Co. They stole the crate because at the time, the XYZ Jewelry Co. was out of gemstones. They had plenty of gold, but could make no decorative jewelry without gems. They have confessed to the crime, but they too don't know the contents of the crate."

"How is that possible?" asked Watson.

"Elementary, my dear Watson," said Holmes. "Obviously they used the contents of the crate, along with their own gold, to make jewelry which has since been sold and cannot be retrieved."

"Precisely," continued Lestrade. "The diamonds were made into rings. I have one here. They shipped out 12 pounds of rings, all exactly like this one. The rubies were made into brooches, all exactly like this one over here. And the emeralds were made into earrings, like those."

"How many pounds of brooches and earrings were shipped?" asked Holmes.

Lestrade answered, "There were 35 pounds of brooches and 33 pounds of earrings."

Holmes declared, "If you will let me take the ring, the brooch, and the earrings, I will solve this conundrum."

"Gladly," said Lestrade, "No one in Scotland Yard has been able to do it."

When they arrived home, Holmes unwrapped the jewelry and examined it carefully. "Observe, Watson," he said, "this ring contains one diamond." He brought out the balance on which he usually weighed his cocaine. "The ring weighs 1.297 ounces and the diamond..." He carefully pried it out, "...weighs 0.110 ounces. Now let us look at the brooch. You see, the brooch is in the shape of a circle, and the rubies make a design in the middle. There are eight rubies, displayed like this:"

```
    *   *   *
  *           *
    *   *   *
```

"What does that look like to you, Watson?"

"A fish," Watson replied. "This must be a clue!"

"Ah," said Holmes, "but what kind of fish is it?"

"How should I know that?" exclaimed Watson.

"Oh, Watson, you're so stupid!" said Holmes. "What color are rubies?"

"Of course," said Watson, "I should have known."

"Now, the brooch weighs 2.629 ounces." Again he carefully pried out one of the gems. "And a ruby weighs 0.203 ounces. Now let us look at an earring. This weighs 1.231 ounces and has three emeralds in it. Each emerald weighs....0.142 ounces. Now Watson, I want you to take these gems to a jewelry store and have them appraised. Also, find out the current price of gold."

Watson soon returned with his report. "The diamond is worth $320. The ruby is worth $37. The emerald is worth $22. And gold is going for $120 an ounce."

"Excellent!" exclaimed Holmes, "Send for Lestrade!"

When Lestrade arrived, Holmes announced, "The insurance company owes ABC Jewelry Co. $169,322."

"You are amazing," said Lestrade. "How did you do it?"

1. How did he do it?

2. What does this have to do with chemistry?

3. What kind of fish was it?

4. What did Holmes do wrong? If you were Holmes, what would you have said?

Take some time now to try to answer these questions before you continue reading.

Holmes was faced with exactly the problem you were faced with earlier: he knew the weight of rings made, and from that he could easily determine the weight of the diamonds in the original crate. Likewise with emeralds and rubies. How did he determine the amount of gold in the original crate? Simple: everything that was not diamonds, rubies, or emeralds must have been gold. You add up what you know, subtract it from the original weight, and there is your weight of gold (oxygen). You should be able to do the same calculations Holmes did, and get the same answer, except that you, being smarter than Holmes, will not make the mistake he did, which was to get so caught up in the calculation that he forgot to remember that he did not know the answer nearly as accurately as he reported it. Since the weight of the rings is presented as "12 pounds," we can assume that there is uncertainty in the last digit and it might have been 11 or 13. Of course this throws our number of diamonds off by up to 12 and our value off by up to $3840. Thus, the closest Holmes could have come to the value of the shipment is $170,000.

Incidentally, what kind of fish was it? Detective fiction aficionados know that the general term for a clue that is there simply to throw the reader off the track is a "red herring."

Chemistry Again

OK, back to our chemistry problem. Here it is again:

An unknown contains carbon, hydrogen, and oxygen. A sample weighing 14.2 mg is burned completely in oxygen and the products are collected and weighed. The CO_2 collected weighs 30.04 mg and the H_2O weighs 14.75 mg. What is the empirical formula of the unknown?

From this data you can calculate easily that there was 6.83×10^{-4} mol of carbon atoms and 1.639×10^{-3} mol of hydrogen atoms in the original unknown. It should be an equally easy task to calculate what these atoms must weigh, but be careful: we have calculated moles of hydrogen *atoms*, not molecules, so we must use 1, not 2, as the weight of each. You should determine that the hydrogen weighs 1.639 mg and the carbon weighs 8.196 mg. Between them we have accounted for 9.835 of the original weight. The rest must have been oxygen, so there must have been 4.365 mg of oxygen in the sample. This we can convert into moles of oxygen atoms (again, atoms, not molecules: use the correct weight). Ultimately we should find 2.73×10^{-4} mol of oxygen atoms to go with our 6.83×10^{-4} mol of carbon atoms and 1.639×10^{-3} mol of hydrogen atoms in the original unknown.

Now all we have to do is play with these numbers to make them pretty. Remember the technique: divide all by the smallest number, then multiply all by whatever you like to get whole numbers. You should arrive at a formula of $C_5H_{12}O_2$.

As with every other aspect of chemistry (and life, except golf), this stuff becomes quite easy after you practice a lot.

Problems

(Answers are provided at the end of the chapter for italicized problems.)

1. How many moles of atoms are in 3 g of carbon? How many moles of molecules are in 3 g of methane (CH_4)? How many moles of atoms are in 3 g of methane? How many moles of atoms are in 3 pounds of methane?

2. Copper comes in two isotopes, 63 and 65. 2.51 g of copper contains 2.37×10^{22} atoms. Using only this information (i.e., without looking at your periodic table), what is the natural isotopic distribution of copper?

3. *Given the following isotopic distribution of krypton, calculate the number of atoms in 0.00300 g of krypton. Don't look anything up: use only these data (and Avogadro's number, which you know).*

Abundance	Mass
0.3%	77.92 amu
2.3%	79.91
11.6%	81.91
11.5%	82.92
56.9%	83.91
17.4%	85.91

4. Zinc has four naturally occurring isotopes, ^{64}Zn (48.6%), ^{66}Zn (27.9%), ^{67}Zn (4.1%), and ^{68}Zn (18.8%).

    a. How many atoms are in 4.67 g of naturally occurring zinc?

    b. Predict the appearance of the mass spectrum of diethyl zinc, $C_4H_{10}Zn$. Draw a picture and label the locations and heights of the peaks.

5. What is the molarity of a solution made by dissolving a weight of 7.6 lb of C_4H_6O in enough ethanol to make 7.6 quarts of solution?

6. The approximate population of earth is 5.3 billion people. If all these people were picking peas at the rate of one pea per second, 24 hours a day, 365 days a year, how long would it take to pick one mole of peas?

7. *If you count all the protons and all the neutrons in the body of a person weighing 165 pounds and add these sums together, what is the combined total? This is not a trick question and it does not require lots of complex or difficult calculations. It does require a good understanding of atomic structure and a little thought.*

8. How much water would you have to add to 1.70 qt of ethanol (C_2H_6O) to make up a 3.65 M (moles/liter) solution? Assume that no volume is gained or lost when the liquids are mixed. Ethanol has a density of 0.789 g/mL.

9. Phosphoric acid (H_3PO_4) and calcium hydroxide ($Ca(OH)_2$) react to form calcium phosphate ($Ca_3(PO_4)_2$) and water.

    a. Write the complete, balanced equation for the reaction.

b. If you start with 5 moles each of phosphoric acid and calcium hydroxide, which is in excess? By how much (in mol)? How much calcium phosphate is produced (in mol)?

10. You have a butane (C_4H_{10}) cigarette lighter with 50.0 g butane left.

   a. How many moles of butane are in the lighter?

   b. Write the complete, balanced chemical equation for the combustion of butane.

   c. What mass of oxygen gas is needed for complete combustion of the butane?

   d. At standard temperature and pressure, 1 mol of any gas occupies a volume of 22.4 L. What volume of oxygen gas would be required under standard conditions (0 °C and 1 atm pressure—a clear, frosty evening at sea)?

   e. What mass of water is produced?

   f. What mass and volume of carbon dioxide gas is produced?

11. From the combustion of a mixture of 4.66 g of butane and 11.1 L of oxygen (see prob 10), 12.7 g of CO_2 were collected. What is the percent yield?

12. Octane, one of the many components of gasoline, burns in excess air by the reaction

$$2\ C_8H_{18} + 25\ O_2 \rightarrow 16\ CO_2 + 18\ H_2O$$

   a. Calculate the volume of oxygen gas needed to react with 2.27 mg of octane, given that the density of oxygen is 1.43 g/L under the conditions of the experiment.

   b. What volume of air is required in a, given that air is 21% oxygen by volume?

   c. *How many pounds of CO_2 will be emitted for every mile a car drives? Assume a typical car gets 22 miles to the gallon, and that gasoline has a density of about 0.7 g/mL.*

13. Phosphorus tribromide can change ethyl alcohol to ethyl bromide. The chemical equation for the reaction is:

$$3\ CH_3CH_2OH\ +\ PBr_3\ \text{----------}>\ 3\ CH_3CH_2Br\ +\ H_3PO_3$$

   a. How many grams of phosphorus tribromide would react completely with 50.0 mL of ethyl alcohol (density 0.789 g/mL)? How many grams of ethyl bromide would you expect?

   b. Suppose you started with 50.0 g of ethyl alcohol and 75.0 g of PBr_3. What reactant will be the limiting reagent? How many grams of ethyl bromide should be produced?

   c. Suppose you started with 50.0 mL of ethyl alcohol and 25.0 mL of PBr_3 (density 2.85 g/mL). What is the limiting reagent? What weight of ethyl bromide should you get? If you obtain 40.0 mL of ethyl bromide (density 1.46 g/mL), what is your percent yield?

14. *PF₃ reacts with XeF₄ to give PF₅ and Xe (an inert gas).*

   a. *What surprises are in the above sentence?*

   b. *How many grams of PF₅ can be produced from 100.0 g of PF₃ and 50.0 g of XeF₄?*

15. Sodium metal (Na) and chlorine gas (Cl₂) react to make table salt, sodium chloride.

   a. What weight of chlorine does it take to react completely with 4.6 g of sodium?

   b. What weight of sodium does it take to react with 1.42 g of chlorine?

   c. If 10.0 g of sodium and 10.0 g of chlorine react with each other, and 15.2 g of sodium chloride are obtained, what was the yield of the reaction?

16. a. Balance the following equation:

   ___ $(C_2H_3CO)_2O$ + ___ H_2O → ___ $CH_2OHCH_2CO_2H$

   acrylic anhydride   hydroacrylic acid

   b. If 10.0 g of water is mixed with 50.0 g of acrylic anhydride, what is the maximum weight (in grams) of hydroacrylic acid that could form?

17. For each of the following reactions, calculate the theoretical yield for 10.0 g starting material (the organic compound). Reminder: first balance the reaction.

   a.

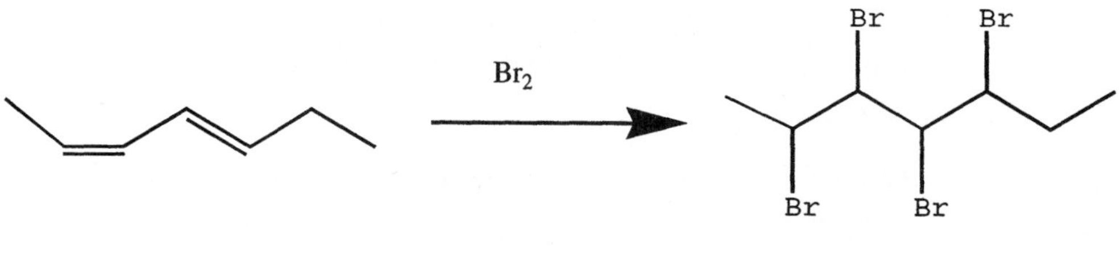

   b.

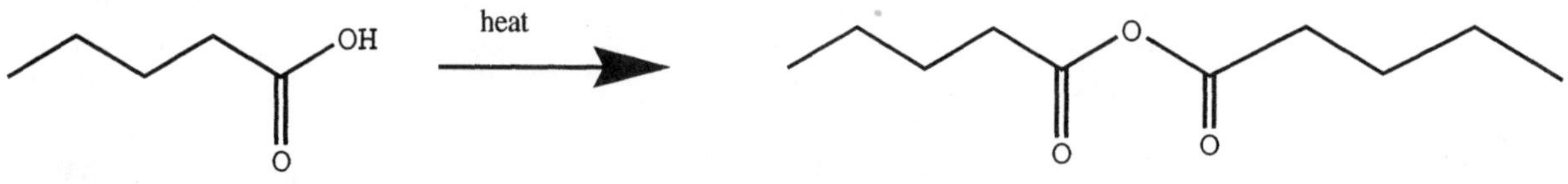

   Note: H_2O is also produced, but not usually shown.

c.

Caution: Water is also produced but typically isn't shown.

If you start with 10.0 g each of the 2 organic starting materials, one will most likely be present in excess and the other will be limiting. Be sure to determine which is which before continuing the problem.

   d. If a lab technician reports carrying out reaction A with a 72% yield, how much product has been produced?

18. *A vitamin C tablet was analyzed to determine whether it really contained 1.0 g of the vitamin, as claimed on the bottle. A tablet was dissolved in water to form a 100.00 mL solution, and 10.0 mL of it was titrated with I_2. It required 10.1 mL of 0.0521 M I_2 to reach the end point. Given that 1 mole of I_2 reacts with 1 mole of vitamin C in the reaction, is the manufacturer's claim correct? Vitamin C has a molecular weight of 176 g/mol.*

19. Yeast can ferment sugars to alcohol according to the following chemical equation:

$$C_6H_{12}O_6 \longrightarrow 2\,CH_3CH_2OH\ +\ 2\,CO_2$$

   a. The density of ethanol is 0.789 g/mL. What volume of ethanol can be produced for every 100.0 g glucose?

   b. Wine is typically 12% by volume alcohol. What mass of glucose is needed per liter of grape juice to produce this concentration of alcohol? (Note that you are calculating a theoretical yield.)

   c. How concentrated (in glucose) must the starting grape juice be to produce 12% wine? Either mass/L or moles/L is acceptable.

   d. A product of 9% alcohol represents what yield from the concentration of glucose calculated in part C?

20. *A substance is analyzed and found to contain 63.2% carbon, 12.3% hydrogen, and 24.6% nitrogen by weight.*

   a. *What is the % composition of the substance by number of atoms?*

   b. *What is the simplest chemical formula you could write for this substance?*

21. In combustion analysis, a sample is burned in excess oxygen in such a way that you can rely on the fact that all of the carbon in the original sample is converted into CO_2, all of the hydrogen into H_2O. A 0.395 g sample of nicotine, which contains only carbon, hydrogen, and nitrogen, was analyzed by combustion analysis, and 1.072 g of CO_2 and 0.307 g of H_2O were obtained.

   a. What is the simplest possible formula for nicotine (empirical formula)?

b. If the molecular weight of nicotine is known to be 162 g/mol, what is the real formula (molecular formula) for it?

22. On the planet Reingoldus, things burn in hydrogen instead of oxygen. When this occurs, all the carbon in a compound turns into CH_4, all the oxygen becomes H_2O, and all the nitrogen becomes NH_3. On this imaginary planet, an unknown sample containing C, H, N, and O and weighing 123 mg was burned in hydrogen to give 108.7 mg of CH_4, 11.55 mg of NH_3, and 24.46 mg of H_2O. What is the empirical formula of the unknown substance?

23. *In the early 1800s, before they knew about atoms, chemists knew that carbon could combine with hydrogen in many different ways. Two of them are described below.*

A. Methane contains 3 g of carbon for every gram of hydrogen.
B. Ethane contains 4 g of carbon for every gram of hydrogen.

John Dalton speculated that having two different substances composed of carbon and hydrogen could be explained by atoms, little tiny balls of substance, that combined with each other in different ratios. Of course, he had no idea how big or heavy the various balls were, but the concept still works, as long as each kind of atom always weighs the same, and is different from other kinds. Place yourself in his shoes, ignoring what we now know about atomic weights, and speculate with him:

a. *If hydrogen units (atoms) weighed 3 times as much as carbon units, could this be consistent with statement A? Statement B? What formulas for methane and ethane would be necessary?*

b. *Still referring to the data in statements A and B above (but not the speculation in part a), if the formula for methane were CH, what would the formula for ethane have to be?*

24. Still referring to statements A and B from problem 23, if carbon were 36 times heavier than hydrogen, what would be the formula for methane and ethane?

25. "Carbonic oxide" contains 3 g of carbon for every 4 g of oxygen. "Fixed air" contains 3 g of carbon for every 8 g of oxygen.

a. Using only the above information, ignoring what we now accept to be true, what would be the formula for each of these substances if carbon atoms weighed 6 times as much as oxygen atoms?

b. Using only the above information, ignoring what we now accept to be true, what would be the formula for each of these substances if oxygen atoms weighed 6 times as much as carbon atoms?

c. If fixed air had the formula CO, what would the formula of carbonic oxide be?

26. "Sulfuric oxide" contains 2 g of sulfur for every 3 g of oxygen. "Sulfurous oxide" contains 1 g of sulfur for every 1 g of oxygen. Answer the following problems from the point of view of John Dalton. Each question is independent of the others, but all relate to the information in this paragraph.

a. What would be the formula for each of these substances if sulfur atoms weighed 6 times as much as oxygen atoms?

b. If sulfuric oxide had the formula SO, what would the formula of sulfurous oxide be?

c. Could sulfurous oxide have the formula SO? Why or why not?

d. Coming back to the present (that is, using the real atomic weights for S and O), what is the real formula for the above two compounds?

Selected Answers

Internal Problems

Test Yourself 1. You have already established that you have 3.74 gajillion grains of salt, and you will use only 2.14 gajillion of them, so there will be 1.6 gajillion left over. Salt weighs 1 ton/gajillion, so this is 1.6 tons of salt left over.

For pepper:

$$10.69 \text{ gajillion rice} \quad \frac{1 \text{ gajillion pepper}}{50 \text{ gajillion rice}} \quad = \quad 0.374 \text{ gajillion pepper used}$$

$$\frac{\begin{array}{l} 0.5 \quad \text{gajillion pepper have} \\ -0.374 \text{ gajillion pepper used} \end{array}}{0.126 \text{ gajillion pepper left over}} \quad \frac{2 \text{ tons}}{\text{gajillion}} \quad = 0.252 \text{ tons pepper left over}$$

Test Yourself 2. 11.77 g NaCl   1.03 M HCl.

End of Chapter Problems

3. 2.16×10^{19} atoms. To get this, you must first calculate the average mass of a krypton atom, then figure out how many moles are in the given weight, and then convert this to a number of atoms.

7. 4.51×10^{28}. How do you get that? Every atom in your body has protons, neutrons, and electrons. We have already determined that the electrons are negligible in terms of weight. Thus your weight consists of only protons and neutrons, each of which weighs 1 amu, or 1 g/mol. So calculate the weight of this person in grams, convert that into moles of atomic particles at 1 g/mol, and convert moles into number using Avogadro's number.

12. c. About 1 lb CO_2 for every mile driven!

14. a. XeF_4??? Xe is an inert gas!! It does not want to react with anything. Why does it make compounds at all? (Find out by taking inorganic chemistry.)

b. 60.48 g

18. No: there is 0.926 g of vitamin C in the tablet.

10.1 mL of I_2 solution contains 5.26×10^{-4} mol of I_2. This will react with 5.26×10^{-4} mol of vitamin C, which weighs 0.0926 g. But we analyzed only 10% of the tablet (10 mL out of the 100 mL solution), so there must be 10 times that much in one tablet.

20. a. 63.2% of the weight of any sample of this substance is carbon. To make working easy, it would be useful to have a particular weight, so we simply pick one. The easiest weight to pick is 100 g, since with that weight the percent of any element is also its weight, but any other weight would work equally well. So, imagine 100 g of this substance. 63.2 g is C, 12.3 g is H, and 24.6 g is N. Now we count:

$$63.2 \text{ g } \frac{\text{mol}}{12 \text{ g}} = 5.27 \text{ mol of C atoms}$$

Likewise we calculate 12.3 mol of H atoms and 1.75 mol of N atoms. For these calculations you use weights of 1 g/mol and 14 g/mol because we are counting atoms, not molecules. The total adds to 19.32 mol of atoms, of which 5.27 mol, or 27%, are C, 64% are H, and 9% are N atoms.

b. We have already calculated a number of moles of each atom: 5.27 mol of C atoms, 12.3 mol of H atoms, and 1.75 mol of N atoms. Divide them all by 1.75 and you get a ratio of 3 mol C to 7 mol H to 1 mol N. C_3H_7N.

23. a. Yes; Yes; C_9H and $C_{12}H$

Details: If we ignore everything we know now, we can define things any way we want. Let's make it easy for ourselves: if H weighs three times as much as C, let's define C as weighing 1 amu and H as weighing 3 amu. Now consider methane, which has 3 g of C and 1 g of H.

$$3 \text{ g C } \frac{\text{mol}}{1 \text{ g}} = 3 \text{ mol C} \qquad 1 \text{ g H } \frac{\text{mol}}{3 \text{ g}} = .33 \text{ mol H}$$

Divide through by 0.33 and you get 9 mol C for 1 mol H, or C_9H.

b. C_4H_3

Details: If the formula for methane is CH, it is obvious that each C must be 3 times heavier than an H. For convenience, let's claim that a C weighs 3 and an H weighs 1 amu. It follows that a *mole* of C atoms weighs 3 g and a mole of H atoms weighs 1 g. Thus 3 g/mole becomes a conversion factor for weight into count for carbon, and 1 g/mole does the same for H.

We know that in ethane, there are 4 g of carbon for every 1 of hydrogen. To get a formula, we need to know not what weights are involved, but how many atoms. So let's convert the weight into count. 4 g of carbon x (1 mole/3 g) = 1.33 moles of C atoms.

$\qquad$ 1 g of hydrogen x (1 mole/1 g) = 1 mole of H atoms.

So in ethane we have 1.33 moles of C atoms for every 1 mole of H atoms. To make a useful formula, we have to adjust these to be whole numbers. It should be clear that multiplying both numbers by 3 will result in the same relationship, but this time as whole numbers: 4 moles of C atoms for every 3 moles of H atoms. Thus ethane must be C_4H_3.

Chapter 7
Alkanes and Cycloalkanes

In this chapter we will take our first extended look at organic compounds. We will concentrate on the family that contains only carbon and hydrogen, and only single bonds. We will learn

- How to name them

- What shapes they take

- How to draw pictures of these various shapes

- How uncomfortable certain rings are

- About equilibrium constants and energy diagrams

- About the different kinds of positions on six-membered rings.

Compounds that contain only carbon and hydrogen are called "hydrocarbons." Double bonds and triple bonds are often referred to as "unsaturation," and hydrocarbons that contain them are unsaturated hydrocarbons. Hydrocarbons with no double or triple bonds are called saturated hydrocarbons; another word for saturated hydrocarbons is "alkanes." Thus, alkanes are the most plain of organic compounds: they have no functional groups, and therefore undergo very few reactions of interest to us (although an entire industry, petroleum refining, is based on exactly these reactions). Nevertheless, there are still some interesting features associated with alkanes, and we will spend this chapter discovering some of them.

The first order of business is to learn how to communicate with each other. Although organic chemists rely heavily on pictures, it is not always convenient to draw them. Sometimes you simply need to be able to communicate in words. This means we need to create words for molecules, that is, we need to learn how to name things. The general term for this is "nomenclature."

NOMENCLATURE

To be perfectly honest, nomenclature is one of the more boring aspects of organic chemistry to a practicing chemist. We like to make stuff, and naming it is a necessary evil. But it *is* necessary, for you as well as for us: for one thing, it will be on standardized exams you probably have in your future, such as the MCAT or the GRE. For another, if you expect to find anything in the "literature," where all science is published, you need to be able to call chemical compounds by their appropriate names, and to recognize when the word you are looking at refers to the substance you have in mind.

There are two directions to nomenclature. You have to be able to read a word and write down the structure it refers to, and you have to be able to look at a structure and write down a word that identifies it. We will learn how to do both, for relatively simple compounds. The rules for doing this fills several books, and we will only scratch the surface, but a little effort should enable us to communicate reasonably well. In future chapters we will spend progressively less time on this subject, but we will still have to at least introduce how to name each functional group in its turn. For now, we will concentrate on the base structures.

First of all, you need to memorize a list of fundamental words. There are specific words for strings of carbon atoms of every possible length, but we will restrict ourselves to the first ten, plus a few important other cases:

Word	Structure
Methane	CH_4 (1 carbon)
Ethane	CH_3CH_3 (2 carbons)
Propane	$CH_3CH_2CH_3$ (3 carbons)
Butane	4 carbons
Pentane	5 carbons
Hexane	6 carbons
Heptane	7 carbons
Octane	8 carbons
Nonane	9 carbons
Decane	10 carbons

Notice that all these words (except the last one) consist of a prefix followed by "-ane." In chemistry the prefix we use for "unspecified" is "alk-," so a hydrocarbon of unspecified length is an "alkane," while one of specified length is "prefix-ane." Now most of these are probably Greek to you, and well they should be: most of the prefixes are derived from Greek. Some of the prefixes should sound familiar: you know about the Pentagon, for example, which is a five-sided building. The prefixes for 5–10 are standard Greek prefixes for their respective numbers. The first four prefixes, however, do not follow that pattern, and I cannot explain why these words were chosen. They just were, and you need to learn them.

One other name you should learn at this point is benzene:

Please note that benzene is a six-membered ring with three double bonds in it. If it did not have these double bonds, it would not be benzene. We will learn more about benzene later, and discover the special nature of these double bonds. Of course, benzene is not an alkane, but I am including it here because it will show up frequently and you will need to know its name.

Simple compounds like these are easy to name. More complicated alkanes are named by combining pieces of these words. The general plan for naming any substance is to choose a basic carbon-chain piece of the substance, name it, and then describe what is attached to this basic frame. Of course, there are rules to follow each step of the way.

It is worth spending a little time explaining why we bother with all the rules you are about to learn. Chemists often need to convey information by words, and it is important that when any chemist speaks a word, every other chemist will assign to it the same meaning. So the name for a molecule needs to be unique to that molecule: another chemist must always call up a picture of that particular molecule *and only* that molecule. We will see in a minute how this simple requirement forces lots of complication into our lives.

Let's consider, for example, the first several alkanes. What is methane? CH_4. There is no other way you can arrange these atoms, so we are OK. Ethane is CH_3CH_3. and propane is $CH_3CH_2CH_3$. But what is butane? Obviously $CH_3CH_2CH_2CH_3$, right? OK, so what should we call the following?

$$CH_3$$
$$H_3C\text{-}CH\text{-}CH_3$$

This still has four carbons, so "butane" could still be a reasonable name. But it clearly is not the same substance as the first thing we called butane; it is an isomer, and we can't let the same word refer to both compounds. Well, we can deal with this. Let's call the first one "normal" butane (*n*-butane, for short), and the new one, the isomer, "isobutane." That solves the problem, because there is no other way to put four carbons together.

Moving on to pentane (C_5H_{12}), what is its structure? There are three ways to put five carbons together. I show you these on the next page, accompanied by their corresponding stick figures, which we need to start getting used to seeing.

$$CH_3CH_2CH_2CH_2CH_3$$

$$\begin{array}{c} CH_3 \\ | \\ H_3C\text{-}CH\text{-}CH_2CH_3 \end{array}$$

$$\begin{array}{c} CH_3 \\ | \\ H_3C\text{−}C\text{−}CH_3 \\ | \\ CH_3 \end{array}$$

Clearly the first one is "normal," meaning straight chain, so we can call it *n*-pentane. The second one is an isomer, so we can call it isopentane. But what do we do with the new one? It is also an isomer, but we can't call it isopentane, because we have already used that word. So we call it neopentane, where "neo" is the Greek prefix for "new." The kinds of names we have just constructed are called "trivial" names.

I think you can see that this approach is soon going to run out of steam. There are five different versions of hexane, nine versions of heptane (can you find them all?), and rapidly increasing numbers of isomers as you get to larger and larger structures. It is unreasonable to continue dreaming up and assigning new names for each modification, let alone trying to remember what they all mean. It is time to end this approach and adopt a more systematic one.

The emperor of chemical nomenclature is the International Union of Pure and Applied Chemistry, IUPAC for short, and the rules I am about to describe are the IUPAC rules. As I mentioned above, the first order of business is to identify a basic unit, and then describe what is attached to it. The basic unit is generally the longest continuous chain you can find. But you need to recognize that this is not always easy to see. It might not be *written* in a straight line. Nevertheless, as long as you can walk from one end to the other by a series of direct attachments, it qualifies as a base. For example, the base unit in the following compound is 12 carbons long. Can you find it? (I know, we do not know a word for 12 carbons, but you can still find it.)

Once you have identified your base, you have to tell your listener what is attached to the base, and where. The "what" is done with a series of words related to the words you learned earlier; the "where" is done with a series of numbers. Groups that are attached to some other base unit are called "substituents." Every substituent has to have a name and a position. Hydrocarbon *substituents* are named just like alkanes, except that the suffix "-ane" is replaced by "-yl." So a one-carbon substituent is called methyl, two is ethyl, and so on. *Positions* are described by numbering the base chain, from one end to the other, and then indicating where the substituents are. This brings up the first complication, since there are two ends to every base chain, and the numbers you get by starting at one end will usually be different from the numbers you get by starting at the other. So this leads to our first rule: always number from the end that results in the lowest number for the first branch point that would be different if you started from the other end. The weird compound I just showed will actually serve as a good example. Below I show the same thing, with the 12-carbon chain identified. Since we do not know the word for 12 carbons, let's call it 12-ane. If you start numbering at the top left, the first black line is on the third carbon; if you start at the bottom left, the first black line is on the second carbon. So we should number from the bottom left in this case. Go ahead, number the chain; it will help.

Now we have to identify the substituents. I like to create a list:

2-methyl, 3-methyl, 5-methyl, 6-propyl, 7-butyl, 8-ethyl, 10-methyl.

Then list them alphabetically:

7-butyl, 8-ethyl, 2-methyl, 3-methyl, 5-methyl, 10-methyl, 6-propyl

and group them by adding the Greek prefixes "di" for two, "tri" for three, "tetra" for four, "penta" for five, and so on. So we get

7-butyl, 8-ethyl, 2,3,5,10-tetramethyl, 6-propyl 12-ane. (Note that the prefixes such as di-, tri-, and tetra- do not count in the alphabetization.)

Convention has it that words are always set off by hyphens while numbers are separated from each other by commas. And the official word for 12-ane is dodecane. So this compound is officially called

7-butyl-8-ethyl-2,3,5,10-tetramethyl-6-propyldodecane

Any chemist in the world, given that name, would write down exactly this structure. Of course, he or she would almost certainly not *write* it exactly like this. More likely it would get written something like the following. But it is still the same molecule. (Convince yourself of this!)

I am sorry to tell you that we are not yet done, though, because there can be complications that did not arise in this structure that you would still need to be able to deal with. For example, let's focus in on the propyl substituent. For simplicity, let's call the rest of the molecule "R" (for "Rest of the molecule") and ignore it:

This is "propylR." What would you call it if it looked like this?

Well, there is still a three-carbon thing attached to the base chain, so it *is* a propyl group, but it is not the same as the first one, so we can't call it the same thing. We could get around this by calling the original propyl group "normal," or "*n*-propyl," and calling this new thing "isopropyl." Indeed, that is what has been done, in the "trivial" system. But as we saw above, the "trivial name game" rapidly gets out of hand. Nevertheless, the various trivial names for propyl and butyl groups are used often enough that you need to know them. Later I will show you how to name things so that this problem disappears.

To explain the way the trivial names for butyl substituents come about, we have to digress for a moment and talk about several categories of carbon atoms that exist in alkanes. Carbons are often classified according to how many *other* carbons are directly attached to them. Look at the following example:

$$H_3C-\overset{H_2}{C}-\overset{|}{\underset{CH_3}{CH}}-\overset{H_3C\quad CH_3}{\underset{CH_3}{C}}-CH_3$$

Each of the CH$_3$ groups has only one other carbon directly attached to it. These are described as "primary" carbons. Hydrogens attached to a primary carbon are called primary hydrogens. If something else, say an alcohol group (an –OH group, remember?), were attached to a one of these carbons, the compound would be called a primary alcohol. Not all the CH$_3$ groups are identical, but they do all share this particular characteristic and so they are all primary. The CH$_2$ group has two carbons directly attached so it is called "secondary." The CH has three carbons attached and is called "tertiary," and the last one (C) has four carbons attached and is called "quaternary." Chemists have developed a cute shorthand for these words:

primary	1°
secondary	2°
tertiary	3°
quaternary	4°

These are pronounced exactly like the spelled out word. The phrase "1° carbon" is read aloud as "primary carbon." If you say "first-degree carbon," this is like fingernails on a chalkboard to your professor. Since you clearly do not want to irritate your professor, you surely will never do this.

Now to get back to naming: let's look at the four ways that four carbons (a butyl group) can be attached to something:

In the second example, the butyl group is attached at a secondary position. Further, this is the only case in which this is true. So calling this a "secondary butyl" (abbreviation: *sec*-butyl) group completely specifies what it is, and that is what the group is called in the trivial system. Likewise, the term "tertiary butyl" (*tert*-butyl or *t*-butyl) is unique for the fourth example. However, in both the first and third examples the butyl group is attached at a primary position, so calling this a "primary butyl" group will not be specific enough. On the other hand, the first one comes naturally by the name "normal," or "*n*-butyl," so the only one left can get the term "isobutyl." You should learn these substituent names, as well as the ones for the propyl group. Anything bigger we will deal with in a more formal way.

I need to point out one unfortunate consequence of what we just did. Consider 5-*sec*-butylnonane, shown below.

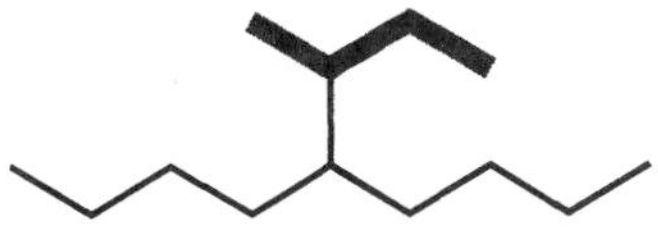

The *sec*-butyl group, shown in bold, is called *sec*-butyl because it is attached to the longer chain at the secondary carbon *of the butyl fragment*. But now that it is attached to the other chain, this is no longer a secondary carbon, it is tertiary! I am really sorry that this is the case, but I cannot change it. It *is* a tertiary carbon in this molecule, but the molecule is still called 5-*sec*-butylnonane. This is one of those things you will just have to put up with.

What would you call the left molecule below?

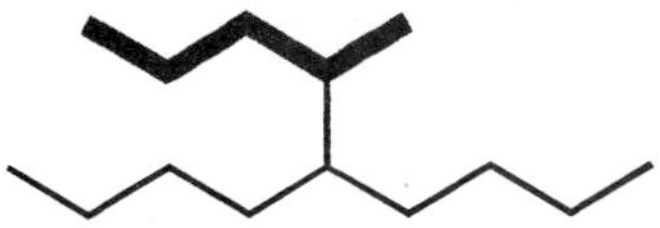

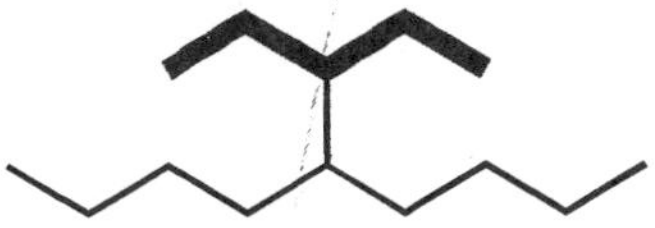

Now we are really in trouble. The bold group is a pentyl group, and it is attached at a secondary carbon of the pentyl group, but there are two different ways we could have done this (see right-hand molecule), so calling the substituent *sec*-pentyl will not do. (Remember, this is considered to be attached at a secondary position, because we don't consider the bond to the main chain as part of the substituent.) It is at this point that we abandon trivial nomenclature (while still remembering the words you just learned) and fall back on the systematic IUPAC approach.

Let me repeat. The names we have just learned are the unofficial, trivial names You do not need to know anything about trivial names for substituents larger than four carbons. For anything larger we will use the IUPAC system described below. The IUPAC system can also be used for the propyl and butyl groups, but enough chemists still use the trivial names that you need to know what they are. Thus, isopropyl, isobutyl, *sec*-butyl, and *tert*-butyl are the only trivial substituent names you need.

To name any substituent strictly by the IUPAC rules, you no longer look at the totality of what is attached to the base unit. Instead, you start counting *from the connection to the base unit* and find the longest chain of the attachment. Repeat: you start counting at the base unit. Even though the bold piece (right structure, previous page) has five carbons in a row, you cannot get to five by starting where it is attached. If you start at the attachment point and start walking in any direction, the best you can do is *three* carbons before you come to the end. So this is defined as a propyl group attached to the base unit, but now this propyl group has something attached to *it*! Now we have to name *that*, and specify its location. You see, the IUPAC system breaks all substituents down into substituents on substituents on substituents, until everything is accounted for. The rule for a complicated substituent is that you must always start *numbering* at the point of attachment, the same place we started counting. Thus, the propyl group attached to the nonyl group is numbered as shown below, and it in turn has an ethyl group attached to the number 1 position of the propyl group. So the IUPAC name for this substance is 5-(1-ethylpropyl)nonane. Notice that the whole group attached at the 5 position of the nonane is put in parentheses, so that the reader won't get confused and think we are referring to something else on the nonane chain.

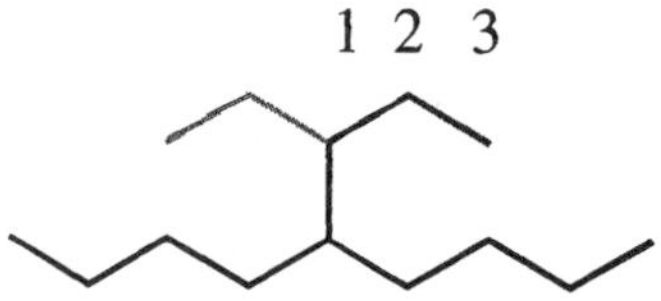

1 2 3

You should convince yourself that the IUPAC name for *sec*-butyl is 1-methylpropyl, for isobutyl it is 2-methylpropyl, and for *tert*-butyl it is dimethylethyl.

Let's do one complicated example and then move on to more interesting things. Name the following:

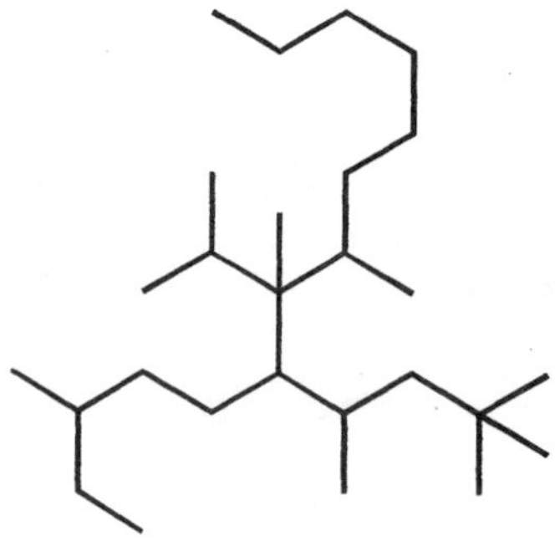

First we need to identify the longest chain. I see one of fourteen carbons. Do you? Find it! Once again we do not know a word for this, but that's OK, we can still deal with the problem by calling it "14-ane." (The official word is tetradecane.) Now we need to number it. Which end should we start with? You should find the decision easy in this case. Now we need to catalog what is attached. Some of the things are pretty simple, others we will have to think about. I see methyl groups at carbons 3, 7, and 8, something complicated at carbon 6 and something less complicated at carbon 7. Are you with me? The thing at carbon 7 you might recognize as an isopropyl group. Technically we are not supposed to mix IUPAC with trivial nomenclature (although people do it all the time), so let's see how to name this formally. The longest attached piece is two carbons, ethyl, and there is a methyl on the number one carbon of the ethyl group. So this attachment is called (1-methylethyl). [Think about this for a second: is there such a thing as 2-methylethyl? What would that look like? Isn't that just propyl? The only time we would ever consider saying methylethyl is when it is 1-methylethyl, so the number is a bit redundant here. We can actually get away with leaving it off.] On to the weird thing on the number 6 carbon of the tetradecane. How do we deal with this? Same as always. We find the longest chain of the attachment, starting at the point of attachment. This is four carbons, butane. We number the butyl group from the point of attachment, and then inventory what is attached to *it*: three methyl groups, on positions 1,3, and 3. So this is a (1,3,3-trimethylbutyl) group. Putting it all together, we have

3,7,8-trimethyl-6-(1,3,3-trimethylbutyl)-7-(methylethyl)tetradecane

Whew!

By the way, if you have multiple identical groups in parentheses, the official prefixes are bis, tris, and tetrakis instead of di, tri, and tetra. You'll see how this works in problem 11 at the end of the chapter.

There are two more terms you ought to know, and again, I have to apologize for them. You learned the name for benzene, and you naturally would now assume that a benzene group, attached to something else, should be called benzyl. This would be perfectly logical, but unfortunately it is not the case. Benzene, when attached to something else, is called a "phenyl" group. The word "benzyl" does exist, but it refers to a phenyl attached to CH_2 attached to something else. You should learn these new terms also.

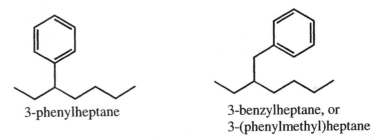

3-phenylheptane

3-benzylheptane, or
3-(phenylmethyl)heptane

To save space, chemists have developed two-letter abbreviations for a few simple substituents. It would be a good idea to learn these also:

Me	Methyl
Et	Ethyl
Pr	Propyl
Bu	Butyl
Ph	Phenyl

Once again, *practice this stuff*—a lot!

ACYCLIC ALKANES: CONFORMATIONAL ANALYSIS

Now that we have words to describe them, let's consider the properties of the alkanes. First, a chemical property: alkanes burn. Of course, you already knew that. Methane is the chief component of natural gas; propane is bottled gas; butane is used in cigarette lighters; gasoline is a mixture of alkanes, often 5–10 carbons long. Other than that, there are no significant chemical reactions that we will worry about in this course, although, as I said previously, armies of chemists in the petroleum industry do nothing but these reactions! To be fair to petroleum chemists, I should also point out that almost everything we learn in our later studies would be impossible without them, since they are the ones who take alkanes and decorate them with functional groups that allow us to do all the other chemistry we will learn. Almost all the man-made chemicals in the world ultimately originate as alkanes in petroleum. Incidentally, I should also mention that not only can chemists do very little with alkanes, the same applies to your body: biological systems cannot metabolize alkanes. This is why mineral oil, a mixture of long alkanes, is such a good laxative: it sails through your intestines without being touched, and is still the same stuff, a nice, lubricating oil, when it gets to the other end.

Physically, you already know that alkanes are non-polar, and therefore they dissolve in non-polar solvents and try to avoid polar solvents like water. Short alkanes (up to butane) are gases at room temperature, longer ones are liquids, still longer ones are solids. So much for observable properties.

The most interesting things about alkanes is not what you can see, but what you can't see. For the time being, you'll have to take my word for what is to follow, since we have not yet considered *how* we could know such things. But believe me, there is plenty of evidence for what I will tell you.

The question we are addressing here is, what do these compounds *look* like? Not color or state (solid, liquid, gas), things that are properties of a visible sample that contains zillions of molecules, but the appearance of single molecules: if you could zoom in and inspect the molecules one at a time. what would you see? Let's consider the alkanes one at a time.

If you don't have a model set (or can't borrow one), go get one now. Personally, I prefer the Framework set by Prentice-Hall, because that kit forces you to deal with hybridization and builds models that are more consistent with the way chemists think about them. On the other hand, they are harder to learn to use, and most students prefer models that have actual balls and sticks for atoms and bonds, such as the HGS set by Freeman. If you are not going to become a chemist, either set will do fine for our purposes.

Methane

Methane is the most boring of compounds. Not only does it undergo few reactions of interest, it also has a boring close-up. As you should expect, it is of course tetrahedral, with bond angles of 109°. Other than that, there is nothing much to talk about.

Ethane

Like methane, ethane has 109° angles, more or less. But ethane has a new feature. To see this, it would be a good idea for you to get out your model kit right now and build a model of ethane.

I will be showing you some pictures of ethane, but I am restricted to two dimensions, while ethane is a three-dimensional molecule, so now is a good time to start trying to see the relationship between flat pictures and real molecules. You need to practice this so that whenever I show you a picture on a page, a 3-D rendering of it appears in your head. This will not occur unless you spend lots of time with your models, and the sooner the better.

There are two carbons in ethane, each with bonds to three other hydrogens. The molecule can *rotate* around the carbon–carbon bond, leading to an infinite number of possible shapes ("conformations") for this molecule. I will show you two different traditional ways to draw ethane on paper; at the same time, you should be looking at your model to see how the pictures relate to the real shape of the molecule.

Sawhorse and Newman Projections

Here is what is called a "sawhorse" projection of ethane:

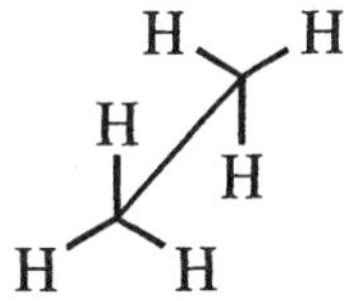

Hold your model so that it looks like this. There is a problem with this representation. If you look at the picture long enough, you will see that you can believe that either the lower carbon is closer to you, or that the upper one is. It's an optical illusion. Although it is irrelevant here, it will be important later on that we agree as to what such a picture means. Since I always see these with the lower carbon as the closer one, let's agree on that. It would be better, perhaps, to draw it as follows, to make the matter clear, but that is seldom done.

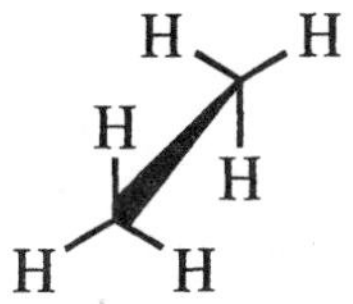

Anyway, if you grab one of your carbons and rotate it relative to the other one, pretty soon you will come to another distinctive conformation, illustrated below:

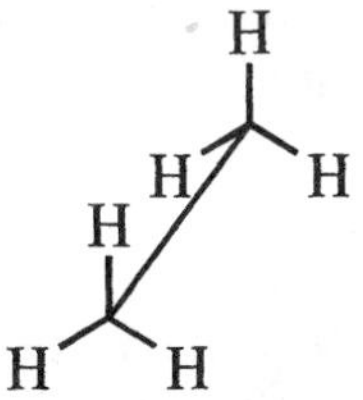

Let's agree (permanently) that the lower left carbon is in front. Do you see the difference between this new conformation and the previous one? And is it also clear that there are many other possible arrangements partway between these two extremes? The two extreme conformations have been given names. The bottom one, where the H's on the front carbon are directly in front of H's on the back carbon, is called the "eclipsed" conformation. The top one, where the H's on the front carbon are exactly between two H's on the back carbon (and *vice versa*), is called the "staggered" conformation. We do not have names for other conformations.

Now take your model, and again hold it so that it looks like the staggered conformation. If you turn it slightly (the whole molecule, not rotation around the bond) so that you are looking directly from one carbon to the other, the back carbon will disappear completely. Could you draw a picture from this perspective? The following attempt to do so is called a "Newman" projection. Sometimes it is more convenient to discuss conformations with Newman projections, and other times with sawhorses, so we should be able to use either. Notice that, in a Newman projection, the point where three lines meet is considered the front carbon, whereas a large circle depicts the back carbon (which is completely invisible when you hold your model this way). Also, notice that the bonds to the front carbon go all the way to the middle, while the bonds to the back carbon stop at the edge of the circle. That's how you know which thing is attached to which carbon.

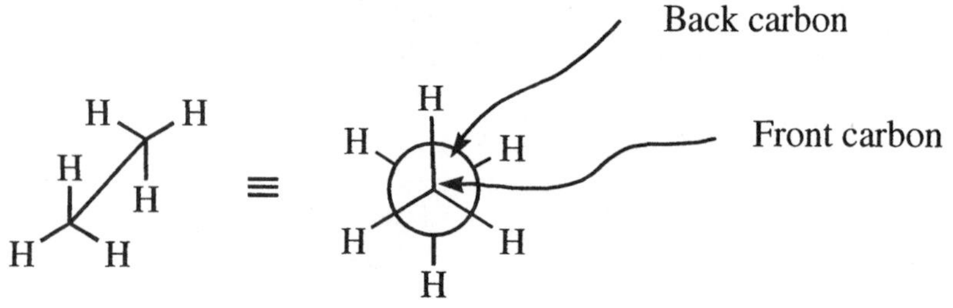

Now rotate the back carbon until you get an eclipsed conformation, and again look straight down the bond. This will be a tricky picture to draw. Not only can you not see the back carbon, you also cannot see the back H's, because they are hidden by the front H's! How on earth could you draw this? Here's how: we cheat. The Newman projection of an eclipsed conformation is rotated slightly away from the true eclipsed position, so that you can see what is attached where, but you are supposed to *imagine* that everything is perfectly lined up. Here's an example:

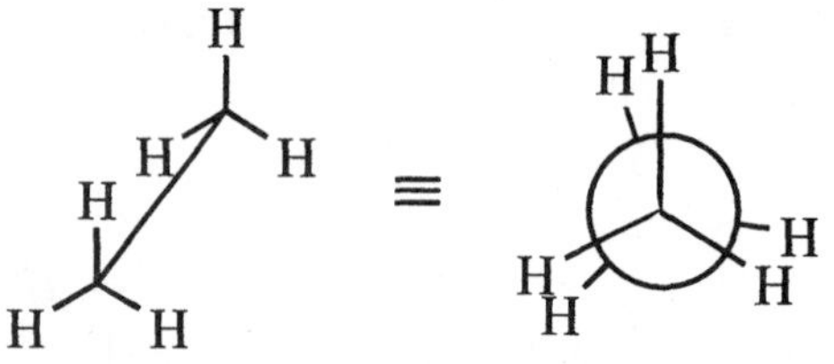

Now, for the important question: does any of this really matter? Does the molecule care whether it is staggered or eclipsed? After all, there is no obvious problem, looking at your model, with either one. But it turns out that the molecule *does* care. Not a lot, but there is a slight difference between these two forms, and the staggered conformation is better. Why? Most likely, because the tubes or sticks in your model that connect the carbons with the hydrogens are, in reality, electrons, and electrons repel each other. The electrons in the front C–H bonds are nearer to the electrons in the back C–H bonds in an eclipsed conformation than they are in a staggered conformation. The eclipsed conformation is therefore a bit higher in energy. We call this discomfort "torsional strain."

I used some pretty nebulous terms in the last paragraph. I said that the staggered conformation is "better," but that this was "not by a lot." Can we be any more precise than this? Yes we can.

When we speak in chemistry of a molecule's comfort level, or happiness (which of course is total nonsense), we are really talking about energy. In chemistry, energy is bad. As we saw earlier with atoms filling up the periodic table, electrons always get put into atoms in such a way as to give the lowest possible energy. The same applies to molecules: they strive to minimize their energy. Low-energy molecules are "better off" than high-energy ones.

Staggered ethane is "better off," "happier," than eclipsed ethane by about 3 kcal/mole. Time out! What are you supposed to think about this number? Is it big or small? How does it relate to other numbers you may have heard about before?

First I should confess that the powers that be (IUPAC) have declared that calories (cal) are out of favor. The preferred energy units are Joules. A calorie is about 4 Joules; obviously, a kilocalorie (kcal) is about 4 kJ. Most organic chemists defy IUPAC and continue to measure things in kcal, and I will follow suit. By the way, the Calorie that you worry about so much when you are on a diet is actually a kilocalorie, and it is written with a capital C to distinguish it from a real calorie. We will use kcal in this book.

So what should you think about 3 kcal/mol? This turns out to be a pretty small amount of energy. A mole of typical carbon–carbon bonds is worth nearly 100 kcal. For reference, this is about the amount of energy it takes to boil 1.5 quarts of water. For a compound just hanging around at room temperature there is usually enough energy available to do most things worth 15 or fewer kcal/mol. So 3 kcal/mol is pretty trivial. Ethane can rotate around its single bond without any significant problem.

Energy Diagrams

Nevertheless, it is still instructive to try to illustrate the energy situation in the course of this bond rotation process. Chemists like to draw what are known as "energy diagrams," pictures in which you follow some process from one point to another and illustrate how the energy changes as you go. Let's construct an energy diagram for rotation around the single bond of ethane. We start by drawing a set of axes. The horizontal axis will be some measure of progress in the process we are illustrating. In the case of bond rotation, it is easy to decide that the angle of rotation is a good measure. Later we will encounter cases where the measure of progress is far less clear. On the vertical axis we'll plot energy. We don't need numbers here, because we have no idea where to start on this axis, but we do know how far we are going. You'll see what I mean in a minute.

Energy

Angle of rotation

Now we start with some randomly chosen conformation of ethane, and place it at some random place near the left side of the graph. I picked an eclipsed form, but I wouldn't have needed to, as we will see. We'll label the angle at this point "0°." I draw a horizontal hash mark to indicate the energy of this conformation. Since the energy axis is not labeled, it doesn't matter where I put it. For convenience, I am leaving off the H's on the ends of the lines, but we all know they are there.

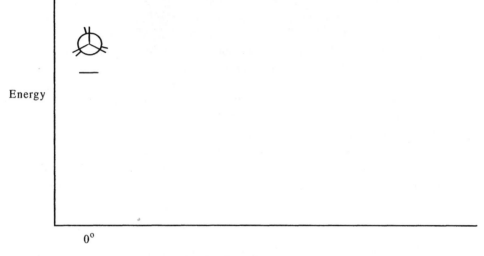

Now we grab one carbon, say the front one, and hold it steady, while we rotate the other one, the back one, in some direction. All of this is random. I'll rotate counterclockwise. What happens to the energy as I do this? We know the answer to this: the molecule becomes more comfortable, lower in energy. Its energy gradually falls until it gets to a staggered conformation. How far has the back carbon rotated at this point? 60°. So the next identifiable point on our graph will be at 60°. How far has the energy gone down at this point? 3 kcal/mol. How do you know? Because I told you so several paragraphs ago. Where do we put the next hash mark vertically on the graph? It doesn't matter, as long as we put it below where we started, because the vertical axis has no scale on it yet. Indeed, the act of putting this new hash mark on our graph actually establishes a scale for the *y* axis.

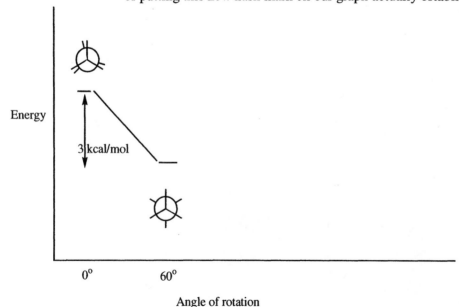

If we continue rotating now, in the same direction, the energy starts to go back up. Eventually, at 120°, it should be right back as high as when we started. So we can add another point to the graph.

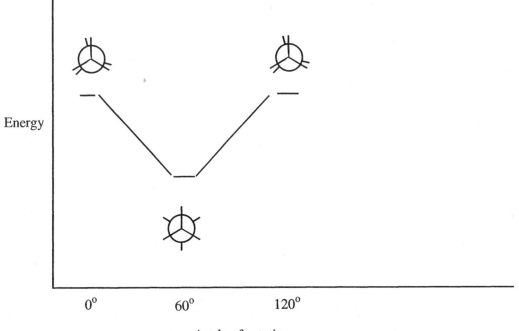

And so on. Now we can pretty up our graph by recognizing that energy changes tend not to make sharp corners. All changes are gradual. So we smooth out the curves a little and eventually arrive at the generally accepted form for an energy diagram showing rotation around the single bond of ethane:

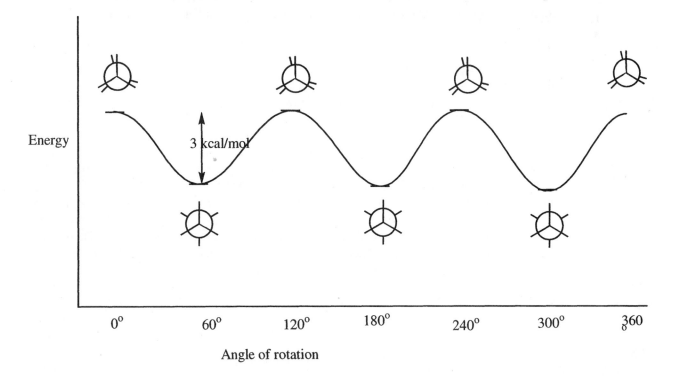

Maybe I didn't need to go through everything in such detail, but with our first energy diagram I wanted to be certain that you understood exactly what was going on. And recall, the amount of energy is a drop in the bucket for a molecule, so while ethane is marginally better off staggered than eclipsed, it still has plenty of opportunity to spin. We say there is "free rotation" around typical single bonds.

ORM: If you have access to the computer program "Organic Reaction Mechanisms," go to "Alkanes" and then "Conformational Analysis" for some nice animations concerning the previous section.

Propane

So what can we say about propane? If you look down either C–C bond in this molecule, here is what you should see:

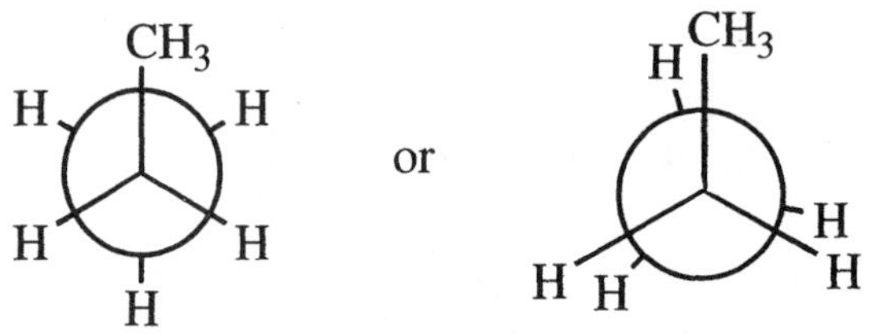

or something in between. And if you keep rotating, you will keep coming back to the same kind of structures. So the energy diagram for rotation around the C–C bond of propane will look exactly like that of ethane, with one exception. Can you figure out what it is? Whereas before, in the eclipsed conformation, hydrogens were lined up with each other, here on one side there is a methyl group lined up with a hydrogen. A methyl group is bigger than a hydrogen. You might think, therefore, that this eclipsed conformation is a bit worse off than the eclipsed conformation of ethane, and you would be right. So while the general shape of the energy diagram is the same, the distances are a bit different: the energy barrier for rotation around the C–C bond of propane is about 3.3 kcal/mol, a shade more than in ethane.

Butane

Build a model of butane, and examine it for rotation. You should be able to see that the end C–C bonds will be almost exactly like those in propane, while the middle C–C bond will be different. Let's focus in on this one.

You should notice that the staggered forms of butane are not all the same: there are two different types of staggered conformations. As you rotate one carbon relative to the other, you come to one of these two forms. We give them names. During a full 360° rotation there are two occurrences of "gauche" (or "methyls adjacent") forms, one where the back methyl group is on the right, as shown, and another where it is on the left. "Gauche," by the way, is a French word pronounced "gohsh" (it rhymes with the first syllable of social). There is also a staggered conformation called "anti" in which the two methyls are "opposite" each other.

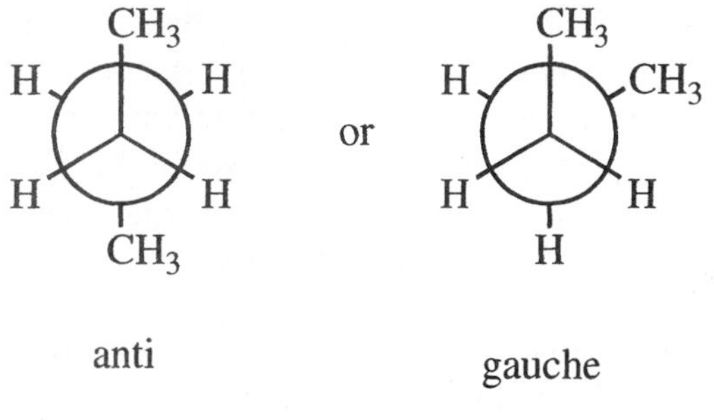

anti   gauche

Do you suppose the gauche and the anti conformations have the same energy? In a gauche conformation, the methyl group on the front carbon is near the one on the back, while in the anti conformation they are pretty far away from each other (in fact, as far away as they can be). To the extent that these groups might rub together, it stands to reason that the molecule would be happier in the anti form. This is correct. The difference is about 1 kcal/mol. So a gauche form is a little worse than the anti form, but still not as bad as any eclipsed form, which must be at least 3 kcal (and probably more) worse than the staggered form. The "rubbing together" of various parts of a molecule is called a "steric interaction."

Now let's look at the situation for eclipsed forms of butane. Again there are two possibilities. Any eclipsed model you build should look like one of the following. We do not have names for these.

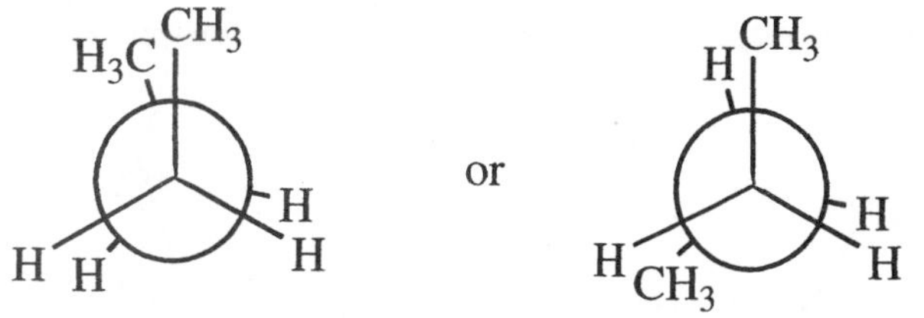

In the form on the left the two methyl groups are really pushing up against each other. In the other form both methyl groups instead push against hydrogens. You should expect that both eclipsed forms are worse than the corresponding eclipsed structures in ethane or propane, and you would be right. You might also guess that the one with two methyl groups banging into each other is even worse than the other eclipsed form, and you would be right again. Both eclipsed forms are worse than any of the staggered forms, but one is better than the other. Both staggered forms are better than either of the eclipsed forms, but one is better than the other. Can we draw an energy diagram for this process? Try it. Here are the steps:

First, draw any conformation you like. Then, holding one of the carbons still, rotate the other. Draw a picture every 60°; that is, whenever you get to an eclipsed or staggered form. The first one you can label 0°, and you should automatically come back to it after 360°.

Inspect these pictures: There are three eclipsed forms and three staggered forms. Eclipsed forms are always at the tops of hills in your diagram, and staggered forms are always in valleys. Which is the worst eclipsed form? It should be the one with the two methyl groups bumping into each other. This will have the highest energy. What about the other eclipsed forms? These will still be bad (at the tops of hill) but not as high on your diagram as the first one. They will also be identical in energy (this is not always the case, but it is here). Look at the three staggered forms and figure out which will be in the lowest valley(s), and where any other valleys should be relative to this.

Draw a set of axes as we did before, labeling the *x* axis "angle of rotation" and the *y* axis "energy." Look at the picture you drew for 0° and put a line near the left side of the graph to represent its energy. Where should you put this vertically? Well, if it is a staggered form, you must be in a valley, and you'll be going up from here, so you probably want it somewhere in the bottom half of the diagram. If you chose an eclipsed form as 0°, you will soon be going down, so put the mark in the top part of your diagram.

Now draw lines that represent the other forms, using the first line as your reference point. Be sure all the hilltops are higher than any valley, and all the valleys are lower than any hilltop. Also, be sure that any two structures with the same energy are at the same height. Then connect your lines with smooth curves.

Here's what I got. Yours might look different if you started from a different conformation, but it should show the same process. Once again, I have left off the hydrogens in my pictures to avoid clutter.

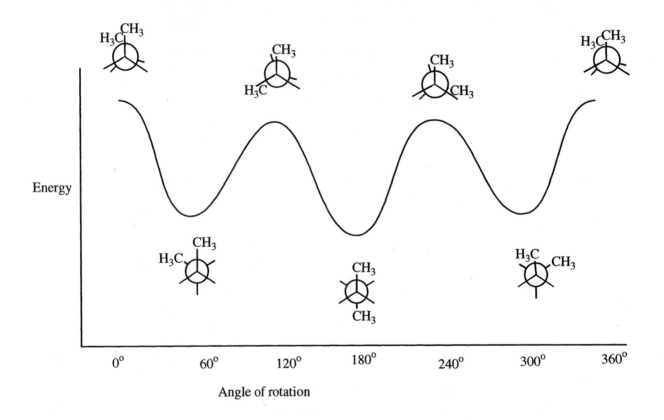

As before, the amounts of energy required here for bond rotation are trivial for a molecule, so we can say there is essentially "free rotation." Nevertheless, given a choice, the molecule is most comfortable in the anti conformation, and that's the way it spends most of its time.

ORM has an animation for butane as well. Check it out, it will help you solidify this section.

Pentane and Higher

It should be clear that the terminal bonds (the ones on the ends) of pentane will be essentially equivalent to those in propane and butane, while the central bonds will behave more or less like the central bond of butane, rotating freely but preferring anti conformations. This is why we often draw such molecules as stretched-out, zigzag pictures. Their chains can still bend around at will, as we will see later, but they like zigzag conformations best. Notice that when you draw the carbon chain as a zigzag on the paper, the hydrogens end up coming out of the paper and going back, as I have tried to indicate in this picture.

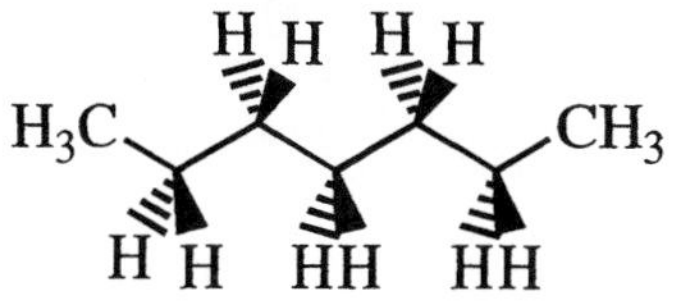

Test Yourself 1

Draw an energy diagram for rotation around the 2–3 bond of 2-bromobutane. To do this correctly you need to know that a bromine atom is considered to be quite a bit larger than a methyl group.

CYCLOALKANES

Compounds with rings are very much like their acyclic (ringless) counterparts: they do very little chemistry of interest to us, because they have no functional groups. No surprise there: taking a long-chain alkane and hooking its ends together really ought to result in no special change. Rings are also named just like their acyclic counterparts, except that you put a "cyclo" in front of the word designating the size of the ring. Rings often are used as the base for a compound name even if there is a longer chain attached, but sometimes it is useful to name the ring itself as a substituent on something else, in which case you change the "-ane" suffix into "-yl" (e.g., "cyclopropyl"), as in 2-cyclopropylhexane, shown below.

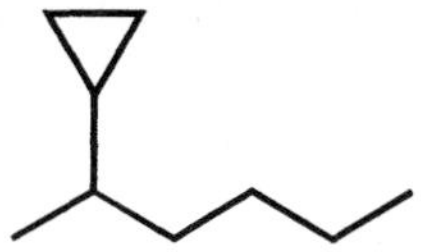

In spite of little chemistry of interest, there are many physical aspects of cyclic compounds that are important, and it is easiest to learn them now when we do not have to worry about reactions also.

Cyclopropane

The smallest cyclic alkane is cyclopropane, with three carbons forming a ring. Try to build a model of this. You may have some difficulty, depending on your model set. Some sets are flexible enough to permit this, some are not, so either you will be unable to build cyclopropane or you will succeed only with some difficulty. The molecule itself exists, but (as your models suggest!) not without some protest. This is definitely an unhappy camper.

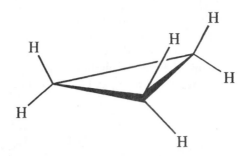

There are actually two different reasons for the unhappiness of cyclopropane. The first is that all hydrogens in the molecule have to be eclipsed. If you look down any carbon–carbon bond you will see this. We can estimate the cost of such eclipsing: eclipsed ethane is 3 kcal/mol worse than staggered ethane, and this is caused by three separate H–H interactions, so each interaction must be worth about 1 kcal/mol. In cyclopropane there are 6 H–H interactions, so this torsional strain is probably causing about 6 kcal/mol of pain.

Angle Strain

But that is not the main problem. It certainly is not what made it difficult to build the model. The problem you had building the model is that you had to bend some bonds. Cyclopropane, being a triangle, requires angles of 60°, but your model kit is designed to produce angles of 109°. So, in fact, are real atoms, and they, like your model, do not fit comfortably as corners of triangles. You have to bend and squeeze the parts to make a three-membered ring, and the same is true with real atoms.

How uncomfortable is cyclopropane? Think about this for a minute. We are claiming that cyclopropane is more uncomfortable than it ought to be; that it has more energy pent up in it than it should have. Can we measure how much more? Perhaps we should start by measuring the total amount of energy in cyclopropane. Can you suggest a way to do this?

Heat of Combustion

One approach might involve burning it. As you know, burning a substance releases heat. We could measure how *much* heat is released by burning cyclopropane and perhaps use that as a measure of how much heat cyclopropane ordinarily *contains*. Of course, thinking of this idea and actually pulling it off are two different things. *How* could we measure the amount of heat released when cyclopropane burns? This is not a trivial exercise: we'd have to make sure that all the cyclopropane we put in our apparatus actually burned, and that it burned completely; we'd also have to make sure that all the heat released was accounted for in some way, that it did not escape into the surroundings; and finally we'd have to figure out a way to measure how much heat this was. The standard way to proceed is called calorimetry. The whole experiment is done in a chamber submerged in a pool of water (a "calorimeter"), and one measures how much the temperature of the water goes up when the sample is burned. If you do the right math (which isn't hard!), you can then calculate the energy content of your sample.

So let's run such an experiment. Oh, there is another experimental detail we need to consider. How much cyclopropane should we burn? Remember, the chemist's unit is the mole, so we eventually want to know how much heat is released by burning a *mole* of cyclopropane (in kcal/mol). So should we burn a mole? Actually, that's pretty wasteful, and unnecessary. In fact, it doesn't matter a bit how much we burn, as long as we know how much we do in fact use and this is enough to measure the heat accurately. How come?

Because we know how much a mole is. Cyclopropane has the formula C_3H_6, so it has a molecular weight of 42 amu per molecule, or a weight (mass) of 42 g/mol of molecules. So if we burn less than 42 g, we simply measure the amount of heat given off by this particular sample, and then adjust the number so that it corresponds to how much heat we would have gotten if we *had* used 42 g. A simple ratio. Mathematically we do this by dividing the amount of heat we get by the number of moles we use, and "kcal/mol" falls right out.

When we do such an experiment, we discover that cyclopropane releases 500 kcal/mol in the course of burning (this is called its "heat of combustion," ΔH_{comb}). So what? Is this how uncomfortable cyclopropane is? No, not at all: after all, other alkanes release energy when they burn, too—even methane, which is presumably not uncomfortable at all. In burning cyclopropane, we did more than just release the strain in the ring, we also destroyed all the bonds, and we made some new chemicals (water and carbon dioxide). There is lots of energy involved in that process too. What we really want to know is how much *extra* energy there is in cyclopropane, compared to how much would be there if it weren't all cramped up. We want to compare the real molecule to "strain-free cyclopropane." Of course, there is no such thing, because in order to be cyclopropane, the molecule has to be cramped.

This is a problem that occurs over and over again in the study of organic compounds, and there have been lots of discussions in various cases as to what the proper reference points or reference materials should be. What should we compare cyclopropane to? Ideally, the same molecule without strain. Many students suggest propane as a suitable model, but there is a problem with this. Propane is C_3H_8, while cyclopropane is C_3H_6. Propane has more atoms and more bonds, so of course it gives off correspondingly more energy when it burns. We need something with the same number of the same bonds. How about propene? That has the formula C_3H_6, just like cyclopropane. But propene has a carbon–carbon double bond, and three hydrogens attached to those double bonded carbon atoms. Cyclopropane has none of that. We need a reference compound that has only what cyclopropane has: three CH_2 groups. Since no such thing exists (other than the strained cyclopropane itself), we have to make a guess as to how it might behave. There are several possible approaches. Perhaps the easiest to understand is one where we find a different compound that *contains* three CH_2 groups in it, but isn't strained. One such compound is pentane, $CH_3–CH_2CH_2CH_2–CH_3$. Of course, pentane has other atoms in it too, namely, two CH_3 groups. But we can guess that the two CH_3 groups in pentane are responsible for the same amount of heat as the two CH_3 groups in ethane (which contains nothing *but* two CH_3 groups). So if we measure the heat of combustion of pentane and then subtract the heat of combustion of ethane, we will be left with the heat of combustion of three CH_2 groups—*with no strain*!

Heat of Combustion

$CH_3–CH_2CH_2CH_2–CH_3$	845.2 kcal/mol
– $\underline{CH_3–CH_3}$	$\underline{372.8}$
$CH_2CH_2CH_2$	472.4

Another approach would be to figure out the energy contribution of *one* CH_2 group, and multiply that by three. To do this you might subtract ethane from propane ($CH_3CH_2CH_3 – CH_3CH_3 = CH_2$) , or propane from butane, or octane from nonane, or a whole bunch of these and average them. When you do the problem this way you get a value of 157.4 kcal/mol for each CH_2 group, and multiplying that by three leads to 472.2 kcal/mole as a best guess for the combustion heat that would be produced by three unstrained CH_2 groups. So either way, we get the same result: about 472 kcal/mol as a guess for the energy content of three CH_2 groups not bent out of shape. Previously we measured 500 kcal/mol as the combustion energy of real cyclopropane, three CH_2 groups that *are* bent out of shape. So what is the "cost" of the bending? Apparently, 28 kcal/mol. This is the "strain energy" of cyclopropane. The molecule contains (and therefore gives off!) 28 kcal/mol *more* energy than we would have expected based on three "comfortable" CH_2 groups. Recall that we attributed about 6 kcal/mol of this to eclipsing that cannot be avoided. The rest must be from what we call "angle strain," the effort it takes to form three bonds at the terrible angle of 60°. Carrying this further, *each* of the three strained angles is apparently contributing about 7.3 kcal/mol of strain energy.

To summarize: We claimed that the amount of energy pent up in cyclopropane consists of the amount of energy ordinarily found in the group of atoms $-CH_2CH_2CH_2-$, plus some extra energy called "strain energy" to account for the fact that the molecule is forced into an uncomfortable shape.

$$E_{total} = E_{normal} + E_{strain}$$

If you know any two of these you can easily figure out the third. We measured E_{total} by burning the actual compound cyclopropane; we guessed at E_{normal} by figuring out how much a CH_2 group was worth and multiplying it by 3; and therefore we were able to calculate E_{strain}.

Cyclobutane

Like cyclopropane, cyclobutane has bad bond angles. A perfect square would have angles of 90° rather than the desired 109°. It would also have perfectly eclipsed hydrogens along each side. Cyclobutane has a total strain energy of about 26 kcal/mol.

Test Yourself 2

If you understood the last section, you should be able to figure out a heat of combustion for cyclobutane.

The strain in cyclobutane is not too different from that in cyclopropane, but this time the 26 kcal/mol is spread over four carbons instead of three, so each must be somewhat less uncomfortable than in cyclopropane. It turns out that cyclobutane distorts itself a little from a perfect square in order to reduce some of its strain, but there is not too much it can do about it.

Cyclopentane

A perfect pentagon would have 109° angles. But it would also have completely eclipsed hydrogens. The molecule flexes a bit to avoid some of the eclipsing, at the expense of a slight increase in angle strain. Overall, however, cyclopentane is pretty happy: it has only about 6 kcal/mol of strain, and most of that is from eclipsing interactions.

Cyclohexane

A perfect hexagon has 120° angles. Not only would there be a problem due to angle strain (too *large* an angle this time), but on top of that all the hydrogens would be eclipsed, which would add about 12 kcal/mol of strain. However, unlike the smaller rings, cyclohexane (and larger) rings are not forced to be flat. Indeed, cyclohexane can flex itself in such a way that all its angles are 109°, and, simultaneously, all the hydrogens turn out to be perfectly staggered. In this conformation, called a "chair" for reasons shown in the following picture, there is no strain at all.

Chair conformation

Cycloheptane and Higher

Larger and larger rings would require larger and larger angles if they were flat, but of course they are not. All of them have enough flexibility to adopt many different conformations. Some conformations have less angle strain but more torsional strain, some the reverse. Sometimes hydrogens start bumping into each other from across the ring. Until you get to rings of about 14 carbon atoms, none can have strain-free conformations. Rings of 7–13 carbons have about 6–12 kcal/mol of strain energy. After 14 carbons the rings are flexible enough to find strain-free conformations.

Cyclohexane Revisited

Because cyclohexane has no strain, six-membered rings crop up all over the place, and there are enough important factors related to their shapes that we need to spend some more time considering this kind of ring. Stop reading and go build such a ring with your model kit right now.

Boats vs. Chairs

Although the chair form of cyclohexane is the best conformation possible, your first model probably did not automatically pop into this shape. This is because your model, unlike a real molecule, is not vibrating all the time, sampling different conformations and then deciding which is best. In order to get from one conformation to another, you have to do something with your hands to cause the conversion. If by accident you built something other than a chair, it would stay that until you changed it. One other ring conformation also has no angle strain, and there is a good chance that's the one you first constructed. This form is called the "boat" conformation, for fairly obvious reasons:

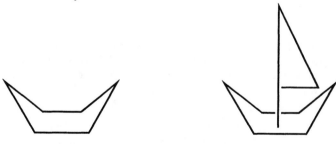

Boat conformation

The boat is not as good as the chair in two ways. First, if you look along the sides, where the oars should be, you will find that all the H's are perfectly eclipsed. Second, if you look across the top, there are two H's across the ring from each other that nearly collide. These are called the "flagpole" hydrogens.

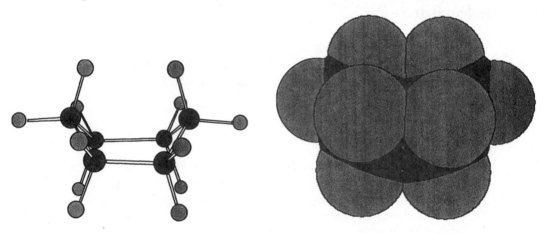

Eclipsing   Flagpoles

The boat, in fact, is about 6 kcal/mol less stable than the chair. So why is it just as easy for your model to come out a boat as a chair? The problem is that model sets are not smart enough to notice either of these problems. As you have already discovered, staggered conformations are more stable than eclipsed conformations, but a model set cannot tell the difference. Further, most model sets are not lifelike enough to show the "bumping" of the flagpole hydrogens. Below I show two pictures of the flagpole problem. The second one shows atoms that are life like in size, and here you can clearly see the problem. But at the same time, models built to scale in this way ("space-filling models") hide many of the bonding features we are usually interested in, so we do not use them very often.

Equilibrium Constants

Since cyclohexane is happier as a chair than as a boat, does this mean that it always exists as a chair? No, it doesn't. Why not? Well, you are more comfortable lying down than standing up, right? Does this mean you are always lying down? Even if you did not have to get up to eat and go to classes, etc., would you spend your whole life lying down? (I realize this is a risky question to ask college students.) No, because at some point you would probably *want* to get up and move around, even though it would cost you energy to do it. The same is true with molecules: they are not always in their most comfortable state, because sometimes they get bored and want to explore other possibilities. The extent to which they do so depends on how uncomfortable the alternative possibilities are.

Imagine you are in a blimp flying over Fort Lauderdale during spring break. You take a photograph of the beach as you pass by. What do you see? A huge crowd, to be sure. Most of the people, maybe 70%, are lying down, with maybe 20% sitting and 10% standing, swimming, or exercising in some way. Thirty minutes later you pass the beach again you take another snapshot, and you see the same thing. But it would be very surprising if exactly the same people were doing the same things: the people swimming now are almost certainly not the same ones who were swimming before. Further, you could never figure out by inspecting the first picture *which* people would be swimming the next time you look. Still, you can safely guess that there will continue to be about 70% lying down, 20% sitting, and 10% other. This is what we call *dynamic equilibrium*: the description of the system as a whole does not change, even though the individuals do. Of course, if a minute before your second picture, several buses had arrived and unloaded thousands of people on the beach, there would be more people standing than expected, but at this point the system would not be at equilibrium: after 30 minutes, these new people will fit the old pattern, and the distribution will be 70:20:10 as usual. What *could* change the description of the system at equilibrium would be a change in temperature. If it got hotter, more people might be swimming than before, for example.

All this applies to molecules too. The study of equilibrium in chemistry is called "thermodynamics." In a system capable of passing freely between two (or more) states, some individuals (molecules) are always in one state and some in the other. You cannot predict which ones will be in which state, but you *can* predict *how many* will be in each state, as a percentage of the whole. Such a system is easily described mathematically by inventing a term called an "equilibrium constant," K_{eq}, which is simply a ratio of the number in one state to the number in the other state. The "numbers" are traditionally described in "concentrations": people per cubic yard, or, in chemistry, moles of particles per liter, abbreviated mol/L or M (molar). The symbol for a chemical set in square brackets means the concentration of that chemical in the unit mol/L. Thus, if A is in equilibrium with B, we show it this way:

$$A \;\rightleftharpoons\; B \qquad\qquad K_{eq} = \frac{[B]}{[A]}$$

Notice that the equilibrium constant is defined as the concentration of the stuff on the right ("product") divided by that of the stuff on the left ("reactant" or "starting material" but NOT "starting product"). Also notice that we could have written the original equation backwards; it would still refer to the same equilibrium process, but now the equilibrium *constant* would be the inverse of the first one:

$$B \;\rightleftharpoons\; A \qquad\qquad K_{eq} = \frac{[A]}{[B]}$$

Both statements are correct, and both of the equilibrium constants are correct, but the numbers are different. Each refers to its own equation. In other words, the equilibrium constant, that is, the number that characterizes the equilibrium concentration ratio, is different depending on whether you write the reaction one way or the other. Like vectors in physics, these particular constants are not just numbers, but, in some sense, directions as well. An equilibrium constant requires you to know the equation to go with it or else it is impossible to interpret.

Every equation has an equilibrium constant associated with it. At a given temperature, the equilibrium constant stays constant, always the same number. Does this mean that the ratio [B]/[A] must always be the same? No it doesn't. It means that [B]/[A] is always the same *if the system is at equilibrium*! Remember the beach: when the bus unloaded, the system was not correctly described by the expected numbers, but after the newly arrived people got settled down (which probably took a while!), it was. Molecular systems are the same: the equilibrium constant works, provided the system has settled down, that is, it has reached equilibrium. But if you encounter it when it is out of equilibrium, any concentrations are possible.

What determines the ratio of B to A (i.e., what will K_{eq} be)? Basically, as you might imagine, this is a question of how comfortable the molecules are in state A compared to state B, or, more specifically, the *difference in energy* between the two states. Molecules prefer a more comfortable state, the one with the lowest energy, but there are always some molecules in any large collection (and all collections of molecules are large!) with enough energy to be in a less comfortable, higher energy state. The ratio always depends on both the energy difference between the two possibilities and also the amount of energy available: that is, the temperature. Specifically, the formal relationship is the following:

$$\Delta G° = -RT \ln K_{eq}$$

where R is a constant one can look up, T is a temperature on the absolute scale (Kelvin), "ln" means the natural logarithm, and $\Delta G°$ is our symbol for a particular kind of energy difference. You probably already know that Δ (pronounced "delta") is the scientific shorthand for a difference or change; we use G here instead of E because we are dealing with a special kind of energy, called the "free" energy, which is a combination of energy pent up in the molecules (technically called enthalpy, symbol H) and also something called "entropy" (symbol S) which, crudely put, is a measure of the "randomness" of the system. The little ° symbol after the G indicates that this energy difference has to be measured under special circumstances; we can ignore it for now. By the way, energy changes are traditionally described as the energy of the product minus the energy of the starting material. So if the product has lower energy than the starting material, $\Delta G°$ will always be negative. We call such reactions "exothermic" (better: "exergonic") since they release energy; reactions that sop up energy by creating materials more energetic than the starting material are called "endothermic" ("endergonic") and they have positive $\Delta G°$ values. It should be clear (from the equation but more importantly because it makes sense!) that exothermic reactions will have K_{eq}'s greater than 1, while endothermic reactions will have K_{eq}'s smaller than 1 (but still positive—all K_{eq}'s are positive, because they are ratios of concentrations, which clearly can never be negative).

We will worry more about some of these things later, but for now you should still be able to use this equation. Be sure you know how to take the natural log of something on your calculator, and that you also know how to work with this equation in the backwards sense: given a $\Delta G°$, could you calculate a K_{eq}? The equation above is entirely equivalent to the one below. If you don't see this, get some help.

$$K_{eq} = e^{-\Delta G°/RT}$$

Be sure you can make your calculator do the appropriate math. Here's a test: Given a $\Delta G°$ of –2.6 kcal/mol, $R = 1.987$ cal/(mol K), $T = 25°C$, what is K_{eq}? Do this now, before you read any further. The answer is the square root of 6515. [I give answers in code so that the actual answer will not be staring you in the face.] Did you get that? If not, check for these common mistakes: $\Delta G°$ was given in kcal/mol, but R is in cal/(mol K), so there is a factor of 1000 that has to be taken into account. Also, the temperature was given in °C, but in the formula it needs to be in K. Finally, $-\Delta G°/RT$ has to be calculated completely before you raise e to that power. And on some calculators, e^x is listed as $\ln^{-1}$. If you still have trouble, get some help.

Let us return now to cyclohexane. I have described the equilibrium shown below, and indicated that the chair is better than the boat by 6 kcal/mol.

chair   boat

Several things should immediately pop into your head. First, at equilibrium, there will be more chairs than boats. Second, the reaction (written in this direction) is endothermic, that is, the molecules are going up in energy from left to right. Third, quantitatively, $\Delta G° = + 6$ kcal/mol. We know this has to be a positive number for two reasons: first, it is defined as the energy of the product minus the energy of the reactant, and we have already determined that the product has more energy, so a larger number minus a smaller one will be positive; second, because *all* endothermic reactions have positive $\Delta G°$ values. What is the equilibrium constant for this reaction? Using the math above, you should get a value of

$$K_{eq} = 3.97 \times 10^{-5} = \frac{[\text{boat}]}{[\text{chair}]}$$

How should you interpret this number? How much of the compound is in the boat form and how much is in the chair? Let me show you three approaches; you pick whichever one works best for you.

A: You will agree, I hope, that the number 3.97×10^{-5} can be divided by 1 without changing it.

$$\frac{[\text{boat}]}{[\text{chair}]} = 3.97 \times 10^{-5} = \frac{3.97 \times 10^{-5}}{1}$$

Now we can multiply the top and bottom of this quotient by some arbitrary number until the result is one we like better. Multiply by 10 until the top becomes something you can handle. If you use 10^5, the top becomes nearly 4, and the bottom becomes 100,000. That says the ratio of boat to chair is about 4 to 100,000, that is, about 4 molecules in 100,004 are in the boat form, while 100,000 are in the chair form. You can put this in percentage form by saying that 100,000 are chair, out of 100,004 total, so the percent in the chair form is 100,000/100,004 x 100% or 99.996%; boats represent 0.004%.

B. We start off the same way,

$$\frac{[\text{boat}]}{[\text{chair}]} = 3.97 \times 10^{-5} = \frac{3.97 \times 10^{-5}}{1}$$

but now we can go directly to percents by recognizing that 3.97×10^{-5} represents the concentration of boats, and 1 represents the concentration of chairs, so the sum, $(1 + 3.97 \times 10^{-5})$, represents the sum of the two concentrations together. Thus, the percentage of chairs is 1/sum and the percentage of boats is 3.97×10^{-5}/sum, expressed as percents. Doing the math, we again get chairs = 99.996%, boats = 0.004%.

C: For those of you who like formulas, try this (where C = chair, B = boat):

$$[C] + [B] = 1$$

$$[B] / [C] = 3.97 \times 10^{-5}$$

Solve these simultaneous equations for the fraction of molecules in each form, which can then be converted to percents by multiplying by 100.

What if we had written the equation the other way around?

We're still talking about the same system, so we had better get the same answers. Let's see. First, it is still the case that the chair form is better than the boat form by 6 kcal/mol, so there will be more chairs than boats. This time the reaction is exothermic as written, and $\Delta G°$ is – 6 kcal/mol. Do the math:

$$K_{eq} = e^{-\Delta G°/RT} = 25{,}159$$

$$25159 = \frac{25159}{1} = \frac{[\text{chair}]}{[\text{boat}]}$$

So there are 25,159 chairs for every 1 boat. The percentage of chairs is 25,159/25,160 = 99.996%, and the percentage of boats is 1/25,160 = 0.004 %.

Flipping Process

Let's see if we can follow what happens physically as a molecule changes from a chair into a boat and back again. Get your model and put it in a chair form. Include the hydrogen atoms. Depending on what kind of model you have, it might resemble one of the following.

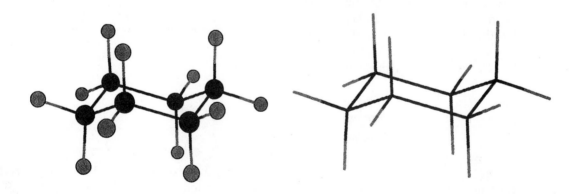

Hold your model so that it looks just like one of these pictures, where the chair points down on the left side and up on the right side. Now grab the left carbon and swing it upward, keeping everything else the same. You will have to apply a little force to get this to happen; it does not flip without resistance. You should also notice that as you do this, there is a point in the motion where you no longer have to apply force: after you have pushed it just so far, it sort of pops into place now as a boat. Go back again to the chair, and you will notice the same thing: after some initial pushing, it just pops into the chair. There is clearly some point between the chair and the boat that is worse off than either of them, and the molecule breathes a sigh of relief as it pops from there in one direction or the other into a boat or a chair.

Let's illustrate this process in an energy diagram. We know the chair is lower in energy than the boat, by 6 kcal, so we can locate two corresponding energy levels on the y axis. The x axis is described by one of those nebulous terms, "reaction progress," perhaps displacement of the left carbon. It doesn't really matter all that much for the moment exactly what it refers to. Here is what we have so far:

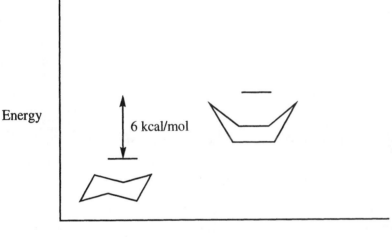

Now, how do we connect these two horizontal lines on our diagram? Do we just draw a smooth curve between them, as we did before in the staggered/eclipsed diagram? No, not this time. Why not? When you are rotating single bonds, as we did before, eclipsed is as bad as you can get. Any motion away from eclipsed makes you better off, so the eclipsed form is at the top of a hill. Here, however, the boat is not at the top of a hill. How do we know? We *felt* it. You felt it when you flipped your model. You had to work to get it out of a boat. Clearly, the molecule was getting worse before it got better. As we decided before, there is something between the boat and the chair that is worse than either of them. The energy diagram has to show this. Here's how.

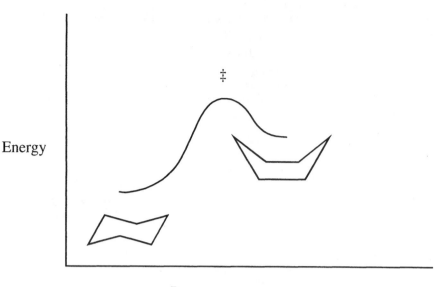

Energy

Progress

So what is this thing I labeled ‡? What does it look like, and what is its energy, relative to the other known structures? It is called a "transition state." A *transition state* is the high point on an energy diagram between two lower, more stable states. In the rotation of ethane, the transition state was the eclipsed form. That case is actually relatively unusual, in that we know exactly what the corresponding transition state looks like. In this case, we don't. You can manipulate your model, and I can manipulate mine, and we'll both come up with something pretty similar as the worst point between boat and chair, but they probably won't be *exactly* the same. The fact is, no one really knows exactly what the transition state for this process looks like, but we know there is one. Further, its actual energy level is known, although for our purposes it is not important what it is, only that it is higher than the level of either of the known structures, but not so high that the molecule cannot get from one "stable form" to the other. In case you are interested, this particular transition state ("TS") is about 11 kcal/mol higher than the chair.

You got from a chair to a boat by popping the left carbon of the chair upward, which corresponds to moving from left to right along the *x* axis of our energy diagram. You went from the boat back to the chair by popping the same carbon back down, moving from right to left on our diagram. But look at the boat: it is symmetric. That is, you can't really tell its bow from its stern. You could also get from boat to chair by taking the *right* carbon and popping it downward. What would you have then? Another chair, one pointing down on the right and up on the left. This is shown in the following new diagram. "But wait!" you say. "My new chair is exactly the same as my old chair! Why bother showing both?" Excellent question. If you build two models, one of the chair on the right and one of the chair on the left, it is true that you will not be able to tell them apart. But try this. Take your model and put a piece of tape on one of the hydrogens (or make one hydrogen a different color). It doesn't matter which one. Flip to a boat, then back to the *other* chair. You will find that now the marked hydrogen is in a new position. In other words, although *you* cannot tell which chair is which, *the atoms can*!

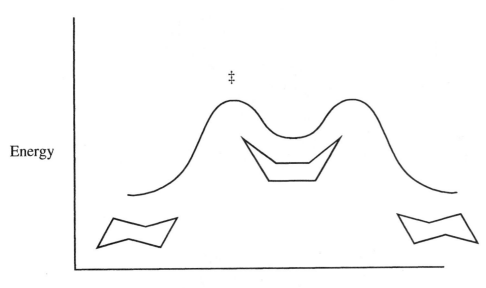

Energy

Progress

Axial vs. Equatorial Positions

The point is that there are two different types of positions for hydrogens in a chair. The ones that are lighter in the left picture below point more or less perpendicular to the general plane of the ring, and are called "axial." The darker ones in the picture are more or less parallel to the plane of the ring (though not quite), and are called "equatorial." Notice that when you flip from one chair to the other, these hydrogens all switch positions; that is, everything that was axial in one chair is equatorial in the other, and *vice versa*.

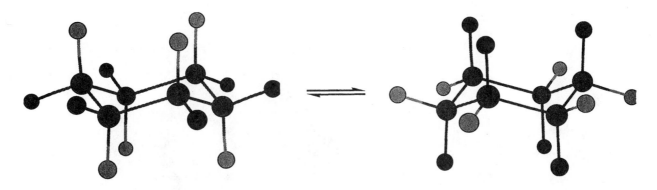

Does a molecule care which chair form it is in? Not this molecule. But consider methylcyclohexane. Simply pick a random H on your model and replace it with a CH_3 group. You will have built one of the two forms shown on the next page, and if you flip from one chair (*via* the boat) to the other, you will obtain the other form shown. Now a molecule can tell the difference between these, and so can you. Further, the molecule cares. Unfortunately, your model does not do justice to the true structure of this molecule, and therefore you probably don't see any big difference between the two that would lead the molecule to prefer one over the other. This is another case where we are deceived by models with atoms that are too small. The following pictures show side and top views of the axial form of methylcyclohexane, one a traditional view and one a "space-filling" view. Hopefully you can see that in real life, the methyl group is crashing into other atoms when it is in an axial position. Indeed, you might notice that the methyl group actually leans a little bit away from vertical in order to avoid this interaction.

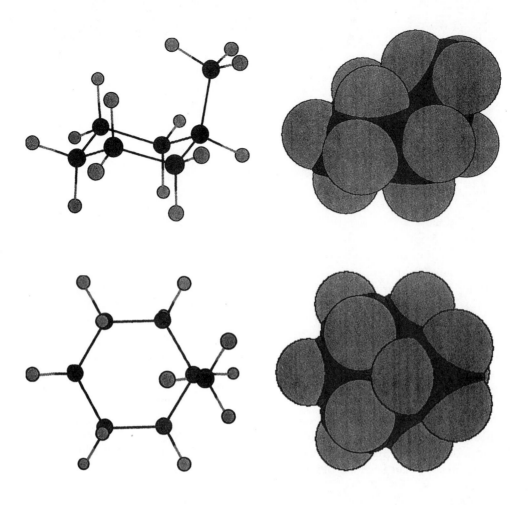

On the other hand, there is no such problem when the methyl group is equatorial:

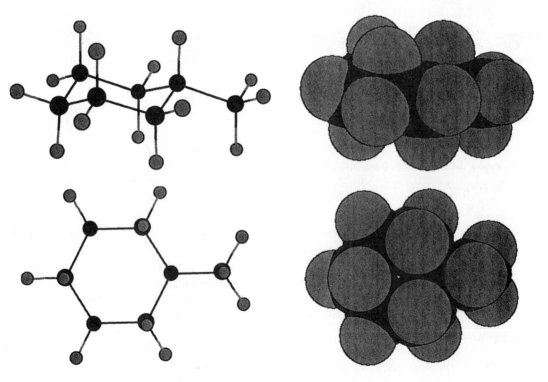

It is important to notice *which* hydrogens the methyl group is crashing into. The problem is not with the ones nearest it on the ring. The axial methyl group bumps into other hydrogens that are also axial, attached to carbons that are two atoms away! If the carbon holding the methyl group is labeled as number 1, the carbon holding the hydrogen bumping into it is at a number 3 position (either one, depending on which direction you start counting). These collisions are referred to as "1,3-diaxial interactions," and they are responsible for a molecule preferring an equatorial to an axial conformation for its substituents.

Methylcyclohexane is in the equatorial conformation 95% of the time at room temperature (25°C). What does this tell you about the energy difference between the two forms? You should be able to use the equations we learned above to calculate this. First you need to figure out K_{eq}, but this is not hard: since 95% are equatorial and 5% axial, the ratio must be 95/5, and the equilibrium constant must be 19. Of course, as soon as we say that we are specifying a particular direction for the reaction. Since we wrote K_{eq} as 95/5, and this represents products over reactants, we must be referring to the equation written below:

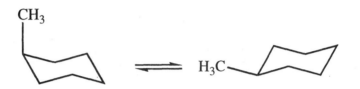

If we had written it the other way, the K_{eq} would be 5/95.

So what is $\Delta G°$ for this process? Using our energy equation you should get –1.7 kcal/mol. (It would be +1.7 kcal/mol if you wrote the reaction equation the other way.)

What do you suppose would happen if we put an ethyl group on cyclohexane instead of a methyl group? Ethyl is bigger than methyl, so the preference should be a bit stronger. In general, the larger the group you hang on a cyclohexane, the greater the preference for equatorial over axial. A *tert*-butyl group has such a strong preference for equatorial that it stays in that form about 99.99% of the time. This is sometimes referred to as being "locked" in a single chair form because it so rarely visits the other.

Disubstituted Rings

What do you suppose would happen if there were *two* methyl groups on our cyclohexane ring? The answer is, that depends on where they are. But first we need to digress briefly to discuss how one names rings with multiple substituents.

As you know, you name a ring by putting "cyclo" in front of the name of the corresponding alkane. If there is one thing attached to a ring, say a methyl group, we can name this "methylcyclowhatever," because every position on a ring is the same as any other. As soon as you put a second group on the ring, however, you need to tell your reader where the second one is relative to the first. You start numbering the ring (your "home base" for naming) where one of the substituents is, and head in the fastest way possible toward the next; that is, you go the short way around the ring. Thus, the following is called 1,3-dimethylcyclopentane and not 1,4-dimethylcyclopentane.

But this is not all. Unfortunately, there are two different versions of this compound! (Actually three, but we'll get to the third later.) The following two compounds are not the same. The one on the left, shown by two different drawings of the same molecule, has both its methyl groups on the same side of the general plane of the ring. We call this *cis*. The other one has the two methyl groups on opposite sides of the plane of the ring. We call this *trans*. So there are two *different* 1,3-dimethylcyclopentanes that can be interconverted only by breaking bonds, and to be complete we have to specifiy which one we are referring to: *cis*-1,3-dimethylcyclopentane is the one on the left, and *trans*-1,3-dimethylcyclopentane is the one on the right. The same applies to rings of other sizes. Isomers like these, with the same connectivities but differing in the directions of their attachments, are called *stereoisomers*.

Now is a good time to review the different kinds of isomers we know. First there are *constitutional* (or *structural*) isomers, molecules with the same constitution (that is, formula) but put together differently; examples would be pentane vs. 2,2-dimethylpropane, or cyclohexane vs. 2-methyl-2-pentene, or 2-methylpentane vs. 3-methylpentane. Each of these molecules has the same formula as its mate, but the pairs are distinguishable in terms of how the molecules are put together, what is attached to what and where. Constitutional isomers are different compounds, and you cannot change one into another without breaking bonds, both in the molecule and your model of it. Structural (constitutional) isomers also— necessarily!—have different names. Second, there are *geometric* isomers, or *configurational* isomers, or *stereoisomers* (these all mean the same thing). These are molecules with the same formula *and* the same attachments, meaning the same atoms are attached to the same other atoms. The only difference is in the relative *direction* of attachment. Examples are the *cis* and *trans* dimethylcyclopentanes above. These, too, are different compounds, and again you cannot change one into the other other without breaking bonds, both in the molecule and your model of it. Stereoisomers tend to have essentially the same names, with special *prefixes* to distinguish them. Third, there are *conformational* isomers, like the staggered and eclipsed forms of ethane, or the boat and chair forms of cyclohexane, or the axial and equatorial forms of methylcyclohexane. These are not really isomers at all: they are just different shapes for the same structure. I prefer to call them conformations rather than conformational isomers, but you may hear the isomer term used. The important point here is that you *can* change one conformation into another without breaking bonds, both in the molecule and your model of it, so the molecules themselves would be constantly changing from one conformation into another, and you normally can't stop this from occurring. Most molecules have zillions of possible conformations, and most conformations have no names associated with them at all unless there is something special about them that makes it important to refer to that particular shape fairly often.

Consider playing with dolls. If you take a Barbie and move the arms so that they are raised rather than down by her side, it is still the same doll you started with, just in a different pose. That is like a conformation. If you take off her arms and reattach them to the wrong shoulders, you have made a configurational isomer, a stereoisomer. There is still an arm on each shoulder, but this doll is identifiably different from the original, and you cannot get back to the original without a screwdriver. You still might call her Barbie, but maybe "inverted Barbie" or some such to distinguish her from the original. Finally, if you put her leg where her head should be and *vice versa*, now you have a different beast altogether, but still made out of the same parts. This is a structural isomer.

So what about dimethyl cyclohexane? There are four *constitutional* isomers to consider: 1,1; 1,2; 1,3; and 1,4-dimethylcyclohexane. The 1,1 compound is uninteresting because there is only one way to build it. But the others are quite interesting: each has two *stereoisomers*, *cis* and *trans* (different compounds), and each of these in turn has two chair forms to consider (different conformations of the same compound). Let's focus on 1,2-dimethylcyclohexane.

First we draw the two stereoisomers. At the same time, you should be building models of these.

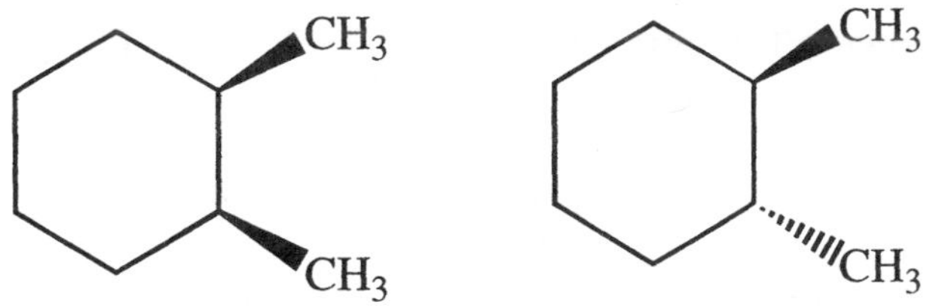

This is not as easy as it sounds. How do you know where to put the groups? One approach is to squash your model flat on a table, like the pictures above. Then it is clear where to put the methyl groups. Another is to notice that even when the model is not squashed, but just sitting on a surface in a chair form, each carbon has one hydrogen pointing relatively up and one pointing relatively down. True, one of them is pointing very much up or down, while the other is pointing only a little up or down, but they still have directions. Choose the "up" hydrogen on one atom to turn into a methyl group, and the up hydrogen on the adjacent carbon to become another methyl group, and you will have made *cis*-1,2-dimethylcyclohexane. Switch either methyl group to the other position on the same carbon to get *trans*.

What have you made? The *cis* compound looks like this:

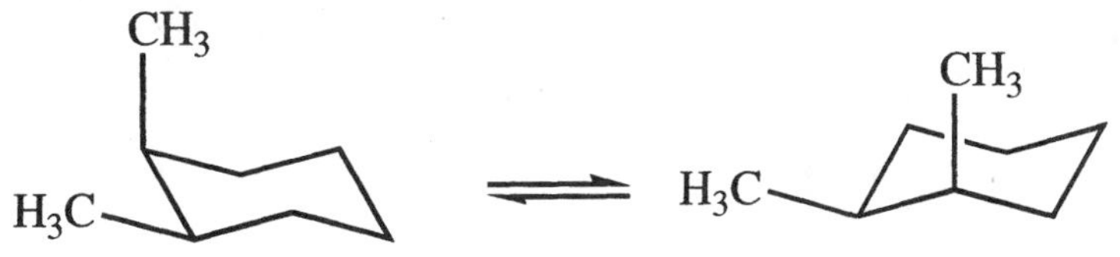

One of the methyl groups is equatorial (good), and the other is axial (bad). Flip to the other chair, and what do you have? The one that was good is now bad, and *vice versa*. Each conformation is just as good or bad as the other, and the molecule flips rapidly back and forth between them, spending half its time in each conformation.

What does the *trans* look like?

In one of the chairs, *both* methyl groups are axial, therefore both of them are unhappy. In the other chair, both are equatorial, a lower energy state. The molecule, being sensible, spends most of its time in the one it likes, namely the diequatorial form. It rarely checks out the diaxial form, and when it does it jumps back as fast as it can. We indicate this unequal distribution of conformations by using unequal equilibrium arrows, with the long arrow pointing to the preferred conformation.

The *trans* form of 1,2-dimethylcyclohexane as a whole is more stable than the *cis* form, because it *can be* in a conformation which is low energy. This does not mean that the *cis* isomer will decide to turn into the *trans* isomer. It probably would if it could, but there is no way you can change a model of the *cis* compound into the corresponding *trans* compound without taking it apart and putting it back together in a different way, and the same is true of the substance itself. *Cis* and *trans*-1,2dimethylcyclohexane are not in equilibrium with each other. They are different compounds, stored in separate bottle in the stockroom.

COMMON MISTAKES: DON'T LET THIS HAPPEN TO YOU!

Many students claim that *trans*-1,2-dimethylcyclohexane is more stable than the *cis* isomer because the methyl groups are farther apart. They think the methyls bump into each other in the *cis* form but not in the *trans* form. NOT TRUE!! The problem with the *cis* form is not that the methyl groups bump into each other, it is that one of the methyls bumps into hydrogens that are two carbons away, the 1,3-diaxial interactions we discussed earlier. Further, there is nothing the molecule can do to avoid such steric crowding. The *trans* isomer also suffers from this problem, and to an even greater extent, in one of its conformations, but it simply chooses not to *be* in this conformation, instead choosing the conformation where there are *no* 1,3-diaxial interactions. It is not the relative closeness of the methyl groups that determines stability in these compounds, it is whether the groups have to be axial or equatorial.

To make sure you understand all this "conformational analysis," consider the 1,3 and 1,4 dimethylcyclohexanes. It is not always true that *trans* is better than *cis*. Draw each molecule carefully, consider the possible chair conformations of each, and draw your conclusions. Use your models to help you with your drawings. Eventually you should work toward being able to do all this *without* models, but for the time being it is not bad to rely on them heavily.

WHO CARES? YOU DO!!

This whole chapter may seem purely academic to many of you. What difference does it make whether you can produce the correct name for a compound? How much strain there is in a small ring? Whether a substituent is axial or equatorial? Whether two substituents are cis or trans? Let me try to address these questions.

Like it or not, you are in the business of communication. Experiments do not become science until they are communicated to other scientists; doctors will not last long if they cannot communicate with their patients. One aspect of communication is using appropriate language. Doctors use (or should use) different language when communicating with patients than when they communicate with each other. With each other, they have special words that mean specific things; words they use because they are more descriptive than the more common words lay people might use for those same conditions. Likewise chemists have words that mean specific things to other chemists. If you are going to communicate with chemists, or understand when they communicate with you, you need to know the lingo. More importantly, information is cataloged in a vast body of writing called, collectively, "the literature." If you expect to obtain any information from printed works, you need to understand how you go about finding that information, and more often than not, this involves naming something by its accepted name.

Strain is important because molecules that have high energy often react in such a way as to lose that extra energy. In Chapter 17 we will discover some special reactivity of three-membered rings containing oxygen, some of which is believed to lead to certain cancers. In Chapter 16 we will learn that compounds react differently depending on whether a substituent is axial or equatorial. Finally, we will learn in many contexts that biological systems are exquisitely attuned to the shapes of molecules, discriminating on even more subtle differences than you have yet seen. Shape is a key factor in determining which biological reactions occur and which do not. Any attempt to understand biology is doomed without a prior understanding of shape.

Drawing Chairs

It may seem silly to devote an entire section of this book to art, but remember: you are also in the communications business. If you join the Peace Corps, and go to a remote country and describe in great detail exactly how they should plant their crops, but you do so in a language they cannot understand, you will accomplish nothing. If you have information you wish to communicate, it is important that you learn to speak the language of the people you are speaking to. That is why we have spent some time learning nomenclature, and that is why you must learn to draw accurate pictures. In chemistry, more information is conveyed through pictures than with words. Organic chemists cannot speak to each other without drawing pictures, and it is critical that both drawer and viewer understand the meaning of the pictures.

One picture that needs to be drawn frequently, and that students often at first have great trouble with, is the chair form of cyclohexane. Below is a picture that we need to study in detail:

Several things you should notice:

- The carbon atoms around the ring alternate between "up" and "down" relative to the general plane of the ring

- Opposite edges of the ring are parallel to each other.

- There are no vertical or horizontal lines within the ring itself.

- Of the lines to the hydrogens, six of them are dead vertical, three up and three down. The "up" ones come from the carbons that are themselves up, and the "down" ones come from the carbons that are themselves down. These six lines represent the "axial" bonds.

- The other six lines to hydrogens are *not* horizontal. Each is just a little up or down from horizontal. They are slightly down if the corresponding axial hydrogen is up, and *vice versa*. Further, of these "equatorial" bonds, three are headed to the right and three to the left. The ones headed to the right are on the right side of the molecule; the ones headed to the left are on the left side of the molecule. (By the way, the word "equatorial" comes from "equator." There is no "i" in the second syllable.)

- Most importantly, you should notice that the drawing of the molecule looks like the model that you built of it. You **must** build models of these things and manipulate them. The pictures you draw must be capable of conveying on paper an image that you have clearly in your head. Such an image cannot get into your head if you do not play extensively with your models. And if you do have an image of a real object in your head, it will be much less likely that you will draw a picture that makes no sense. (I've seen a lot of these from students.)

There are also several other things you should discover from models of this molecule, in particular, things that are *not* true. Again, it is critical that you realize that the things I am saying apply to the object, the model. They apply to the drawing precisely *because* they apply to the model:

- There are no vertical lines coming straight up from carbons that are down, and *vice versa*. Some students draw lines this way on exams. Look at your model (if you aren't holding one right now, go get one) and you will see: there aren't any such bonds. Therefore, a line pointing in this direction has no meaning! Worse, it implies *wrong* information. Whether you *intend* to convey axial or equatorial is irrelevant: if you draw such a line, you are conveying nonsense, and you will be graded appropriately.

- There are no lines going left from the right side of the molecule, and *vice versa*. Again, these lines convey nonsense, and they strongly suggest that a student does not understand the material.

Well, it's one thing to look at a pretty, computer-generated picture, and quite another to make one yourself. Eventually you will develop your own style, but for now, here is a method that should work every time:

1. Start in the middle. Draw two lines that are angled slightly from horizontal, one on top of the other. The bottom line should have slipped a little downhill from the upper one. In other words, they are not directly above one another:

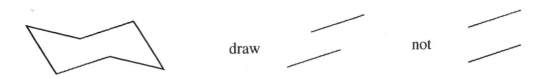

draw   not

2. Connect the uphill ends of these lines with a V angling <u>down</u>. Not too steep.

draw   not

3. Connect the downhill ends of these lines with a V angling <u>up</u>. Again, not too steep.

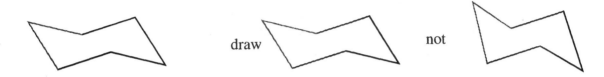

draw   not

Now you have the carbon skeleton. The next task is to fill in the peripheral bonds. Once you get good at this, it will not be necessary to fill in all of them, just the ones you want to show. But it *is* important to put all lines where they belong. And for the next three weeks at least, you should never draw a cyclohexane chair without showing all the bonds.

4. An axial bond, at any carbon, is dead vertical. Remember to make sure it goes *up* from an up carbon and *down* from a down carbon. By the way, this is why we did not make our first two slanted lines directly above one another. If we had, the vertical lines going up would be right on top of each other, making the picture hard to read.

NOT

5. Equatorial bonds go out and slightly up or down: down if the axial bond was up, up if the axial bond was down. If for some reason you did not actually draw the axial bond, imagine it there, and then put the equatorial bond correctly. Extra good pictures will have the equatorial bonds approximately parallel to the carbon–carbon bonds that they are *not* attached to.

Never draw equatorial bonds headed into the ring:

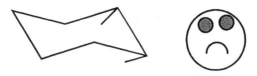

Above all, **PRACTICE, PRACTICE, PRACTICE!!!**

Flipping Rings

Again, it is important that you realize that everything I say in this section is true not because I say it is, but because you believe it is, and you believe it because you can see it with your own eyes on the model you are holding. Build methylcyclohexane by putting a methyl group on an equatorial position. Hold it so that the methyl group points to the right, and that the carbon it is attached to is down.

Now grab the carbon with the methyl and twist it upward, keeping the rest of the molecule where it was. Then grab the opposite carbon and twist it downward. When you get done, you will have a new chair, in which the methyl group is now axial, not equatorial.

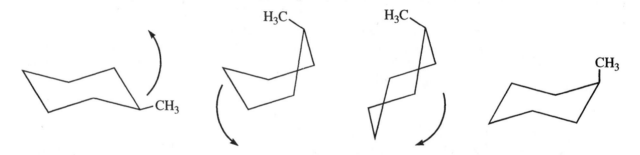

But look carefully: the carbon it is attached to is now "up." It will not do to draw this as if the methyl simply moved to an axial position and nothing else changed:

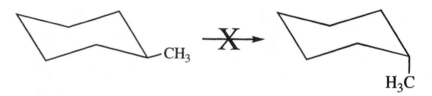

The picture above implies that the methyl swivels on its ring carbon atom without any other changes in the molecule. But that is not what happened! The chair actually inverted! Damn! Does that mean you have to learn how to draw a chair slanted to the left, and another one slanted to the right? Yes, it does, but that's OK, because you do it exactly the way I described above: Start in the middle and work outward, but start with lines slanted in the opposite direction. *You must learn how to draw both kinds of chairs!* The only accurate way to show a ring flip is to indicate that the new chair is the opposite of the old one. It is true that all axial groups have become equatorial, and vice versa, but you cannot simply switch them.

It is equally incorrect to draw this process as if the methyl moved to a different position:

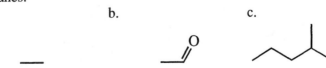

Although both of these wrong drawings show the correct product, they do not do the process justice, and if we are to be clear in our thinking, we have to show the process the way it really occurs. Wrong drawings are the product of laziness, a refusal (or perhaps inability) to draw the other chair. Get over it!

It is true that the three pictures of axial methylcyclohexane above are all identical, just rotated relative to one another, so who cares how they are drawn? First, it is important that you visualize the true process, and it is unlikely that you are visualizing it correctly if you draw the results in the second or third way. But more important, the results are *not* always equivalent! Consider the same process with *trans*-1,2-dimethylcyclohexane. Shown below are two versions of the ring flip, the first one (the correct way) acknowledges that the ring has flipped, and the second one (the "lazy" way) in which the artist simply switched positions for the groups.

It is unlikely that any of you will see it at first (and it's not important to us just yet), but if you very carefully build models of the two products shown, you will discover that no matter what you do, you cannot superimpose one on the other. They don't represent the same substance! (We'll get into this phenomenon later.) If you draw things the lazy way, you could be conveying seriously wrong information.

Problems
(Answers are provided at the end of the chapter for italicized problems.)

1. Write out the structures, showing all bonds and non-bonded pairs for the following. Name the alkanes.

    a.   b.   c.

d.   e.

2. Draw the line diagrams for

   a. *1-methyl-3-isopropyl-5-sec-butylcyclohexane*

   b. *3,3-dimethyl-4-(2-ethylpentyl)-5-(1-methyl-2-ethylbutyl)decane*

3. Write 10 *different* isomers of the formula $C_5H_{13}N$.

4. Provide a name for the following compound.

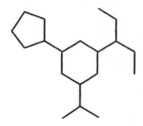

5. a. Name the following compound, using the longest chain as your base.

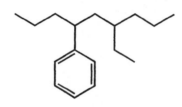

   b. Name it again, using the ring as your base.

   c. Show the structure of the following compounds.

      1-isobutyl-2-*sec*-butyl-4-*tert*-butylcyclohexane

      5- (2-methylpropyl)-6-(1,2,2-trimethylpropyl)decane

6. *Draw structures for the following compounds. If the name given is not the proper one,
   rename the compound properly.*

       4-Ethyl-3-methyl-5-(2-methylbutyl)octane

       1-Chloro-3-(3-chloro-1,2-dimethylpropyl)cyclohexane

       2-Methyl-3-(1-isopropyl-1,2-dimethylpropyl)hexane

7. Draw structures for the following compounds. Rename them if they are named incorrectly.

cis-1,4-Diethylcyclohexane

4-Ethyl-5,5-dimethylpentane

2-Isopropyl-4-methylpentane

trans-1-Isobutyl-3-methylcycloheptane

all the isomers of C_6H_{14}

8. Show structures for the following compounds, then name them correctly.

   a. 3-sec-butyl-4-(1-methyl-2-isopropylpropyl)hexane

   b. 1-tert-butyl-1-sec-butyl-1-isobutylethane

9. Name the following compounds:

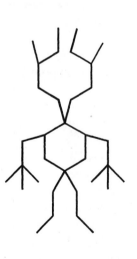

10. Name the following compounds.

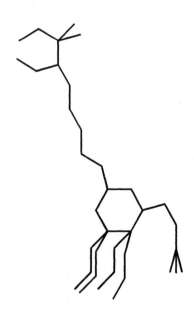

11. Each of the following structures represents _____ of

A. a different isomer
B. a different conformation of a molecule
C. a different configuration of a molecule
D. both A and B
E. both A and C
F. different representations of a single molecule

a.

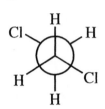

b.

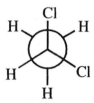

c.

12. Draw the Newman diagram for the following conformations of 1-chloro-2-bromoethane.

   a. gauche
   b. anti

13. *Draw an energy diagram for rotation around the C–C bond in 1,2,2 tribromoethane.*

14. Draw an energy diagram for the rotation about the 2–3 bond of 2-bromobutane. Draw Newman pictures of each peak and valley of your diagram. Note that bromine is considered to be larger than methyl.

15. *Draw an energy diagram for the rotation around the central bond of 2-chlorobutane. Note: in order to answer this question you have to make an assumption. Specify your assumption, then answer the question.*

16. To determine the energy due to strain in a cyclobutane ring, you need the combustion energy for what series of compounds?

17. *How would you analyze the above combustion energies to get a value for the strain energy per $-CH_2$ in cyclobutane?*

18. *Cyclobutane has 26 kcal/mol of strain energy. If 3.7 g of cyclobutane is burned in a calorimeter, how much energy will be given off? You will have to use data given in the chapter to answer this.*

19. A 1.05-g sample of cyclooctane was burned in a calorimeter and was determined to release 11.8 kcal. Given the following data, what is the strain energy of cyclooctane? (Note that these numbers are not exactly correct—I have adjusted them to make the calculation easier. This means that numbers recalled from previous problems or found in the chapter will not work here. Use only the data here to do this calculation.)

Compound	Heat of combustion
Ethane	374 kcal/mol
Propane	530
Butane	686
Pentane	842
Hexane	998
Heptane	1154
Octane	1310

20. Cycloheptane has a strain energy of 6.0 kcal/mol and a heat of combustion of 1105 kcal/mol. Ethane has a heat of combustion of 374 kcal/mol. What is the heat of combustion of pentane? (The answer will not be the same as some numbers you may have seen previously, because I doctored the numbers in this problem.)

21. *From the following (incorrect) data, calculate the heat of combustion of ethane:*

    Cyclobutane has a strain energy of 20 kcal/mol and a heat of combustion of 528 kcal/mol. Pentane has a heat of combustion of 681 kcal/mol.

22. Draw the stable chair form for *trans*-1,4-dimethylcyclohexane. Draw a circle around the equatorial hydrogen atoms and a box around the axial hydrogen atoms.

23. For *cis*-1,3-dimethylcyclohexane,

    a. Draw the most stable chair conformation.
    b. Draw the least stable chair conformation

24. Which is more stable, *cis*- or *trans*-1,4-dimethylcyclohexane? Why? Illustrate your explanation with *clear* drawings.

25. Draw and label (*cis* or *trans*) the two isomers of 1-ethyl-4-methylcyclohexane. For each isomer, draw the two possible chairs and indicate which of the two will be preferred and by how much (a lot or a little). Which of the four is the most stable (label as "best")? Which of the four is the least stable (label as "worst")? Which of the two isomers is most stable?

26. *The following compound exists in two isomers, the* cis *and the* trans. *Further, one is more stable than the other. Predict which isomer is more stable, and suggest a reason for it.*

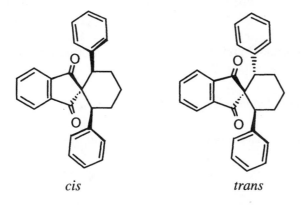

cis   trans

27. Decahydronaphthalene (also known as "decalin") exists as two isomers: *cis* and *trans*. Both have their six-membered rings in chair conformations. Explain, using models. Then attempt to draw the two structures.

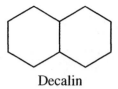

Decalin

28. *Ethyl cyclohexane prefers the equatorial conformation to the axial conformation by 2.2 kcal/mol.*

    a. *Draw each.*

    b. *Calculate the equilibrium constant for the conversion of equatorial into axial at 25° C.*

    $$\Delta G = -RTlnK_{eq} \qquad\qquad R = 1.987 \; cal/mol \; K \qquad K = \;°C + 273$$

    c. *What percent of the molecules are in each form at that temperature?*

29. Draw the two isomers of 1-ethyl-3-methylcyclohexane. For each isomer, draw the two possible chairs and indicate which of the two will be preferred, and by how much (a lot or a little). If in the *trans* compound there is a ΔG of 0.650 kcal/mol between the two chairs, what is the percentage composition of this substance at equilibrium at 20 °C?

30. At 20 °C a particular derivative of cyclohexane exists as 20% boat form and 80% chair form.

    a. What is the K_{eq} for the "flip" from chair to boat?

    b. What is the ΔG for the "flip"?

    c. If ΔG remains constant and the temperature increases to body temperature (37 °C), what is the K_{eq}? What are the percentages of the compound in the boat and in the chair form?

31. As you know, ethane prefers the staggered conformation over the eclipsed conformation by about 3.0 kcal/mol. At room temperature (25 °C), what percent of the molecules will be in each form (assume that only these two forms are possible, i.e., ignore all the other in-between forms)?

32. Isopropylcyclohexane prefers the equatorial conformation with an equilibrium constant of 42 at room temperature (25 °C).

    a. *What percent of molecules are in the axial form at room temperature?*

    b. *What is ΔG for this process at room temperature?*

    c. *Assuming that the ΔG value does not change with temperature (it actually does), calculate the percent of each form at 200 °C.*

d. *Explain why this molecule has such a strong preference for an equatorial isopropyl group. Your answer should include at least two pictures and clearly show why there is a problem in one of them.*

33. a. Below are the two "anomers" of glucose. (One is the chief constituent of starch, the other of cellulose.) Which of the two do you think is more stable (lower in energy, "happier"), and why? Show specifically the problem one of them has that the other does not.

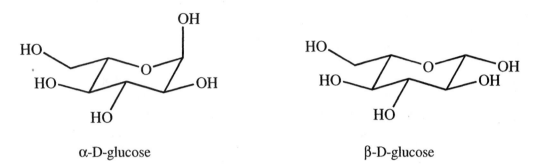

α-D-glucose   β-D-glucose

b. Imagine that the energy difference between the two forms is 1.2 kcal/mol. Assuming that there is an easy way for the molecules to convert from one to the other (this is NOT simply a conformational change), and that they can therefore achieve equilibrium, what would the equilibrium constant be at body temperature (38 °C)?

c. Using your answer to part b, determine the percent distribution of the anomers of glucose in the body, that is, what % is in the α form and what % in the β form?

34. a. At room temperature (25°C) cyclohexane can exist as a boat or a chair; 99.99075% of the time, it is a chair. What is the energy difference ($\Delta G°$) between boat and chair?

b. What is the strain energy of boat cyclohexane? (This does not require any numbers I have not given you.)

35. *Trans*-2-butene is more stable than *cis*-2-butene by 1.1 kcal/mol. Suppose that somehow equilibrium could be established between the two. How much *trans*-2-butene would be present in the product mixture, expressed as a %, at 25 °C? How much at 200 °C?

36. If a *tert*-butyl group on a cyclohexane ring prefers to be equatorial 99.99% of the time at 25 °C, what is the energy difference in kcal/mol between equatorial and axial *tert*-butylcyclohexane?

37. *Here's a tough one. There are two different substances named trans-1,2-dimethylcyclopentane. Make molecular models of them and see if you can find the relationship between them. We'll explore this kind of isomerism next semester.*

Selected Answers

Internal Problems

Test Yourself 1. See answer to problem 15.

Test Yourself 2. In the last section we determined that a normal, unstrained CH_2 group is worth 157.4 kcal/mol. Cyclobutane has four CH_2 groups, so it should give off 4 x 157.4 = 629.6 kcal/mol *if it had no strain*. But it does have 26 kcal/mol of strain, so it will give off 629.6 + 26 = 655.6 kcal/mol. This is the heat of combustion of cyclobutane.

End of Chapter Problems

2. a.

b.

(Not actually the correct name.)

6.

Name should be 3,7-dimethyl-4-ethyl-5-propylnonane

Name should be 2,3-dimethyl-3,4-diisopropylheptane

10.
1-pentyl-1-(1,1-diethyl-3-propylhexyl)cyclohexane = "supermane"

1,1-bis(3-methylpentyl)-2,6-bis(2,2-dimethylpropyl)-4,4-dipropylcyclohexane = "humane"

1,1,2,2-tetrapropyl-3-(3,3-dimethylbutyl)-5-(6-ethyl-7,7-dimethylnonyl)cyclohexane = "giraffane"

13.

15. You have to assume something about the relative size of a chlorine and a methyl group. In fact they are similar, but if we were to guess that chlorine is bigger, the diagram would come out like this:

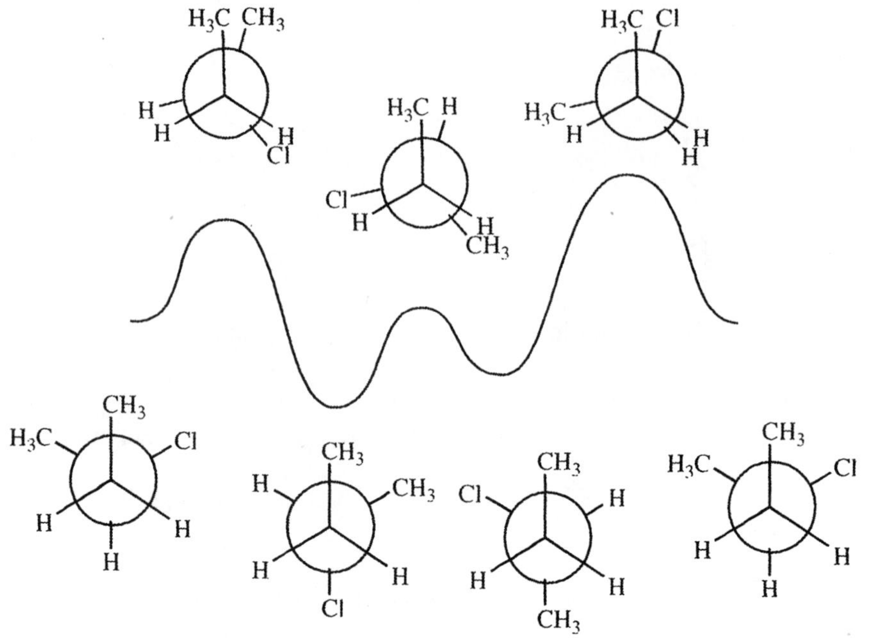

18. 43.3 kcal

To get this, you first figure out the amount of energy in 4 CH$_2$ groups: 157.4 x 4 = 629.6 kcal/mol. Then add 26 to this to get the heat content of a mole of cyclobutane: 655.6 kcal/mol. We are burning less than a mole, specifically, 3.7 g (and a mole is 56 g).

$$3.7 \text{ g} \quad \frac{\text{mol}}{56 \text{ g}} \quad = 0.066 \text{ mol} \times 655.6 \text{ kcal/mol} = 43.3 \text{ kcal}$$

21. 300 kcal/mol

26.

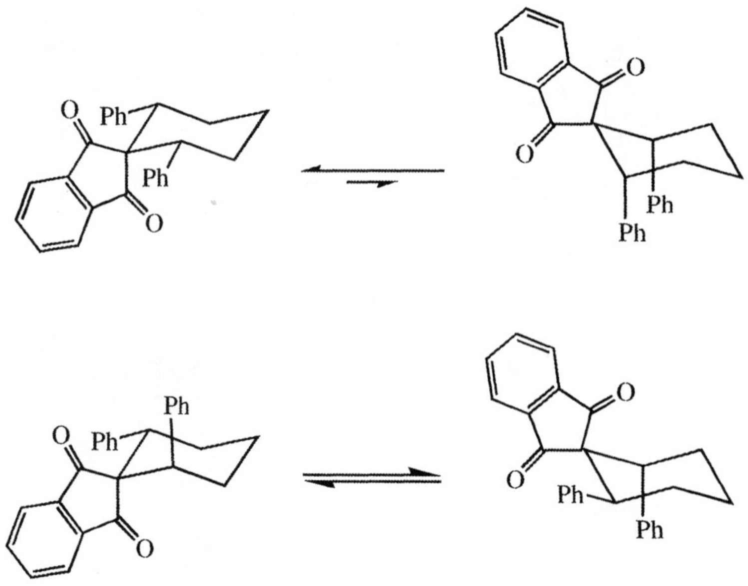

Cis can be diequatorial and happy. *Trans* has no choice but to make one of the phenyl groups axial, where it is unhappy. Thus the *cis* isomer is preferred.

28. a.

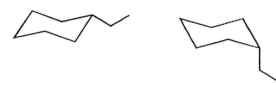

    b. 0.024

    c. 2.3% axial, 97.7% equatorial

32. a. 2.3% axial

    b. axial to equatorial, −2.2 kcal.mol

    c. 91.3% equatorial; 8.7% axial

    d.

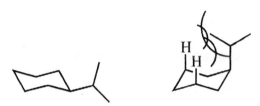

              major steric problems here

37. They are non-superimposable (i.e., not identical) mirror images.

Chapter 8
Acids and Bases

One of the most important reactions in organic chemistry actually has little to do with carbon. It is what we call "proton transfer," the moving of a proton, H^+, from one atom to another. In this chapter we will

- Define acids and bases

- Learn about conjugate acids and bases

- Learn how to measure the strength of acids and bases

- Learn how to use the numbers that result from this measurement

- Learn why certain acids and bases are stronger than others

- Have our first encounter with resonance.

In previous chemistry courses you have probably encountered acids and bases. You may have learned that acids are compounds that release H^+, and bases are compounds that release OH^-. This is a legitimate approach, called the "Arrhenius" definition of acids and bases, but it is too narrow for our purposes. For organic chemistry, it is more useful to adopt the "Bronsted–Lowry" definition. Under this definition, an acid is still something that *donates* a proton, but the concept of base is broadened to include anything that *accepts* a proton. Thus, a proton transfer reaction is by this definition an acid–base reaction, and it is necessary to take some time out now to try to understand acids and bases.

DEFINITIONS

Acids

An acid is something that loses a proton, H^+. Any molecule that has a hydrogen in it could potentially lose H^+, leaving behind the electrons that formerly kept it attached. Now is as good a time as any to start using a special kind of arrow to illustrate how reactions proceed. You can think of these as cartoon instructions for how to get from here to the next step. The most important thing to remember is that by convention, the arrows are *always* instructions for the electrons, *not* for the atoms themselves. Here's how to do it:

Start by drawing the substance on which you want to operate: H—Cl

Then draw an arrow that shows what happens to the electrons when the hydrogen detaches itself from this molecule. It is important to consider whether the hydrogen keeps with it any of the electrons it is currently sharing. In this case, it will not, because it is leaving as H^+. So both electrons in the H—Cl bond will be deposited onto the chlorine atom. We use a curved arrow to indicate this: H—Cl .

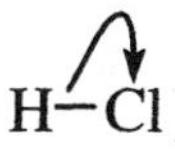

Now we draw a *new* picture to show the result of this action. *Never* draw the result of your action on the same picture as the instructions themselves! You will get hopelessly confused if you do that. The result of the above arrow is that there is no longer a bond between the H and the Cl. Further, the H now has no electrons, and is therefore H^+, and the Cl now has 8 electrons of its own, so it is now Cl^-. Here is how the whole sentence should look:

$$H\text{—}Cl \longrightarrow H^{\oplus} + Cl^{\ominus}$$

COMMON MISTAKES: DON'T LET THIS HAPPEN TO YOU!

Many of you will try to draw this reaction by showing how the H atom moves, as shown below:

$$H\text{—}Cl \longrightarrow H^{\oplus} + Cl^{\ominus}$$

This is WRONG. Do not do it. The arrow MUST show the movement of the electrons, not the atom. Why? Tradition! There is no necessary property of atoms that demands that arrows be used in this way, but chemists have agreed that the arrows will show how *electrons* move. Once that has been said, it is essential that arrows always mean the same thing; otherwise true communication becomes impossible. It is not a law of nature that *mesa* should mean "table" in Spanish, but it does, and if you choose instead to use *agua* for "table" you will have trouble communicating with a Spaniard. To communicate with chemists, always use their language: *arrows show the movement of electrons.*

Bases

A base is something that accepts a proton. To do this it must have some electrons to which an H⁺ can attach. This is most often (though not always) an unused pair of electrons, usually called a "lone pair." Any atom that has a lone pair of electrons can in theory act as a base. It may not be a good base, but it is potentially able to pick up a proton. Some atoms with lone pairs also have minus charges, such as OH⁻. Others do not, such as H_2O. Both types can act as bases, though, by using their lone pairs to accept protons:

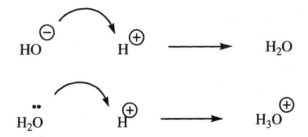

Note here that the arrow again shows what the electrons will do. It would probably be better if the arrow were shown finishing in the space between the oxygen and the hydrogen (as shown below), since that is where the electrons will in fact end up, as a new bond, but for some reason chemists often show the electrons going all the way to the atom that will receive them. This is another one of these unfortunate conventions that we have to get used to.

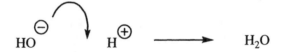

Free Protons

In both the acid and base examples given above, I have implied that there are protons floating around somewhere as H⁺. I showed them detaching themselves from an acid, departing into space, or just wandering around and waiting to be captured by a base. Although chemists often carelessly depict protons this way, in fact it is generally agreed that there are no "free" protons. Protons are nearly always attached to something. When they detach from an acid, it is because some other base has come along to take them. When a base acquires a proton, it is picking it up from some other acid. Thus, it is uncommon for an acid or a base to act by itself: there is usually some base or acid involved acting as its partner. That is why such reactions are called acid–base reactions. So the following expressions are more accurate versions of the above examples:

$$H_2O \quad H\!-\!Cl \longrightarrow H_3O^{\oplus} \;+\; Cl^{\ominus}$$

$$HO^{\ominus} \quad H\!-\!OH_2^{\oplus} \longrightarrow H_2O \;+\; H_2O$$

Conjugate Acids and Bases

Not only does an acid nearly always have a base serving as its partner (something to accept the proton it is releasing), it also has a base that could be described as its relative. Consider this: an acid is something that releases a proton. The proton must have been attached to the acid by a pair of electrons. After the proton is gone, that pair of electrons is now sitting around. The result is, by definition, a base: the new substance has a place where a proton could attach, namely, the place the other proton just left. Any substance that acts as an acid creates a base. The base created by an acid releasing a proton is called the **conjugate base** of the original acid. All acids have conjugate bases. *The conjugate base of any acid is the same substance, minus a proton* (note that this also changes the charge). The conjugate base of HCl is Cl^-. The conjugate base of H_2O is OH^-. The conjugate base of H_3O^+ is H_2O. The conjugate base of OH^- is O^{-2}. (I did not say forming such a base would necessarily be easy or desirable, but by definition, a conjugate base is the same substance minus a proton.)

The same applies to bases. A base is something that accepts a proton. If it did not have one before, the new substance certainly has a proton now, the one that just arrived, and it therefore could potentially be given away again. Any substance that acts as a base in the process creates an acid. The acid created by a base accepting a proton is called the conjugate acid of that base. All bases have conjugate acids. *The conjugate acid of any base is the same substance, plus a proton* (note that this also changes the charge). The conjugate acid of OH^- is H_2O. The conjugate acid of H_2O is H_3O^+. The conjugate acid of H_3O^+ is H_4O^{+2}. (I did not say it would necessarily be easy or desirable, but by definition, a conjugate acid is the same substance plus a proton.) Trick question: what is the conjugate acid of H_4O^{+2}? Answer: If you said H_5O^{+3}, you fell for the trick. There *is* no conjugate acid of H_4O^{+2}! Not because you can't put three plus charges on an atom, although that is something to be avoided if possible (at least in organic systems), but because by this point you have run out of lone pairs. Draw the Lewis structure of H_4O^{+2}. You will see that there are no lone pairs left, so this substance has no possibility of accepting another proton.

You should see now that any acid-base reaction will have one acid and one base on the left side of the equation, and a new acid and a new base on the right side. The acid on the right side of the equation is the conjugate acid of the base on the left, and *vice versa*; the base on the right side of the equation is the conjugate base of the acid on the left, and *vice versa*.

HCl	+	H_2O	$\rightarrow$	H_3O^+	+	Cl^-
Acid		Base		Acid		Base
Conjugate acid of Cl^-		Conjugate base of H_3O^+		Conjugate acid of H_2O		Conjugate base of HCl

STRENGTHS OF ACIDS AND BASES

The most important point of this chapter, apart from the definitions established above, is that we would like to be able to know whether a given acid–base reaction will occur with any pair of partners. For example, the equation written above could equally well be written backwards:

$$H_3O^+ \quad + \quad Cl^- \quad\quad \rightarrow \quad\quad HCl \quad + \quad H_2O$$

Which way will it in fact go? This question can be phrased in many ways, all of which mean the same thing:

1. Which of the two acids (HCl or H_3O^+) is stronger, i.e., which one will do a better job of casting off its proton? The stronger acid should succeed in getting rid of the proton. As the reaction goes forward, the stronger acid will become its conjugate base, while the weaker acid will be stuck with the proton.

2. Which of the two bases (Cl^- or H_2O) is stronger, i.e., which will do a better job of snaring a proton? The stronger base should succeed in gaining the proton. As the reaction goes forward, the stronger base will become its conjugate acid, while the weaker base will be stuck with its lone pair unused.

3. Which of the lone pairs (one on the Cl^- or on the H_2O) is a better home for a poor, lonely H^+? The H^+ will end up at the place where it is most comfortable, derived from the better base of the two; in landing there the proton will convert that base into its conjugate acid.

4. You can picture the reaction as involving an H^+ suspended between two partners, engaged in a tug of war for it, sort of a *menage a trois*: $Cl^- \cdots H^+ \cdots OH_2$. To continue the analogy, imagine two guys each trying to dump the same woman on the other (version 1), two guys fighting for the love of the same woman (version 2), or a woman choosing between two suitors (version 3). In any case, the partnership corresponding to the most overall attraction (or the least repulsion) is likely to win. Be my guest and reverse the genders if it helps you identify with the analogy!

Any way you look at it, a reaction like this will proceed away from the stronger acid and base toward the weaker acid and base; *the reaction will favor the side of the weaker acid and base*. So in order to answer the original question, we need to know how strong various acids and bases are. Fortunately, there is an easy way to find out: In a series of reactions, keep one partner the same and vary the other, and then measure how far the reaction goes in each case. What you get from this process is a measure of relative strength for the partner that changed. Consider, for example, the following series of reactions:

H–Cl	+	H_2O	→	H_3O^+ +	Cl^-
H–F	+	H_2O	→	H_3O^+ +	F^-
H–SH	+	H_2O	→	H_3O^+ +	HS^-
H–NH_3	+	H_2O	→	H_3O^+ +	NH_3

In this series we have progressively changed the acid on the left but kept the base (water) the same. The reaction that goes farthest to the right in this list will be the one involving the strongest acid of the four on the left. The next best reaction will correspond to the next best acid, and so on. Indeed, the strength of any acid on the left could be defined as the extent to which its reaction with water goes to the right. That is precisely how these acids are ranked. But before I show you numbers, we must establish a few more concepts.

What Do We Mean By "Extent"? Equilibrium Constants

All the reactions shown above, indeed all the reactions in this chapter, are technically equilibrium reactions, which means simply that they can go both forward and backward. In other words, the proton always gets to choose where it wants to be. Equilibrium reactions should be shown with equilibrium arrows,

$$\rightleftharpoons$$

so all the equations above are technically wrong. Equilibria can always be characterized by their inherent tendency to go more toward the left or right in the equation *as it is written*, and this tendency is in turn expressed by an "equilibrium constant," K_{eq}. The equilibrium constant is, by convention, the amount of material that ends up on the right divided by the amount of material on the left, where "amount of material" is expressed in terms of concentration, moles of compound per liter of solution. Recall that the shorthand for "the concentration of X" is [X]. Thus, for the first equation in the list above,

$$K_{eq} \;=\; \frac{[H_3O^+][Cl^-]}{[HCl][H_2O]}$$

This equilibrium constant really *is* a constant, and it will not change; it will have the same value whenever these four components are present together (at a given temperature) and the reaction has reached equilibrium, i.e., the mix has figured out where it wants to sit. For this particular reaction, K_{eq} has been measured to be 200,000. Clearly, the numerator must be a lot larger than the denominator; in other words, this reaction sits pretty far to the right. *Reactions that tend toward the right have large equilibrium constants.* HCl has a strong tendency to come apart: HCl is a strong acid.

Look at the fourth equation in the list. For this,

$$K_{eq} \;=\; \frac{[H_3O^+][NH_3]}{[NH_4^+][H_2O]} \;=\; 2 \times 10^{-11}$$

Clearly, to get a tiny value like this from the fraction, the numerator must be a lot smaller than the denominator; in other words, here we have a reaction that lies far to the left. *Reactions that tend toward the left have small (but still positive) equilibrium constants.* NH_4^+ has a low tendency to come apart: NH_4^+ is a weak acid.

It should be clear to you that a reaction with no strong preference for one side or the other will have a numerator and denominator of similar magnitude, so K_{eq} will be near 1. Notice that all K_{eq}'s are positive, but they can range from extremely small to extremely large positive numbers.

A "Simplification": K_a

All four of the reactions above are usually perceived as happening in water as a solvent. This is the H_2O on the left side of the equation. You add a bit of the acid and see what happens. One of the things that happens is that a tiny bit of the water disappears, to become H_3O^+. (By the way, H_3O^+ shows up often enough that you ought to know its name: "hydronium ion.") But even if *all* of the acid comes apart and protonates the water, there will still be only a negligible effect on the total amount of water present, assuming (as we have) that water is there in large excess. Nobody could measure the difference between the amount of water present before the reaction and after. In other words, the amount of water in the "system" does not significantly change, and we can get away with treating it as a constant. K_{eq} does not change either (as we established earlier), so the product of K_{eq} and $[H_2O]$ is essentially yet another constant. Therefore, the equation:

$$K_{eq} = \frac{[H_3O^+][Cl^-]}{[HCl][H_2O]}$$

can be rearranged to

$$K_{eq}[H_2O] = \frac{[H_3O^+][Cl^-]}{[HCl]}$$

This new constant, the product $K_{eq}[H_2O]$, is called the "acid dissociation constant," or the "acidity constant," K_a. So in general,

$$K_a = \frac{[H_3O^+][\text{conjugate base}]}{[\text{conjugate acid}]}$$

K_a is basically the same as K_{eq}, discussed above, but it differs from it by a (constant) factor of 55.6, the concentration of water in moles of water per liter of water. Like K_{eq}, K_a ranges from very large numbers to very small numbers, all positive, and the interpretations of the two are essentially the same as well. So K_a is a number that expresses how strong an acid is with respect to the base water. Using this new constant, HCl has a K_a of 1×10^7, still large, and NH_4^+ has a K_a of 1×10^{-9}, still small.

Another "Simplification": pK_a

It is a pain to write numbers like 1×10^{-9} and 1×10^7 as expressions for how far a reaction goes. Chemists have found it easier to work with the *logarithms* of these numbers. The logarithm, you may recall, is roughly the exponent on the 10's above. Since most acids have K_a's that are quite small (i.e., their reactions with water do not proceed very far) using the logarithm approach would lead to most acids being described by negative numbers. For convenience, then, chemists also switch signs and define

$$pK_a = -\log K_a$$

This probably does not seem like much of a simplification to you, and I agree, but it is nevertheless what we are stuck with. HCl therefore has a pK_a of –7, which means it is a strong acid, and NH_4^+ has a pK_a of 9, meaning that it is a weak acid. We are just playing games with numbers here. All you really need to remember is that low pK_a numbers represent strong acids, high pK_a numbers represent weak acids, and each increment of one in such a number actually represents a factor of 10 in the corresponding equilibrium constant. In summary, pK_a is not a new concept, it is simply a different way of expressing the K_a, which a measure of how strong an acid is, so pK_a, likewise, is a measure of how strong an acid is.

How Do We Measure "Extent"?

How did I get these numbers? Look at the equation:

$$H{-}Cl \;+\; H_2O \;\rightarrow\; H_3O^+ \;+\; Cl^-$$

If we start with plain old water and add HCl, every time an HCl molecule comes apart it will create one H_3O^+ and one Cl^-. It will also leave behind one less intact HCl molecule than was there before. Thus, after a while, the amount of H_3O^+ and the amount of Cl^- present will be the same (call it X), and the amount of HCl left over will simply be the amount we added minus X. (Be sure you understand this! Many students think the amount of HCl left over should be the amount we started with $- 2X$. This is wrong!) Thus, by measuring X we get all the numbers we need to calculate K_a. What is X? It is, among other things, the concentration of H_3O^+, and many of you will remember from high school that we can easily measure the amount of H_3O^+ in a solution. (We play the same kind of "number game" we played above, taking the negative log of this concentration and calling it pH, but the point is that we can measure the H_3O^+ concentration easily.) So with one simple type of measurement we can calculate pK_a's for a wide variety of acids.

Test Yourself 1

A sample of 6.00 g of acetic acid (CH_3COOH) is diluted with enough water to make 1 L of solution. After waiting for equilibrium to become established, the pH is measured to be 2.87. What is the pK_a of acetic acid? Do this now. Be sure you understand the answer at the end of the chapter before you proceed.

Test Yourself 2

Can you do one backwards? Now that you know that the pK_a of acetic acid is 4.73, figure out what will happen if you put 60.0 g of acetic acid in enough water to make 1.0 L. (*Hint*: One minus a very small number is still pretty much one. The answer is given at the end of the chapter. Do not look at until you have your own.

Partnerships

Only one more point before we get to the numbers we're interested in. Imagine an acid, call it HA, trying to get rid of its proton. This process will create the conjugate base A^-, which will immediately try to get the proton back. To say that an acid is strong, that it is eager to dump its proton, is also to say that it is stable without its proton; in other words, A^- is happy as is, and is not eager to get the proton back. So A^-, as a base, is weak. *Strong acids always have weak conjugate bases.* By the same token, if HA is a weak acid, very reluctant to give away its proton, it stands to reason that if it loses its proton it will fight very hard to get it back: *Weak acids always have strong conjugate bases.* In general, the number we apply to an acid also provides a description of the conjugate base, in reverse. Now we have to be very careful about this, and make sure we are using the right partnership: The number we get for NH_4^+ as an acid will also describe NH_3 as a base, but it tells us nothing about NH_3 as an acid. The number we get for NH_3 as an acid also describes NH_2^- as a base, but this does *not* describe NH_3 as a base. Students are constantly mixing these things up. Part of the reason, I think, is that most tables of the type that follow include only the first two columns, since the third is obvious to someone with experience. The problem with this approach, of course, is that if you want to look up the strength of some base, you have to look on the table in a column that is not there! My students have always had a terrible time with this, so I am including both acids and bases in my table.

Note that there is a danger here. A pK_a is an "acid dissociation constant" and refers to the ease or difficulty of removing a proton from the substance identified. For example, the number -1.7 properly describes H_3O^+ as an acid. It is the pK_a of H_3O^+. The same number properly describes H_2O as a base, but it is *not* the pK_a of H_2O. It is still the pK_a of H_3O^+, and the pK_a of H_3O^+ is what you need to describe the basicity of H_2O. Get it? The pK_a of H_2O does not describe the base power of H_2O, it describes the *acid* power of H_2O, and you will find that farther down in the table at 15.7, partnered with the conjugate base, OH^-. As if this weren't confusing enough, chemists often get careless: to describe the base power of NH_3, one *should* say that the pK_a of NH_4^+ is 9, but chemists often refer to the pK_a of ammonia as 9 even when they are considering it as a base. In fact, the pK_a of ammonia is actually 38, which by definition refers to ammonia as an acid. I cannot change the fact that chemists tend to be careless and can only urge you to be perceptive in your listening, and always take the time to figure out whether a speaker (writer) is referring to the substance in question as a base or an acid. I hope the following table will help.

Table

Conjugate Acid	pK_a	Conjugate Base
H_2SO_4	−10	HSO_4^-
HI	−10	I^-
HBr	−9	Br^-
	-7	
HCl	−7	Cl^-
$PhSO_3H$	−6.5	$PhSO_3^-$
H_3O^+	−1.7	H_2O
HNO_3	−1.5	NO_3^-
HF	3	F^-
CH_3COOH	5	CH_3COO^-
HOCOOH	6.5	$HOCOO^-$
HSH	7	HS^-
PhS–H	7	PhS^-
$H_3N–H^+$	9	NH_3
HCN	9	CN^-
	9	
$HOCOO^-$	10	CO_3^{-2}
PhOH	10	PhO^-
	11	
	13	
CH_3OH	15.5	CH_3O^-

Acid	pKa	Conjugate Base
H_2O	15.7	OH^-
CH_3CH_2OH	16	$CH_3CH_2O^-$
(cyclopentadiene structure)	16	(cyclopentadienyl anion structure)
(acetamide structure)	17	(acetamide anion structure)
(acetaldehyde structure)	17	(acetaldehyde enolate anion structure)
$(CH_3)_3COH$	18	$(CH_3)_3CO^-$
(acetone structure)	19	(acetone enolate anion structure)
(methyl acetate structure)	25	(methyl acetate anion structure)
$N\equiv C\text{--}CH_2\text{--}H$	25	$N\equiv C\text{--}CH_2^-$
$H\text{--}C\equiv C\text{--}H$	25	$H\text{--}C\equiv C^-$
(N,N-dimethylacetamide structure)	30	(N,N-dimethylacetamide anion structure)
$Ph_3C\text{--}H$	31.5	Ph_3C^-
$H\text{--}H$	35	H^-
$H_2N\text{--}H$	38	H_2N^-
$(iPr)_2N\text{--}H$	40	$(iPr)_2N^-$
$PhCH_2\text{--}H$	41	$PhCH_2^-$
$H_2C=CHCH_2\text{--}H$	43	$H_2C=CHCH_2^-$
(benzene structure)	43	(phenyl anion structure)
$H_2C=CH\text{--}H$	44	$H_2C=CH^-$
$H_3C\text{--}H$	48	H_3C^-
$CH_3CH_2\text{--}H$	50	$CH_3CH_2^-$
(cyclohexane structure)	51	(cyclohexyl anion structure)

USING THE NUMBERS

OK, you already know that low numbers mean strong acids. So the strongest acids are on the top of this table, in the left-hand column, and the weakest acids are at the bottom, again in the left column. You also know that strong acids correspond to weak bases, so the weakest bases are at the top of the table, in the *right* column, and the strongest bases are at the bottom, again in the right column. Any time you want to consider a substance acting as an acid, you need to find it (or a close relative of it) in the *left* column; any time you want to consider a substance acting as a base, you need to find it (or a close relative) in the *right* column.

Now we are in a position to look at specific reactions. Consider this one:

$$HF \; + \; NH_3 \; \rightleftharpoons \; NH_4^+ \; + \; F^-$$

There is an acid and a base on the left, and a different acid–base pair on the right. The acid on the left is HF, with a pK_a of 3. The acid on the right is NH_4^+, with a pK_a of 9. Which is the stronger acid? HF is. So HF will lose its proton and give it to NH_3, which will turn into NH_4^+. The reaction will go from left to right.

Now try this one:

$$HF \; + \; H_2O \; \rightleftharpoons \; H_3O^+ \; + \; F^-$$

Using the same logic, we discover that H_3O^+ is a stronger acid than HF, so this reaction will go from right to left. In other words, HF does not come apart very much in water. Notice that in these and all other examples, *reactions prefer the direction of the weaker acid and base.*

Here is another way to use the table: protons will always move down and to the right. An acid will donate a proton to any base lower on the table than its own conjugate base, and the extent of the reaction will depend vaguely on the slope of a line you might draw connecting the acid and the base. Thus, HCl will react violently with CH_3^-, but CH_3COOH will not react very much with H_2O.

$$CH_3COOH \; + \; H_2O \; \rightleftharpoons \; H_3O^+ \; + \; CH_3COO^-$$

Now, we need to be careful. This does not mean that CH_3COOH will not react *at all* with H_2O; it simply means that the equilibrium written above lies far to the left. How far? Again, this depends on the difference in pK_a values between the two acids involved. If the line connecting the acid and the base slants steeply upward then effectively nothing will occur (like, perhaps, the reaction between CH_4 and Cl^-), but if the species are reasonably close in acidity then there will be some detectable reaction. How close is "reasonably close"? Say, 15–20 pK_a units. The above reaction between CH_3COOH and H_2O, involving acids that differ by about 7 pK_a units, will lie far to the left, but the stuff on the right will still be detectable. We will see later that sometimes even a trace of ionized material can be very important.

So far we have used the table to compare one acid with another, but we can also consider the relationship between a single substance and the surrounding medium. Look back to the original equation defining K_a:

$$K_a = \frac{[H_3O^+][\text{conjugate base}]}{[\text{conjugate acid}]}$$

If you take the log of both sides and then multiply by −1, you get the following:

$$pK_a = pH - \log\left(\frac{[\text{conjugate base}]}{[\text{conjugate acid}]}\right)$$

This is called the Henderson–Hasselbalch equation, but you do not need to remember that. When the values of [conjugate base] and [conjugate acid] are equal, their quotient is 1 and the log of that is 0, so $pK_a = pH$. Thus, pK_a can be thought of as the pH necessary to get half of an acid to dissociate (come apart). A corollary of this is that if the pH of the medium is lower than the pK_a of the acid, log ([conjugate base]/[conjugate acid]) has to be a negative number, and there must be more acid form present than there is conjugate base. Conversely, if the pH of the medium is higher than the pK_a of the acid, log ([conjugate base]/[conjugate acid]) must be a positive number and there must be more conjugate base present than acid. Did that make sense? Perhaps not, if your logarithmic math is a little weak. Think of it this way. When the pH is the same as the pK_a, you have the same amount of the conjugate acid and conjugate base. If the solution is more acidic than that, there is more of the acid form; if it is more basic than that, there is more of the base form. OK? You can also turn this thinking around and figure out what ratio of conjugate acid to conjugate base you would need to introduce into a solution in order to *force* the pH to have some particular value. This is the basis for buffer problems, which you may have seen in the past and will almost certainly see in the future.

Why Are These Numbers What They Are?

Electronegativity

The table clearly states that HF is a stronger acid than H_2O, which is in turn stronger than NH_3, which is itself stronger than CH_4. What the table does not state is why. So why is it the case? We are looking at stripping a proton off of each species, leaving behind a minus charge on the respective atoms, F, O, N, or C. Does this give you a hint? We learned some time ago that electronegativity increases as you go up and to the right in the periodic table. Thus, it should be no surprise that F, the most electronegative of elements, is more comfortable with a negative charge than is O, which in turn is happier than N, and C should feel the worst. If F is the best place for a minus charge in this group, this anion should be the easiest to make, so HF should be the strongest acid of the group. In other words, to some extent we can even predict the relative placement of entries in the table above.

In the same sense, it should not surprise us that H_3O^+ is a stronger acid than H_2O. In both cases, when we rip off a proton, we are leaving electrons on an oxygen atom, but in the former case we are creating the neutral, perfectly happy molecule H_2O, while in the latter case we are forcing a negative charge on the species in the form of OH^-.

There are other trends we are not yet equipped to understand. Why, for example, is HF a weaker acid than HCl, even though F is more electronegative than Cl? The traditional explanation has to do with size: Cl⁻ is larger than F⁻, since it is lower in the periodic table and has a full shell number 3 instead of shell number 2. Likewise, Br⁻ is larger yet, and so on. Each of these would suffer somewhat from a minus charge, but the ones lower on the periodic table could spread the charge over a larger area, so no area would be afflicted with very much charge. As a general rule, spreading a charge over a larger area makes ions considerably happier. This factor apparently is more important than electronegativity when dealing with atoms in a given column of the periodic table. Since I⁻ is more stable than Br⁻, it is easier to make it, so HI is a stronger acid than HBr (by a little bit).

Here's another thing that doesn't seem to make sense: why is there such a range for the acids with OH bonds? H_2SO_4, CH_3COOH, ArOH, CH_3OH, $(CH_3)_3COH$ all appear to leave behind electrons on a negatively charged oxygen. (I'll bet you didn't find that many. You need to be able to look at the formula H_2SO_4 and recognize that there are two OH bonds. Practice your Lewis structures!!) Shouldn't they be pretty similar? Yet the pK_a's in this series run from −10 to +18, 28 orders of magnitude! Although we will not be able to predict numbers exactly, we can at least try to learn why the trends are in the direction they are. We will end up using the same concept we just introduced: namely, that spreading charge over a larger area is helpful. But we will use the principle in a different way.

The first thing to do in contemplating why one reaction might be better than another is to draw all the relevant species. It is not sufficient simply to examine the two chemicals that are being compared: you must also examine the chemicals they are becoming. It is very often the structures of the products, rather than the starting materials, that hold the answer. Let's pick two acids that are somewhat different in acidity, CH_3OH (pK_a 15.5) and CH_3COOH (pK_a 5). First we have to draw all the structures accurately, both of the acids and their conjugate bases.

$$CH_3OH \quad \rightleftharpoons \quad CH_3O^{\ominus}$$

methanol   methoxide

acetic acid   acetate anion
                             a carboxylate anion

You probably do not see any significant difference in these structures, either for the two acids or for their conjugate bases on the right. But there is one, and it is time we learned what that difference is. The concept involved is called "resonance," and it is one of the more important concepts in organic chemistry. When your teacher asks you on an exam, "Why is compound A more something than compound B?", often the answer has something to do with resonance (the next best guess is steric factors).

Resonance

Here's the deal: the bottom anion, called a "carboxylate anion," is special, because the Lewis structure we drew for it is not unique. Keeping all the atoms frozen in position, and without breaking or forming any new partnerships between atoms, there is another way we could have distributed the electrons to make a sensible structure. Instead of

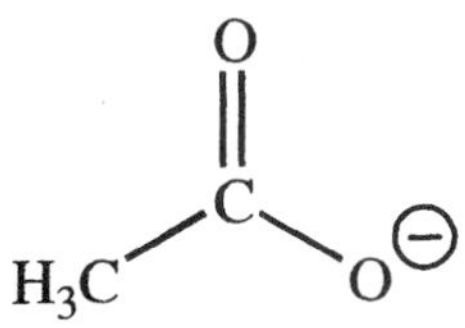

we could have drawn

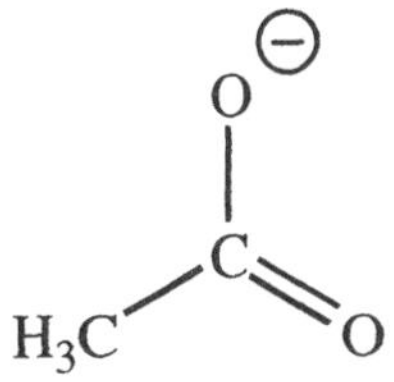

Further, we can draw a sensible pathway by which the electrons can achieve this new distribution:

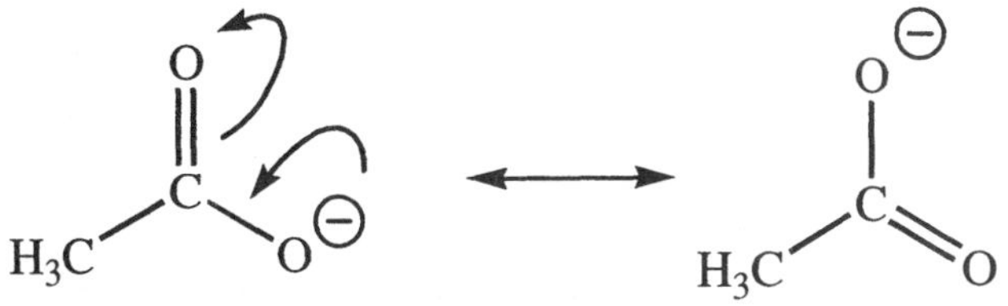

When you can redistribute electrons without moving any atoms and without making or breaking any attachments, you are in a funny twilight zone called resonance, and the molecule must be interpreted in a special way. Instead of imagining the molecule popping back and forth between its two structures, we imagine it as a *cross* between the two, a *hybrid*. Indeed, the structure is often referred to as a "resonance hybrid." It is *not* A, it is *not* B, it is something in between. It is as wrong to think of this as a "jumping back and forth" as it would be to think that if you looked quickly into the barn, you might catch your mule at one instant being a horse and at the next being a donkey. It is neither: it is something in between, neither horse *nor* donkey. Notice the special double-headed arrow used in the above picture: an arrow like this is reserved to express resonance, and it means "something between these structures." You must always use such an arrow to show resonance, and never when resonance is not involved.

But wait: aren't A and B exactly the same thing? If I gave you one of each and asked you if they are the same, you would say yes, and you would be right. But in this context, we are not comparing two molecules to see if they are the same. We are looking at two pictures of the same individual molecule. Furthermore, we have locked the molecule in space: no bond rotation is allowed. The question are, "Is this electron distribution unique? Could we attach the same atoms to the same partners using a different electron distribution?" If there is another electron distribution that maintains all the same attachments, then you have resonance.

COMMON MISTAKES: DON'T LET THIS HAPPEN TO YOU!

Many students look at the two structures shown for the carboxylate anion and see that the double bonds are in different places Quite naturally, they imagine that the double bond has shifted from one place to the other, and show it like this:

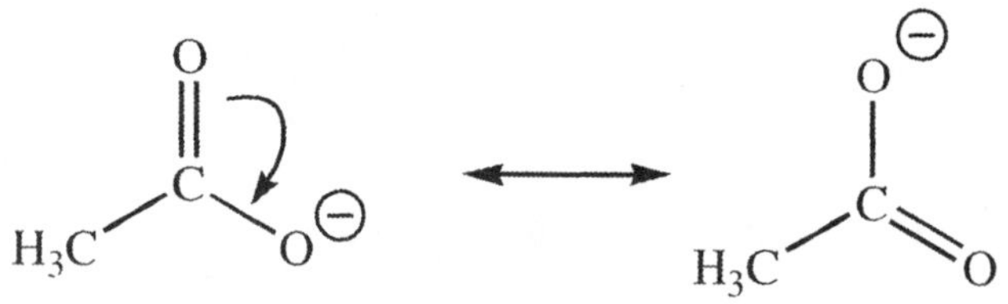

AARGH!! Don't do this!!! Let's look carefully at what this misguided student actually did. To make it clear, I will include all the electrons:

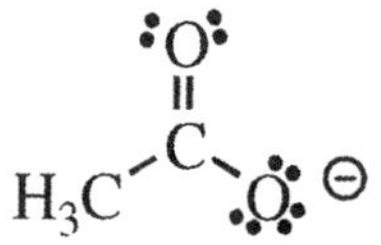

The arrow the student drew instructs you to take two of the electrons in the double bond and put them between the carbon and the other oxygen. What would you get? The following, if you really followed these instructions:

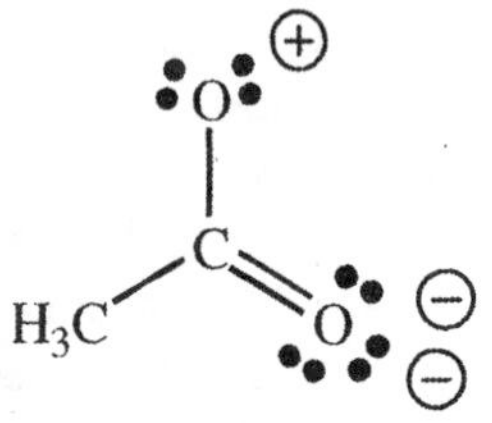

The upper oxygen now has only 6 electrons around it, and only 5 of its own, so it has a positive charge. The bottom oxygen now has 10 electrons around it (a capital offense by itself) of which 8 are owned by the oxygen, giving it two negative charges. This is a total disaster!

Arrows show the movement of *electrons*. You should never show electrons moving toward a negative charge, or away from a positive charge. If there is a negative charge shown, the arrow should *start* there. If there is a positive charge shown, the arrow should head there. The most common mistakes on organic exams are caused by students moving arrows in all kinds of nonsensical ways. Don't let this happen to you!

So what's the big deal about resonance? If the carboxylate anion we drew is really a cross between two structures, then there can't be a full minus charge on the bottom oxygen—or the top oxygen either; there is really only part of a charge on each. Indeed, sometimes chemists try to represent the situation by drawing the following:

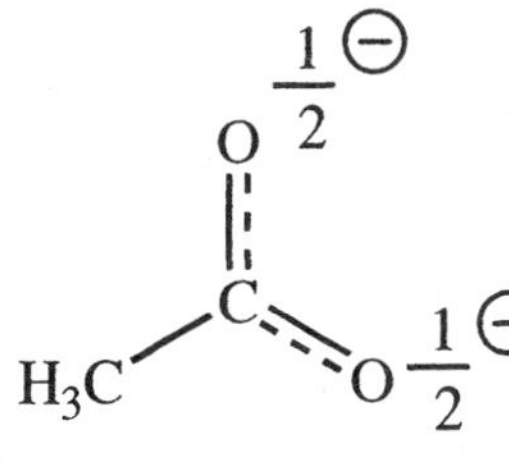

$$O \overset{\frac{1}{2}\ominus}{}$$

Remember what we decided about halide anions? That having a charge spread out over the largest possible area is important? Well, here we have it again! Resonance is a way of spreading out charge, making a structure more stable than you would expect based on a single structure, because the charge is not really stuck on a single atom but shared between two (or more) atoms. A species for which you can write resonance structures is almost always more stable than you would otherwise predict, sometimes by quite a bit. In this case, for example, the ability of the minus charge to spread out over two oxygens in the carboxylate anion makes the ion significantly happier than methoxide, where a minus charge is stuck on a single oxygen. For this reason acetic acid is more acidic than methanol by 10 pK_a units, a factor of 10^{10} in equilibrium constants.

What does one of these things really look like? We believe that there is a p orbital on each atom of the three that are communicating, and the electrons spread out over this system, as shown below:

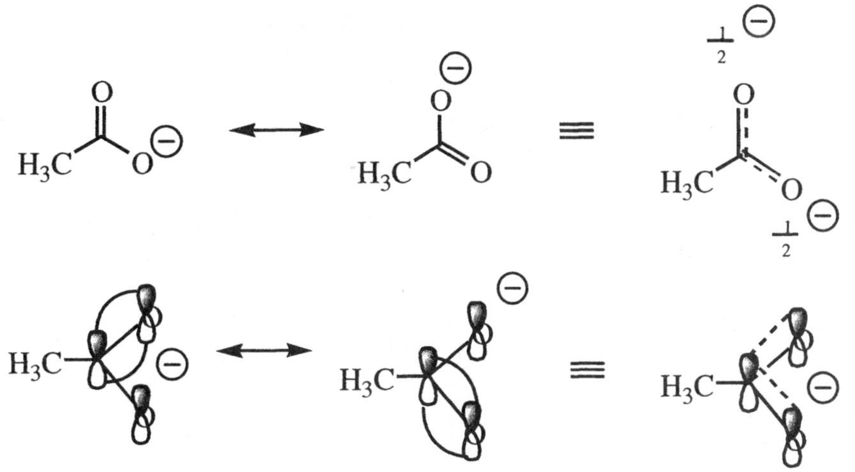

What about H_2SO_4? Let's look at its structure, and also at the structure of its conjugate base:

The first thing you should notice is that I have drawn sulfur with 6 bonds, and there apparently are 12 electrons around it. If you ever do this with a second-row element like carbon or oxygen you will get your hand severely slapped, but an "expanded octet" is permissible with third row elements due to the availability of empty d orbitals. Don't worry about this for now, just accept that sulfur and phosphorus and other similar elements are allowed to violate the octet rule in the sense of having too many electrons around them. The more important thing to notice is that, having drawn the anion shown on the right, one can now rearrange electrons, without moving anything else, to create not one but two new structures for the anion:

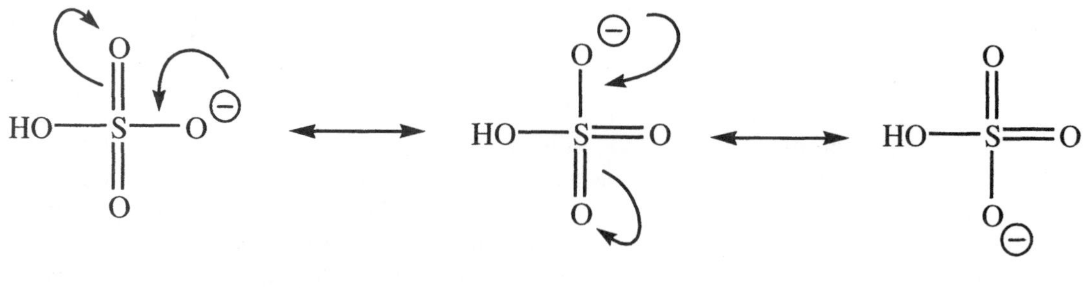

The real ion is a compromise among all three resonance structures, meaning that the charge is spread over three oxygens, and each oxygen really holds only 1/3 of a charge. This anion is so stable that sulfuric acid is more acidic than any other compound in our table.

How do you find resonance structures? This is a matter that will become much easier with practice, but here is one hint. For now, let's focus on anions, as in the above cases. Always start your thinking with the negative charge. This represents an excess of electrons, electrons that potentially could be shared. To generate a new resonance structure when a pair of electrons is present, these electrons must be able to go somewhere else within the same molecule—otherwise we'd be breaking the rule that no new partnerships can be formed. Since there are no places in the molecule that are short of electrons, the only reasonable place to put excess electrons is somewhere from which electrons can move away, again without breaking any partnerships. If you think about this you will realize that the "electron sink" we are looking for has to be a double or triple bond so some of the electrons holding the atoms together can be displaced and there will still be some left over. Therefore, you should look for nearby double bonds, and then move your minus charge in that direction, allowing electrons in the double bond to retreat away from the place where the new electrons are coming from. More thought should convince you that such a double bond has to be at a place exactly one and two atoms away from the charge you are trying to accommodate. Otherwise it can be of no help.

We will be encountering resonance frequently in the next several chapters, so even though it seems weird to you now, you will get more used to it as time goes on.

Using These Concepts

It's one thing to read the acidity table, but it is quite another, more important thing to be able to look at some problem and see how its solution is related to data in the table. Here are some examples.

1. Is water a strong enough acid to protonate fluoride ion? We need to be able to take a sentence like this and turn it into an equation. What it asks is, does the following reaction proceed significantly to the right?

$$F^- \quad + \quad H_2O \quad \rightleftharpoons \quad HF \quad + \quad OH^-$$

To answer the question we look on the table and discover that HF is a stronger acid than H_2O, so this reaction prefers the components on the left side and will not proceed to the right. Alternatively, we see that to get from the acid water to the base fluoride is an uphill process on the table, so the reaction will not occur significantly.

2. Is sodium propoxide ($CH_3CH_2CH_2ONa$) a strong enough base to remove a proton from HCN? Again, we must turn the question into a chemical equation. First we must recognize that when sodium is part of a molecule, it almost always exists as Na^+, and its role is to hang around and watch what happens but not get involved ("spectator ion"). If the sodium is Na^+, then the other part of the molecule must have a minus charge, making it $CH_3CH_2CH_2O^-$. This is what we should put in our equation.

$$CH_3CH_2CH_2O^- \quad + \quad HCN \quad \rightleftharpoons \quad CN^- \quad + \quad CH_3CH_2CH_2OH$$

Now we have a problem: we can find HCN on the table, but not propoxide. What do we do next? We need to make an estimate. This is the true art involved in reading such a table. The table cannot contain every species you would ever want to look up. It can only show representative examples of various *types* of species that are important. So we have to find an entry on the table that is similar to the thing we want to know about. We want to know about the acid strength of propanol or the base strength of propoxide. The reaction we looked for (unsuccessfully) on the table was removal of a proton from the OH group of propanol. Clearly, now we should try looking for other entries on the table that entail removing a hydrogen from an oxygen. There are quite a few of these, so we have to narrow the search down a bit: The oxygen bearing the H should have no charge and after the H leaves the oxygen should have a – charge. This helps, but there are still entries with pK_a's of –10, –6.5, –1.5, 5, 10, 15, 16, and 18 that meet this criterion. So let's look more closely: in propanol, the OH is attached to an alkyl group. So we should look for an OH that is attached to an alkyl group. Now we can narrow our choice down to 10, 15.5, 16 and 18. (Do you see the ones I mean in the table? Look for them!) But the one at 10, PhOH, has double bonds nearby. (Did you notice that? You need remember what Ph stands for, a benzene ring.) So now we need only choose between 15.5, 16 and 18, and we can therefore be fairly confident that we are dealing with a pK_a in the upper teens. This is close enough to allow us to answer our original question: propoxide *is* a strong enough base to remove a proton from HCN.

By the way, what happened to the Na^+? It was hanging around the $CH_3CH_2CH_2O^-$ when we put it in; now that the reaction has taken place, it is hanging around the new negative charge, CN^-. If you like, you can write this as NaCN, but chemists often find it easier simply to ignore the spectator ions and deal only with those that actually take part in the reaction. The resulting equation is called a "net ionic equation."

3. How strong an acid is H_3S^+? Hmm, this is not on the table, and there is nothing much like it. Let's see what *is* on the table:

NH_3	38	NH_4^+	9
H_2O	16	H_3O^+	–2
H_2S	7	H_3S^+	??

It would clearly be easier to remove a proton from a positively charged species than from the corresponding neutral species, so we can be certain that the pK_a of H_3S^+ is considerably lower than 7. However, the actual difference between a positively charged species and the corresponding neutral species does not seem to be predictable: with nitrogen, the difference is about 30 pK_a units, and with oxygen it is only about 20. If sulfur were in a direct line from these two (N→O→S) in some sense, we might make a guess of 10, but sulfur is below oxygen on the periodic table, not next to it, so there is no reason to believe that the 30 → 20 trend should continue. Our pK_a table contains no third-row elements from which to estimate the difference between a positively charged species and one with the same neutral atom, so we are left to guess. Since H_2S is a stronger acid than H_2O, it seems likely that H_3S^+ will be a stronger acid than H_3O^+, though perhaps not by the same amount. Therefore I would guess that H_3S^+ probably has a pK_a of something less than –5, possibly as low as –10. (More complete tables show it to be about –7.)

4. Lysine is one of the 20 or so naturally occurring amino acids that make up most of the protein in your body. What will lysine look like at pH 12? At pH 7? At pH 0?

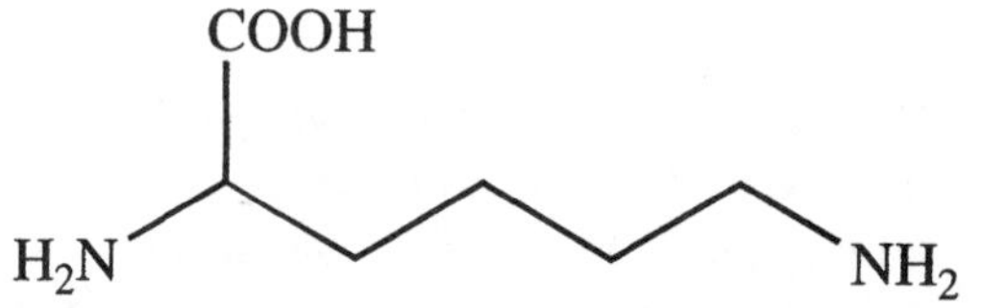

COOH

H₂N   NH₂

Huh? What is this question asking? As usual, the first step is to write down some reactions. There are three functional groups in this molecule: two amines and one carboxylic acid. In real life these do not act independently, but we have no way of accounting for that with our table, so for our purposes, we will assume that each group acts as if it were the only group in the molecule. So let's start by looking at the carboxylic acid group, which can undergo the following reaction:

$$H_2O \ + \ RCOOH \ \rightleftharpoons \ RCOO^{\ominus} \ + \ H_3O^+$$

Lysine is not on your table, but there is a carboxylic acid, which has a pK_a of about 5. The question asks, will this equilibrium lie mostly to the right or to the left in solutions of various pH's? To answer this you need to remember the relationship of conjugate base to acid as a function of pH. This is the Henderson–Hasselbalch equation given earlier in this chapter and repeated below, which is actually nothing more than a rearranged version of the K_a equation:

$$pK_a \ = \ pH \ - \ \log \left(\frac{[\text{conjugate base}]}{[\text{conjugate acid}]} \right)$$

Recall that this means that for any conjugate acid–base pair, if the pH is below (more acidic than) the pK_a, there is more of the conjugate acid, while if the pH is above (more basic than) the pK_a, there is more of the conjugate base.

The carboxylic acid group has a pK_a of about 5. At pH of 12, we are a lot more basic than pH 5, so there must be a lot more of the base form than the acid form at this pH; indeed, we can claim that essentially all the molecules will look like the conjugate base, $RCOO^-$. At pH 7 we still have more base than acid (you can demonstrate from the equation that there would be 100 times more base than acid; do it!). At pH 0, however, we are below pH 5 and the acid form will predominate. At this pH nearly all the molecules look like RCOOH.

What about the amine group? There are two reactions it can undergo:

$$H_2O \; + \; RNH_2 \; \rightleftharpoons \; RNH^{\ominus} \; + \; H_3O^{\oplus}$$

$$H_2O \; + \; RNH_3^{\oplus} \; \rightleftharpoons \; RNH_2 \; + \; H_3O^{\oplus}$$

The top one has a pK_a value of about 38. No matter which of the three specified pH's we choose, the solution will be *much* more acidic than pH 38 (which, in reality, cannot exist). It is no exaggeration to claim that all the molecules will favor the conjugate acid form, RNH_2, over the base RNH^-.

The bottom reaction has a pK_a of about 9. At pH 12, we are above that. The molecules will look like RNH_2 rather than RNH_3^+ (by a factor of 1000). At pH 7, the molecules will prefer RNH_3^+ by a factor of 100. At pH 0 they will all be RNH_3^+.

So what will lysine look like at these pH's? At pH 12 the molecule will look like this:

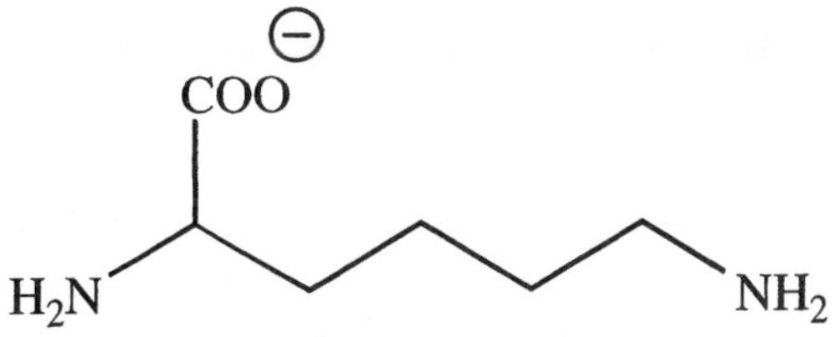

At pH 7 we are above the pK_a of the carboxylic acid but below that of the protonated amines. Thus, the amine groups will be in the form corresponding to the acid side of their equilibrium, while the carboxylic acid will still be present as its conjugate base:

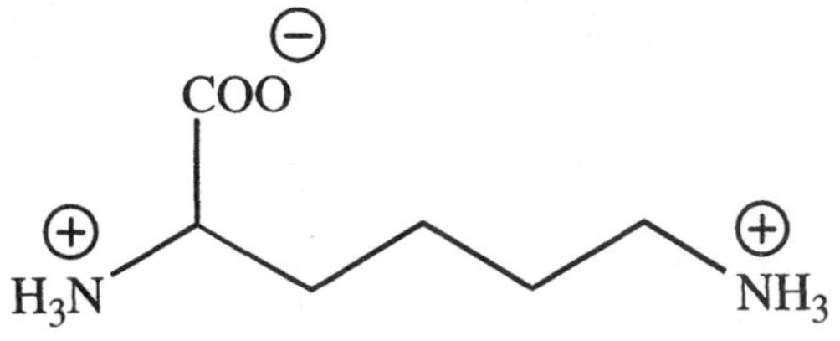

Notice that pH 7 is neutral, which is pretty close to the pH that organisms maintain. Thus, under physiological conditions the amino acids all have their NH_2 groups protonated, becoming NH_3^+ groups, and their COOH groups deprotonated, becoming COO^- group, as shown above.

Finally, pH 0 is below all the pK_a values, so all three groups will have the form shown on the conjugate acid sides of the respective equilibria:

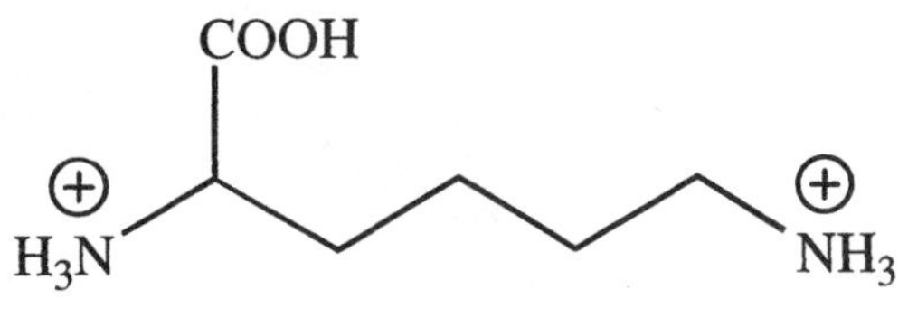

By the way, the true pK_a values for the groups in lysine are not exactly what we estimated, but they are close.

LEWIS ACIDS AND BASES

So far we have been discussing acids and bases in the context of proton transfers, where an acid is defined as a proton donor and a base as a proton acceptor. The official name for this is the "Bronsted–Lowry" definition of acids and bases. Much of the time when we refer to acids and bases in organic chemistry, this is what we are talking about. But there is a more general approach that will prove useful to us occasionally, which instead makes use of the Lewis definition. Lewis simply inverted the point of view to arrive at his definition: if a base is a proton acceptor, and it accepts a proton by providing a pair of electrons for the proton to bond to, then a base must also be an electron pair donor. This in fact became Lewis's definition of a base: an electron pair donor. What does that accomplish for us? Not much. Anything that was a base before is still a base; anything that was not still is not. What this new definition does do is broaden the context within which we can use the word. Whereas before we could only call something a base if it donated its electrons to a *proton*, we can now call it a base if it donates them to something else. In a future chapter we will learn another term for electron donation, but "Lewis base" will remain an acceptable term for any species donating its electrons for use by any partner.

So the Lewis concept does not allow us to define as a base anything that we were not already prepared to think of as a base. But the same is not true for acids! According to Bronsted–Lowry theory, an acid is a proton donor. What is it about a proton, H^+, that allows it to accept a pair of electrons? It has room in its outer shell for two more electrons, and it would be relatively happy to accommodate them. Thus, *any* species with room for two more electrons in its outer shell, and no objection to having them, might be expected to behave in the same way. Such things are what we call Lewis acids, electron pair acceptors: think of them as imitators of H^+ without *being* H^+. For example, BF_3 has only six electrons around boron, two short of an octet. Presumably it would be delighted to pick up two more to complete the octet. BF_3 is in fact an eager Lewis acid. Likewise, any other neutral compound containing an element from the third column, like $AlCl_3$, will be a Lewis acid. We will encounter Lewis acids later on in our study.

Lewis acids and bases are extremely important in organic chemistry. Nearly all the reactions we will study could be described as Lewis acid–base reactions in some sense. But 95% of the time we will instead refer to them by other names, so the concept does not arise *explicitly* that often. I will remind you about this when it seems appropriate.

COMMON MISTAKES: DON'T LET THIS HAPPEN TO YOU!

It is not true that *any* atom with vacant orbitals in its outer shell is a Lewis acid. Na^+, for example, has no electrons in its number 3 shell (indeed, it used to have one, but got rid of it, which is why it now has a plus charge). The problem is, Na^+ does not *want* two electrons in that shell. To be a Lewis acid, an atom ought to already be pretty close to having a full shell so that the extra pair of electrons will do it some good. Na^+ is perfectly happy as it is. For our purposes, Na^+ never does anything other than pal around with nearby negative charges. Many organic students naively claim that Na^+ plays an important role in some reaction they have shown, often using it to initiate an important process. Although there are certain reactions where so-called counterions are important, for our purposes we will ignore them. Na^+ should be regarded strictly as a "spectator ion," which means it stands on the sidelines and watches the action (often running up and down the sidelines to be *near* the action), but it does not participate in the game.

Problems

(Answers are provided at the end of the chapter for italicized problems.)

1. Identify the conjugate acid and conjugate base of each of the following. Write "none" if there is none.

Conjugate Acid		Conjugate Base
	CH_4	
	NH_3	
	CH_3OCH_3	
	CCl_4	

2. *Using the pKa values you have available to you, predict the products of the following reactions. (If there is no reaction, write NR).*

   a. $CH_3COOH + NH_3 \rightarrow$

   b. 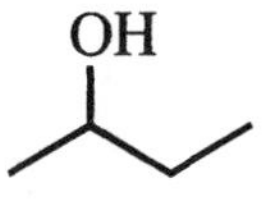 $+ NaNH_2 \rightarrow$

   c. 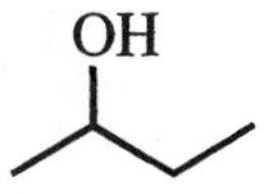 $+ HCl \rightarrow$

3. In which direction do the following reactions lie? Explain your choice.

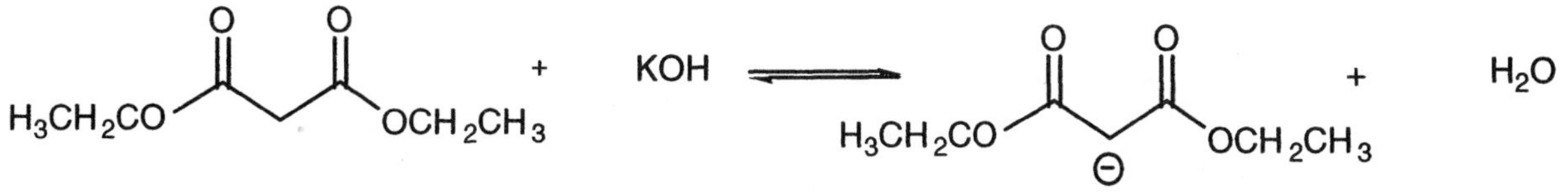

$$CH_3CH_2NH_2 + H_2S \rightleftharpoons CH_3CH_2NH_3^{\oplus} + HS^{\ominus}$$

4. Use your pK_a tables to suggest an appropriate base to accomplish the following reaction:

$$R-C \equiv C-H \longrightarrow R-C \equiv C \ominus$$

5. *For the following list of compounds:*

   a. *Circle the strongest acid.*

   b. *Put a box around the strongest base.*

   c. *Show the conjugate acid and base of each compound listed. Write none if there is none.*

   d. *Is the first compound a strong enough acid to protonate the third? (A yes or no answer will not do here. Show your reasoning.)*

   Conjugate Acid   Conjugate Base

   CH_3OH

   NH_3

   H_2S

   SH^-

6. *Explain why (not how do you know) phenol (benzene with an alcohol group attached) is more acidic than methanol.*

7. In each of the following pairs, which will be more acidic, and *why*? (For most of these, your pK_a tables will be of no use.)

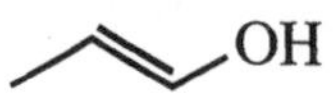

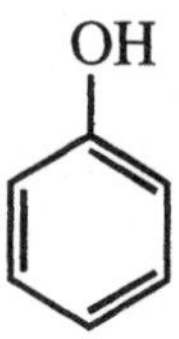

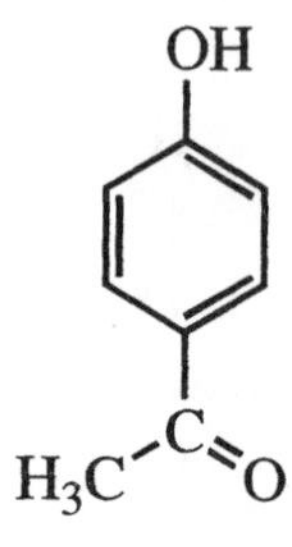

8. *The following compound can be protonated on any of the three nitrogen atoms. Determine which nitrogen atom is the most basic and why. (Hint: Think resonance.)*

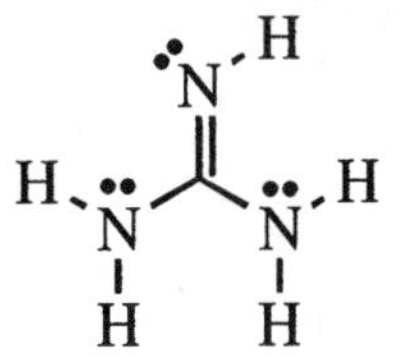

9. Acetic acid, CH_3CO_2H, has a pK_a of 4.73. Calculate the value of K_{eq} for the equilibrium

$$\text{acetic acid} + H_2O \rightleftharpoons \text{acetate anion} + H_3O^{\oplus}$$

10. *425 mL of an aqueous solution containing 13.0 g of Reingoldic acid (MW 142) is found to have a hydronium ion concentration of 0.000743 mol/L. What is the pK_a of Reingoldic acid?*

11. a. The pK_a of NH_4^+ is reported to be 9. What should be the concentration of hydronium ions in a solution in which 72 g of NH_4Cl has been dissolved in enough water to produce 350 mL of solution?

    b. Since "pH" is the negative logarithm of $[H_3O^+]$, what would be the pH of the above solution?

12. For each of the following reactions, would you expect the equilibrium constant to be greater than one or less than one? Why?

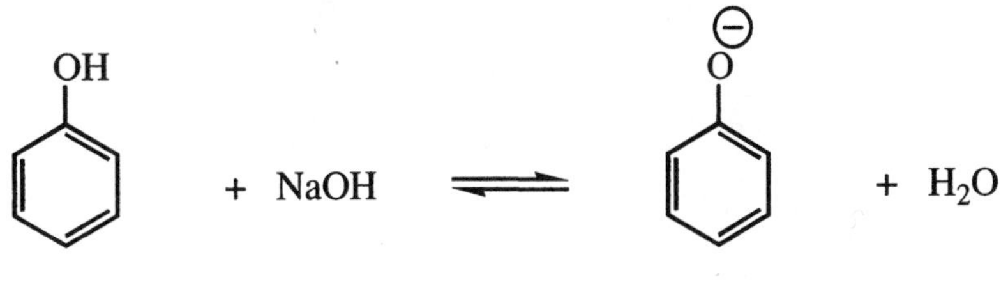

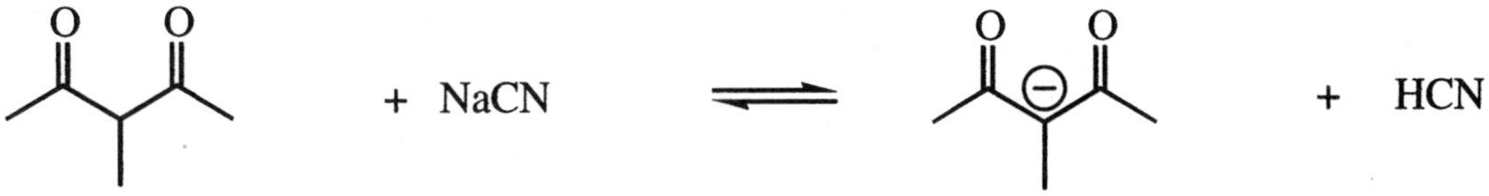

$$SO_4{}^{2\ominus} \quad + \; HI \quad\quad \rightleftharpoons \quad\quad HSO_4{}^{\ominus} \quad + \; I^{\ominus}$$

13. What will each of the following molecules look like at a pH of 3, 8, and 13?

14. *In what pH range will the following molecule have a net charge of –2? –1? 0? +1?*

15. Show the product of the reaction between BH_3 and NH_3. Show any formal charges that are required.

Selected Answers

Internal Problems

Test Yourself 1: First we write the equation

$$CH_3COOH + H_2O \rightarrow H_3O^+ + CH_3COO^-$$

and the expression for the acidity constant:

$$K_a = \frac{[H_3O^+][CH_3COO^-]}{[CH_3COOH]}$$

Then we write down what we know: before the reaction we had 6.00 g of acetic acid in 1 L of solution, and we can easily calculate that this is 0.100 M. *After* the reaction we have a pH of 2.87. Since

$$pH = -\log [H_3O^+]$$

we can calculate that $[H_3O^+]$ *after* the reaction has come to rest is 0.00135 M. If you can't do this, get some help with your calculator (or your math). In order to calculate K_a we need to know $[CH_3COOH]$, $[CH_3COO^-]$, and $[H_3O^+]$ after the reaction. We now know one of them. Can we get the rest? You should. Consider this: if you had 100 Bic pens with caps on them, and you put them in a box and shook for a while, some of the caps would come off. If you could count all the loose caps afterward, don't you think you could figure out how many capless pens there were? And how many intact pens-with-caps? Clearly, every time a cap comes off, you lose one pen-with-cap and create one cap and one capless pen. So the number of capless pens must be the same as the number of caps, and the number of pens-with-caps must be what we started with (100) minus that number. We'll do the same here: every time an H comes off, you lose one CH_3COOH and you gain one CH_3COO^- *and* one H_3O^+. So the number of CH_3COO^-'s must be the same as the number of H_3O^+'s, and the number of intact CH_3COOH's must be the number we started with (1M) minus that number. So $[CH_3COO^-]$ is 0.00135 and $[CH_3COOH]$ is 0.09865. Plug these into the K_a equation and you should get 1.84×10^{-5}. The pK_a is the log of this, with the sign changed, namely 4.73.

Test Yourself 2.

First we write the equation

$$CH_3COOH + H_2O \rightarrow H_3O^+ + CH_3COO^-$$

and the expression for the acidity constant:

$$K_a = \frac{[H_3O^+][CH_3COO^-]}{[CH_3COOH]}$$

Then we write down what we know: We know that the pK_a of acetic acid is 4.73, so the K_a must be $10^{-4.73}$ or 1.86×10^{-5}. We also know that before the reaction we had 60.0 g of acetic acid in 1 L of solution, and we can easily calculate that this is 1.0 M. As the reaction proceeds, some of the acetic acid comes apart. Each time one molecule comes apart, we lose one CH_3COOH and gain one CH_3COO^- *and* one H_3O^+. If this happens x times, then at equilibrium the concentrations will be

$$CH_3COOH \ + \ H_2O \ \rightarrow \ H_3O^+ \ + \ CH_3COO^-$$

Before	1 M	0	0
At equilibrium	1 M – x	x	x

We can plug these into the K_a expression and solve for x. If necessary, you can use the quadratic equation to do this, but often you can simplify the equation by recognizing that since K_a is very small, not very much acetic acid will come apart. That is, x is real, but insignificant compared to 1, and the concentration of acetic acid *after* you lose x of them is still pretty much 1. [Mathematically, this is that same as saying that you can ignore x in (1 – x).] So the K_a expression becomes

$$x^2 = 1.86 \times 10^{-5}$$

x = .0043 M (which you should confirm is pretty small compared to 1 M). This is the concentration of H_3O^+ at equilibrium (and also CH_3COO^-, although we do not care about that).

$$pH = -\log[H_3O^+] = -\log(.0043) = 2.37$$

Notice that we began this problem with 10 times as much acetic acid as the previous problem in the text, but the amount of H_3O^+ produced was *not* 10 times as much as before. This is a consequence of having two concentrations in the numerator and only one in the denominator. But the equilibrium constant *does* remain constant.

End of Chapter Problems

2. a. $\quad CH_3COO^- + NH_4^+$

b.

What happened to the Na? It is Na^+, and it's hanging around the O^- we made. It is not wrong to show it there, but for simplicity's sake it is often left off.

c.

5.

Conjugate Acid		Conjugate Base

$$H_3C-\overset{\oplus}{\underset{H}{O}}\!{}^{H} \qquad\qquad H_3C-OH \qquad\qquad H_3C-\overset{\ominus}{O}$$

$$H_3C\overset{\overset{\oplus}{O}-H}{\diagup\!\!\diagdown}CH_3 \qquad H_3C\overset{O}{\diagup\!\!\diagdown}CH_3 \qquad H_3C\overset{O}{\diagup\!\!\diagdown}\overset{\ominus}{CH_2}$$

$$H-\overset{\overset{H\oplus}{|}}{\underset{\underset{H}{|}}{N}}-H \qquad \boxed{NH_3} \qquad \overset{\ominus}{:}\overset{..}{N}-H \atop \underset{H}{|}$$

$$H-\overset{\overset{H}{|}}{\underset{\oplus}{S}}-H \qquad \bigcirc\!\!\!\!\!H-\overset{..}{S}-H\!\!\!\!\!\bigcirc \qquad \overset{\ominus}{:}\overset{..}{S}-H$$

$$H-\overset{..}{S}-H \qquad \overset{\ominus}{:}\overset{..}{S}-H \qquad \overset{\ominus}{:}\overset{..}{S}\overset{..}{:}\overset{\ominus}{}$$

Note that the bottom species is not on the table as an acid. How is one to guess how strong an acid it is? Clearly (I hope) you would expect it to be harder to remove a second proton, making a double minus charge, than a first; thus this must be *less* acidic than H_2S. So it cannot be the strongest acid.

Both CH_3OH and H_2S are missing from the table as bases. We can guess that CH_3OH as a base is pretty similar to H_2O, so that is no problem. What do we do with H_2S? We know that HS^- is a weaker base than HO^-, so we can guess that H_2S is a weaker base than H_2O. So it cannot be the strongest base.

d.

$$H_3C-OH \quad + \quad NH_3 \quad \rightleftharpoons \quad H-\overset{\overset{H\oplus}{|}}{\underset{\underset{H}{|}}{N}}-H \quad + \quad H_3C-\overset{\ominus}{O}$$

$pK_a = 15.5$ $pK_a = 9$

Weaker acid: reaction favors this side.

Answer: NO!

6.

Bottom anion is *much* more stable due to resonance!

$$\text{H}_3\text{C}-\text{OH} \quad \xrightleftharpoons \quad \text{H}_3\text{C}-\text{O}^{\ominus} \qquad \text{p}K_a = 15.5$$

8. If the top nitrogen becomes protonated, there are four resonance structures that describe the cation.

But if either of the bottom nitrogens are protonated, all the charge is stuck on a single atom:

10. Let's call Reingoldic acid "HR."

$$\text{HR} + \text{H}_2\text{O} \quad \xrightleftharpoons \quad \text{R}^{\ominus} + \text{H}_3\text{O}^+$$

The *initial* concentration of HR can be calculated from the data given as 0.215 M. The *equilibrium* concentration of H_3O^+ is given as 7.43×10^{-4} M; clearly the equilibrium concentration of R^- must be exactly the same, since whenever one forms, so does the other. The equilibrium concentration of HR is 0.215 minus this small number—for all intents and purposes, still 0.215. So we can plug all these numbers into the K_a equation:

$$K_a = \frac{\left[R^{\ominus}\right]\left[H_3O^+\right]}{\left[HR\right]} = 2.57 \times 10^{-6}$$

Now take the log of this number and change the sign: $pK_a = 5.6$

14. –2: anything over 9.
    –1: about 5 through 9.
     0: Never (actually *at* pH 5 it will have a charge of 0, but this is tough to see.)
    +1: anything under 5.

Chapter 9
Reaction Intermediates

The study of the shapes of organic molecules is fascinating, and some chemists devote their careers to it, but it must be acknowledged that *most* organic chemists find that the fun part of chemistry is reactions. We love to figure out how to make molecules (and to do it), and to discover new ways of accomplishing certain transformations. We like to make new materials and see what their properties are, sometimes anticipated and sometimes not. (Teflon, for example, was made entirely by accident the first time, and its refusal to dissolve in anything or stick to anything was a source of great frustration at first, until people realized how useful those properties could be!) You have successfully negotiated most of the preliminary material we needed to cover, and we can now move into the realm of organic reactions. In this chapter we will

- Discuss the symbolism for writing organic reaction equations

- Discuss the kinds of pathways organic reactions tend to follow

- Learn about cations, radicals, and anions as intermediates

- Revisit resonance and learn about other stabilizing factors.

Consider a generic organic reaction:

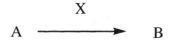

This means that A is converted to B by using X to make it happen. X is often called a "reagent." There are several things that chemists studying this reaction would do. First, they would catalog it in their memory banks alongside many others they already knew: this is one reaction (of probably many) that compounds of type A undergo; this is one way (probably among many) of making compounds of type B; this is one (of probably many) transformations that reagent X is capable of effecting. When we get to real examples, by the way, I will show you how to catalog the reactions you are learning. But most important for the moment, our chemists would want to try to understand *how* this reaction happens. *How* does A change into B? What does X do to make this happen? Does it happen all at once or are there some resting points in between A and B? What is the relationship between this reaction and the many others that are known?

The *how* of a reaction is called a *mechanism*, and the study of reaction mechanisms will occupy most of our time from here on out. Why do we worry about mechanisms? Because, without them, organic chemistry would be nothing more than a catalog of facts. Indeed, this is how organic chemistry used to be taught, and that is why it developed such a fearsome reputation as a killer course that required tons of memorization. I cannot deny that there is a lot you will have to remember, but mechanisms, understood properly, will give you a framework on which to hang all these reactions. Once you understand how reactions work, you will recognize that there aren't really all that many fundamental types. Most of the reactions that students of 40 years ago struggled to memorize are really minor modifications of the same thing, and once you know the general reaction type, it becomes obvious what would happen in a slightly different situation.

Most of the reactions we will study are already reasonably well understood. However, we should note two things: first, everything we "know" about mechanisms is a guess. It tends to be a good guess, an educated guess, one consistent with lots of experiments designed to test our guess, but the fact remains that no one has ever watched a molecule go through a reaction. We have to infer what is happening from indirect evidence. Later we will look at some of the methods people use to make their guesses. Second, almost all the reactions I will describe are actually somewhat more complicated than I will pretend. The reason for oversimplifying at this stage is that unless you plan to become a professional chemist, it is unlikely that you will ever need to know all the gory details of these reactions, and learning them will (in my judgment) confuse you more than help you. If you do plan to become a professional chemist, or even a non-chemist who does a lot of chemistry, you should consider taking a follow-up organic course that covers some of the topics left out of this book and includes many details I have glossed over.

A typical pattern for a reaction is the following: a molecule is sitting happily around, minding its own business, when some troublemaker shows up and does something to it to make it uncomfortable. The molecule then goes through one or more contortions ("steps") in an effort to become comfortable again. One of those possible steps, most of the time, simply undoes whatever the troublemaker did and takes the molecule back where it came from; but that is not always possible, and even when possible, if the troublemaker is still hanging around, the process may repeat itself until the molecule devises a new solution to the problem. Sometimes it takes a few operations to find a new happy state, other times the process is pretty straightforward. Below is an example of a typical reaction and its mechanism. (You should not expect to understand what is written there; we will study this in detail in the next chapter.)

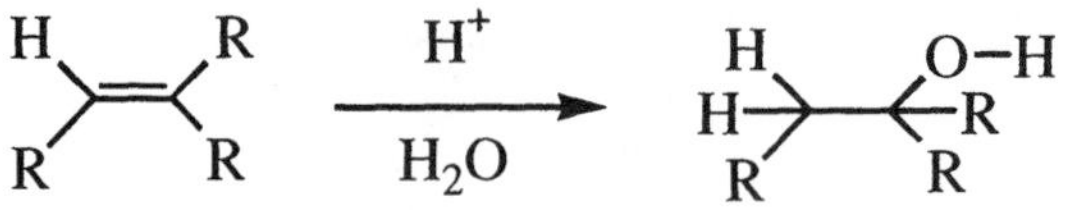

Mechanism:

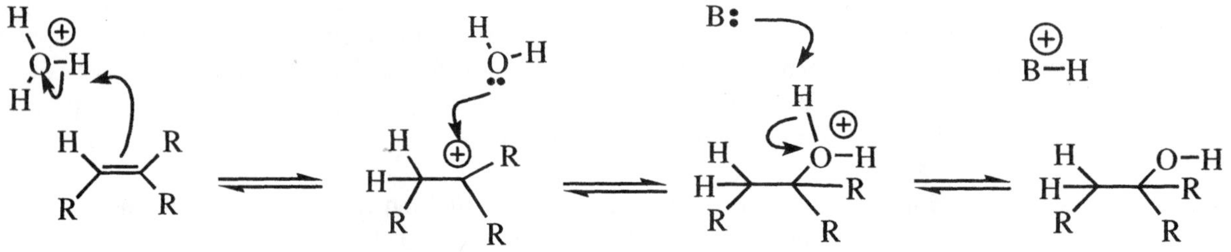

The key to learning organic chemistry is becoming familiar with the patterns. There is a limited number of types of unhappy states that a troublemaker can create, and there is a limited number of ways the molecule can react to them. Once you know these, you will find that most (though not all) of the reactions we study fall into one of a few familiar categories.

Intermediates

There are three common types of "unhappy states" that organic molecules typically find themselves subjected to. All of them relate to some deviation from the normal resting state of carbon in molecules, namely, that of four covalent bonds. Normal carbon has four bonds, each representing two electrons, so carbon experiences an octet. Uncomfortable carbon has three bonds. In place of the fourth bond there might be zero, one, or two electrons but no partner.

If there are no electrons at the fourth position, the carbon owns only three electrons. (Remember how to count by ownership? If you don't, review the section on formal charge!) Therefore this carbon has a positive charge, which we indicate by a plus sign in a circle. This is a cation: specifically, since the charge is on a carbon, a "carbocation." Carbocations are high in energy, because they have this charge, and also because the carbon lacks an octet. If there is one electron in the fourth position, the carbon owns four electrons and has no charge. It is nevertheless high in energy because it does not see an octet. This is called a "radical" or sometimes a "free radical." We will spend very little time discussing radicals in this text. If there are two electrons in the fourth position, the carbon now owns five electrons and has a negative charge, which we indicate by a minus sign in a circle. We may also show the pair of electrons, as in the picture above, but that is not necessary, because if there is a minus charge, the pair of electrons *must* be there, and they can be (and sometimes are) assumed. By the way, the converse is not true: although writing an unshared pair of electrons clearly requires that there be a minus charge, it is not considered correct to draw the electrons without the charge. If there is a charge you are required to show it. Recall that a negative ion is called an anion; this being on a carbon, it is a "carbanion." Carbanions have octets around carbons, but they are nevertheless high in energy because they also have charge.

Let's consider each of these in more detail.

Carbocations

Carbocations suffer from a positive charge. They have three bonds to other partners and nothing else. Like all other atoms with three areas of electron density, they adopt a trigonal geometry and are sp²-hybridized. Later on it will be important to remember that carbocations are planar.

A carbocation's prime mission in life is to stop being a carbocation. There are two approaches to this: it can somehow get rid of the positive charge, or it can gain a negative charge. Both strategies involve providing the missing electrons to the poor starving carbon. As we will see, there are other things a carbocation can do, but they only delay the inevitable: eventually, one of the above two things has to occur.

It is important that you be able to distinguish a carbocation from other kinds of cations you have seen and will see. A carbocation has a positive charge because it is short of electrons, which means that the carbon atom itself is capable of housing two more. That makes it a Lewis acid, if you remember our earlier definition. It also makes a carbocation different from other cations frequently encountered. The oxygen atom of H_3O^+, for example, also has a positive charge, but if you examine its structure carefully, you will see an important distinction between this species and a carbocation: H_3O^+ has a complete octet. The oxygen atom has a positive charge, *not* because it is short of electrons, but because it is sharing too many of the ones it has. The solution to this problem is *not* to add more electrons, but to eject a partner. Both kinds of cations, those with too few electrons (carbocations) and those with too many partners, play important roles in the reactions we will soon be learning.

Inductive Effect

Carbocations are not all alike. You will recall from our discussion of nomenclature that we classify carbon atoms as primary, secondary, and tertiary, depending on how many other carbons are attached to them. These same terms apply to carbocations. You couldn't predict this ahead of time, but it turns out that the more substitution there is on a carbocation, the better off it is. Don't get me wrong: essentially all carbocations are unhappy and desperate to stop being carbocations. Nevertheless, tertiary carbocations are *less* unhappy (more stable) than secondary carbocations, which in turn are more stable than primary carbocations. Later on we will see some data that should persuade you that this is true. For now, simply believe me.

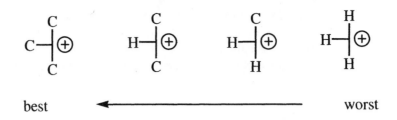

best ⟵————————————————⟶ worst

What can we conclude from this? What follows is not fancy chemistry to be memorized (though I hope you will remember it) but simple logical deduction from facts: every time we replace a hydrogen with a carbon on a carbocation, the cation gets better. Carbon must therefore be a better neighbor for a positive charge than hydrogen. What would make for a good neighbor to a carbon atom that is poor and destitute, suffering from a shortage of electrons? Clearly, generosity. If you were poor, you would benefit from rich neighbors who shared their stuff, rather than other poor neighbors who kept everything for themselves. Compared to hydrogen, carbon must be generous. Carbon apparently shares its electron density with a neighbor more freely than hydrogen does. The chemical term we use to describe this is "**electron donating.**"   Carbon is electron donating relative to hydrogen, so tertiary carbocations are better off than secondary, etc. There are other neighbors that are even more generous than carbon, as we shall soon see.

The general term applied to this effect of a group stabilizing (or destabilizing) its neighbor by electron donation or withdrawal through σ bonds is called the "**inductive effect.**"

Resonance Effect

There is another very important factor that can make one carbocation better than another, and that is resonance. We discussed resonance in the previous chapter in the context of anions, but it applies equally well to cations. Any process that serves to spread out charge will make a species more stable than the analogous structure with concentrated charge. Resonance, you will recall, involved shifting electrons in a structure without moving any atom and without creating or destroying partnerships. In the last chapter, we looked closely at anions and tried to find someplace we could push the extra electrons, someplace that could accommodate them, namely, an adjacent double (or triple) bond. In the case of cations, we again begin our thinking with the source of the discomfort, the atom with the charge, and try to help it out without involving the outside world and also without breaking any partnerships.

If the cation is of the type that has too many bonds, the only way to solve its problem is to *break* a bond. If all its bonds are single bonds, you are out of luck; you cannot break a single bond without destroying a partnership, and that would no longer be resonance. However, if the cation is involved in a double bond, you could break one of the two without destroying the partnership. *In the process, be sure to bring the electrons toward the atom that needs them!!!!* Here is an example:

Notice that in the process of solving the second oxygen's problem, we had to take electrons away from carbon, creating a problem for it, so that carbon now has a positive charge. This should not surprise us: the molecule began with a positive charge, and we did not change the makeup of it, so it must still have a positive charge. All we did was pretend the charge was somewhere else. *Pretend*? Yes, pretend: remember, resonance structures have no reality. The positive charge is not really on the oxygen, nor is it on the carbon. It is partially on each. The charge is shared between them. The fact that we can write two resonance structures means that the structures themselves are poor approximations to the actual structure, which is really something in between them. Such an ion is also better, happier, more stable than it would be if it were forced to be either of the structures we wrote rather than a hybrid of the two.

If the cation is of the type that is short of electrons, such as a carbocation, then we try to solve its problem by bringing in more electrons. Again, for resonance purposes we are forbidden to look outside the molecule to do this, and we are forbidden to break any single bonds. The only potential sources of help, therefore, are nearby electrons not currently being used for anything (lone pairs) or electrons involved in double or triple bonds. A suitable lone pair needs to be on an atom immediately next to the cation; if a double bond is to be of help, it needs to be between the next atom and the one after that. The following examples should illustrate the point.

The first example shows how to use a lone pair to help a nearby cation. In the process of doing this, of course, the atom with the lone pair ends up sharing it and thereby acquires a positive charge of its own. As before, this should not surprise us, since the structure is known to be positively charged; it is just a matter of where we pretend the charge is. You might also notice that the example I used here is exactly the same one I used in the previous paragraph. The point is that no matter which of the two resonance forms you think you have, or which one you write first, they are in fact the same thing, with a charge that is neither on the carbon nor the oxygen exclusively but shared by both. Resonance stabilization of a carbocation by a neighboring oxygen (or nitrogen) is extremely important, but, in my experience, it is also the thing students miss most often in describing reactions. You should train yourself to notice situations in which there is a lone pair on an atom next to a positive charge and then show the resonance structures involved.

The second example shows how a double bond can assist a carbocation. Again, in pursuing this we find that we are taking electrons away from another atom—the one at the far end of the double bond—thus saddling it with a positive charge. But again, neither carbon actually has to put up with the entire charge; it is shared between them, and the result is therefore more stable than any charge localized on a single carbon (more stable even than a tertiary carbocation).

In both of these cases, notice that the electrons (arrows) are moving *toward* the positive charge, which only makes sense, since arrows always show the movement of electrons, and the positive charge represents an atom that needs electrons.

ORM: Check out the section on "Ion Stability" in the computer program "Organic Reaction Mechanisms." There are some nice animations to help explain the inductive effect and the resonance effect.

Radicals

We will not be discussing radical reactions much in this particular book, but it will still not hurt to spend a few minutes learning what radicals look like and how they behave. Radicals share with carbocations the fact that they lack an octet, and although they are not charged, they have many properties in common with cations. For example, radicals, like carbocations, are sp^2-hybridized and have trigonal planar geometry (though it may not be obvious why). Also like cations, tertiary radicals are better than secondary ones, which are better than primary. That is, substitution is good for a radical. We can explain this in the same terms we used before (carbon is electron donating relative to hydrogen), because even though a radical is not positively charged, it still wants one more electron and appreciates generous (electron-donating) neighbors. Finally, radicals, like carbocations, are improved by resonance, although the way chemists draw electron arrows for radicals is different from what we have seen before. I will not show you the details now, because in my experience, once students learn how to draw radical-type arrows, it is difficult to get them to stop, and in truth there are very few circumstances where they are needed.

A radical's discomfort stems from the fact that it has an open bonding position containing one electron. To fill this up, the radical needs a partner that brings with it a single electron to join with the electron already present, creating a bond. There are two standard ways to accomplish this: the radical can find another radical to join up with (which does happen, but fairly seldom, since the chances of one radical finding another one are slim), or it can attack some molecule that is complete and forcibly tear off a partner with one electron, leaving the victim molecule with its own radical problem. This victim then becomes vicious and does the same to some other innocent bystander, and the process continues until eventually two radicals do encounter each other and cancel each other out. The process of a radical reacting with a whole molecule to create a new whole molecule and a new radical, which in turn does this again, is an example of a *chain reaction*. When you encounter radical reactions in some later course, you will learn more about chain reactions.

I should mention that radicals have been implicated in the aging process, and many of the vitamin supplements people eat, such as vitamin E, are for the purpose of intercepting radicals so that they will do you less harm.

Carbanions

The third possible type of intermediate is a carbanion. Carbanions have four areas of electron density around them, and so ought to be sp^3-hybridized and tetrahedral. Further, since we have already established that carbon is an electron-donating substituent, we ought to conclude that carbon would be a *bad* neighbor for a carbanion, which already has excess electron density, and the stability order ought to show that primary carbanions are more stable than secondary, which are better than tertiary. While most of the above statements turn out to be true, they are also largely irrelevant, because ordinary carbanions are so unstable that they do not play a significant role in the chemistry we are about to study. Look on the pK_a table in the previous chapter: carbon with a minus charge is at the very bottom of the base column, meaning that these species are among the most energetic anions that can possible exist and will be exceedingly difficult to make. Thus, they do not show up often.

So why are we discussing carbanions at all? It turns out that carbanions *do* play an important role in the chemistry we are about to learn, but only when there is something about them that makes them more stable than ordinary carbanions. Such as? The first thing that ought to pop into your head at this point is resonance. We have seen the important role resonance plays in stabilizing both anions and cations in other contexts, so surely resonance should also be important in stabilizing anions on carbon. Sure enough, the majority of carbanions we will encounter are stabilized by resonance. By far the most common example is what we call the "enolate anion," an example of which is shown below.

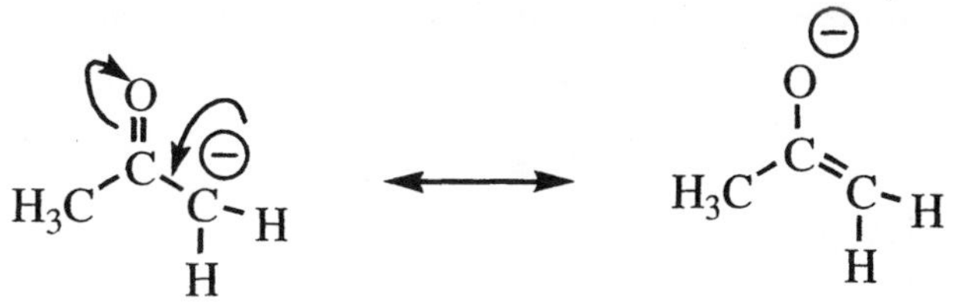

We should take some time here to analyze this particular species. If I asked you to draw a picture of just the structure on the left, you would say that the carbon of the double bond is sp^2-hybridized and planar, while the carbon with the anion is sp^3-hybridized and tetrahedral. On the other hand, the structure on the right clearly has a double bond between the two carbons, so both must be sp^2 hybridized and planar.

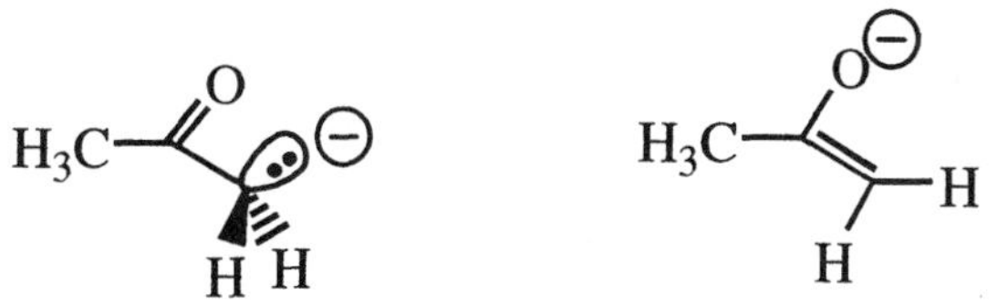

This brings up a problem. Recall rule one of resonance structures: no atom can move. Yet these two structures appear to have different geometries around the rightmost carbon. It turns out that the picture I drew on the left, with an sp^3 carbon for the anion, is wrong. Clearly, we cannot get around a planar geometry for this carbon in the right structure, and since nothing can move, that must be its geometry in the left structure as well. So a more accurate picture of these resonance structures is the following:

Recognizing the true geometry reminds us of an aspect of resonance structures we have covered before. Between the carbon–oxygen double bond and the anion associated with a p orbital on the carbon atom, we can see that we have a continuous array of three p orbitals among which four electrons are shared. These resonance structures are one feeble attempt at showing this phenomenon:

It will become important later on to remember that any anion resonance-stabilized in this way is actually flat.

We will encounter two other important varieties of carbanions in our study. One is readily understandable; the other is not. Take a look at the anions with a charge on a carbon in the pK_a table in Chapter 8. Ignoring the ones that are resonance stabilized, there are really only three types left: those with charge on sp^3 carbons, sp^2 carbons, and sp carbons. Here is what you should have found:

	pK_a
$H-C\equiv C-H$	25
	44
	50

Can we explain this trend? As I've told you before, when you are trying to explain something, be sure you look at the whole picture. Here we are trying to explain why one reaction is better than another, but we have not actually drawn the complete reactions yet. We'd better do that:

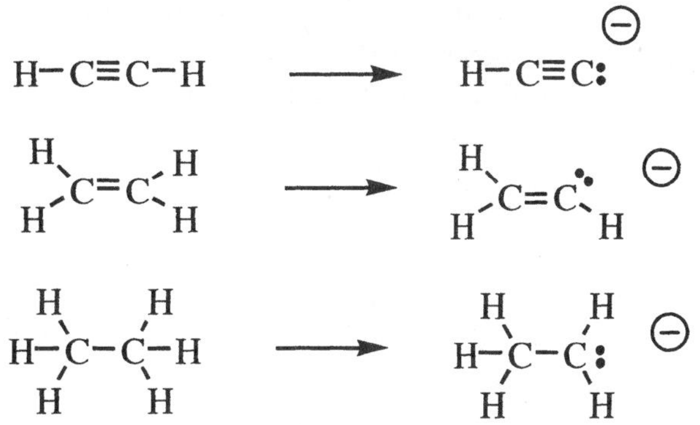

$$H-C\equiv C-H \longrightarrow H-C\equiv C:^{\ominus}$$

$$\begin{array}{c} H \\ \diagdown \\ H \end{array} C=C \begin{array}{c} H \\ \diagup \\ H \end{array} \longrightarrow \begin{array}{c} H \\ \diagdown \\ H \end{array} C=\ddot{C} \begin{array}{c} \\ H \end{array}^{\ominus}$$

$$\begin{array}{ccc} H & & H \\ | & & | \\ H-C-C-H \\ | & & | \\ H & & H \end{array} \longrightarrow \begin{array}{ccc} H & & H \\ | & & | \\ H-C-\ddot{C}: \\ | & & | \\ H & & H \end{array}^{\ominus}$$

In each case we are removing a proton, H^+, and leaving behind electrons. Where are those electrons stuck? In one case (the bottom one) in an sp^3 orbital, in the next an sp^2 orbital, and in the top case in an sp orbital. What do we know about these things? Reaching back to Chapter 4, you should recall that an sp^3 orbital is derived from one s orbital and three p orbitals, all mixed up together. It is ¼ s and ¾ p. An sp^2 orbital is 1/3 s and 2/3 p, while an sp orbital is ½ s and ½ p. There is something else: all the way back in Chapter 2, we learned that at any energy level, the s orbital is lower in energy than the p orbitals. Therefore, the more s-like a hybrid orbital is, the lower its energy will be. An sp orbital, being ½ s, is lower in energy than an sp^2 orbital, which in turn is lower in energy than an sp^3 orbital. It follows that unshared electrons dumped into an sp orbital are lower in energy than electrons in an sp^2 or sp^3 orbital. This is what we observe in the trend shown above: it is easier to remove a proton from the tail of a triple bond, leaving electrons in an sp orbital, than it is to do the same on a double or single bond. The effect is rather dramatic: it is *a lot* easier to make an anion on a triple bond than on a double or single bond. In fact, it is easy enough to make an anion on a triple bond that we can relatively readily prepare such things and use them in organic chemistry. The other types are generally inaccessible.

The third approach to making anions you could not predict. Indeed, they are not really anions at all, in the true sense, but we are going to treat them as if they were anyway. If you take almost any compound containing a carbon–halogen bond, where the halogen is chlorine, bromine, or iodine (for various reasons fluorine does not show up nearly as often in organic systems), and treat it with magnesium metal in the right way, a magic reaction occurs that results in a magnesium atom jumping in between the carbon and the halogen.

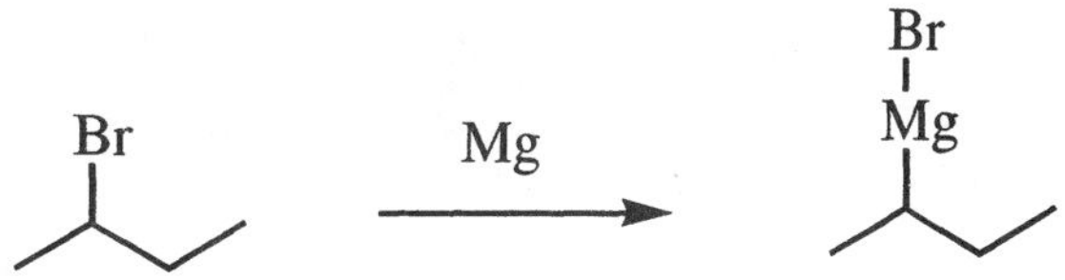

There is no way you would be able to understand what actually happens here, or why, but you can still understand some important consequences of this weird transformation. Focus your attention on the carbon–bromine bond. You already know that bromine is more electronegative than carbon, so this bond is polarized such that the carbon is partially positive, and the bromine is partially negative. Thus, the carbon should behave slightly like a cation and be attractive to (and attracted to) negative charges. On the other hand, magnesium is in the second column of the periodic table and is relatively electropositive; carbon is *more* electronegative than magnesium. So in the right-hand compound, the carbon has a partial *negative* charge. It turns out that this negative charge is pretty substantial, and many of the reactions of compounds like this can be understood pretty well if we simply pretend that there is a real negative charge on the carbon atom attached to the magnesium, and ignore the magnesium altogether. These compounds are called "Grignard reagents" (say "Grin-yard") after their discoverer, Victor Grignard, who won the Nobel Prize for discovering this transformation and figuring out some neat stuff the compounds do.

Problems

(Answers are provided at the end of the chapter for italicized problems.)

1. Write complete sets of resonance structures for the following species. Be sure you draw the arrows that show the movement of electrons as you proceed from one structure to the next. *Answers to the first and second structures are given.*

2. *Explain* why *(not how do you know)* ammonia is more acidic than diisopropyl amine.

3. The following compound can be protonated on any of the three nitrogen atoms. Determine which nitrogen atom is the most basic and why. (*Hint*: Think resonance.)

Selected Answers

1.

2.

$$pK_a = 38$$

$$pK_a = 40$$

Alkyl groups are electron donating. Both anions are suffering from an excess of electrons, but the lower anion is a bit more unhappy than the upper one because of the extra donation from the alkyl groups. Thus this anion is harder to form, so diisopropylamine is less acidic.

Chapter 10
Alkenes I

Now for our first real foray into the realm of organic reactions. We will start by learning about addition reactions of alkanes. In this chapter we will

- Learn how to name alkenes, including geometric designations

- Learn about the addition of hydrogen halides (HX) to double bonds

- Discover Markovnikov's Rule, why it works, and why it sometimes doesn't

- Start systematically keeping track of the reactions we encounter

- Learn about the addition of water to double bonds, which is a way to make alcohols

- Learn how to make ethers

- Apply our knowledge of reactions by working some synthesis problems.

NOMENCLATURE

Compounds with double bonds are called *alkenes* or *olefins*. Double (and triple) bonds are often described as "positions of unsaturation," and compounds containing them are referred to as "unsaturated hydrocarbons."

Alkenes are named by exchanging the "-ane" ending of an alkane for "-ene." (Trivial name alert: ethene is usually called "ethylene," and propene is sometimes called "propylene." Most other alkenes are named according to the rules.) The base name of a compound with a double bond is not necessarily the longest chain in the molecule; it is taken from the longest chain *that has the double bond in it*. Two new problems also arise, however. The first is that you have to indicate where along the chain the double bond is. This is done with numbers, as before. Of course, the double bond occupies *two* carbons in the chain, but since they are necessarily adjacent, it is customary to specify only the first of the two (the carbon with the lower number). As in choosing the base name of the compound, so also in the numbering; the double bond takes priority over other things, so the chain is numbered so as to give the double bond the lowest possible number. If more than one double bond is present, you include as many as possible in the base chain, and the name of the compound becomes "something-diene" or "something-triene" (recall di- and tri- are prefixes for two and three), again specifying with numbers where along the chain you find the double bonds. For example, the following compound is 8,9,9-trimethyl-4-pentyl-1,4,6-decatriene.

But we are not done. Recall that a double bond consists of a σ bond and a π bond, and that the π bond requires side-to-side overlap of p orbitals, so the bond cannot rotate without the overlap (and thus the "second" bond) being broken. So the two compounds below are *different* compounds, separable from each other. Each can be placed in a bottle and will remain indefinitely what it is and not convert into the other. Thus, we need to be able to name these things as different chemicals.

We use the same trick and the same names we applied to cyclic compounds. If two substituents are on the same side of the double bond, we call the compound *cis*. If they are on opposite sides, we call it *trans*. Thus, the compound on the left is called *trans*-2-butene, and the one on the right is *cis*-2-butene. Notice that by "opposite sides" we are referring to sides of a wall that *includes* the two carbons of the double bond, not a wall that bisects the double bond. (Of course, trying to base a name on the latter would do no good, as the methyl groups in both the above compounds are on opposite sides of that wall, so we still would not have differentiated the two.)

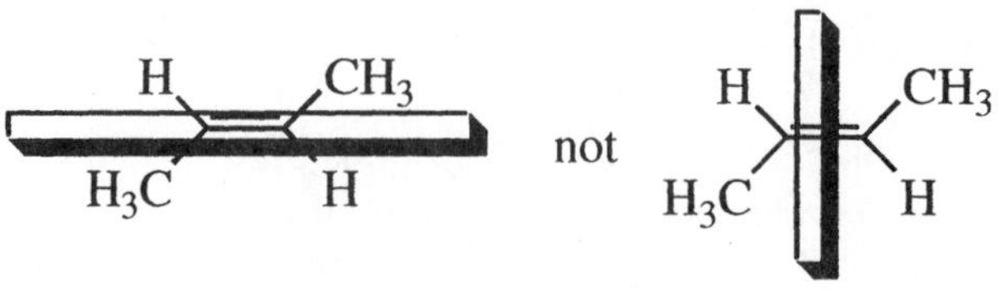

Naming compounds *cis* and *trans* works in lots of cases, but occasionally you run into trouble. For example, consider the compound we started with, the decatriene. Clearly the double bond at the number 1 position does not need to be labeled for geometry, because it has two of the same thing (H's) on one end, and therefore it does not matter what you put where on the other end. The double bond at number 6 can be labeled *trans* without any ambiguity. But how should we label the double bond at number 4? Is this *cis* or *trans*? Well, it's *trans* if you are worried about the relationship of the right part of the chain with the piece on the left that contains the double bond, but it would be *cis* if you focused your attention on the part on the left that is longer. We need some more guidelines as to what to call this one.

Chemists have created such guidelines, and created a system of nomenclature that lets you know they are using the guidelines. When in doubt, we don't use the words *cis* and *trans*, but instead we use the letters *E* and *Z*, which stand for *entgegen*, the German word for "opposite," and *zusammen*, the German word for "together." Thus, *Z* sort of replaces *cis*, and *E* replaces *trans*. It is never wrong to use these letters, but many chemists continue to use the words *cis* and *trans* when there is no ambiguity as to what they mean. When there *is* ambiguity, such as in our case, this new system removes it by telling us exactly what relationship to specify. Here's how:

We first divide the double bond *through* the double bond, as in the right-hand picture above, and at each end we determine an order of priority for the two (guaranteed two!) substituents present. The main rule is that higher atomic numbers take priority over lower ones. This particular set of rules is called the Cahn–Ingold–Prelog system, after the chemists who invented them.

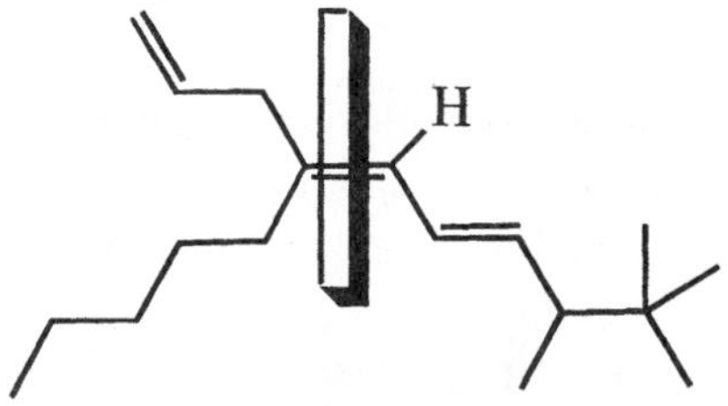

Let's now apply the rules to our example. On the right side it's easy: there is a bunch of stuff, starting with a carbon, on the lower branch, and nothing but a hydrogen on the upper branch. Carbon has atomic number 6, and hydrogen is 1, so the lower branch on the right takes priority over the upper branch. On the left, things are not so easy. Both the upper and lower branches begin with carbon. The rules say, if you can't decide, keep going until you can. Stepping onto each carbon, we look around and see two hydrogens and another carbon. Still can't decide, so we take another step, onto the next carbon. Now these two are not the same. On the lower branch, the second carbon has two hydrogens and another carbon (in addition to the carbon we just came from), while on the upper branch we find one bond to hydrogen, one to carbon, and another to carbon (the double bond to carbon is treated for this purpose as if it were two single bonds to carbon). So, according to our rules, the upper branch on the left takes priority over the lower branch *even though the lower branch as a whole is bigger and heavier.*

In summary: on the right, the lower branch takes priority over the upper branch. On the left, the upper branch takes priority over the lower branch. We then determine the relationship of the high-priority branches to each other. If they are opposite each other, as in this case, we call the compound *E*. If they had been on the same side of the wall (the other wall, the one that includes both carbons of the double bond) we would have called it *Z*. So now to name our compound. We shouldn't mix *cis-trans* with *E-Z* nomenclature, so we'll use *E-Z* for everything. Here goes. Number 1 double bond does not get labeled (remember, there is no need for a label here); number 4 is *E*; number 6 is *E*. This compound is officially 8,9,9-trimethyl-4-pentyl-1,4(*E*),6(*E*)-decatriene.

The naming may seem complicated, but it is necessary to specify these things unambiguously. Trust me, it will get easier with practice.

Test Yourself 1.

Name the following compound. Chlorine as a substituent is called "chloro-," not "chloryl-."

PHYSICAL PROPERTIES

As with alkanes, there is little to say about the physical properties of alkenes, except that small ones are gases, medium ones are liquids, and big ones are solids. Because the double bonds detract from the regularity of the structures, alkenes do not pack together quite as efficiently as alkanes, and therefore they have somewhat less tendency to aggregate than alkanes of the same size. This is part of the reason why animal fats, which have lots of long, saturated alkane chains in them, tend to be solids, while vegetable oils of the same length, but containing one or more double bonds, tend to be liquids. On the other hand, if you do a magic reaction (that we will learn about later) and convert the double bonds into single bonds by adding hydrogen, you can make oils solidify. This is the basis of the production of margarine and other products referred to as "partially hydrogenated vegetable oils."

Alkenes tend to be very non-polar, and therefore they dissolve in non-polar solvents like hydrocarbons but do not dissolve in polar solvents like water.

Because the corresponding atoms are held together by two bonds rather than one, the measured length of a carbon–carbon double bond is less than that of a single bond. Single bonds tend to be about 1.54 Å (Å is the abbreviation for Angstrøm, which is 10^{-8} cm or 10^{-10} m), while double bonds tend to be closer to 1.33 Å. Although the difference is very small in terms of actual distance, it amounts to about 13% of the distance between the two atoms, which is significant.

ENERGETICS

Recall that a double bond consists of bonds of two different types: a σ bond and a π bond. The σ bond consists of an end-on interaction of two orbitals pointed directly at each other along the line connecting the two nuclei. The π bond, on the other hand, is the side-by-side overlap of two p orbitals that are actually directed perpendicular to the internuclear axis.

This is a less efficient form of overlap, and it should not surprise you to learn that a π bond is weaker than a σ bond. Can we measure this? Yes we can: we can measure the amount of energy it takes to rip apart any bond (but let's not worry about how at the moment). The actual strength of a bond varies a bit with the structure of the compound it is in, but a typical carbon–carbon, carbon–hydrogen, or hydrogen–hydrogen (there is only one kind like that!) bond is worth between 85 and 105 kcal/mol. About 100 kcal/mol is a reasonable ballpark estimate to keep in your head. A typical carbon–carbon double bond is worth something on the order of 170 kcal/mol. So we can estimate that the second (that is, the π) bond alone is worth about 70 kcal/mol, or about 2/3 to 3/4 what a single bond is worth. So it is not surprising that when alkenes react, it is the π bond that is involved in almost every case.

Advanced question: When you look at bond strengths in more detail, you find that ordinary single bonds (for example, in ethane) are closer to 90 kcal/mol, and the π bond of a typical alkene is worth about 66. This means that the σ bond of an alkene must be worth about 104 kcal/mol, substantially more than the σ bond of ethane. Why do you suppose the σ bond of ethylene is so much stronger than the σ bond of ethane? (Hint: Consider the bond length.

It turns out that not all double bonds are alike in this sense. *Trans* alkenes are usually (but not always) more stable than the corresponding *cis* alkenes; this is because in a *cis* isomer there is usually more steric crowding, meaning that the groups are more likely to bump into each other, while in a *trans* isomer the groups tend to be less in each others' way. However, as you already know, under ordinary circumstances *cis* isomers cannot simply decide to change into the corresponding *trans* isomer, so the greater stability of the *trans* isomer often has no particular consequence. Comparative stability only matters when a molecule can choose what it wants to be, and an alkene, once formed, does not have that choice. Thus, the greater stability of the *trans* isomer only has consequences with molecules that do have a choice. This situation will crop up in a few reactions that involve the *formation* of alkenes. In other words, *while an alkene is being formed* it sometimes gets to choose whether to be *cis* or *trans*, and in these circumstances it does matter that *trans* is usually better.

Another aspect of double bond stability is that, in general, double-bonded carbons are happier when their other attachments are to carbon rather than to hydrogen. In other words, there is a stability order like that shown below. We use the term "substituted" to refer to a carbon with anything other than hydrogen attached to it. Thus, the trend shown can be expressed in words by saying that the more substituted an alkene is, the better.

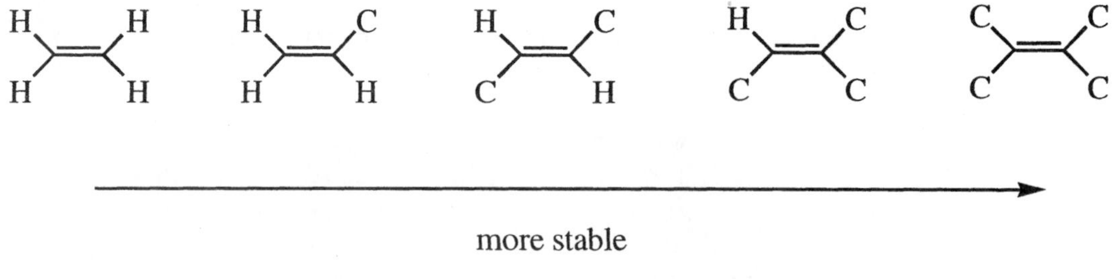

more stable

This again will become important with molecules that have a choice, which tends to be when they are in the process of being formed. Thus, these are considerations that will apply most often when we get to the chapter on the formation of alkenes. For now, we will concentrate on their reactions.

REACTIONS

Addition Reactions

Hydrogen Halides

The important feature that alkenes possess, and alkanes do not, is a π bond. This is a cloud of electrons, partly above and partly below the plane of the molecule, and the π electrons are available to any passers-by who might be particularly interested in grabbing onto electrons. Chemicals that are looking for electrons are called "electrophiles," Greek for "electron-lovers." The one chemical you have encountered so far that behaves like this is the proton, H^+. Of course, as I mentioned earlier, protons do not really float around by themselves; they are always attached to something, but if that something is eager to get rid of protons (i.e., if we have a pretty strong acid), then a proton can leave its current home and drop onto the electrons of the double bond. Consider, for example, HCl. We could show the process in a cartoon like this:

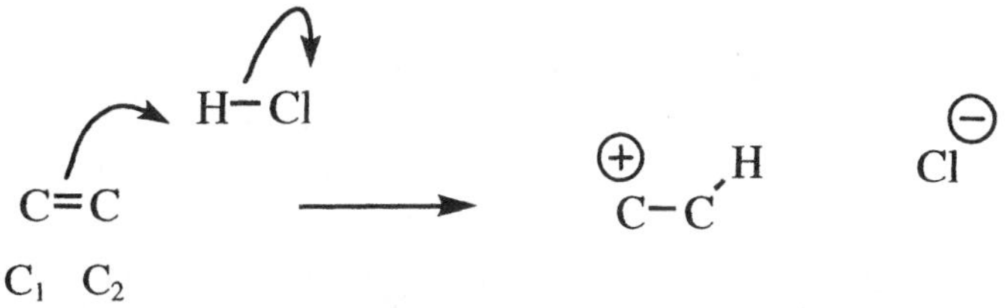

Notice immediately several things.

1. Although we said the proton drops onto the double bond, the action we show in our cartoon is *not* that of the proton moving onto the double bond; we instead show the electrons of the double bond going out to get the proton. This is because, by the convention we insisted on earlier, arrows always show the movement of electrons. An arrow must always start where the electrons are!

2. The *curvature* of the arrow also has an important meaning. In this case, the arrow curves to the right. This means that the electrons currently between C_1 and C_2 are moving to the area between C_2 and the H. They form a new bond between C_2 and the H in the new structure on the right. The electrons are being removed from C_1, and no new electrons come in to replace them, so C_1 ends up with a positive charge. It is a carbocation.

3. The curvature of the arrow is sometimes irrelevant. In the case we just discussed we could just as well have shown the arrow curving to the left, as shown below. If we did, the new bond would be between C_1 and the H rather than C_2 and the H. *In this case* that makes no difference, but in many cases it does make a difference, so usually you have to show the proper arrow to depict the proper reaction.

4. One arrow is not sufficient to show this reaction. If you showed only the arrow going from the double bond to the hydrogen, without the other arrow, you would be showing

electrons attaching to a hydrogen that already is associated with two electrons. Since you do not show those original two electrons retreating onto the chlorine, you would be claiming that the hydrogen ends up with two bonds, four electrons, and you certainly would not want to make a nonsensical statement like that!

5. This actually is a reaction you should have been pretty reluctant to draw in the first place. Why does it happen? After all, we started with two perfectly happy, neutral species, and we created a nasty carbocation and an anion. Would molecules really choose to do this? The answer is, not very often. If we mix a million of the guys on the left, we certainly cannot expect to get a million of the guys on the right. In fact, these molecules, just like you, would be reluctant to undergo such a change. But if you had millions of these guys floating around, every once in a while *one* of them *would* do this. Remember what we said about equilibrium: an equilibrium constant may be very small, but that does not mean that the reaction *never* happens. Given the zillions of molecules present in any noticeable sample, there would still be many instances of this reaction occurring, even though it is an unfavorable one, provided it is not *too* unfavorable.

Markovnikov's Rule

Now, the example we have used is obviously not a real one, because we have not specified what is attached to the carbons of the double bond. We just showed a generic reaction. Let's consider a real case. What would happen if propene were the alkene used? (Why didn't I call it 1-propene?) This immediately brings up a new question related to point number 3 on the previous page. *Which direction this time should the arrow curve?* The way to answer such a question, as with most other questions in organic chemistry (or any other science, for that matter), is to consider all the possibilities and decide which one we (or, more importantly, the molecules) think is better.

Examine the reactions above carefully. What you should notice is that the stuff on the left (what you start with) is the same in both cases. Looking at the products, both reactions make chloride ion (Cl⁻). So the only difference between the top reaction and the bottom one is the structure of the resulting carbocation. Is either carbocation here better than the other? Yes! If you recall from last chapter, 2° (secondary) carbocations are happier (lower energy) than 1° (primary) carbocations. The top reaction generates a 1° carbocation, and the bottom reaction generates a 2° carbocation. If the molecule is smart, it will choose to take the bottom route.

Molecules *are* smart. Therefore, this reaction will proceed as shown in the bottom reaction above, generating a secondary carbocation rather than a primary one. In general, when a reaction like this has a choice, it will always choose to proceed through the better (i.e., more stable) intermediate. We have a name for this preference: it is consistent with "Markovnikov's rule."

But, of course, we are still not done. We have so far taken a happy molecule, propene, and made it unhappy, by reacting it with a strong acid. The unhappy species, the carbocation, must now do something to relieve its unhappiness. As mentioned in the previous chapter, there are two ways it might proceed: it can either discard a piece with a positive charge or gain something negative. Let's consider these possibilities in turn.

Where did our carbocation come from? From propene. How did we get it? A proton was added. Where did the proton land? On a carbon, *not* the one that now has the positive charge, but the one next door. It should make sense that the molecule could reverse this process: throw out the proton that just came in, and thereby become happy again. Below is the cartoon that depicts this process:

Notice several things:

1. When trying to fix a carbocation by throwing out a proton, you need to select a proton from a carbon *adjacent* to the positive charge, so the electrons holding it to its carbon can retreat into the space between the two carbons and make a double bond.

2. Even though we are throwing out the proton, our cartoon does not explicitly show the proton going away. Instead, it shows the electrons on the chloride anion coming to get it. Again, this is a function of our strict convention that our arrows show the movement of *electrons*.

3. Both arrows are necessary. We certainly could not show the arrow on the left without the arrow on the right, because that would again lead to a hydrogen with two bonds. We *could*, and often do, show only the arrow on the right, as shown below. This picture is inaccurate, in that it implies that the proton drifts off into space with no partner, whereas in fact, there is always something present to grab it. However, in many cases, we do not know precisely what that something is, so sometimes it is easier to show the reaction *as if* the proton drifted off alone. I shall try not to do this too much, but you will see it done.

4. Unfortunately, this has not gotten us anywhere, because when we get done with the process we are right back where we started originally. What this means is *not* that such a process does not occur; indeed, it does occur. It simply means that we should probably spend little time worrying about it here, because we get nothing new out of it. It does

indicate one thing, however: the reactions I drew on the previous pages are wrong, in that the reaction arrows I drew (the straight one in the middle, not the curved ones showing the movement of electrons) were all one-directional, when in fact they should have been equilibrium arrows. The material we put in the pot can go to the right, and it can turn around and come back. That is an equilibrium, and should be shown that way.

$$Cl-H \quad \overset{H}{\underset{CH_3}{H_2C=C}} \quad \rightleftharpoons \quad \overset{\ominus}{Cl} \quad \overset{H}{\underset{H}{H-C}}-\overset{\oplus}{\underset{CH_3}{C}}\overset{H}{}$$

The possibility of a carbocation solving its problem by ejecting an adjacent proton is called "elimination"; we will study this reaction in detail in Chapter 16. You need to keep this possibility in mind, because every time a carbocation is generated, elimination is one of several means it has to solve its problem. In this particular example it took us back where we came from, because we *made* the carbocation by adding H$^+$ to an alkene, but there are other ways of making carbocations, and then elimination becomes an important option.

Since throwing out a positive charge did not get us anywhere new, let's consider the other option, which is picking up something with a negative charge. Is there such a thing around? Sure enough! The moment we created this carbocation, we simultaneously created a chloride anion. Perhaps that negative charge could help us out. We draw it this way:

$$\overset{\ominus}{Cl} \quad \overset{H}{\underset{H}{H-C}}-\overset{\oplus}{\underset{CH_3}{C}}\overset{H}{} \quad \longrightarrow \quad \overset{H}{\underset{H}{H-C}}-\overset{Cl}{\underset{CH_3}{C}}\overset{H}{}$$

You should notice several things:

1. A carbocation is always seeking electrons, which makes it an electrophile.

2. The chloride anion has electrons; it is seeking a nucleus to share them with. That makes it a "nucleophile."

3. One arrow is sufficient in this case, because the destination (the carbocation) has a space available for electrons, so when they arrive, nothing else needs to happen.

4. The product is neutral and happy, but it's not the same stuff we started with (propene and HCl). We have carried out a reaction!

Overall, this is what has occurred:

$$\equiv \quad \overset{HCl}{\longrightarrow} \quad \overset{Cl}{\diagup}$$

Now, be careful with that picture. You can easily see the new chlorine atom in the product. What you cannot see without some thought is that there is also a new hydrogen atom in the product that was not there before. There are *two* new bonds in the product, one on each of the carbons that used to be part of the double bond. This is what is called an "addition" reaction, because new stuff has been *added* without anything going away.

Overall, this is *how* it occurred (i.e., this is the mechanism for the reaction);

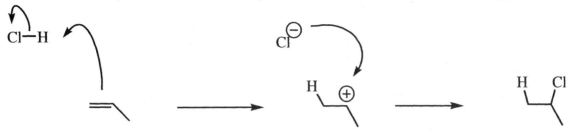

You should again notice several things:

1. The two ends of the original double bond are different. One end (the right end, as drawn) has more carbons attached to it. We call this the "more substituted" end, not because it has more attachments (both ends have two: left has two hydrogens, right has one hydrogen and one carbon) but because it has more *carbon* attachments.

2. Each end got a new partner. The less substituted end got the hydrogen, and the more substituted end got the chlorine (Markovnikov's rule).

3. We understand why number 2 is true. The reaction involves a carbocation, which is formed because hydrogen attaches first, before chlorine. The better carbocation is the more substituted one, so naturally the hydrogen attaches itself to the less substituted end, leaving the chlorine to attach to the more substituted end.

4. The reaction occurs in two steps. Halfway through there is a high-energy, unhappy species, known as an *intermediate* (in this case, a carbocation). An energy diagram (sometimes called "reaction coordinate diagram") for the reaction has to include this. Because the intermediate is high in energy, it is hard to make but eager to react further. The energy diagram shows that a high hill must be climbed making the carbocation, but only a very small one in proceeding on to the product. It also shows that the product, with two new single bonds, is lower in energy (better off) than the starting materials. This is the driving force for the reaction.

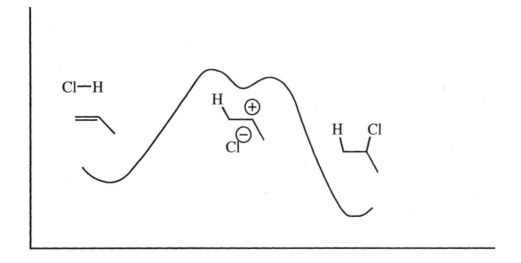

When Markovnikov proposed his rule, he actually expressed it in terms of point number 2 above (some people remember this as "the rich get richer"). Markovnikov had no idea why his rule worked, since mechanisms were not understood in his time. He simply noticed that in a large number of similar addition reactions, the less substituted end of a double bond always got the hydrogen, and the more substituted end got the other piece of the acid (the rule is not restricted to HCl). So he suggested a general observation that has come to be known as Markovnikov's rule. We can do much better than Markovnikov, because we understand *why* hydrogen goes to the less substituted end. So we do not need to memorize any rule at all, provided we simply *understand how molecules think*. Indeed, it is fair to say that my chief aim in writing this book is to teach you how molecules think. If you come away from this course able to predict reactions because you know how molecules think, we will both have succeeded!

ORM: In the program "Organic Reaction Mechanisms," check out the section under Addition Reactions called "HBr to propene" for a nice animation of the last three pages.

We really have a big advantage over Markovnikov: because we know the *reason* for his rule and don't have to memorize his statement of it, we can explain cases in which his rule (as he stated it, but not as we understand it) goes wrong. For example, in each of the following reactions hydrogen in fact goes to the more substituted, rather than the less substituted end of a double bond. These are results that would have baffled Markovnikov, but we can explain them easily. I'll work through the first one with you; then you should do the other two.

Test Yourself 2

Why do the following reactions violate Markovnikov's rule?

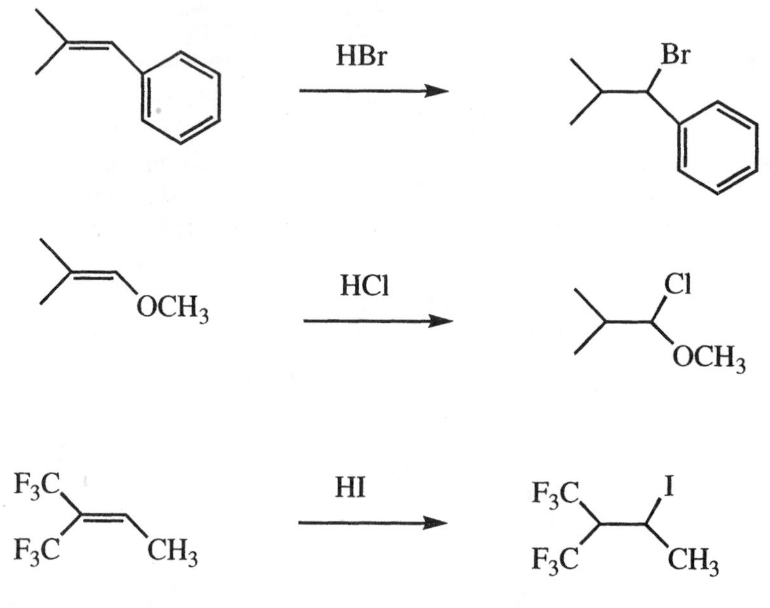

First let me clear up something that ought to be bothering you. In this first example, there are four double bonds shown. Why did the reaction occur at only one of them, and why that one? The answer (which you would have no reason to know before now) is that double bonds in a benzene ring do not behave like other double bonds. Why not? Because benzene has its own special resonance, illustrated as follows:

The "double bonds" we often show in such a ring are not really locked between any pairs of carbon atoms, but can be approximated by either of two structures, which means, of course, that benzene is neither, but something partway in between. For this reason a benzene ring is often shown with a circle in the middle, to emphasize the smeared nature of its bonds. The significant point is that these are *not* true double bonds but more like "1.5 bonds," and therefore they do not do what other double bonds do. The picture on the right emphasizes that the special nature of benzene depends on there being a p orbital on each atom of the ring, so that this cyclic "racetrack" for electrons can exist. We will spend a whole chapter on benzene chemistry later on; for now, just remember that benzene is special and does not undergo any of the reactions we are describing in this chapter.

Getting back to the question, there is really only one double bond in this molecule that is a candidate for addition, and that is where the reaction takes place. But it happens backwards, for this and also for the other two examples shown. Markovnikov's rule says that the hydrogen should go on the less substituted end, but it instead goes on the more substituted end. Why???

The answer lies in the mechanism, and our understanding of the reasons behind Markovnikov's rule. If Markovnikov had known what we know, he never would have stated the rule the way he did. He would have said, as we do, that a reaction like this proceeds *via* the best possible carbocation. Normally that is the more substituted carbocation, but in this case it must not be. Why??? The only sensible way to find out is to write down the possibilities and inspect them. Do this now, and think carefully about what you have written before looking at the answers at the end of the chapter.

Cataloging Reactions

So far we have learned one reaction. It is a very general reaction and applies to almost any double bond (except those in benzene rings). It also applies to any hydrogen halide (except HF, which you would not use anyway because it eats glass, and your container would spring a leak!). It is now time to learn how to enter a new reaction systematically in your memory banks: how to catalog it. Most likely, you will begin by working with note cards, but very soon the reactions should become permanently ensconced in your memory banks.

First, write down the reaction we learned. You'll want to write it in a general form, so that you won't be fooled into thinking that it applies only in the specific case you have written. Chemists have invented a way of writing molecules with unspecified attachments. Whenever we want to indicate the possible presence of a carbon attachment, but we do not care what it is, we use the letter "R." This stands for the word "radical" for historic reasons, but it doesn't matter. You can think of it as "random," or "rest of the molecule." Further, chemists traditionally use the letter X to indicate a random halogen, that is, chlorine, bromine, or iodine. Here is the general form of the reaction we just learned:

You should probably write this on a note card. If you are into drilling yourself, you might consider putting the starting material and the reagent on the front and the product on the back, but as we will see there may be reasons to organize your cards in other ways as well.

What we just wrote does not do this reaction justice, however. For one thing, it does not indicate that if the alkene is unsymmetrical, the reaction has a preference as to which end gets the H and which gets the X. Perhaps we should add a second line,

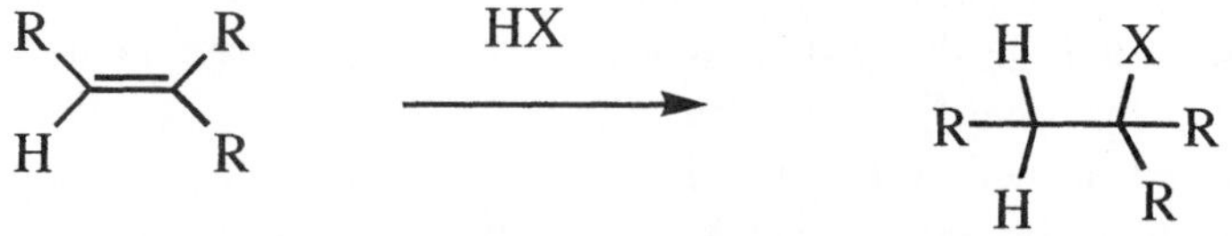

to indicate that fact. I assume that you (the reader) will recognize that there are also many other permutations in which one or more of the remaining R's could be H's.

Still, there is more. What we have written does not tell us anything about the mechanism of this reaction. We need to have that somewhere, because you surely want to remember it. Thus, it *might* be a good idea to write the whole sentence we just wrote on the front of the card, and the mechanism on the back:

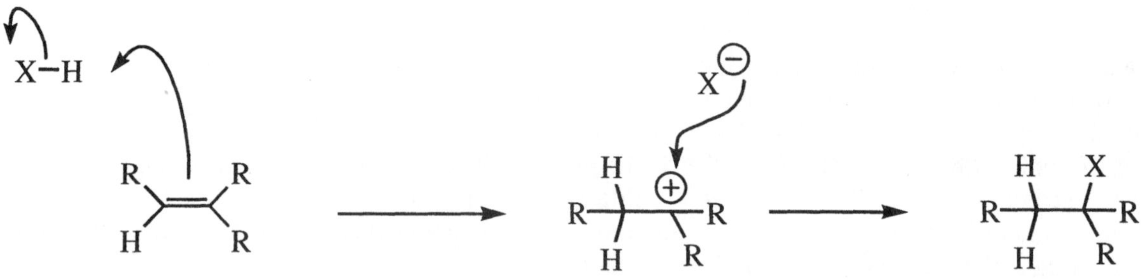

More stable carbocation:
Markovnikov's rule

Now, for the most important question of all, which you probably didn't think was even a question: where should you put this card? Obviously, this is your first card; why does it matter where you put it? I claim that it matters a great deal, because there are at least three ways you will want to remember this reaction.

1. *Reactions of alkenes.*

You should now be able to take any random alkene and show what would happen if you treated it with any of the hydrogen halide acids. For example, what are the products of the following reactions? (Don't just look at such questions, answer them!)

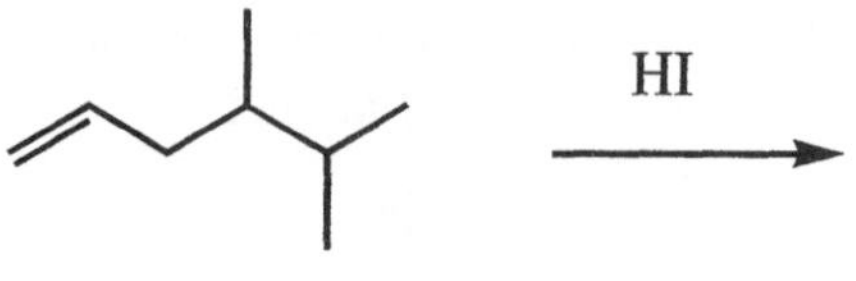

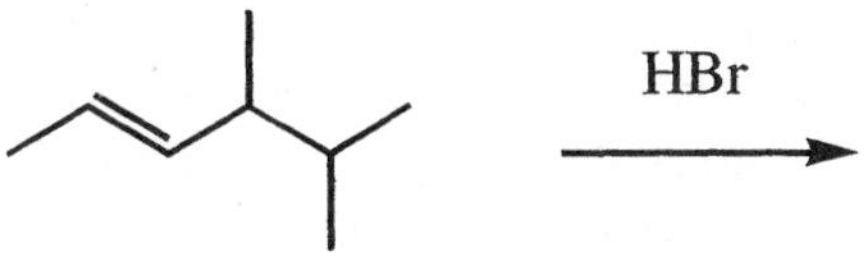

HBr
⟶
2 answers

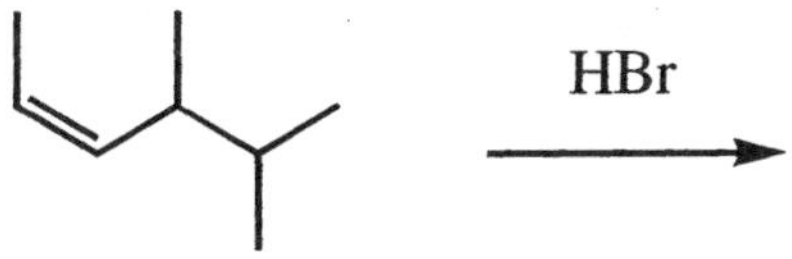

HBr
⟶
2 answers

HCl
⟶

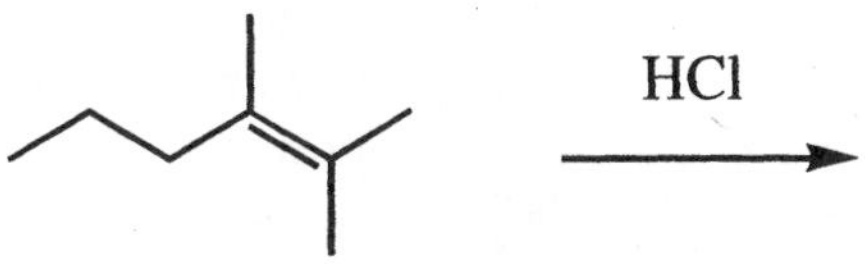

HCl
⟶
2 answers

For every new reaction you learn, you should take the time to write out several examples of it. You don't need to wait for me (or your professor) to write down sentences like the ones above. You are perfectly capable of doing that yourself. Write out three more examples of this reaction.

2. *Ways to make alkyl halides.*

You should be able to take a given alkyl halide and show what alkene you would have to use if you wanted to make it. For example, how would you make the following? NOTE: *Do not blow this off.* This is the hardest part for many students: learning to think backwards. It is the key to synthesis, the process of designing ways to make compounds, which is what most distinguishes chemistry from other sciences. It is imperative that you do these examples, and many others, over and over until it becomes easy for you to write down the answers. The ones below are easy, but you will not be able to do hard ones until you have the simple stuff down. We will be working our way up to harder problems. For now, how might you make the following?

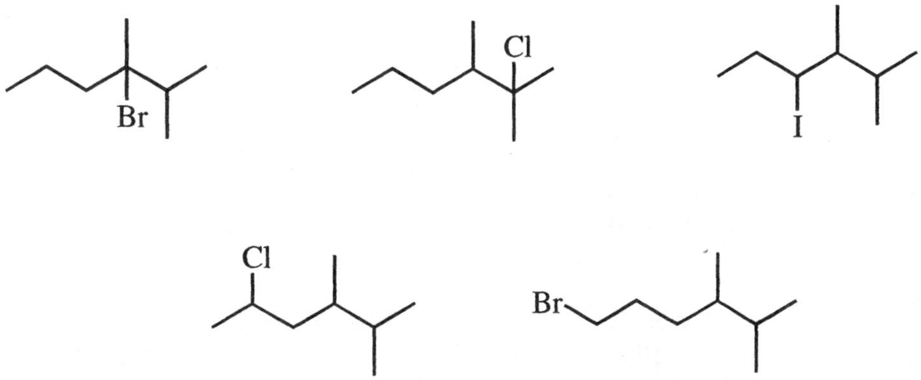

Hint: One of these compounds in fact *cannot* be made using the reaction you have learned. Which one, and why? For one of them, there are four different possible starting materials you might propose, of which three are good choices and one is not. For one of them there could be three different starting materials, of which one is a good choice. In one case there are two different possible starting materials, only one of which is a good choice. And for one there are two reasonable starting materials, but both are pretty poor. *You will not adequately understand this section until you can explain all these statements!*

Since I know this is hard for many of you, let me lead you through the first one. We are trying to make

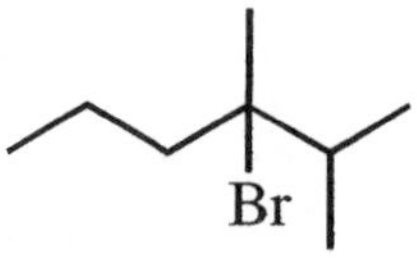

using the only reaction we know, which is the addition of HX to a double bond. I hope it is clear to you that in this case HX will be HBr. The Br in our product molecule got there by adding HBr to some double bond, so the position of the double bond *before* the HBr was added must have included that carbon that now bears the bromine. Clearly, then, the double bond must have been between that carbon, the one that now has the bromine, and one of its neighbors. In other words, we must have added HBr to one of the following three alkenes:

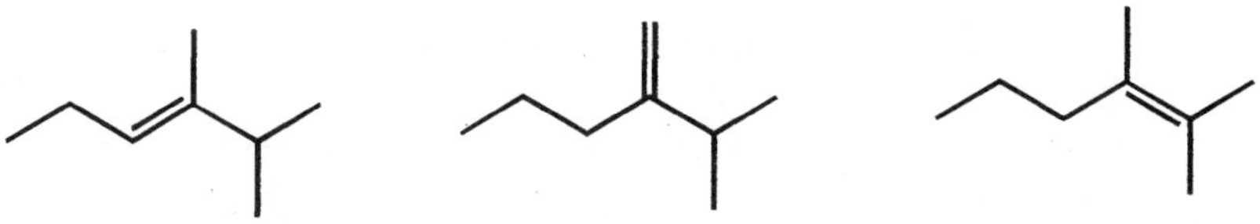

So now we have to look at these possibilities and decide whether adding HBr to them would *really* provide our desired compound. The first one has a double bond that is more substituted on the right than on the left. Markovnikov's rule tells us that when we add HBr, the hydrogen will add to the left carbon and bromine to the right. That is what we want. The middle one also has an unsymmetric double bond, more substituted on the bottom than on the top. Markovnikov says that the top carbon will get the H and the bottom will get the Br, just what we want. The third possibility also has a double bond that is different on the two ends, but both ends are equally substituted (albeit by different things). Either way we add the H, we will get a tertiary carbocation, so Markovnikov's rule cannot make a prediction in this case, and neither can we. Both carbocations should be about equally happy (or unhappy), so we should get some of each, and in this case we would expect to get a nearly equal mixture of our desired product and the following undesired product:

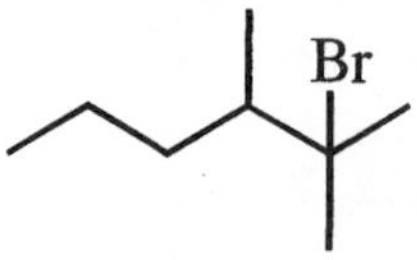

Therefore this third starting material would be a bad choice, since we would expect at best about half of the product to be what we want.

So which one is this? We have identified three possible starting materials, of which two should work well and one should not. This does not correspond to any of the possibilities I listed under the original problem. We must have missed something, and yes, we did. Look closely at the first answer we came up with, 2,3-dimethyl-3(*E*)-hexene. There is a double bond in it, which has an ethyl group, a methyl group, and an isopropyl group attached. The ethyl group on the left is *cis* to the methyl group on the right. This is a different molecule from the Z isomer, in which the ethyl group is *cis* to the isopropyl group, but both would lead to the same product, because after the reaction the double bond is not there anymore. So there are really four different starting materials we could have used for this one, and three of them would work well. None of the other alkenes we suggested as possible starting materials suffer from this ambiguity, because each of them has one end with identical substituents.

3. *Things that HX can do to a molecule.*

Why would you want to remember that? Because sometimes you will be faced with a situation where you have a sample whose structure is unknown, but you do know that it reacts with HX to make something new. If you know the kinds of things HX is capable of doing, you can narrow down the possibilities for what your unknown might be.

Therefore, I would recommend filing your card in three different places, or three different file boxes. How can you do that? Make three of them! Start a file of "Reactions of Alkenes" and put one of the cards in there. Start another file of "Ways to make Alkyl Halides" and put another in there. Make a third file of "Reactions using HX" and put the third one in there. As we learn more reactions—a *lot* more reactions—add to your files, or make new ones, and you will gradually build a nice system for learning organic chemistry.

Water Additions

The reaction we just learned is one of a family of related reactions called acid-catalyzed addition reactions. They all share certain common features: they are two-step reactions, in which the first step is addition of a proton to the double bond, thereby creating a carbocation according to Markovnikov's rule, while the second step is reaction of that carbocation with something in the neighborhood. So far that "something" has been a halide ion that accompanied the proton, but that is not necessary: any source of electrons (i.e., any nucleophile) could do the job. This is true because a carbocation is a very strong electrophile, which means that it is *really* eager to find electrons to react with, and therefore the electron source does not have to be one that is itself particularly eager to react. The carbocation will throw itself at any old nucleophile. One nucleophile commonly encountered in these situations is water.

Imagine this situation: a proton adds to an alkene, as we have discussed, producing a carbocation, but the carbocation cannot locate a halide anion to react with. Why not? Perhaps because there wasn't one—after all, there are other acids we might have used besides HX. Maybe we used H_2SO_4 in water. (HSO_4^- is a lousy nucleophile; when you see H_2SO_4 you can think of it as H^+.) Anyway, let's imagine that this poor cation is floating around desperate for electrons, and all it sees is water. What will it do? Naturally, it will drag in the lone pair from an oxygen atom in a water molecule to take care of its positive charge. Here is a cartoon that shows this:

You should notice several things:

1. I got sloppy. I told you earlier that protons do not really float around by themselves, and yet here I wrote H^+ as the troublemaker who got this whole thing started. Since this reaction is taking place in water (How do we know that? Well, there *must* be water around, because it will be involved in the next step...), the H^+ was most likely attached to a water molecule, and it was actually H_3O^+. Thus, a better picture would be the one below. Nevertheless, a picture like the one I drew first is not uncommon, and you might as well get used to seeing things described that way.

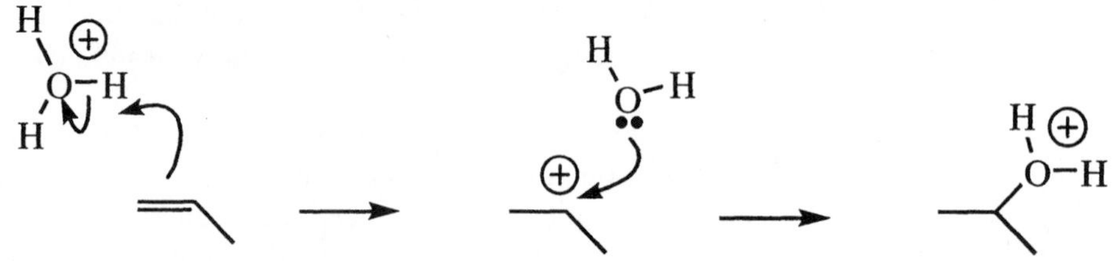

2. I only showed one lone pair on the oxygen of the water. Aren't there two? Of course there are. But the second one doesn't do anything. It's still there in the product on the right, and while it would not hurt to write it in, both in the middle structure and on the right (and on the left, too, for that matter), lone pairs that are non-participants often do not get written, if for no other reason than to reduce clutter.

3. We aren't done. Sure, our carbon atom is relieved not to be a carbocation any more, but now the oxygen is suffering from too many bonds. Recall, this is the kind of cation that cannot be helped by importing a negative charge, because there is not room for another bond. The only help for this oxygen is for it to abandon one of its current bonds, *keeping the electrons for itself.* Only two candidates exist. The oxygen can throw off the carbon it just grabbed, reverting to H_2O and a carbocation, and yes, it does do this—but that only takes us right back where we came from, so all we need to do to indicate it is to add an equilibrium arrow. Alternatively, the oxygen can throw off a proton, as shown below.

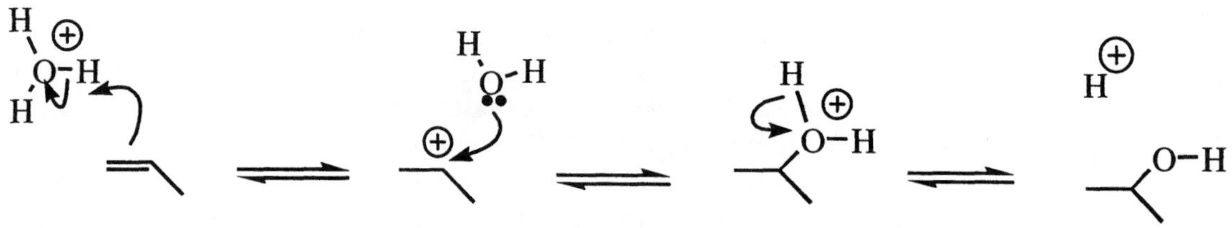

4. I got sloppy again. This time I showed H^+ drifting off by itself. I should have shown something coming to get it. Most likely, that something would be water, and I could show it as below at the left. But the "proton grabber" could also have been a second molecule of the alkene, since the present cation is just as good a source of protons as H_3O^+ is. And there are other possibilities as well. When chemists know that *some* base is coming to take a proton like this, but there are many possibilities (and, in truth, we don't really care which one it is), they sometimes use a catch-all abbreviation for a base, B:. Here the letter B stands for "base," not the element boron, which we emphasize by showing the pair of electrons that it is guaranteed to have (otherwise it could not be a base!). So we might write the process as shown below right. In fact, the sloppy

representation above is probably as good as any, even though we should all recognize it as being sloppy.

5. We are done! We have created a new, happy structure, this time an alcohol. Now it's time to create a new card:

Mechanism:

Energy diagram:

File this card under (1) reactions of alkenes, (2) ways to make alcohols, and (3) reactions of aqueous acid.

COMMON MISTAKES: DON'T LET THIS HAPPEN TO YOU!

In the last step of this reaction a proton is lost from the oxygen cation. Many students do fine up to that point, but show the last step this way:

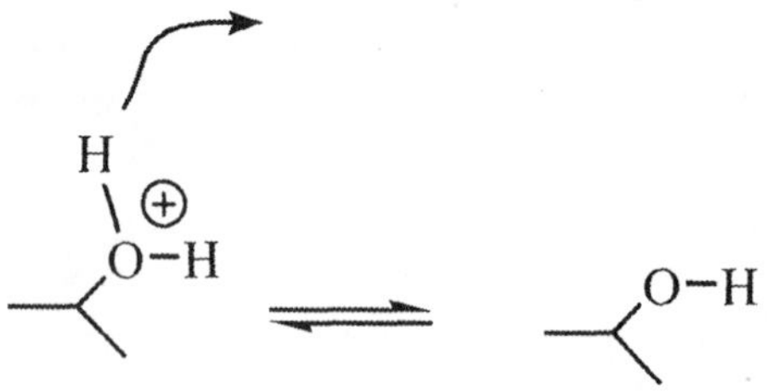

I realize that it is quite natural to want to show the proton going away like this, but it is *wrong*. Remember, arrows, by convention, show the movement of electrons. There *are* no electrons on the outside of this hydrogen atom. The only electrons around are in the bond holding it to the oxygen. The arrow has to start there. Further, these electrons are not moving away from the positive charge: that would make it even more positive. The electrons move toward the oxygen, giving the oxygen atom complete rather than partial ownership of them, and relieving its charge. That leaves the proton with no electrons to drift off. Extra good mechanisms will show a partner arriving (as above, the unnamed base B:) to shepherd the proton out of the picture.

One more thing. Remember, the bad guy that got this started was H^+ (admittedly attached to something else). At the end of the reaction, the last thing we did was to dump H^+ (admittedly onto something else). So we haven't actually used up any H^+ at all in running through the process. Whatever we use at the beginning, we regenerate at the end. The H^+ in this reaction is what is referred to as a "catalyst," something involved in the reaction, allowing it to proceed faster, but not consumed. In our study we will find that many reactions involve catalysts, and that the most common catalysts are acids and bases. Your body is also full of catalysts, called enzymes. Enzymes are much more complicated than simple acids and bases, but much of what they in fact do is provide acids and bases in the right places at the right times. In other words, to a large extent enzymes are simply specialized acid–base catalysts.

ORM: Under Addition Reactions, watch "Water to an Alkene" for a nice animation of this section.

Test Yourself 3

As before, you should now be able to reel off fairly quickly the answers to the following questions. Keep practicing until you can!

What are the products of the following reactions?

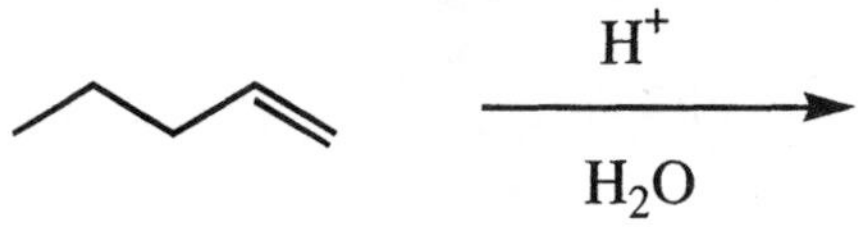

Write three more examples of this reaction.

What should you start with to make the following alcohols?

You do these the same way we did the problems in the last section, on HX additions. The OH is clearly attached to a carbon that used to be part of a double bond. It must have been double-bonded to one of its neighbors. Sometimes there can only be one choice; sometimes there is more than one possible choice. In each such case, you should check to make sure that the particular alkene you've selected will in fact make the product you want (it won't always), and also see if there are any geometric isomers available for that alkene (i.e., *cis* and *trans* isomers). (Note: One of the targets above is impossible for you at this point, two have one plausible answer, one has two answers, and one has three.)

Show the mechanism for each of your reactions—this practice is important!

COMMON MISTAKES: DON'T LET THIS HAPPEN TO YOU!

It is very common and very natural—and very wrong—to show the following:

The problem is in the second step, where you show OH⁻ as the nucleophile. This is a very natural thing to do, because after all, it is OH that you want to have attached to the carbon, so the easiest thing to do is to simply say that it is there and let it come in. But this is *wrong*, for the simple reason that OH⁻ *cannot exist* under these conditions. Remember, the carbocation we make is so hot to trot that its lifetime is essentially nil. It reacts as soon as it is formed. We cannot suggest that we make the carbocation and let it wait around for us to add something new to the pot that we want to become attached. If what we want to attach is not immediately handy, the carbocation will find something else that is. The other thing to remember is that in order to make the carbocation to begin with, we had to use a pretty strong acid. After all, the alkene didn't particularly want to pick up a proton in the first place, and only did so because there were so many of them around trying to find a home. We are carrying out this reaction under strongly acidic conditions. So whatever the carbocation will react with has to be already sitting around in strongly acidic conditions. Does it seem likely to you that much OH⁻ will be present in strong acid? Presumably you once learned that OH⁻ and H⁺ react rapidly to make H_2O. So your gut should tell you that it does not make sense for OH⁻ to be sitting around in strong acid, waiting eagerly to react with a carbocation. But even if your gut does not tell you that, your head can. Consider the following equation:

$$H_2O \quad + \quad H_2O \quad \rightleftharpoons \quad H_3O^+ \quad + \quad OH^-$$

conjugate acid ..... conjugate base

Go back to Chapter 8 and reread the section on how to use pK_a values. There you will find the following equation:

$$pK_a = pH - \log((\text{conjugate base})/(\text{conjugate acid}))$$

What this means is that "if the pH of the medium is lower than the pK_a of the acid, log ((conjugate base)/(acid)) has to be a negative number and there must be more of the conjugate acid form than the conjugate base." In our case,

$$15.7 = pH - \log((OH^-)/(H_2O))$$

If the pH were 14.7, then log $((OH^-)/(H_2O))$ would have to be –1, which means that $(OH^-)/(H_2O)$ would have to be 1/10. In other words, there would be 10 times as much H_2O as OH⁻ at pH 14.7, 100 times as much at pH 13.7, and 1000 times as much at pH 12.7, etc. Under the strongly acidic conditions necessary to protonate a double bond, we must certainly be at pH 0 or below, which means that the amount of OH⁻ available is essentially zero. The chances of our carbocation running into an OH⁻ ion is pretty much nil.

Another way of reaching the same conclusion is to remember (from high school?) that there is another constant, K_w, associate with water. This equation actually falls out of the above one if you manipulate it properly:

$$K_w = (H^+)(OH^-) = 10^{-14}$$

This equation tells you quite directly that if (H^+) is high (low pH), then (OH^-) must be very low.

The message here is that under acidic conditions, the only nucleophiles you can reasonably use are ones that are able to exist in an acidic environment. These are the conjugate bases of strong acids. Look on your pK_a chart. What are they? The halides, resonance-stabilized oxygen anions like HSO_4^- and NO_3^- (which are pretty lousy nucleophiles), and water. *Do not use OH- and other O- nucleophiles in acidic conditions!* They *cannot* be present!

Alcohol Additions

We noted previously that in the reaction shown below, there is one lone pair on the water molecule that is merely along for the ride; one that is so irrelevant, that we usually don't even bother to show it.

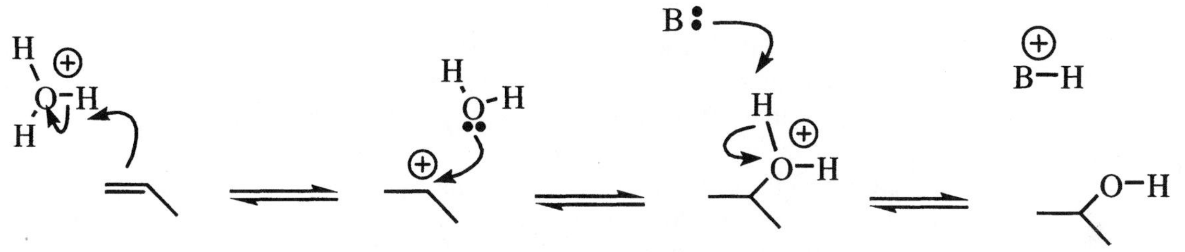

What we did not note at the time is that there is also a hydrogen atom on the water molecule that plays no role in the process. Clearly, one of the hydrogens has an important role: it has to be there in order to be thrown off at the end. But the other one, what does it do? Nothing! Does it even matter that it is hydrogen? Surely not, since it does not have any role. Therefore you should be able to guess that other molecules containing at least one lone pair and at least one hydrogen on the same atom ought to be able to undergo the same reaction. Indeed, you are right.

The most common example of this is when one of the hydrogens on the water is replaced with an alkyl group, in other words, if the reaction is carried out in alcohol solvent instead of in water. Then we can write exactly the same process, with a minor difference:

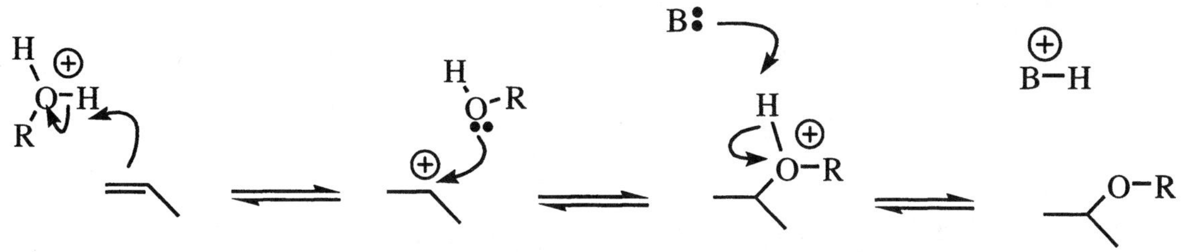

But what is this? Now we have made an ether! Voila! A new reaction.

For example,

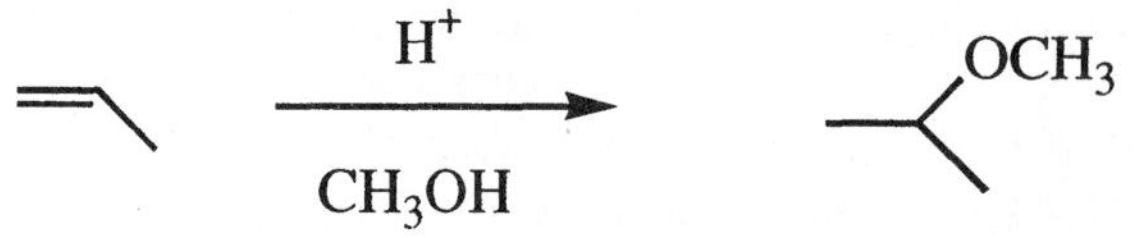

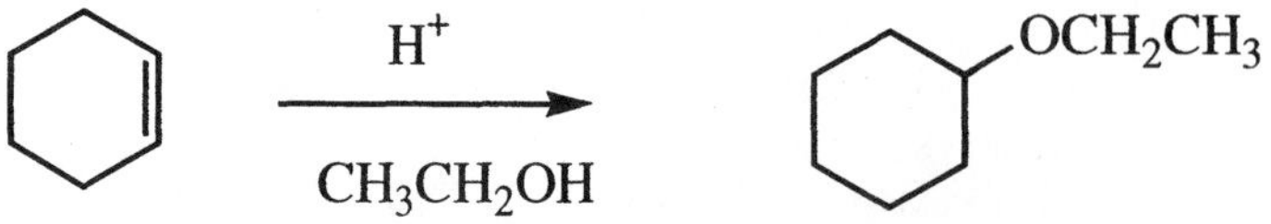

Make yourself some new cards. You have yet another reaction of alkenes, your first synthesis of ethers, and your first reaction in acidic alcohol. Write at least three examples of this reaction.

But wait! Is this really a new reaction? In a sense, all three reactions we have learned so far are variations on essentially the same thing! All you do is protonate an alkene and add to the carbocation some handy nucleophile. If that nucleophile is negatively charged, you are done. If it is neutral, the combination needs to drop a proton and the reaction is done. Big deal.

Congratulations! The sameness of these reactions is precisely what I wanted you to discover. It is what makes them easy to understand and easy to remember. All you need to know is that acidic conditions can protonate double bonds, making the best possible carbocation, and then that carbocation finds the nearest nucleophile and sucks it in, making a new bond to whatever it found.

Nevertheless, some of you will have a little difficulty with this ether business. The reason has to do with the multiple ways a given ether can be formed. Therefore, do not skim over this section, but do each problem seriously. It may cost you a little time now, but it will help you in the long run.

Test Yourself 4

What are the products of the following reactions?

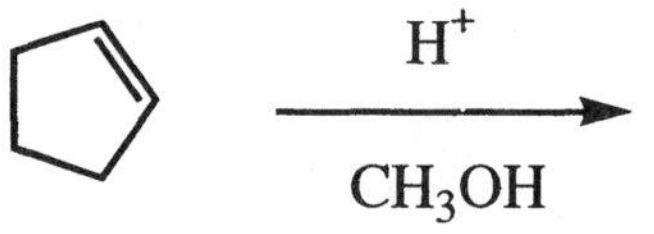

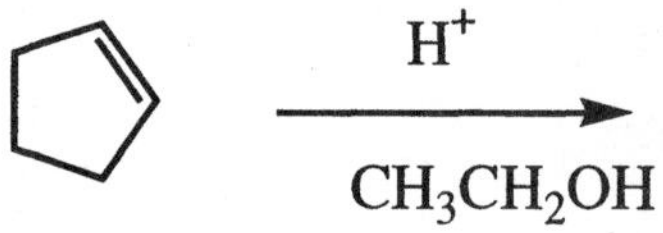

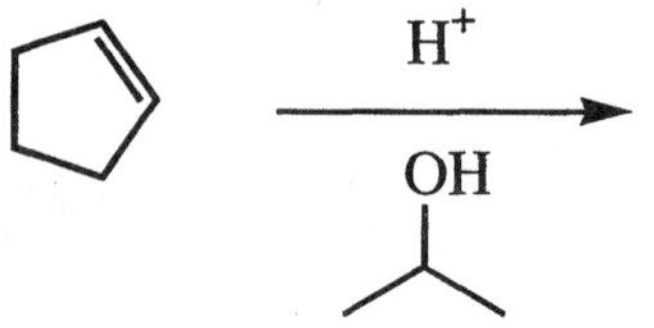

How would you make the following ethers?

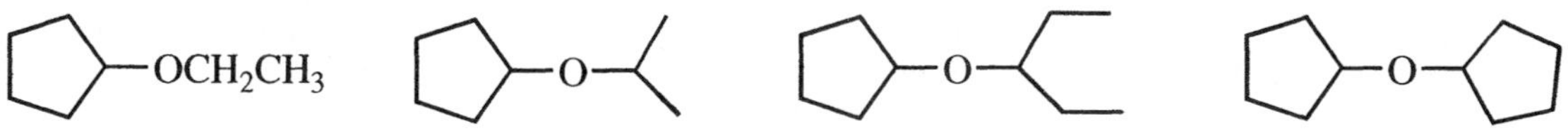

How about these?

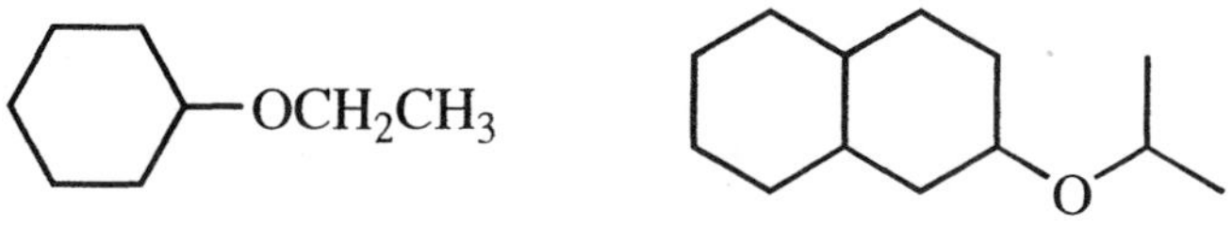

Let's take a look at that last one, because I think you may have missed something. In fact, I'll bet a dollar you did.

You have been doing these, I hope, exactly like the ones we did before when we were using HX and H$_2$O. You identified the piece that came in (boxed in the following picture) and recognized that the carbon it is attached to used to be part of a double bond.

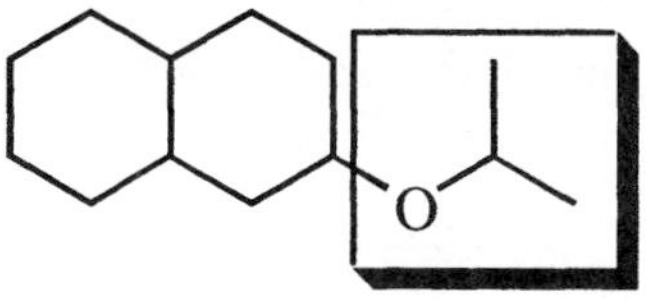

You usually have two choices for that double bond. It could have gone up from the point where the oxygen attaches, or down, meaning that the alkene you started with in this example could have been either of the following:

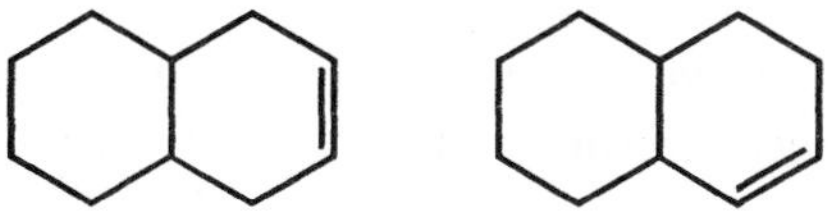

Of course, there are two ends to each of these double bonds, and they are in both cases equally substituted, so the ether could end up on either carbon of the double bond. If you used the one on the right, those two products would be the following:

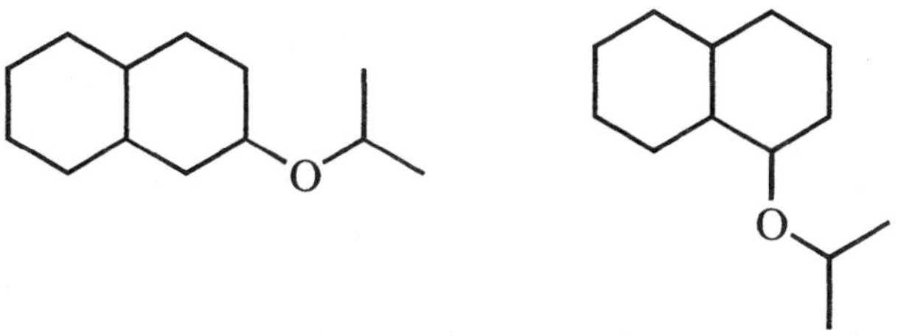

These are not the same thing, and you would get some of each, so this is not an ideal way to make the compound you want. On the other hand, if you used the one on the left, you could get either of the following:

I hope you can see that these *are* both the same thing. So clearly the alkene on the left would be the preferred starting material. What should you add to it? Acid and isopropyl alcohol:

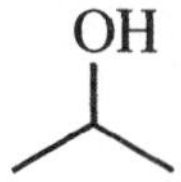

Why wouldn't you add the anion of the alcohol, the one without the proton on the oxygen? Go back and read the "Common Mistakes" section at the end of the last section!

Some of you are feeling pretty good right now, because I just bet you a dollar that you missed something, and so far you got all that. You have my congratulations, because if you got all that, you are doing pretty well but you don't have my dollar yet. Take another look at our target compound. I drew a box around the piece that came in, on the right, and we figured out that the piece on the left had to be an alkene, and we decided which alkene it should be. But unlike the previous two reactions, where we made halides and alcohols, an ether has carbon on both sides of the oxygen, and just by looking at the product there is no way to tell which side started out as the alkene and which as the alcohol. What if we drew the box another way?

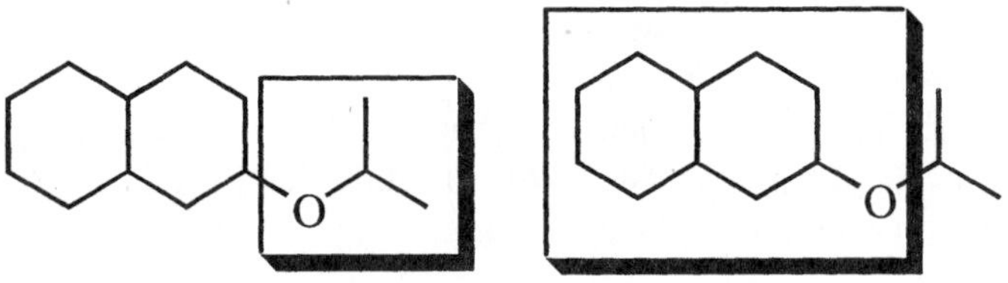

Isn't it also true that if we started with the following compounds we would also get the product we want? Show the mechanism for this.

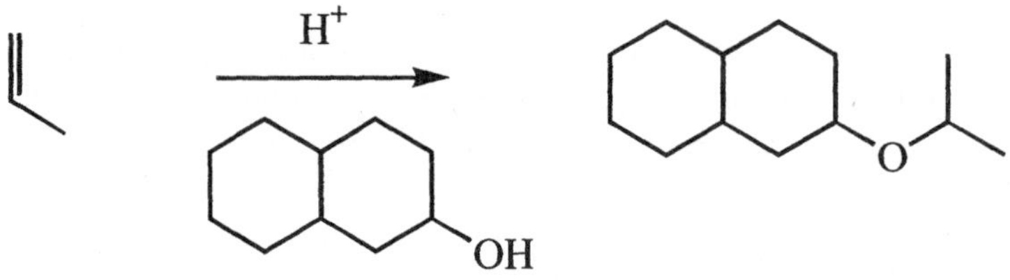

You see, the nuisance (or, depending on your attitude, the opportunity) here is that there are two sides to every ether, and either side could originally have been alkene while the other side was an alcohol. This means there are often (but not always) more ways to make an ether than you might at first imagine, and if one is not ideal often the other way will work better. Go back to the examples you did earlier and look now for more ways to make the same compounds:

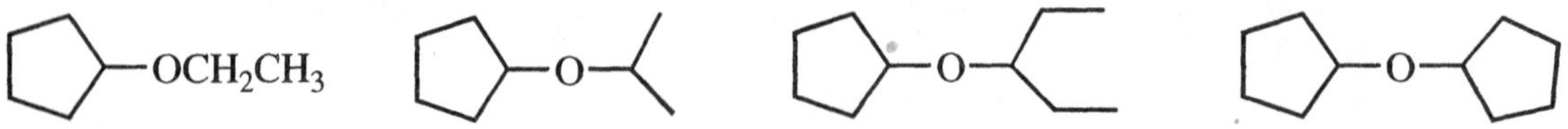

2 ways   2 ways   3 ways, but only   1 way
                                          1 good one

Feel free to send me the dollar you owe me.

Test Yourself 5

If you really understand what we have been doing in this section on addition reactions, you should be able to solve the following problems.

1. Suggest a mechanism for the following reaction. "Suggest a mechanism" means, using only the chemicals shown, put in the curved arrows that show how the starting material is converted into the product. There is no implication that this product would be the only one formed, simply that under these conditions you would get *some* of it, and you want to figure out how that might occur.

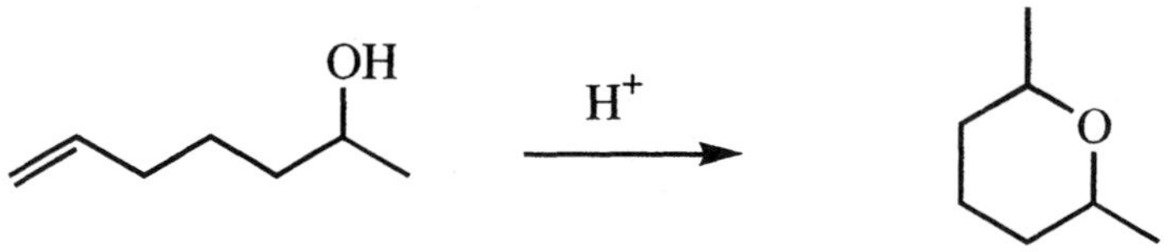

2. When propene is allowed to react with HBr, polypropylene sometimes forms as an undesired by-product, as shown in the following equation. Suggest a mechanism for the formation of this polymer.

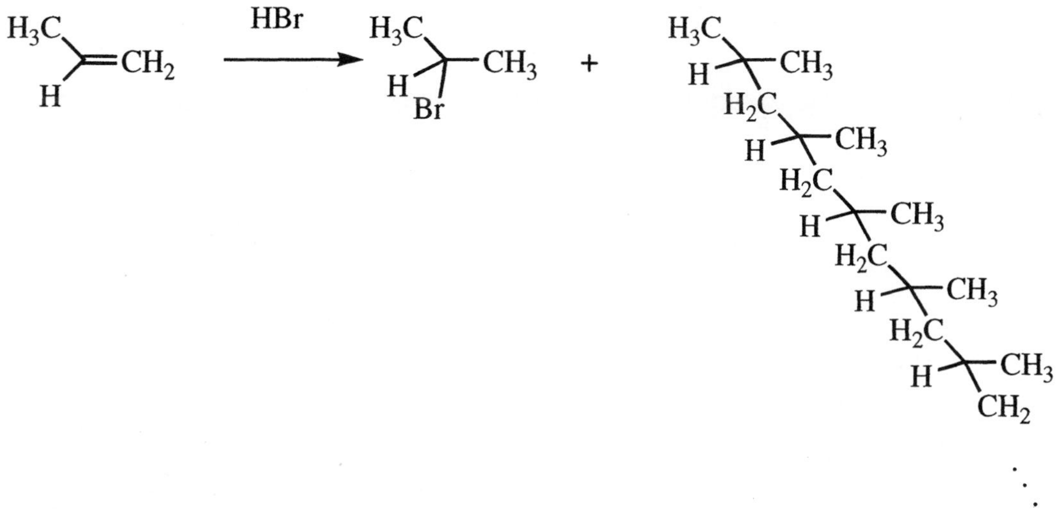

3. Explain why the alkene dihydropyran is exceptionally reactive in addition reactions, and (therefore) why the following reaction is exceptionally rapid.

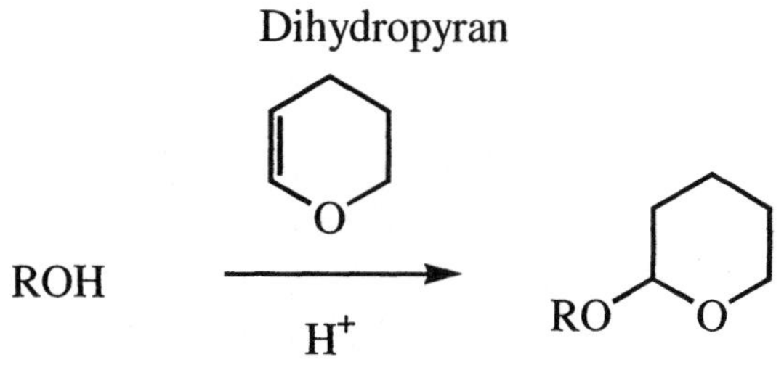

Dihydropyran

WHO CARES? YOU DO!!

Addition reactions may seem pretty far afield from anything you are interested in. But most of you probably are aware of steroids, either from the angle of athletes who take them to boost their performance, or because certain of them, like cholesterol, can be health problems if there is too much in your body. Steroids are complex, polycyclic molecules. Below is the structure of cholesterol, and other steroids are very similar, especially with respect to the size and arrangement of the rings.

Where do these things come from? Most of the steroids in your body are made right there, from small pieces, through a long sequence of reactions. After several of these reactions, a long piece results called squalene.

In the presence of the right enzyme, this compound then folds up in a very special way, and a carbocation is generated near the end of it (later we'll learn how):

Next a series of addition reactions occurs. A nearby double bond donates its electrons to the carbocation, creating a new carbocation:

This then reacts with the next double bond, which in turn reacts with the next, stitching up the steroid structure almost like a zipper.

There is more to the reaction than this, and we will actually come back to it in the next chapter, but for now you should see how the entire steroid backbone arises in one step thanks to a carefully orchestrated series of addition reactions. You should also notice that one of these additions happens backwards, in that it violates Markovnikov's rule (which one?). This is the magic of enzymes. We could never make this happen in the lab, because Markovnikov's rule would prevent it, but the enzyme holds the carbons in such a position that they have no choice but to react in this way. You may also be interested to know that chemists are getting closer and closer to being able to do such things without enzymes, by altering the starting structure in such a way as to make the desired reaction more favorable than undesired ones. You may learn more about such non-enzymic steroid cyclizations in later chemistry courses.

Synthesis

In a sense, the problems you have been doing that say "How would you make the following ethers," for example, can be described as synthesis problems; synthesis being the process whereby you make something from other stuff. However, these are the simplest possible synthesis problems, in that the answer is some combination of chemicals you would mix to make the target compound in one step. *Real* synthesis problems are never that easy. For example, one might ask: how could you make the following ether *using only alkenes as your organic starting materials?*

Many students have incredible trouble with problems of this type, so let me try to make several things clear.

1. You are not being asked to mix particular alkenes with some other chemicals in the hope that the desired product will just pop right out. Yes, you are to make the compound shown, and yes, you are to start with alkenes, but there is nothing in the problem that says that you have to do it all in one step. This is like, perhaps, being asked to build a house using plywood sheets as your only source of lumber. This does not mean that you can't have any 2 x 4's in your house. It simply means that if you want some 2 x 4's, you have to make them from plywood. So be it: you can do that!

2. The worst thing you could possibly do, when confronted with a problem of this type, is write down the structure of some starting alkenes. To choose a different analogy, this problem is like being asked to make a cake using only things that are already in your kitchen—no shopping allowed. You wouldn't just take random things out of your cupboard and start mixing and praying! You would look up recipes for cakes, check to see what you need to make them, and establish whether these things are in your kitchen. Or, using the wood analogy, you do not start randomly nailing pieces of plywood together. You look at the plans for the house, and see what kinds of pieces of wood you should use, and then see if you can make what you need out of plywood sheets. In other words, you start the *thinking* process at the final product (the cake or the house) and see what components go into making it. You work backwards until you have a recipe/plan that starts with materials you are permitted (or have available) to use. Then you start cooking/nailing.

3. You do not need to consider how to turn this molecule into the starting materials requested. Think about the cake again: if I asked you to make a cake using only ingredients that are in your kitchen, you need to think about how to make a cake, what ingredients you need to do so, and check whether you have those ingredients or whether you could make some equivalent from what you do have (for example, did you know that if you have no brown sugar, you can make it by combining regular sugar with molasses?). But at no time in this process do you ever give a thought to how to turn a cake into flour and sugar, do you? By the same token, you should not waste your time thinking about how to turn the target molecule shown into something else. That is not your task. Your job is to turn something else into that target, and to identify what that something else is.

I cannot emphasize this enough. Even Nobel Prize-winning chemists do not look at a molecule and write down starting materials they would use to make it. They consider: what would I ideally mix to make this compound, and then, what would I need to mix together to make *those* materials, and they continue until the answer is something they actually have. The thinking process is backwards. This takes some getting used to.

So we work backwards. Consider the target I proposed: what would we need to mix to make this stuff? Hopefully you have practiced enough already to be able to write down four different pairs of compounds that, when mixed with acid, would give this structure:

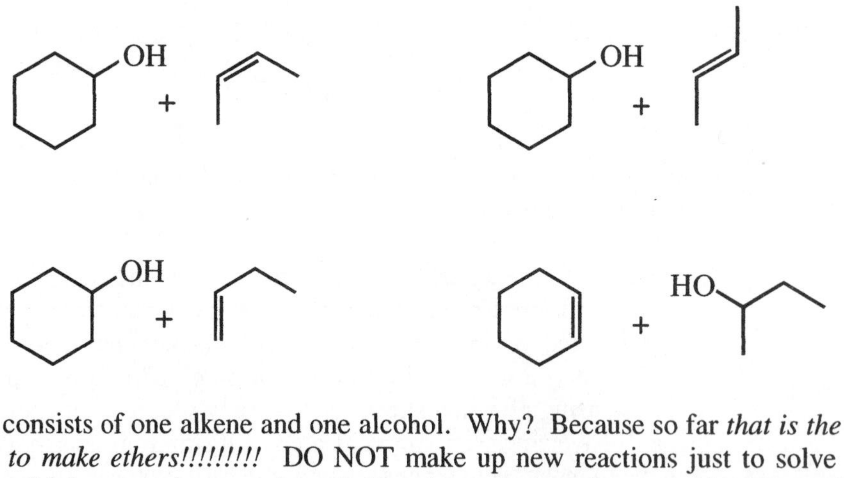

Each pair consists of one alkene and one alcohol. Why? Because so far *that is the only way you know to make ethers!!!!!!!!!!* DO NOT make up new reactions just to solve synthesis problems! Stick to reactions you know! You do not know of any reaction that makes ethers directly from alkenes only. Tough luck. Use the (only) reaction you *do* know, and use it right. Any of the above pairs will serve. This is no different from what we have done in previous exercises. The difference is, in previous exercises we were finished at this point, and this time we are not. In each case, one of the compounds we wrote down is, indeed, an alkene, and that is fine, but the other compound is not. So let us focus in on that other compound and see if we can make *that* from an alkene. Sure enough, this, too, is just like some of our previous examples. Indeed, all synthesis problems require nothing more than simple, known reactions strung together in some sensible order. So one possible answer to our problem would be the following:

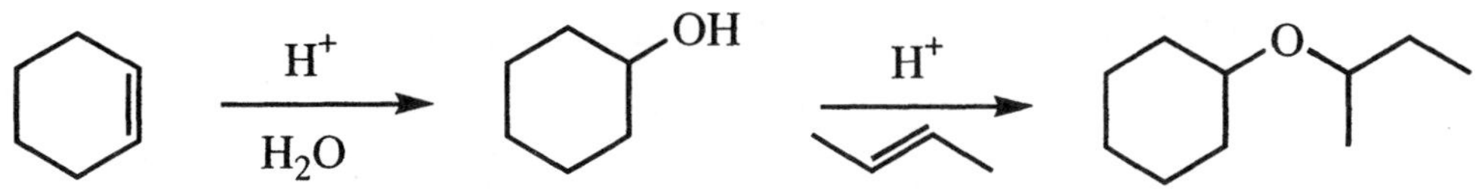

There is one important point about this answer that I would like to impress upon you. Because you are reading it as a line of text in a book, naturally you read from left to right. When professionals write the answer to a problem like this, they do *not* start at the left and write. Just as their thinking starts at the product, so does their writing. First, our professional would write down the structure of the product, then what is needed to make this, and then, if any of the materials fails to qualify under our ground rules, what you need to make *them*. Thus, the answer would take shape more like this, reading from top to bottom:

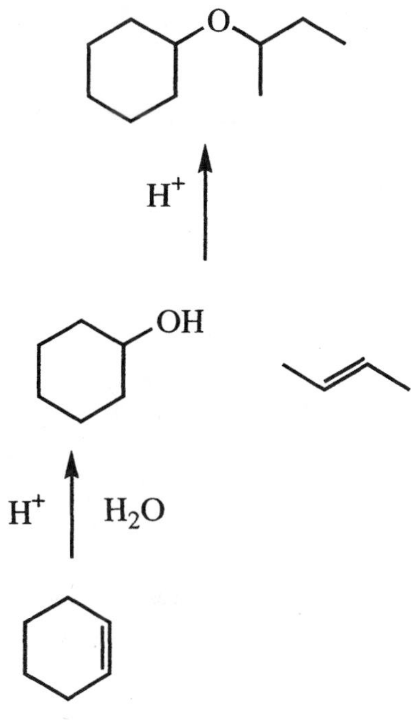

Most real syntheses, of course, are more than two steps long, and some of the best minds in the world spend their time dreaming up long, complicated syntheses of big, complicated molecules. It is an extremely challenging sport, and loads of fun.

WHO CARES? YOU DO!!

Why am I bothering you with synthesis? After all, few of you will actually end up in careers
that require you to synthesize new molecules, but all of you will use the fruits of synthesis.
One important arena in which synthesis plays a major role is the pharmaceutical field.
Virtually all the drugs we take to cure diseases are made by chemical synthesis, and the need

for new drugs will never end, since bacteria are constantly developing immunities to the drugs we have, forcing scientists in turn to develop ever more sophisticated drugs. Although many new drugs are discovered in the natural world, once a promising lead is discovered the need always arises for large quantities of material to test, and, if the tests are successful, even larger quantities to sell. Large-scale syntheses of drugs are designed exactly the way you will learn in this course, although the list of available reactions includes methods we will not cover. Most important, the difficulties and limitations of real commercial syntheses are the same as those you will encounter here. Finally, viewed as a mental exercise, synthesis develops the habit of looking at a problem from more than one starting point, a very useful frame of mind to have.

Problems

(Answers are provided at the end of the chapter for italicized problems.)

1. Show the preferred products of the following reactions (one each). Show the mechanisms for all of them.

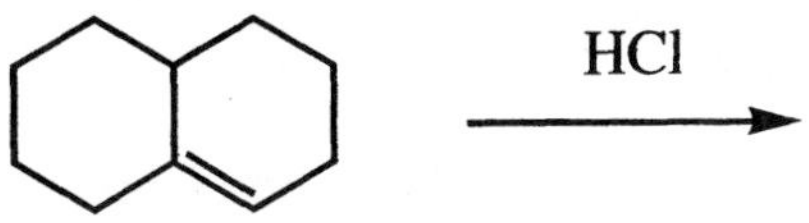

HCl

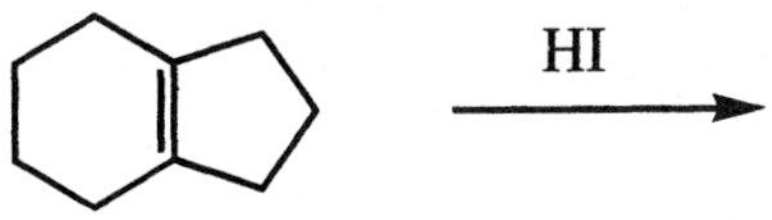

HI

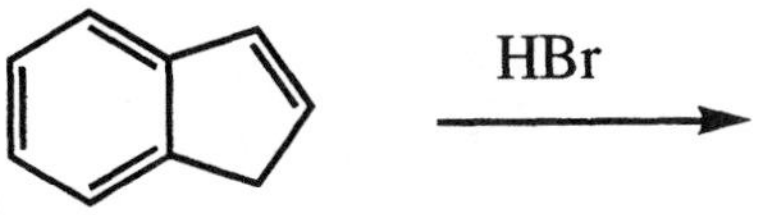

HBr

2. Show what starting materials would be necessary to prepare the following halides.

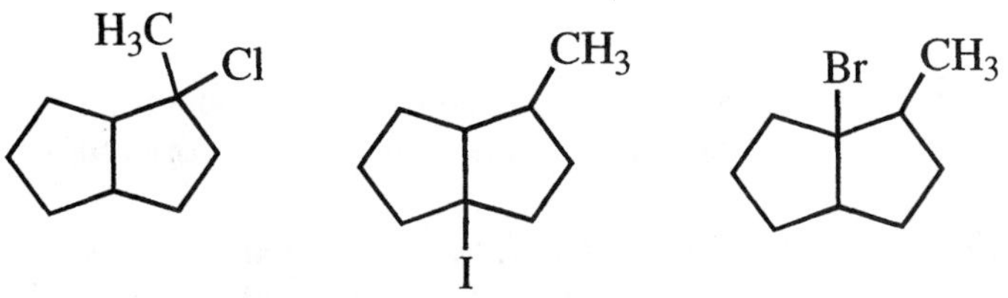

3. When 2-hexene reacts with HBr, two products result, 2-bromohexane and 3-bromohexane. When 3-hexene reacts with HBr under the same conditions, only one product is observed. Why?

4. *When 1,3-butadiene is treated with HBr, two major products are found: 3-bromo-1-butene and 1-bromo-2-butene. Draw the appropriate structures and show a mechanism that explains why both products appear.*

5. Show the mechanism of the following reaction. Show how both products would be formed and indicate which would not be observed and why.

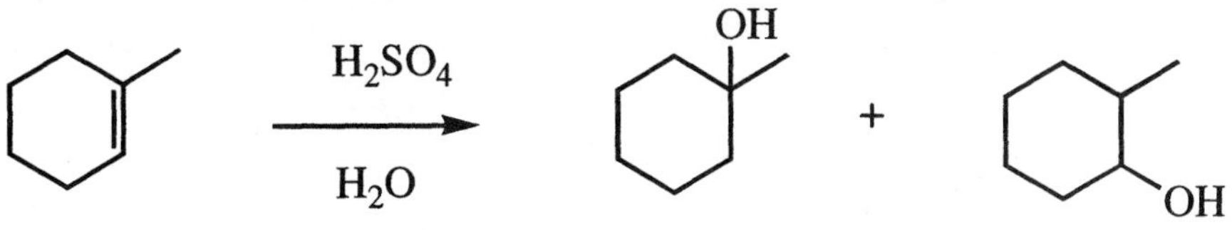

H₂SO₄

H₂O

OH

+

OH

(Don't be thrown by the H₂SO₄ in this equation. It is simply a chemical one might use to make a solution acidic. If it makes you feel better, you can replace "H₂SO₄" with "H⁺.")

6. *Enamines such as the one shown below react exceptionally rapidly with acidic water. Explain why.*

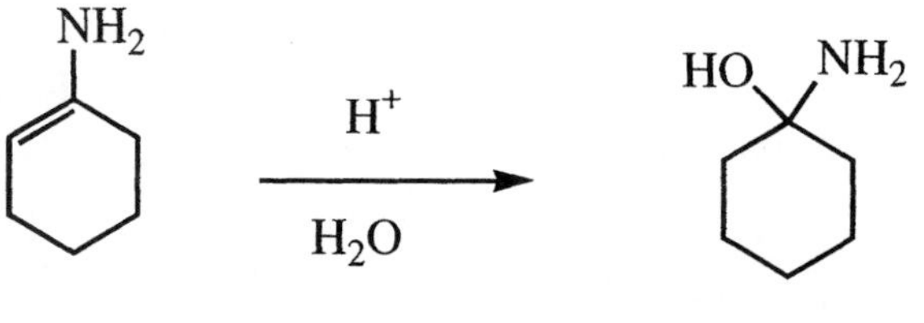

7. The following compound contains an alk*ene* with an alcoh*ol* attached and is called an *ene-ol* or enol. Suggest two different possible neutral products, which are not enols, from the reaction of an enol with aqueous acid. Show the mechanism for the formation of each. *Hint*: pay careful attention to resonance. Carbocations that are directly on carbon–carbon double bonds ("vinyl cations") are generally not very good.

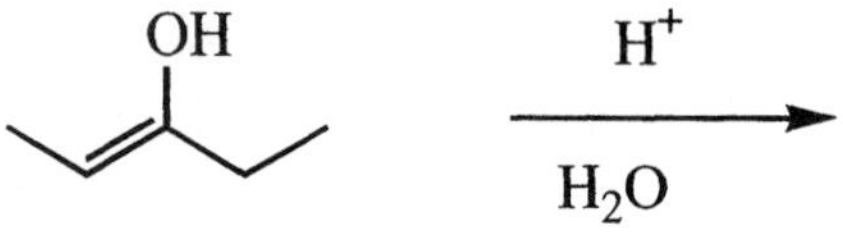

8. a. When propene reacts with acidic water, 2-propanol is obtained. Show a mechanism for this process.

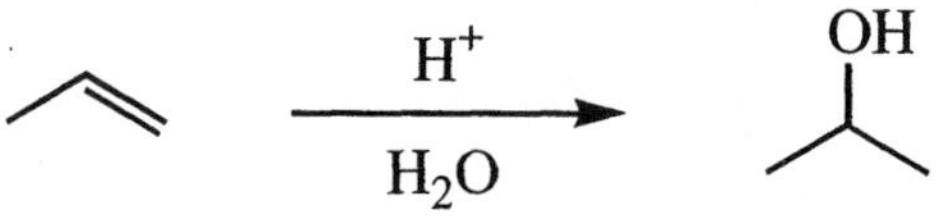

   b. If there is more propene than water, one can obtain an ether in the above reaction. Identify the ether, and show a mechanism for its formation.

9. *The following alkene is exceptionally good at making ethers. Show the mechanism of the acid-catalyzed reaction of this alkene with ethanol. Be sure to specify where the reaction will take place, and explain why this alkene is better than others at this reaction.*

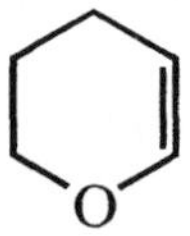

10. Answer the question from problem 9 about this alkene.

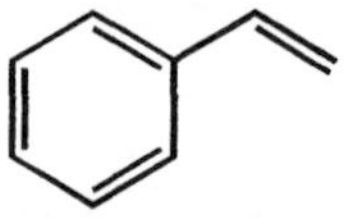

11. Suggest a mechanism for the following reaction.

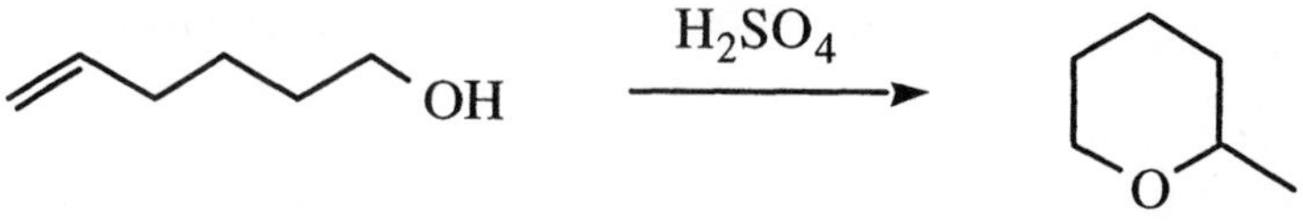

12. *Show a mechanism for the following reaction:*

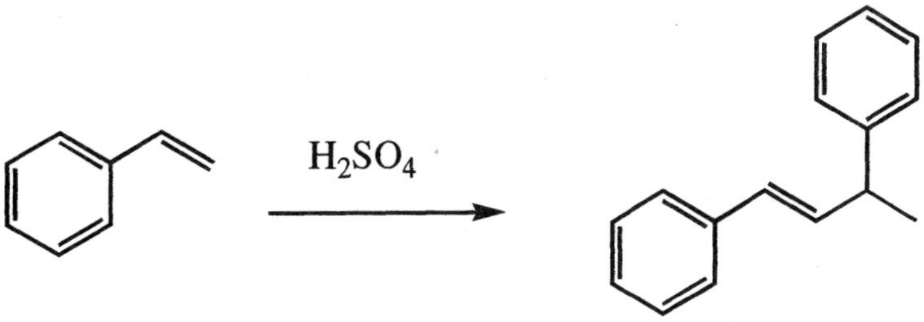

13. Suggest two different pairs of starting materials that would lead to the following product. Show a mechanism for the formation of the product in each case.

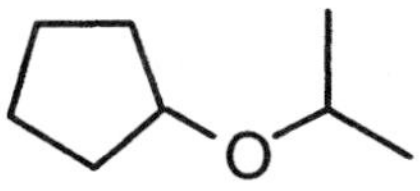

14. Show three different pairs of starting materials that would lead to the following product.

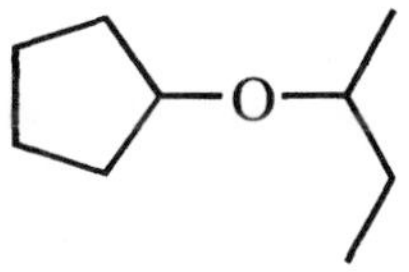

15. Show four different routes for forming the following ether. Which of them would be the best choice for this synthesis? Why? For the other three, why would they not be such a good choice?

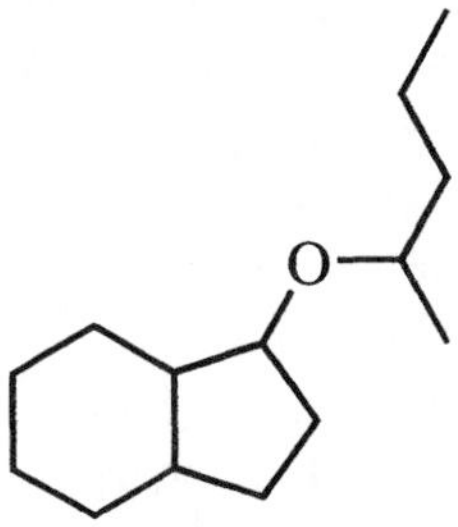

16. *Explain why you could not make the following ether using reactions you know:*

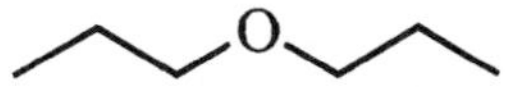

17. Compound **A** is both an alcohol and an alkene. Upon treatment with acid, both of the following products are obtained. Suggest a structure for **A**. There are two reasonable answers.

Selected Answers

Internal Questions

Advanced question, p 219: A double bond is considerably shorter than a single bond. This means that the two orbitals overlapping to make the σ bond overlap *more* in an alkene than in an alkane. More overlap leads to more (i.e., stronger) bonding.

Test Yourself 1.

3,6-dichloro-5-propyl-2(Z),5(E)-nonadiene

Test Yourself 2.

Your gut tells you the top carbocation should be better than the bottom one, since the top one is tertiary and the bottom one is secondary. But the facts tell you that the molecule likes the bottom process better than the top one, and the molecule is not stupid, so there must be something the molecule knows that you don't see. Either the top carbocation is not as good as you think it is, or the bottom one is better than you think it is. Let's look at them carefully and see if we can find something special associated with either. Eventually you should come to recognize that there is something special about the bottom cation. What we have drawn is not a complete description of this cation. The positive charge can be helped by other electrons in the molecule, and this can happen without any atoms moving. What we have here is a resonance-stabilized carbocation. Indeed, there are four locations that can acquire a share in this positive charge. No wonder the molecule prefers to form this carbocation rather than the top one!

Since this carbocation is better (more stable) than the top one, this is the one that forms. The bromide anion then comes in to take care of the positive charge where it finds it, namely, on the less-substituted of the two original double-bonded carbons. Why doesn't the bromide anion attach at any of the other places we showed sharing this positive charge? If it did, that carbon would become sp³-hybridized and no longer have a p orbital with which to communicate with its neighbors. The benzene ring would have lost its "benzene-ness," which is what makes it so special. Benzene is very reluctant to lose its special cyclic communication. It is happy to help out a positive charge next to it, as we showed in the above picture, as long as all the atoms of the ring remain sp²-hybridized. But if bromide were to react at any position other than where it did, the product would no longer contain a benzene ring. Benzene is very selfish of its benzene-ness and insists that bromide stay away.

Many students look at the above structures and claim that in all but one of them, the benzene has lost its "benzene-ness" already. This is not true. Recall that the specialness of benzene requires a racetrack of electrons comprised of the six p orbitals of the ring. Any of the structures shown above, if drawn out, looks like the following (this is, of course, necessary, because the molecule has a shape, and the four structures we drew are only our poor attempt to represent it using the "bonds" we are used to drawing). This retains the racetrack, so the specialness of benzene is conserved, even though the individual resonance structures do not all show three double bonds in the ring. But if the bromide reacted with this cation in the wrong place, the racetrack would be gone.

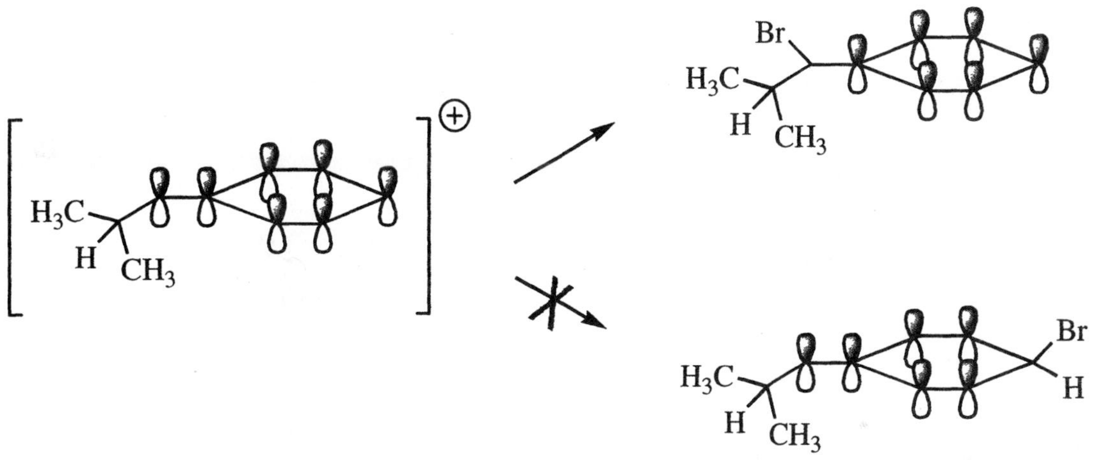

Similarly, in the second example, the "wrong" carbocation is more stable than the "right" one due to resonance:

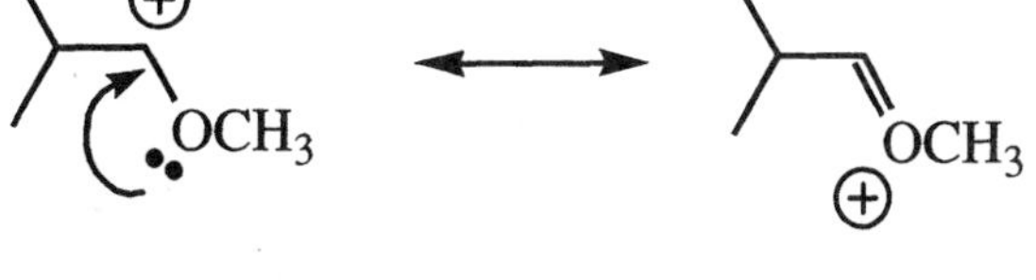

The third example is inverted from the first two, in the sense that instead of the "wrong" carbocation being *more* stable than expected, the "right" one is *less* stable than expected:

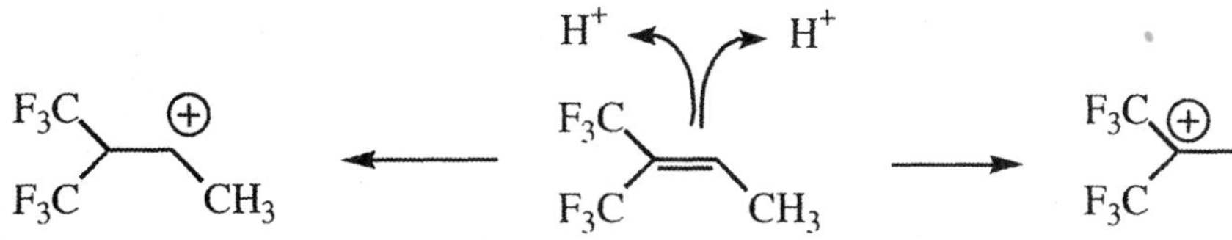

The right one is tertiary while the left one is secondary. Tertiary is *normally* better than secondary because alkyl groups are electron donating. Are these? Each carbon has three electron-hogging fluorines on it, making these carbons themselves electron withdrawing. So these *particular* carbons do *not* help a positive charge next door, they actually hurt it. Thus the secondary carbon is better off.

Reactions of Alkenes, p 227:

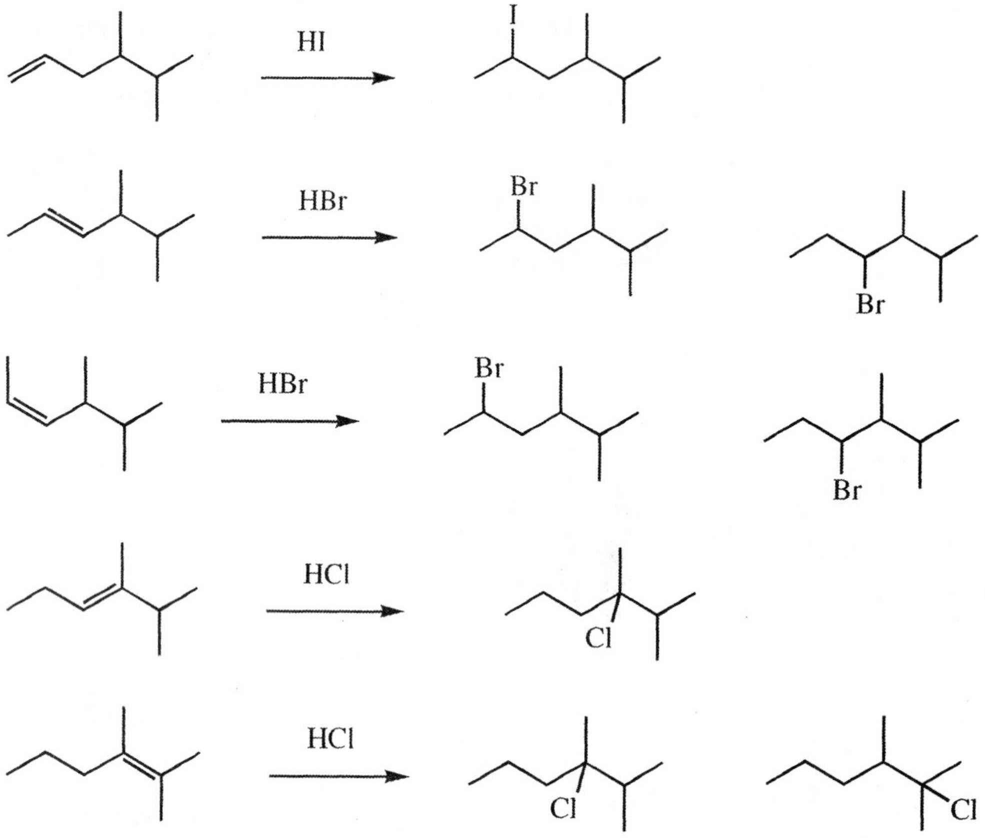

Ways to make alkyl halides:

bad

good

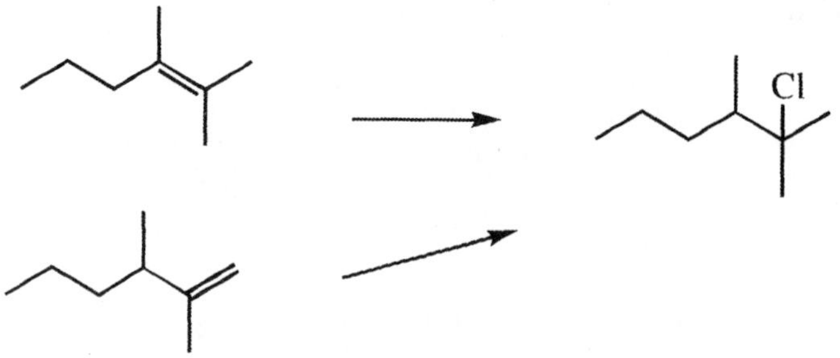

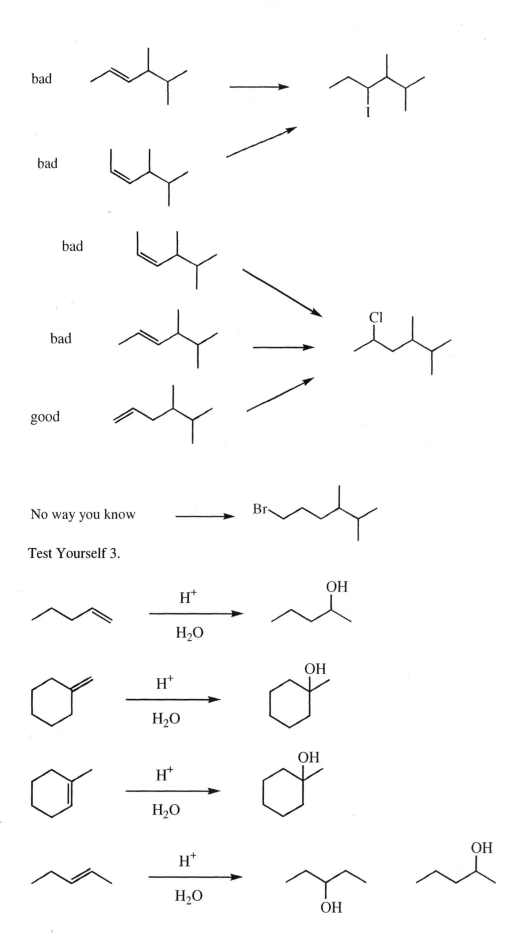

bad

bad

bad

bad

good

No way you know

Test Yourself 3.

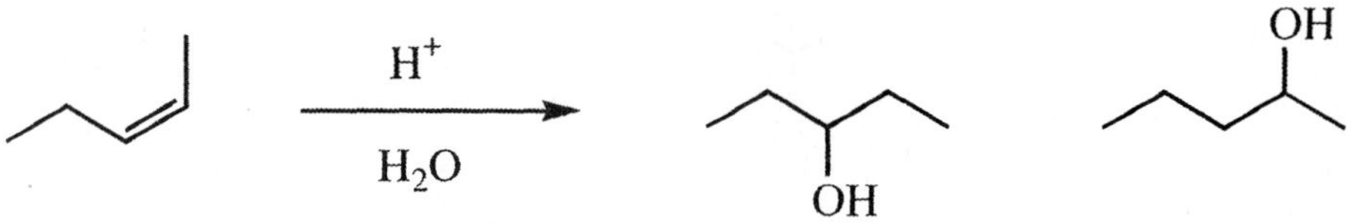

What should you start with?

good

good

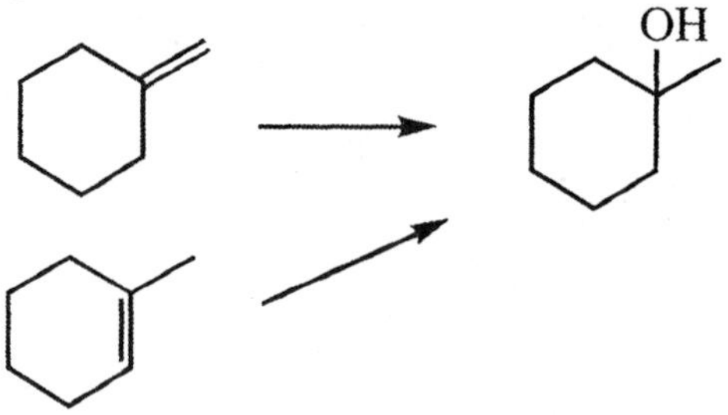

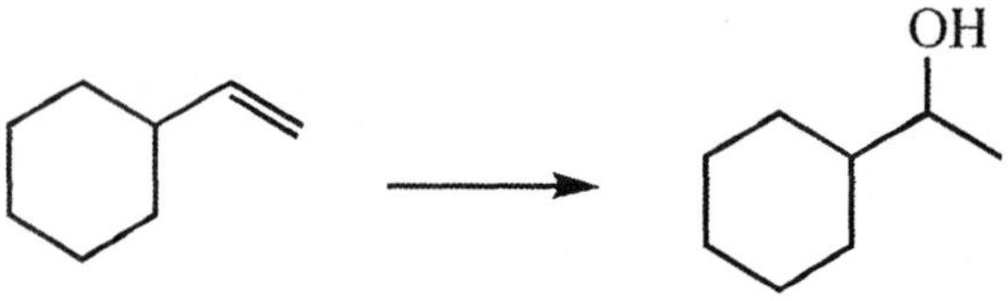

No way that you know

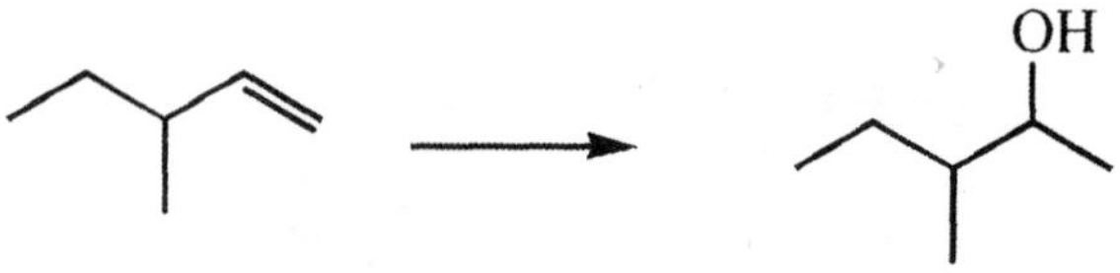

good

good

good

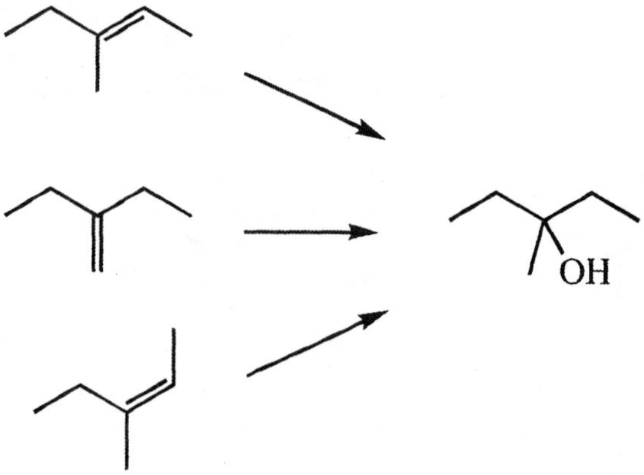

Test Yourself 4.

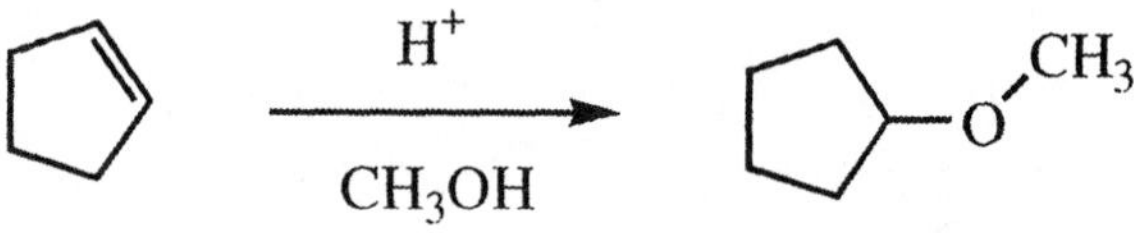

How would you make the following?

How about these?

Also:

but both of these are bad

Test Yourself 5.
1.

2.

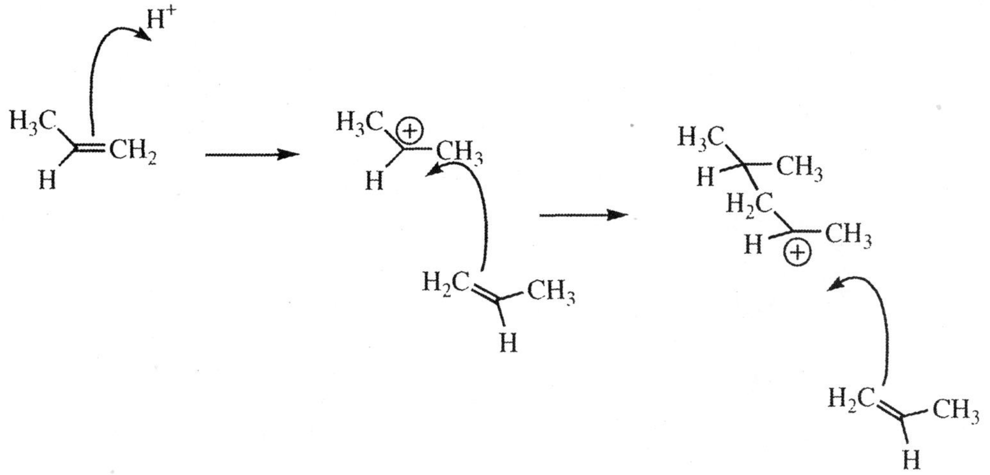

H^+

H_3C — CH_2

$H_3C^{(+)}$ — CH_3 H

H_2C — CH_3 H

H_3C CH_3
H
H_2C
H $CH_3^{(+)}$

H_2C — CH_3 H

etc.

3.

H^+

ROH

resonance-stabilized cation
is easy to form

RO — H

RO —

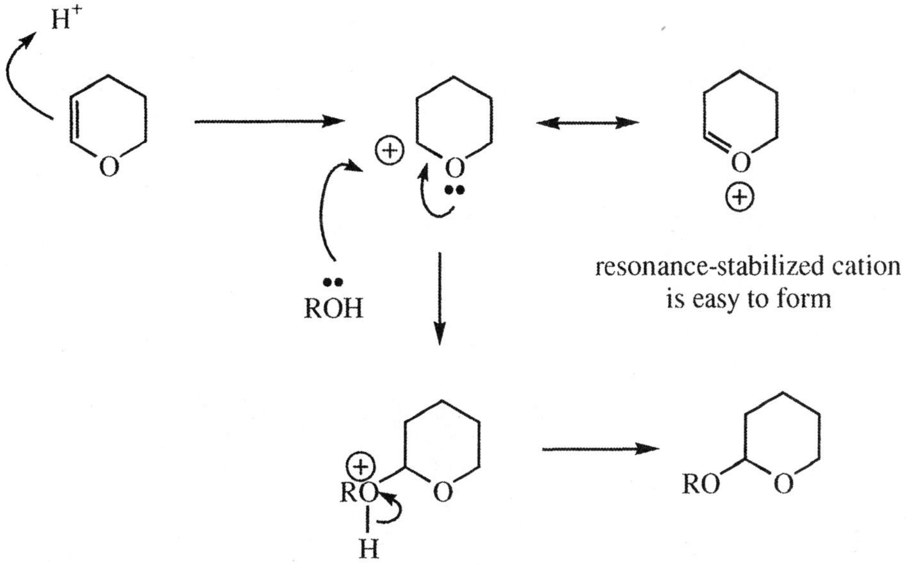

4.

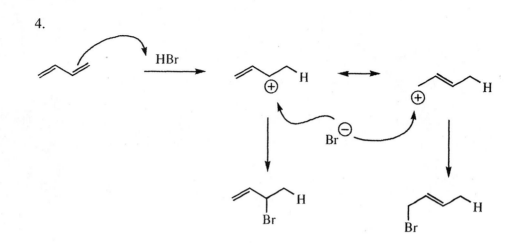

6.

9.

resonance stabilized cation
is easier to make, therefore this
reaction is faster than others

EtOH

12.

3 more resonance structures

16.

We do not know how to make this carbocation!

The only way we know to make carbocations is by protonating double bonds, and you can never get a cation on the end of a chain that way
(unless it is only two carbons long!)

Chapter 11
Alkenes II

In the last chapter we learned the basics of addition reactions to double bonds. In this chapter we will

- Learn some of the complications involved in addition reactions—specifically, rearrangements

- Learn about addition reactions initiated by electrophiles other than the proton—specifically, bromination

- Learn about epoxidation, ozonolysis, and hydrogenation.

REARRANGEMENTS

In Chapter 9 I told you that there are basically two ways a cation can take care of its problem: it can drop a plus charge or bring in a minus charge. We have now seen examples of both. In the addition of HX, the carbocation formed brings in an anion, and everyone ends up happy. In the addition of water and alcohols, the carbocation formed initially attracts a pair of electrons associated with a neutral species, which then acquires a plus charge and itself dumps a proton. In all three reactions, the initially formed carbocation also had the option of ejecting a proton from an adjacent carbon, but that leads back to the starting point, so we did not look at it carefully (we will soon).

There is one other option available to carbocations, and we need to spend some time considering it. Certain carbocations are in situations such that they can become somewhat better off *without* losing a positive charge. They will still be very unhappy and highly reactive, but not *quite* as unhappy as they were before. What they do is reorganize their bonds in order to transform themselves from one cation into another. The process is called "rearrangement." This makes a total of three things that cations can do: grab a nucleophile, eliminate a proton, or rearrange. *Remember these three things!!!*

Let's take a close look at an example of a rearrangement. Consider the following reaction:

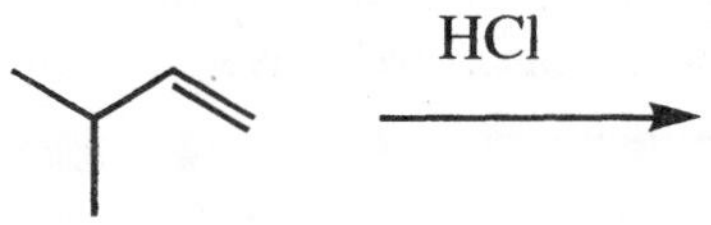

The first thing that happens, as you should expect, is that a proton adds to the double bond, creating a carbocation, which waits around for something with which it can react:

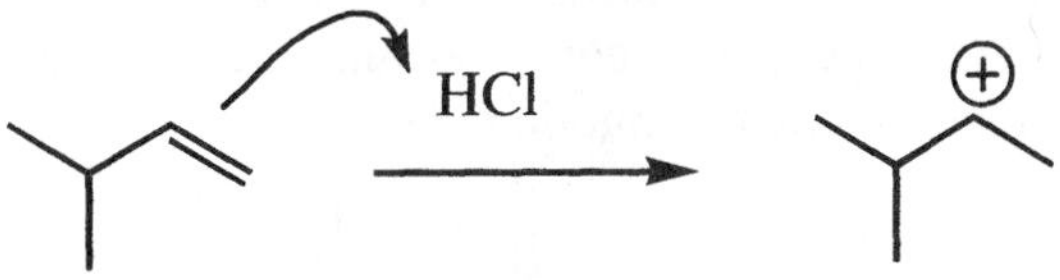

It is certainly true that this intermediate is a highly reactive carbocation, and it does not take long for it to find a partner. Nevertheless, it will have to wait a finite amount of time for a collision with something else to occur. If there happens to be something it can do by itself to ease the pain slightly, often it will choose to do so. Below is a bigger picture of this carbocation, showing the rearrangement process. The carbon with the charge is short of electrons—it has only six and it wants eight. It is also a secondary carbocation, which is not the worst kind but also not the best. In an effort to lessen the pain, it steals some electrons— and a new partner!— from the carbon next door.

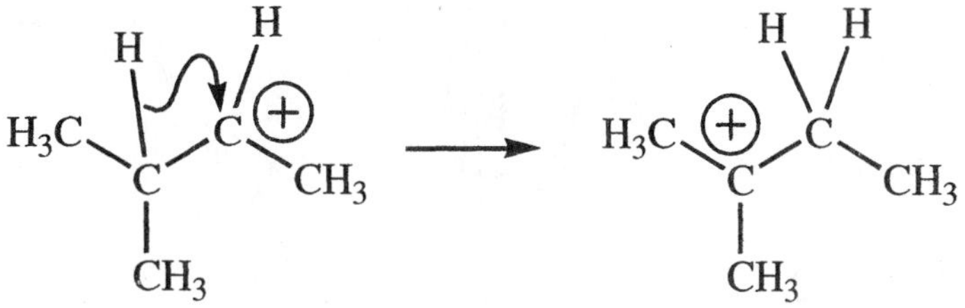

You should notice several things:

1. Both the electrons and the attached partner (the hydrogen) have moved to where the carbocation used to be. Since hydrogen with its electrons is called "hydride," this is described as a "hydride shift." We can indicate it by giving the electron arrow a slight upward curvature at its origin, to signify that the electrons are detaching from carbon and staying with the hydrogen, then plopping down on the next-door carbocation. If we had used a simple downward curved arrow, this would have instead implied that the electrons remained with the carbon and detached themselves from the *hydrogen*, and the result would be a double bond (this does happen, but it is not the process we are discussing at the moment), as shown below:

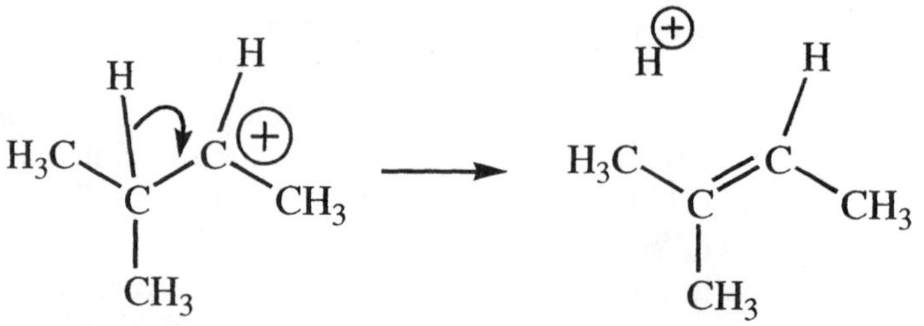

2. The carbocation has no interest in its own bonds. After all, those are the six electrons it already has a stake in. In looking for new electrons, it looks to the bonds between its neighbors and *their* partners. If one of those bonds could move away and leave behind a better carbocation than what we started with, there is a good chance it would happen. In the present case, we have traded a secondary carbocation for a tertiary one, an improvement.

3. There are actually several possibilities to consider here, but only one of them results in an improvement. For example, one of the hydrogens on the methyl group on the right could have moved, but this would have left us worse off than before (primary carbocation), so we will reject that. Or, one of the methyl groups on the left could have moved, but this would have left us no better off than before (things like this sometimes do happen, but most rearrangements result in improvement, so we will not consider rearrangements that do not).

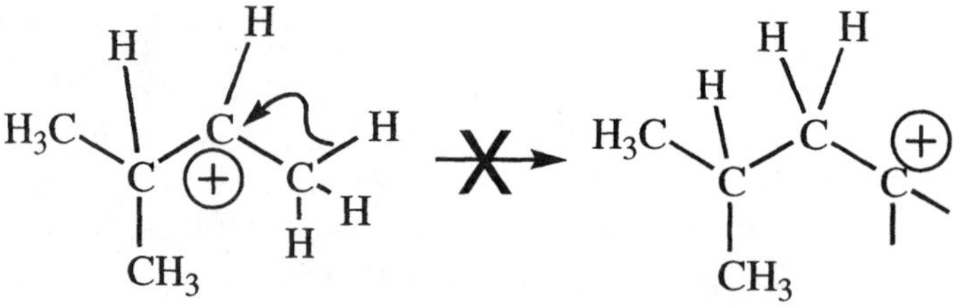

Primary carbocation

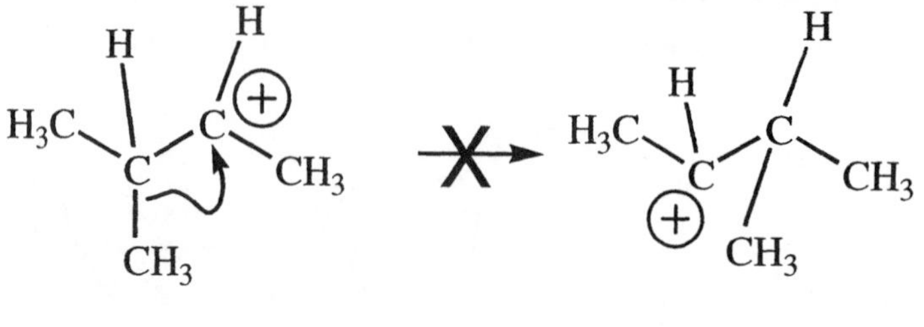

Secondary carbocation

4. What we have shown here is not resonance! We have completely broken one attachment and made another. Further, in doing so, the hydrogen has actually changed its position (remember, one of the ground rules of resonance is that all the atoms must remain frozen in space). The structure of the ion changes during this process.

5. We are not finished. Once a rearrangement takes place, we are still left with a carbocation. This carbocation does what any other carbocation does—namely, tries as hard as it can to stop being one. In this case, it would trap a Cl⁻ that comes floating by.

Overall, then, there are two possible products for this reaction. Molecules that react with chloride before they have time to rearrange would make one product; molecules that rearrange before they react with chloride would make another:

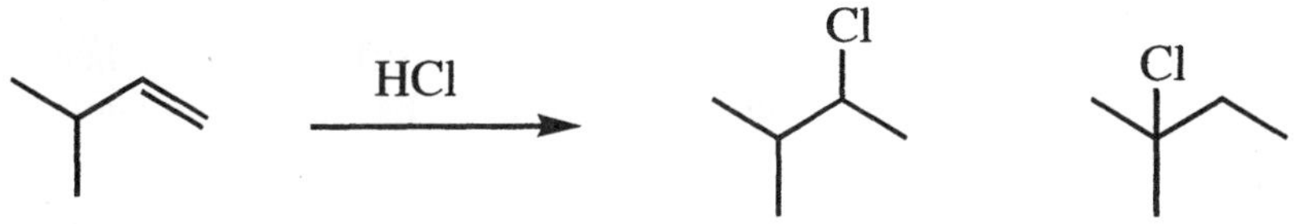

Now don't get confused by this equation. Some students read this to mean that a single molecule of starting material somehow becomes two different things (creating matter in the process, I suppose). The equation instead means that a sample of starting material, containing zillions of molecules, would result in the same number of molecules of product, but some would be of one type and some of another.

What is the mechanism of this process? Here's how we write it:

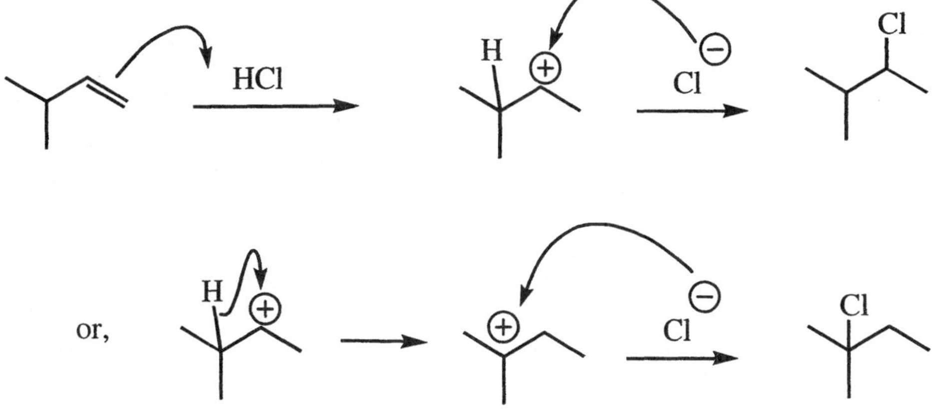

In summary, rearrangements often occur when a molecule can see a way to improve a carbocation by shifting its bonds around. Almost all rearrangements involve carbocations: if you think a reaction involves a rearrangement, you'd better produce a carbocation during that reaction and have the rearrangement happen while the carbocation is still present.

ORM: In "Organic Reaction Mechanisms," under "Addition Reactions," look at "HBr to 3-methyl-1-butene" and "HBr to 3,3-dimethyl-1-butene" for a nice animation of the above section.

It is all very well for me to show you how such a thing happens, but how should you know when to worry about it? How do you know which molecules are likely to do this? First of all, the reactive species has to be a carbocation. That does not help you yet, since all our reactions so far involve carbocations, but this will change soon. Second, once you have a carbocation, you should take a quick look at the carbons next door and see if any of them, if they lost a bond, would lead to a better carbocation than the one you've got. If so, there's a good chance of rearrangement. Don't forget to take into account all the factors you know: tertiary carbocations are better than secondary, etc., but resonance is also important.

Example:

Suggest three reasonable products for the following reaction. Try to get an answer before you look at mine.

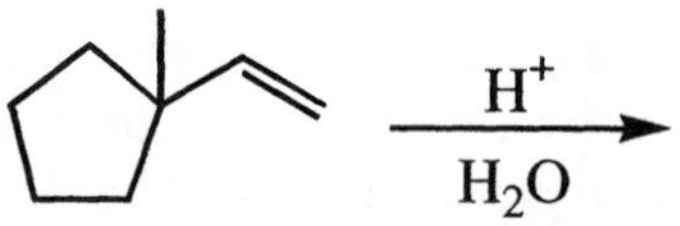

Answer: This reaction should start off like any other. You should never propose any monkey business unless you already have a carbocation, and you don't yet. But you will soon.

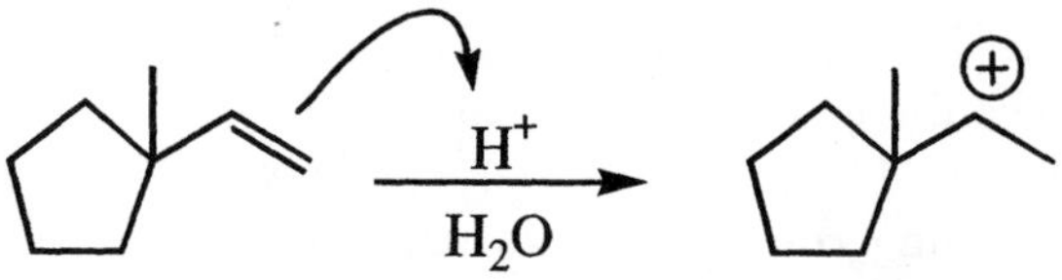

Now we have a carbocation. We know we will get this one because of Markovnikov's rule. There are two possibilities: it can react with water now, or it can rearrange. If it reacts with water right away, we get the following:

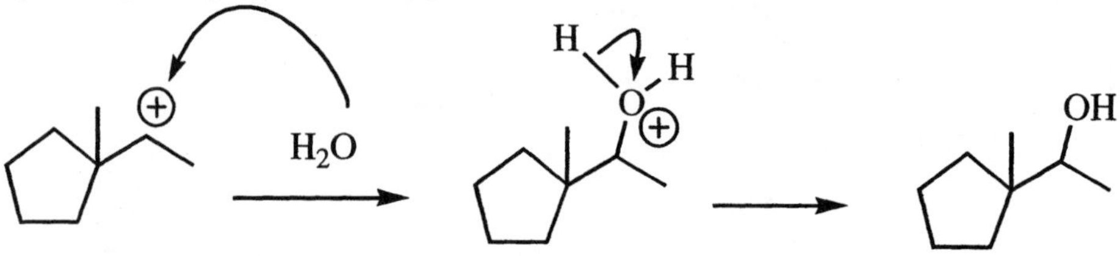

Is there a decent rearrangement for it to undergo? Look at the carbon *next to* the one with the charge. There are four bonds to carbon there (i.e., it is quaternary). If one of them were to disappear, the carbon would be tertiary, better than the secondary carbon where we have a cation now. So it is reasonable to suggest that one of those bonds could move next door. Like this:

Now *this* can react with water to make a different product:

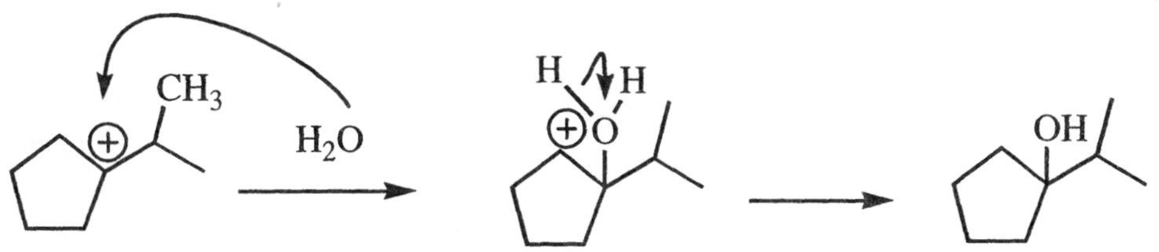

To carry out this rearrangement, we chose one of the bonds between the quaternary carbon and one of its partners (other than the carbocation). Why did we choose this particular bond? Certainly it is as good as any other, but, for reasons I cannot explain, many students have a hard time realizing that it is in fact *no better* than any of the other bonds. There are two other eligible bonds to consider, and the fact that they are part of a ring does not make them any less eligible as sources of electrons. In other words, the following process is just as valid as the first rearrangement we did:

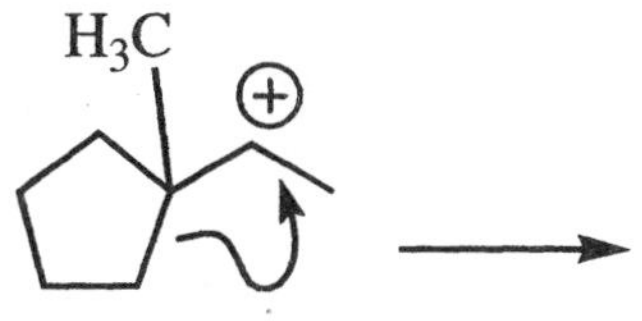

But what does it make? Both beginners and professionals alike often have trouble seeing what is formed in a complex rearrangement like this. My advice is, *don't try to figure it out!* Instead, draw the product with the atoms exactly where they are in the starting material, changing *only* the bond you moved. *Then* look at what you have made. In this case you should be able to see that this weird thing is actually a six-membered ring, so now you can redraw it in a reasonable shape.

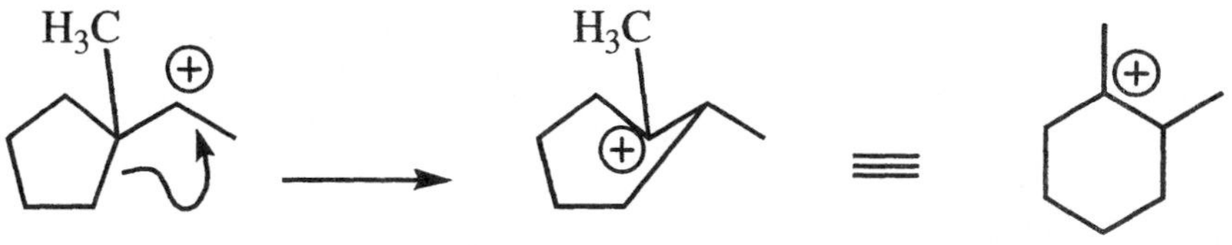

Notice that I did not put an arrow between the second and third structures. This is not a *process* the molecule undergoes, simply a redrawing of the picture for *our* benefit (the molecule knows perfectly well what it is; we are the ones who are having trouble seeing it!). Instead of an arrow there is an equivalence sign, meaning "is the same as." This new ion, of course, can also trap water to make yet a third product.

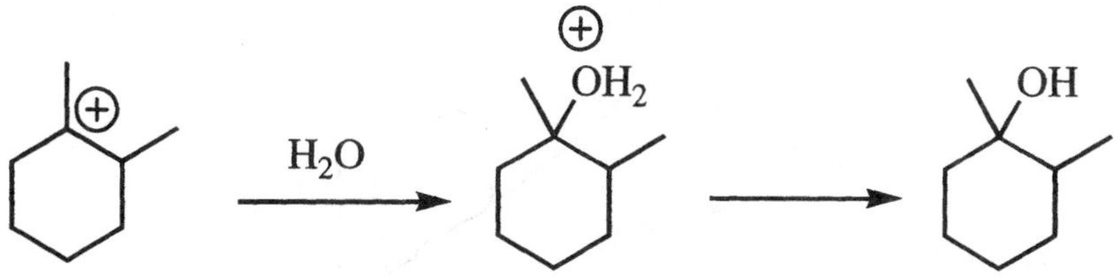

So we have found three reasonable products for this reaction. But wait! Isn't there still another? There is a third bond that could have rearranged; why didn't we consider that? OK, let's do it.

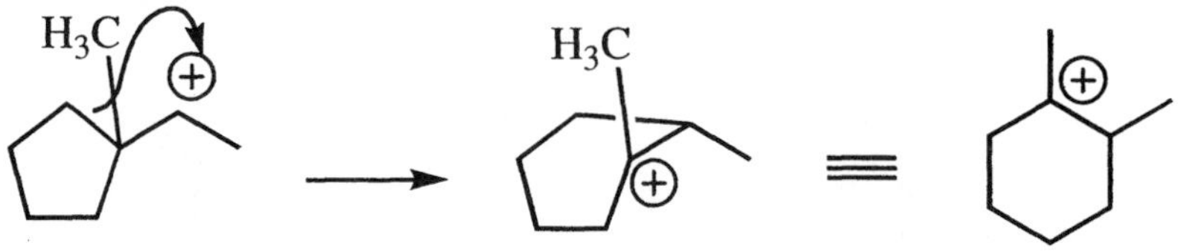

After looking at it, you suddenly realize that this bond also can and does participate in the rearrangement, but it does not lead to anything we haven't already considered (though that is certainly not obvious at first). So these are the three reasonable products we can expect from this reaction.

Nomenclature alert: when a rearrangement involves moving a carbon from one atom to its neighbor, as in the above example, the process can be referred to as a "Wagner-Meerwein rearrangement." You do not need to remember this name, but in case you see it, you now know what it refers to.

Try another example: Show a mechanism for the following reaction and explain *why* the reaction occurs the way it does.

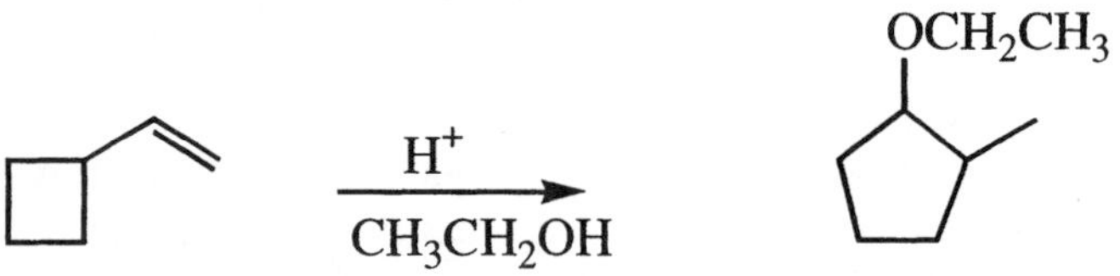

Answer: Clearly there has been a rearrangement. We cannot even think about how this occurs until we have a positive charge somewhere. So we will begin normally:

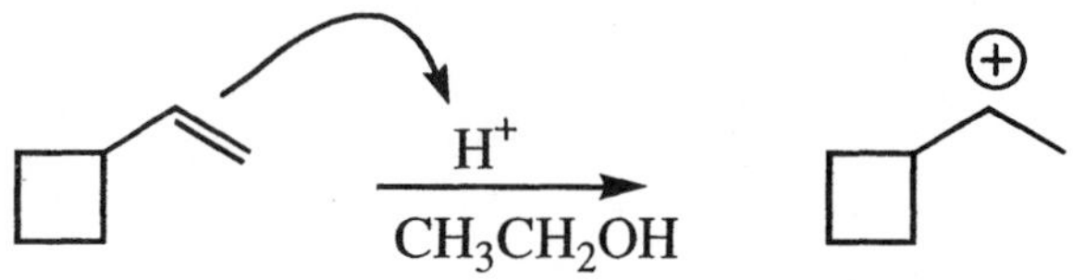

Now we have a carbocation, so we should see if there is a reasonable rearrangement we can do. There are two:

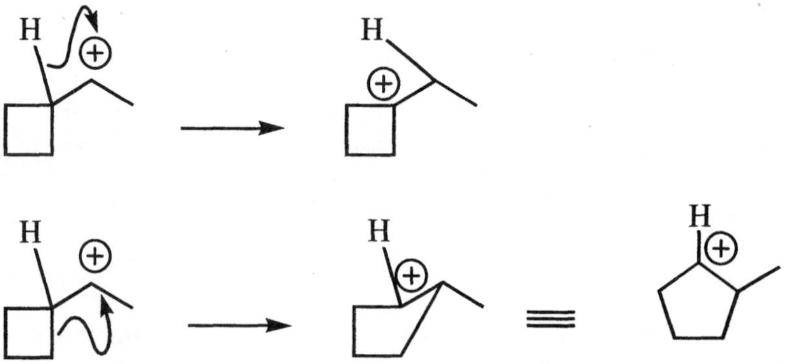

(There is a third that is the same as the second.) The second one clearly is the one we want, since it has changed the four-membered ring into a five-membered ring, so we will continue with it. All that remains is for this carbocation to react with the solvent.

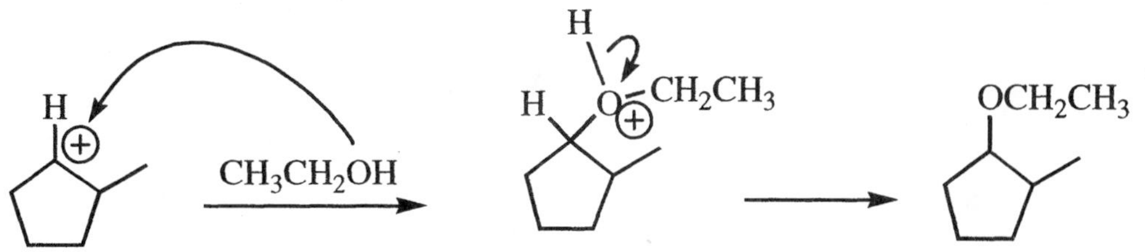

There is still the other part to the question, however: *why* does this occur, and, although the question does not ask this, you should: why does the first rearrangement *not* occur? After all, the first rearrangement converts a secondary carbocation into a tertiary one, while the second simply trades a secondary for another secondary. Let's tackle the second part first. Neither the question nor the answer stated that the first rearrangement does not occur. It simply asks where the product shown comes from. There is no implication that this is the only product. The first rearrangement might very well occur, and there might be a product related to it. The question simply does not address this issue. It states only that a certain product is observed and asks you to justify it. Nothing more. Very often there are many products from a reaction, and one of them might be more difficult to rationalize than others, so that is the one someone might be wondering about.

Nevertheless, in this particular case, one could make an argument that the first rearrangement might not occur. At first glance, of course, it looks *more* likely than the second. However, there is more to molecular stability than just the carbocation. This molecule has a cyclobutane ring, which we have already determined is suffering somewhat because its bond angles are forced to be about 90° while normal sp^3 carbons want angles of 109°, so these carbons are suffering by about 20°. As the rearrangement occurs, one of the carbons of the cyclobutane is becoming sp^2-hybridized, because it is becoming a carbocation with only three partners instead of four. Normal sp^2 carbons want angles of 120°, so this one carbon is becoming *more* strained than it was before. So even though the carbocation is changing from secondary to tertiary, at the same time the molecule as a whole is becoming more strained. It is anybody's guess which factor is more important, but if someone told you that the first rearrangement does not occur, you could now justify that observation. I should point out that in the actual case, I do not know whether the first rearrangement occurs or not, and in a sense it does not matter. If it does, I can show how it happens; if it does not, I can explain why it chooses not to. What I can't do is predict what will actually happen. To know that, we have to run the experiment and find out.

Onward. The second rearrangement *does* happen. Why? Well, at worst, we could point out that the molecule is no worse off after the rearrangement than before, since both carbocations are secondary, so there is at a minimum nothing preventing such a rearrangement. But we can do better. Again, we have to consider the molecule as a whole. The starting material has a cyclobutane ring, which is strained. After the rearrangement, we no longer have a cyclobutane. The ring has enlarged to a cyclopentane, which has far less strain than a cyclobutane. So even though the carbocation center itself is about as happy in one form as the other (secondary), the molcule as a whole has a strong preference for rearranging.

Test Yourself 1

Seeing rearrangements and showing how they occur takes practice. Practice! For example, suggest a mechanism for the following reaction, and explain why it occurs the way it does:

WHO CARES? YOU DO!!

In the last chapter I showed you the cyclization of squalene to the steroid skeleton. We stopped that reaction in midstream in Chapter 10. Now we can take it the rest of the way. You might recall that we started with squalene, generated a cation, and cyclized it:

What happens next is a series of rearrangements, each of them trading one tertiary carbocation for another. First a hydrogen (hydride) moves:

Then another:

Then a methyl, and another methyl, and finally a proton is lost:

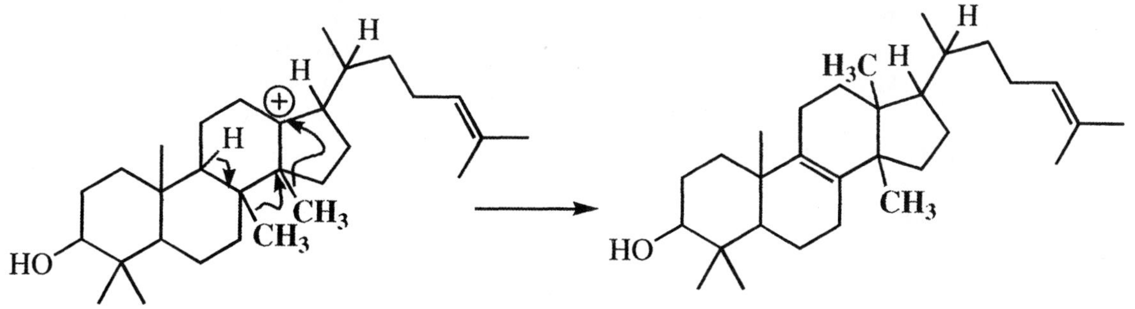

This last compound is called lanosterol, and it is the first cyclized intermediate one can isolate in the steroid sequence. There are still a few reactions to be done before the product becomes cholesterol, but I hope you can see that in this one reaction (from squalene to lanosterol) there has been an incredible transformation!

BROMINATION

All the addition reactions we have covered are two-step reactions, where the first step is the reaction of an electrophile with the double bond. So far, in all our examples the electrophile has been H^+. Thus, additions of HX, water, and alcohols, all begin with H^+ adding to the double bond. There is another important electrophile that often initiates additions to double bonds, and it should come as a surprise: Br_2. (Cl_2 and I_2 do the same reaction, in exactly the same way, so everything we discuss in relation to Br_2 is transferable to those cases.) Why is this a surprise? All our other reactions began with acids, such as HX, in which one partner, the H, was significantly less electronegative than the other, and therefore suffering from a partial positive charge. Clearly, in Br_2, neither partner is more electronegative than the other! On the other hand, we do know that while each bromine atom sees eight electrons in Br_2, each would also be quite happy to have full control of all eight that it sees (to become Br^-), so there is a bit of a tug of war going on here. Since both atoms are equally strong, neither ever wins this tug of war, but both atoms might be happier if some new electrons were to show up. This is where an alkene can help.

We can imagine, as a first guess, that the alkene donates electrons exactly as we imagined before:

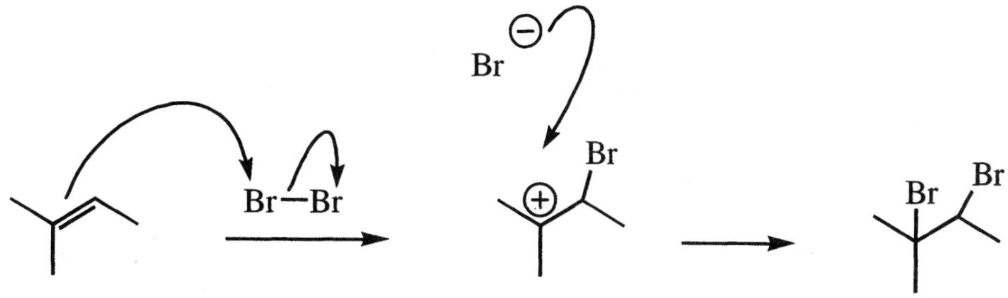

This correctly predicts that the product of the reaction is one in which *two* bromine atoms have added to the double bond, one at each end. Although you could not have predicted this, it turns out to be a very rapid reaction, so rapid that chemists use it as a test for alkenes, as follows. Bromine (Br_2) has a deep reddish-orange color, caused by the bromine–bromine bond. If you mix some bromine with an alkene, the color will rapidly disappear, since the product no longer contains this bond. Thus, you can test to see whether some unknown substance is an alkene by checking to see whether it "decolorizes" bromine.

However, the mechanism I wrote above is not completely correct. How do we know? Because there are aspects of the reaction that are inconsistent with the mechanism as I have written it. You don't see the problem because I haven't told you about it yet. It has to do with stereochemistry. To understand the problem, we have to go back and consider the stereochemistry of earlier reactions we looked at.

Consider the addition of HCl to 1,2-dimethylcyclopentene:

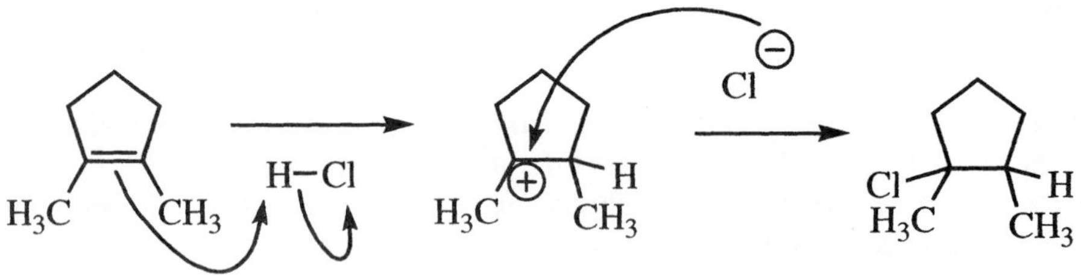

I want to focus on the second step and see if we can predict whether the product will have the two methyl groups *cis* or *trans* to each other. To do this, we have to look carefully at the carbocation. Here is a blow-up of it. You may want to build a model to see in three dimensions what I am talking about.

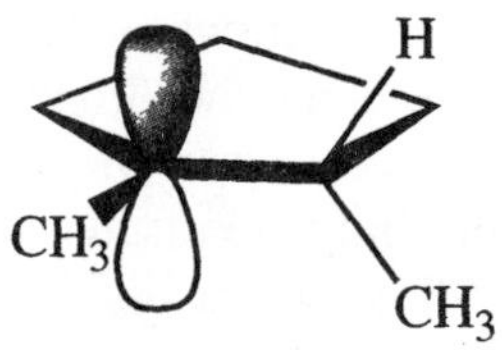

The positive charge is in the p orbital I have shown. Notice that the carbon with the positive charge is flat, and the chlorine is free to come in either from the top (as drawn) or from the bottom. If it comes in from the top, the methyl groups will end up *cis* to each other, whereas if the chlorine comes in from the bottom, the methyls will end up *trans*:

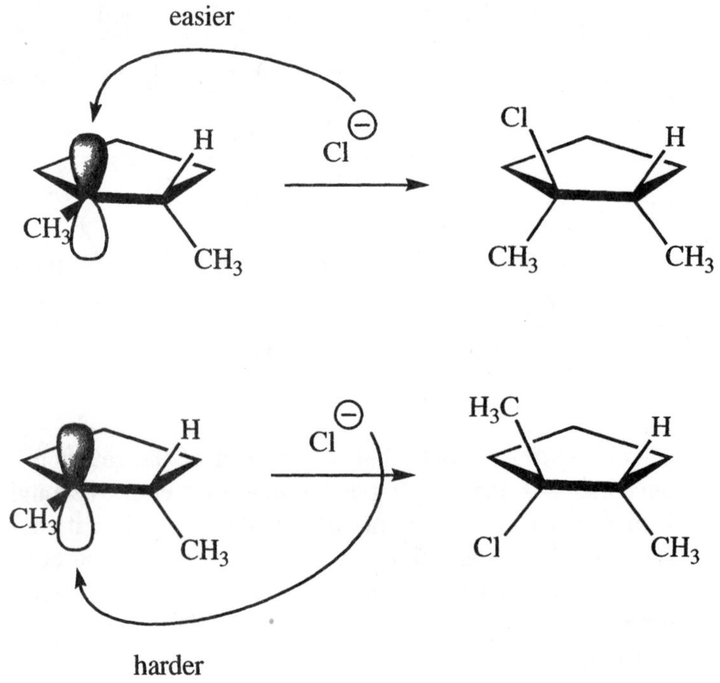

easier

harder

Now it is not quite true that the chlorine is equally free to come in from either the top or the bottom. To enter from the bottom, it has to brush past a methyl group on the adjacent carbon, while from the top the only thing in its way is a small hydrogen. Therefore, we might predict getting *more* of the top compound, with the methyl groups *cis,* than of the bottom one, *not* because it is a better product, but because the process of forming it is easier. However, this preference for *cis* over *trans* methyls will be far from absolute. There is a preference for *cis* product, but some will still be *trans*.

Using the same approach in the reaction with Br_2, if the mechanism were as I wrote it, we would predict a preference for *trans* methyl groups this time, because a bromine atom is larger than a methyl group, and therefore the bottom reaction (below) is easier than the top, but we would still expect some *cis*:

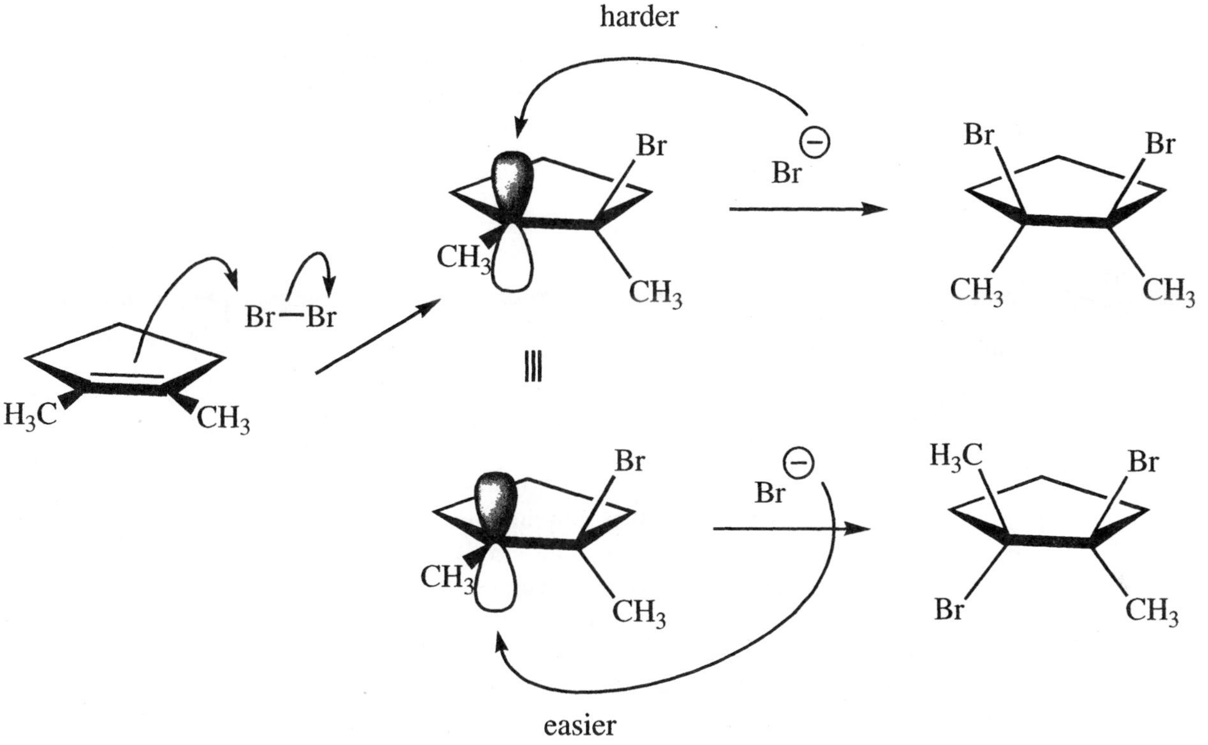

But this is not what we observe. If you do this reaction in the lab, you discover a complete absence of *cis* product. *Trans* is the only product you get. Our mechanism has to be adjusted somehow to account for this observation.

You almost certainly would not come up with the explanation here that chemists have settled on, but I do hope you will be able to see that their explanation accounts for the facts. It has been concluded that the picture I drew of the carbocation, in this case, is inaccurate. In order to account for the second bromine *never* coming in from the same side as the first, chemists are convinced that the first bromine is not off to the side, as we originally pictured it, but directly over the top of the original double bond. If so, it would completely block that face of the double bond and force a second bromine to approach from the opposite side; that is precisely what we see.

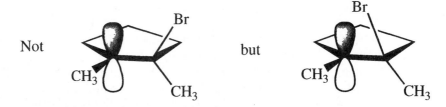

Not   but

But why on earth would the bromine do that? Isn't this a little strange? Yes, it is a lot strange. The notion is that this is no ordinary carbocation. Recall that bromine has lone pairs of electrons around it. With the bromine atom hanging over the top, those electrons could help relieve the positive charge on carbon without having to move, as shown in the second structure below. Further, as this picture suggests, according to this model there is really no difference between the two original carbons of the double bond, and the right-hand carbon can share in the positive charge just as well as the one on the left. In other words, we have here a kind of resonance hybrid with the positive charge spread out over several centers, thus making it better off than an ordinary carbocation. *That* is why the bromine atom sits in the middle. Such a three-membered ring is called a "cyclic bromonium ion," or just a "bromonium ion," and all reactions of double bonds with halogens are thought to proceed through intermediates with an analogous structure.

I must acknowledge immediately that in writing these resonance structures, we have violated one of the rules of resonance I gave you previously, namely, that you may not make or break any single bonds in writing resonance structures. Like most rules in chemistry, they work most of the time but there are occasional exceptions. This is one; if you study chemistry further, you may encounter other things called "non-classical" carbocations that likewise have resonance involving single bonds. For our purposes, this will be one of only two exceptions we will run into.

In writing the structure of a bromonium ion, one really ought to write all three resonance structures, but you will often see just the middle structure written, the assumption being that the reader understands that it is intended to stand for all three simultaneously.

Here is a description of the reaction that might help you make more sense of it. Recall that a double bond consists of a σ bond and a π bond, as shown. The bromine molecule approaches this source of electrons and sits down on top, using electrons from *both* of the carbon p orbitals simultaneously to create a bromonium ion. What we have here is an orbital picture of the resonance hybrid we wrote above. Then the second bromine atom, now an anion, comes in from the bottom—at either carbon—allowing the original bromine to move over and make a full bond with the other carbon.

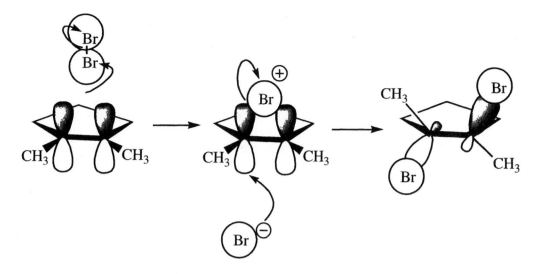

How should you write this on one of your cards? Like this:

Reaction:

$$\underset{R}{\overset{R}{\diagdown}}C=C\underset{R}{\overset{R}{\diagup}} \quad \xrightarrow{\text{Br}_2} \quad \underset{R}{\overset{Br}{\diagup}}\!\!\!\!\underset{R}{\overset{R}{\diagdown}}$$

Bromines *trans* if
stereochem can be
determined

Mechanism:

ORM: Under "Addition Reactions," see "bromine to an alkene" for an animation of the above section. We have only considered the reaction labeled "reaction with bromide ion," but you ought to be able to understand the other animations just as well.

"MAGIC" REACTIONS

Epoxidation

There are three more reactions of alkenes you need to know. These are so complex that it would probably be easier to memorize them than to try to understand what is actually going on. The first involves reagents known as "peracids" (that's the word "acid" with a "per," meaning "extra," in front). Peracids are carboxylic acids with an extra oxygen. They are often written RCO_3H. When you see this you should immediately build in your head a picture like the following:

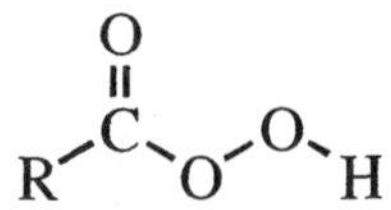

Recall that a carboxylic acid looks almost like this but with one fewer oxygen atom. Carboxylic acids are acidic, because when their proton is lost, the resulting anion is resonance stabilized:

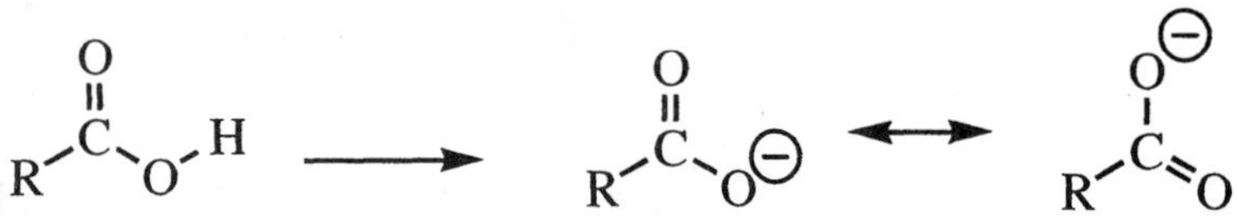

The anion of a *per*acid cannot experience the same resonance stabilization; therefore peracids, despite their name, are not particularly acidic. However, they do undergo an interesting "magic" reaction with alkenes, the purpose of which (from their standpoint) is to get rid of that extra oxygen. What they do is to dump it onto the alkene.

At the end of the reaction, the peracid is no longer a peracid; it is now an ordinary carboxylic acid. The extra oxygen now appears squarely on top of what used to be the double bond. The product has in it a three-membered ring containing an oxygen, which is called an "epoxide" (also "oxirane," although few people other than textbook writers use that term). This reaction, then, is an "epoxidation" reaction. It works with just about any double bond. The most common peracid in use is called *meta*-chloroperbenzoic acid, shown below. This is a mouthful, so most people abbreviate it *m*CPBA.

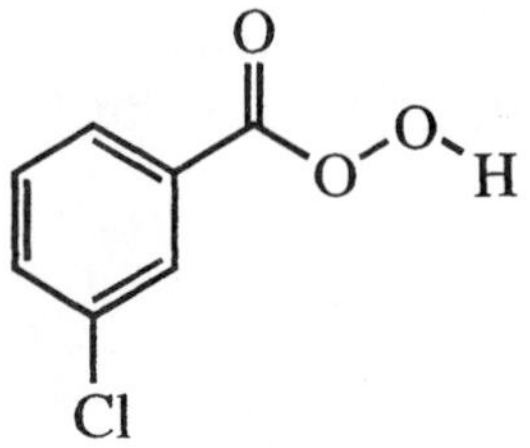

meta-chloroperbenzoic acid

Here are some examples:

WHO CARES? YOU DO!!

Later on we will learn some of the reactions of epoxides, at which point this reaction will prove very useful. For now, it's enough for you to know that epoxides are also very important biologically. Recall that the first step in steroid cyclization is production of a cation within the squalene molecule. It is actually an epoxide that is the immediate precursor of this cation. Epoxides also play an important role in cancer. For example, it has been shown that benzo[a]pyrene, an ingredient in charcoal smoke (and therefore everything you barbecue), is carcinogenic, not because the molecule itself is a problem, but because your body turns it into an epoxide, which is the real bad guy. Of course, biological systems do not use mCPBA to epoxidize molecules, they are much more elegant than that, but the effect is the same.

Ozonolysis

The second "magic" reaction is ozonolysis. This is the official term for a cleavage reaction of a chemical with ozone, the very same gas that is in short supply in the atmosphere over the poles and which is needed to help shield us from harmful ultraviolet radiation from the sun. Ozone is O_3, and it will be well worth your while to take time out here and see if you can write a sensible Lewis structure for it. *Hint*: Ozone is not cyclic, you cannot write any structure in which all the atoms are happy, and several possible structures (resonance structures) exist with varying problems. The mechanism of the reaction of ozone with alkenes is quite complicated, and there is really no need for us to explore it in detail. The result is quite remarkable, however: ozone acts like a pair of molecular scissors, slicing a molecule wherever there is a double bond. The scissors cuts through *both* bonds, separating the molecule into separate pieces, and replaces the former carbon partners with oxygen atoms. Like this:

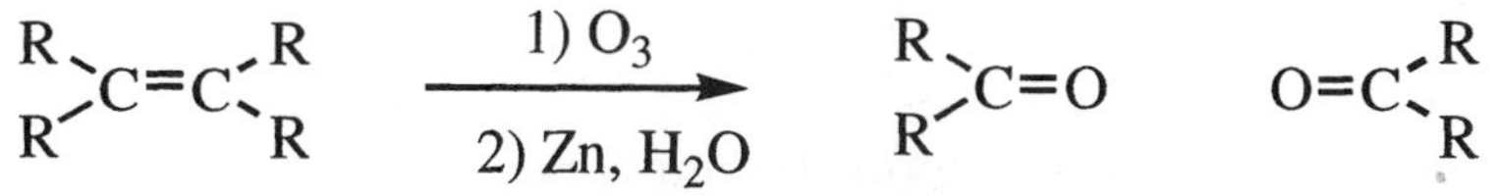

The result is ketones if all the R's are carbons, or perhaps aldehydes if any of them happen to be hydrogens. Notice what is above and below the arrow: it says, first add ozone, then, when that is done, add zinc and water. It does not matter that one instruction is above the arrow and one below. But it does matter that one says "1)" and the second says "2)." This is our shorthand for two sequential operations, and it is important to distinguish sequential operations from a situation in which all the chemicals are dumped together.

Here are some examples:

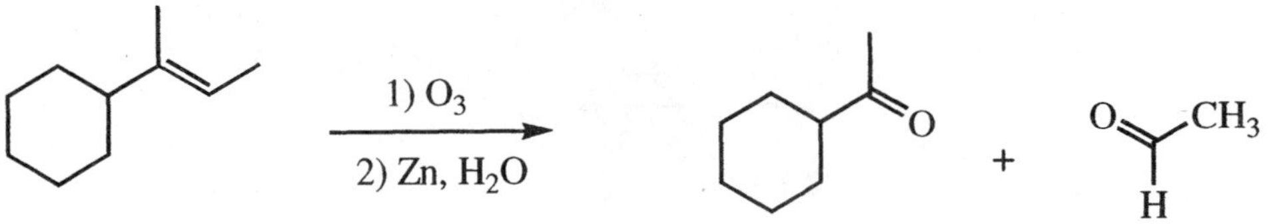

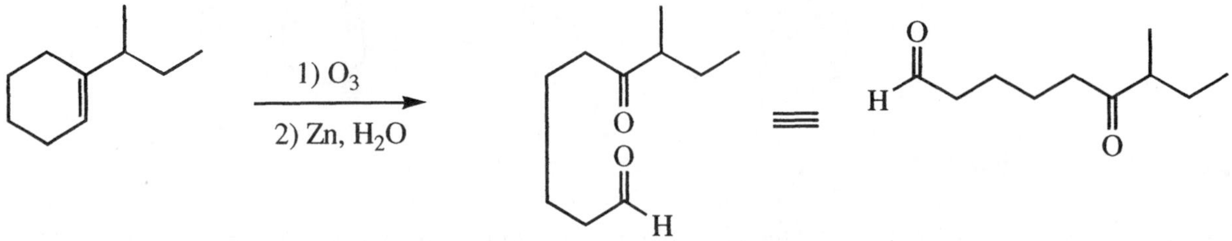

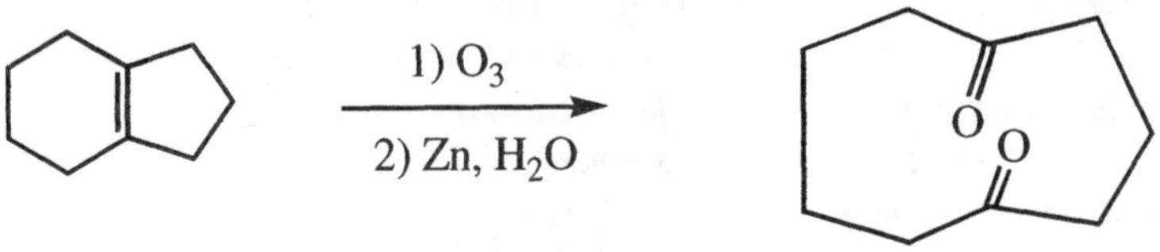

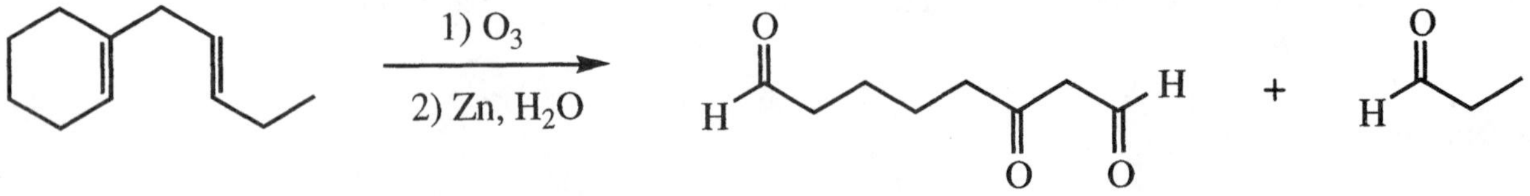

Be sure you take a close look at that last one and really understand it!

Don't forget to make cards for this reaction. It is, after all, the *only* way you know (for now) of making aldehydes and ketones.

Incidentally, this propensity for ozone to react with double bonds has some economic consequences. In addition to the upper atmosphere, another place where ozone is in unusually high concentration is in smog. Automotive tires are made of rubber, a polymer with a significant number of double bonds in its structure. Tires tend to have fairly short life spans in smoggy cities. Apparently the tire business is booming in Los Angeles.

Test Yourself 2

Do you think you can handle ozonolysis reactions? Then try thinking backwards! Say that an unknown compound is treated with ozone followed by zinc and water, and the products shown below are obtained. What was the structure of the unknown compound you started with?

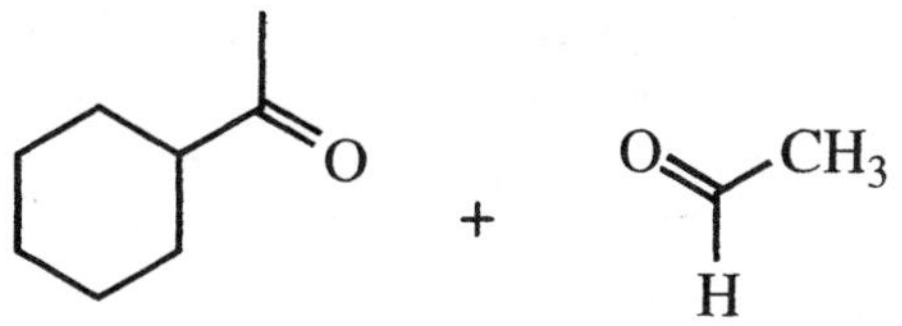

If your answer is the first compound in the previous examples, you only got it half right! The starting material could also have been the *Z* isomer, shown below, which is a different compound. Since ozonolysis slices double bonds, the stereochemistry around those bonds disappears in the products, which means that stereoisomers always give the same products.

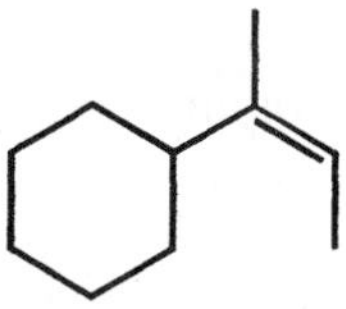

Try again with the following products. One answer is included in the set of examples we just did; however, I can think of *nine others* without much difficulty! Can you? By the way, don't ever suggest putting *trans* double bonds in rings smaller than 10 members, because this would produce tremendous strain. (A few such molecules have been made, but only at great effort.) *Hint*: Two contain a cyclohexene, two contain a cyclooctene, and six contain a cyclohexadiene (two double bonds in the same six-membered ring).

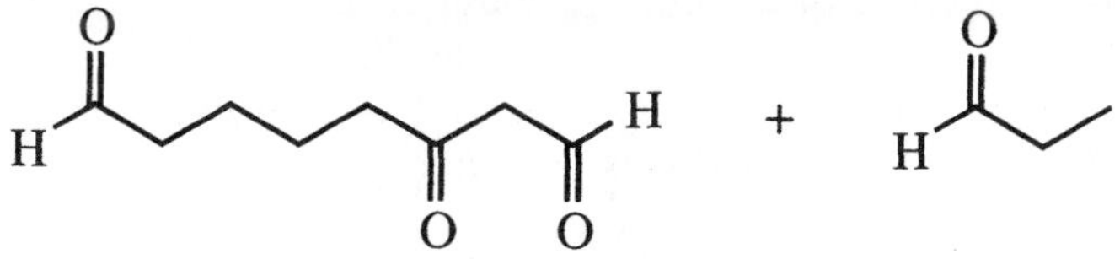

Hydrogenation

The third "magic" reaction is called hydrogenation, which means, "reaction with hydrogen." Many students confuse "hydrogen," H_2, with hydrogen *ion*, H^+. There is a world of difference. H^+ is the business end of an acid: it is short of electrons, and it reacts wherever it can find electrons, creating a positive charge in the process. H_2 is the elemental form of hydrogen, and it is relatively inert. It is pretty happy as is and does not want either to give up electrons or accept them. Indeed, when you mix hydrogen with most other chemicals, nothing at all happens. Even mixing hydrogen with oxygen produces absolutely no reaction, unless you set it off with a spark (at which point there is a very violent reaction to make water!).

Hydrogen gas reacts with alkenes only when pushed to do so. There has to be an appropriate catalyst present, usually a metal of some sort, that magically allows the hydrogen to react with the alkene. In many cases even this is not enough, and the hydrogen has to be really pushed onto the alkene by increasing the pressure of the gas. Nevertheless, by adjusting the conditions, most alkenes can be forced to react with hydrogen. We write the reaction this way:

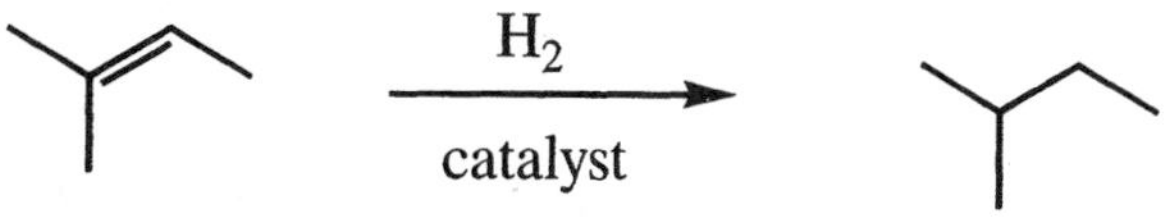

What has occurred? This is hard to tell, because I haven't written in explicitly the new hydrogens that are attached to the molecule. However, if you are comfortable reading these stick figures, you know that there are two hydrogens in the structure on the right that are not present in the structure on the left: one on each carbon where the double bond used to be. (Until you get used to the stick figures, go ahead and write them in!) These are the two hydrogens that used to be part of an H_2 molecule. Reactions in which hydrogen atoms are added to a molecule are often called "reductions."

In spite of what you may have seen in the movies, you cannot buy a chemical called "catalyst." While it is all very well to declare (as in the above equation) that some catalyst is necessary, when you get around to doing this reaction, you will need to know what actually needs to be put in! It turns out that there are a variety of catalysts that work pretty well for this particular reaction, but the most common ones are nickel, palladium, and platinum. Usually these are used (for this purpose) as a very finely divided form of the metal, deposited on powdered charcoal for ease of handling. Thus, it is better, in the above equation, instead of saying "catalyst," to say "Pd/C," which is the chemist's abbreviation for "palladium on carbon."

Hydrogenation reactions are almost always exothermic. You should be able to anticipate that: after all, I told you before that C–H and H–H bonds are worth about 100 kcal/mol, while a π bond is worth only about 70. In this reaction, we are breaking one π bond and one H–H bond (we need to put in about 170 kcal/mol) but we are getting back two C–H bonds, worth about 200 kcal/mol, so we should be better off by about 30 kcal/mol after the reaction is over. This should bring up a question in your mind: if this reaction is so favorable, why doesn't it happen rapidly by itself? When you mix an alkene and hydrogen gas, why does the mixture just sit there unless there is a catalyst present? The answer has to do with rate. The reaction is exothermic, yes, but as reaction begins, the energy of the system goes up until you reach a transition state. In this reaction, the transition state is very high in energy. In other words, in order to make this reaction happen, you have to put in a good bit of energy before you start getting any back. Even though you eventually will get back more energy than you put in, the molecule does not know this up front, and the amount of energy it takes to get things going is larger than the molecules are willing to tolerate. Therefore, no reaction takes place—until you provide a lower energy pathway. This is what the catalyst does. As an analogy, consider a village in a valley surrounded by huge mountains. Even though on the other side of the mountains there might be an even better place to live, the villagers do not know about it, and even if they did, they'd have no inclination to drag all their belongings over the mountain to the new home. On the other hand, if there were a tunnel through the mountain, it would not be so difficult to get to the better place, and many of the villagers would probably move.

The energy diagram below illustrates this. The hydrogenation reaction is exothermic by about 30 kcal/mol, and the presence of the catalyst does nothing to change that. The catalyst simply provides a lower energy pathway for the reaction to follow.

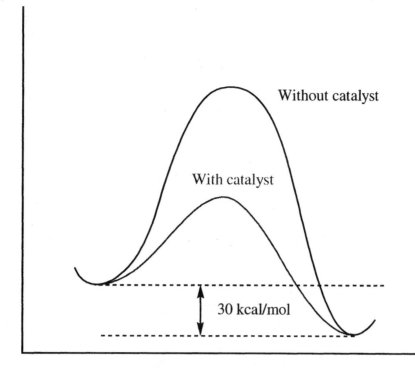

It is interesting to look at the stereochemistry of this reaction. You will recall that the reaction of bromine with alkenes was exclusively *trans*; that is, one bromine went on the top and the other one on the bottom of what used to be a double bond. We explained this by suggesting that the two bromine atoms come in at different times: first one creates a bromonium ion, covering up the top, then the other comes in from the opposite side. Hydrogen does just the opposite: both H's enter from the *same* side. We explain this by claiming that they come in more or less simultaneously. You can imagine the H–H molecule sitting down atop the C=C bond as shown below. The fact that hydrogen adds in *cis* fashion to double bonds gives us some insight into the mechanism: chemists believe that the H_2 molecule sits down on the catalyst surface, which weakens the H–H bond and allows the H_2 to react with the alkene. I have left the catalyst out of the picture below, but it is shown in the ORM movie referenced soon.

Here are some examples.

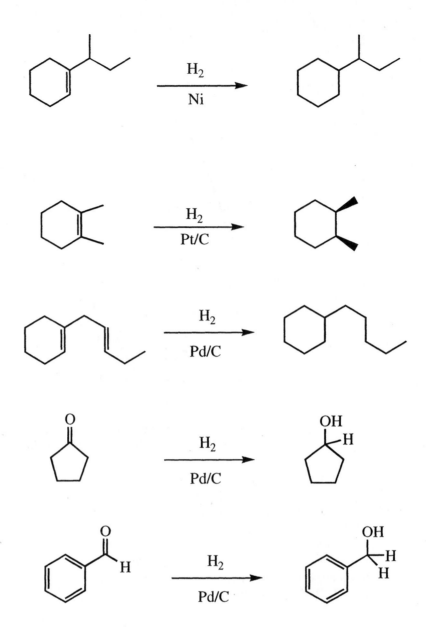

Notice that there is only one example in which I showed stereochemistry. This is not because the hydrogens fail to add *cis* in the other examples, it is simply that this is the only case among all the examples in which there is a stereochemical relationship in the product that we can *see* (all the others permit rotation around the new single bonds).

Notice also the last two examples. The hydrogenation reaction works on carbon–oxygen as well as carbon–carbon double bonds. In fact, one hardly ever reduces carbon–oxygen double bonds this way, because there are better ways of accomplishing the same transformation. When we learn those better ways, we will stop doing it this way, but for now this is the only way we have!

Finally, note that the double bonds of a benzene ring are impervious to this reaction, as they are to most of the other double-bond reactions we have learned.

ORM: Although I have presented this as a "magic" reaction, there is actually a fair bit known about how it proceeds. If you are interested in seeing one suggestion of how to visualize this, look in ORM under "Addition Reactions," "Hydrogen to an alkene."

Don't forget your cards. You may need some extra ones for this reaction, because it can be used to make not only alkanes but also alcohols. In particular, it is the only way you know so far to make *primary* alcohols (see last example).

WHO CARES? YOU DO!!

As you know, saturated hydrocarbons tend to stretch out in zig-zag fashion, as shown below:

It is not hard to imagine that several of these placed near each other would fit together quite well. This is the case in biological substances known as triacylglycerols, shown below:

Molecules like these, with long saturated chains, are the chief components of animal fats, and, properly processed, butter. As you know, these are all solids at room temperature, which is consistent with molecules that fit together comfortably. Vegetables also make compounds like this, but almost always there is some unsaturation in one or more of the long chains:

Obviously, these do not pack together nearly as nicely, and although they are of essentially the same size and weight, they tend to be liquids—vegetable oils, like olive oil. How can you make vegetable oils behave like animal fats, so that they will be solids at room temperature? Get rid of the double bonds! If you look on a package of margarine, you will discover that it contains "partially hydrogenated vegetable oils." They have done our hydrogenation reaction on the oil to make it solid, so they can pass it off as a butter substitute!

SYNTHESIS REVISITED

Let's see if we can put together everything we have learned. Suggest a synthesis of the following compound, starting with any alkenes you like as your only sources of carbon.

How do you go about solving a problem like this? I should probably repeat what I told you in the last chapter: *No one* can look at such a problem and instantly write down the alkenes one should start from. Unless you have been instructed to start from *specific* compounds, it is counterproductive to try to think in terms of the first reaction you would actually carry out. The only sensible way to approach this problem, and virtually every other synthesis problem, is to try to think of the *last* reaction you would perform. What chemicals might you mix to make *this* compound? In order to approach this question, you must have in your memory banks a list of all the reactions that result in various functional groups. If you sift through those vaults, you should find one *and only one* reaction that results in ethers. *Do not try anything else!!!!!*

The reaction you know that results in ethers is the reaction of an alkene with an alcohol in the presence of acid catalyst. Your task now is to figure out which alkene and which alcohol you might mix to pull this off. There are several possibilities. We covered this last chapter, so it should be review.

Before going any further, we'd better check to make sure that each of these "disconnections" will really work the way we have said it will. Sure enough, the precaution was an important one here: if you follow the reactions through, you will find that the one on the right side will work as written, but the one on the left side will not, because when the double bond at the lower left is protonated, it will not give a carbocation at the end (primary) position but instead at the secondary position, according to Markovnikov's rule. So, we must rule out the left approach and focus on the right one. This approach requires one of the two alkenes shown and also the alcohol shown next to them.

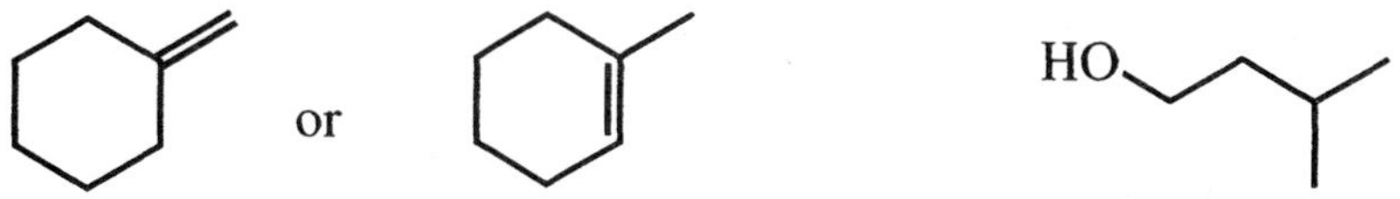

or

Now our rules of the game stated that we had to start with alkenes as our only sources of carbon. Clearly both of the alkenes shown are fair game, so we are done with that part of the problem, but the alcohol we have written is not an allowed starting material. As I told you before, this does not mean we cannot use it, only that we cannot stop here. We have to show where it will come from: ultimately some alkene. However, and this is important, the problem has changed. We no longer care a whit about the ether specified in the problem. That is not the issue anymore; we clearly have found *out* how to make it! Our only concern, at this point, is how we will make the alcohol shown above. Everything else is taken care of.

So we have a new problem: how can we make that alcohol? Again, it is not necessary that we make it immediately from an alkene, only that whatever we use, call it A, itself has to be made ultimately from an alkene. In other words, we take this one step at a time until we discover logically how we can use an alkene as our precursor. *You cannot force synthesis!!*

We approach this problem, like the last one, by asking how we might make the compound in question. Again, if you search your memory banks for ways to make primary alcohols, you will find only one method (at the moment). You *must* use *that method*! The only way you know to make a primary alcohol is by the reduction of an aldehyde:

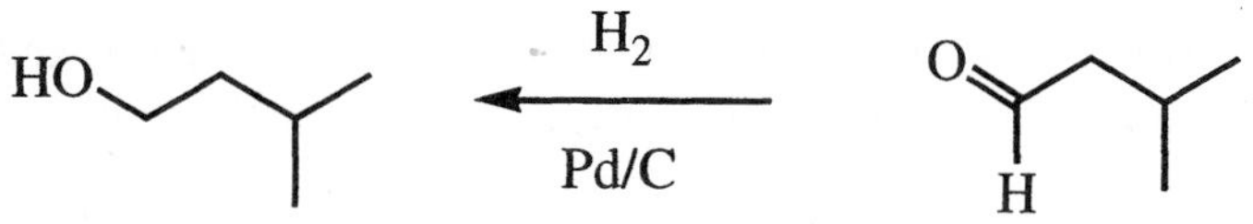

You have no choice in the matter. This is the only method you know for making primary alcohols, and this particular aldehyde is the only starting material that will give you *this* primary alcohol. But it's still not an alkene. Aren't we violating the rules? No, it only means we are not yet done. Once again, we have a new problem. Our problem now is, how can we make the above aldehyde? Another search of the memory banks (or your cards, until they are reliably transferred to your memory banks) will produce one and only one reaction that makes aldehydes. This is the ozonolysis of an alkene. AHA!! An alkene! Which one? There are actually many alkenes that would work here, but only two or three that make sense to write down.

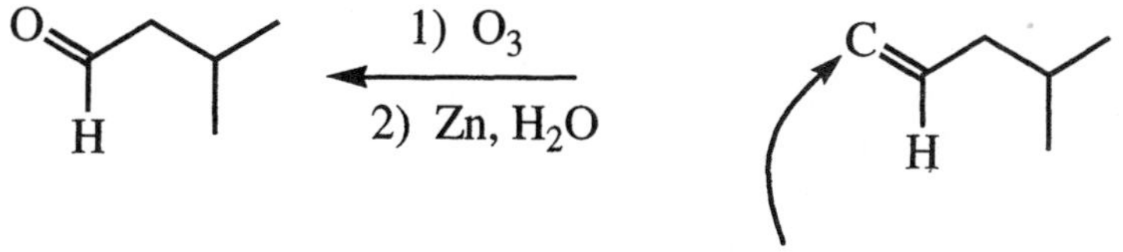

where anything at all could be on this carbon

The three choices that seem most logical to me are, first, the simplest possible answer, where there are hydrogens on the C in question. A disadvantage of this approach is that you will lose a piece of the molecule, but so what! A second possibility is one in which, after slicing, both pieces are what you want; i.e., start with a symmetric alkene. There are two such symmetric alkenes. Thus, I would choose one of the following:

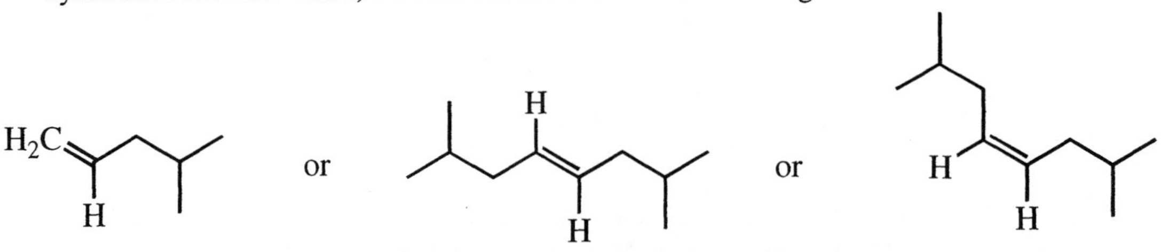

or

or

A well-written answer to the original question would look like this:

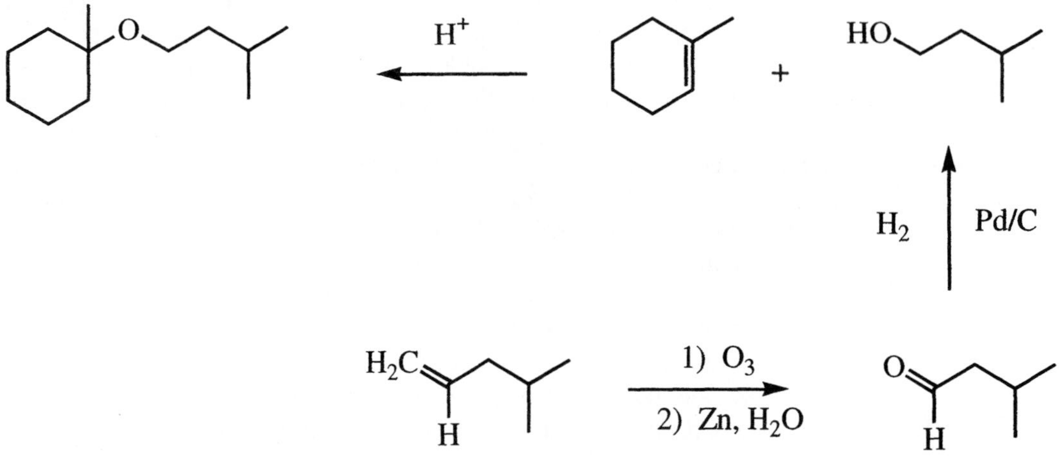

Notice that every carbon that ends up in the final product can be traced back ultimately to a carbon in an alkene molecule, thus fulfilling the requirements of the problem.

Synthesis at this stage of your studies is very easy, because for almost every transformation you might need, there is only one possible choice. Thus, it behooves you to get used to the thought process now by practicing *a lot*, so that later, when there are more choices available to you, you can concentrate on the chemistry and not worry about how one approaches synthesis problems in general.

Problems

(Answers are provided at the end of the chapter for italicized problems.)

1. Show the major organic products of the following reactions.

    a.

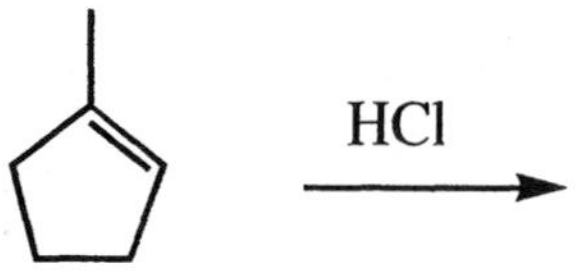

    HCl

    b.

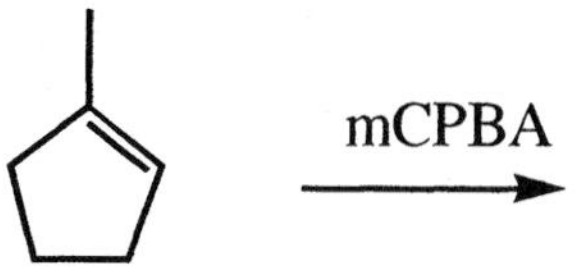

    mCPBA

    c.

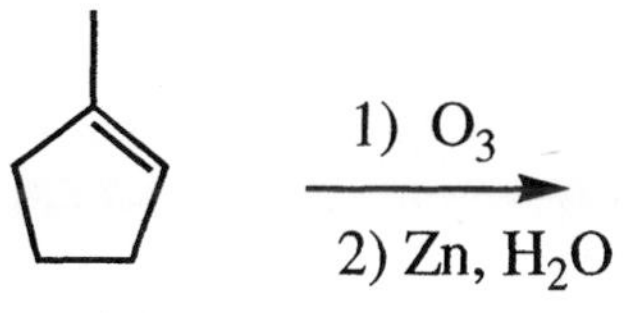

    1) O_3
    2) Zn, H_2O

    d.

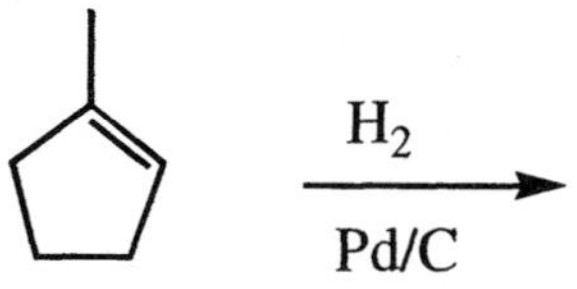

    H_2
    Pd/C

    e.

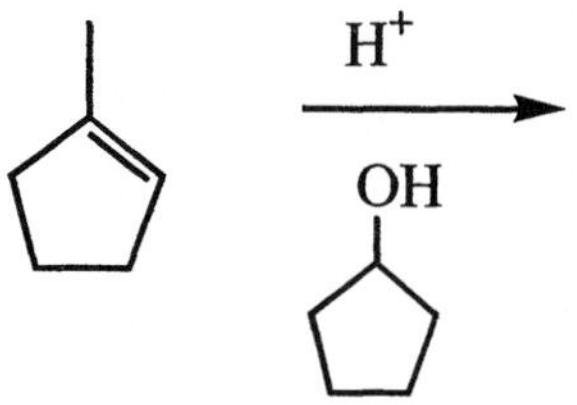

    H^+
    OH

2. Suggest a mechanism for the following reaction.

    H^+

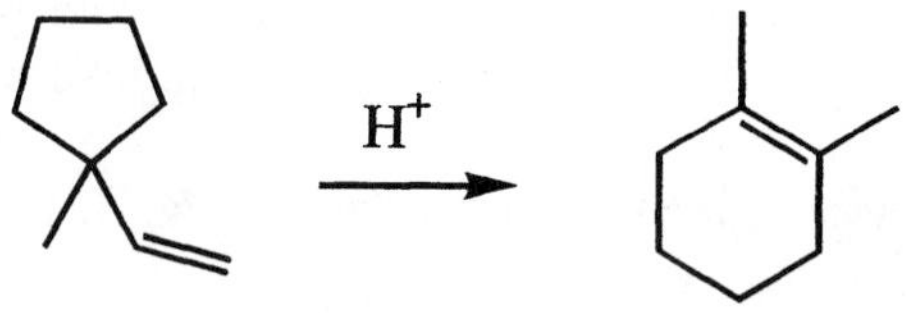

3. Suggest a mechanism for the following reaction, and explain why it occurs.

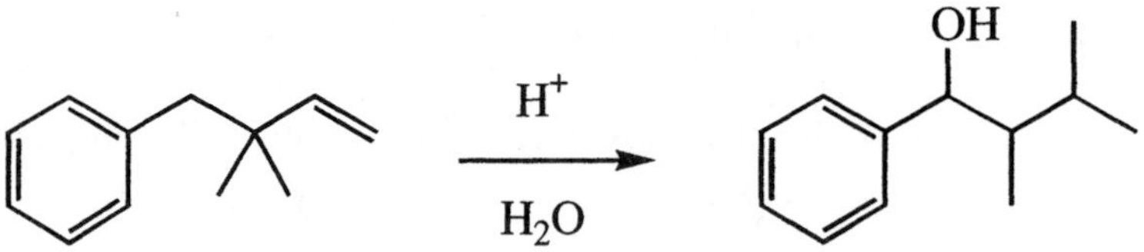

4. *If bromine dissolved in concentrated sodium chloride solution is added to cyclohexene, the product is largely 1-bromo-2-chlorocyclohexane. Explain this, using a reaction mechanism. What stereochemistry do you predict for the product (cis or trans)? Explain why.*

5. A substance **A** has a molecular weight of 136 and contains only C and H. Upon ozonolysis followed by zinc, it gives only the following compound. Show three different possible structures for **A**. (*Hint in answer section.*)

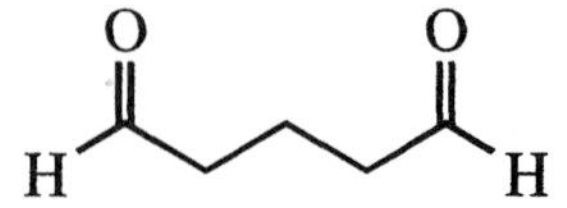

6. *A compound **C** has the formula $C_{12}H_{16}$. Upon ozonolysis followed by zinc, it gives only the following compound.*

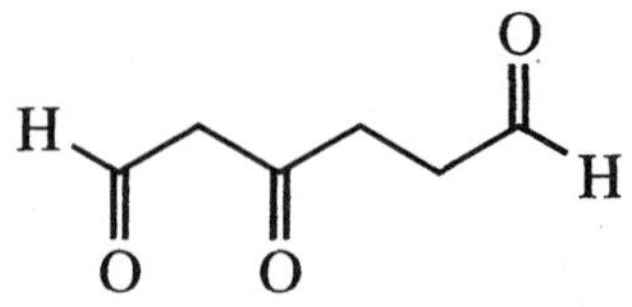

   a. *Show 4 different possibilities for **C** (there are at least 18).*

   b. *When **C** is treated with one equivalent of Br_2, a mixture containing three different products of formula $C_{12}H_{16}Br_2$ is produced. Which of your above answers are still possibilities?*

7. A molecule has the formula $C_8H_{11}Br$.

   a. How many units of hydrogen deficiency does it have?

   b. If 100 mg of this compound is treated with hydrogen and a palladium catalyst, 2.13 mg of hydrogen react with the compound. How many rings are in the compound?

   c. If the product of the reaction from part b is 1-bromo-2-ethylcyclohexane, show 10 possible structures for the original compound (there are at least 30).

8. You are given 4 bottles, labeled **A**, **B**, **C**, and **D**, all with the formula C_4H_8. You do the following experiments:

A reacts with H_2 and palladium to give a new compound, **E**, and 28.7 kcal/mol of heat.
B reacts with H_2 and palladium to give the same new compound **E**, and 27.5 kcal/mol of heat.
C reacts with H_2 and palladium to give the same new compound **E**, and 30.4 kcal/mol of heat.
D does not react with hydrogen under these conditions.

A, B, and **C** all react with HBr to give the same new compound **F**. **D** does not react with HBr.

A and **B** react with Br_2 to give the same new compound **G**. **C** reacts with Br_2 but gives a different new compound **H**. **D** does not react with Br_2.

Identify **A–H**.

9. A 4.1-g sample of compound **A** (molecular formula C_6H_{10}) is found to react with 0.20 g of hydrogen (H_2) in the presence of platinum catalyst.

   a. How many grams of bromine (Br_2) should react with 8.2 g of **A**?

   b. Treatment of **A** with ozone followed by zinc leads to three products: **B**, **C**, and **D**. **B** is found to be a three-carbon ketone, and **D** contains only one carbon atom. Show the structures of **A** through **D**.

10. Suggest syntheses of the following compounds, starting with any hydrocarbons you like. A model will be helpful for the last one. (*Answer to part c given in the answer section.*)

a.

b.

c.

Selected Answers

Internal Problems

Test Yourself 1.

Resonance-stabilized cation is better than secondary.

Test Yourself 2.

End of Chapter Problems

4.

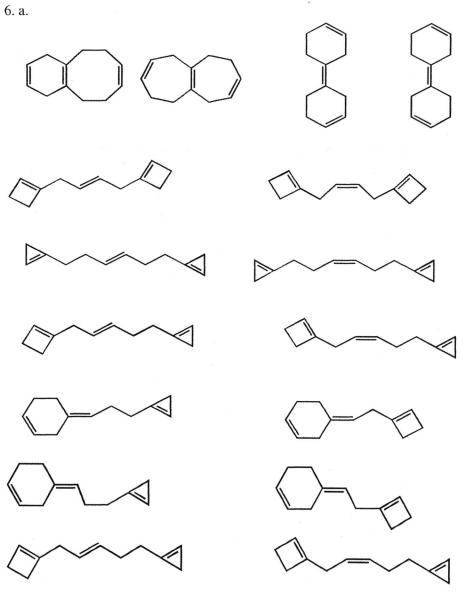

5. *Hint*: Remember stereochemistry.

6. a.

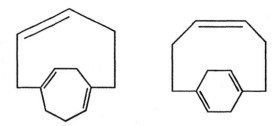

6. b. This rules out any compound that has only two kinds of double bonds. There are eight such structures ruled out: all of the first eight on the list except for the first one, and also the last structure on the list.

10. c. This has proved difficult for many students, so I'll lead you through this one. The goal is to synthesize *cis*-1,2-dibromocyclodecane. I have shown this in two different views below.

We know that when bromine adds to a double bond, the two bromines come in *trans* to each other. Thus, when this compound was first formed, it must not have been in this conformation, but rather the following:

(This will be *much* easier to see with a model.) Thus, it should be clear that the double bond to which bromine added was the one that looked like this:

The only reason we can accomplish this is that the ring is large enough to allow a *trans* double bond to exist in it.

Chapter 12
Alkynes

Alkynes are molecules containing triple bonds. You will recall that a triple bond consists of a σ bond and *two* π bonds. Therefore much of the chemistry of alkynes is similar to alkene chemistry. There are, however, some interesting new wrinkles that we will encounter. In this chapter we will

- Learn the similarities and differences between alkene and alkyne addition reactions

- Discover keto–enol tautomerism and a new way to make ketones

- Learn two ways to reduce alkynes to alkenes, specifying stereochemistry

- Remember that terminal alkynes can be deprotonated.

NOMENCLATURE

Alkynes are named just like alkenes, except that you use the suffix "-yne" to denote a triple bond, rather than "-ene" to denote a double bond. Of course, the problem of geometry (*cis* vs *trans*) does not apply to triple bonds, since the two atoms of the triple bond and the two atoms they are bonded to are all in a straight line. The only special name you need to know is that the simplest alkyne, ethyne, is almost always called "acetylene" instead of ethyne. This is unfortunate, since the "-ene" ending makes you think it is an alkene, but it is not. Since all other alkynes could be thought of as derivatives of acetylene, that is, acetylene with other stuff hanging on it, the whole family is sometimes referred to as acetylenes rather than alkynes.

Here is an example.

This would be *trans*-4-ethyl-4,5-dimethylnon-6-ene-2-yne. It is an alkene and also an alkyne (or an acetylene).

PROPERTIES

You may recall that single bonds are about 1.54 Å long, and double bonds are about 1.33 Å. As you might guess, triple bonds are shorter than double bonds, about 1.20 Å. They are also stronger than double bonds, of course, consisting of three bonds instead of two, but as was the case for alkenes, the π bonds are weaker than the σ bond.

Like alkanes and alkenes, alkynes tend to be non-polar and therefore soluble in other non-polar substances but not in polar ones. Also (again like the others), the simple alkynes are gases at room temperature up to 4 or so carbons, liquids up to about 18–20 carbons, and then solids.

REACTIONS

Most of the reactions of alkynes are simple extensions of what we have already seen for alkenes. In particular, addition reactions are quite common, and indeed, the results are predictable based on what we already know.

Addition Reactions

Hydrogen Halides

In Chapter 10 we learned that when HX adds to an alkene, it does so in two steps, first protonating the double bond to produce a carbocation (and if there is a choice, the better carbocation), followed by a reaction between the halide anion and the new carbocation. The result is an H on one carbon and an X on the other, according to Markovnikov's rule:

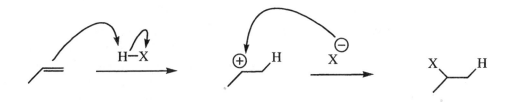

When HX adds to an alkyne, exactly the same thing happens. However, there are a couple of new wrinkles that we need to address.

First, you should notice that there are only three possible substitution patterns with alkynes: unsubstituted, monosubstituted, and disubstituted, as shown.

$$H-C\equiv C-H \qquad H-C\equiv C-R \qquad R-C\equiv C-R$$

There is only one alkyne that is unsubstituted—acetylene itself—so we needn't spend much time on that system. Monosubstituted alkynes, by necessity, have the triple bond at the end of a chain, so they are often referred to as "terminal acetylenes," while disubstituted alkynes can be called "internal acetylenes," since the triple bond is *not* at the end of some chain (the significance of this will become apparent soon).

Acetylene itself and all the internal acetylenes are unaffected by Markovnikov's rule, since both ends of these triple bonds are equivalent in the eyes of a carbocation. Notice particularly the case of internal acetylenes, where the two ends would lead to equally stable *but not necessarily identical* carbocations. For example, 2-pentyne could react with HCl in either of the following ways:

You should notice several things:

1. There are five carbons in this particular starting material. Many students have trouble with the picture as it is drawn, thinking that the triple bond extends for the length of the horizontal line. But if you think about the geometry of an alkyne, with sp-hybridization for the two carbons involved in the triple bond, there *must* be four carbons in a straight line, so the picture is quite logical. You simply have to mentally fill in carbon atoms not only at the ends of the triple bond *but also* at the ends of the lines coming out from them. So the following are two pictures of the same material:

2. When a proton attaches to one of the carbons, the hybridization changes from sp to sp^2, and the geometry from linear to trigonal, as shown. However, the remaining carbon still has only two areas of electron density around it and therefore remains sp-hybridized and linear. The carbon with the positive charge has *two* p orbitals on it. One is involved in the double bond with the partner, while the other is empty and corresponds to the positive charge.

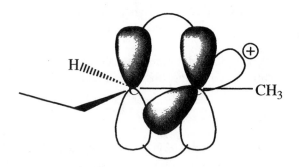

This is actually a double bond, one end of which is also a carbocation, a so-called "vinyl cation." Vinyl cations are less stable than most saturated cations—they are similar to primary cations in stability—and they are therefore harder to form. Addition reactions with triple bonds tend to be slower than the analogous reactions with double bonds, because the intermediate for the triple bond reaction (the vinyl cation) is higher in energy.

3. When a chloride ion attaches to the carbocation, it can come in from the side *cis* to the alkyl group and *trans* to the H, or from the opposite side (*trans* to the alkyl and *cis* to the H). If the chloride comes in *cis* to the alkyl group you get the Z product, but notice that the carbon chain actually runs *trans* in this substance. In the picture I drew of the reaction I showed only the Z product, but in fact there is nothing to prevent the E product from being formed also, so I should actually have shown four different products. We can even predict *which* products should be most abundant. Recall that double bonds cannot rotate on their own, so the result will not be determined by which product *feels* better (is lower in energy) but rather by which one is *made* more easily. You might guess from the picture of the carbocation that a chloride anion would have an easier time entering from the side bearing the H rather than the side where the R group is, so you would then expect to get more *E* product than Z product. This is generally found to be true. (Incidentally, compounds with halides directly attached to double bonds are called "vinyl" halides.)

4. Each predicted product has a double bond in it, and HCl reacts with double bonds. Therefore, we should expect that under the reaction conditions, it is at least possible that we won't actually isolate the products shown above; they continue on to make new products. I'll show the situation for one of them, and you can trace the process with the other:

At this point we need to decide which of the two types of dichloride products shown will actually be formed (or whether, indeed, there will be any preference at all). Again we rely on our understanding of how the reaction works: it is a two-step process with a carbocation intermediate. Like all the others we have seen of this type, the product will be determined by which carbocation is better. Which will it be? As always, the best approach is to look at the options:

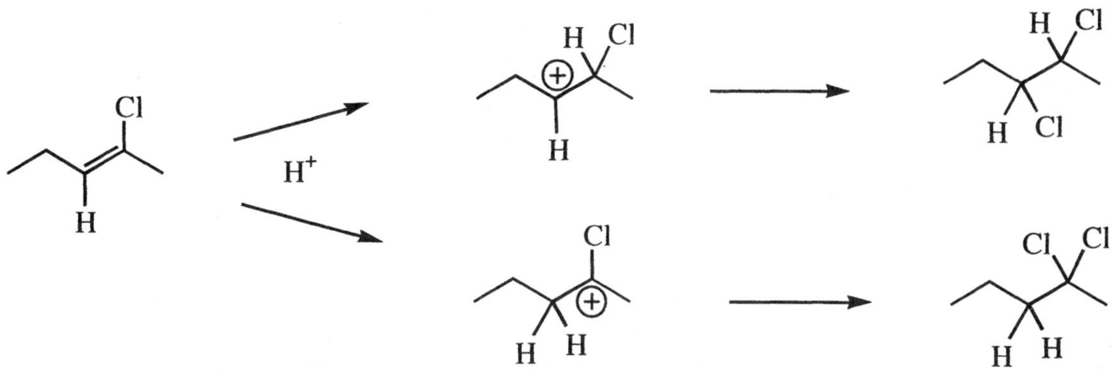

Does either look better or worse than the other? Many students conclude that the bottom carbocation is worse, for good reason: it has a chlorine attached. Chlorine is electronegative and would draw electron density away from a carbon already suffering from a shortage of electrons. This is good thinking, but it goes only part of the way. What this suggestion overlooks is that chlorine also has lone pairs on it, so the bottom carbocation has another resonance structure that can help delocalize the charge, while the upper carbocation does not have such a resonance structure.

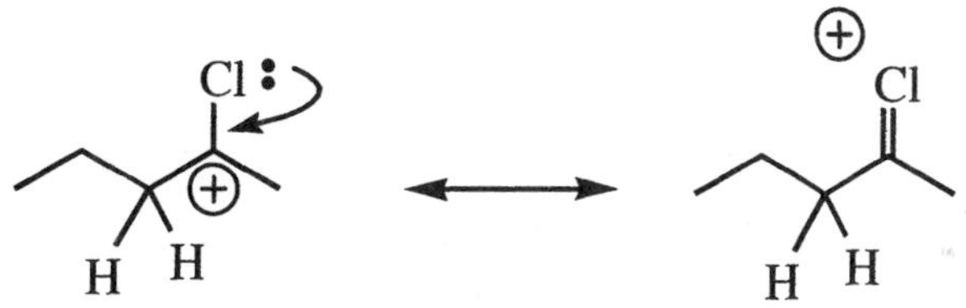

Thus, we have one valid argument saying that the bottom carbocation should be worse due to the electron-withdrawing effect of the chlorine (this is called an "inductive" effect), and another argument (also valid) saying that the bottom carbocation should be *better* due to the electron-donating lone pairs (resonance effect). We have actually seen this problem before, you may recall (in Chapters 9 and 10), and as we discovered there, we also find here that resonance is more important than an inductive effect. In other words, in real life (and we could not know this for sure without running the experiment and *observing* what really happens), the bottom route is the one the molecules choose. So if HCl adds *twice* to an alkyne, the second chlorine always goes on the same carbon as the first one, leading to what is known as a "geminal" dichloride. (Remember that the alternative, a "vicinal" dichloride, could be made by adding Cl_2 to a double bond.)

Monosubstituted (terminal) alkynes react with HX exactly as you should expect: first the proton adds, leaving the carbocation on the more substituted end, and then the halide takes care of the positive charge. If HX adds twice, both X's again end up on the same carbon, the one that was substituted in the original alkyne.

As you should also expect, water adds in exactly the same way HX does, by analogy to what happens with an alkene. Terminal alkynes add specifically in one direction, while internal alkynes can produce two different products.

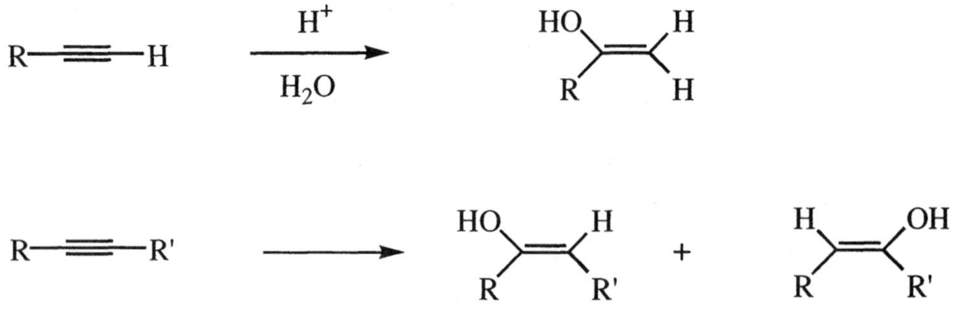

R≡H $\xrightarrow[\text{H}_2\text{O}]{\text{H}^+}$ [HO and H on the alkene carbons, R and H]

R≡R' ⟶ [enol: HO, R, H, R'] + [enol: H, R, OH, R']

You should have no trouble writing a mechanism for this, one that exactly mirrors what you did for alkenes. Don't forget that what finally joins with the cation is not OH⁻ but H_2O.

Keto-Enol Tautomerism

Don't make cards yet, because we are not done. There is an important difference between this alkyne reaction and the preceding one: whereas it is possible to stop the addition of HX to a triple bond and obtain a vinyl halide, it is not possible to stop the addition of water to a triple bond to obtain a vinyl alcohol (vinyl alcohols are called "enols"). In other words, the products shown above are *not* the products you actually isolate from this reaction. What do you get? The clear implication from what I said is that the process goes farther, and "farther" most likely means more addition to the new alkene. Let's examine the situation The next step will be:

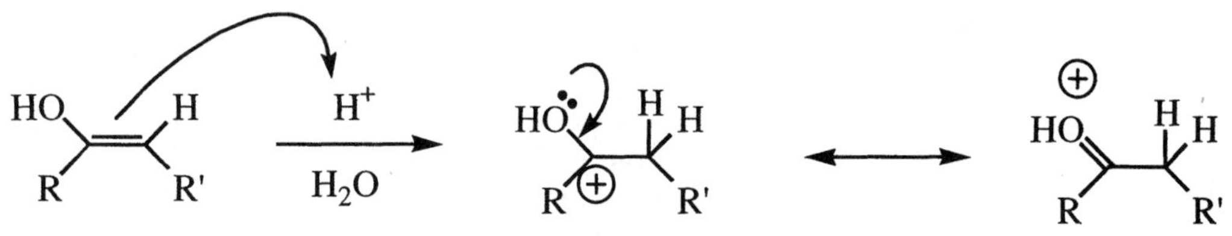

[mechanism scheme: enol with curved arrow, $\xrightarrow[\text{H}_2\text{O}]{\text{H}^+}$, carbocation intermediate with resonance, ⟷, oxocarbenium resonance structure]

Notice that I've paused once again to show a resonance structure. It is critically important that you begin to see this kind of resonance for yourself, because it crops up all the time. One of the consequences of such resonance is precisely the circumstance we are dealing with now: one cannot stop this reaction at the enol stage because the second protonation happens faster than the first, and it happens so quickly because the cation that arises is happier than most carbocations, meaning that the reaction has a smaller hill to climb. Another consequence of this resonance is that the new carbocation finds an alternative way to get rid of its positive charge. Not only can it go back where it came from or pick up a nucleophile (such as a water molecule), but it can also eject a proton on oxygen to regain neutrality.

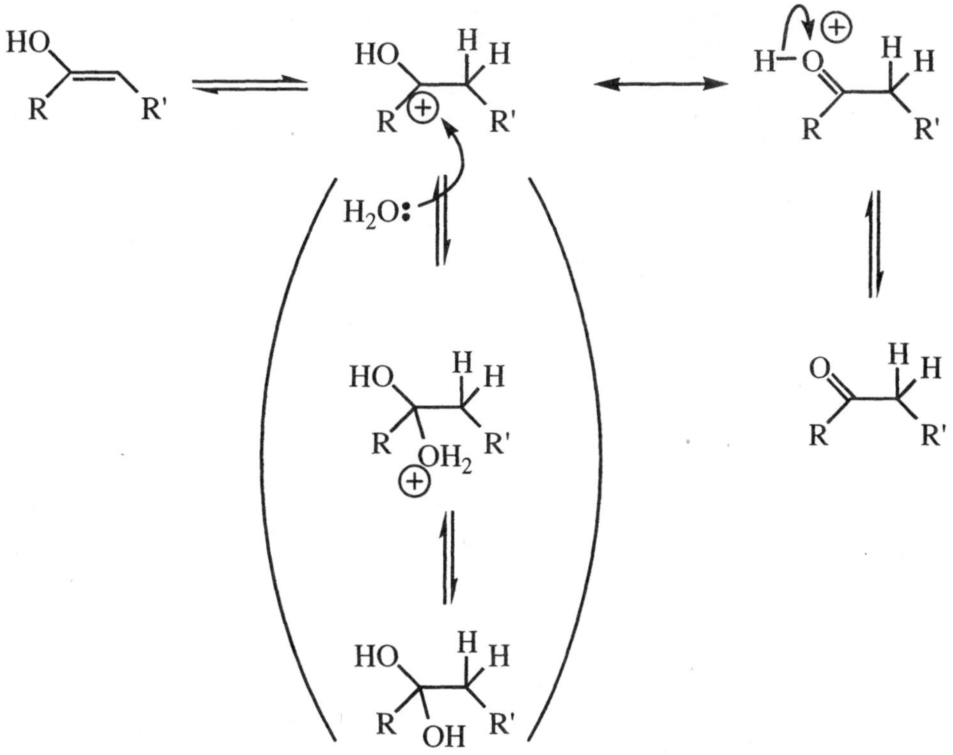

HO–C(R)=C(R')H ⇌ structure with HO, H, H, R, R' and (+) carbocation ⟷ resonance structure H–O(+)=C(R)–C(R')(H)(H)

H_2O:

(middle route)

HO, H, H / R, OH_2(+), R'

⇌

HO, H, H / R, OH, R'

(right route)

O=C(R)–C(R')(H)(H) (ketone)

If it chooses the right-hand route, the product is a ketone. If it chooses the middle route, the product is a geminal dialcohol ("geminal" is the same word we used above to refer to two groups on the same carbon). You have no way of knowing it at this point, but in fact a ketone is generally a lower energy species (happier) than a geminal dialcohol. Now, this would be an irrelevant consideration if the molecules had no way of jumping back and forth; that is, reactions lead to the best (most stable) product only when there is an opportunity for choice, and "choosing" means that the process can try out all the possibilities without getting stuck anywhere. You should notice that the picture above is full of equilibrium arrows, which means all the reactions are reversible. So in this case, a molecule *can* choose which product it prefers to be, and it does. The product of this reaction is always a ketone and not a dialcohol.

The conversion of an enol to a ketone (and back; the latter conversion does occur occasionally even though the molecule is happier as a ketone—remember equilibrium constants) is a member of a special class of reactions called *tautomerisms*. Ketones and the corresponding enols are called tautomers. The distinguishing feature of tautomers is that they are related to each other by the shift of a double bond and a proton.

(enol, C=C with O–H) ⇌ (ketone, C=O with CH)

The double bond moves to an adjacent position, and a hydrogen moves from the end of the new double bond to the end of the old double bond. No other change occurs. This is tautomerism. Now we have also seen *how* this happens, so we know that this latest description of the phenomenon is misleading. In particular, we know that the H shown in the enol is *not* the same atom as the H shown in the ketone. Recall: the way the change occurs is that, first, the double bond reacts with H^+ (from the solvent, not the hydrogen from the OH of the enol), and in a second step the H on oxygen is lost. Nevertheless, the net result is that hydrogen is now in a new position, and the double bond has moved. Tautomers. We will see these again.

You have just learned a new way of making ketones. You should make a card for this important reaction to help you remember it. The overall transformation looks like this:

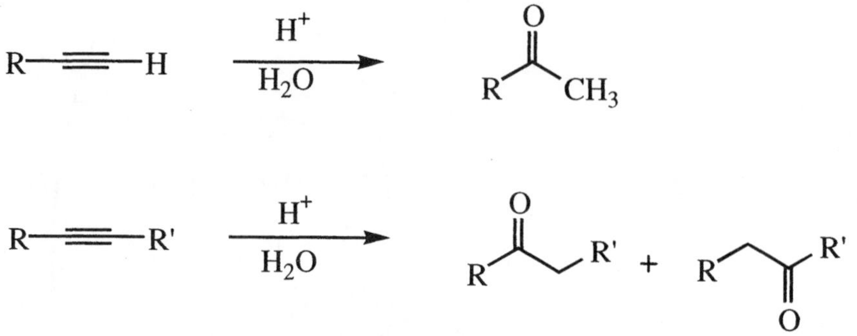

It should be clear to you why I have shown only one product for the top reaction and two products for the bottom reaction. If the reason is not clear, you need to read this section again, working out in detail for yourself all the mechanisms referred to. It should also be clear that this method is a very nice way of making methyl ketones, that is, ketones that have nothing but a methyl group on one side and almost anything you want on the other. But it is not so hot for other ketones, because when you start with an internal alkyne, there is no way to steer the reaction toward one carbon to the exclusion of the other.

Here is a complete picture of the mechanism for the top reaction, from start to finish:

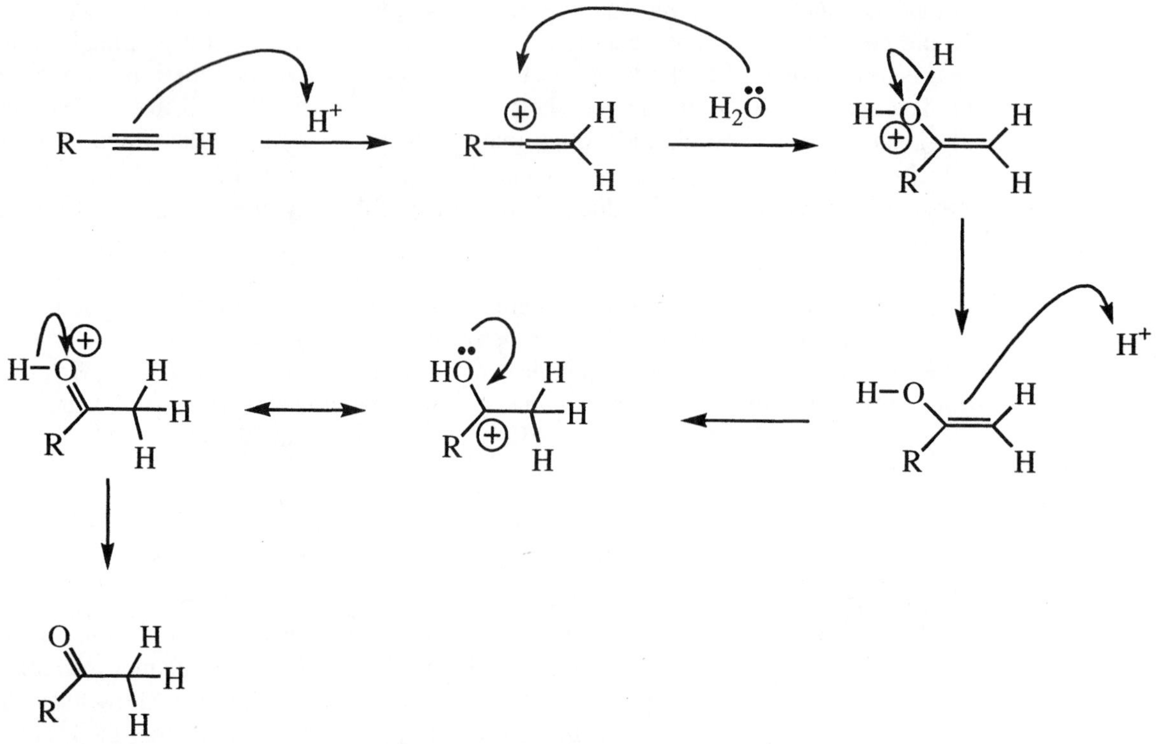

Notice that at the appropriate place I used a resonance arrow instead of an action arrow. It is wrong to use a one-headed arrow when you need a resonance arrow, and *vice versa*.

There is one more technical detail about this reaction that I should mention. I told you earlier that vinyl cations are less stable than ordinary cations and, therefore, electrophilic addition reactions on triple bonds are hard to get started (since the very first thing you must do is *make* a vinyl cation). Occasionally, these reactions need an extra kick in the pants to get going. This particular reaction, that of an alkyne with acidic water, is such a reaction. Although many books (including this one) gloss over the details, this reaction is almost always run in the presence of a mercury salt, usually mercuric sulfate, $HgSO_4$. The Hg^{2+} cation apparently acts as a Lewis acid and somehow serves to get the process started. The precise way it does so is a bit confusing, and I prefer not to discuss it. So for our purposes you are free to write the mechanism the way I have shown you. The only change you need to make is that in the overall reaction statement you should add this mercury salt as an ingredient:

Be sure to make this change on the cards you just made.

Bromine

Bromine adds to triple bonds exactly the way it adds to double bonds. The reaction proceeds *via* a cyclic bromonium ion, and the product contains two bromines, this time separated by a double bond. The two are specifically *trans* to each other. The reaction is much slower than that with an alkene, however, because the intermediate bromonium ion is a three-membered ring with a double bond present in the ring. This is considerably more strained than a three-membered ring containing only single bonds, and it is therefore harder to make, so, as I have said, the reaction proceeds more slowly.

The above considerations imply that it should be difficult to make a dihalide from an alkyne, since the product dibromoalkene should react more rapidly with bromine than the alkyne itself, becoming a tetrabromide. It is easy to make tetrabromides this way, by using plenty of bromine, but apparently if you limit the amount of bromine you *can* get the dibromoalkene. This fact appears to be inconsistent with the statement that alkynes react more slowly with bromine than alkenes do. Both statements appear in many books, but I have not been able to find any source that reconciles the two. I believe it is because the first two bromines are electron withdrawing, making the resulting dibromoalkene much less nucleophilic than an ordinary alkene, and apparently even less reactive than an alkyne.

Ozonolysis

Ozone, O_3, reacts with triple bonds as well as with double bonds. The mechanism is even more complicated than that for the alkene reaction, so again we will not go through it. Again the reagent slices right through the molecule, this time through all three bonds of the triple bond, leaving a carboxylic acid at both ends. For example,

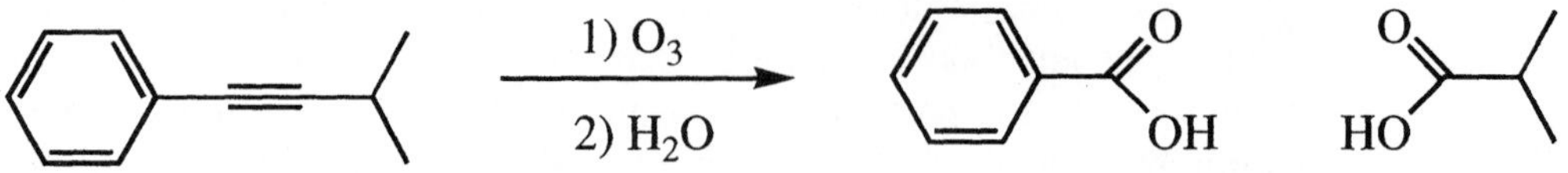

Notice that in this case no zinc is needed in the second step, unlike the reaction with alkenes. This may be somewhat difficult for you to remember, since I never explained what the purpose of the zinc was in the first place. It serves an important function in the alkene reaction, and if it were missing there would be different products. It is not *necessary* that you omit it from the alkyne reaction, by which I mean that if you put it there anyway, it will not mess up the reaction, it simply will waste some zinc. Later I will explain all that, but for now, just try to remember the distinction. To be honest, it won't be so terrible if you forget.

Reductions

To cis *Alkenes*

You will recall that one of our magic reactions of alkenes was the hydrogenation reaction, in which we added hydrogen gas with a catalyst. The result was that two hydrogen atoms are added, one to each carbon of the double bond. We can do the same thing with triple bonds, under the same conditions:

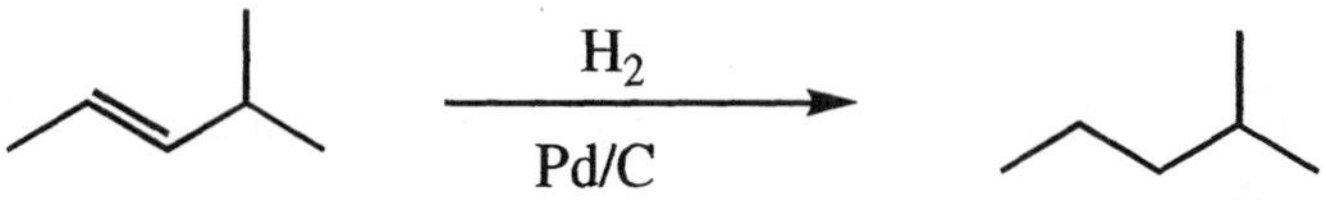

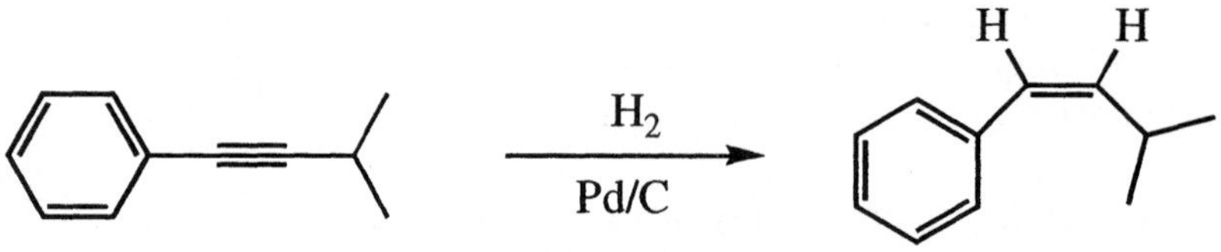

You should also remember, in the case of alkenes, our establishing that the two hydrogen atoms always come in from the same side of the double bond, suggesting that they both attach more or less simultaneously. The same is true with triple bonds, so you always get a *cis* double bond from this process. However, this raises a problem: The product is an alkene, and alkenes themselves react with hydrogen and the same catalyst already present in the reaction mixture. Can we actually stop the reaction at the alkene, or will it always continue on to add another hydrogen molecule and generate a saturated product (an alkane)? This is a question we cannot answer just by looking at the molecules: it requires an experiment. We might, for example, run this reaction with a limited amount of hydrogen and see what happens. Before we do this, however, it is worth thinking about the factors that would determine what we see.

Consider the general reaction shown below.

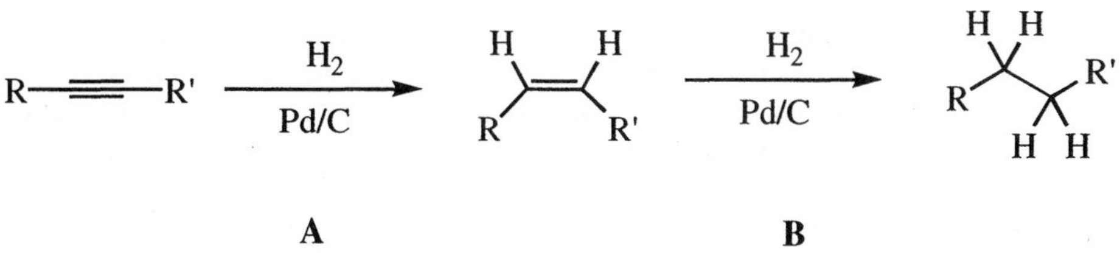

$$R\!-\!\!\!\equiv\!\!\!-R' \xrightarrow[\text{Pd/C}]{H_2} \text{(alkene A)} \xrightarrow[\text{Pd/C}]{H_2} \text{(alkane B)}$$

A   **B**

Imagine that reaction A is faster than reaction B. What would happen if we used only a limiting amount of hydrogen: say, 1 mole of H_2 per mole of alkyne (that is, only enough to reduce each molecule once, but not twice). At the beginning, of course, some alkyne is certain to get converted into alkene. But after a while there will be both alkene and alkyne present in the reaction mixture, the two competing for a limited amount of hydrogen. If reaction A is faster than reaction B, the alkyne will win the competition, and to a first approximation all the alkyne will be converted to alkene, but none will go on to alkane (because we will run out of H_2). So the only product will be an alkene. But what if reaction B is faster than reaction A? As soon as we convert some alkyne into alkene, that alkene will react with more hydrogen, and we will get alkane. Of course in order to do this we use up two molecules of H_2 and only one of alkyne. Reaction will continue until we run out of hydrogen. What will be left? Half the alkyne will have been converted into alkane, and the other half will not have reacted at all (because we ran out of H_2). Thus, the question we asked above (can we stop at the alkene?) really depends on whether the addition of hydrogen to alkynes is faster or slower than the addition of hydrogen to alkenes. As it turns out, alkynes react faster.

Thus, by carefully limiting the amount of hydrogen we use, it is in fact possible to reduce alkynes to *cis* alkenes. But measuring gases carefully is an experimental pain, and chemists have figured out a better way to accomplish the same thing. Consider: if reaction A is easier than reaction B, but neither occurs at all without a catalyst, it is at least conceivable that there exists a catalyst good enough to make reaction A happen but not good enough to make reaction B happen. Indeed, there is; in fact, there are several of them. All of them are based on the traditional metal catalysts we have been using, but each has been "doped" with a substance (called a "poison") that reduces its reactivity. The precise nature of the poison varies with the recipe chosen, and we do not need to get into the details here. Suffice it to say that when you need to do this, you can look it up. Catalyst with poison goes by the generic term "Lindlar catalyst." With a Lindlar catalyst it is not necessary to limit the amount of hydrogen used, because no matter how much is present, it will never react with a double bond, only with a triple bond. Thus, you can easily reduce triple bonds to *cis* double bonds with hydrogen and Lindlar catalyst. We'll write it this way:

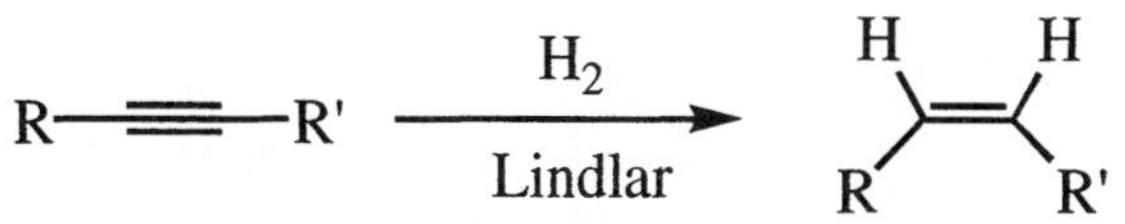

$$R\!-\!\!\!\equiv\!\!\!-R' \xrightarrow[\text{Lindlar}]{H_2} \text{(cis alkene)}$$

You do not really need to know this, but to satisfy your curiosity, among the "poisons" used to make Lindlar catalyst are barium sulfate ($BaSO_4$) and quinoline in appropriate amounts.

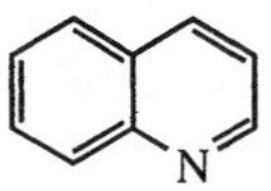

quinoline

If you want a double bond in a particular place in a molecule, and if you want it specifically *cis*, reduction of an alkyne with hydrogen and Lindlar catalyst is the perfect way to proceed. But what if you want the double bond to be *trans*? Catalytic hydrogenation will not work, because the two hydrogens always come in from the same side. Fortunately, there is an alternative reduction method for alkynes that does result in *trans* alkenes, so in fact you can make any disubstituted alkene you want starting with an alkyne. Since (as we will discover later) alkynes are fairly easy to make, this opens up a lot of possibilities.

The reduction of alkynes to *cis* alkenes is a consequence of the *process*: the two hydrogens come in at the same time from the same side. There is no possibility of making a *trans* alkene *even though the trans isomer is usually better*! If you want to make the product *trans* instead of *cis*, you take a different approach and allow the molecules to choose for themselves. The distinction I am making here is one that will come up frequently, and it has a special name: kinetic *vs.* thermodynamic control of a reaction process. When the products are determined by the *process* of making them (i.e., whichever one is made preferentially is what you get), the reaction occurs under kinetic control, meaning that the *rate*, or speed, of competing processes determines what you get. If a reaction gets to try out all the various possibilities and decide which product it *likes* better (which requires that it be reversible!), then the reaction is under thermodynamic control, meaning that the stability or energy of the products dictates the nature of the product mixture. In this case, we want a *trans* double bond; we want the better isomer, so we can let the molecules choose by running the reaction under thermodynamic conditions. Of course, that is easy to say, but how can we in fact do it?

Alkenes never jump back and forth between their *cis* and *trans* forms, so chemists have had to come up with a process that is functionally equivalent. The approach is to feed electrons to the alkene, and do so in the presence of a proton donor. The reagent used to feed electrons is sodium metal, Na, and the proton donor is ammonia, NH_3, in the liquid state. The reaction is written as follows:

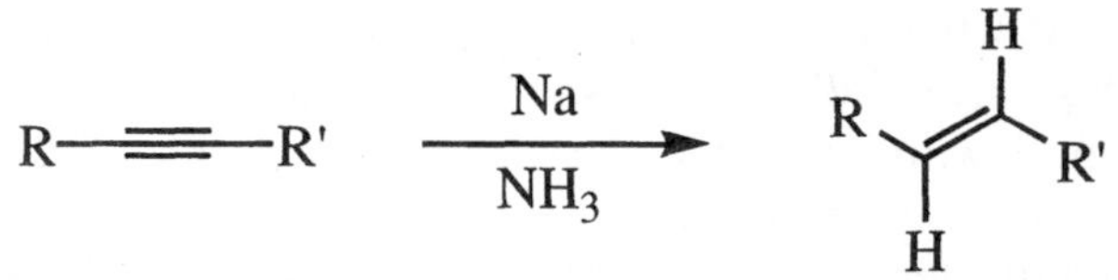

COMMON MISTAKES: DON'T LET THIS HAPPEN TO YOU!

Many students confuse the reagent Na in NH_3 with a different reagent, $NaNH_2$. To the untrained eye they look different by only one trivial number, but there is a world of difference between them. $NaNH_2$ (which we will encounter later in this chapter) is like NaOH, in that it is an ionic substance composed of Na^+ and NH_2^- ions. The Na^+ part, as always, does nothing at all, just sits around and watches the action (a "spectator ion," remember?). The NH_2^- part is a strong base, as you should know not only from the pK_a chart, but also because you know that anions on nitrogen are not particularly happy, certainly less so than anions on oxygen (like OH^-, which is no slouch of a base itself). Na in NH_3 is totally different. The Na does not have a positive charge on it—it is the neutral metal itself, which, we learned in Chapter 3, is dying to lose that electron and *become* Na^+. This is why we use it: its function is to dump that electron, and the alkyne we put in the solution will pick it up. The NH_3 is completely separate from the Na, a liquid solvent that will also serve as a source of protons when we need them.

This reaction need not be regarded as a magic reaction: we can easily understand it. The Na gives up an electron, and the alkyne accepts it. We need now to learn how to show such a process, and also how to picture the product. First, the process: so far all the arrows we have drawn in our reaction mechanisms have shown the movement of electron *pairs*, that is, two electrons at a time. In this case we want to show only *one* electron moving, and to do that chemists have agreed to use an arrow with only one barb on its head, a so-called fishhook. Here's how we write it:

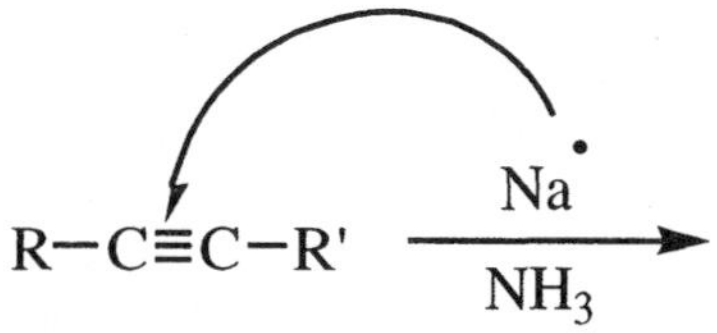

$$R-C\equiv C-R' \quad \xrightarrow[\text{NH}_3]{\text{Na}^{\bullet}}$$

So what does this make? Each carbon already has 8 electrons around it and cannot hold more. How can we simply add another? We solve this just as we would with any other Lewis structure in Chapter 3. Count electrons and distribute them as well as we can. Between the two carbons in the middle (i.e., ignoring the bonds to the R groups, which we will not touch), there were initially six electrons, so now there are seven electrons to distribute. We can do this in two ways, which differ only in electron placement and are therefore resonance structures:

$$R-\overset{..}{\underset{}{C}}=\overset{\bullet}{\underset{}{C}}-R' \quad \longleftrightarrow \quad R-\overset{\bullet}{\underset{}{C}}=\overset{..}{\underset{}{C}}-R'$$

But we are not done, because we have not yet assigned charges. The sodium lost its electron and acquired a positive charge; therefore somewhere there must be a negative charge. Sure enough, in each resonance structure there is one carbon atom that owns more electrons than it should, so it has a negative charge. We need to indicate this:

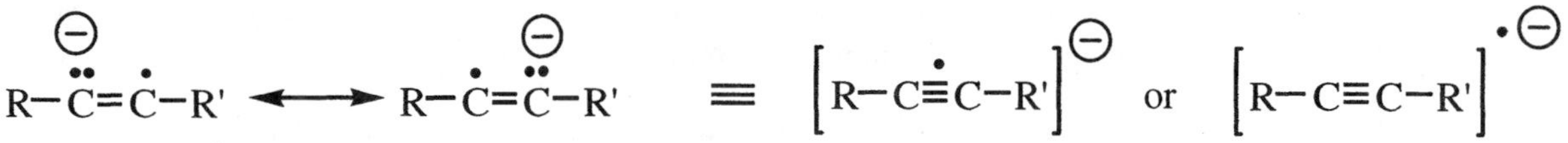

$$R-\overset{..}{\overset{\ominus}{C}}=\overset{\bullet}{C}-R' \longleftrightarrow R-\overset{\bullet}{C}=\overset{..}{\overset{\ominus}{C}}-R' \quad \equiv \quad \left[R-C\overset{\bullet}{\equiv}C-R'\right]^{\ominus} \text{ or } \left[R-C\equiv C-R'\right]^{\bullet\ominus}$$

Sometimes this is indicated through one of the special structures on the right—sort of a shorthand for the resonance hybrid. In any case, the species has an unpaired electron, so it is a radical; it also has an extra electron, so it is an anion. These things are called "radical anions." Clever, right?

In either resonance form, there is a negative charge (and an unshared pair) on a doubly bonded ("vinyl") carbon. If you look on your pK_a chart, you will find something resembling this in the base column at pK_a 44. In other words, you should expect the radical anion to be a very strong base, capable of yanking a proton from any species in the acid column higher than 44 on the chart. Is there such a species in the reaction mixture? Indeed there is: ammonia, NH_3, with a pK_a of 38. So our radical anion ought to pull a proton off NH_3:

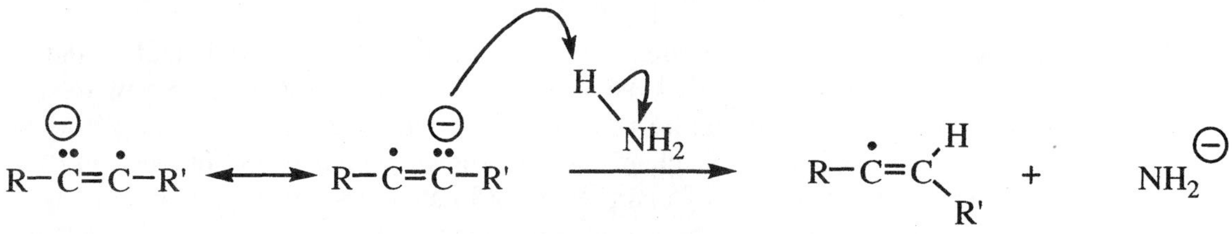

Notice several things:

1. The geometry and hybridization of one of the carbon atoms changes in this process, while the other one remains sp-hybridized and linear.

2. We made NH_2^-, and we already had Na^+ from the first step. Hanging around each other, these constitute $NaNH_2$! The substance we mentioned earlier is an accidental by-product of this reaction.

3. The organic product so far is a radical—still unhappy—and the carbon with the unpaired electron does not have an octet.

4. We can fix the octet problem by giving the system another electron (just one, so we use a fishhook again) from another sodium atom (*not* an ion!):

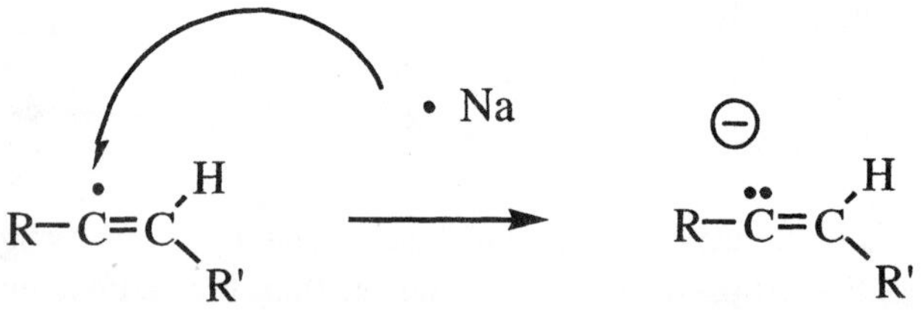

5. This time the anion is *not* resonance stabilized. The unshared electron pair is stuck in its own orbital on a particular carbon atom. That carbon atom has *three* areas of electron density around it and is no longer linear but sp^2-hybridized and trigonal. So the picture above is not quite accurate, and should be drawn as follows:

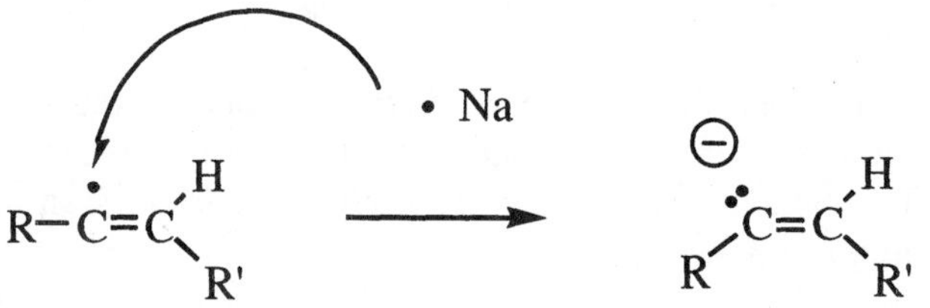

Here is where the geometric trick comes into play. We need to look at this anion in greater detail.

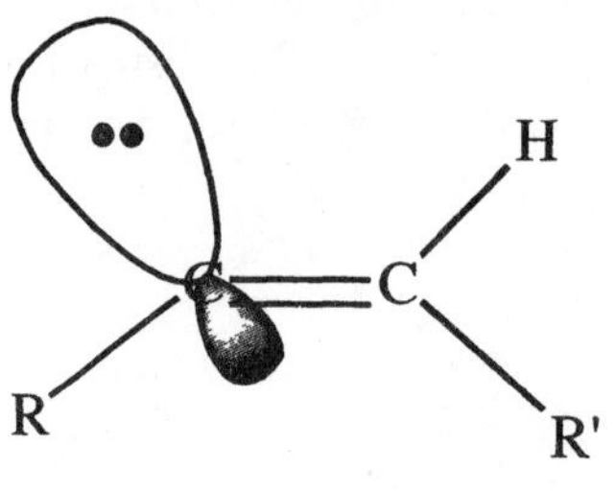

You may recall that I told you a long time ago that these orbitals (in this case, the sp^2 orbital pictured) have a small tail that normally does not matter, because some partner keeps the orbital distorted in the direction of the partner. But here the hybrid orbital does not have a partner, and there is nothing that forces it to stay pointed in the direction it is now pointing. Orbitals like this undergo a process called "inversion" (ammonia does the same thing), in which the tail becomes the big end and the big end becomes the tail, permitting the anion to flip back and forth between two geometries:

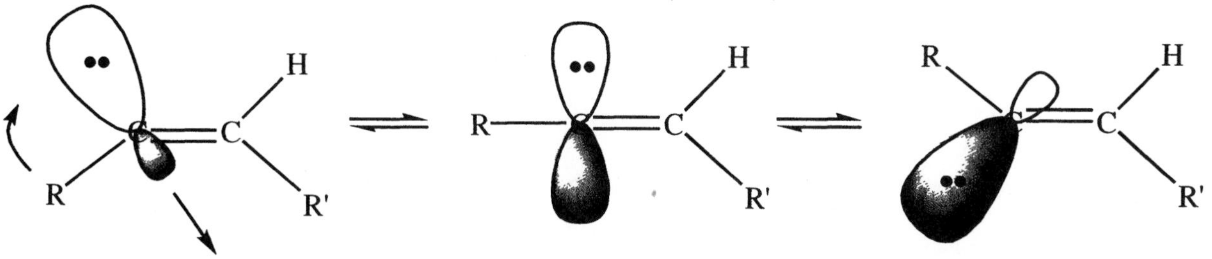

Note that this is not resonance: physical motion occurs—the R group actually wags back and forth as shown—so we use equilibrium arrows rather than double-headed resonance arrows. The important point here is that *at this stage in the process*, there is the opportunity for the anion to adopt whichever conformation it prefers, since it can freely jump back and forth between the two possibilities. For steric reasons, the anion almost always prefers the conformation on the right, where the R groups are *trans* to each other, so in this mixture, most of the anions will have the *trans* geometry.

Of course, we are not done yet, because we still have an anion, and a pretty bad one at that: it is a strong base with a pK_a (for the conjugate acid) of 44. Once again, our ion will rip a proton from whatever it finds nearby (another molecule of ammonia), and once it does the geometry of the double bond will be locked in place:

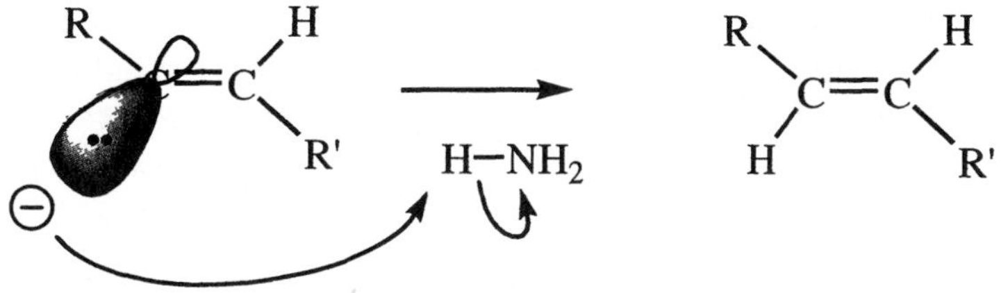

Overall, this is a clever way to allow the target molecule to choose a preferred geometry even though the molecule itself does not have that option: you let the system choose *just before* it becomes product, under circumstances where it makes the same choice a real molecule would.

Overall, the reaction looks like this:

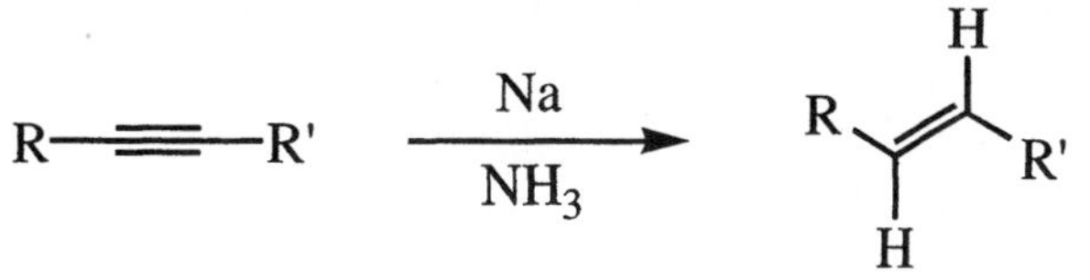

And the mechanism, gathered all together, looks like this. If any part of this does not make sense, go back and read about that step in the previous section.

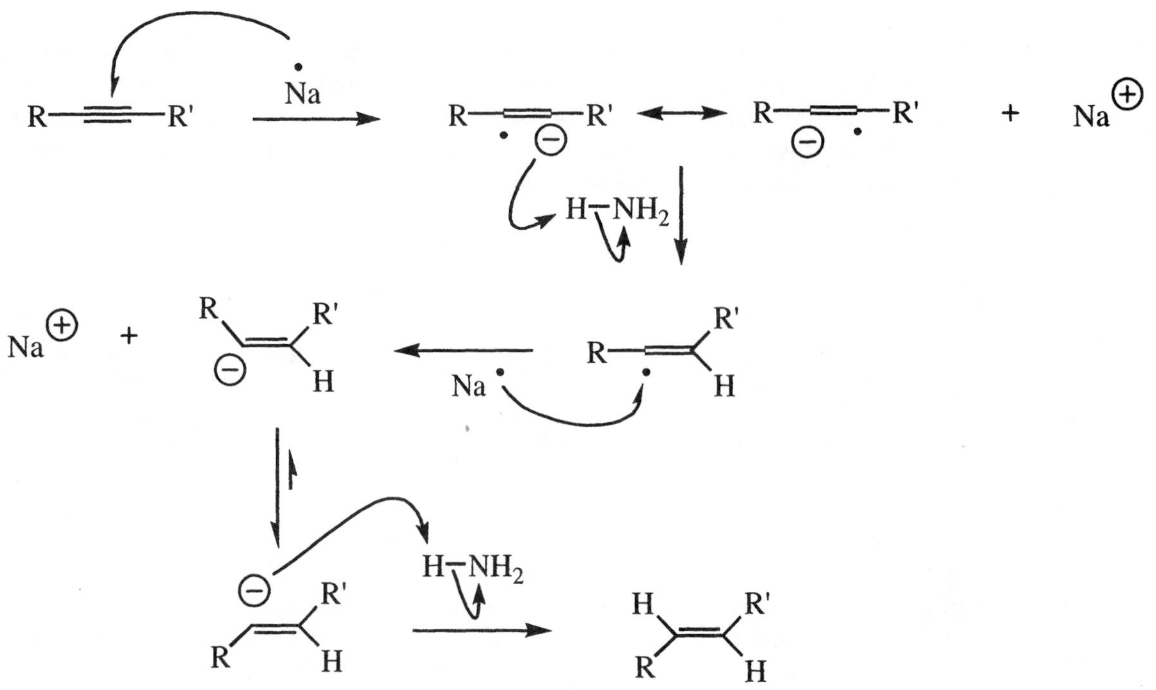

In summary, then, given any alkyne, we can reduce it at will to either a *cis* or a *trans* alkene. To make the *cis* alkene, use hydrogen and Lindlar catalyst; to make the *trans* compound, use sodium and ammonia instead.

FYI: When sodium is dissolved in ammonia, something very strange happens. The electron in fact detaches itself spontaneously from the sodium atom, leaving Na^+ and a free electron. The electron is not really free: it is surrounded by ammonia molecules. These are referred to as "solvated electrons," and they are actually the source of the electrons involved in the sodium–ammonia reduction of alkynes. For reasons that are not at all obvious, electrons solvated in ammonia are a deep blue color, not unlike the color of this type (if you have a color version of this!); running sodium–ammonia reductions is an immensely satisfying aesthetic experience! Interestingly, if you allow the ammonia to evaporate, you get back metallic sodium.

Anion Formation

There is one more reaction of alkynes we should discuss. We do not need to spend much time on it, because we already introduced it at the end of Chapter 9. Due to the fact that alkyne carbons are sp-hybridized, if there is a hydrogen attached to one (i.e., with a terminal alkyne or acetylene itself), a strong base can remove that hydrogen. The pK_a for a terminal alkyne is only about 25, so any base below 25 on the table will do the job. A common one is NH_2^-, which generally comes as our old friend $NaNH_2$ (*not* Na in NH_3). $NaNH_2$ is called sodium amide ("AMM-mid," or "am-MID," or "am-MIDE"), or sodamide ("SO-da-mid") for short. You can buy this and weigh it out and use it, if you are careful. Why do I say "if you are careful"? Consider the following reaction:

$$NaNH_2 \quad + \quad H_2O \quad \rightleftharpoons \quad NH_3 \quad + \quad NaOH$$

Look on your pK_a chart and decide which direction this reaction will go. Don't wait for me to do it for you, you can do this!

First of all, in order to use the pK_a chart, you have to imagine stripping away the extraneous counterions, and then look at the stripped-down form of the reaction: what we call a "net ionic equation," in which all the non-participants have been removed:

$$\overset{\ominus}{N}H_2 \quad + \quad H_2O \quad \rightleftharpoons \quad NH_3 \quad + \quad \overset{\ominus}{O}H$$

All these species are on the pK_a chart, so you should easily be able to determine that the equilibrium lies far to the right, because NH_2^- is a much stronger base than OH^-. So what? There is water in the air!!! If you weigh $NaNH_2$ in an ordinary room, the chemical on the balance will begin to react with water vapor in the air, and if you wait very long it will no longer be what you took out of the bottle, but instead a mixture of ammonia and sodium hydroxide (which is *not* basic enough to pluck a proton from an alkyne, and which stinks besides). So you have to take special precautions when you use sodamide, but it can be done. The reaction can be shown like this:

$$R-C\equiv C-H \quad \xrightarrow{\text{NaNH}_2} \quad R-C\equiv C^{\ominus}$$

Note that this is technically an equilibrium, but it lies so far to the right that it is basically a one-way street.

Shortly, we will use anions like this (called "acetylide anions") to make all kinds of alkynes, so file this process away for future use.

Problems
(Answers are provided at the end of the chapter for italicized problems.)

1. Show the major organic products of the following reactions. Write NR if there is no reaction.

a.

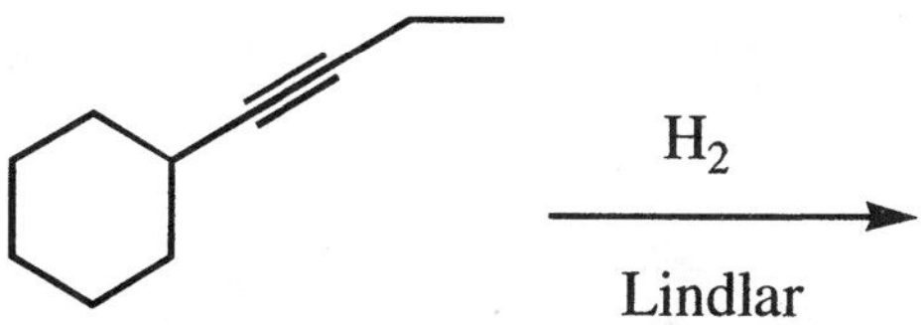

$$\xrightarrow[\text{Lindlar}]{\text{H}_2}$$

b.

$$\xrightarrow[\text{2) H}_2\text{O}]{\text{1) O}_3}$$

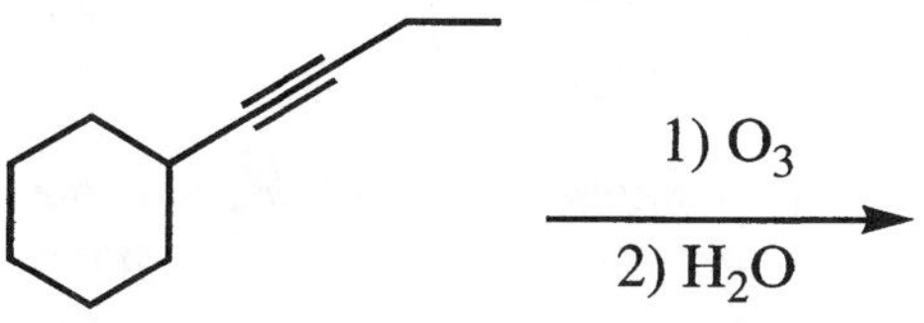

c.

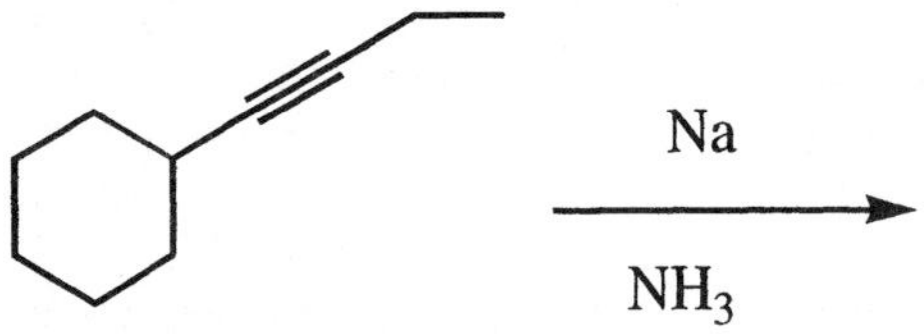

$$\xrightarrow[\text{NH}_3]{\text{Na}}$$

d.

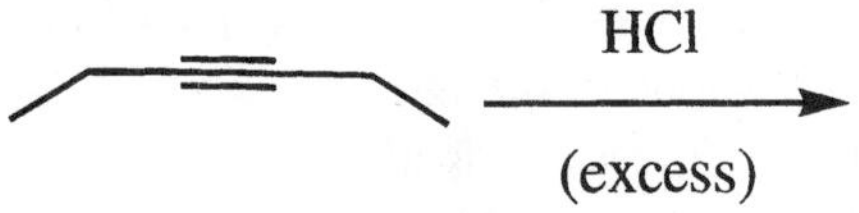

$$\xrightarrow[\text{(excess)}]{\text{HCl}}$$

e.

$$\xrightarrow[\substack{\text{H}_2\text{O} \\ \text{HgSO}_4}]{\text{H}_2\text{SO}_4}$$

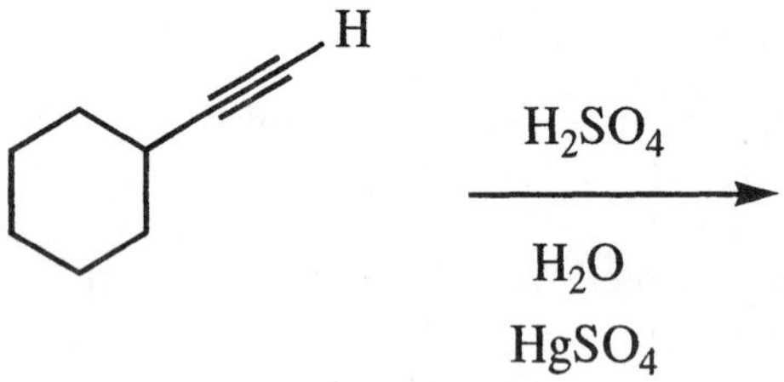

2. Based on the information in the chapter, do you think it *ought* to be possible to stop the addition of HCl to an alkyne at one equivalent, or should two HCl's add to every molecule? Based on the information in the chapter, do you think it *is* possible to stop at one equivalent? Defend your conclusions.

3. When 2-pentyne is treated with aqueous sulfuric acid in the presence of mercuric sulfate, a mixture of two products is obtained. When 3-hexyne is treated under the same conditions, only one product is obtained. Show the three products, write mechanisms for their formation, and explain why only one product is formed in the latter case.

4. *Imagine there were a way to add water to a double bond in anti-Markovnikov fashion, as shown below. (Such reactions exist, although we will not encounter them in this book.) Do not worry about how this occurs—consider it magic.*

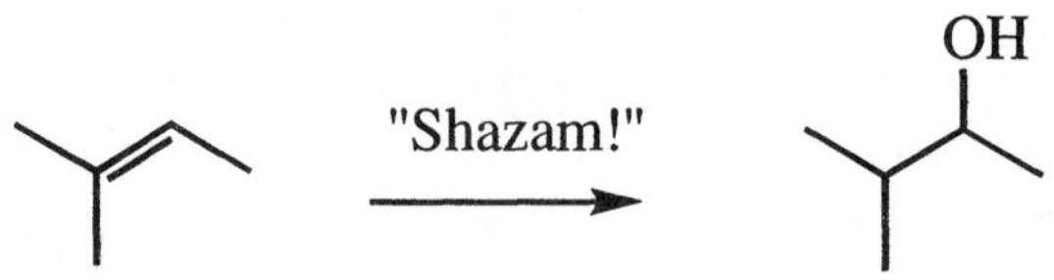

Based on the chemistry you learned in this chapter, predict the outcome of the use of Shazam reagent on the following alkyne. You may assume that there is aqueous acid present under the reaction conditions. Show a mechanism for the reaction, to the extent you can, but remember, you do not know how Shazam reagent works.

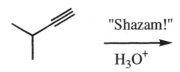

"Shazam!"
$\xrightarrow{H_3O^+}$

5. A 5.0-g sample of compound **A** with the molecular formula C_8H_9Br reacts with 13 g of Br_2. Treatment of **A** with Na/NH_3 gives **B**, with a molecular weight of 187. Treatment of **A** with ozone produces **C** and **D**. **C** has the molecular formula $C_6H_8O_4$. **D** has a molecular weight of 139.

   Interpret all the evidence, and suggest reasonable structures for **A–D**.

6. *A compound **B** has the formula $C_{17}H_{18}$. Upon treatment with hydrogen and palladium on carbon, 165 mg of **B** absorbs 7.43 mg of hydrogen. If Lindlar catalyst is used instead, 125 mg of **B** absorbs 2.25 mg of hydrogen. How many rings, double bonds, and triple bonds are in **B**?*

7. Propose a synthesis for each of the following, starting in each case with alkynes:

8. Suggest syntheses of the following molecules, starting with any alkynes you like. A model will be helpful for the last one.

Selected Answers

4.

6. 4,1,2.

There are nine units of hydrogen deficiency. 165 mg of **B** is 7.43 x 10^{-4} mol. 7.43 mg of H_2 is 3.714 x 10^{-3} mol, or 5 times as much. So each mol of **B** reacts with 5 mol of H_2. This tells us that 5 of the units of hydrogen deficiency involve multiple bonds. 125 mg of **B** is 5.63 x 10^{-4} mol. 2.25 mg of H_2 is 1.125 x 10^{-3} mol, or twice as much. This tells us that there are two triple bonds. Since two triple bonds account for four units of hydrogen deficiency, and five of them were due to multiple bonds, one must have been due to a double bond. The rest must be rings.

Chapter 13
Substitution Reactions

Now that we know how to make some alkyl halides and alcohols, it seems sensible to consider a few reactions of such things. In this chapter we will

- Define leaving groups, and learn what makes one good

- Define nucleophiles, and learn what makes one good

- See how kinetics helps us investigate reaction mechanisms

- Learn about S_N2 and S_N1 reactions

- Discover how to adjust leaving groups, so that we can substitute alcohols

- See how all this greatly expands our synthetic abilities.

Before we do any of this, however, we ought to take time out to learn how to name halides and alcohols. I won't spend long on this, because you have probably already figured out how this is going to work.

NOMENCLATURE OF HALIDES AND ALCOHOLS

Both halides and alcohols can be named in two different ways. The functional group (halide or alcohol) can be the base of the name, in which case the rest of the molecule is named as a substituent. This is just what we have done before, except that here we separate the functional group as a separate word. For alcohols, there are two alternatives within this model: you can either say "alkyl alcohol," in two words, or you can contract the name to one word by using the name of the corresponding alkane and changing the "-e" to "-ol." The examples below should illustrate this.

The other approach is to use the alkane chain as the base of the name and consider the functional group to be a substituent. In that case, again we name these compounds just the way we have named everything else, but when halogen is a substituent it is called "halo-," and the functional part of an alcohol is called "hydroxy-."

Examples:

H_3C-Cl

chloromethane
methyl chloride

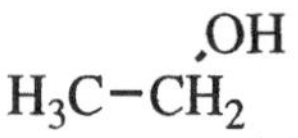

H_3C-CH_2 OH

ethyl alcohol
ethanol
hydroxyethane
(drinking alcohol)

OH

2-propanol
2-propyl alcohol
2-hydroxypropane
isopropyl alcohol
isopropanol
(rubbing alcohol)

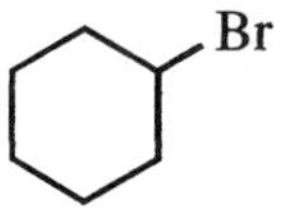

Br

cyclohexyl bromide
bromocyclohexane

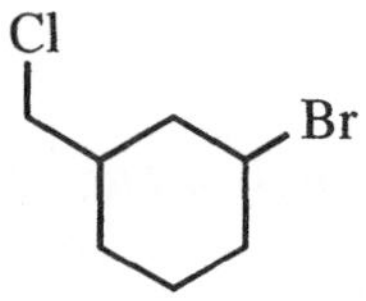

1-bromo-3-chloromethylcyclohexane

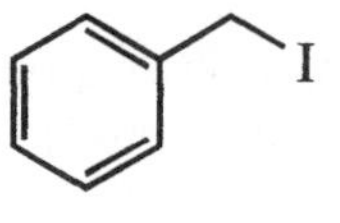

iodomethylbenzene
phenylmethyl iodide
benzyl iodide

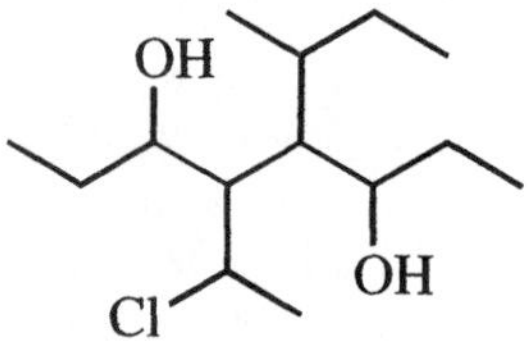

4-(1-chloroethyl)-5-(1-methyl-propyl)-3,6-dihydroxyoctane

4-(1-chloroethyl)-5-*sec*-butyl-octane-3,6-diol

COMMON MISTAKES: DON'T LET THIS HAPPEN TO YOU!

Even though the –OH group is called "hydroxy-," an alcohol is not a hydroxide. There are no alcohols that come apart the way sodium hydroxide does, to make OH- floating around in solution. Don't misinterpret the name to imply that they do, and don't ever show this happening!

$$NaOH \longrightarrow Na^{\oplus} \quad + \quad OH^{\ominus}$$

$$ROH \overset{\times}{\longrightarrow} R^{\oplus} \quad + \quad OH^{\ominus}$$

REACTIVITY

Halides and alcohols share the characteristic of having an especially polar bond in the molecule: the bond from carbon to halogen or oxygen is polarized in the direction of the **heteroatom**, giving this particular carbon a partial positive charge. (The term "heteroatom" is used to refer to atoms other than carbon and hydrogen in organic molecules; you know the prefix "hetero-" means "different" from the term heterosexual.) The partial positive charge is responsible for most of the chemistry we will observe.

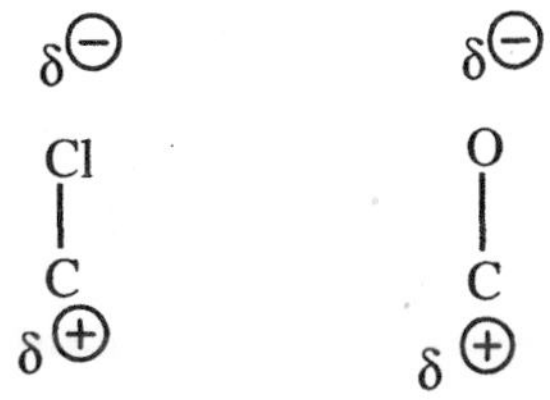

$$\overset{\delta\ominus}{Cl} \qquad \overset{\delta\ominus}{O}$$
$$| \qquad |$$
$$\underset{\delta\oplus}{C} \qquad \underset{\delta\oplus}{C}$$

Carbons bearing these substituents are electrophilic: not as electrophilic as carbocations, but still sufficiently electrophilic to be attractive to nucleophiles. The special reaction we will study in this chapter is one in which a nucleophile is drawn to this carbon. Unlike reactions we have already studied, where the carbon electrophile was a full carbocation and therefore extremely attractive to nucleophiles—and able to accept the nucleophile's electrons without difficulty—we are talking here about electrophilicity in neutral species. The carbon in question already has four bonds (the other three are not shown but assumed in the above pictures) and cannot accept a new bond without breaking one of the bonds it already has. In other words, we will be talking about a reaction in which carbon trades one of its current bonds for a new one, a "substitution" reaction. Below is an illustration of a generic substitution reaction.

$$Y^{\ominus} \qquad \overset{R}{\underset{R}{R-C-X}} \longrightarrow \overset{R}{\underset{R}{R-C-Y}} \qquad X^{\ominus}$$

| Nucleophile | Starting material (Substrate) | Product | Leaving Group |

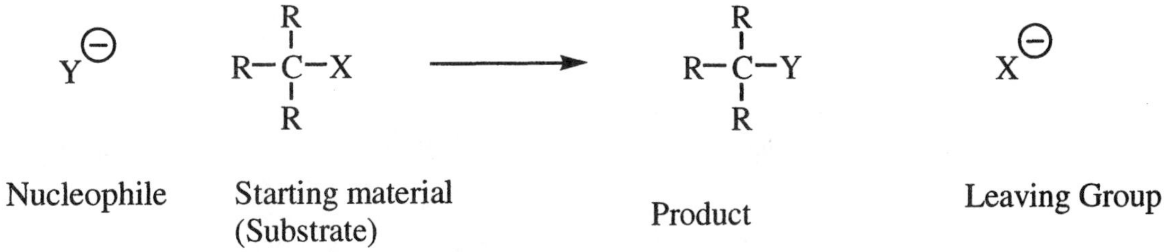

Let's meet the players.

Starting Material

There is really nothing to say about this other than a word about words. For reasons I have never figured out (it certainly did not come from me!), many of my students always refer to starting materials as "starting products." This expression is guaranteed to set your professor's teeth on edge. A product is what you have *after* a reaction. You cannot have a product before a reaction. Learn to say "starting *material*." "Substrate" is the generic term for "that which is acted upon," and is a term used frequently in biochemistry to describe molecules operated upon by enzymes. Recall, by the way, that the carbon to which X is attached could be primary, secondary, or tertiary (that is, the R's in the above structures could be either alkyl groups or hydrogens).

Leaving Groups

A necessary aspect of a substitution reaction is that something has to leave, and since the nucleophile is bringing its own electrons with it, whatever leaves has to take *its* electrons with *it*. An atom or group of atoms leaving a molecule and taking along a pair of electrons is a recurring theme in organic mechanisms, and the departing species is so important that it has its own name: a "leaving group."

Be careful with this: a leaving group is not just anything that leaves—after all, in some sense the carbon is leaving X also. By definition, a "leaving group" has to *take its electrons with it*. Further, it does not have to be small, or negatively charged. In the following examples, the leaving groups are highlighted in boxes.

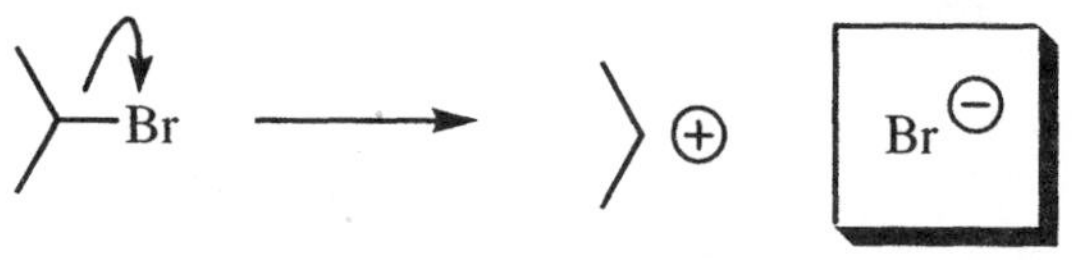

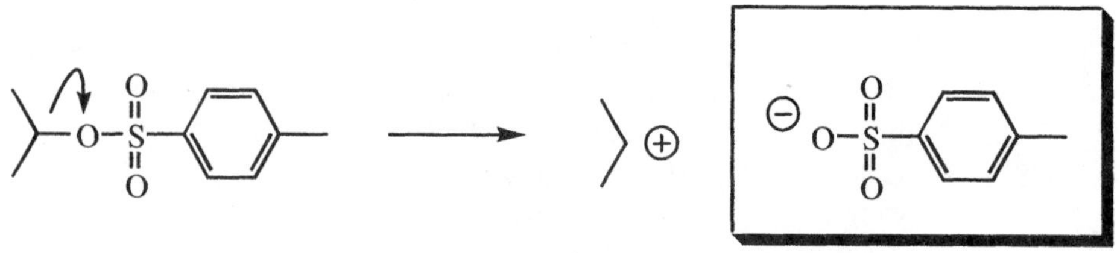

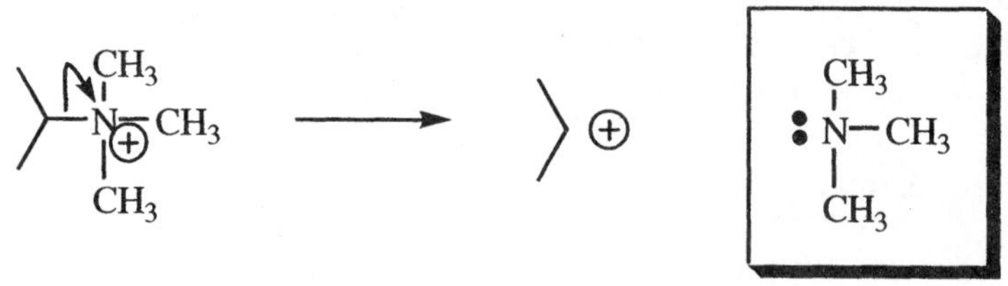

The leaving group is *whatever leaves with its electrons*, the entirety of it. Now, an important question for us to examine is the following: how will we know when a group is likely to leave? In other words, what makes something a *good* leaving group? It turns out that we are already much closer to an answer to this than you might realize.

A leaving group is something that is currently attached to a *carbon*, and we are asking about its propensity to break the bond with carbon, with both electrons being kept by the group in question. We have actually asked this question before, phrased slightly (but only slightly) differently. Look back to Chapter 8, where you will find a long list of acids and bases in ranked order. A strong acid is something that has a bond that readily breaks between a *hydrogen* and its partner, with the electrons remaining with the partner. A strong acid is also a substance whose conjugate base (obviously a weak base) is quite happy with the excess electrons it got from that bond. I suggest that whatever factors allow a partner to take off with electrons in Chapter 8 might apply here equally well. If a conjugate base is happy with excess electrons in Chapter 8, it should also be happy with them in Chapter 13. In other words, *good leaving groups are weak bases.* We can identify good leaving groups directly from our pK_a chart! All those bases near the top of the chart are good leaving groups; molecules that have one of these groups attached to a carbon are in a position to lose it.

So what are examples of some good leaving groups? Among the best and most commonly encountered are the halide ions, the $ArSO_3^-$ ion (middle picture, previous page), and H_2O. Further, considering only the halides, iodide is best, followed by bromide, followed by chloride. What would be a bad leaving group? Anything that you would consider a strong base. This includes, in most cases, OH^- and OR^-, and certainly anything below them on the base side of the pK_a chart.

Nucleophiles

A nucleophile, as you already know, is something that has electrons it is willing to share. It is also, by virtue of having these electrons, a base. We call it a base if it uses its electrons to grab a proton (H^+), and we call it a nucleophile if it uses the same electrons to go after anything else (usually carbon). What we have not yet discussed is how to rate nucleophiles, because in Chapters 9–12, our electrophiles (carbocations) were so potent that any nucleophile in the vicinity was good enough to bond with them. But now we need to know what makes something a *good* nucleophile. And again, I suggest to you that most of the answer we are looking for is to be found in the pK_a chart, which is, after all, a ranked list of precisely these species, arranged in the order of how well they can grab a partner proton (from bottom to top, right column). Surely something that has a strong attraction for a proton should also have a strong attraction for a positive (or partially positive) carbon. This is generally an accurate statement, but the correspondence with pK_a is not quite as direct as it was in the case of leaving groups.

It is certainly true that the general trend holds up: the stronger something is as a base, the stronger it is as a nucleophile. But there are two areas where this trend falls apart, and they ought to be mentioned here. The first is a steric consideration. Hydrogens are always on the outside of molecules, and a species trying to grab a hydrogen does not have to dig very deep to do so. On the other hand, carbons are almost always surrounded by other junk (including lots of hydrogens), and any species trying to go after a carbon atom has to brush aside a bunch of other stuff to get there. The consequence of this is that small, strong bases are indeed good nucleophiles, but some large, good bases are quite bad nucleophiles, not because they do not want to react with carbons but simply because they cannot reach them. Thus, within the category of oxygen nucleophiles, for example, all the anions below have similar basicities, but the last one (called *tert*-butoxide; or *t*-butoxide, for short), which is actually the best base of the lot, is quite a poor nucleophile. It is usually sold as its potassium salt, "potassium *t*-butoxide."

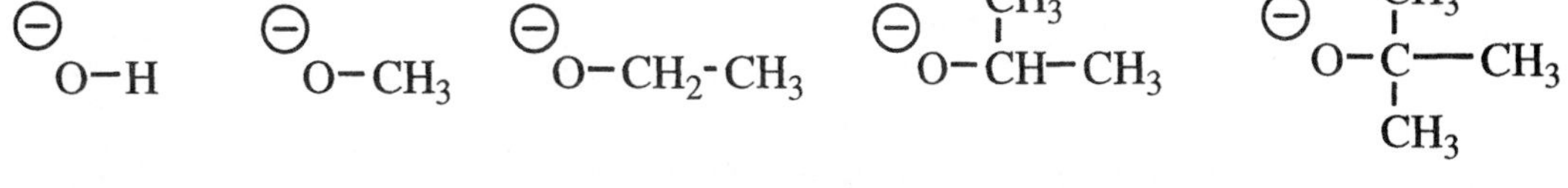

$$\overset{\ominus}{O}-H \qquad \overset{\ominus}{O}-CH_3 \qquad \overset{\ominus}{O}-CH_2-CH_3 \qquad \overset{\ominus}{O}-\overset{\overset{\textstyle CH_3}{|}}{CH}-CH_3 \qquad \overset{\ominus}{O}-\underset{\underset{\textstyle CH_3}{|}}{\overset{\overset{\textstyle CH_3}{|}}{C}}-CH_3$$

Among the nitrogen anions, which are considerably more basic than oxygen anions, there is again one that is similarly so bulky that it virtually refuses to act as a nucleophile, even though it is a terrific base. This is the compound shown in the next picture, normally used as its lithium salt, called "lithium diisopropylamide." That name is a mouthful, so the substance is known as LDA for short.

Potassium *t*-butoxide and LDA are two members of a limited family called "non-nucleophilic bases." We will later encounter occasions when we need a reagent that acts as a base but not as a nucleophile, at which point we'll take these down off the shelf.

The second case of nucleophiles deviating from the general trend (better base equals better nucleophile) also is related to the fact that the carbon being attacked is buried inside the substrate, and the nucleophile has to squeeze past some junk to get there. The facts this time are counterintuitive. You already know that I⁻ is a better anion (and therefore a weaker base) than Cl⁻, because it is bigger, and there is more area within which it can spread out its charge. It would stand to reason, then, that I⁻ should also be a weaker nucleophile, *both* because the anion is a weaker base, *and* because it is bigger. Unfortunately, I⁻ turns out to be a *better* nucleophile than Cl⁻. Why? The term used to rationalize this observation is "polarizability": Bigger things are said to be more polarizable. I⁻ is a better nucleophile than Cl⁻ because it is more polarizable. What does this mean? The best explanation I have been able to come up with is "squishiness." Look at it this way. Chloride is kind of like a baseball: hard and compact. Iodide is kind of like a nerf basketball. It's bigger than a baseball, but if you needed for some reason to get it through a chain link fence, you could do it, because you can deform the ball and squeeze it through, while a baseball would never go through even though it is smaller. Similarly, large, squishy atoms have an easier time squeezing into tight places than small, hard ones. For this reason bigger atoms tend to be more nucleophilic than smaller atoms, other things being equal. That last phrase is important, by the way! This comparison only works within a given column of the periodic table, and also when one is comparing atoms of the same charge. For example, SH⁻ is a weaker base but a stronger nucleophile than OH⁻, because it is more polarizable. Likewise, H_2S is a better nucleophile than H_2O. But we could not go so far as to say the H_2S would be a better nucleophile than OH⁻, because OH⁻ has a negative charge that H_2S does not have. So this polarizability business only enters in when other factors have been taken into account.

Another case of deviation from the general trend of basicity vs. nucleophilicity has to do with solvation, but I don't want to get into that here.

You should note that the halide ions, being among the weakest bases, ought to be pretty bad nucleophiles as well, but in fact they are rather good, and iodide is one of the better nucleophiles available. The halides have the rare property of being good nucleophiles but not good bases. Later we will discover that substitution reactions are complicated further by competing reactions in which the nucleophile acts instead as a base, leading to "elimination" reactions. The halides tend not to get involved in this complication, so they are excellent nucleophiles to use in studying the nature of substitution reactions.

SUBSTITUTION REACTIONS OF HALIDES

We have met the actors, so now let's watch the play. So far what I have told you is who is on stage at the beginning, and who is there at the end, but this doesn't make for a very lively evening at the theater. We want to see the plot. What happens between the beginning and the end? In our case, how does the starting material (remember: not "starting product") turn into product? Unfortunately, this particular play takes place entirely behind a curtain, with actors so small we can't see them, so we are left to infer the plot of the play from various pieces of evidence we collect. How do chemists actually figure these things out?

The first order of business is to write down all the possible ways we can imagine of getting from beginning to end. We are dealing with a very simple reaction here, a substitution. There are only a limited number of ways we could pull this off. Here is a typical reaction:

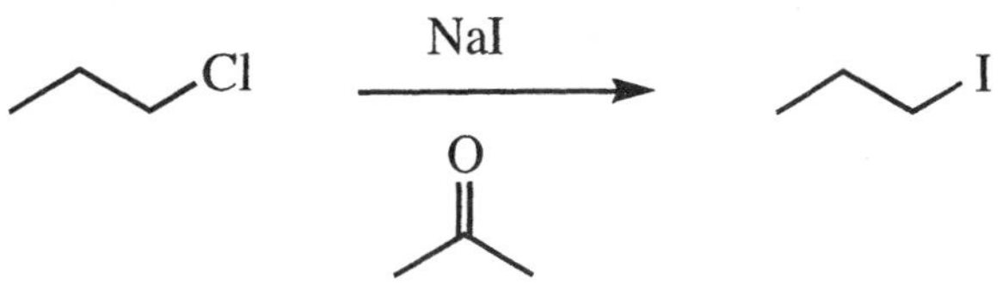

Don't be distracted by the extra stuff you see. You know that NaI will come apart in an appropriate solvent into Na$^+$ and I$^-$. Thus, there must be a solvent around, and the other substance shown (acetone) must be the solvent. That's about the only purpose it serves. So really, the business part of this equation can be summarized as:

Two things happen in the course of the process: iodide comes in, and chloride leaves. There are three (and only three) ways to imagine this combination of events, at least in terms of timing.

1. Iodide enters first, then chloride leaves.

2. Chloride leaves first, then iodide enters.

3. Both things occur at the same time.

If anyone can come up with a fourth possibility, I'd like to hear about it. The three possibilities are illustrated below.

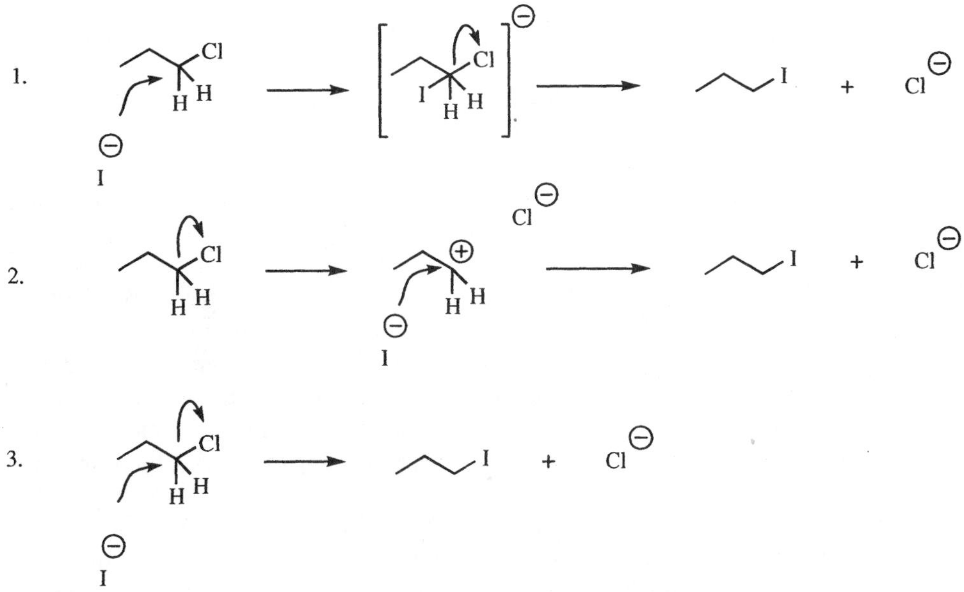

The next step is to examine these timing possibilities for chemical reasonability. You should immediately recognize that pathway 1 involves a species (the one in the middle) with five bonds to the same carbon, which requires ten electrons surrounding an atom that can hold only eight. As written, this is being proposed as a genuine substance with some finite lifetime. Hopefully, by now you are so well attuned to what is proper for carbons that you would reject this structure out of hand, deciding that whatever the mechanism is it cannot possibly involve such a ridiculous species.

Now we are down to two possibilities. Possibility 2 suggests formation of a carbocation. We know that carbocations can and do exist, although they are uncomfortable species that react rapidly. We also know that there are different grades of carbocation, and the one shown here, a primary carbocation, is among the least stable. Nevertheless, even though we may be prejudiced against this mechanism (nothing wrong with that!), we have no grounds for rejecting it completely.

What about the third mechanism? It's something we have not seen before, but it does not seem to involve anything that would cause us to reject it immediately.

Kinetics

The next consideration is how we should go about figuring out what *really* goes on behind that curtain. Is this a play with two separate acts (mechanism 2) or only a single one (mechanism 3)? Our chief way of dealing with questions like this requires measuring the speeds of certain reactions. The official term for speed in chemistry is "rate," and the subspecialty that deals with rates is called "kinetics."

Measuring rates is a way to find out information about the slowest step of any reaction. Why? What is so special about the slow step? The slow step of any reaction is special, because that is the only one we can really measure. Consider an assembly line at a Ford plant making carburetors. There are ten people on the line, and each can do his or her job in two to three minutes, except for the klutz at the number 4 position, who takes an hour to put on his piece. How fast will carburetors come off the line? One an hour, right? It makes no difference how many steps there are before or after this slow guy: production will always be limited to his speed of one an hour. If the number six worker suddenly starts to do his job poorly, taking 30 minutes to do something that should be accomplished in 3, we would never know it: there would still be one carburetor produced every hour. All steps before or after the bottleneck are invisible to someone outside the plant counting finished pieces. The same is true with chemical reactions: all steps before or after the slow one are invisible from a reaction rate standpoint. On the other hand, you can turn this realization around and recognize that by investigating the overall rate of a reaction, you can learn a great deal about the slowest step.

How can we learn things about a reaction by measuring rates? Here's another analogy that might illustrate the point. Imagine that you are running a trucking company plagued by accidents. Drivers keep returning from their routes with damaged trucks. They insist they are being hit by other cars, and that it is not their fault, but you suspect your people are bad drivers and are running into things (not necessarily cars) on their own. Without following them around, could you figure out what is going on? With the right data, you could.

Suppose you keep track of how often accidents occur (the rate, get it?). You send out 40 trucks, and 2 return with damage; you send out 80, and 4 are damaged. Does this surprise you? It shouldn't: the rate *should* depend on how many trucks are out on the road. More trucks, more accidents. But now, check for a relationship with how many *other* vehicles are on the road. You send out 80 trucks on a day with light traffic, and 4 come back damaged. Now send out 80 trucks on a day with heavy traffic, and again 4 return damaged. Have you learned anything? If other cars were smashing into your trucks, wouldn't you expect more accidents when there are more other cars around? There should have been more than 4 accidents on the second day, if your drivers were telling the truth. On the other hand, if your drivers are crashing into loading docks, for example, it shouldn't matter how many other cars are out there. In other words, you can tell how many vehicles are involved in creating the damage by determining the relationship of the damage rate to the quantity of other vehicles.

The same is true with molecules. You can tell how many molecules are involved in a reaction by determining the relationship of the reaction rate to the quantity of other molecules present.

Let's look back at our proposed mechanisms and see what we might expect to learn by measuring rates. Mechanism 2 requires two steps. The first step involves taking a neutral, comfortable molecule and breaking it apart into a cation and an anion. This is likely to be pretty difficult. On the other hand, the second step involves taking a very uncomfortable species, a carbocation, and combining it with an anion to create something neutral and happy. I think you should have no trouble predicting that for this mechanism, the first step should be more difficult (slower) than the second. (Later we will examine some experimental evidence for this.) So any rate information we collect will involve only the first step. Notice that in the first step the nucleophile is not in any way involved. The only thing that happens is that the leaving group falls off. This does not require any other characters: the starting material just stands on stage and disintegrates. This is like the truck drivers crashing into loading docks: no other vehicles are involved, and the rate should depend only on how many bad truck drivers are out there. Since only one molecule is involved in the slow step of the proposed reaction, it is called a "unimolecular" reaction. And since this is a substitution reaction in which the new partner is a nucleophile, we call this reaction "Substitution, Nucleophilic, Unimolecular," or S_N1 for short.

On the other hand, mechanism 3 occurs in one step. The starting material doesn't do a thing until the nucleophile appears on the scene. These two must collide with each other in order to make the reaction happen. This is like a two-vehicle collision, in which the rate should depend not only on how many trucks are out there, but also how many cars. Nucleophilic substitution of this type would involve two molecules colliding in the slow step, so it is "bimolecular," and the reaction is a "Substitution, Nucleophilic, Bimolecular," or S_N2 for short.

We've proposed two possible mechanisms, so let's measure some rates and find out how our reaction really takes place. This brings up a logistical problem. How can we measure rates? The easiest way is to measure the amount of one of the ingredients present at various intervals. If we measure how much *product* there is, the answer will of course be zero at the beginning, and the concentration will then increase with time and serve as a direct measure of how fast the reaction is going. Alternatively, we could measure the amount of starting material present. This will have some definite value at the beginning and (gradually or rapidly) go *down*. The rate of the reaction is therefore the difference between the amount present at time zero and the amount at some other time *t*, divided by the amount of time that has elapsed.

The S$_N$2 Reaction

OK, let's make measurements for a specific case. Here are some data for the reaction of 1-chloropropane with sodium iodide. (Note: All of the numbers on this page are made up and oversimplified in order to help you learn the concepts.)

Initial Conditions	Time (min)	[1-chloropropane], mol/L
	0	0.37
[RCl] = 0.37 M	10	0.36
	20	0.35
[I$^-$] = 0.10 M	30	0.34
	40	0.33
Rate = _____		

What is the rate? The amount of starting material changes by 0.01 mol/L (M, or "molar") during each 10-minute interval, so our rate is 0.01 M/10 min = 0.001 M/min. Have we learned anything? Not yet. Let's run a second experiment, this time starting with an 0.74 M concentration of 1-chloropropane. Make a prediction: what do you think the rate will be?

Initial Conditions	Time (min)	[1-chloropropane], mol/L
	0	0.74
[RCl] = 0.74 M	10	0.72
	20	0.70
[I$^-$] = 0.10 M	30	0.68
	40	0.66
Rate = _____		

As I hope you predicted, the rate doubled to 0.002 M/min. Have we learned anything? Still no! We knew that would happen!

Both of the above experiments were done with [I$^-$] = 0.10 M. Let's do a new one with [I$^-$] increased to 0.20 M.

Initial Conditions	Time (min)	[1-chloropropane], mol/L
	0	0.37
[RCl] = 0.37 M	10	0.35
	20	0.33
[I$^-$] = 0.20 M	30	0.31
	40	0.29
Rate = _____		

The rate is again 0.002 M/min. Now the question is, with what do we compare this number? The easiest comparison to make will be one that involves only one change. Therefore, we should compare this 0.002 M/min result to the original 0.001 M/min value obtained from the first experiment, where the concentration of 1-chloropropane was the same as it is here. And we see that, holding everything else the same, doubling the amount of I⁻ doubles the rate. This tells us something. It tells us that I⁻ must be colliding with our 1-chloropropane during the slow step (the only one we can measure) of this reaction. If I⁻ were not involved in the slow step of the reaction, changing its concentration would have no effect on the rate. What does this tell us about the mechanism of the reaction we have run? Clearly, our alkyl chloride and the I⁻ have to collide to make anything happen. In an S_N1 reaction, the slow step involves the chloride falling apart with no help from anyone else. In an S_N2 reaction, a nucleophile collides with the chloride to get the process going. This reaction we have just studied must have an S_N2 mechanism. We have determined that the rate depends on both the substrate and the nucleophile: in mathematical terms,

$$\text{rate} = k\,[\text{RCl}]\,[\text{I}^-]$$

Notice that the equation above is simply a shorthand way of stating the sentence that precedes it: the rate increases if you increase *either* [RCl] or [I⁻]; the k is simply some constant that makes the numbers work out. This k is called the "rate constant," and every reaction has one.

We have just acquired a piece of evidence indicating that the play we are "watching" is a one-act play, where everything happens at once as shown in the mechanism we labeled 3, which we have now learned to call S_N2. The evidence so far is that, based on the kinetics, there must be a collision between the nucleophile and the substrate in order for the reaction to happen. Can we accumulate some more evidence? What if, instead of using 1-chloropropane as the substrate, we used 1-*bromo*propane? Make a prediction: should the reaction go faster or slower?

Chlorine in our reaction is functioning as the leaving group. Changing the leaving group should affect the reaction rate *if the leaving group is actually leaving during the step we are measuring*. Well, we believe that this reaction only has one step, so obviously changing the leaving group should change the rate. In which direction? Well, is bromide a better or worse leaving group? Look on your pK_a chart and find out (or remember from before that atoms lower in the periodic table make better leaving groups). Bromide is a better leaving group, so this change should make the reaction go faster; that is, the rate constant should be larger for this reaction when bromide is the leaving group than when chloride is the leaving group. Sure enough, it is: below are data for three S_N2 reactions done under identical conditions, changing only the leaving group. Of course, if mechanism 2 (S_N1) were correct, this change would also make the reaction go faster, so our new experiment really does not confirm our choice of mechanism, but at least it does not prove it wrong!

Leaving Group Effect on S_N2 Reaction

Leaving Group	Relative Rate
Cl⁻	1
Br⁻	42
I⁻	79

What if, instead of using I⁻ as the nucleophile, we used Br⁻, keeping the substrate as 1-chloropropane? Changing the nucleophile should affect the reaction rate *if the nucleophile is actually entering during the step we are measuring*. Well, we believe that this reaction only has one step, so obviously changing the nucleophile should change the rate. In which direction? Bromide is a worse nucleophile than iodide (remember that squishiness makes bigger nucleophiles better even though they are worse bases), so the reaction should get slower. Sure enough, it does: below are data for three S_N2 reactions done under identical conditions, changing only the nucleophile. What if mechanism 2 were happening? In the slow step of mechanism 2 (the first step) the nucleophile is not involved! Changing the nucleophile should have no effect at all on the rate of an S_N1 reaction. So this time we *have* obtained further evidence that we made the correct choice of mechanisms.

Nucleophile Effect on S_N2 Reaction

Nucleophile	Relative Rate
Cl⁻	1
Br⁻	26
I⁻	1119

There is more. Notice that in order to get from starting material to product, all in one step, there must be a point in the reaction where the nucleophile has begun to attach itself, and the leaving group has begun to leave. The transition state for this reaction must have a structure something like the one shown in brackets below. Also shown is an energy diagram that illustrates the idea.

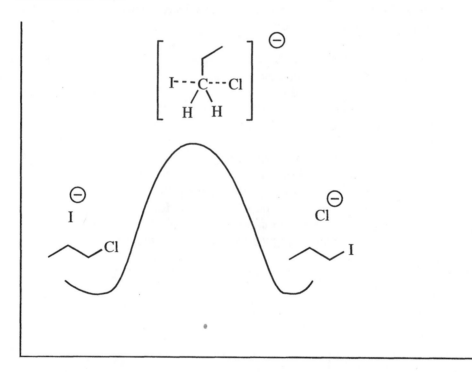

In the transition state shown there is more junk than normal around the carbon atom at the center of the action: five other atoms instead of the normal four (though only three of them are really "bonded"). This looks pretty crowded. The particular molecule in question has no difficulty handling the clutter, because two of the atoms are tiny hydrogens; that is, the starting material is a primary chloride. But what if it had been secondary? Now in the transition state there would be five things, four of which are pretty big. This suggests trouble, and we find out in fact that secondary substrates are somewhat slower than primary ones in undergoing S_N2 reactions. And consider tertiary substrates! Here the transition state would be so crowded that its energy is prohibitively high. The rate of S_N2 reactions on tertiary substrates is so slow that we can legitimately claim that they simply do not happen.

Substrate Effect on S_N2 Reaction

R	Relative Rate
CH_3-	1
Primary	0.03
Secondary	0.0008
Tertiary	No reaction

So now we have three different types of evidence supporting our choice of mechanism, all related to kinetics. First, the rate dependence on concentrations tells us that there must be a collision between the substrate and the nucleophile involved. Second, when we change either the nucleophile or the leaving group, the rate changes. Third, primary substrates react faster than secondary, and tertiary ones do not react. All these findings are consistent with the hypothesis that the reaction is proceeding by mechanism 3, which we are now calling S_N2.

ORM: In "Organic Reaction Mechanisms," under "Substitution Reactions," look at "Classic Reaction, S_N2" for a nice animation of the above section. You can watch it either with or without the solvent.

Below is a typical case of such a reaction and the mechanism for it.

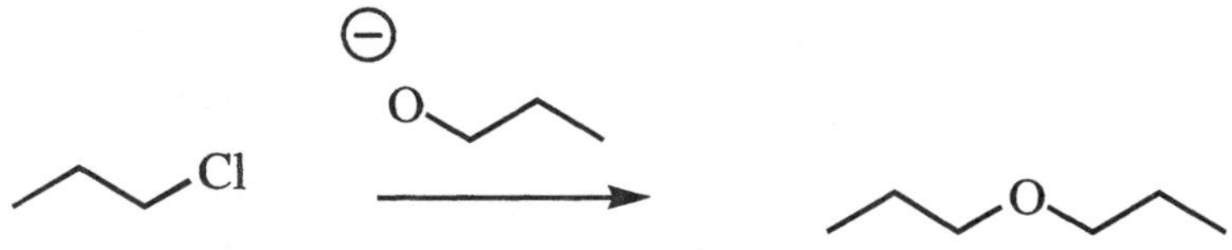

Mechanism:

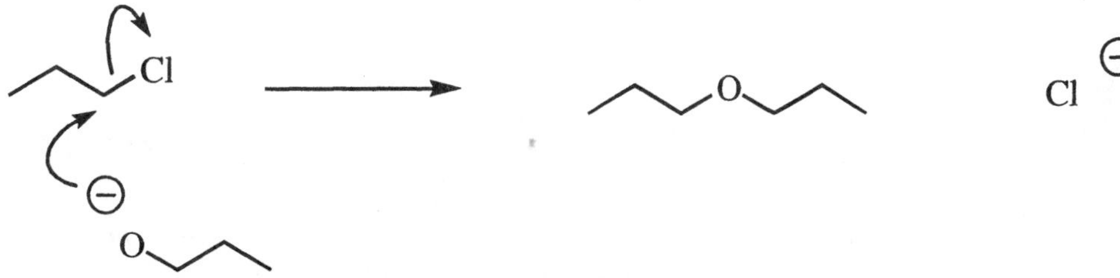

There is something important about the mechanism I just wrote that you can't actually see. The good news is that if you do it wrong, your professor won't be able to see it either, but the bad news is that it could muddle your thinking and make this stuff more difficult, so we might as well nip the problem in the bud. There are two curved arrows in the mechanism I wrote. What order should you write them in? (Of course, once they are written, no one knows in what order you wrote them, so in that sense it does not matter; what I am about to tell you relates to a thinking error that can only be detected by *watching* someone draw the mechanism.)

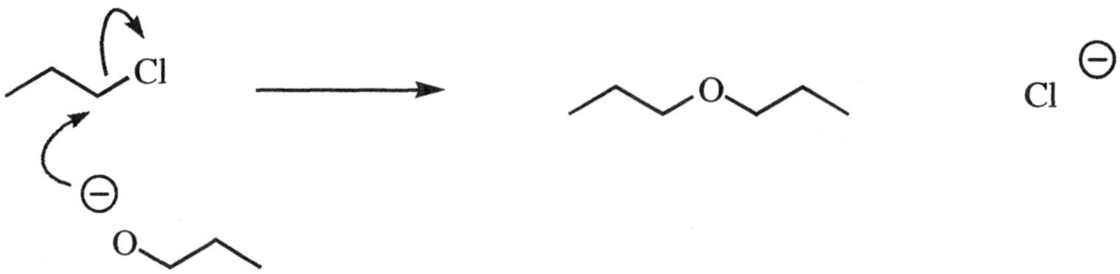

Most students, without instruction to the contrary, draw the top arrow first, and then the bottom one. This is completely backwards. It implies (not on paper, but in your head) that the chloride leaving is what gets this thing started, and then the nucleophile falls into place. But we have already shown that the reaction does not occur unless there is a collision between these guys, so the nucleophile must be what gets it started. You can think of the nucleophile as the aggressor in an S_N2 reaction. The proper way to show an S_N2 mechanism is to draw the bottom arrow first, then the top one, so that in your head you are thinking, "the nucleophile is what makes this happen."

You should notice, by the way, that this product is precisely the one we said (at the end of Chapter 10) that we could not synthesize using reactions we knew earlier. Now we can. This might be worth remembering (hint, hint). The only problem is that what it uses as a reagent (the thing over the arrow the first time we wrote it) is something you have not seen before. What the heck is this, and where did it come from? Let's take a look.

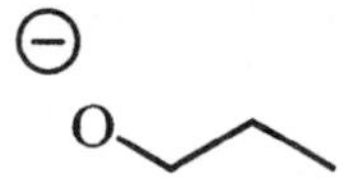

This is an anion, and presumably it must have a cation associated with it: one not mentioned in the equation, but one we know must be there. The most likely candidates are Na^+, K^+, Li^+, etc. We can assume that such a cation is hanging around somewhere; we can also assume that it doesn't matter what cation it is, or even, for that matter, that it is present at all. The real question is, how can we get the necessary anion? As is the case with most anions, the easiest way to get it is to remove a proton from the corresponding conjugate acid. Imagine that the thing that's now an anion once had another proton on it—in other words, that it was a perfectly normal compound, propyl alcohol—and that we had ripped off that proton:

Here we are looking at part of an ordinary acid–base reaction, and you should be able to use your pK_a chart to figure out how to carry it out. Simply find propyl alcohol (or a close relative) on the acid side, and then choose a base from below that on the base side. One base you might notice (because we have seen it before) is $NaNH_2$, which you know you can buy and use. This should do a fine job. Later we will learn a more common way to accomplish the same thing.

By the way, you just learned a new way to make ethers. This is called the "**Williamson ether synthesis**." Make up some cards, right now before you forget! And take some time to write down at least three examples of this new reaction.

Speaking of cards, when is the last time you reviewed your *old* cards? You know, the ones you made up from Chapters 10 and 11. You should make a point to scan through your decks of cards at least every other day. Right now that is not so tough, but as we go on your decks will get bigger and bigger. The better you know the old reactions the easier it will be to add some new ones. If you let this slide, at some point you will be faced with having to recall 50 reactions for some test. It is far better to already know 45 of them cold, and only have five more to add.

Here's another example of a substitution, using the "acetylide" anion we learned about in Chapter 12.

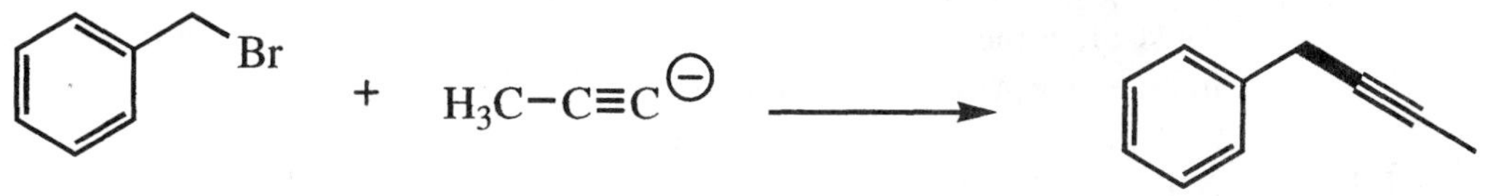

Ooooh, now this is a good one. Look at what we have done: we have created a new carbon–carbon bond, the bold one in the product. What's the big deal? Up until now, if we wanted to make something containing five carbons in a row, we had to start with something containing five carbons in a row. We had no way of making longer chains of carbons using the reactions we had learned. But now we can. Further, we also know how to make the anion shown above (see the end of Chapter 12). And if we wanted to, we could carry out this process twice with the same alkyne. Watch:

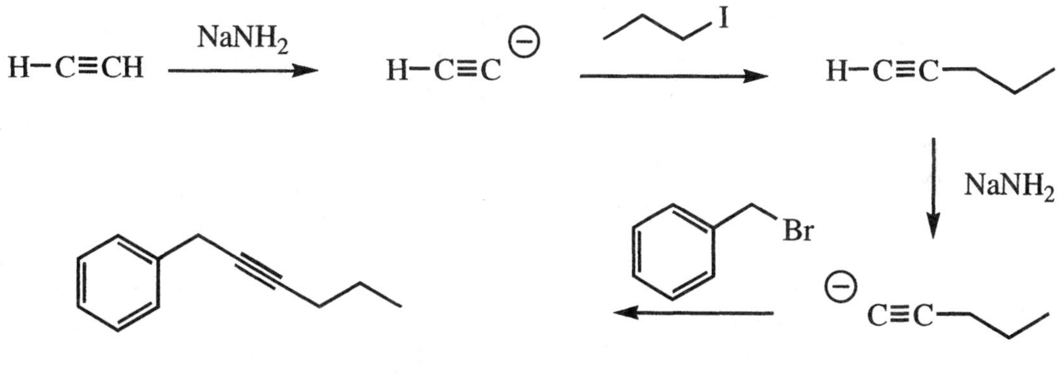

Using this approach we can build up almost any alkyne we want! And further, we know how to reduce alkynes to alkenes (either *cis* or *trans*, if we happen to care), and we know lots of reactions of alkenes. This opens up a huge door to making a wide range of compounds from simple starting materials. This definitely calls for some new cards.

Three points of caution.

1. This is the *only* carbon–carbon bond forming reaction you know. Later you will learn others, but for now *do not try to make a carbon–carbon bond using any other reaction!!!* If you need to make a carbon chain longer than what you are told to start with, this reaction has to be incorporated somewhere in the mix!
2. Don't lose carbons. It is very easy to draw pictures and forget that both the carbon with the minus charge and the carbon with the leaving group must be present in the product.

Until you get used to doing this, it will be a very good idea to count the carbons before and after each step.

3. S_N2 reactions work best on primary substrates, next on secondary, and not at all on tertiary. Further (and I haven't told you this before), they do not happen at unsaturated centers. That is, if you have a leaving group directly on a double or triple bond (including a benzene ring), it cannot be substituted in this way. Don't try it!

ORM: In "Organic Reaction Mechanisms," under "Substitution Reactions," look at "Alkylation of an Alkyne, S_N2" for a nice animation of the above section.

We will return to all of this at the end of the chapter when we practice more syntheses.

To summarize the S_N2 reaction: Nothing happens until there is a collision. A nucleophile initiates the reaction by actively attacking a carbon atom bearing a leaving group, driving the leaving group out. This requires an aggressive nucleophile, one that wants to form a bond. Most often the nucleophile is negatively charged. The carbon bearing the leaving group has to be either primary or secondary, as well as saturated.

The S_N1 Reaction

And yet, below is a known substitution reaction that seems to violate many of the rules:

Recall that "Et" is our abbreviation for "ethyl," so it's ethyl alcohol ("ethanol") under the arrow, our solvent for this reaction. Clearly, iodine has taken the place of bromine in this molecule, so there has been a substitution; but we suspect that this cannot be an S_N2 reaction, because it is happening at a tertiary center. What can be going on here? Of the three possible mechanisms we listed originally, we rejected the first as being unreasonable and identified the third as S_N2, which this is not. The only other possibility is the second proposed mechanism, which we labeled S_N1. Could this be an S_N1 reaction? Let's find out.

Again, our tool is kinetics. If this is an S_N1 process, then no matter how much I^- is present, the starting material should disappear at exactly the same rate. This in fact turns out to be the case. Further, if this is an S_N1 reaction, and we were to change the starting material to a chloride instead of a bromide (worse leaving group), the reaction should slow down, but if we changed the nucleophile from I^- to Cl^- (worse nucleophile), that should have no effect at all on the rate. Both of these predictions also turn out to be correct. Mathematically, for an S_N1,

$$\text{rate} = k \ [RCl]$$

Finally, in an S_N1 reaction there is a carbocation intermediate, and we already know that it is easier to make tertiary carbocations than secondary, and primary cations are pretty bad (unless they are resonance stabilized). So we would expect this type of reaction to occur more readily at tertiary centers than secondary, and hardly at all at primary centers. Also true. So we surely have found an S_N1 reaction, and here is its mechanism:

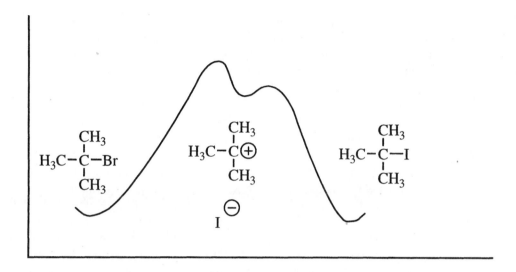

We have here a two-step reaction with a carbocation intermediate. Thus, there should be several features of it that are already familiar to us.

1. The energy diagram should look just like the energy diagram for an addition reaction, and in fact it does:

2. Carbocations can react with alcohols. The one formed here should also. Indeed, we generated it in alcohol solvent, so it ought to react with the solvent even more often than with the nucleophile (I⁻) we added. It does, and the major product of the above reaction is actually not the iodide but the corresponding ether:

The ether-forming reaction works just fine with no iodide present at all. The important message here is that it does not matter how a carbocation is generated: once present, it will do what all carbocations do, given the circumstances they find themselves in. It should be clear to you that the same reaction carried out in different alcohols would lead to different ethers, and that in water the product would be an alcohol. Notice that the nucleophile here is quite poor—a neutral molecule of solvent. Because an S_N1 reaction produces a carbocation, which is a super electrophile, you do not need to provide it with a super nucleophile. It will suck up anything that passes by.

3. One of the other things that carbocations do is rearrange. Carbocations generated *this* way ought to rearrange also. Sure enough, they do.

Test Yourself 1

Try writing a mechanism for the following reaction:

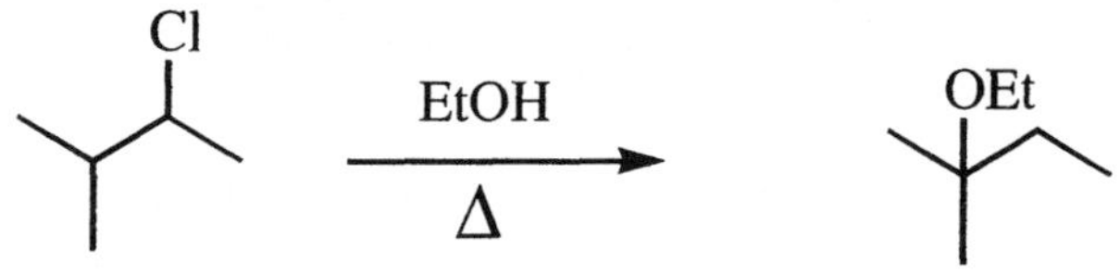

(Note the symbol Δ, which is the Greek capital letter "delta." In chemical contexts it stands for heat. Heat typically makes reactions go faster, and many reactions would be so slow that you have to heat them up if you want to make something happen this year. You can typically ignore the role of heat in mechanisms.) Go ahead, work on the problem now.

There is another important factor that applies to S_N1 reactions, but not so much to S_N2 reactions, namely, the solvent. Why? We have taken a perfectly good, neutral molecule and suggested that it simply disintegrates, all by itself, into a cation and an anion. If the cation is decent, and the anion is decent, this should have a chance, but there is one more important criterion to be met: the molecule must be sitting in a medium that can support charges (that is, a polar solvent). Otherwise, creating charge separation will be prohibitively energetic. You need a polar solvent to lower the energies of these ions. Indeed, the vast majority of S_N1 reactions are carried out in a polar solvent like water or alcohol *with nothing else present*! The starting material simply falls apart and reacts with the solvent. Such a reaction is called a "solvolysis," from the Greek "lysis," meaning cutting, by the solvent. If the solvent is water you can be more specific and call it a "hydrolysis," "cutting by water." As a general rule (though there are some important exceptions to this), if the solvent does not contain an –OH group, it is probably not polar enough to support an S_N1 reaction.

ORM: In "Organic Reaction Mechanisms," under "Substitution Reactions," look at "Unimolecular, S_N1" for a nice animation of the above section.

I need to remind you again that substitution reactions, either S_N1 or S_N2, will not occur with compounds in which the leaving groups are attached to unsaturated centers. Nucleophilic substitution reactions are specific for saturated substrates.

Motivation

A natural question at this point is, "Why would a molecule ever do an S_N1 reaction?" After all, the molecule is perfectly content (no charge, everyone has an octet), and suddenly, without warning, it just elects to fall apart into highly energetic ions! This does not seem like a sensible thing to do!

This is a very fair question. And as I have pointed out, it takes a special set of circumstances for the reaction to happen. Both the leaving group and the cation have to be pretty good. Further, the solvent has to be polar, so that the interaction between the solvent and the new ions can help to compensate for the lost interaction between the leaving group and the carbon it used to be attached to. The reaction often takes heat, which suggests that it is not entirely spontaneous: molecules have to acquire a certain amount of excess energy before they will just bust apart. This same point is also intrinsic in the energy diagram we write, which shows a significant increase in energy as the molecule comes apart. Energy does exist in the surroundings, and not all moelcules have the same energy. Some have more and some have less than the average, and some of those that have more will have enough to come apart. Finally, there is one more factor you have probably not considered: entropy. Taking a single molecule and splitting it into two pieces substantially increases the freedom of movement (entropy) in the system, and, as we discussed in Chapter 5, increasing entropy is a significant factor in determining what molecules are able to do.

S_N1 and S_N2 Summary

In summary: The overall result of both types of reaction is the same: one nucleophile replaces another. It is only the *mechanisms* that differ.

S_N2 reactions go best at primary carbons, then secondary, and they fail at tertiary carbons. They also require an aggressive nucleophile to push out the leaving group. On the other hand, they do not require a polar solvent. S_N1 reactions go best at tertiary carbons, then secondary, but won't work with primary carbons (unless the carbocation can be stabilized by resonance). They require a polar solvent but do not require a good nucleophile.

	S_N2	S_N1
Nucleophile	Must be good—it initiates the reaction	Any nucleophile will do; often solvent
Leaving Group	Must be good	Must be good
Solvent	Does not need to be polar	Must be polar: you are forming ions
Carbocation	None formed	Must be good: therefore...
Substrate Structure	Nucleophile must have access: primary better than secondary; tertiary no good	Tertiary better than secondary; resonance stabilized also very good. Simple primary is very bad

The classic conditions for an S_N2 reaction are treatment of a primary or secondary alkyl chloride or bromide with NaI in acetone. Iodide is one of the best nucleophiles we have and is definitely good enough to push out a good leaving group, which is also present. Acetone is a solvent that is generally not polar enough to allow alkyl halides to ionize, so we expect no competition from S_N1 reactions. One wonderful thing about this particular set of conditions is that it just so happens that NaI is soluble in acetone, but NaCl and NaBr are not. Of course, it is essential that the NaI be soluble in order for the reaction to proceed at all, because that's the only thing that makes I^- available. What is unusual, and very nice, is that the corresponding chloride and bromide are insoluble. This means that at the beginning of the reaction we have Na^+ and I^- floating around. As the reaction proceeds, some of the I^- reacts with substrate, creating alkyl iodide and Cl^-. The latter gets together with Na^+ and discovers that the combination is insoluble in the solvent, so it crashes out of solution. As the reaction proceeds you will see a white precipitate developing. The beauty of this is that you can see something has happened, and there is never any Cl⁻ floating around in solution, so you do not have to worry about the reaction going backwards.

The classic approach to an S_N1 reaction is to take a tertiary alkyl halide and simply dissolve it in alcohol or water *with no additional nucleophile*. With time, sometimes with some heat, reaction takes place all by itself.

SUBSTITUTION REACTIONS OF ALCOHOLS

So far all the substitution reactions you have seen have been done on compounds with good leaving groups, namely, halides. Often one wants to substitute a leaving group that is not so good; specifically, it would often be useful to be able to turn an alcohol into a halide.

$$R-OH \xrightarrow{\quad ? \quad} R-Br$$

Both of the substitution mechanisms we have discussed, S_N1 and S_N2, require that the leaving group leave in the slow step. What is currently attached to this carbon is an OH group, which, if it left, would do so as OH⁻, something we have already identified as a bad leaving group (a fairly strong base, the anion of a rather weak acid). What can we do? We can help it leave. After all, what is attached to the carbon right now is OH, but it is not necessary that OH⁻ be the thing that leaves. Our chief concern is the carbon–oxygen bond. What is attached to the oxygen is irrelevant to us once the bond breaks. So it might be useful to think of some carbon–oxygen bond that *would* be easy to break.

This is simple. We look for good leaving groups based on oxygen, and as I told you before, we find good leaving groups near the top right of the pK_a chart. One that you should quickly find there is H_2O. If H_2O is a good leaving group, then carbon attached to H_2O should be able to eject it. But a carbon attached to H_2O is not an alcohol, it is an alcohol with an extra proton; i.e., a thing with a positive charge:

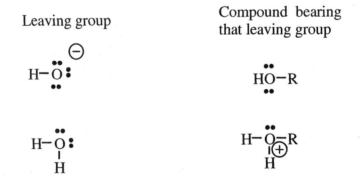

Leaving group   Compound bearing that leaving group

What this tells us is that if we could figure out a way to put another proton onto an alcohol, *then* the carbon would be attached to a good leaving group, and the carbon–oxygen bond would readily break. How can we accomplish that? You should be able to figure this one out. If we want to add a proton to something, we are talking about an acid–base reaction. The base is the alcohol, so you find an alcohol, or something resembling it, on the base side of the pK_a chart. The closest you can come is H_2O, which is actually a very close relative of an alcohol in terms of the likelihood of the lone pair accepting another proton. To choose an appropriate acid, choose one higher than this on the pK_a chart, obviously on the acid side. What you find there, among a few others, are HI, HBr, and HCl.

So what happens if you treat an alcohol with one of these?

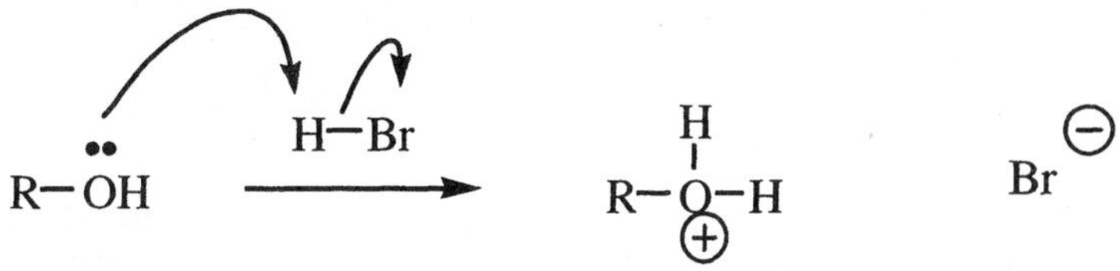

$$R\text{--OH} \quad \xrightarrow{\quad H\text{--Br} \quad} \quad R\text{--}\overset{+}{\underset{\,}{O}}\text{--H} \quad\quad Br^{\ominus}$$

Now what? What happens next depends on the structure of the alkyl group R. If R is tertiary, surely there cannot be an S_N2 reaction, but the conditions are right for an S_N1 process. So that is what happens: the leaving group (water) departs, leaving a carbocation, which then picks up bromide (or something else) to form the product. If R is primary, we surely will not get an S_N1 reaction, but conditions are right for an S_N2 process, so that is what happens: bromide anion attacks and pushes out the good leaving group water. If R is secondary, either reaction (or really both!) might occur. In any case, the result is that you end up with a halide where the OH used to be.

So the net reaction is the following:

$$R\text{--OH} \quad \xrightarrow{\quad H\text{--X} \quad} \quad R\text{--X}$$

Later we will learn some better ways of accomplishing the same result, at least in certain cases, but for now this is the only way you know for turning an alcohol into something else—namely, a halide. Don't forget to catalog this reaction along with your growing list of others, and write several examples of it.

ORM: In "Organic Reaction Mechanisms," under "Alcohols and Ethers," look at "rx alcohol with HBr" for a nice animation of the above section.

Will this work with other nucleophiles? What if we wanted to do the following?:

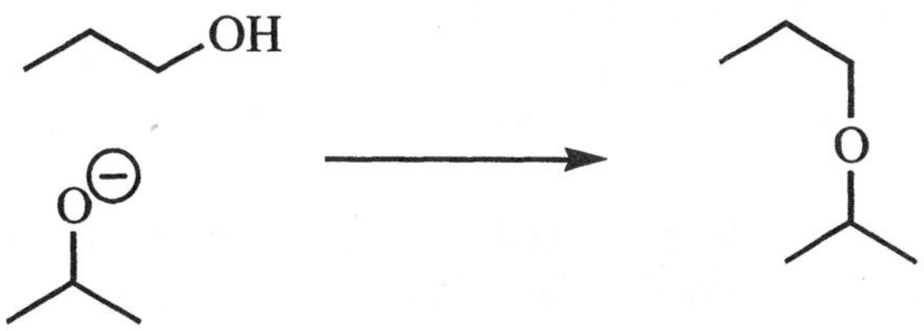

Couldn't we just mix these things and treat the mixture with an acid to make the OH into a leaving group? It looks wonderful on paper:

But there is a serious problem with this. We cannot tell an H^+ to react only with what *we* want. It will react with what looks best to *it*. It will find the best base and go there. What is the best base? The anion, by a mile (look on your pK_a chart, or, better yet, just think about it). So if you add acid to this mixture, you will only put a proton onto the O^- (from which you probably just removed it!). Well, what if we protonate the alcohol first, and then, in a second step, add the anion?

Still trouble. As soon as you add the anion, which is a strong base, it will pick up a proton, either from the medium (which you made acidic in order to protonate the alcohol) or directly from the protonated alcohol itself. Proton transfer is one of the fastest reactions in chemistry. If some simple proton transfer is possible, it will probably occur before any other process has a chance to occur.

The bottom line here is that the only anion you can use to displace a protonated alcohol is one that is compatible with acidic conditions. And, according to your pK_a chart, about the only anions that can exist in acidic solution are the halides, so this reaction is useful only for converting alcohols into halides. But don't despair! Once you have a halide, *then* you can convert it by another substitution into whatever else you want!

SYNTHESIS REVISITED, YET AGAIN

Now we are in a position to do some really cool synthesis problems. Here's one.

Suggest a route for synthesizing the following compound, starting with any compounds you like containing three or fewer carbons as your only sources of carbon. You may also use any inorganic reagents you are familiar with.

Yet again I must remind you that synthesis *cannot* be done by starting at the beginning, especially in a problem like this, since you have no clue what the "beginning" should be. You start at the end and work your way backwards. Here goes.

This is an ether. You know two ways of making an ether. The first one we learned, back in Chapter 10, cannot be used to make an ether like this one, because it relies on addition to a double bond, and a double bond at the end of a chain would be subject to Markovnikov's rule. Therefore, the ether function would wind up at the second carbon. Thus, the *only* way to make *this* ether is by a substitution reaction. A substitution reaction means that something comes, and something goes. The carbon undergoing substitution must have a leaving group on it, and there must be a nucleophile available to come take its place. The nucleophile becomes a part of the product, so it must be part of the compound we are already looking at; the leaving group left, so it is *not* there in the compound we are looking at. It should be clear that the nucleophile in this case must have included the oxygen, and this oxygen is attached on both sides to primary carbons, so it must have displaced something else at a primary carbon, and the only way to do that is with an S_N2 reaction. S_N2 reactions require strong nucleophiles, so the oxygen must have had a minus charge when it came in and displaced some leaving group, most likely a halide. This gives us two options for making the required compound:

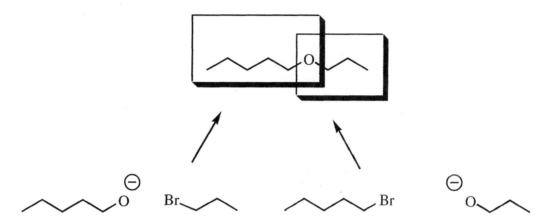

Apparently we have to choose either the left route or the right route. It actually makes no difference in this case, although for some problems one route will be a lot easier than another. Let's choose the left route. The ground rules said that we can use anything we want that has three or fewer carbons. Propyl bromide meets this rule, so we are finished with that part. The only problem left is how to make the left anion. That is easy: we remove a proton from a molecule that has one more proton. We must also specify how, and we know that too:

$$\text{\raisebox{0pt}{$\diagdown\!\diagup\!\diagdown\!\diagup$}OH} \xrightarrow{\text{NaNH}_2} \text{\raisebox{0pt}{$\diagdown\!\diagup\!\diagdown\!\diagup$}O}^{\ominus}$$

Unfortunately, this starting material, pentanol, does not meet our requirements, so we must figure out where it will come from. *One step at a time!!!!* It is a primary alcohol, and there is only one way you know for making primary alcohols, and that is reduction of an aldehyde. So do it:

$$\text{\raisebox{0pt}{$\diagup\!\diagdown\!\diagup\!\diagdown$}=O} \xrightarrow[\text{Pd/C}]{\text{H}_2} \text{\raisebox{0pt}{$\diagdown\!\diagup\!\diagdown\!\diagup$}OH}$$

There is also only one way you know for making aldehydes. So do it:

$$\text{\raisebox{0pt}{$\diagup\!\diagdown\!\diagup$}=CH}_2 \xrightarrow[\text{2) Zn, H}_2\text{O}]{\text{1) O}_3} \text{\raisebox{0pt}{$\diagup\!\diagdown\!\diagup\!\diagdown$}=O}$$

Whoa! We are trying to make something that has five carbons, and we eventually have to show how it comes from something with three carbons. But this compound has six carbons! Aren't we going in the wrong direction? Yup, but you have no choice. Later we will learn more efficient ways to do these things, but for now, every reaction has been the only one possible.

There is only one way you know for making alkenes. So do it:

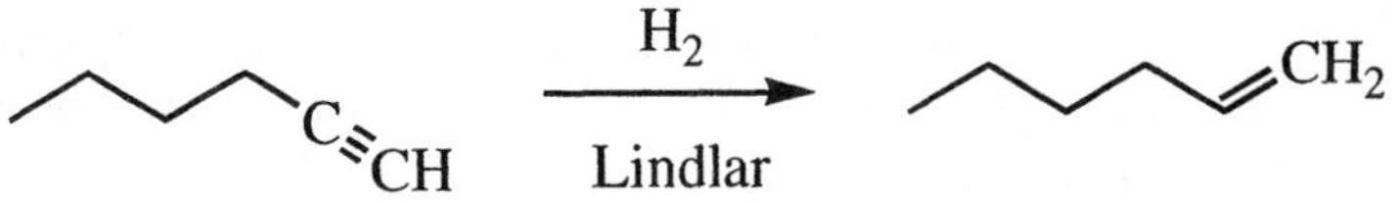

Now, at last, we have come to a molecule that can be made from smaller pieces. Remember, the only way you know for making carbon–carbon bonds is reaction of an acetylide anion with some primary or secondary alkyl halide in an S_N2 reaction. Here is a place we can use it:

One of these compounds, acetylide anion, has fewer than three carbons, so that can be a stopping point. It turns out that the anion of acetylene can actually be bought, as its sodium salt, so you could go straight to a stockroom and get some. Ordinarily, though, it is a good idea to tell your reader how to make charged species starting from neutral ones:

So we are done with that piece. The other piece has four carbons, and we still have to get down to three. Where could the butyl bromide come from? There are two reactions you know for making alkyl halides, but addition reactions never put the halide on the end carbon, so this one must have been made by the other route: substitution of an alcohol:

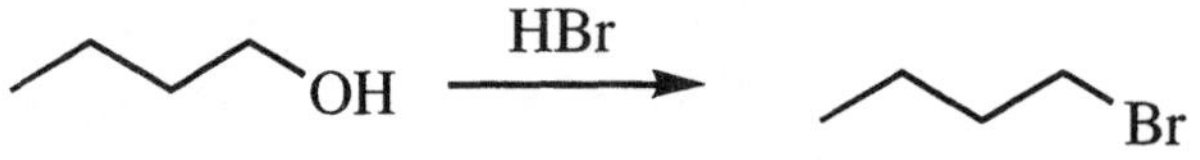

This requires a four-carbon alcohol, and we have just shown how, by a long sequence of reaction, we could turn it into a five-carbon alcohol. The same sequence, then, should convert a three-carbon alcohol into a four-carbon alcohol. Here is how it would look on paper:

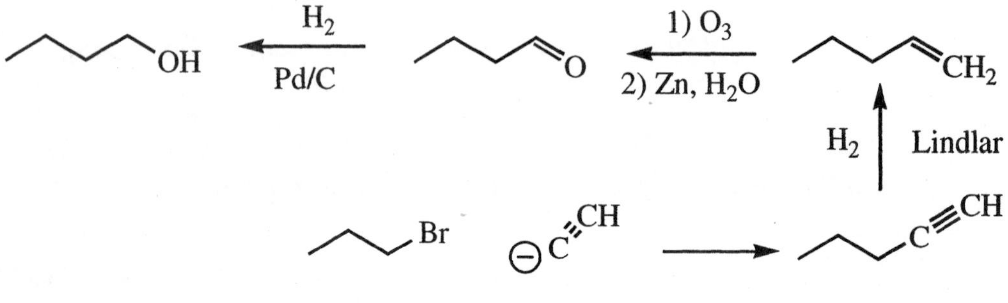

Wow, we even saved a step, because the three-carbon bromide is a legitimate starting material.

So here is a complete answer to the synthesis problem, written as you might produce it on an exam:

Practice, practice, practice! This should become quite easy after a while. Have fun!

Problems

(Answers are provided at the end of the chapter for italicized problems.)

1. Show the major organic products of the following reactions. Write NR if there is no reaction.

a.

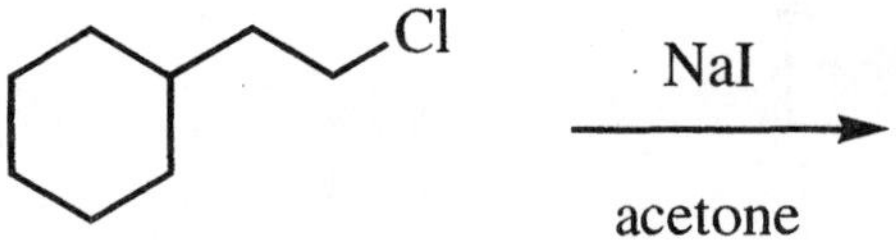

$$\xrightarrow[\text{acetone}]{\text{NaI}}$$

b.

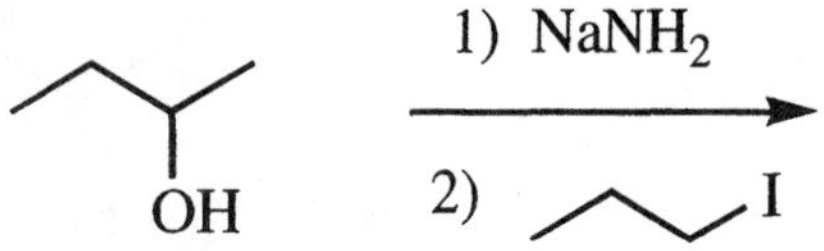

1) NaNH$_2$

2)

c.

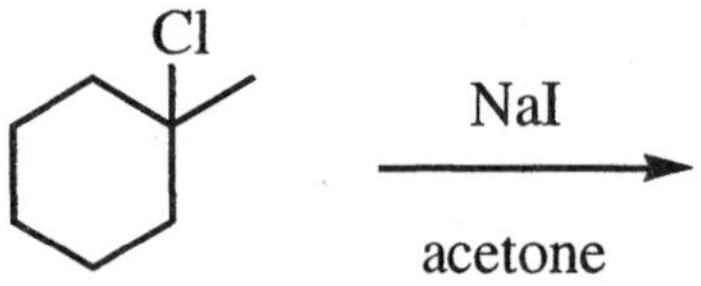

$$\xrightarrow[\text{acetone}]{\text{NaI}}$$

d.

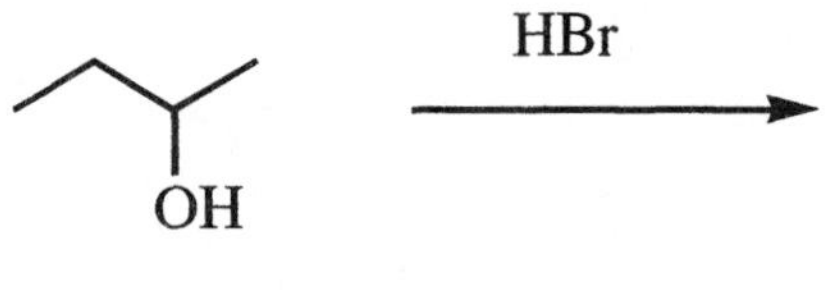

$$\xrightarrow{\text{HBr}}$$

e.

$$\xrightarrow[\triangle]{\text{EtOH}}$$

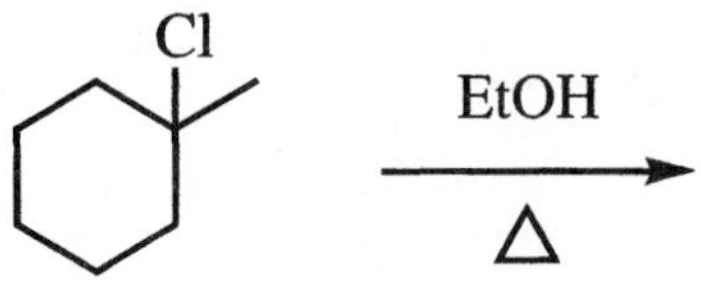

2. Which S_N2 reaction in each of the following pairs is faster? Explain.

3. The following data were obtained for the reaction described by the equation below. "Initial rate" means the rate of the reaction at the very beginning. Of course, the rate changes as time goes on. $[X]_0$ means the concentration of X at time 0 (i.e., the beginning of the reaction). This also obviously changes with time.

$[A]_0$, mol/L	$[I^-]_0$, mol/L	Initial rate, mol/L-sec
1.5×10^{-2}	1.5×10^{-2}	3.88×10^{-4}
3.00×10^{-2}	1.5×10^{-2}	7.74×10^{-4}
1.5×10^{-2}	3.00×10^{-2}	3.85×10^{-4}
3.00×10^{-2}	3.00×10^{-2}	7.75×10^{-4}

Is this an S_N2 reaction? If not, propose a mechanism consistent with the above facts.

4. A reaction produces the following data:

$$RCl \; + \; NaI \; \rightarrow \; RI \; + \; NaCl$$

Time 0		Time 1 minute
[RCl], mol/L	[NaI], mol/L	[RCl], mol/L
3.75×10^{-2}	4.83×10^{-2}	3.50×10^{-2}
3.75×10^{-2}	9.65×10^{-2}	3.25×10^{-2}
7.51×10^{-2}	4.84×10^{-2}	7.00×10^{-2}
7.50×10^{-2}	9.97×10^{-2}	6.50×10^{-2}
1.50×10^{-1}	1.00×10^{-1}	???
3.75×10^{-2}	???	3.00×10^{-2}

    a. Is this an S_N1 or an S_N2 reaction?
    b. Fill in the missing numbers.
    c. Suggest a structure for R that would be consistent with this information.
    (*Hint*: Create a fourth column labeled "rate.")

5. Answer the same questions from problem 4 for the following reaction. (Hint: Again, create a fourth column for rate. To fill in the last row, you will have to use the rate equation on page 315 or 320, whichever is appropriate.)

Time 0		Time 10 minutes
[RCl], mol/L	[NaI], mol/L	[RCl], mol/L
3.75×10^{-2}	4.83×10^{-2}	3.50×10^{-2}
3.75×10^{-2}	9.65×10^{-2}	3.51×10^{-2}
7.51×10^{-2}	4.84×10^{-2}	7.02×10^{-2}
7.50×10^{-2}	9.97×10^{-2}	7.00×10^{-2}
1.50×10^{-1}	1.00×10^{-1}	???
???	???	3.00×10^{-2}

6. Draw energy diagrams for an S_N2 reaction as well as an S_N1 reaction, including pictures describing each peak and valley.

7. In general, primary halides almost always undergo S_N2 reactions rather than S_N1 reactions. An exception to this generalization, however, is the case of benzyl halides, which can undergo either S_N2 or S_N1 reactions. Explain why these particular primary halides are willing to undergo S_N1 reactions.

8. The following reaction is called the pinacol rearrangement. Suggest a mechanism for the reaction, and explain why it occurs the way it does.

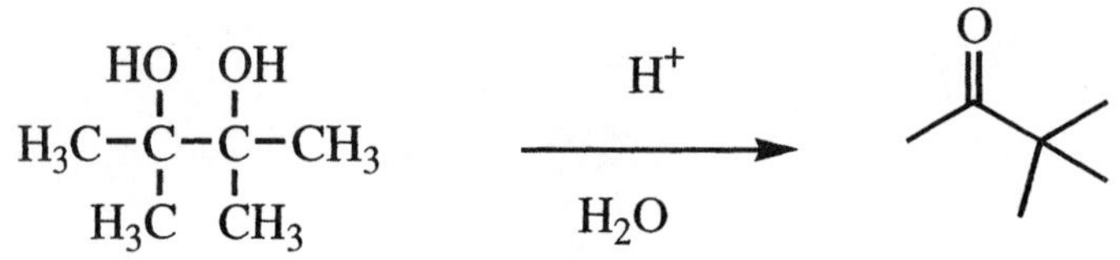

9. Suggest a mechanism for the following reaction, using curved arrows. *Be sure to indicate any resonance structures that are relevant.*

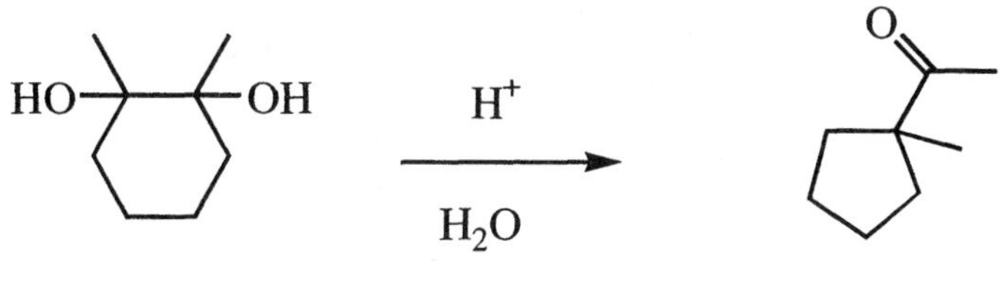

10. Suggest a mechanism for the following reaction.

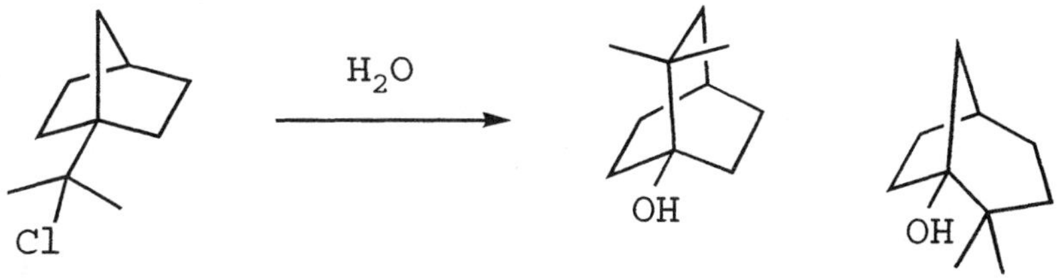

11. a. Suggest a mechanism for the following reaction.

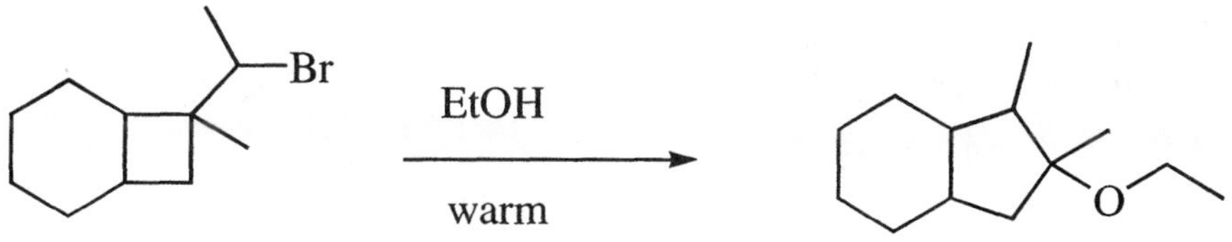

b. Show other likely products from this reaction. I can think of six obvious ones, and many more arising through complicated processes. Show mechanisms for obtaining each of the new products.

12. a. *Show at least three products that could be expected to form upon heating 2-iodo-2-methylpropane together with methanethiol (CH_3SH) in the solvent ethanol.*

b. *To what extent should the rate of disappearance of 2-iodo-2-methylpropane be affected by doubling the concentration of methanethiol? Assume all other conditions remain the same as in part a). Explain briefly why the rate effect should be as you stated here.*

c. *What effect, if any, should doubling the concentration of methanethiol have on the yields of the various products proposed in part a)?*

d. *Speculate on what product would predominate (and why) if the substrate in part a) were changed to 1-iodopropane.*

13. You have learned how to make an ether using an alcohol and an alkene. One can also make an ether by reacting an alcohol with itself in the presence of acid. Show a mechanism for the following reaction:

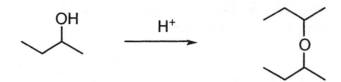

14. *Show how to synthesize the following compounds, starting from anything you like containing three carbons or less.*

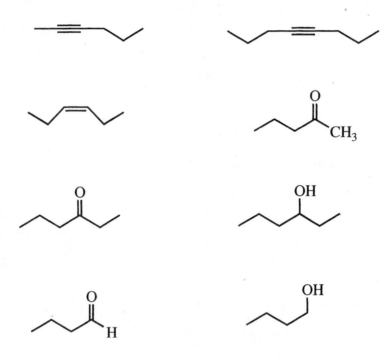

15. a. Suggest a synthesis of *cis*-4-octene starting with organic compounds containing three or fewer carbons.

b. Suggest a synthesis of $CH_3CH_2CH_2OCHCH_2CH_3$ starting with organic compounds containing three or fewer carbons. $\quad CH_3$

c. Suggest a synthesis of $CH_3CH_2CH_2OCH_2CH_2CH_3$ starting with organic compounds containing three or fewer carbons.

d. Suggest a synthesis of $CH_3CH_2CH_2OCH_2CH_2CH_2CH_3$ starting with organic compounds containing three or fewer carbons.

16. The following compound could be prepared by a Williamson ether synthesis in either of two ways. Show the two ways and explain why one would be preferable to the other.

17. Suggest a synthesis of the following compound, starting with anything you like containing three carbons or less.

18. Starting from anything you like containing three carbons or less, suggest syntheses of the following molecules.

19. *Suggest a synthesis of the following alkanes starting from anything you like containing three carbons or less.*

Selected Answers

Internal Problems

Test Yourself 1.

End of Chapter Problems

5.

	Time 0		Time 10 minutes
	[RCl]	[NaI]	[RCl]
	3.75×10^{-2} M	4.83×10^{-2} M	3.50×10^{-2} M
	3.75×10^{-2} M	9.65×10^{-2} M	3.51×10^{-2} M
	7.51×10^{-2} M	4.84×10^{-2} M	7.02×10^{-2} M
	7.50×10^{-2} M	9.97×10^{-2} M	7.00×10^{-2} M
	1.50×10^{-1} M	1.00×10^{-1} M	1.40×10^{-1} M
	3.2×10^{-2} M	any value	3.00×10^{-2} M

For this last line, you need to solve the rate equation, rate = k [RCl], where the rate is ([RCl] at time 0) – ([RCl] at time 10 minutes), all divided by the time, 10 min. We abbreviate this $(C_0 - C_{10})/10$.
From any of the first lines we can conclude that rate = .0067 [RCl]

Thus, in the last line $(C_0 - C_{10})/10 = .0067 \, C_0$

Since we are given C_{10}, we can solve this for C_0.

Note: This is actually oversimplified. Since the rate depends on the amount of RCl, and the amount of RCl is constantly going down, so is the rate. To solve this kind of equation accurately requires a bit of calculus. In other words, the real rate is not $\Delta[RCl]/\Delta t$, but $d[RCl]/dt$. If you integrate the real rate equation, you get the following:

$$\ln [RCl]_t = \ln [RCl]_0 - kt$$

To get k properly, you must plot $\ln[RCl]_t$ vs. time and take the slope. For this course, you do not need to know this indented material.

S_N1; *tert*-butyl

7.

A resonance stabilized carbocation is more stable than a tertiary one.

8.

tertiary carbocation

resonance stabilized cation--even better

12. a.

b. No effect. This is an S_N1, depends only on concentration of substrate.

c. Second product will increase, because carbocation will more often encounter methanethiol. Of course, 1st and 3rd will therefore decrease.

d. 1-iodopropane cannot do an S_N1, so it will have to undergo an S_N2. The best nucleophile present is methanethiol. Therefore, almost all the product will be ⌒⌒SCH$_3$.

14.

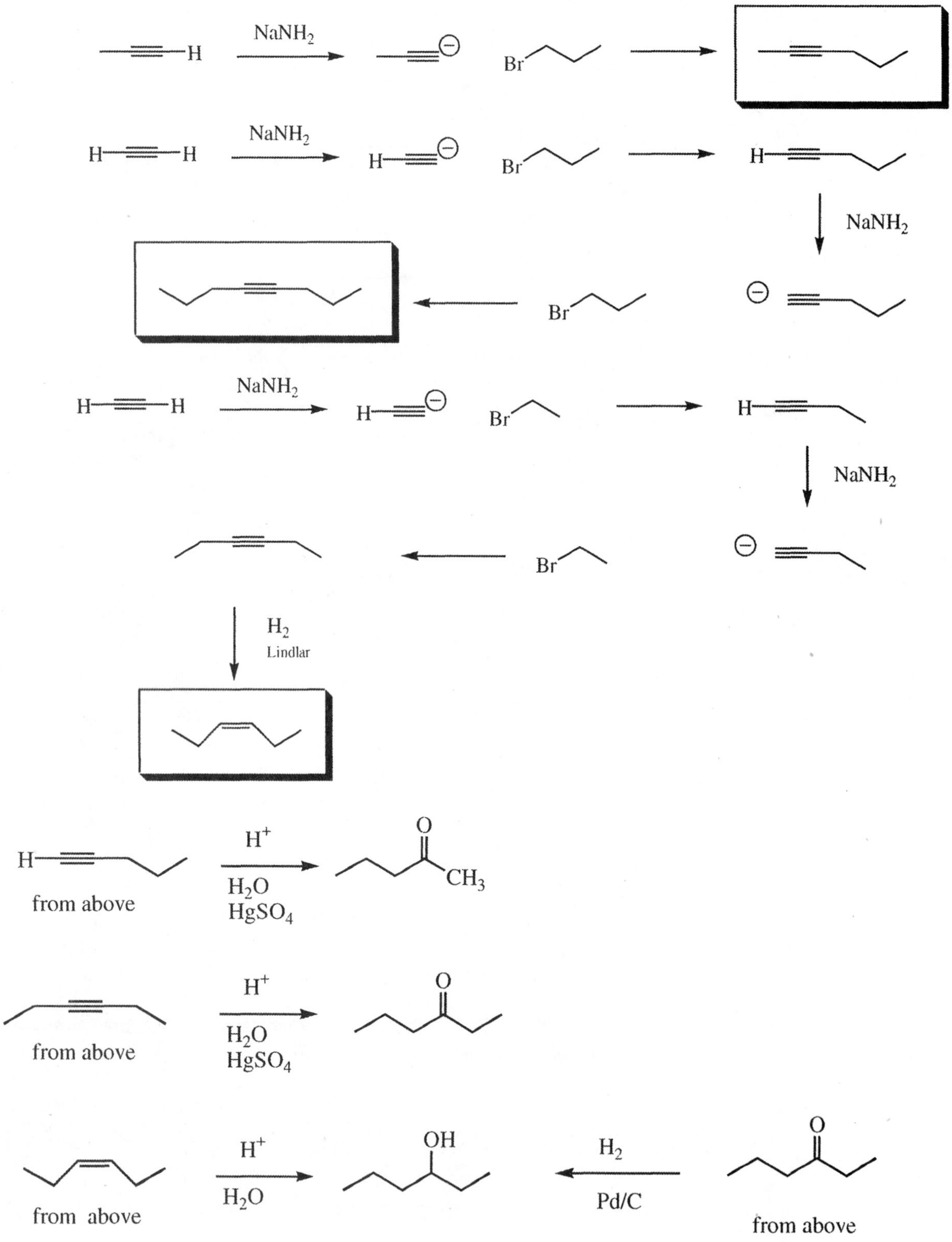

NaNH$_2$

NaNH$_2$

NaNH$_2$

NaNH$_2$

Br

Br

Br

Br

Br

NaNH$_2$

NaNH$_2$

H$_2$
Lindlar

H$^+$
H$_2$O
HgSO$_4$

from above

H$^+$
H$_2$O
HgSO$_4$

from above

H$^+$
H$_2$O

from above

H$_2$
Pd/C

from above

CH$_3$

OH

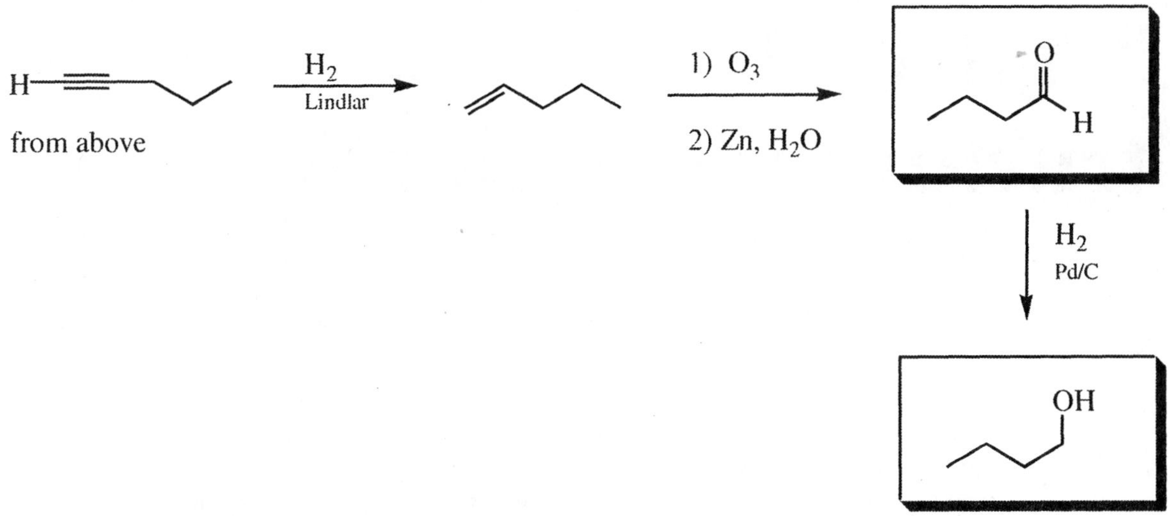

H—≡— (from above) →[H₂ / Lindlar] ⇒⟍⟍ →[1) O₃; 2) Zn, H₂O] | O‖ butanal (H) |

→[H₂ / Pd/C] | ⟍⟍⟍OH |

19.

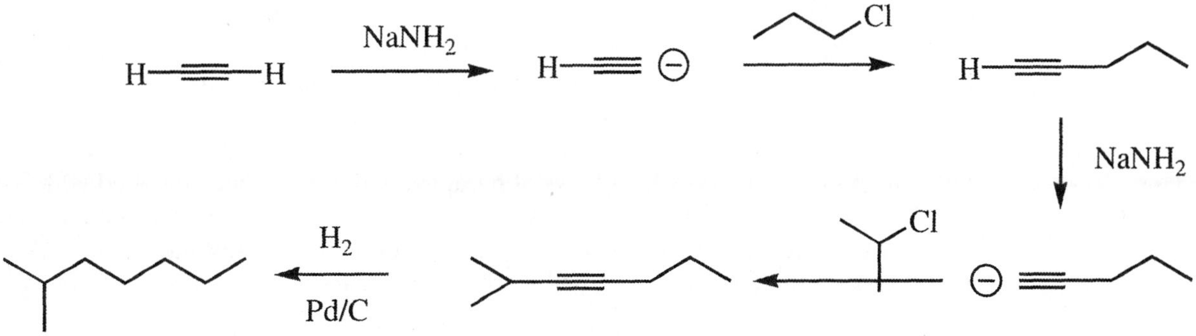

H—≡—H →[NaNH₂] H—≡—⊖ →[propyl chloride] H—≡—⟍⟍

→[NaNH₂]

⊖—≡—⟍⟍ →[isopropyl chloride] branched alkyne →[H₂ / Pd/C] branched alkane

⟍⟍⟍OH →[HBr] ⟍⟍⟍Br

From Prob 13

H—≡—H →[NaNH₂] H—≡—⊖ →[butyl bromide] H—≡—⟍⟍⟍

→[NaNH₂]

⊖—≡—⟍⟍⟍ →[isopropyl chloride] branched alkyne →[H₂ / Pd/C] branched alkane

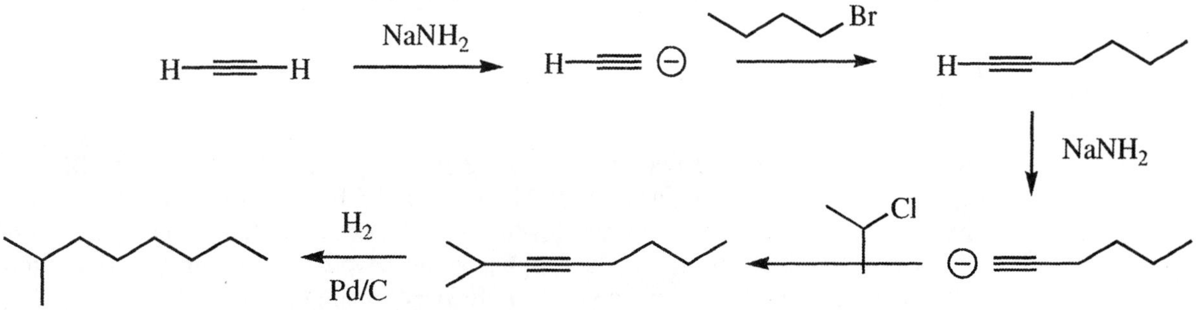

Chapter 14
Structure Determination

Ever since we started talking about reactions, I have been telling you that such-and-such a reaction makes these products, and this other reaction makes these other products, etc. Some of you may have wondered at many points along the way, "How do you know that?" What evidence do we have that molecules are put together the way chemists claim? How do we know organic structures?

There is a great deal of history associated with questions like these, but I will ignore most of it and dive straight into the field called spectroscopy. In the old days, there was a lot of fascinating slicing of molecules in an attempt to end up with something recognizable—the ozonolysis reaction, for example, was one of the common tools for structure determination. These days almost all structures are determined by analyzing the interaction of a compound with various kinds of electromagnetic radiation (of which "light" is a tiny portion). The ultimate in this business is X-ray crystallography, which zaps the compound with X-rays and in effect takes a photograph of exactly which atoms are where. This technique is revealing wonderful information about the structure and function of proteins, but it is beyond the scope of this course. In this chapter we will discuss the four basic kinds of spectroscopy:

- Mass spectrometry

- Ultraviolet-visible (UV-Vis)

- Infrared (IR)

- Nuclear magnetic resonance (NMR).

There are scientists who spend their entire careers mining the details of all of these techniques, and each provides a wealth of information, but we will only be able to scratch the surface.

Acknowledgment

Most of the spectra shown in this chapter were downloaded from the SDBS website, operated by the National Institute of Materials and Chemical Research in Japan. This site has an extensive collection of spectra that may be accessed free of charge, as long as you don't take too many at once. The chapter would have been difficult to write without them! Take a look at http://www.aist.go.jp/RIODB/SDBS/menu-e.html

MASS SPECTROMETRY

Mass spectrometry is unique in several senses. For one, it generally does not involve an interaction of molecules with "light" of any sort, so it is not formally spectroscopy at all. (You may have noticed that it is called spectro*metry* rather than spectro*scopy*. The "spectro" part means a spreading out. The suffix "scopy" implies light.) Mass spectrometry involves a spreading out of particles, ions that are traveling through a magnetic and/or electric field, as we already discussed in Chapter 1.

To review: You arrange for your sample to be zapped with electrons, which results in a *loss* of electrons from your molecule, therefore creating positive ions with essentially the same mass as the original molecule itself. These ions are accelerated down a tube and through a magnetic field, where their paths are bent by an extent that depends on their mass. If you place a detector at the other end of the zone that registers where ions are landing, you can get a report of how many particles arrive and what their masses are.

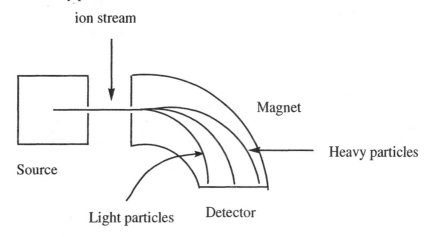

If we put a pure substance into this instrument, the output will tell us several things. First and foremost, we can assume that the particles of heaviest mass probably correspond to exactly what we put in (called the "parent ion" or "molecular ion"). So generally the first thing you do with a mass spectrum (mass spec, for short) is look for the heaviest signal detected. That usually equals the molecular weight (more accurately, mass) of your molecule. What good is that? Well, obviously, if your molecule has a mass of 120 amu, it can't have more than 10 carbons in it! And if it had 10, there would be no room for other atoms, so it probably has less than 10. Let's see, it could have 9, weighing 108 amu, and then perhaps 12 hydrogens, so C_9H_{12} is a possibility. Could it be C_8H_{24}? No, because 8 carbons hold at most 18 hydrogens. But there might be something other than C or H; perhaps there is an O. If there is an O, weighing 16 amu, the rest of the molecule has to weigh 104 amu: this might be C_8H_8. Or maybe there are two O's. And so on. Nevertheless, note that the molecular weight all by itself constrains your possibilities significantly.

Isotopes

The obvious question at this point is this: if you put in a pure substance, why do you see anything *other* than the parent ion? There are really two important answers. One has to do with isotopes. Carbon is mostly ^{12}C, but about 1% of the carbon in the world is ^{13}C. In any random collection of carbon atoms, about 1% of them will be heavier than the others by 1 amu. We have put millions of molecules into this instrument, enough so that we can rely on statistics. Every carbon in our molecule has about a 1% chance of being ^{13}C, so we should *expect* to find a few molecules that are *heavier* than the calculated weight of our molecule based on ^{12}C. For example, consider 1-phenylethanol, shown below.

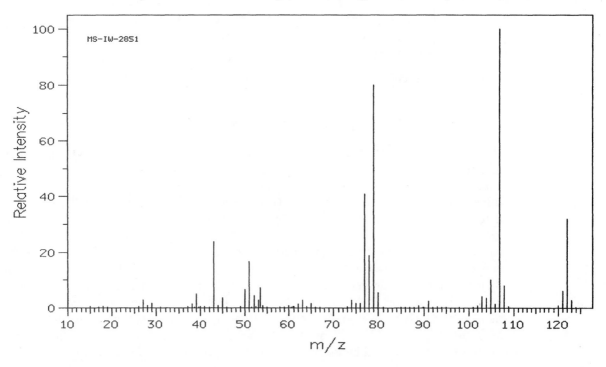

If you calculate its molecular mass using normal values (12 for carbon, 1 for hydrogen, 16 for oxygen), you should get 122 g/mol. So we should expect to see a peak in the mass spec at 122. But occasionally a molecule will show up weighing 123, because occasionally one of the carbon atoms present will be a ^{13}C. This will actually occur more than 1% of the time, because it occurs 1% of the time for *each* carbon, and there are 8 of them in this molecule. Any one of them might cause an increase in the weight. So actually about 8% of the molecules should weigh 1 unit more than "normal." Below is a real mass spec for this molecule, and you should notice the small peak at 123. It is difficult to tell from the picture whether this peak is exactly 8% the size of the 122 peak, but it is certainly about that, and if we got better information from the instrument it would turn out right. (The *x* axis is labeled *m/z*, which is the mass-to-charge ratio. For our purposes we can consider this to correspond to the molecular mass of the ion. The reason it is not labeled "molecular mass" is that the curvature of the particles depends not only on their masses but also their charges. If some particle had a charge of more than one, it would curve twice as much as normal. Fortunately, almost all particles have a +1 charge, so we can ignore this complication.)

So it turns out that I was not quite telling the truth when I told you that the peak of highest mass corresponds to the molecular weight of your molecule. There are peaks with slightly higher masses, indicating the presence of heavier isotopes of carbon, hydrogen, or oxygen. However, as you can see from this example, they are also very small, and frankly, in many cases they are ignored. The molecular weight is the heaviest *major* peak, clearly 122 in this mass spectrum, and dribblings one and two mass units higher just reflect isotopic variants. These variants can be used for fancy interpretations, but we can ignore them except in two cases, and these are the cases we started with in Chapter 1. Recall, the world's bromine comes as two isotopes, 79 and 81, in a 1:1 ratio; the world's chlorine comes as both 35 and 37, in a 3:1 ratio. So molecules containing chlorine or bromine will have quite significant *pairs* of peaks at the upper end of the mass spec. Look at the spectrum of methyl chloride shown in Chapter 1 for an example of this.

Fragmentation

Obviously there are more peaks in this spectrum than the one for the whole molecule and its isotopic variants. There are lots of peaks at masses lower than 122. What are these doing here? We answered this question back in Chapter 1, although you may not remember it: the process of obtaining mass spectral information involves so much energy that some molecules do not survive the ordeal; instead they break into smaller pieces. The resulting pieces ("fragments") also show up in the mass spec if they retain the positive charge (see below), and by analyzing the pieces we can often learn more about the molecule. Indeed, mass spectra have so much detail in them that a professional can identify many molecules using this information alone! Further, the pattern of peaks, including the isotopic clusters and the fragmentation pattern, can vary so much that no two molecules will give exactly the same spectrum. Under standard conditions, however, a given substance will always result in substantially the same spectrum. In other words, the mass spectrum can be used as a highly reliable fingerprint of a molecule, and when you compare your pattern to a library of hundreds of thousands of stored patterns, if you obtain an exact match you can positively identify an unknown. Of course, no one wants to sit down and actually make the comparison with thousands of spectra, but computers do not mind doing so, and there are now excellent computer programs for this purpose. Mass spectra have become quite important in the legal arena, where they are accepted as evidence of the identity of samples obtained at crime scenes, such as illicit drugs.

But you will not always have a computer with you, much less one with thousands of spectra in it for comparison purposes! What can you as an individual scientist make of a mass spectrum? Besides looking at the whole molecule (the "parent"") and deducing its molecular weight, you can look at some of the fragments and try to guess what pieces they might represent. After all, these pieces must have been part of the original molecule.

As it turns out, it is often easier to look at differences rather than the fragments themselves. Huh? The pieces that fall off a molecule tend to be somewhat small, like methyl, ethyl, –OH, –OCH$_3$, etc., weighing less than about 40 amu. Because there is no such thing as a perfect system, there is always air in a mass specrometer, and air contains water, nitrogen, oxygen, carbon dioxide, argon, and other atoms and molecules with low masses. This part of the mass spec tends to be somewhat crowded and hard to interpret, and indeed, many mass spectrometers are set so they don't even look below 50 amu. So what good are these small pieces? Consider some molecule weighing 150, from which a methyl group (–CH$_3$, weighing 15) falls off. This would create a piece weighing 15, but also a piece weighing 135. The methyl group may get lost in the low-weight part of the spectrum, but the 135 peak should stick out like a sore thumb. Further, if you see a peak at 150 and another at 135, you can easily figure out that something that weighs 15 has fallen off. Since the upper end of the mass spec tends to be less crowded than the lower end, it is usually easier to recognize fragments by the difference between the parent and what's left rather than by observing the small fragments themselves. (Another reason to concentrate on the upper end is that most mass spectrometers can only "see" positive ions, and when an ion falls apart into two pieces, only one of them is charged. It is usually the larger piece that has the charge, and is therefore recorded in the mass spec. The smaller pieces sometimes fly away straight and invisibly.)

Look again at the spectrum of 1-phenylethanol: You see the parent ion at 122. The next significant peak below this (in mass) is at 107. The difference is 15. There is only one simple organic piece that weighs 15, and that is a methyl group. So a methyl group must have fallen off. What's more, it must fall off pretty easily, since the peak at 107 is actually larger than the one at 122! Of course, we already knew there was a methyl group on this molecule, because I told you its structure, but if this were an unknown, we could now conclude that 1) it weighs 122, and 2) it has a methyl group in it somewhere.

Anything else? There is also a peak at 105. This is 17 away from the parent. The most common piece that weighs 17 is –OH. So perhaps there is also an –OH group in this molecule. (Again, we already knew there was, but if we didn't we could now guess it.)

What else? There are significant peaks at 77 and 79. The 77 peak is due to a phenyl group, C_6H_5, and most molecules with a phenyl group will have a peak at 77. You are not yet equipped to explain the 79 peak. But there it is. The message is that we do not need to be able to explain every peak in a mass spec, but useful information can often be obtained from the parent peak and a few at lower mass. [Actually, it turns out that the triplet of peaks at m/z 77, 78 and 79 is typical for alkylbenzenes and their derivatives, indicating the formation the phenyl cation, ionized benzene, and protonated benzene. So these "unexplained" peaks do have explanations, if you study the field sufficiently!]

A great deal of sophistication can go into mass spectral interpretation, but for our purposes all we need to be able to do is pick out the parent, look at the first few fragmentations, and see if we can make some good guesses about the structure. In certain cases, you may actually be able to completely assign a structure based on a mass spec alone. The example above is a good one: pretend you didn't already know the answer. What could you learn from this mass spec?

1. The stuff has a molecular weight of 122.

2. It has a methyl group.

3. It has a phenyl group.

4. It probably has an –OH group.

Add the weights of these pieces up and they come to 109. There are only 13 amu's you have not identified, and it is hard to imagine any way to get 13 amu's other than with CH. Further, there is only one way to put all these pieces together. So you could actually identify this particular compound simply from its mass spec. This is a rare piece of good luck, though.

About the only other thing you need to know about mass spectral interpretation, at this level, is what pieces are most likely to fall off. The best rule of thumb I can give you is that single bonds between atoms easily break when one of the pieces can stabilize the positive charge remaining on that fragment. So carbonyl groups and benzene rings often cleave neighboring bonds because the cations produced (phenyl or "acyl") are relatively stable, as these things go. Most favorable is the cleavage of a C-C bond one away from a C-C double or triple bond or an aromatic ring, because allyl- and benzyl-type cations are particularly stable. Also, anything you might identify as a leaving group, such as chlorine, bromine, or iodine, is a prime candidate for leaving in a mass spectrometer.

For example, predict the mass spectrum for the following molecule. Do it!

The first thing to do is add up what the whole thing weighs. When you do this be *sure* to use the masses for the most abundant isotopes (e.g., C = 12), *not* the masses listed in the periodic table (such as 35.5 for chlorine). In this case you should get 178. So we predict a peak at 178. Then guess what pieces might fall off the thing. I told you that bonds adjacent to carbonyl groups are good candidates for breaking, so we should expect bonds 1 and 2 to break (see below for labels). The molecule would lose 45 amu (leading to a piece weighing 133) if bond 1 broke, and breaking bond 2 would give us a biggest piece weighing 105. Bond 3 is in the category of bonds one away from a benzene ring, and that would leave a piece weighing 91. (A peak at 91, due to C_7H_7, is very common for compounds with the phenyl–CH_2 group, called benzyl, because radicals and ions based on the CH_2 group are resonance-stabilized by the benzene ring. When you see a large 91 peak, think "benzyl.") Finally, we expect the phenyl group to fall off, weighing 77.

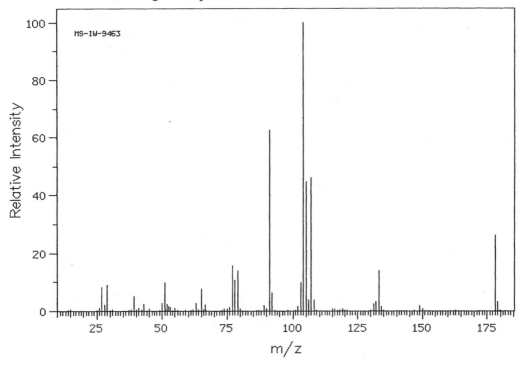

So we predict peaks at 178, 133, 105, 91, and 77. Below is the real mass spec of this. It has peaks at 178, 133, 105, 91, and 77. You may notice that in addition to the 77 peak, there are also peaks at 78 and 79, which were referred to earlier. And there are peaks that we did not predict, including big ones at 107 and 104; indeed, the peak at 104 is the biggest one of all! There are explanations for these peaks, but they are not straightforward. Interested students will study mass spectrometry in greater detail in later courses. Overall, however, I'd say we did a pretty good job guessing what the spectrum would look like, even with our elementary understanding of the process.

Test Yourself 1

Now try one in the other direction. Can you guess what gave the following spectrum?

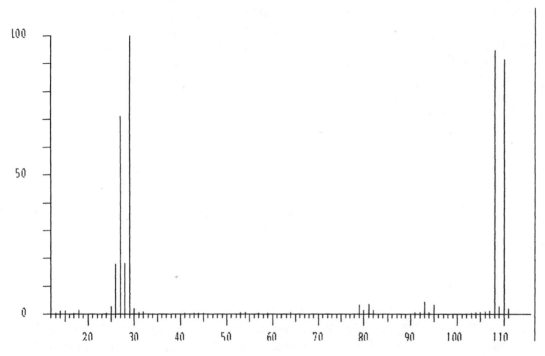

Hint: there is no obvious way to lose two amu from a molecule, so *both* of the heavy peaks must come from parents, involving different isotopes.

SPECTROSCOPY

As mentioned above, spectro*scop*y involves the interaction of molecules with *light*. We detect an interaction of light with molecules by shining light through a sample of the molecules and checking to see whether any is absorbed. The instrument we use is called a *spectrometer*. The diagram below is for a generic spectrometer, which can be thought of as either a UV-vis spectrometer or an IR spectrometer. The only difference between the two is the kind of light emitted from the source.

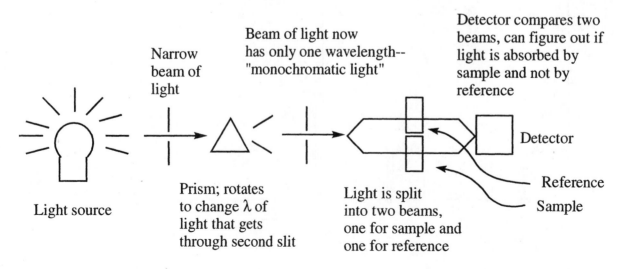

Narrow beam of light

Beam of light now has only one wavelength-- "monochromatic light"

Detector compares two beams, can figure out if light is absorbed by sample and not by reference

Detector

Reference

Sample

Light source

Prism; rotates to change λ of light that gets through second slit

Light is split into two beams, one for sample and one for reference

A clever feature of this spectrometer is the way it splits the light beam into two identical beams. The reason for this is that you want to discover what wavelengths of light are absorbed by the molecule you are interested in. But you cannot just lay a chunk of the stuff all by itself in the light beam. It has to be dissolved in some solvent, and contained in some vessel. Both the solvent and the vessel are made of chemicals, and could potentially absorb light, but you are not interested in that. So you put an identical vessel containing solvent but *not* sample in the "reference" beam. The instrument compares the two beams, and registers absorption only when light is absorbed by the sample and *not* by the reference. By this trick you can be sure it is your molecule doing the absorbing, and not the solvent or the vessel.

UV-Vis Spectroscopy

Why do molecules interact with light? For the same reason that atoms do (see Chapter 2): they have energy levels, and light of the proper energy (v, λ) can make them jump from one energy level to another, while light with the wrong energy is ignored. In the case of atoms, discussed in Chapter 2, electrons jump between the main energy levels of isolated atoms; we used this information first to deduce that such energy levels exist, and then indirectly to explain the periodic table based on our understanding of the significance of these energy levels. In this section we will examine electron energy levels in *molecules*. It turns out that there are some, and that light in the energy range of visible light or slightly higher (ultraviolet light, "UV") can make electrons jump between them. Recall that visible light covers the region of about 400–700 nm; UV light corresponds to about 200–400 nm.

Why are there energy levels in *molecules*? First, they are mainly associated with bonds. You recall that we have two chief kinds of bonds (there are others, but we will not deal with them in this course): σ and π. A bond is a sharing of electrons between two atoms. If the orbitals that are sharing the electrons (sp^3, sp^2, sp, s) are pointed at one another along the internuclear axis, we call the result a σ bond. If they are parallel to each other (always p orbitals, in our experience), they form a π bond. What I did not tell you earlier is that every time you form a bond—which is an orbital shared by two atoms and into which you may place two electrons—you also form another, higher energy orbital that usually has *no* electrons in it. (Whenever you *combine* two orbitals you have to *get* two orbitals—just as we did in hybridization.) This higher energy orbital (called an "antibonding orbital") normally is of no interest to us, and we can ignore it, as we have so far; but it is there, and with the right amount of energy, you can pop an electron from a bonding orbital into the antibonding orbital. This will register as an absorption of energy if you look for it properly.

The business of antibonding orbitals is confusing and really not necessary for you to understand now, so we need not dwell on it. Here is all you *really* need to know:

1. In organic molecules, UV and visible light are absorbed only if there are π bonds present. This means double and triple bonds.

2. A double or triple bond all by itself is not enough. In order to be detected by our normal instruments, multiple bonds need to be bunched together in a special way, called conjugation. Conjugation is more accurately a lining up of the p orbitals from two or more double bonds. You remember that a π bond comes from a side-by-side overlap of p orbitals. If you bring up a second π bond so that *its* p orbitals are lined up with the first ones, you get a string of 4 aligned p orbitals:

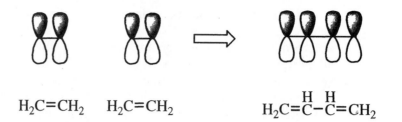

$$H_2C{=}CH_2 \qquad H_2C{=}CH_2 \qquad\qquad H_2C{=}\overset{H}{C}{-}\overset{H}{C}{=}CH_2$$

Note that this is not implying that a chemical reaction has accomplished some transformation; we are simply noting that two double bonds in a molecule could result in a linear array of four p orbitals. In the molecule shown above, 1,3-butadiene, the two p orbitals on the left are involved in one double bond, and the two on the right are involved in a different double bond. According to our picture, the p orbital on carbon 2 supposedly interacts with the p orbital on carbon 1 and ignores the p orbital on carbon 3. But how does he know to do that? It's like if you go to a dance and two equally attractive partners are vying for your attention, it would be hard to choose one and ignore the other. The p orbital on carbon 2 has a roving eye and will actually interact with the attractive orbitals on both sides. This raises his energy (as it might yours!) and, of special consequence to us, reduces the energy gap between a bonding orbital and an unoccupied, antibonding orbital where a burst of light might send him. *Conjugation always reduces the bonding/antibonding energy gap; more conjugation reduces it more.*

3. In order to be conjugated, double (or triple) bonds have to be in just the right relationship to each other. If they are too close together or too far apart, they won't be conjugated. Conjugated multiple bonds are always in the relationship double-single-double-single-etc.

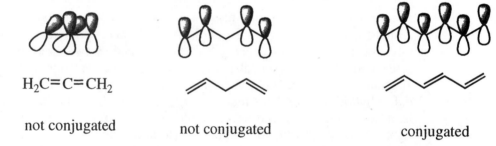

$$H_2C{=}C{=}CH_2$$

not conjugated   not conjugated   conjugated

4. Here is the most important point: the longer the conjugated system, the higher the wavelength (and the lower the energy) of light that is absorbed. This makes sense, given what I told you earlier: more conjugation reduces the energy gap over which an electron has to jump. If the jump involves a smaller amount of energy, then light of lower frequency (longer wavelength) can do the job. For example, here are some data:

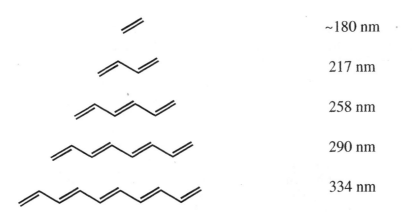

	~180 nm
	217 nm
	258 nm
	290 nm
	334 nm

Remember I said that an isolated double bond is not enough to be detected by our normal instruments. This is true: normal instruments cannot see below about 200 nm, so ethylene does not really count. For the rest, you see that there is a significant increase in wavelength as the conjugated system gets longer. It is not a perfect arithmetic sequence (meaning that the wavelength does not increase by exactly the same amount for each new double bond), but clearly each extension in the conjugation makes the appropriate wavelength longer.

The key point of this whole section is not so much that this kind of spectrum helps you figure out a structure, because frankly there are few chemists left who use UV-vis spectroscopy for this purpose. What I really want to do here is to remind you that you are carrying a spectrometer around with you right now: your eye is an extremely good spectrometer that happens to be sensitive to light in the wavelength range 400–700 nm. All the molecules shown above are completely invisible to you (colorless), because all *visible* light would be ignored by the compounds. However, you should anticipate that by stringing together a few more double bonds, you would get compounds that *do* absorb some visible light, and therefore appear colored to us humans. Indeed, highly conjugated organic compounds really are colored, and I have dragged you through this long discussion largely to state that fact. Below is "β-carotene," a highly conjugated and (surprise!) orange substance found in carrots (hence the name).

$\lambda_{max} = 455$ nm

You are probably aware that light from the sun or a light bulb, so-called "white light," contains all the visible colors, and it looks white. When light bounces off an object, if all the light is reflected, the object looks white. But if the object absorbs some of the light, what is reflected is missing light of that wavelength, so it looks colored. Conversely, any light that looks colored is missing some wavelengths of light. Look at your clothes. Is there any color on them? If so, your clothes are absorbing certain wavelengths of light. Why? Most likely because they have conjugated organic compounds in them. Most dyes used in clothing are organic, and colored organic compounds are highly conjugated. Indeed, many early organic chemists spent most of their time making various conjugated compounds in a search for dyes with various colors and compatibility with clothing, and many fortunes were made in this field.

Although we are not going to use UV-vis spectroscopy significantly for structure determination, there are several aspects of light absorption that might be of interest to you. For example:

- Bleach. How does bleach work? Does it really make clothes whiter? Does it really make clothes cleaner? Most bleach is essentially chlorine, Cl_2, dissolved in water. It reacts with double bonds in very much the way we discussed in Chapter 11, although there are a few differences. The point is, imagine you have some stain on your shirt that is colored. If it is organic, it is colored because there are lots of double bonds in it, conjugated with each other. Now imagine a chemical comes along and reacts with one or more of these double bonds. Presto! It is no longer colored. The chemical is still *there*, just no longer colored. How does bleach work? It does not clean your clothes, it simply makes the dirt invisible! (This is an oversimplification, but there is a fair amount of truth to it!)

- Many of you have done titrations using that awful stuff that no one can pronounce, phenolphthalein, as an indicator. You may remember that the indicator is pink in basic solution and colorless in acidic solution. Why? Below is the structure of phenolphthalein in both acid and base.

acid form   base form

In the acid form there is lots of conjugation, but each benzene ring is an island of conjugation unto itself: it cannot communicate with the other two rings, because one carbon in the middle has no p orbital. Thus, the longest string of conjugation present in the acid form is four double bonds (left ring plus a C=O). But in base, the molecule unfolds in such a way as to put a p orbital on that middle carbon. Now there is a p orbital on every atom in the molecule, and they are all in communication with each other. This produces an intensely pink-colored compound, and the transition from colorless to pink and back is an excellent indicator of when a solution changes from acid to base and back.

A very similar test has been used by forensic scientists who find a spot of dark substance at the scene of a crime and wonder if it is blood or paint. Blood contains an enzyme that can catalyze the following conversion, producing the same colored compound we saw in the phenolphthalein reaction. Using the compound on the left, investigators can tell whether an unknown substance is blood or not.

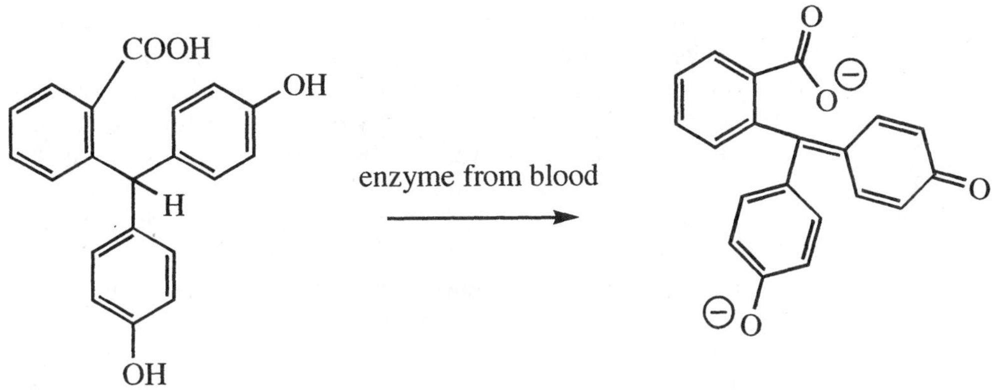

enzyme from blood

- How do your eyes work? Why do they serve as a detector for visible light? They, too, have conjugated compounds in them. The specific substance in your eyes that does the job is called "retinal," shown below. It is half of carotene, so it makes sense that eating carrots is good for your vision! The version sitting in your eye has a *cis* double bond partway along the chain. This is important, because it gives the molecule a shape that allows it to nestle down into the pocket of an enzyme called "opsin." When light hits this molecule, the light is absorbed due to extended conjugation, and the extra energy allows the *cis* double bond to become *trans*. But in the *trans* form, retinal no longer fits into the pocket of opsin, so it comes out. This acts as a signal to the enzyme that a beam of light has been received, and the enzyme begins the process of sending a signal to the brain. Eventually the *trans*-retinal is converted back to its *cis* form and it can sit down in the pocket again, ready to receive another photon of light.

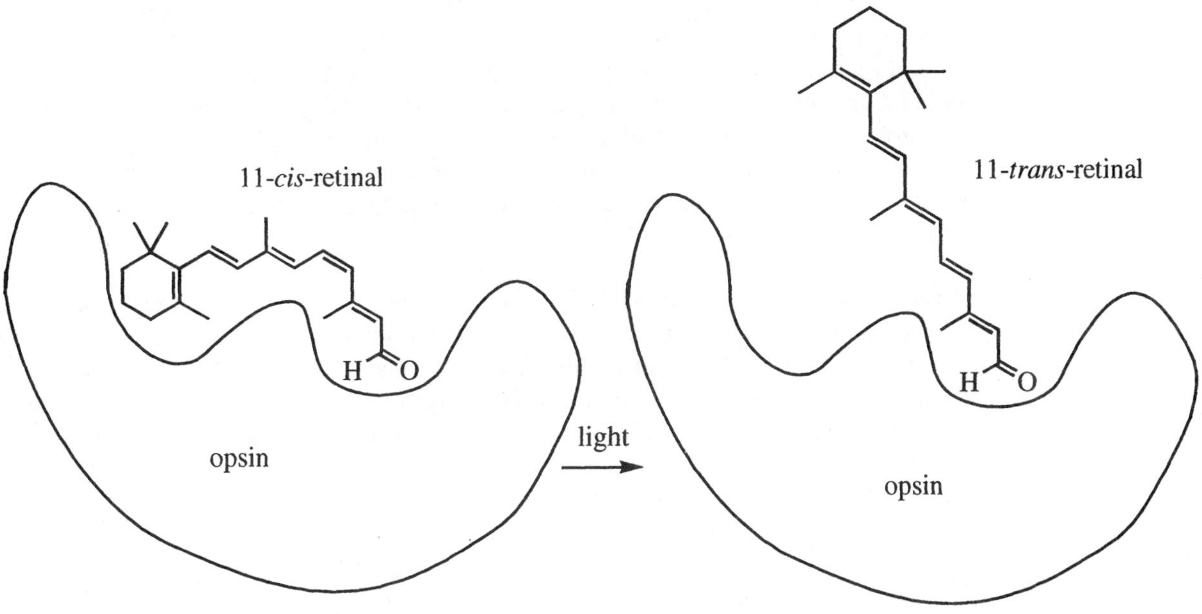

- The height of a peak in a recorded UV-vis spectrum is directly proportional to how much of the absorbing compound is present. This is usually expressed in the form of Beer's law:

$$A = \varepsilon \, c \, l$$

where A is an observed absorbance (the height of the peak) from the spectrometer, c is the concentration, l is the length of the sample cell, and ε is a proportionality constant, different for each substance and characteristic of the substance, called the "extinction coefficient" or "molar absorptivity." The usefulness of this is that if you know the extinction coefficient for a substance, and you measure the absorbance of a sample containing it, you can find out *how much* of that substance is in your sample. In other words, this technique is very useful for *quantitative* determination of materials. This also lends itself extremely well to doing kinetics, since you can watch some peak get larger or smaller as a reaction proceeds, and thus at any time during the reaction have a precise measure of the concentration of the stuff causing that peak.

UV-vis spectroscopy has many important uses, and much can be learned from the study of the electronic properties of molecules, but we will end our discussion of it here.

Infrared Spectroscopy

In attempting to identify an unknown substance, one of the important pieces of information one would like to know is, what functional groups are present? Is this an alcohol? A ketone? A carboxylic acid? An alkyne? Although there is much more information than this available from IR spectroscopy, identification of functional groups will be our chief use of it.

How do we get the information? How can light tell us what bonds are present? (Remember: it's the *bonds* that *define* a functional group!) For this purpose it is useful to think of bonds as springs. Atoms that are bonded together are like balls attached by springs. Although we have discussed bond lengths as if there were fixed distances between the atoms, in fact there are not. There is a measurable *average* distance between any two atoms, but the atoms are actually jiggling and vibrating all the time. They have a certain amount of energy as they sit around jiggling, and by feeding them extra energy (light), you can get them to absorb some and end up with more energy than they had before. And guess what? Like everything else at the atomic level, this kind of energy is quantized. In other words, individual bonds are associated with energy levels, just like the atoms we described in Chapter 2, and by studying what kind of light (wavelength, frequency) a bond can absorb, we can tell something about its nature.

A typical IR spectrum is shown below. The first thing you need to know is that this spectrum is upside down compared to most other spectra. While most spectra have the output running along the *bottom* of the paper if the sample is not absorbing light, resulting in an absorption "peak" that goes up, IR spectra are almost always shown with the line running along the top when nothing is happening, and an absorption "peak" is actually represented as a dip. The particular spectrum below has a group of peaks near 3000, a large peak at about 1725, and a whole bunch of peaks to the right of 1500.

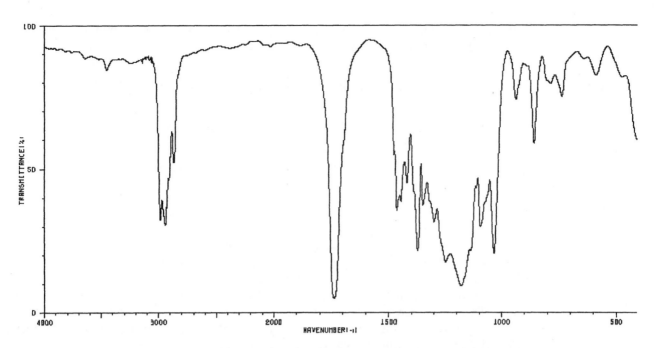

What are the axes on this chart? The *y* axis is called "transmittance," a measure of how much light is getting through. (When little light comes through the sample, it is absorbing a lot, so transmittance is inversely related to absorbance, and a dip in an IR spectrum represents the absorption of light.) The *x* axis is labeled cm^{-1} ("inverse centimeters" or "reciprocal centimeters"). What the heck is this? The word normally used is "wavenumber," and it is the inverse of the wavelength. Why plot in the *inverse* of the wavelength, instead of the wavelength? Actually some instruments do plot in wavelength, and some in both, but most chemists think about IR spectra in terms of the unit cm^{-1}, because that is what they have become used to. Recall from Chapter 2 that

$$ E \;=\; \frac{hc}{\lambda} \;=\; hc \;\frac{1}{\lambda} $$

Since *h* and *c* are both constants, energy is therefore directly proportional to $1/\lambda$, wavenumbers, so by plotting in wavenumbers we are getting a picture related to the energy involved. Higher numbers along the *x* axis represent higher energies.

The wonderful thing about IR is that each type of bond absorbs energy in a relatively narrow window on the IR scale. For example, carbon–hydrogen bonds all show up in the vicinity of 3000 cm^{-1} (more accurately, 2850–3100). Carbon–oxygen *double* bonds are always between 1650 and 1800 cm^{-1}. Carbon–oxygen *single* bonds are always near 1100–1300 cm^{-1}. Unfortunately, there are several other types of bond that also appears in the 1200–1300 range, so seeing a peak there does not tell you a whole lot. Indeed the whole spectrum below about 1500 cm^{-1} has so many peaks in it that that part is rarely used for interpretation; on the other hand, because it is complex, it is like a fingerprint (this is called the "fingerprint region" of the spectrum), and can be used to determine if two substances are the same. Like mass spectrometric evidence, this has become useful in court cases. But also like mass spectrometry, unless you carry a library of thousands of compounds' IR spectra around with you, this fingerprinting feature is of little use to you, a chemistry student.

What *would* be of use to you would be to recognize some areas in the spectrum that are exclusive: places where certain types of bonds show up but no other types do. If you could find such regions, seeing a peak there would be a tip-off of a certain type of bond in the molecule. There are four important regions like this in the spectrum, and we will be able to make good use of three of them.

$$O-H\left(\text{also } N-H\right) \qquad 3200\text{-}3600 \text{ cm}^{-1}$$

$$C-H \qquad 2850\text{-}3300$$

$$C\equiv N, \quad C\equiv C \qquad 2100\text{-}2300$$

$$C=O \qquad 1650\text{-}1800$$

Why can we use only three of them? Consider the region near 3000 cm^{-1}. Imagine you have an unknown organic compound, you acquire an IR spectrum, and you see peaks near 3000. What have you learned? That the stuff has C–H bonds. What was the last organic compound you saw that did *not* have C–H bonds? You've *never* seen one (although a few do exist). This really provides no *useful* information, since just about every organic compound has C–H bonds. The other three regions are useful, however, especially the alcohol and carbonyl regions. If you see an IR spectrum with a large peak near 1700 cm^{-1}, you can virtually guarantee that there is a carbonyl group in your molecule. Further, if you are working on an unknown substance and you propose a structure that has a carbonyl group in it, you can look at the IR to see if you are right. For our purposes, there will be a large peak between 1650 and 1800 cm^{-1} *if and only if* there is a carbonyl group in the molecule.

The compound producing the IR spectrum shown above has a carbonyl group in it. So now you know what a carbonyl peak looks like.

Below is the IR of 2-phenyl-2-propyl alcohol.

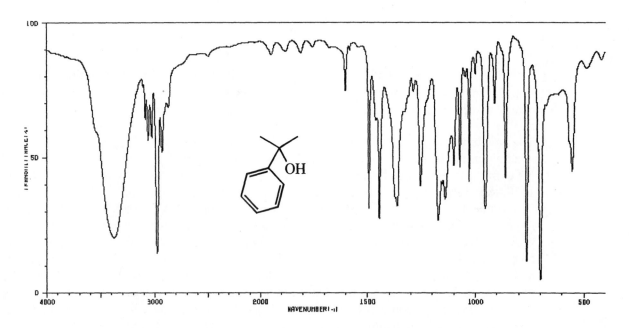

So now you have seen what an alcohol usually looks like in the IR. OH peaks tend to be somewhat broader than other peaks, as the one here is. You should also notice that although the peak is broad, it is pretty much finished by the time it gets to the C–H bonds at 3000 cm^{-1}.

Below is the IR of phenylacetylene.

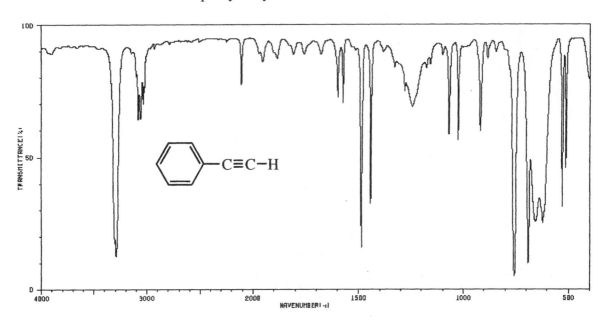

There are two things to notice here. One is that the peak I promised you between 2100 and 2300 cm^{-1} is pretty small. It's there, but it doesn't jump out at you like the others we have seen. Carbon–carbon bonds tend to be pretty faint in the IR for reasons we will not discuss (in a nutshell, polar bonds show up better than non-polar ones). On the other hand, there is quite a large peak at 3300 cm^{-1}, large enough to be an OH peak, but of course it can't be because there is no OH in this molecule. This is at the tail end of allowable C–H bonds, so it must be one of them, and since we have never seen anything in this area before, it is probably due to the special C–H bond in this molecule, namely, the one on the triple bond. This is true, and you will probably never see C–H bonds past 3100 except for this kind.

You may be wondering, why is *this* C–H bond so different from the others we have seen, so that it vibrates with such a different energy? It turns out that it is part of a trend that becomes clear if you look closely at precisely where certain bonds show up. In particular, within the C–H region, it turns out that alkanes show up at about 2850–2960 cm^{-1}, alkenes tend to be between 3000 and 3100, and alkynes are near 3300 cm^{-1}. Recall that this cm^{-1} scale increases with the energy of the light: 3300 cm^{-1} represents stronger photons than 2900. This tells us that it is harder to stretch a C–H bond on a C–C triple bond than one on a C–C single bond—the spring is stronger. Why? The bond in the alkyne is composed of an sp orbital from carbon interacting with the hydrogen; on an alkane, it would be an sp^3 orbital. And an sp orbital is lower in energy than an sp^3 orbital (because an s orbital is lower than a p, and an sp orbital is ½ s while an sp^3 orbital is only ¼ s). Thus, the electrons in the C–H bond on an alkyne are at lower energy than those in an alkane, the bond is stronger, and it takes more energy to stretch it. IR can actually detect this subtle difference.

The last IR we'll look at here is of butyric acid. A carboxylic acid has both a carbonyl group and an OH group. We should expect to see evidence of both in the IR spectrum: the carbonyl should give a peak near 1700 and the OH near 3400.

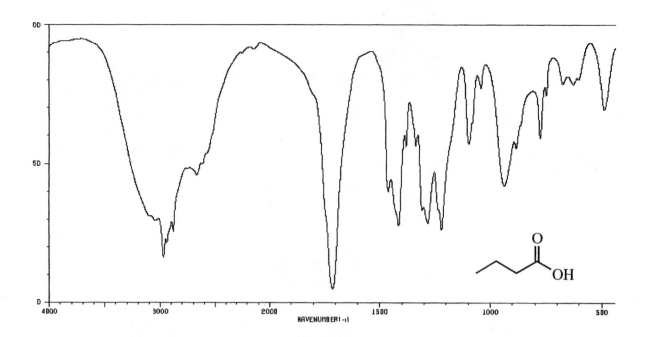

Sure enough, the carbonyl is there, very close to 1700 cm^{-1}, but what is that mess on the left? If you look closely, you might be able to see that this mess is actually the superposition of a *very* broad peak that runs all the way from 3600 to 2400, with some stuff on top of it near 3000. The stuff near 3000 is clearly due to the C–H bonds; the huge broad peak is characteristic of the OH of a carboxylic acid. Recall that the OH of a normal alcohol finishes *before* the C–H's kick in, as in the spectrum of 2-phenyl-2-propyl alcohol above. The difference is that, in carboxylic acids, the OH is broader and far enough to the right to plow right through the C–H region. When you see this (in combination with a carbonyl group) it is a tip-off that you have a carboxylic acid.

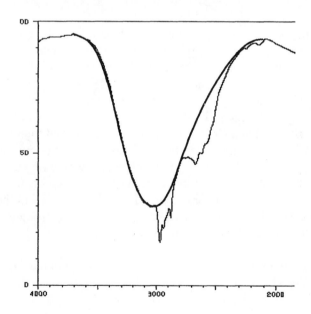

Test Yourself 2

See if you can figure out the structure of an unknown compound from just its mass spec and IR, shown below. *Hint*: Learn what you can from the IR. Then look at the molecular weight, and guess a reasonable formula, keeping in mind the IR information. Look at some mass spec fragments, and try to piece all this information together.

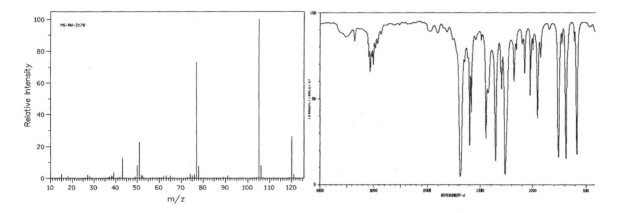

Test Yourself 3

Propose reasonable structures for the isomeric compounds that give rise to the following five sets of spectra. *Hint*: proceed as above. Once you have a formula, there are only a few ways you can put this together. Write all the reasonable structures, then see if you can match structures to spectra.

a.

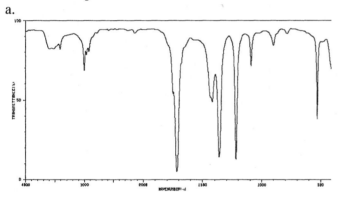
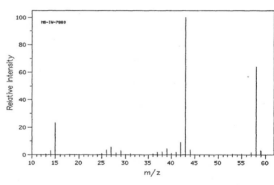

b.

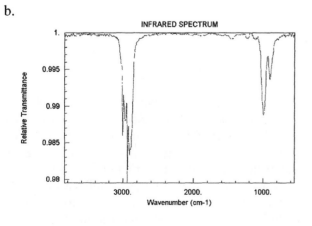
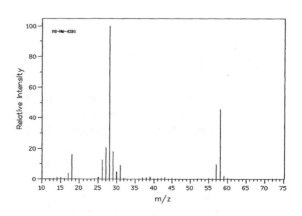

c. **Note: the strong IR band in this case is at about *1600* cm^{-1}.**

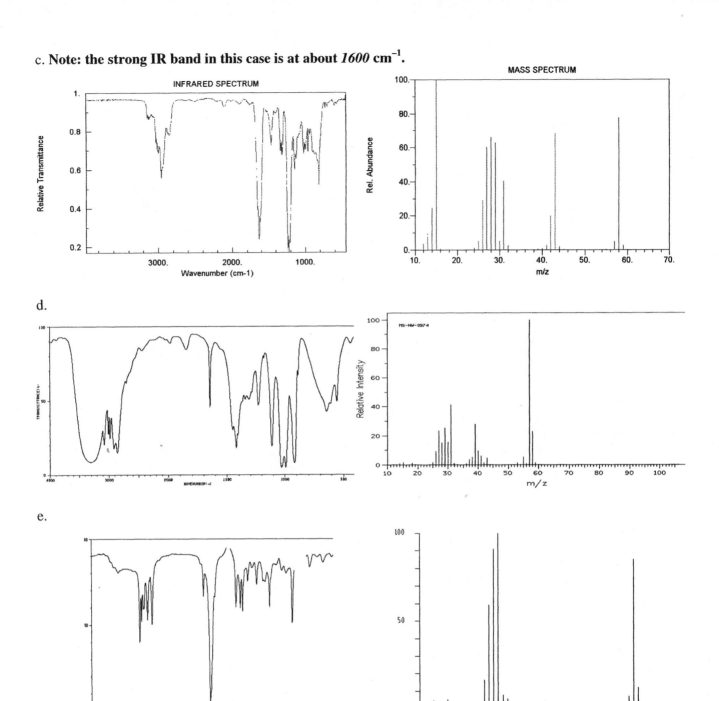

d.

e.

NMR Spectroscopy

So far we have learned techniques that allow us to determine the weight of an unknown molecule along with a few pieces it might contain (mass spec), and also to identify functional groups that may be present (IR), but most of the molecule does not involve such tags. How do we get information about the fundamental backbone of a molecule—the way its carbon chains are put together? The most powerful technique for routine structure determination in organic chemistry is nuclear magnetic resonance (NMR) spectroscopy. Unlike all the other spectroscopies, this one looks past the electrons right into the nuclei of various atoms. It turns out that some nuclei, like electrons, also have energy levels, and that these nuclei can be made to jump from one level to another with electromagnetic radiation. The nuclear energy levels are so close together that the amount of energy necessary to induce the corresponding jump is minuscule—all it takes is radio waves, just like the kind broadcasting your favorite music. Nevertheless, with a sensitive enough instrument, the absorption of radio waves can be detected, and a great deal of information becomes available from the results of such an experiment.

As noted above, NMR means "**n**uclear **m**agnetic **r**esonance." The "nuclear" part is obvious: we are looking at nuclei. The term "resonance" means that the radio waves must have the right frequency to induce a jump. Where does "magnetic" come in? It turns out that different states available to a nucleus have exactly the same energy (and therefore there is no gap to detect by putting in energy) *unless* the sample is inside a magnetic field. In other words, this technique only works with a sample hanging in the middle of a big magnet. Once the sample is in a magnet, *then* there will be distinct energy levels, and we can see differences between them by scanning with various radio waves until we hit the right ones. The stronger the magnet, the farther apart the energy levels get, and the more energetic the radio waves we would need to induce a transition. This can be illustrated by the following diagram, which is simply a graphical representation of the previous sentence.

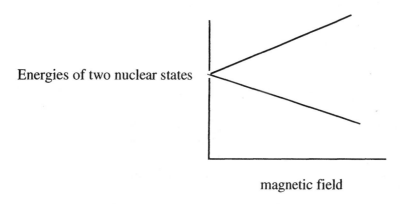

Energies of two nuclear states

magnetic field

Thus, the key to NMR is getting your sample inside a magnet, the bigger the better, and then, in effect, playing music to it, changing the station until you find the kind of music the sample likes. Modern NMR actually shortcuts this process. Imagine you have a new friend. You could try to find out what kind of music she likes by playing all your CDs, one at a time, and seeing whether she prefers rap, or folk, or reggae, or heavy metal, or whatever; but it would be easier to show her your collection all at once and let her choose what she wants to listen to. A better analogy is that if you wanted to find the resonance frequency of a bell, you could hook it up to a frequency generator and play all the frequencies one at a time until you find one that makes the bell vibrate; alternatively, you could hit the bell and listen to the tone it gives off. All modern NMR is carried out in the latter way, by hitting the nuclei and seeing what frequencies they emit, a technique called "Fourier transform" or "FT-NMR," but this is a detail that need not concern you, at least as far as spectral interpretation is concerned.

It is useful in doing NMR to work with the biggest magnet possible. To create a strong magnetic field, scientists typically use electromagnets, but the amount of electricity it would take to keep a huge magnet running would be prohibitive if it were not for superconductivity. Certain materials actually can conduct electricity with no resistance at all, so by building electromagnets out of these materials, scientists manage to create tremendous magnetic fields without using much electricity. Basically, you plug the magnet in, let it start up, then unplug it—and it keeps going! Currently, superconducting materials only work at extremely low temperatures, near absolute zero, so these magnets need huge thermos bottles around them filled with liquid helium to keep them cold. The technology necessary to do all of this has been perfected for some time, so all we need to do is learn how to interpret the output.

WHO CARES? YOU DO!!

Many of you have heard of MRI, a fancy medical technique that allows doctors to learn things about your insides without cutting you open. It complements X-rays, which see through soft tissues and show hard things, like bones and tumors. MRI focuses on soft tissues like the brain and other organs. How does it work? It is nothing but NMR. The big machine is a big magnet, and while a person is inside it they play radio waves, just like music, and the nuclei in the body sing out their favorite frequencies. By analyzing these frequencies and other properties of the nuclei, doctors can learn a great deal about what is going on. The technique is completely harmless, no different from standing in front of a large magnet. But when it was first introduced, no patient would give permission for it. Why? Originally it was called NMR imaging, because doctors could get an image of organs by use of NMR. But patients wanted to know what the "NMR" stood for, and when doctors told them it was "nuclear magnetic resonance," they all refused. Why? The name has the word "nuclear" in it, and patients had visions of being put in a nuclear reactor and getting bombarded with radioactivity. Nothing could be farther from the truth, but as soon as the word "nuclear" was on the table, patients refused. The solution? Eliminate the word. The technique is now called MRI, "magnetic resonance imaging." The procedure happens to look at nuclei, as it always did, but no one objects anymore!

Not all nuclei respond to this technique. In particular, nuclei with even numbers of both protons and neutrons are invisible to NMR. The rest can all be observed, if you look properly, but in this book we will concentrate on only two nuclei: ^{13}C and ^{1}H. These are especially simple because they have only two possible states available to them, called "up" and "down," and therefore only one energy gap to worry about.

^{13}C NMR

The first thing you should remember about ^{13}C atoms is that there aren't very many of them. Only about 1% of the carbons in the world are this isotope, so most of your sample will be invisible to carbon NMR. It takes a fancy instrument to see that tiny 1% of your sample (or else a very large sample). Nevertheless, modern instruments can do it.

There is a great deal to ^{13}C NMR that I will gloss over in this section. For our purposes, ^{13}C NMR is simply a technique for counting carbons. Every different "kind" of carbon in the molecule will make its own line on the spectrum. No more, no less. The end.

Huh? I can illustrate this best with an example. Consider the four different compounds with the formula C_4H_9Br. (Can you draw all of them without looking? Can you name them? If not, time to review!) Their structures are shown below.

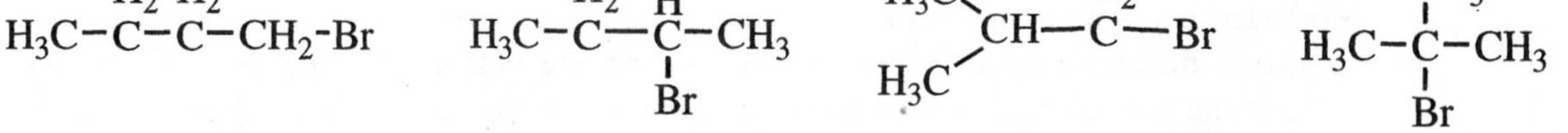

$$H_3C-\underset{H_2}{C}-\underset{H_2}{C}-CH_2\text{-Br} \qquad H_3C-\underset{H_2}{C}-\underset{\underset{Br}{|}}{\overset{H}{C}}-CH_3 \qquad \underset{H_3C}{\overset{H_3C}{\diagdown}}CH-\underset{H_2}{C}-Br \qquad H_3C-\underset{\underset{Br}{|}}{\overset{CH_3}{C}}-CH_3$$

The one on the left, n-butyl bromide, has carbons in four different environments ("four kinds of carbon"). Be careful here: we are not simply looking at whether an atom is primary, secondary, or tertiary; in that case you might claim there are only two "kinds" in this molecule. While it is true that there are only two different *kinds* of carbon in this sense, each carbon present is identifiably different from any of the others. Look at them: one has a bromine attached. One (and only one) is one carbon away from there, one is two carbons away, and one is different by being part of a CH₃ group. All four can be described differently. Likewise, the second strucure (2-bromobutane, or *sec*-butyl bromide) also has four distinguishable carbons. The third structure (1-bromo-2-methylpropane, or isobutyl bromide) has only three different kinds of carbons. Yes, of course, there *are* four carbons, but two of them, the two methyl groups, are completely indistinguishable. There is nothing you can say about one that is not true of the other. Clearly, *as written* they are different—one is down on the page, and one is up—but by now you should recognize that the picture on the page is but a feeble attempt to represent the actual molecule, and if you build a model, you will see that these two carbons are completely identical. So there are only three different carbon environments in this structure. And in the fourth structure (*tert*-butyl bromide, or 2-bromo-2-methylpropane), there are only two. Below are the ^{13}C spectra for these four compounds. Can you tell which is which?

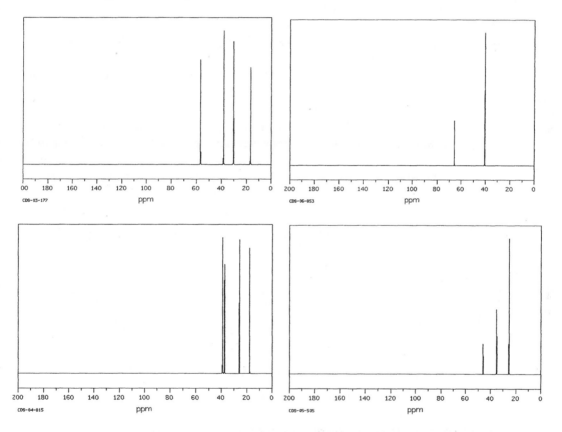

Even though I have told you next to nothing about how to interpret these things, it should still be pretty obvious that the top right spectrum, with two peaks, is from the *tert*-butyl compound, and the bottom right one, with three peaks, corresponds to isobutyl. The left ones both have four peaks, and either one could be *n*-butyl while the other would be *sec*-butyl. Based on this information alone, we cannot tell these two apart.

You should notice several other things about the spectra:

1. The plural of "spectrum" is "spectra." Do not say "spectrums." And do not say "a spectra."

2. The *y* axis has no scale. It is simply "height." The *x* axis is labeled ppm, which stands for "parts per million." The atoms we are talking about are different from each other, but not by much: the radio frequencies it takes to get them to "sing" differ from each other by only a few hertz out of several million. This is like trying to tune your radio to a station at 104.7539 and not 104.7538. Clearly there must be very fancy radio tuners in this instrument.

3. The height of a peak does not correlate directly with how many carbons are responsible for that peak. In both spectra on the left there is one carbon for each line, but the various peaks are not exactly the same height. In *tert*-butyl bromide, there are three carbons making one peak and one carbon making the other. Clearly one is larger than the other, but not quite by a factor of three. Most obviously, in isobutyl bromide there are two identical carbons, and two others different from each other. Thus, we might expect one peak to be twice as high as the other two, which should in turn be about the same height. This is not the case. The message here is that we cannot attach much meaning to the height of a peak in ^{13}C NMR (this will be different when we get to hydrogen).

4. All the peaks you see show up toward the right side of the spectrum, at 70 ppm or less. The location of a peak on the *x* axis is called its "chemical shift," δ, and it is reported in how far away it is, in ppm, from a standard, which has been chosen to be tetramethylsilane, always abbreviated TMS. Saturated carbons like these, sp^3-hybridized, almost always show up at a δ between 0 and 70 ppm. Carbons involved in carbon–carbon double bonds, sp^2-hybridized, usually appear between 100 and 150 ppm. This includes carbons that are in benzene rings. Sp^2 carbons involved in carbon–oxygen double bonds (carbonyl groups) are typically between 160 and 220 ppm. Sp carbons are usually between 80 and 100 ppm. This is about all the interpretive data we will need.

Type of Carbon	Chemical Shift Range
sp^3	0 – 70
sp^2	100 – 150
sp	80 – 100
Carbonyl	160 – 220

You may be wondering, why do different kinds of carbons appear in different places? Aren't they all in the same magnet, and therefore seeing the same magnetic field? Shouldn't they all require the same amount of energy, and therefore the same frequency, to jump from one level to the next? Excellent question! If they all were *experiencing* the same magnetic field, they *would* all absorb the same frequency of radiation. And yes, they are all in the same magnet, so all are *exposed* to the same magnetic field. The linguists among you may notice the careful hedging of the last two sentences. All the nuclei are *exposed to* the same magnetic field, but they are not *experiencing* the same magnetic field. This is because each nucleus is buried inside a cloud of electrons, its own electrons, and these shield it somewhat from the applied magnetic field. The effectiveness of the shielding is a very sensitive function of the electron density around that particular carbon atom. Since you know that some substituents are electron donating and others are electron withdrawing, it should not surprise you that each carbon atom might be surrounded by a slightly different electron density from all the others, unless two of them are really identical. I do mean slightly: remember, the frequencies necessary to cause jumps differ from each other by only a few parts per million, but that is enough difference for this instrument to detect easily. The practical result, then, is that every different carbon atom in a molecule shows up at a different spot on the spectrum, so we can simply count the lines. What's more, we now also know something about what *kind* of carbon (hybridization) each one probably is.

A little terminology here. Recall the field–energy diagram I showed you before:

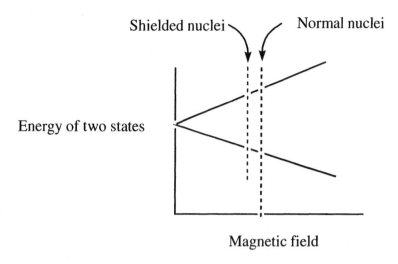

Imagine your instrument is operating at some random magnetic field represented by the black (right) dotted line. The two possible energy states of the nucleus ought to be separated by the energy gap indicated by where this line crosses the diagonal lines, and "light" of appropriate frequency ought to cause this transition. However, nuclei that experience significant shielding will not feel quite this much magnetic field. They will feel a somewhat smaller field, represented by the left dotted line above, and therefore the two states of their nuclei will be separated by a smaller energy gap. That means that "light" of slightly lower frequency will cause the transition. More shielding means lower frequency. OK? The frequency scale is actually the *x* axis of an NMR spectrum, and it is backwards, meaning that frequency increases as you move to the *left*. Thus, the more the shielding it experiences, the further to the *right* an atom will appear in the spectrum, at lower frequency. This is also usually referred to as "upfield" in the spectrum, while peaks toward the left are said to be "downfield" (and "deshielded"). If you want to know why, read the next paragraph. If you don't care, skip it.

Unfortunately, early NMR instruments were set up in a curious way. I have described the experiment as holding the magnetic field steady and slowly scanning the frequency of the radio waves until there is a match. Such an instrument would work, but it is equally true that if you kept the frequency constant, you could slowly change the magnetic field a bit. This would slowly change the energy difference between the allowed states of the nucleus until there was a match with the radio waves, and you would end up with the same information. It turns out that it was easier to build instruments the second way, so that's the way they were built. Now consider again the field–energy diagram above. If the instrument is set at the precise frequency that corresponds to the energy gap at the black dotted line ("normal"), an atom that is shielded will not have the proper gap to absorb that energy. Its gap is too small, because the magnetic field it feels is smaller than what we have put in. In order to get it to absorb the prescribed frequency of radiation, we would have to increase the gap between the states, which means that we have to increase the magnetic field. More shielding, higher field. Get it? Peaks to the right in the NMR spectrum represent nuclei that are shielded, and appear "upfield." Peaks to the left are deshielded, and appear "downfield." Unfortunately, downfield corresponds to higher, not lower, ppm values. Sorry, that's the way it is.

Test Yourself 4

Below are eight ^{13}C-NMR spectra of molecules, all with the formula C_9H_{12}, and all containing a benzene ring. You won't be able to tell them all apart, but you should be able to identify some. Try it! *Hint*: It might be a good idea to start by writing down all the structures you can think of with this makeup, and predicting their ^{13}C-NMR spectra. How? Count the number of unique carbons, and estimate where each would appear in the spectrum. Recall that the double bonds in a benzene ring are smeared out; it will help you to see which atoms are identical if you simply draw the ring with a circle in the middle. As I see it, there is only one compound that you should expect to identify uniquely; there are two different pairs of possible answers and one group of three that you cannot tell apart from this information alone. Try to match groups of spectra with the appropriate groups of compounds.

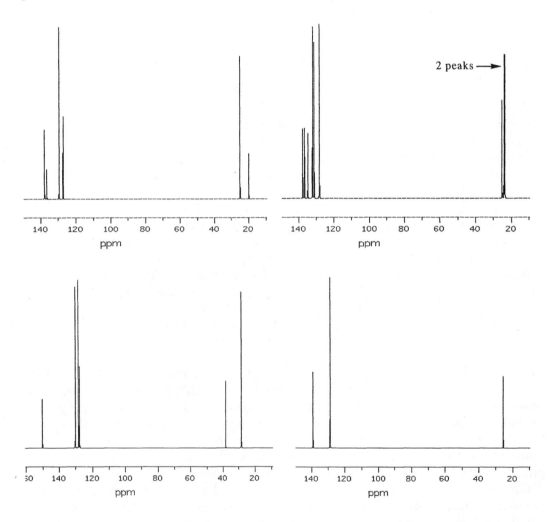

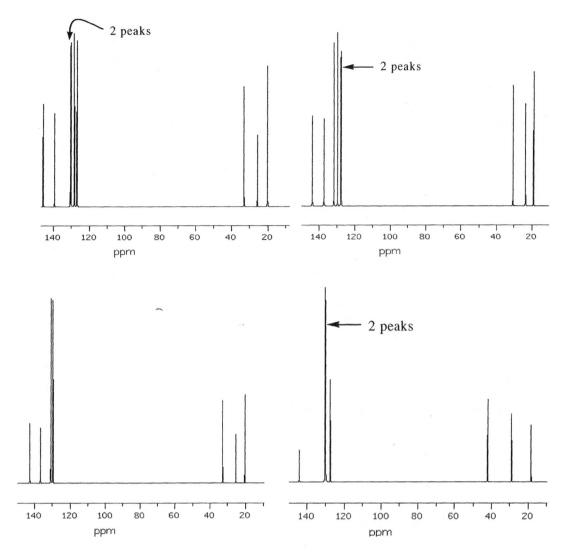

Probably the hardest part of this exercise is deciding how many peaks are shown in each spectrum! The problem is, I copied and pasted these several times, shrank them, and doctored them up to get them onto the page. Further, the lines one sees are not always perfectly smooth, even right off the instrument. As a result, many of the lines above have slight shoulders that might appear to you as separate lines. I've tried to indicate those cases where two "real" lines overlapped. Otherwise, if you can see a space between two lines, you should consider them separate; if not, they are probably part of the same line.

^1H NMR

The other nucleus we will consider for NMR is the proton. Nearly all hydrogens are the isotope ^{1}H, so this technique is a lot more sensitive than ^{13}C NMR—100 times more sensitive just from the standpoint of natural abundance. There are also special aspects of ^{1}H NMR that make the spectra both more complicated and more informative. In particular, we can tell from a hydrogen spectrum not only what kind of carbons comprise a compound's backbone, but *where they are with respect to each other*. In other words, ^{1}H NMR allows us to establish connectivity, without which we could never specify a structure for sure.

The NMR experiment for protons is just like the experiment for carbons. There has to be a magnetic field, preferably large, and radio signals are transmitted to determine which frequencies the nuclei like. (More accurately, these days all frequencies are broadcast simultaneously, allowing all the nuclei to absorb energy at once, then the radio waves are turned off and you "listen" to detect which frequencies are re-emitted by the excited nuclei.) As with carbons, each *different* proton experiences a slightly different magnetic field, depending on its level of shielding, with more heavily shielded protons appearing upfield (to the right) in the spectrum and deshielded ones appearing downfield. Again the differences are minute, parts per million—indeed, with protons, the differences are even *more* minute, since the carbon scale encompassed about 200 ppm and the proton scale typically covers only 10–12 ppm—but the instrument is capable of detecting *very* small differences. As before, nuclei associated with saturated carbons tend to appear upfield in the spectrum (to the right), and those on unsaturated carbons are usually downfield. However, with only a few ppm in which to distribute themselves, it is far less certain that each different nucleus will appear in a unique place; that is, there are more cases, with proton NMR, of atoms that are supposedly different showing up by accident in the same place.

With ^{13}C NMR, there is essentially one type of information available from the spectrum: the location of each peak (chemical shift). We expect each different carbon to give its signal at a different place. We cannot tell from the spectrum how many carbons of each type are present, nor what is attached to what. We can learn both of these things with proton NMR.

There are *three* types of information available from an HNMR spectrum. These are:

- Chemical shift, which tells you about the electronic environment of the protons being observed;

- Integration, which tells you how *many* protons are being observed; and

- Spin-spin coupling, which tells you about the *neighborhood* of the protons being observed.

We will consider each of these in turn.

Chemical Shift

In the ^{13}C spectrum we did not get very specific about where to expect various carbons, except that saturated ones tend to be upfield and unsaturated ones downfield. Within these broad guidelines there can still be significant variation. It is indeed possible to be a bit more specific about where certain carbons will appear, but small factors can have large effects, and it is difficult to analyze the spectra in detail. Proton NMR is much easier in this respect, in that there are relatively narrow ranges associated with various types of protons, and the effects are (to a first approximation) additive.

Recall that an exact chemical shift is determined by the electronic environment of the nucleus, since electron clouds are what shield nuclei to a greater or lesser extent from the applied magnetic field. Electron-withdrawing substituents reduce the electron density and therefore reduce the shielding, so hydrogens that have electron-withdrawing substituents nearby will appear downfield from those that do not. If you know where a "normal" proton should appear, then each electron-withdrawing group will move it downfield (to the left) by a predictable amount:

Hydrogen Type	Chemical Shift (δ)
CH$_3$–R ("normal")	~1
CH$_3$–Cl	3.1
CH$_2$Cl$_2$	5.3
CHCl$_3$	7.25

You should notice several things:

1. The additive relationship for the deshielding effect is not perfect, but it is not far off: every time you add a chlorine, the chemical shift seems to move by something close to 2.1 ppm.

2. To figure out where you would expect CH$_2$Cl$_2$ (if I hadn't already told you), you cannot look at the value for CH$_3$Cl, which is 3.1, and double it. The reason is that the value for CH$_3$Cl is not a direct measure of the effect of one chlorine. It is the *sum* of the effect of one chlorine *plus* some base value. The *effect* of one chlorine is about 2.1 ppm, so the effect of two chlorines ought to be 4.2, and you should add *that* to the base value (1, in this case) to get a predicted chemical shift for CH$_2$Cl$_2$ of about 5.2, which is pretty close to what we see.

Many students find this quite complicated. They really shouldn't, though, because they deal with this kind of thing all the time in everyday life. For example, suppose you were in the market for a car, and you wanted tinted front and rear windows. The single car available on the lot has only its front windows tinted, and it costs $17,500. No one would jump to the conclusion that to get tinted front *and* rear windows it would cost you $35,000. Further, if you found a car without any tinted windows for $17,000, you could figure out that the tinted front window was worth $500, and you could then guess that a tinted rear window would also cost about $500, so a car with both would be around $18,000. We are doing nothing different here. A stripped-down proton, one hanging around on a saturated carbon with nothing special about it, shows up near 1 ppm. Everything else is relative to that.

3. Almost every atom we deal with routinely in organic chemistry is more electronegative than carbon, so almost all protons we see will appear at 1 ppm or higher. It is very rare to find protons to the right of 0.9, although it does occur. Some even show up at δ values as low as –4 or more, but you are very unlikely to encounter them.

Below is a short list of where certain types of protons can be expected. Be careful: this shows where they show up, the sticker price, if you will. The *effect* of a group is the difference between this value and the base value of about 1. Notice that the chart is listed in H environment; although there are some slight differences, for our purposes it does not matter whether you are looking at a CH$_3$, a CH$_2$, or a CH.

Hydrogen Environment	Chemical Shift (δ)	Effect of Group
R–CH ("normal")	~1	
[acyl] R–C(=O)–CH	2 – 2.5	1 – 1.5
[phenyl]–CH	2 – 2.5	1 – 1.5
X–CH (X = halogen)	3 – 4	2 – 3
R–O–CH	3.5 – 4	2.5 – 3
[ester] R–C(=O)–O–CH	4 – 4.5	3 – 3.5

Try an example: Where would you expect peaks from the following molecule?

$$H_3C-O-CH_2-C(=O)-CH_3$$

This molecule has three kinds of hydrogen in it. Two of the guesses should be easy. The methyl group on the left, for example, is attached to an oxygen, so it should show up at δ 3.5 to 4, according to the chart. The methyl on the right can also be read directly from the chart, and should be between 2 and 2.5. But what about the CH_2 group? Where should these two hydrogens be? There are three ways we can approach the problem, all equivalent. We can think of this group as an ordinary ("normal") group that is influenced by both an oxygen and a carbonyl; we can think of it as a group next to a carbonyl that is further influenced by an oxygen; or look at it as a group next to oxygen that is further influenced by a carbonyl. (Remember the cars: It would be like a stripped-down car to which you want to add front and rear tinted windows; one with tint in the front to which you add tint in the rear; or *vice versa*. You should get the same result whichever you do.) If you go through any of the processes above you should conclude that this CH_2 group ought to show up somewhere between 4.5 and 5.5.

Below is the actual NMR spectrum of this compound. You can see that two of the peaks are right where we expected them; one must be a bit off from our prediction, but not too far.

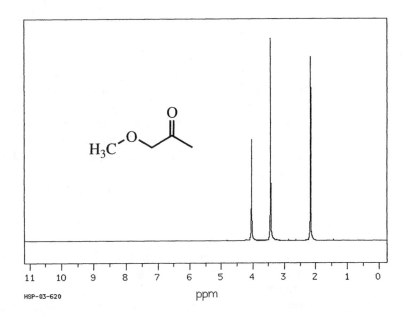

HSP-03-620

ppm

Be aware that the above list of chemical shifts and effects applies only to hydrogens on saturated carbon. Hydrogens attached to atoms other than saturated carbons have their own "normal" locations, and they tend not to vary as much with substitution. Here are a few:

Hydrogen Type	Chemical Shift (δ)
R₂C=CHR (vinyl)	4.5 – 7
aromatic C–H	7 – 8
aldehyde RCHO	9 – 10
carboxylic acid RCO–OH	11 – 13
R–O–H	1 – 6 (variable)

There is something else you ought to notice about the spectrum we looked at above. Besides the fact that the peaks showed up more or less where we expected them, they are also not all the same size. In particular, one of them is not nearly as big as the other two. We know that the large peak at about 2.0 must represent three protons: the methyl group next to the carbonyl. The other two peaks, a large one and a small one, represent three hydrogens and two hydrogens. Hmmm...Might there be some correlation between the size of a peak and the number of protons that give rise to it? We discovered in ^{13}C-NMR that this was not the case. But it turns out that in ^{1}H-NMR, there *is* a connection.

But this still doesn't seem right. The small peak is about half the size of the other two, and if it comes from the CH_2 group it should be 2/3 the size. Further, the two CH_3 groups should be the same size; they look close, but still not quite the same. What gives? The key here is that it is not the *height* of the peak that tells the story, it is the *area* underneath it. If one peak is a little fatter than another, the two could encompass the same areas but not be the same height. Indeed, when you measure the areas under these three peaks, you do come out very close to a 2:3:3 ratio.

How do we measure such areas? Some of you may have solved a problem like this before by actually cutting out peaks from the paper and weighing them. Fortunately this is not necessary anymore. The instrument can do it for us. You will see this kind of information reported in two ways. One shows actual numbers above the peaks, as in the left spectrum below. Each such number refers to a *relative* area under the specified peak. The second way is to show what is called an "integral trace," a line on the paper that rises by an amount related to the area under the curve, as in the right spectrum below. To interpret these, you consider the squiggles to be steps, and you measure (or estimate) the height of each step. The two steps on the right are about 1.5 times as high as the one on the left, a 3:2 ratio. (You don't need an actual ruler to check this: create your own, using a piece of paper with some roughly equally spaced lines, or the length of your pencil lead, or anything else handy.)

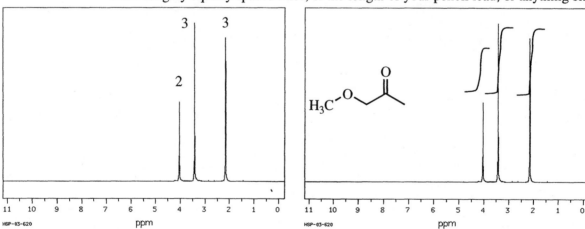

Spin-Spin Splitting

So far it looks like we can determine something about the electronic environment of a particular kind of proton by the location of its peak, and we can determine (at least on a relative basis) how many of each kind are present. The best part of all is that it turns out we can also determine if there are other protons nearby, and if so, how many. The reason for this is that nearby protons also affect the local magnetic field, so each proton in a molecule "feels" a slightly different field depending on what its neighbors are doing. I'll try to explain with an example.

Remember the basis of the NMR phenomenon: there are two states available to a nucleus, and in a magnetic field they have different energies. The property of the nucleus that we are observing is called "spin," and it is usually described as being "up" or "down." Don't worry about precisely what this really is (*I* don't even know!). A normal nucleus can have a random up spin or down spin. It doesn't care, because they have the same energy. In a magnetic field the two do not have the same energy, however, and the NMR experiment is an exercise in taking an up-spin nucleus and flipping it to down, or *vice versa*. Whatever this thing called "spin" is, it interacts somehow with magnetic fields.

Consider some molecule with two different, but adjacent, hydrogens. A generic example is shown below.

$$H_A \quad \quad H_B$$
$$X \underset{X}{\overset{}{\big|}} \quad \underset{Y}{\overset{}{\big|}} Y$$

For the sake of argument, let's imagine that H_A has a chemical shift of 3 ppm, and H_B has a chemical shift of 2 ppm. Let's also imagine that in the particular molecule we are observing, the spin for H_A is down. This means that this particular H_A is shielded from the external magnetic field by some amount, such that when the spectrometer gets to 3 ppm it thinks it is the right time for it to jump from spin down to spin up. But wait! That is what would happen if there were no H_B around, but there is, and H_B in this molecule also has a spin, either down or up. Since spin, whatever it is, interacts with magnetic fields, it must influence the magnetic field felt by H_A. The spin of H_B adds (or subtracts) a little bit from the field felt by H_A, with the effect that in this particular molecule, the line we see for H_A won't quite be at δ 3, it will be either a tad higher or lower due to the spin of H_B. That is the story for this one molecule, but of course there are millions of molecules in our sample, and in about half of them the spin on H_B will be up, and in the other half it will be down. So for about half of our molecules the line for H_A will be just a shade to the left of 3 ppm, and for the other half it will be just to the right.

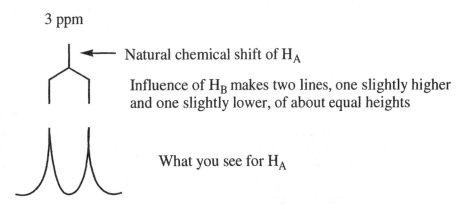

3 ppm

← Natural chemical shift of H_A

Influence of H_B makes two lines, one slightly higher and one slightly lower, of about equal heights

What you see for H_A

What happens when you get in the vicinity of 2 ppm, where H_B should absorb? Exactly what you would now predict: the influence of H_A causes the signal for H_B also to split into two peaks. So in this imaginary compound, you would see one "doublet" in the spectrum at 3 ppm and another doublet at 2 ppm. Below is a real spectrum that has such a pair of doublets, one at 7 and one at 8 ppm.

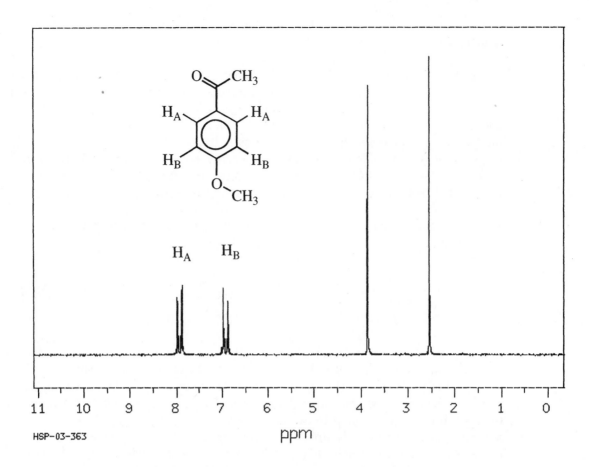

This effect of the spin of one nucleus on another one nearby is called spin-spin splitting (or coupling).

Let's expand on this idea: What if we had one proton on one carbon, and *two* protons next door?

$$H_A \quad \overset{H_A}{\underset{X \quad \underset{Y}{Y}}{\diagdown}} \quad \overset{H_B}{\diagup}$$

We can analyze this situation just like the last one. Again, let's assume that H_A has a chemical shift of 3 ppm, and H_B has a chemical shift of 2 ppm. What do you see at 3 ppm? As above, the signal for H_A ought to be there, but because H_B can be either up or down, and in some molecules it is up and in others it is down, what you will see is *two* peaks, one slightly higher than 3 ppm and one slightly lower. Exactly what you saw before. But what do you see at 2 ppm? This is tougher: there are *two* other protons (H_A) in the neighborhood, and each one, independently, could be either up or down. If both are up, the peak will be off from 2 ppm by some amount; if both are down it will be off by the same amount in the other direction; and if one is up and one is down, their effects will cancel each other out and the peak should be exactly where it would be in the *absence* of H_A. Note that there are two ways this latter circumstance could occur.

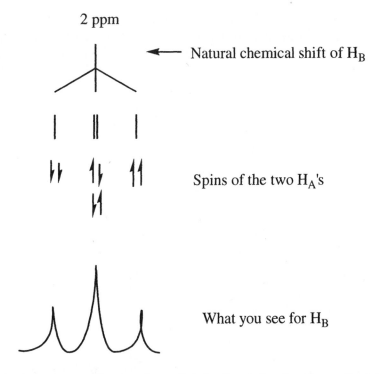

2 ppm

← Natural chemical shift of H$_B$

Spins of the two H$_A$'s

What you see for H$_B$

Notice that this time you get *three* peaks, a "triplet," and the sizes (areas!) of the three peaks are in a 1:2:1 ratio. Below is the real spectrum of one molecule of this kind, 1,1,2-trichloroethane. You can see the large doublet for the CH$_2$ group, split by one proton, and the smaller triplet for the CH group, split by two neighboring protons.

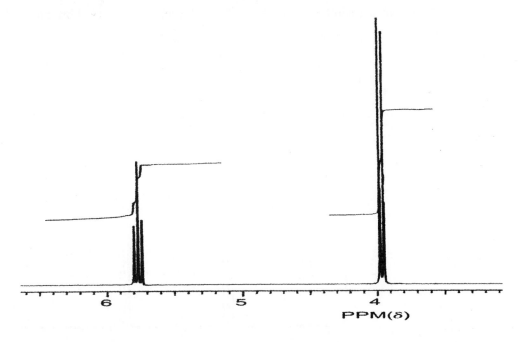

PPM(δ)

So one neighbor makes a peak split into two; two neighbors make a peak split into three—are we seeing a pattern here? Indeed we are. The number of peaks you can expect for a particular proton is given by the number of neighbors plus 1, often called the "n + 1 rule." Further, each group has a standard intensity pattern: we have seen that doublets come as 1:1

peaks; triplets come as 1:2:1 peaks. What should we expect for quartets and higher? The expected pattern is given by something called Pascal's triangle, which is reproduced below. Each entry is given by the sum of the two entries above it (including the non-entries adjacent to the triangle).

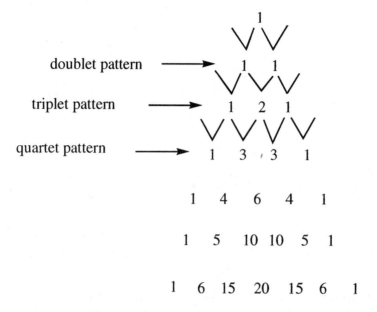

doublet pattern ⟶

triplet pattern ⟶

quartet pattern ⟶

$$1$$
$$1 \quad 1$$
$$1 \quad 2 \quad 1$$
$$1 \quad 3 \quad 3 \quad 1$$
$$1 \quad 4 \quad 6 \quad 4 \quad 1$$
$$1 \quad 5 \quad 10 \quad 10 \quad 5 \quad 1$$
$$1 \quad 6 \quad 15 \quad 20 \quad 15 \quad 6 \quad 1$$

Notice that all multiplets are symmetric, big in the middle and fading away toward the outside.

Now consider another example: below is the spectrum of ethyl bromide, CH_3CH_2Br.

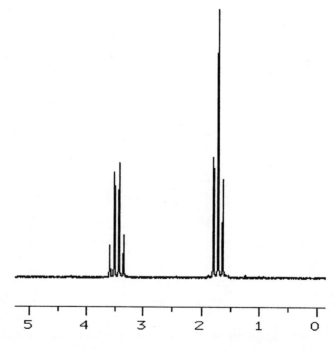

This is a pattern you should burn into your brain: an ethyl group shows a triplet for the methyl, adjacent to two neighbors, and a quartet for the methylene, with three neighbors, in a 3:2 overall integration ratio. The triplet is of necessity near 1 ppm, since it has nothing special attached to it, and the quartet is downfield from it, at a chemical shift that depends on what exactly is attached on the other side. There are a couple other things you might notice about this spectrum:

1. The triplet for the CH₃ group is not really at 1 ppm. The electronegative bromine has the effect of pulling electron density away from the CH₂ group, which is why the latter is at 3.5 ppm. This bromine is certainly not attached directly to the CH₃ group, but its effect is still not zero. Being only one carbon away, it continues to have an "attenuated" effect on the shielding of the methyl group, whose signal is pulled a little (but not nearly as much as the CH₂ group) downfield. In general the effect of any substituent falls off rapidly with distance, but it can still have a minor effect when it is two carbons away. That is why the CH₃ group is a little to the left of where you would have expected it.

2. The two inner peaks of the quartet are supposed to be the same height. Likewise the two outer peaks for both the quartet and the triplet. They are close, but not quite the same. Indeed, the triplet and the quartet seem to be leaning toward each other a little bit. This is a common phenomenon—groups that are "coupled" to each other almost always lean toward each other. There is a long, complicated explanation for this that you don't want to know.

Let's look at another example. This one is 1-propanol, CH₃CH₂CH₂OH

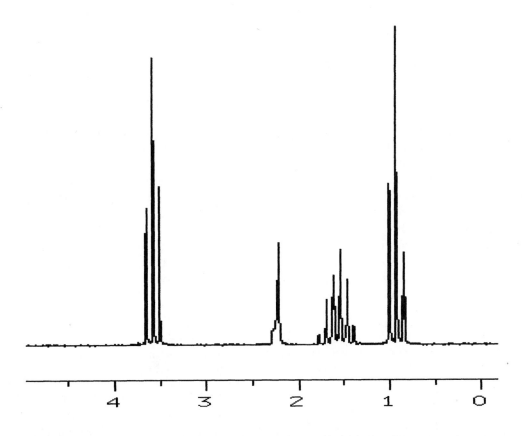

The CH₃ on the end is clearly the peak at 1 ppm. It has nothing special attached to it, so it belongs near 1 ppm; further, it has two immediate neighbors, so it should be a triplet. It also has two other neighbors further away; nevertheless, we see a pretty clean triplet. The apparent (and generally correct) conclusion is that spin-spin coupling operates only at short distances. The general rule of thumb is that the information can travel through three bonds but not four.

H–C–C–H   not   H–C–C–C–H

This means that it is generally only the hydrogens on the carbons immediately next to the one you are looking at that matter.

The next group is the first CH_2. We expect to find it also near 1 ppm, but perhaps a bit to the left because it is only one carbon removed from an oxygen. It has two neighbors on its right (as I wrote it) and three on its left, for a total of 5 neighbors; and $5 + 1 = 6$. We expect 6 lines, and according to Pascal's triangle they should be in a 1:5:10:10:5:1 ratio. Lo and behold, at about 1.5 ppm we find 6 peaks, big in the middle and small at the outsides. Not quite symmetric, the group leans a bit to its right, but it's not too bad. And all together, the group signal appears to be about 2/3 the size of the methyl triplet. (If we had an integral on this spectrum we could demonstrate that.)

On to the second CH_2 group. This has a CH_2 on its left (again, as I wrote it) and an OH on its right. Since it has an oxygen directly attached we expect this to be quite downfield, near 4 ppm. It has two protons on its left and one on its right, for a total of three, so it should be a quartet. But it is quite clearly *not* a quartet, it is a triplet! Apparently this CH_2 group is aware of only two of its three neighbors. This, too, turns out to reflect a general phenomenon: hydrogens attached to oxygens do not count as "neighbors" for splitting purposes.

Why not? Think again about the reason that neighbors have any effect at all: while you are observing H_A, its neighbor, H_B, will be either up or down. In those molecules where H_B is up, you get one line; in those where it is down, you also get one line but in a slightly different place; the spectrum of a large collection of such molecules shows what looks like a doublet. The cause of the effect is the fact that each neighbor has some particular spin that changes the overall magnetic field by a little bit. But what if H_B were constantly changing back and forth between being up or down? Then H_A would experience an *average* of the two, as if H_B were not there at all. This is exactly what happens with OH protons. They hydrogen bond with each other, so that each oxygen is always bonded to one H and partially bonded to another, and likewise each H is bonded to one O and partially bonded to another. Frequently they swap partners, and this happens so often that the NMR experiment cannot keep up with it. In other words, NMR can't tell whether a particular OH proton is up or down. So OH protons do not affect their neighbors by spin-spin coupling, and for the same reason the neighbors do not affect the OH protons. It is as if there were a wall between the OH group and the rest of the molecule, inhibiting communication of spin information.

Actually this swapping is not always as fast as I have described. Sometimes you'll see an NMR spectrum in which the OH protons *do* couple with their neighbors. The swapping (officially, "exchange") rate depends on solvent, temperature, concentration, and purity, among other factors. One factor that is important is the presence of *extra* protons: a trace of acid makes the exchange very rapid. If you have a sample where the OH is coupling, wave your sample over the opening of a bottle of concentrated HCl. The gas floating there will get absorbed into the sample, and that trace amount will usually be enough to speed up the exchange and eliminate the coupling.

Some of you may have been wondering why identical protons do not couple. There are three protons in a methyl group, for example: why is each not split by the other two? The answer is the same as the explanation given above for why OH protons do not split their neighbors. OH protons are constantly changing back and forth between being up or down, and do not stay put long enough to influence their neighbors. CH protons do not change back and forth like that all the time, but they *do* change back and forth when the right amount of energy is available, that is, when the energy of the radio waves matches the energy difference between the two states. This, of course, is the same for all identical protons, so while you are observing any one of the protons of a methyl group, the others in that methyl group are madly hopping back and forth at the same time, so they are not able to influence each other. But protons on neighboring atoms, which are not in resonance at the same frequency, *can* influence their neighbors, as we have seen.

Another consequence of this rapid exchange of OH protons is that the exact magnetic environment of an OH proton is less well defined that that for other types of protons, because specific OH protons are not always attached to the same molecules! Thus, you find that the signals for OH protons in an NMR spectrum tend to be on the broad side rather than sharp lines. Also, if you add a drop of D_2O to a sample containing an alcohol, the signal for the OH hydrogen will immediately disappear. Almost all the H gets swapped for D thanks to the same hydrogen bonding mechanism, and deuteriums are invisible in the NMR, so the signal disappears.

Getting back to our propanol spectrum, we expect that the CH_2 group on the right (the signal on the left!) *should* be a triplet, because it is split by the CH_2 group to its left but it ignores the OH on its right.

The last thing in the spectrum is the OH proton itself. We don't know quite where to expect this—the chart says it could be anywhere between 1 and 6 ppm—but we expect it won't couple with its neighbors, therefore it will be a singlet, and it should be somewhat broad. Sure enough, just such a peak appears at about 2.3 ppm.

Finally a word about integrations. From left to right in the spectrum above, the groups we see should have integral ratios of 2:1:2:3. The web site where I found this did not have integrals, but I am confident that if we ran this sample on a spectrometer that measured integrals, we would get ratios very close to the expected values. The reason I bring this up is that if you just look at the spectrum, the triplet on the left and the triplet on the right look very nearly the same size even though they can't be. The message is that you often cannot estimate integral ratios by eyeballing the peaks themselves. Apparently the triplet on the right has somewhat fatter peaks (we cannot see it in this picture) so that if we integrated, it would in fact integrate to 50% more area than the triplet on the left.

If you understand what this is about, it should not be too tough to look at some structure and predict a spectrum for it. Let's try one: Predict the proton NMR spectrum for 2-chlorobutane. Do it now, before I walk you through it.

Of course, the first order of business is to write down the structure. If you can't do this by now, it is time to drop the course!

$$\underset{\displaystyle H_3C-\underset{\displaystyle |}{\overset{\displaystyle |}{C}H}-CH_2-CH_3}{\overset{\displaystyle Cl}{}}$$

Now we identify how many different kinds of protons we find here. I see four: the methyl on the left, the methyl on the right, the CH, and the CH_2. Next we should decide *where* each of these ought to be in the spectrum, that is, we should guess at the chemical shifts. The methyl on the right is ordinary in every respect: it should be near $\delta = 1$. The CH group has a chlorine attached, so it should be between 3 and 4 ppm, according to our chart. Finally, the methyl on the left and the CH_2 group are both pretty ordinary, but one carbon away from one with a chlorine. These are likely to be a shade to the left of 1 ppm.

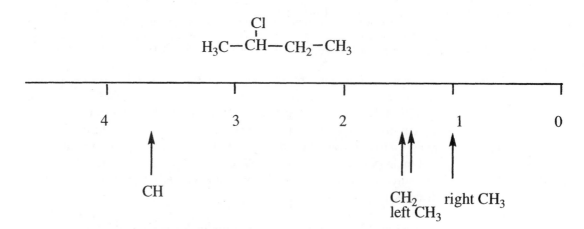

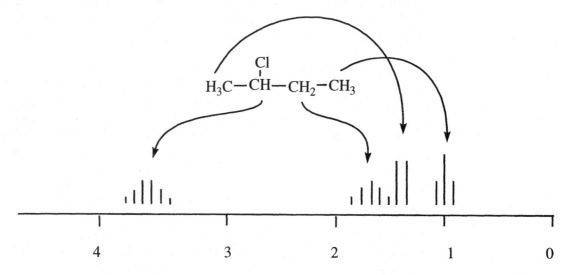

Now we can ask about the patterns each will have. The right CH_3 has two neighbors so it should be a triplet. Triplets come naturally in a 1:2:1 ratio. The left CH_3 has one neighbor so it should be a 1:1 doublet. The CH_2 has four neighbors (three on one side and one on the other) so it should be split into five lines, and the CH should be split into six lines.

Now we have to adjust these groups for size. The two CH_3 groups should be the same size, but be careful: "size" here refers to the area under all the lines that make up the group, put together. The methyl group at 1 ppm is split into three peaks, while the one at 2 ppm is split into two peaks. Each of the two peaks of the doublet represents half the area of the CH_3 group; the three peaks of the triplet represent 1/4, 1/2, and 1/4, respectively, of a CH_3 group. So the largest peak of the triplet should be about the same height as either peak of the doublet (assuming they are all about the same width, which is not always the case). These will be the largest peaks in our spectrum, so let's draw them to fill the page:

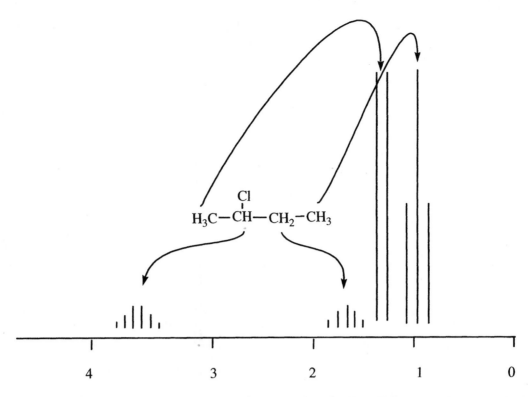

Now we have the two little groups to fix. The CH₂ group, all put together, should be about 2/3 the size of the CH₃ groups; the CH group. All put together, should be about 1/3 the size of the CH₃ groups. We could calculate exactly how big each peak should be, but NMR is not usually that accurate. Let's just eyeball these. The CH₂ group clearly needs to get a lot bigger than it is, while the CH group is not too far off. So here is my final prediction:

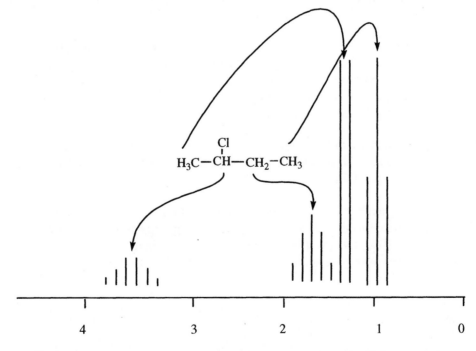

This assumes that we get real lucky and the peaks from the CH₂ group do not overlap with the peaks from the left CH₃ group. If they do, things will be much messier between 1 and 2 ppm.

Below is the real spectrum for this substance. As you see, we did pretty well. The heights of the peaks are not quite in the relationship we predicted, but the integration is OK and each pattern looks pretty much the way we predicted.

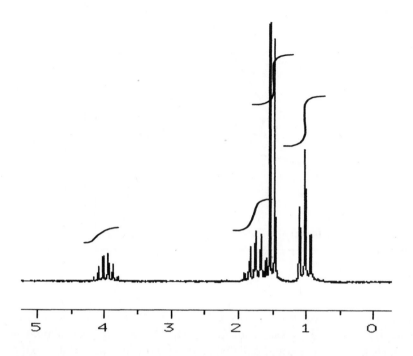

What we have done so far is actually the easy part. What is much tougher, but in fact 99% of the use of NMR, is to look at a spectrum and figure out a structure from *it*. You have to move from stage A to stage B, and now is the time to do it. The *way* to do it, like everything else in this book, is to practice: a lot.

First I want to give you a series of hints and potential complications. Then we can work on some problems. There are actually many, many possible complications: NMR is very much more complicated (and also more powerful) than I have let on. Those of you who are chemistry majors will learn some of the complications, and the power, in later courses. For now we will just scratch the surface. Advice:

1. Honor thy father and thy mother. Woops! Wrong list. But speaking as a father, this is good advice nonetheless. When was the last time you called home? Maybe this would be a good time.

2. Remember the ethyl group, and keep it holy. When you see a triplet near 1 ppm and a quartet to the left somewhere, an ethyl group should jump into your head. Maybe once in 1000 times you will be wrong (after all, there are exceptions to everything), but those are pretty good odds. Burn this pattern into your brain:

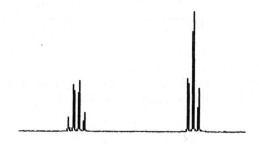

3. Coupling is communicated through bonds. Thus, if H_A makes H_B become a doublet, separated by 10 Hz in the spectrum, then H_B will make H_A become a doublet, also separated by the same 10 Hz, because the information has to travel through the same set of bonds to get from one to the other. The size of the split is called the "coupling constant" (or "splitting constant"), J_{AB}. So when you see an ethyl group, for example, you should discover that the distance between the peaks of the quartet is the same as the distance between the peaks of the triplet. (Coupling constants are *not* reported in ppm because they do not change with the strength of the magnetic field, unlike chemical shifts.)

This brings up our first complication: recall the last example we did, 1-propanol. The middle CH_2 group appeared as a sextet, due to its five neighbors, three on one side and two on the other. However, the spin information traveled to this CH_2 group from the CH_3 group over one set of bonds, and from the other CH_2 group over a different set of bonds. Luckily, both sets were about equally efficient, and the pattern of a sextet remained. However, if one path turned out to be more efficient than the other (this often happens), then you would *not* get a sextet for the middle CH_2 group. The CH_3 group would make it a quartet, and each peak of the quartet would be further split into its own triplet by the other CH_2 group, with a different coupling constant. Result? One huge mess. Not all patterns can be dissected as simply as the ones we have looked at so far. Sometimes you just have to shrug your shoulders and say, "There's a bunch of protons here." Such a signal is usually described as a "complex multiplet."

4. Another possible complication is related to this one. Most parts of an organic molecule, those that are not attached to something electronegative, appear between 1 and 2 ppm, but there is a limited amount of space in that region. Further, if there are other protons nearby, each peak will be split into several peaks by spin-spin coupling. I think you can imagine that with several complicated multiplets in the same region of the spectrum, it will not be uncommon for them to overlap with each other, at which point it becomes very difficult to determine which line goes with which multiplet. Things can get very messy indeed. Below is the spectrum of 1-pentanol. Figure out what you think the spectrum *ought* to look like, and compare it with what you see.

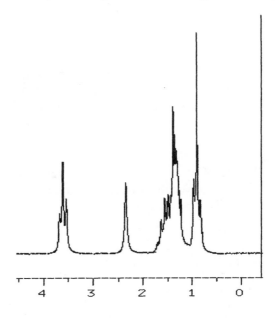

Three of the four CH_2 groups are very close to each other, and after splitting, they overlap so much you cannot distinguish them.

Chemistry departments these days are spending big bucks to buy "high-field" NMR spectrometers. Why? The above spectrum was taken on what is known as a 90-MHz instrument. This means that the magnet is of such strength that it takes about 90-MHz radio waves to make the protons sing. Different protons have "notes" that differ by parts per million of this 90 MHz, about 90 Hz. In other words, the distance between 1 and 2 ppm on the scale is 90 Hz. Meanwhile, these protons are coupled to each other with values on the order of 5–10 Hz. A triplet will be spread out over 20 Hz or so; a quartet over 30 Hz or so. Put a couple of these into a 90-Hz region and you cannot help but suffer from overlap. On the other hand, imagine building the same instrument with a huge magnet, one that would require 400 MHz to make the protons resonate. The protons still couple to each other by 5–10 Hz, so a quartet will still occupy 30 Hz or so (remember that coupling constants do not change with magnetic field strength), but now there are 400 Hz between 1 and 2 ppm. So there is much less chance of overlap. Below is a spectrum of the same substance that we just looked at, 1-pentanol, but this time taken on a 400-MHz instrument.

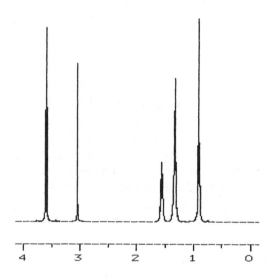

I hope you can see that there is much better resolution in this spectrum. Of the three CH_2 groups that were overlapped at 90 MHz, one has separated completely from the others, at about 1.6 ppm. Further, the triplets at 3.6 and at 0.9 ppm are much cleaner than they were at 90 MHz. All in all, this spectrum is much easier to deal with. This is one of the reasons that high-field instruments are in such demand.

5. Beware of disappearing edges. Consider the following spectrum, of isopropylbenzene. Write the molecule's structure and then predict its spectrum.

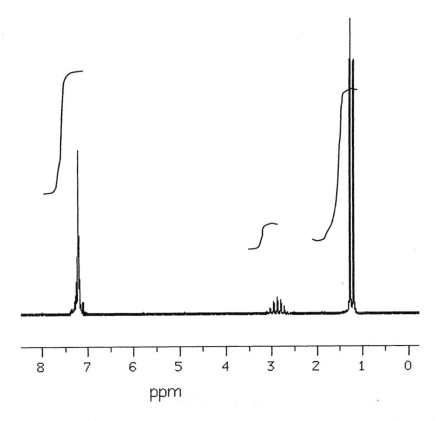

You should have predicted a large doublet, worth six H's, at about 1 ppm, just what you see; and a septet (7 lines) between 2 and 2.5, worth 1 H. Ours is a little to the left of that, but acceptable. However, in order for the huge doublet to fit on the page, the septet has to be pretty small. Further, the center line of this small pattern is 20 times higher than the outer lines. Where does this leave the outer lines? Pretty much invisible, that's where. If you didn't know this was a septet, could you really count seven lines in it with any confidence? I couldn't. But if you have a large doublet that integrates to 6 H's, and a small multiplet that integrates to 1 H, it is hard to imagine any way to put them together that does not result in a septet. So you look at this spectrum and say, "It *must* be a septet." (Of course, if you had a real spectrometer in front of you, you could increase the size of the peaks to be sure those outside lines were there, but you do not have that luxury.)

6. Consider the region between 7 and 8 ppm. Nearly all benzene-type (officially, "aromatic") protons appear here. Many compounds contain benzene rings (and their relatives), but there is only a limited amount of space between 7 and 8 ppm. Many aromatic protons accidentally appear in the same place. This should not surprise you: consider the protons labeled H_A and H_B in the following two molecules:

In the first molecule these protons are identical (*all* the protons in benzene itself are identical). In the second molecule (toluene) they are technically not identical (after all, you can describe one as being in a different position relative to the methyl group from the other), but how different are they to the instrument? The only difference between them is that there is a methyl group some distance away, and methyl groups replacing hydrogens have relatively little effect on chemical shifts anyway. It turns out that on most instruments, one cannot distinguish H_A from H_B in toluene; indeed, *all* the protons on the ring of toluene are pretty much in the same place, unless you use a very high field instrument, as shown below.

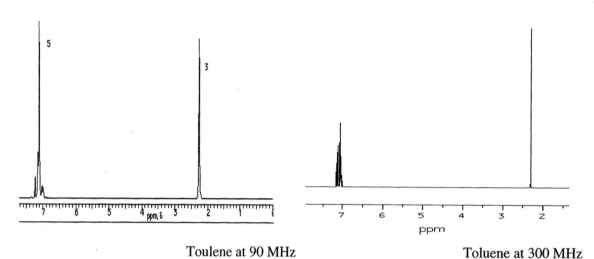

Toulene at 90 MHz   Toluene at 300 MHz

The reason the ring protons of toluene are in about the same place is that the group on the ring (methyl) is not all that different from the group in the same position of benzene itself (hydrogen, obviously). On the other hand, if the group on the benzene is electronically quite different from the hydrogen it replaces, then it is often easy to distinguish different protons on the ring. Examples are acetophenone and anisole, shown below.

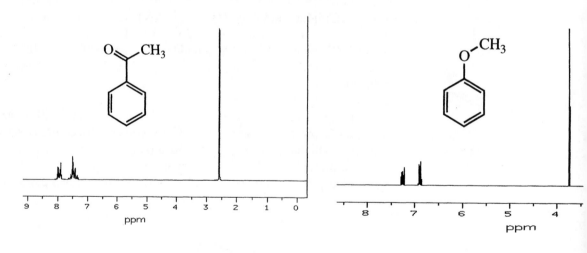

You might notice in these two examples that the five remaining ring hydrogens seem to be in two groups, a group of three and a group of two. Later on we will be able to explain why this is the case.

There is one more aspect of aromatic protons that can help you significantly. Consider a ring with *two* substituents on it. There are in fact three different ways the two substituents might be arranged, and we will discuss this in more detail when we get to the subject of aromaticity:

"ortho"   "meta"   "para"

All I want you to notice at this stage is that in both the "ortho" and the "meta" substances, all four remaining hydrogens are different from one another. I haven't drawn them in, but clearly there is one hydrogen at each position of the ring that does not have X or Y. And each one of them can be described in a way that distinguishes it from the others. So all four H's could have different chemical shifts, and they could couple to each other, and in general, we should expect a pretty big mess from this. On the other hand, in the "para" compound, there are only two different kinds of H's, those adjacent to X and those adjacent to Y. The ones next to X are identical to each other, likewise the ones next to Y. Further, these two kinds of H's are neighbors to each other, and ought to split each other. So for this molecule, we ought to see a fairly simple pattern of two doublets.

A word of warning, however: remember I told you that patterns coupled to each other tend to lean toward each other. The closer they are to each other in chemical shift, the more intense this leaning is. Protons that are widely separated in the spectrum (not in the molecule—they obviously have to be *near* each other in the molecule in order to couple at all, right?) might have a slight lean, like this:

But protons that are near to each other will have an accentuated lean:

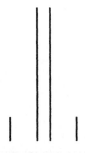

Aromatic protons are already constrained to be between 7 and 8 ppm, so they often look more like the latter pattern more than the former. And every once in a while, by sheer accident, you will find protons exactly the right distance apart, and with the right coupling constant, so that they look for all the world like a quartet:

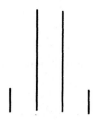

You will almost never see a *real* quartet between 7 and 8 ppm (why not?), so it shouldn't fool you there, but this kind of pattern can occur elsewhere, and you might be fooled into thinking something is a quartet when it is really two doublets. This probably won't happen often, but it is worth keeping in mind as a possibility.

7. Only one more warning before we try to solve some spectra problems. Most NMR instruments require that the sample be dissolved in a liquid solvent—and most solvents have protons. Since there is a lot more solvent than sample, you would see essentially nothing of value if you did not correct for this somehow. The standard approach is to use a solvent in which the hydrogens have all been replaced with deuteriums. Deuterated solvents dissolve things just as well as their normal counterparts, but the deuteriums are invisible in the NMR. There are only two problems associated with this. One is that deuterated solvents are expensive, but we put up with that; the other is that you cannot get perfect solvent, in the sense that there will always be a few molecules left in which the proton(s) did not get replaced by deuterium, so there will always be a small signal where that hydrogen shows up. The most common NMR solvent is deuterochloroform, $CDCl_3$, and some residual chloroform, $CHCl_3$, will be in it. This shows up in the spectrum at δ 7.27. If you see a small singlet there that does not seem to belong to your molecule, it probably does not. You can ignore it. You will also occasionally encounter a spectrum in which a *reference compound* (tetramethylsilane, or "TMS") has been added to produce a standard signal at δ 0. This is not so common anymore, but you will likely encounter it now and then. Any peak at 0 can safely be ignored.

Now let's look at some spectra.

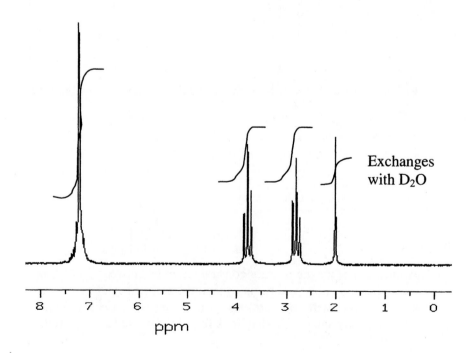

Exchanges with D_2O

First, spend at least 10 minutes trying to solve this one yourself before reading on. *Hint*: the formula is $C_8H_{10}O$, but if you are really good you could solve the problem without that information.

When I look at a problem like this, the first thing I do is to try to catalog the things I can tell about the substance, and only later do I try to put the pieces together. What do we know about this stuff from its NMR spectrum? One thing that should jump out at you right away is the peak on the far left. This is at about δ 7.2. Although there are occasional exceptions, by far the most common thing to appear between 7 and 8 is a benzene ring (or something similar). When you see significant peaks in this region you should immediately write down a benzene ring. Is it plain old benzene? Obviously not—if it were, there would be nothing else in the spectrum. Clearly something is attached to our benzene ring. It is often useful to try to figure out *how many* things are attached. How do you do that? You know that benzene has six hydrogens on it. If there is one attachment, there will only be five hydrogens. If there are two attachments, there are four hydrogens left, and so on. Look at the integration curves (the step curves, remember?). They appear to be in about a 5:2:2:1 ratio. You can tell this by using your fingers, a ruler, a portion of your pencil, anything you want, to estimate the height of the steps. We know that there are 10 H's in this molecule, but even if we did not, it would be a pretty good guess that we are looking at 5, 2, 2, and 1 hydrogens. If there are 5 hydrogens in the aromatic region, it is a pretty safe guess that we have a benzene with one substituent attached. That one substituent must contain all the rest of the hydrogens from the molecule. So we can write the following to indicate what we know so far:

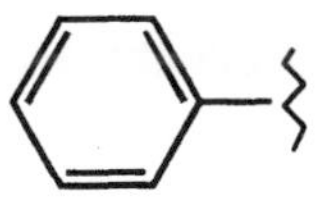

Turning our attention to the rest of the spectrum, we see groups of 2 H's, 2H's, and 1 H. Further, the 1 H exchanges with D_2O, a dead giveaway that it must be an OH proton. So we must have an alcohol here, and the one H is the hydrogen on the alcohol. We can write

–OH

as another piece of the puzzle.

Now we have two groups of 2 H's. Further, each one is a triplet, meaning that it is *next door* to two H's. How can you get a group of two H's adjacent to two H's, and a second group of 2 H's that are adjacent to 2 H's? It should be pretty obvious that the easiest way to do this is to have a $-CH_2CH_2-$ group. That is the third piece of the puzzle. Can we put these together? Certainly! In fact, there is only one possible way! The compound must be

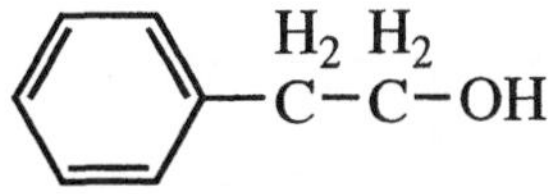

Now pay careful attention here: we are not done. We have made some reasonable guesses and come up with a reasonable structure. But there may be other equally reasonable structures that also account for the information we have considered so far. Most importantly, there is other information we have not considered yet, and if this structure is not consistent with the rest of the information, one of the other structures (not yet thought of) might prove to be better than this one. We have to make sure that this proposed answer is consistent with the whole spectrum we see, not just part of it. This is like checking math problems when you get preliminary answers (a process too few of you probably engage in!)

The best way to check your answer is to start with the structure and predict its spectrum, and see if you come up with something pretty close to what is there. So let's do it.

Start with the benzene part. It should give a signal between 7 and 8 ppm, and that signal should be worth 5 H's. So far so good. What should it look like? There are three different kinds of hydrogen on the benzene ring, and they all have neighbors, so they should be split into a terrible mess; if we saw a mess we could accept it. But all the protons in this case are very similar to each other, so it might also be that they will appear as a broad singlet. That is what we see. We can deal with that. On to the next group, a CH_2 group. This is next to a benzene ring, so it should be at about δ 2–2.5. But it is also influenced by the oxygen atom on the end, dragging it a bit farther downfield. It should be a triplet, due to the two adjacent protons, and it should integrate to two H's. We find this at about δ 2.9. Next, another methylene group, adjacent to an oxygen so it should be near δ 4. We see this at about δ 3.8. It should be a triplet, because it is coupled to the two H's next door (but *not* to the OH) and it should integrate for 2 H's. Finally the OH proton should be a singlet, possibly broad, anywhere from δ 1–6, and it should integrate to 1 H. Further, it should exchange with D_2O. We have that also, at about δ 2. And there is nothing else in the spectrum. Oh yes: we were told that the formula is $C_8H_{10}O$, and our structure agrees with that too. *Now* we are done.

We'll do one more together. Then you have to practice a lot on your own. First, try this yourself before you look at the answer.

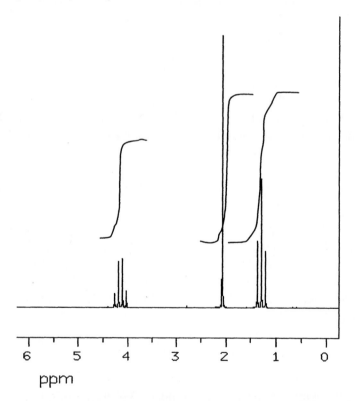

Uh-oh, he didn't give us a formula this time. Well, we'll just have to do the best we can. What do we see? The first thing that should bite you in the nose is an ethyl group: a triplet and a quartet, in 3:2 ratio, at δ 1.2 and 4.1, is a dead giveaway for an ethyl group. Further, we can be confident that this ethyl group is attached to something pretty electronegative if the quartet has been dragged all the way downfield to δ 4.1. The normal candidates for such a thing are halogens, like chloride, or maybe oxygen. If we put a halide on an ethyl group, there can be no other pieces to the molecule, but there *is* another peak in the spectrum, so the compound cannot be a halide. The logical guess is that we have an ethyl group attached to an oxygen. Write it down:

$$-OCH_2CH_3$$

What else? The only other peak here is a sharp singlet at δ 2.0 that integrates for three protons. A singlet worth 3 H's is almost always a methyl group. Write it down:

$$-CH_3$$

Can this methyl group be attached directly to the OCH_2CH_3 group? If it were, where would that methyl be? Near 4, right? But it is not. There must be something between these two pieces, something invisible to the NMR but capable of leaving the methyl group near 2 rather than 4 ppm. What could that be? Looking at your chemical shift chart, you see that two things that place a methyl group between 2 and 2.5 are a carbonyl group and a benzene ring. It is unlikely we have a benzene ring in this molecule (although it is possible, if all six positions were substituted), but a carbonyl would be invisible in proton NMR. Perhaps there is a carbonyl next to the methyl group:

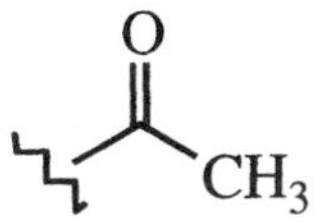

Could this be attached to the OEt group? Yes it could, giving us the following possibility:

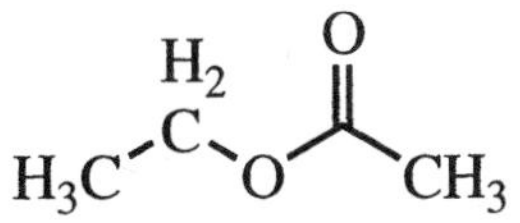

Don't forget, though, check the whole thing! Is this answer really consistent with everything in the spectrum? Yes, it turns out that it is. Is it right? Who knows? It *might* be right. It is a reasonable possibility for the spectrum we see. It is, perhaps, the *simplest* possibility. But it is far from the only possibility. Below are five other molecules that are equally consistent with the spectrum: each would give a spectrum very much like the one we see, and with the current information you cannot prove that one of these is any more likely than another, or yet many others we haven't written or even thought of.

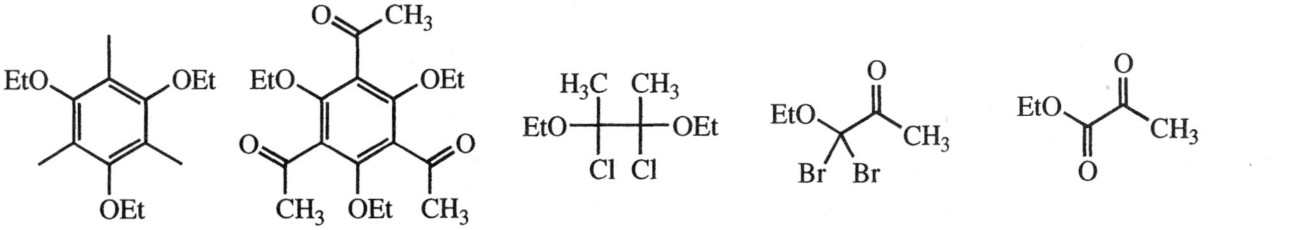

Of course if we had an IR of the substance, or a ^{13}C NMR, or a mass spec, or all three, we could decide which of these, if any, is the right answer, but without more data you simply cannot tell. The best answer under these circumstances is generally the simplest (and it happens to be the correct one in this case).

Now it's your turn. Below are some problems to practice on. Do them!

In addition, if you want more, here are several web sites with further information on spectroscopy and more practice problems:

http://www.nd.edu/~smithgrp/structure/workbook.html
http://homework.chem.uic.edu/NEXT.HTM#
http://www.chem.ucla.edu/~webspectra/

Problems

Identify the following compounds based on the spectra given.

1.

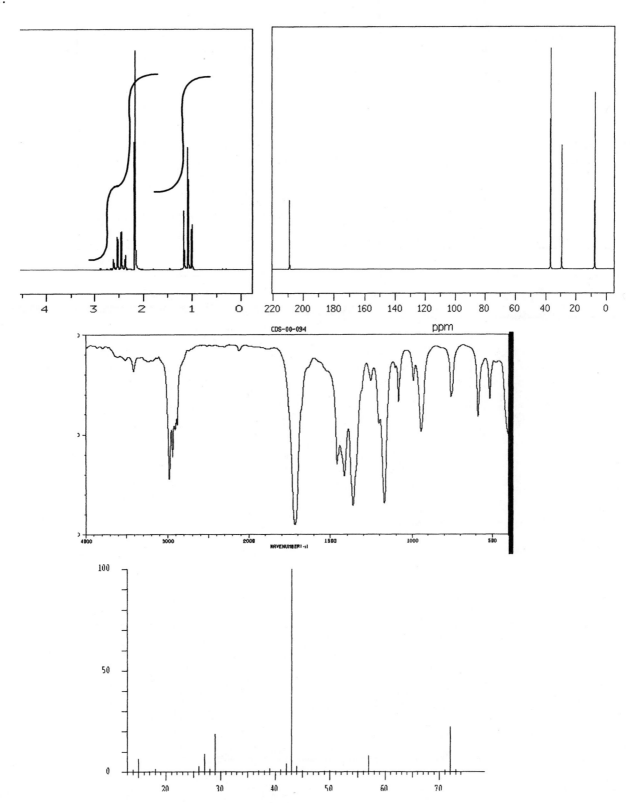

CDS-00-094

2.

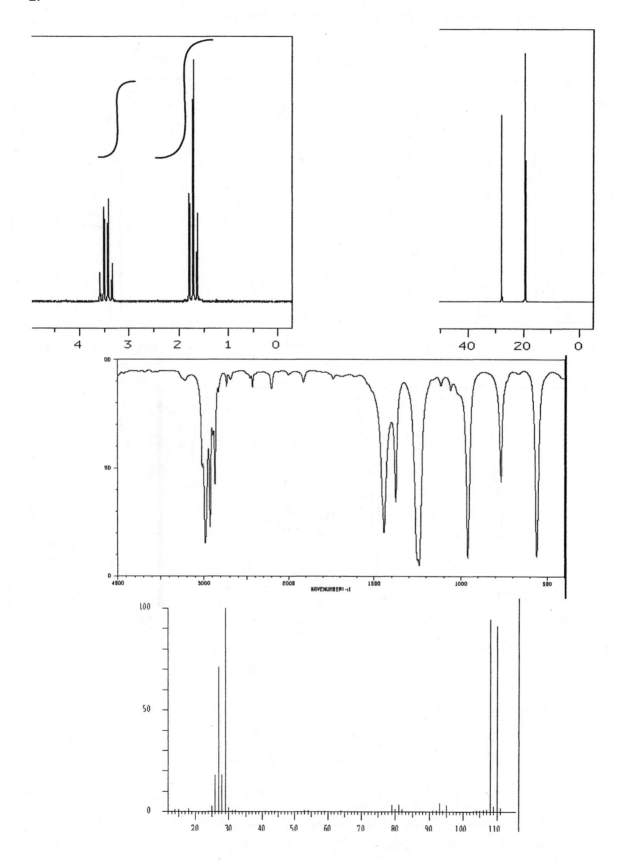

3.

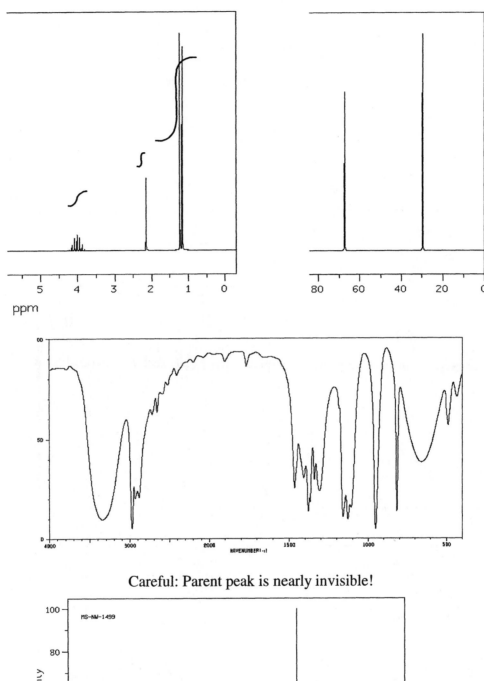

Careful: Parent peak is nearly invisible!

4.

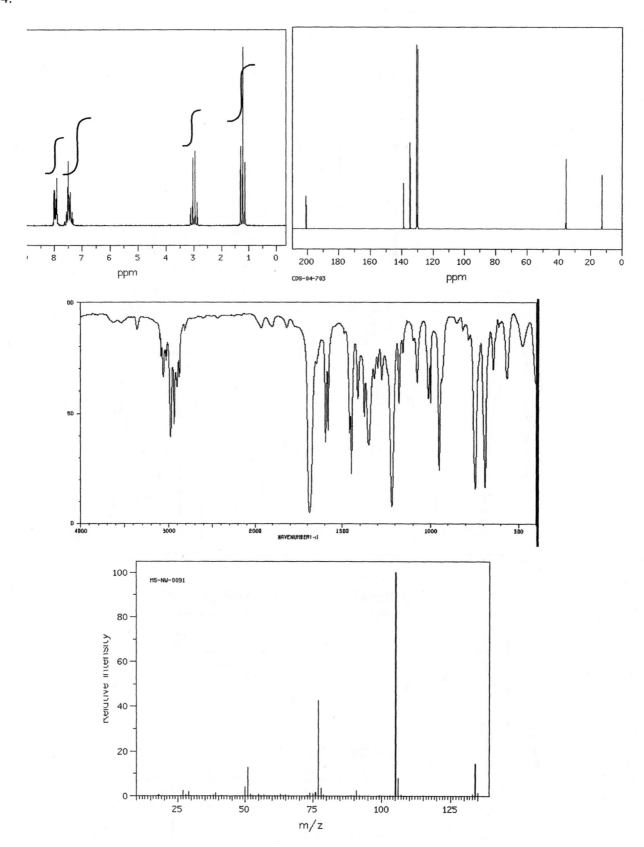

CDS-04-703

MS-NW-0091

5. *Hint:* There is one nitrogen in this compound. *Hint 2:* Beware of duplicate groups!

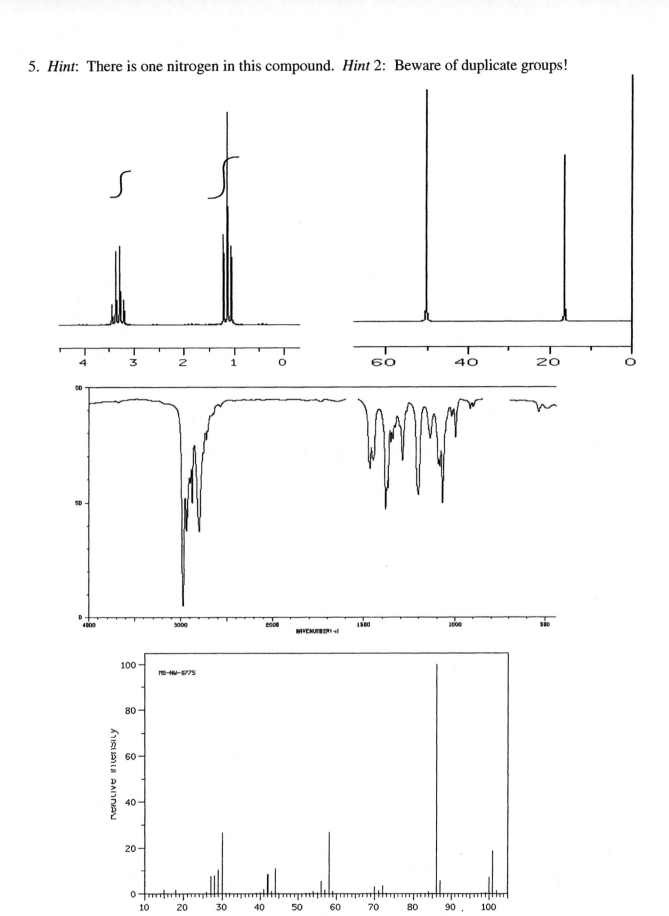

6.

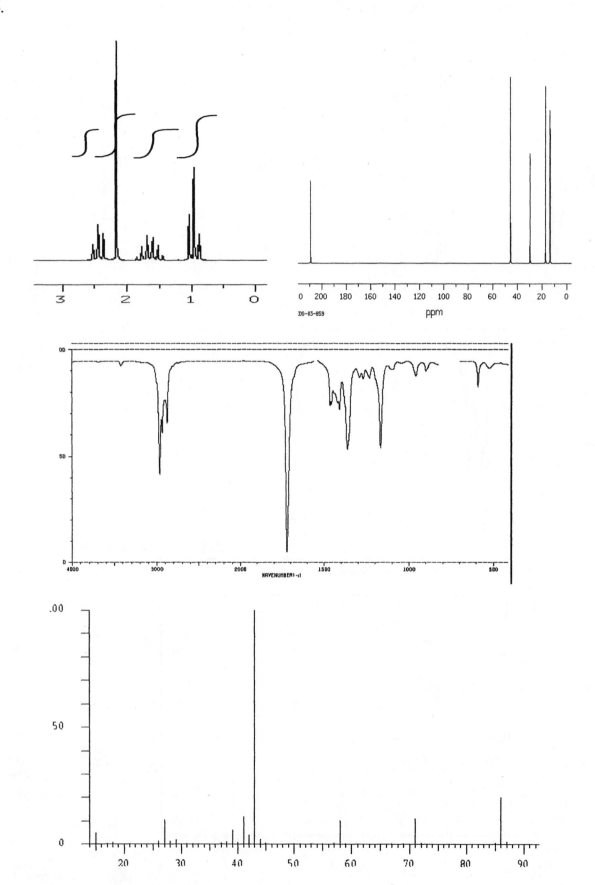

7.

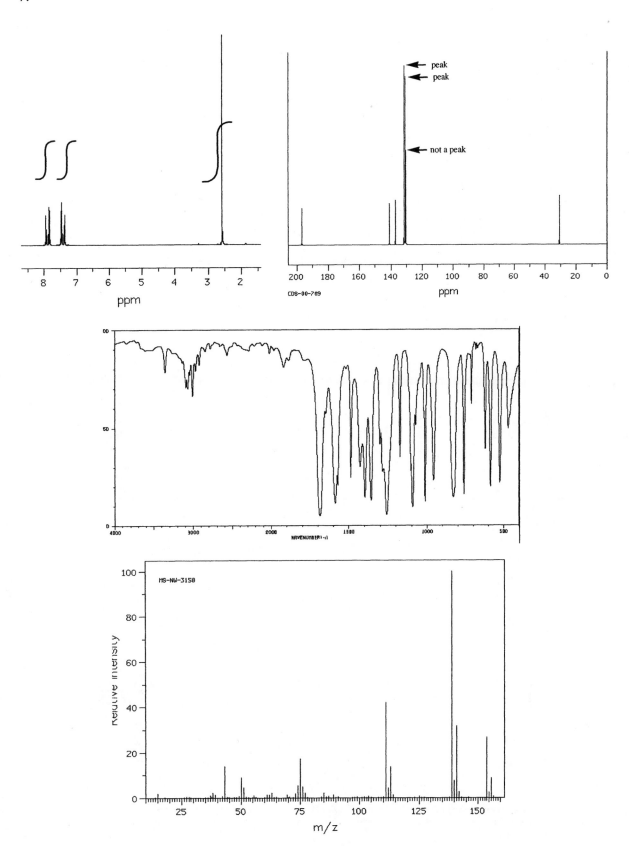

8.

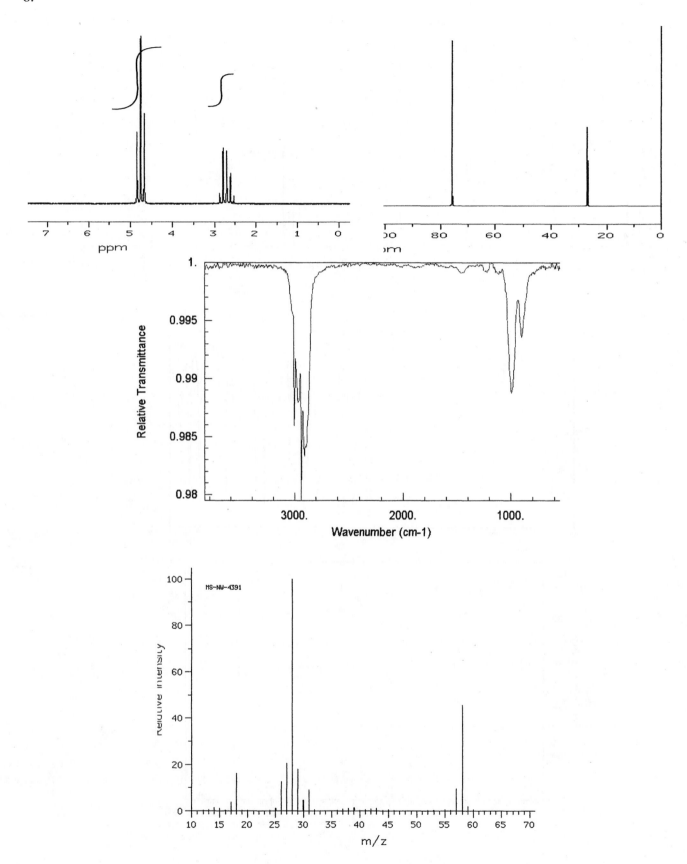

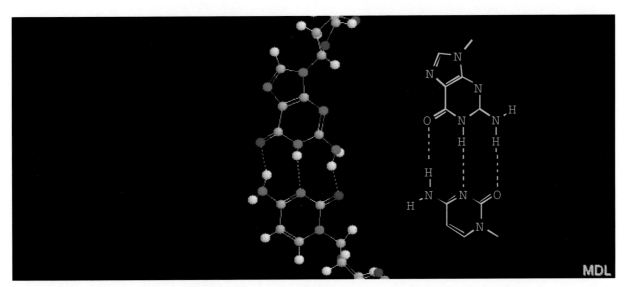

Page 91

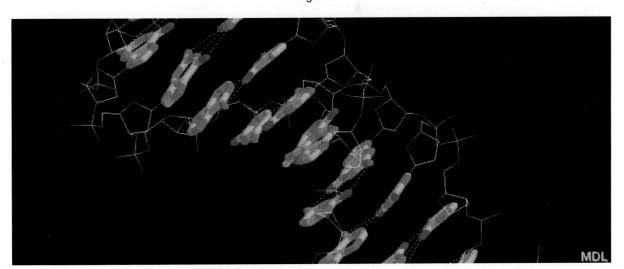

Page 91

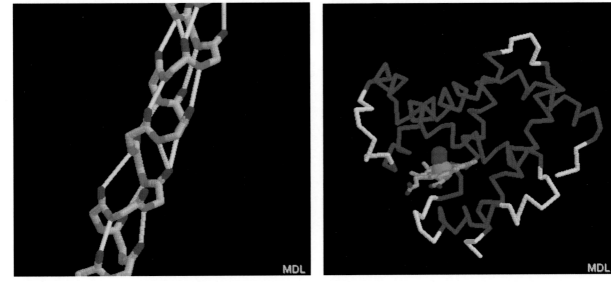

Page 93

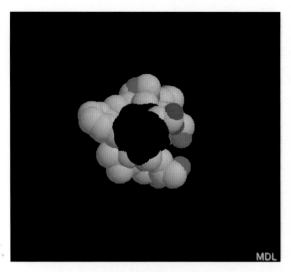

Page 93

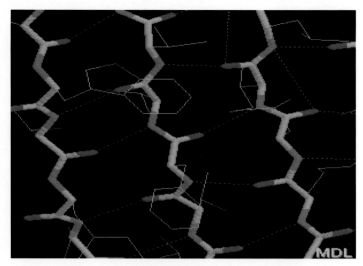

Page 721

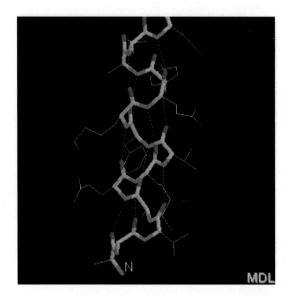

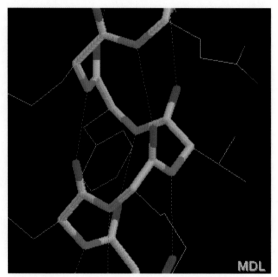

Page 720

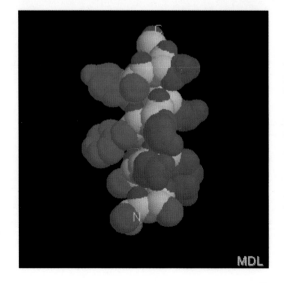

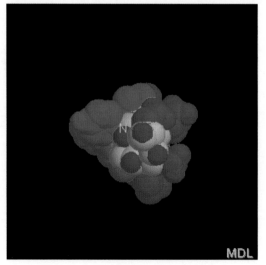

Page 721

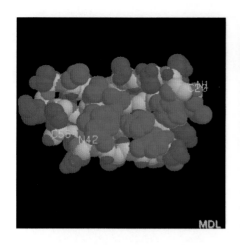
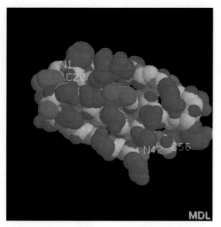
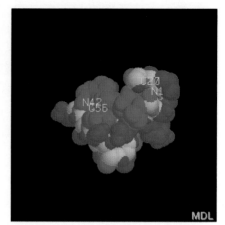

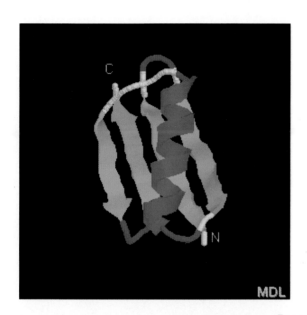
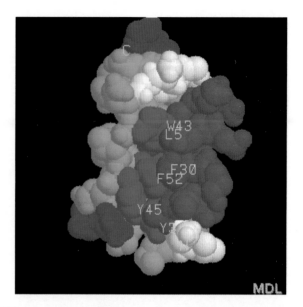

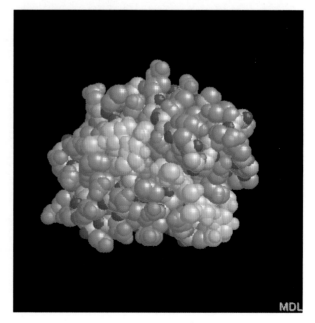

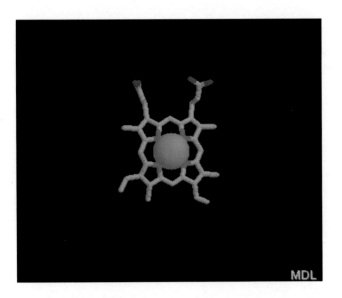

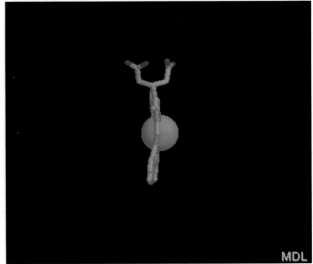

Page 768

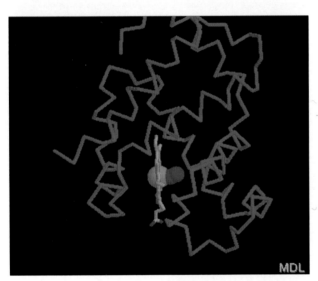

Page 768

Page 769

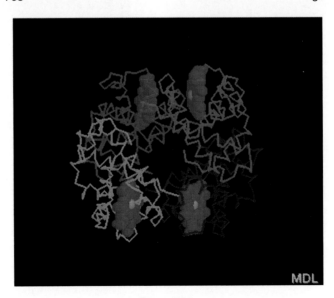

Page 769

9. This actually is not related to the material in this chapter, but serves as a preview to the next. Get out your model kit and build a model of 2-bromobutane. Now get together with three classmates who have done the same, and compare your models. Are they all absolutely identical? Could you superimpose each on all of the others? You are free to twist and turn the models in any way you wish, but keep them intact for purposes of comparison (i.e., don't break any "bonds"). If two models cannot be superimposed, they must be different! What relationships can you find among the models you have built?

10. Do the same with a molecule corresponding to the formula below. (You'll want to agree on what colors to use to represent which elements.) Again compare your models. How many different structures do they in fact represent? Test this *empirically* by trying to superimpose one model upon another. Are there other unique models that can be made corresponding to this arrangement of atoms? How many compounds are actually represented by the formula below? Can you detect any systematic relationships among the various molecules?

Selected Answers

Internal Problems

Test Yourself 1. Ethyl bromide

Test Yourself 2.

Test Yourself 3.

a   b   c   d   e

Test Yourself 4.

6 sp^2 3 sp^3 2 sp^2 1 sp^3

4 sp^2 3 sp^3 4 sp^2 2 sp^3

Chapter 15
Chirality

A mirror image is what you see when you hold something up to a mirror. All objects have mirror images. (Well, almost all—vampires do not!) Some objects are exactly the same as their mirror images, others are not. In this chapter we will

- See how the concept of mirror images applies to chemistry

- Learn the words that apply to mirror image isomers and stereoisomers in general

- Discover how to name mirror-image isomers

- Find out how to differentiate mirror-image isomers

- Figure out how to separate mirror-image isomers

- Study the stereochemistry of substitution reactions.

Warning: this chapter is full of new concepts and the related vocabulary. I will make it as easy as I can, but there are some things that you will just have to work at learning. Much of the material will make no sense at all unless you build models! This chapter is involved heavily in the three-dimensional shapes of molecules, and the untrained eye is very unlikely to be able to interpret the necessary three-dimensional picture based on a two-dimensional drawing. Later you will be able to do this, but for now, build models of everything! There is no substitute for becoming intimately familiar with your model set.

PLANES OF SYMMETRY AND CHIRALITY

Let's start by dividing the world we see into two classes of objects: those that are identical to their mirror images and those that are not. By identical, we mean that you can pick up the mirror image and, by manipulating it properly, completely superimpose it on the original object. Objects that fall into this category include a table, a chair, a fork, a drinking glass, a plate, a baseball, an undershirt, and many others. All of these share a common characteristic: they have a plane of symmetry. A what? A plane of symmetry is an imaginary plane with which you slice an object in half, such that whatever you see on one side of the plane is exactly mirrored on the other side. In other words, a plane of symmetry acts like an internal mirror. Below is a feeble attempt to show a chair and an undershirt with their planes of symmetry.

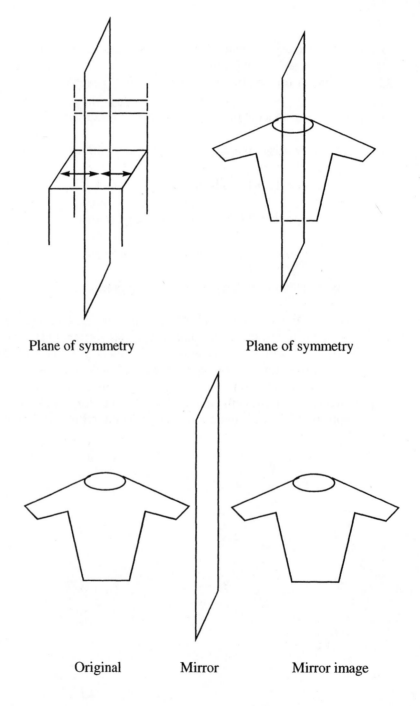

Plane of symmetry   Plane of symmetry

Original   Mirror   Mirror image

A plane of symmetry is a mirror that slices through an object, but it does not create a mirror image of the entire object. To do that you need to hold a mirror *next to* the object. Objects that have an internal plane of symmetry, as described, *also* have the property that they are superimposable on their mirror images. Hopefully you will agree that if you could reach into a mirror and pull out the reflection, each of the mirrored objects listed before would be identical to the original, and if I handed you one randomly, you would not be able to tell whether it was the original or the mirror image. I have attempted to illustrate this also for the undershirt. Objects in this category are called *achiral* (pronounced ay-kai-rill) meaning not *chiral*, where chiral is the word we use to describe the other category, coming up next.

The other category, then, is made up of those objects that *cannot* be superimposed on their mirror images. Examples include each of your hands, a glove, a screw, an undershirt with the tag in it, a T-shirt that says "Tommy" on it, a shoe, a coffee cup with a picture of Juan Valdez on it, and many other objects. Below I've attempted to show this situation also. All such objects share the property that they do *not* have a plane of symmetry. Objects in this category are called "chiral," from the Greek for "hand." The property is often referred to as "handedness," which could be considered a synonym for "chirality." For objects in this category, say a glove, the word by itself does not constitute a complete description: it needs an adjective, such as *right* glove, to complete the description of a particular item. That works for some objects, those that we are used to describing in such terms; but the two shirts shown below we would have trouble describing as "right" and "left." If we needed to specify them, we would have to invent new words.

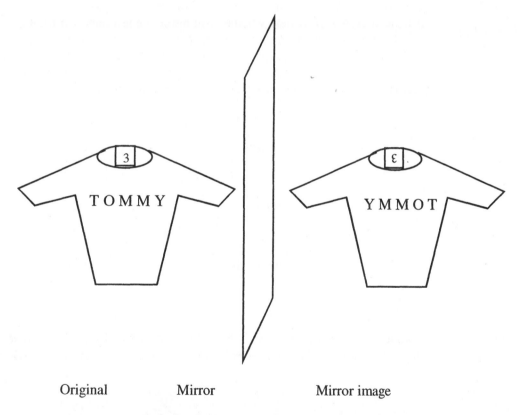

Original   Mirror   Mirror image

Notice that these pictures look like they might be superimposed simply by flipping one on top of the other, but that is because they are pictures. Real objects would not superimpose because their fronts and backs would not match up. The pictures *do* superimpose (and the pictures are not chiral), because they *do* have a plane of symmetry, namely, the plane of the paper on which they are printed.

Molecules are just like objects. There are achiral molecules that have planes of symmetry and can be superimposed on their mirror images, and chiral molecules that do not have planes of symmetry and cannot be superimposed on their mirror images. We will not discuss the rare exceptions to these statements. Most (but not all) chiral molecules have what is called a stereogenic center (abbreviated *stereocenter*, sometimes called *chiral center*), which is a carbon atom (or other tetrahedral atom, but almost always carbon) that has four *different* substituents attached to it; most (but not all) achiral molecules have no such stereocenter. We *will* discuss the exceptions to these statements, but first we have to establish what they mean!

Repeat: there are no vampire molecules. *All* molecules have mirror images. What we are worrying about in this chapter is not whether there *is* a mirror image (of course there is!), but whether a particular mirror image is distinguishable from the corresponding original.

Let's first take a minute to explain why we are bothering with this topic. It turns out that virtually all molecules of biological importance are chiral. What is more, they exist in nature in only one of the two possible mirror-image forms: exclusively one-handed gloves, if you will. Thus, if you are in the business of understanding biological systems, and are dealing with a particular molecule, you not only need to know the structure of the molecule but also which mirror-image form you have—whether it is the left-handed one or the right-handed one. More important, because the existing parts of your body are already chiral, a new chemical introduced into the body (a medicine, perhaps) will interact differently with your system depending on which "handedness" the molecule has (if your body is built entirely of left-handed gloves, it matters whether you introduce left hands or right hands). For example, the two mirror-image forms of the substance "carvone" smell different: one smells like spearmint, the other like caraway. Many cases are known where one version is beneficial but its mirror image is harmful. Thus, it is important for us as scientists interested in natural systems to be comfortable with the issue of mirror images in chemistry.

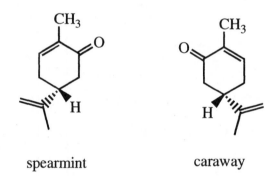

spearmint   caraway

STEREOCENTERS

So what is a stereocenter? As I said above, it is usually a carbon that has *four different* things attached to it. They could be four different atoms, as in the following,

$$\text{Cl} \quad \text{Br}$$
$$\text{H} \quad \text{F}$$

or four different carbon chains, as in the following,

or any combination thereof. The key is this: stand on the suspected carbon and start walking in any direction, taking snapshots along the way, and then go back to the beginning and strike out in a new direction, again taking pictures along the way. If there is any difference at all in the two sets of pictures, those two directions are different. If there is no difference, then they are the same, and that center can't be a stereocenter.

For practice, identify all the stereocenters in cholesterol, below. You should find eight of them.

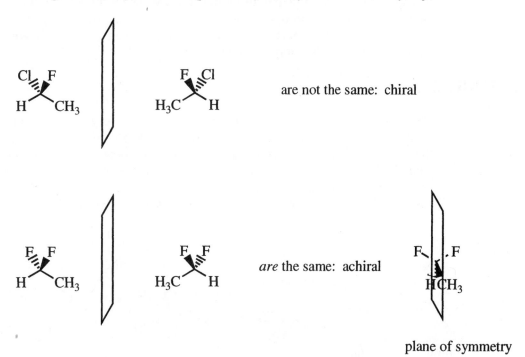

What's the big deal about stereocenters? Get out your model kit. Build any random molecule with a stereocenter. Then build its mirror image. Now study them carefully. They aren't the same, are they? *Any molecule containing one stereocenter is chiral.* Now build a molecule with no stereocenter (take the first molecule and adjust it so that two of the substituents are the same). Convince yourself that this time the mirror image *is* the same as the original. Finally, locate the plane of symmetry in the molecule you just built.

are not the same: chiral

are the same: achiral

plane of symmetry

The two molecules you built in the first case are as different from each other as your left hand is from your right. Life's experiences have already trained you to recognize the difference between left and right hands, feet, gloves, etc., but not left- and right-handed molecules. We need to train ourselves to see these peculiar differences, and we need to invent ways to draw them, name them, and distinguish them. We will do all of that. For now, we need to introduce some relationship words.

ENANTIOMERS

If the mirror image of a molecule is superimposable on the original, we do not need a name for the mirror image. It is simply another copy of precisely the same thing. The molecule is, of course, achiral in this case. The mirror image of a *chiral* molecule is (by definition) *not* the same thing. We call it the **enantiomer** ("en-ANT-ee-oh-mer"). An enantiomer of a molecule is a mirror image of a molecule *that is not identical to the original*. Note carefully: an enantiomer is not defined as merely a mirror image. A mirror image that is the *same* as the original (that is, the mirror image of an achiral molecule) is *not* an enantiomer of it. An enantiomer is always the mirror image of a *chiral* molecule. Further, the word enantiomer is a relationship word, like "bigger." You do not describe something as being bigger, unless there is something else under discussion. Likewise, a chiral molecule must be an enantiomer *of another molecule* (its mirror image).

What is the difference, really, between two enantiomers? After all, your right hand can (in principle) do anything your left hand can do, as long as it does not involve an asymmetric object (like a right glove) or an asymmetric operation (like turning a screw). For normal operations, like holding a baseball, a hammer, etc., for anything *symmetric*, either hand is equally good. Although you have probably become better at throwing a ball with one hand than with the other, it is not because one hand is better configured to do the operation, but simply because you have practiced more with it. The same is true with molecules: anything you try to do with molecules in a symmetric environment, such as a laboratory flask—procedures such as measuring melting points, boiling points, or solubilities, acquiring spectra of various types, and running all the reactions we have studied—will occur identically with either enantiomer. Indeed, if I handed you two containers, one containing one enantiomer, and one containing the other enantiomer of the same substance, you could not tell them apart with any ordinary laboratory operation.

Racemic Mixtures

There is an important corollary to this: if the laboratory environment cannot distinguish enantiomers, it follows that when a reaction is run in which enantiomers are made (such as the addition of HCl to 1-butene, as shown below), both enantiomers *must* appear in equal amounts. A mixture that contains equal quantities of the two enantiomers of a chiral substance is called a **racemic mixture**, or racemate ("ruh-SEE-mic," "RASS-um-ate"). Under ordinary conditions, organic reactions always give racemic products if the starting materials are achiral.

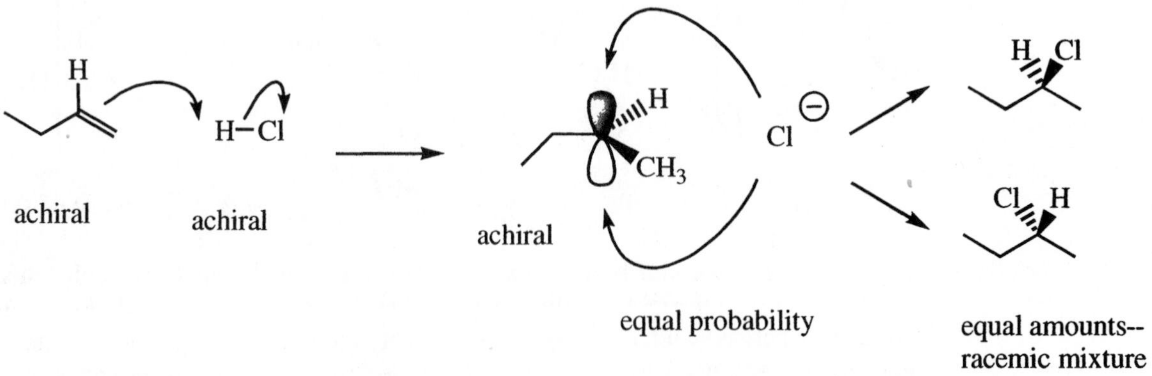

There is another important corollary to this: if you start with a racemic mixture, and the two enantiomers behave exactly the same way in everything you do to them, then it must be impossible to separate them from each other.

Hmm, think about this: you cannot make one enantiomer without making the other, you cannot separate them, and even if you could, you could not tell them apart. So why are we bothering with this? As I told you before, *biological* systems have no trouble at all making one and not the other, or distinguishing between enantiomers. There is a famous case of a drug called thalidomide, which was used as a sedative and a morning-sickness remedy in the 1960s. It was (and is) a terrific drug and had wonderful effects on the people who used it. Unfortunately, unbeknownst to the doctors, the patients, or the company that made it, it was only one enantiomer of the drug that was having these beneficial effects, even though the drug was sold as an equal mixture of both enantiomers (a racemic mixture, because of the way it was synthesized). This in itself is not necessarily a problem—people were simply getting half the intended dosage, because only half the molecules were doing the job. The real problem was that the *other* enantiomer, the one that was not acting as a sedative, was instead causing major birth defects when taken by pregnant women. Many children were born in the 1960s with flippers for arms and/or legs, and eventually the cause was traced back to thalidomide. Thalidomide was then banned, but not before many people were affected.

Thalidomide:

Draw the two enantiomers of thalidomide.

By the way, thalidomide was never approved for use in the United States, because the drug-approval process here is more stringent and time consuming than it is in other countries. This fact has frequently been the cause of great frustration to suffering people, who see patients in other countries receiving apparently useful drugs that are unavailable in the United States, and many of them go to great measures to obtain such drugs. But authorities point to cases like thalidomide to justify the extra care. Thalidomide recently *has* been approved for certain uses in the United States—just not by women who are or might become pregnant.

Optical Activity

I told you earlier that you could not tell two enantiomers apart with any ordinary laboratory operation. This is true, but fortunately there is one *un*usual laboratory technique available that does distinguish between the two enantiomers of the same substance. Once again, it relates to light, but to a property of light that you are probably unaware of.

Light, you will recall, can be thought of as a wave, typically drawn as a sine wave:

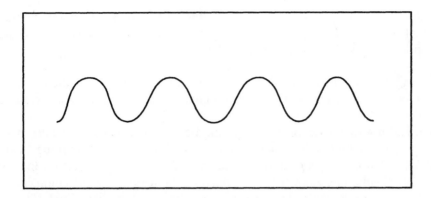

Of course, this picture depicts the wave as an oscillation in a particular plane—the plane of the paper. But with real light, there is no reason why the oscillation should not be occurring in other planes as well, indeed, in all planes. A beam of light should contain light oscillating in the plane perpendicular to this one, shown dashed below, and in every other imaginable plane. Looking at it end-on, you should see waves coming toward you in all possible planes.

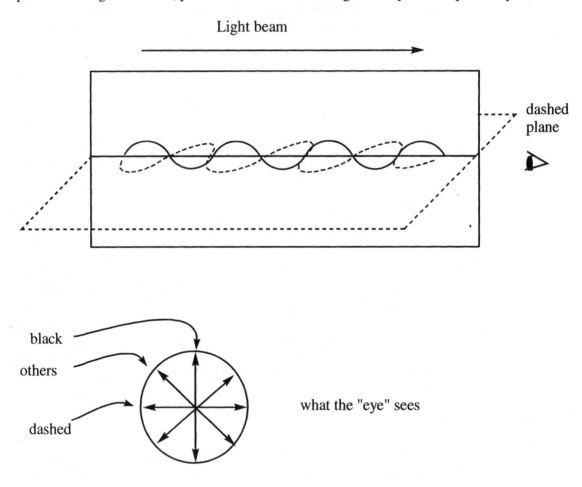

what the "eye" sees

Plane Polarized Light

A *polarizing* lens is a piece of material that can take ordinary light, oscillating in all directions, and allow only one plane of light to get through, say the vertical plane. You may have seen polaroid sunglasses before; they eliminate much reflected light by selecting a particular plane to let the light through. The light that comes through such a lens is called "plane-polarized light," meaning that the light oscillations are restricted to a single plane. Of course, if the polarizing lens were rotated from what is shown in the picture, some other plane, perhaps the dashed one, or one in between, would be let through. It all depends on how the polarizing lens is held. In other words, a polarizing lens has a direction associated with it.

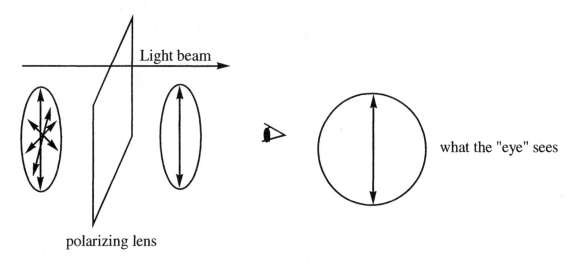

Light beam

what the "eye" sees

polarizing lens

Polarimeter

Your eye, however, does not know that it is now seeing light in only one plane. All it knows is that the light it was seeing has gotten dimmer. But if you put a second polarizing lens in the path, you can easily tell what has happened. If the second polarizing lens is lined up with the first, the same light will get through. But if the second one is perpendicular to the first, *no* light will get through, and you see blackness. (Have you seen the little discs they sometimes sell with polarized sunglasses? They are for encouraging you to do precisely this experiment.)

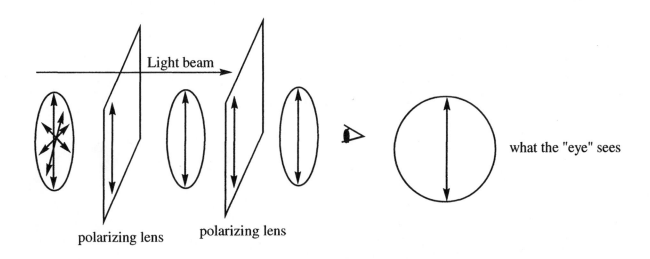

Light beam

what the "eye" sees

polarizing lens   polarizing lens

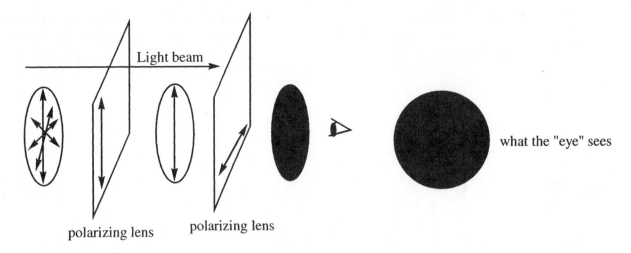

Light beam

polarizing lens   polarizing lens

what the "eye" sees

By rotating the second polarizing lens, you can determine the plane of polarization of the light coming into it. Does this make sense? As you rotate the second lens, the one closer to your eye, the light gets dimmer and dimmer, and when the "window" goes black, you know the second lens is perpendicular to the plane of the light coming at it. That plane, of course, is determined by the plane of the first lens.

Or is it? What I have described so far is accurate if there is nothing between the two polarizing lenses, or if the gap is filled by air, water, or anything else that is not chiral. However, if you make a solution of thalidomide, cholesterol, sugar, or anything else that is chiral (and enantiomerically pure, meaning you have only one of the mirror-image forms of it), something weird happens. The plane of polarization of the light rotates! If you do this inside a system like the one we have shown above, with two polarizing lenses, the second lens will now have to be rotated by a *different* amount to make the light disappear. By measuring how much you have to rotate the second lens, compared to what was necessary with a blank sample, you can tell how much the polarizing plane of the light was rotated by the sample.

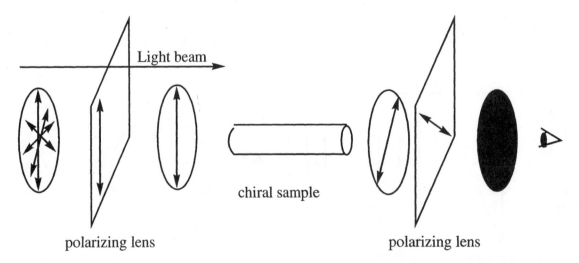

Light beam

polarizing lens

chiral sample

polarizing lens

The instrument we have just designed is called a polarimeter. You use it exactly as I described: inserting a blank (an empty sample-holder), you rotate the second lens until you reach some clearly determined point. This can be the point at which the *most* light comes through, or the *least*—it doesn't matter, as long as you can find that point reproducibly. Then you put your sample in and again find the corresponding point. If the sample is not chiral, the "characteristic point" should require the same amount of rotation of the lens as before. If the sample is chiral, however, the lens will need to be rotated by a different amount relative to the blank sample. The extent to which a sample rotates the plane of polarized light is called the *optical rotation*, and it is given the symbol α. A rotation is reported as both a number and a direction—after all, the light could be rotated either clockwise or counterclockwise from the point of view of the eyeball in the pictures. Clockwise is by convention reported in terms of positive numbers and counterclockwise by negative numbers. A clockwise rotation is also referred to as "dextrorotatory," or *d*, and a counterclockwise one as "levorotatory," or *l*, with the consequence that (+) and *d* are synonyms, and (–) and *l* are synonyms.

COMMON MISTAKES: DON'T LET THIS HAPPEN TO YOU!

These symbols, (+) and (–), are *not charges!!!!* They are simply indications of which direction the light was rotated by a particular sample. They are a way of reporting the results of an experiment, nothing more; (–)-carvone refers to that enantiomer of carvone that rotates the light in a counterclockwise direction. It could also be called *l*-carvone (that's the letter el, not a one).

Substances that rotate the plane of polarized light are called "optically active." Obviously, in order to be optically active a substance must be chiral. Thus, the terms "optical activity" and "chirality" are often used interchangeably, although, as I will explain later, to do so is not exactly correct.

Here's the interesting thing: enantiomers rotate light by equal amounts, but in opposite directions. If natural cholesterol rotates light clockwise, its enantiomer (unnatural cholesterol) will rotate light by exactly the same amount, but counterclockwise. Voila! We now have a way to distinguish enantiomers in the lab!

Specific Rotation

We have to be very careful with this measurement, however, both in terms of the amount of rotation and its direction. First let's worry about the *amount* of rotation. I said that unnatural cholesterol would rotate light by exactly the same amount in the opposite direction, but shouldn't the amount of rotation depend on how much sample is in the light beam? If molecules are making the rotation occur, however it is they do it, more molecules should lead to more rotation. So the observed rotation α should be proportional to the concentration of the solution you make, and indeed it is. In a sense it is also dependent on the sample container: if you have two tubes (like test tubes) filled with exactly the same concentration of sample, but in one case you shine the light through in the long direction, and in the other you shine it in the short direction, don't you suppose that the long tube will result in more rotation? After all, even though the two solutions have the same number of molecules *per unit volume*, the first experiment is allowing the light to pass through more volume and therefore more molecules, so it should result in more rotation. The key dimension here is the *length* of the path the light travels through (with sample in it). This is the same thing we encountered with UV-vis (and IR) spectroscopy.

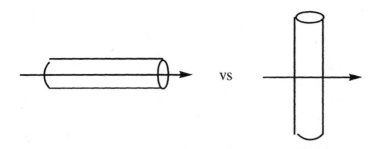

Thus, the observed rotation α depends on the product of the concentration and the length of the tube: this product corrects for the number of molecules the light beam sees. If you take an observed rotation and divide by the product of the concentration (*c*, traditionally expressed in g/mL) and the length (*l*, traditionally reported in this particular application in dm, decimeters, and almost always with a value of 1.0 dm), what you are left with is a measure of the intrinsic ability of this kind of molecule to rotate light. We call this the compound's *specific* rotation, [α].

$$[\alpha] = \frac{\alpha}{c \cdot l} \qquad \text{or} \qquad \alpha = [\alpha] \cdot c \cdot l$$

The observed rotation, α, is likely to vary each time you measure it, depending on exactly how you make up your sample, but the specific rotation, [α], should always be the same, because that corrects for the experimental variables. It is specific rotation that you will see reported in the scientific literature, because anyone, anywhere, measuring the rotation of cholesterol should come up with the same value for a given specific rotation. Specific rotation is a property of the substance, and not dependent on the experiment, just like melting point or IR spectrum. It is, in a real sense, the built-in ability of the substance to rotate light, and it varies from compound to compound.

I should caution you that even the specific rotation is not perfectly invariable: it does depend on the temperature, and especially on the wavelength of light used. Thus, a complete report of specific rotation also reports these variables. What you see in the literature might look like this:

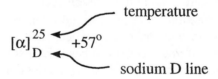

The "sodium D line" is a particular wavelength of light that sodium lamps give off—it's the yellow-orange light you sometimes see from street lights—and most polarimeters are outfitted with the proper light for making this measurement. For our purposes I will omit the superscripts and subscripts from specific rotations, but when you see them in some other book or in the research literature, you now know what they mean.

A more subtle issue has to do with the *direction* of the rotation observed in a given measurement. Imagine you are the eyeball in the picture I drew, and you discover that in order to blacken the light, you have to make the second polarizing lens horizontal with no sample, as in the following picture on the left, but with the sample, it needs to be at an angle, as in the right picture. What rotation did you observe?

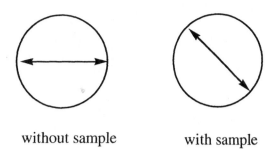

without sample   with sample

You had to rotate the lens 45° clockwise to eliminate the light. But what if you had turned it counterclockwise by 135°? The second polarizing lens would still be perpendicular to the plane of the light, and you would still see darkness. You could have done either. Has the light been rotated clockwise or counterclockwise? One or the other has occurred, *but you can't tell which*! Rotating the lens in either direction eliminates the light, so how can you know whether the light rotated +45° or –135°? For that matter, how do you know the light was not rotated by +225°? Or by –315°? Or any other value corresponding to 180° more or less than these? The answer is, you can't: not with this single measurement. There is simply no way of knowing what the light did, because every 180° of rotation of the lens brings you back to the same observation, namely, darkness.

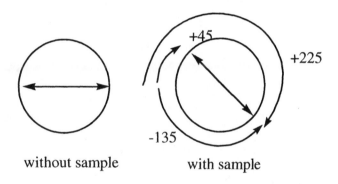

without sample   with sample

Because the instrument will give you the same reading with every 180° of rotation, it only needs to have 180° worth of numbers on it. Most polarimeters are labeled between +90° and –90°. They leave it up to you to recognize that a reading of +45° might not really represent a rotation of +45°.

A single measurement leaves you with many different possible values for α, the observed rotation, and you cannot calculate a specific rotation until you know what the true observed rotation is. Indeed, with a single measurement you cannot even tell whether the light rotated clockwise or counterclockwise. So what to do?

The answer is simple, if you think about it. The amount of rotation you *observe*, α, depends on the concentration of the sample. If you make up a different sample, using the same substance in, say, slightly higher concentration, the light should rotate a little more. The amount more should correspond exactly to the amount by which you increased the concentration. Let's imagine that the first measurement we made, described above, was for a concentration of 1.5 g in 10 mL of solvent, or 0.15 g/mL. Let's now make up a new sample using 2.0 g in 10 mL of solvent, or a concentration of 0.20 g/mL. This new concentration is 1.33 times as large as the first (0.20/0.15), so the rotation should also be 1.33 times the first. Measure it! Here is what you see:

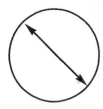 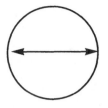

First sample   Second sample

Based on only the first measurement, we concluded that the sample might have rotated the light +45°, or + 225°, or –135°, or –315°, or many other amounts. Perhaps it would be useful to make a table of possible values. The second measurement resulted in the second lens needing to be horizontal (exactly where it started from, by coincidence!). To get there it might have rotated not at all (0°), or 180°, or 360°, or –180°, or –360°, etc. Let's make a table of these possibilities too. Both tables consist of a single number read from the instrument, and then a series of other numbers that are increments of 180° more or less relative to the previous one. Both lists could in principle go on forever in both directions, but for practical reasons we can stop after the first several entries.

First Measurement	Second Measurement
.	.
.	.
.	.
+405	+540
+225	+360
+45	+180
–135	0
–315	–180
–495	–360
.	–540
.	.
.	.
	.

Recall that the second sample was more concentrated than the first by a factor of 1.33. So the second reading should also be larger than the first by the same factor of 1.33 (assuming we used the same sample cell, and therefore the length was the same). If the first reading was really +45°, then the second reading should have been +60°. It wasn't. Therefore, the first reading was *not* +45°, *even though the instrument said it was!!!!* In saying that the first reading is +45°, what the instrument is *really* saying is that it is "+45° plus or minus some multiple of 180°. I can't tell; you figure it out!"

So we can eliminate +45° from our list of possibilities. What about –135°? If the first reading was really –135°, the second one should be 1.33 times that, or –180°. Sure enough, that *is* one of the possibilities shown in the second column. In other words, if our substance has rotated the light by –135° at the first concentration and by –180° at the second concentration, we should observe exactly what we did observe. So –135° is a possibility for the rotation at the first concentration.

Are we done? We have an answer! But before we settle on that, perhaps we should check to make sure it is the *only* possible answer. After all, we also had answers after one measurement; we just did not know which one was correct. We check the rest of the possibilities just the way we did the first two: pick a reading in column 1, multiply by 1.33, and see if the answer appears in column 2. If you do this, you will discover that there is indeed a second possible answer. That is, *another* number from column 1 has a mate in column 2, such that if these were the *real* readings, we would observe exactly what we did observe. Find it!

So now we still have a problem. There are two possibilities left, and we still don't know which one is correct. But we are better off than before; after one reading we had listed six possibilities (there are actually many more), and now we have eliminated several of them, and there are only two left (there would be more if the original list had been longer). What now? No problem! We eliminated some of the possibilities from the first list by repeating the experiment at a different concentration. We should be able to delete more possibilities by doing so again. See for yourself that by making a third concentration (say, 2.25 g in 10 mL) we could distinguish our two remaining possibilities. Try to do this without looking at the answer that follows. Do it now.

If you use 2.25 g /10 mL or 0.225 g/mL, this is 1.5 times the concentration of the original solution. So the reading in this case should be 1.5 times the original reading. From the first two readings, we have two possibilities left: $-135°$ or $+405°$. If the original reading was really $-135°$, then the third reading (this new one) should be 1.5 times that, or $-203°$. Now of course if the light is rotated by $-203°$ the instrument won't know that—as I told you before, it reports the smallest equivalent number. The instrument cannot distinguish this reading from any other reading that differs from this one by a multiple of 180°. So $-203°$, to the instrument, is indistinguishable from $-23°$, or $+157°$, or many others. The reading you will get from the instrument will be the one value from this set that falls between $+90$ and -90, namely, $-23°$. So if the $-135°$ possibility were correct, the third reading would be $-23°$. Now prove for yourself that if the $+405°$ possibility were correct, the third reading would be $+68°$. If we actually make the third measurement, then, we will get one of these readings ($-23°$ or $+68°$), and we will know which of the original readings was the right one. From that we could calculate a specific rotation.

Test Yourself 1

A compound X is optically active. You take 3.0 g of X and dissolve it in 10 mL of ethanol (a solvent, nothing more) and obtain a polarimeter reading of $-74°$. If you start with 4.0 g of X in 10 mL of ethanol, you obtain a reading of $+81°$. If you start with 4.5 g of X in 10 mL, you get $+69°$. What is the specific rotation of X?

WHO CARES? YOU DO!!

The material we have covered on the last several pages is very unusual for an introductory organic text. Indeed, I have not been able to find any other introductory book that acknowledges that a clockwise rotation is indistinguishable from a counterclockwise rotation with a single measurement. Most books stop at the discussion of specific rotation. So why have I bothered you with this? I think it is fair to say that there are very few students taking this course out of general interest. Nearly all of you expect to be scientists of some kind. Scientists deal with data, and more and more frequently these days data is spit out from instruments. Most modern instruments (like our cars, as well) have become highly computerized, so that an answer appears on a digital screen, and the experimenter writes it down. The instrument is treated as a magic box that knows the answer, and the answer is treated as a gift from God. It is important that scientists understand, to the extent possible, what is going on inside their black boxes, and it is especially important that scientists (and laypeople as well) recognize what instruments can and cannot do. Since this material is mathematically fairly simple, involving nothing more than addition and some ratios, I thought it would be useful to go through the process of analyzing what information is (and is not) present in a reading from a polarimeter. You should take similar care with data presented to you from other sources as well, whether it be a blood test or a chemical analysis. All scientific data should be treated as suspect: data can be ambiguous, as in the above case, or misleading, or off by a little, or dead wrong. People who recognize and accept this will have a huge advantage over those who don't, in both science and life.

Using a polarimeter, we can distinguish between the two enantiomers of a chiral substance. One will rotate the light clockwise: we call this the (+) or *d* (for dextrorotatory) enantiomer. The other will rotate the light counterclockwise and we call that the (–) or *l* (for levorotatory) enantiomer. Notice that we can't know which is which until we do the experiment. Even if we drew accurate pictures of each, and we knew that bottle A contained material with a particular structure and bottle B contained its mirror image, we still would not know which was (+) and which was (–) until we ran the experiment. Although some progress is being made using fancy calculations, in general it remains true that *there is no simple correlation between structure and optical rotation.*

One thing is certain, however. If you know what direction one enantiomer rotates the light, you can be sure the other one rotates it by the same amount in the opposite direction. Further, an equal mixture of the two (a "racemic" mixture, remember?) will not rotate light at all. As often as the light hits a molecule that rotates it clockwise, it will hit another molecule that rotates it back again. Light will emerge from the other end of the tube exactly as it came in. A racemic mixture, then, is *not* optically active, even though each molecule in it is chiral. Optical activity is an observable bulk property, whereas chirality is a structural property. Thus, "optically active" and "chiral" should not be used as synonyms, even though many people make this mistake.

Depicting and Naming Chiral Molecules

Using optical activity as a technique, we can now distinguish a sample of a chiral substance from a sample of its enantiomer, assuming for the time being that we are somehow able to create separate bottles of the two. One bottle we can label (+), or *d*, and the other we would label (–), or *l*. But we still have a problem. We can also draw pictures of the two enantiomers, but we don't know which bottle corresponds to which enantiomer. And even if we did, we do not yet have a way of referring to the pictures without drawing them. In other words, we need a way to *name* the two bottles differently (since, after all, they are different substances) in such a way that each name leads to the correct *structure*. I will leave until later the problem of determining which structure applies to which substance. For now let's assume that we know. We still need to agree on ways to show these things, and especially on ways to name the structures.

Here are several different pictures of 1-bromo-1-chloroethane. Can you figure out which ones are the same, and which are mirror images? This is a very tricky business, and I confess I would have some trouble with this question if I didn't know some tricks.

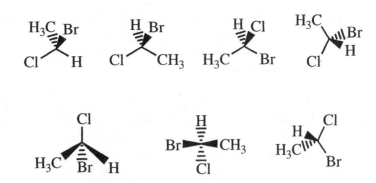

Try to make some assignments. Then build models and figure out whether you were right. Build a model of the first one and label it in some way to keep track of it. Then, one at a time, build the others and determine whether they are identical to or the mirror image of the first one. It really will be worth your time to stop and do this now!

What did you find out? I found four pictures of one enantiomer and three of the other. Did you? We need to be able to visualize molecules like these in three dimensions and to make these kinds of assignments without always carrying models around with us. To be quite honest, I did not answer this question by mentally picking up each molecule and rotating it to see if it was the same as one of the others (which is presumably what you did with your models). I *can* do that, if forced to, but at this point you probably cannot, and the reason I can is that I have already worked for years with models and pictures. More important, I have learned how to name the things, and any two things with the same (complete) name are identical. So I solved this particular problem by naming each picture and then seeing which ones were the same based on their names. Still, even doing it this way requires mental gymnastics you are not yet used to.

Think about this: how would you describe your left hand to someone with no sense of left and right, characterizing it in such a way as to distinguish it from your right hand? What can you say about one of your hands that is not equally true of the other? This is no trivial question. Go on, ponder it for a minute.

One possibility you might have thought of depends on the fact that to get from the thumb through the other fingers to the little finger on the left hand requires moving clockwise, but counterclockwise with the right hand. Of course, before you can use this fact in distinguishing the two hands, you must also specify that the palm must be turned towardsyou. In other words, you can successfully distinguish the two by telling your partner 1) how to hold the object, and 2) that in moving along a specified path you would need to go clockwise or counterclockwise depending on which hand it is. (This is cheating a little, because someone who does not know right from left probably also cannot distinguish clockwise from counterclockwise, but we will ignore that point!) We will apply exactly the same strategy to molecules.

Cahn–Ingold–Prelog Nomenclature

The following approach to naming is called the Cahn–Ingold–Prelog system, after the chemists who invented it. We have actually seen part of it before, back in Chapter 10, when we were designating double bonds as *E* and *Z*, remember? The trick there involved establishing a priority order for the substituents attached to a particular carbon. We need to do the same thing here. Each stereocenter has four things attached to it. We need to establish a priority order for them.

The rules are the same ones we used before: higher atomic numbers take priority over lower ones, and if you can't decide, keep going until you can. It would be worth looking back at Chapter 10 and reviewing how the rules work. There is no point in my repeating the whole song and dance here.

Back in Chapter 10, we had to distinguish only two substituents at a time. For each end of a double bond, we needed to designate a substituent of high priority and one of low priority. Now we are dealing with a stereocenter characterized by four substituents, so we need to establish a high, medium, low, and bottom priority. This is not always easy, but by following the rules, you can do it. Let's start with an easy case, the example on the previous page. The stereocenter has directly attached to it: a hydrogen, a carbon, a chlorine, and a bromine. Clearly, the bromine ranks high, the chlorine medium, the carbon low, and the hydrogen bottom. (Many books use 1, 2, 3, and 4 here, but I always get confused as to which is high!)

Now for the naming rule. Remember the steps: hold the object as instructed and then move in a specified path. The holding instructions are simple: hold the molecule so that the bottom priority substituent is directly *away* from you. Note carefully, this is probably *not* how the molecule has been drawn on the page. You have three options: you can build a model based on the picture and then pick it up and hold it properly; you can adjust the picture until it is the way you need it to be; or (here's where the mental gymnastics come in) you can imagine yourself inside the picture, standing in the right position. Although to a beginner the latter is by far the hardest of the three, it is in fact the way most real chemists view these pictures, and you should work toward developing that ability. Once the molecule is held (or imagined) properly, you then need to move along some specified path and determine whether you are going clockwise or counterclockwise. The specified path is also simple: from high to medium to low. If this route is clockwise, we call the stereocenter *R* (from the Latin *rectus,* meaning right). If it's counterclockwise, we call it *S* (from the Latin *sinister,* meaning left).

Take the first structure I drew:

If you build this with your model kit, and then hold it so that the hydrogen (bottom priority) is directly away from you, the hydrogen will be invisible behind the carbon, and what you will see is shown above on the right. If you now trace the path from high to medium to low (namely, from Br to Cl to CH$_3$), it goes clockwise, so this stereocenter should be labeled *R*. This substance is (*R*)-1-bromo-1-chloroethane.

Can you do this without building a model? Here's the way I do it: I imagine myself in the picture, where the eye is below. What the eye sees is precisely what is in the right-hand picture above: bromine on the top right, methyl on the top left, and chlorine below. *In my head,* I then proceed from high to medium to low and recognize the route as clockwise or *R*. It may be a while before you can do this, but keep trying.

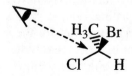

There is one secret alternative to this. Hopefully, you will agree that my method would be easiest to use if the hydrogen were in the position occupied by the methyl in the above picture: *already* going away from you. So simply put it there! The trick is this (and you should convince yourself that it works by playing with your models): every time you switch two substituents at a stereocenter, *any* two substituents, you invert that center's configuration, meaning you change it from *R* to *S* or *vice versa*. In the above picture, if you want the hydrogen where the methyl is, just switch them:

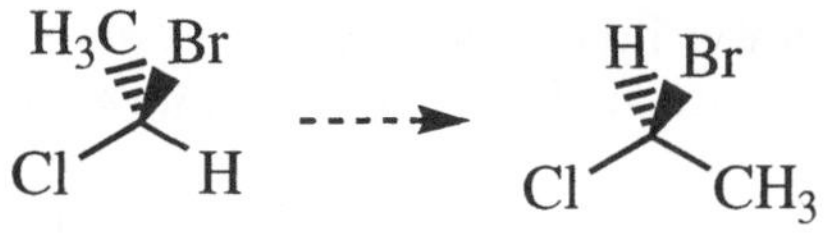

Now you have a choice. One possibility is to name the thing you just drew, but recognize that what you now have is the *opposite* of the structure you were asked to name. The new one is *S*, so the original must have been *R*. Or, you can switch another pair of substituents, recognizing that the third thing you will draw is guaranteed to be the same as the first, since you made *two* switches. So name your third substance.

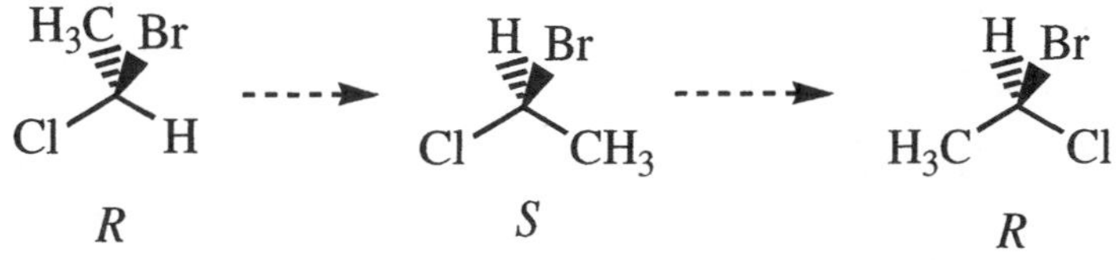

Now label the other six pictures in the original example. Try to do it without your models, and then check yourself with the models. You should end up with three *R*'s and four *S*'s.

Here's a harder example to practice on. Label all the stereocenters in the following molecule. Try it yourself before I show you how.

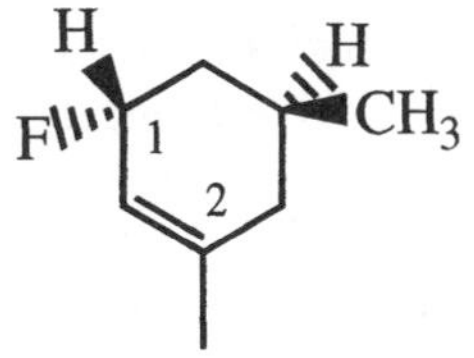

First you need to *find* the stereocenters. Hopefully you had no trouble identifying two stereocenters in this molecule. Start with the one bearing the methyl group. We have to prioritize the four things attached. Clearly the hydrogen is bottom, but the other three atoms are all carbons. No problem; just look further. The methyl group has nothing but H's on its carbon. The other two have two H's and one C. So the methyl group has low priority. The other two groups are still tied. The branch going to the left continues with yet another carbon; the branch going down does the same. Are you with me? Now there is a difference. On the left branch, what is attached to this carbon (the one labeled "1")? One H, one C, one F. And on the lower branch (labeled "2")? Woops, there is a double bond! How do we deal with this? The rules say you count the carbon of the double bond twice. So carbon 2 is attached to three C's, besides the carbon we came from. Two of the three C's I am referring to are actually the same one counted twice. OK? So now we have to decide which of the carbons (1 or 2) has higher priority? The decision is based on the best direction you can go from here, not the sum of what is attached. Even though the three carbons, taken together, on carbon 2 add up to higher atomic number than the H, F, and C on carbon 1, carbon 1 wins this game, because it has the single highest atomic number substituent (F). So the branch in direction 1 is high, the branch in direction 2 is medium, the methyl is low, and the H is bottom. Hold the molecule so that bottom is away from you (it already is) and count high to medium to low. This is counterclockwise, so the stereocenter with the methyl is labeled *S*.

Now we go to the stereocenter with the F. Clearly the H is bottom and the F is high. The other two are both carbons so they are tied for the time being. Keep going! One of these has 2 H's and a C; the other has 2 C's (the double bond counted twice) and an H. The H cancels the H, and the C cancels the C, but the lower branch has a C where the upper branch has an H. The lower branch wins. So this stereocenter is *R*. Did you get that? Don't forget, you have to turn the molecule around before you count, so that the H is away from you.

The full name for this molecule, then, is 3-(*R*)-5-(*S*)-3-fluoro-1,5-dimethylcyclohexene.

Test Yourself 2

For a real challenge try cholesterol:

I get four *R*'s and four *S*'s.

As you know, professors love to turn things around on you. If I have shown you how to look at a structure and give it a name, your professor will obviously try to invert this: given a name, what is the structure? I should not have to tell you this, nor should I have to create a problem for you to solve along those lines—this is the sort of thing you should be doing automatically as you read and study. But let's work through one. Draw (*S*)-2-chlorobutane.

Start by figuring out what this thing is:

$$H_3C-\underset{\underset{H}{|}}{\overset{\overset{Cl}{|}}{C}}-CH_2\text{-}CH_3$$

Obviously, the stereocenter we need to draw correctly is carbon-2. It has four different groups, and we can order them easily. Lowest is hydrogen, and the others go chlorine to ethyl to methyl, from high to low. So let's draw a picture and put the things in the right places. Obviously, we want the hydrogen back, so put it there, as in the middle picture, and then add the other three in counterclockwise direction.

Alternatively (and again I should not have to say this), you could simply draw one and name it. If it is the wrong one, switch two groups, and you'll have the right one.

ABSOLUTE STEREOCHEMISTRY

We still have a problem. Let's take a particular example, glyceraldehyde:

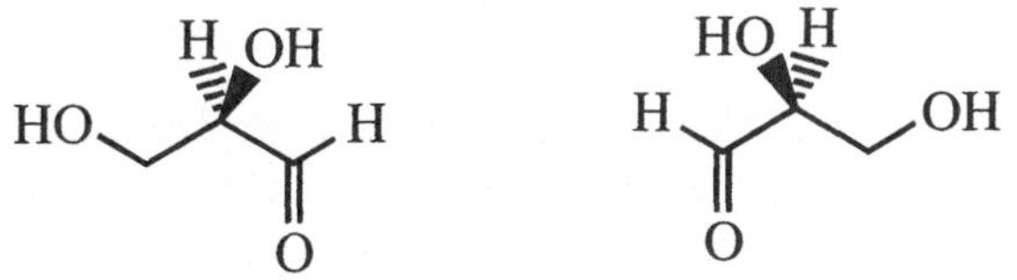

There are two versions of this, both shown above. One of them is *R*, one is *S*. (Which is which?) Also, if we had samples of these, separate from each other, we could determine which is (+) and which is (–) by putting them in a polarimeter. But we have two bottles [one labeled (+) and one labeled (–)] and two pictures (one labeled *R* and one labeled *S*). Which picture goes with which bottle? *We don't know!!!* There is absolutely no way of knowing this unless we take a detailed photograph of the molecule labeled (+) and figure out which picture it corresponds to. Then we would know what is called its **absolute stereochemistry** (or absolute configuration). Chemists did not figure out how to do this until the 1950s, using X-ray crystallography, but they recognized that there were two forms and had available to them separate samples of each over 50 years earlier. Since there was no way to tell which sample had which rotation, but needing to put some label on each bottle, they guessed! It was this very molecule that they guessed with. ("They" was actually Emil Fischer, a great chemist at the end of the 19th century.) The sample with (+) rotation was arbitrarily assigned the left structure above, what we now call the *R* configuration, while the (–) isomer was assigned the *S* structure. Many, many other compounds were similarly labeled based on direct (or indirect) comparison with these. There was of course a 50:50 chance that all the assignments were wrong, so that when the structures were finally determined for real, all the structures might have to be reversed. By great good luck, it turns out that they guessed right. However, we were left with one unfortunate legacy. The labels *R* and *S* had not been invented at the time. The *R* structure for glyceraldehyde was originally designated by the label D and the *S* structure by the label L. These capital letters refer to the *absolute* stereochemistry of a molecule, while the small letters *d* and *l* refer to a direction of rotation. The small and capital letters were chosen to correspond for one particular molecule, glyceraldehyde, but there is no intrinsic connection between them. You *cannot* expect D compounds always to have *d* rotations.

The D and L convention today is applied almost exclusively to biological molecules, in particular to carbohydrates and amino acids, so we will put this topic on the shelf until we need it. For our current purposes you only need to know about the *R*, *S* nomenclature. And what you need to know is that even though you can label a picture as *R* or *S* by looking at it, you cannot label the compound as (+) or (–) by inspection. This is an experimental measurement, and you have no way of knowing ahead of time whether the *R* form or the *S* form will turn out to be the one that is (+). Of course, once you determine which is (+), it is obvious that the other one is (–).

COMPOUNDS WITH MULTIPLE STEREOCENTERS

So far we have dealt primarily with molecules that contain only one stereocenter. But of course many molecules have more than one. Consider the following pictures:

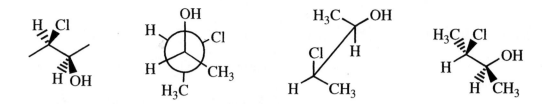

How many of these representations are identical? Which depict mirror images of each other? Are there any that have a relationship not yet covered? Are there any possible stereoisomers *not* shown above? Take some time right now to work on these questions.

Diastereomers

As in the previous examples, I take the easy way out: instead of trying to manipulate the pictures in my head and see if they are the same or mirror images or something else, or, alternatively, building them with my model kit, I just name them. I know that whenever I hold a molecule up to a mirror, every center labeled *R* in the original becomes *S* in the mirror, and *vice versa*. Any molecule that comes out labeled the same as another must *be* the same (assuming I named it correctly). Any molecule in which every label is the opposite of every label in some other molecule is the enantiomer of that other one. Any molecule in which some labels are the same and some are different is a new kind of beast. We have a name for this, too: **diastereomers** ("dye-ah-STER-ee-oh-mers"). These are still stereoisomers, but not enantiomers.

Here's what I get for the four molecules above (from left to right), where the first label applies to the carbon with the chlorine and the second to the carbon with the OH: *RR, SR, RR, SS*. Clearly the first and third pictures must be identical, and the first and fourth (and therefore the third and fourth) are enantiomers. The second is a diastereomer of all the others.

Further, we can easily see by this method (but much less easily by building or visualizing) that there must still be one more version of this. Clearly, the carbon with the chlorine can be either *R* or *S*. For each of these, the other carbon could be either *R* or *S*. So there should be 2 x 2 = 4 different isomers. The pictures above include *RR, SR,* and *SS*, but not *RS* (the enantiomer of #2).

Indeed, we can generalize from this example to much larger ones. *Each* stereocenter in a molecule can be either *R* or *S*, so the total number of stereoisomers possible for any random molecule should be 2 x 2 x 2 x 2, etc., for however many stereocenters there are. If there are n stereocenters, there are at most 2^n possible stereoisomers. Consider cholesterol, with its eight stereocenters. There are 2^8 or 256 different ways to put this together, all differing only in stereochemistry. One (and only one) of these is cholesterol; another is the enantiomer of cholesterol. The remarkable thing is that your body is so exquisitely tuned to these things that it can tell the difference among all of them and will make and process only the right one of all the possibilities.

Unfortunately, there are some special cases we have to consider where this 2^n rule breaks down. Occasionally, you will encounter a molecule in which two of the possible isomers turn out not to be different after all. For example, consider the possible isomers of 2,3-dichlorobutane. Here they are: *RR, RS, SR, SS*. I have drawn these for you.

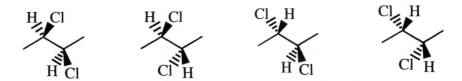

But focus on the middle two: if you take either of these and flip it end for end, you recognize that it is exactly the same as the other!

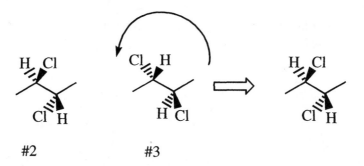

#2   #3

There aren't four of these, there are only three! We built all the possibilities, and yet two of them turned out to be the same. Why did this happen? Because the molecule is the same on both ends. It cannot tell *RS* from *SR*. Put another, more general way, this *RS* compound contains a plane of symmetry. It does?? Where? Well you can't see it because you have to rotate the molecule internally to find it, but the molecule *can* rotate internally without becoming a different molecule. If there is any allowed conformation of the molecule that has a plane of symmetry, then the molecule has one. And of course, you already know that a molecule with a plane of symmetry is achiral and has no enantiomer (it has a mirror image, but no enantiomer, because the mirror image is the same as the original).

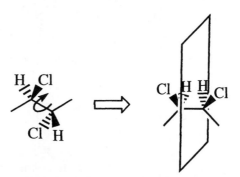

So here is a strange molecule: it has two stereocenters, but the molecule is not chiral! I told you earlier that there were exceptions to the statement that every molecule with a stereocenter is chiral: here is one of them. It is still true that every molecule with *one* stereocenter is chiral. But with more than one, and if the molecule has an internal plane of symmetry, it is possible to build molecules that are not chiral even though they have stereocenters. These molecules turn out to be important enough that we give them a special name: a ***meso* compound** is a molecule containing stereocenters, but is nevertheless achiral.

Test Yourself 3

At this point you should be able to answer questions like the following.

For the following molecules:

a. Label all the stereocenters (as *R* or *S*).
b. Indicate any relationships that exist among the structures (enantiomers, diastereomers, identical, etc.).
c. Indicate the optical rotation of each, if you can tell [(+), 0, or (−)].
d. Label any *meso* compounds.

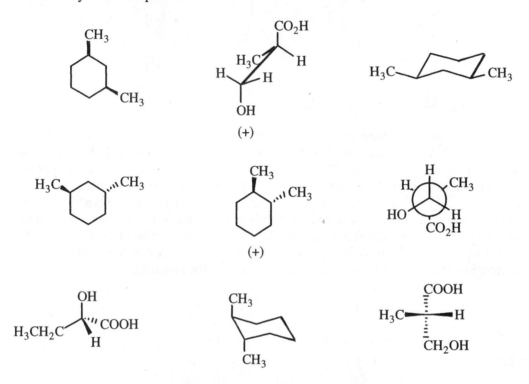

GENERATING PURE ENANTIOMERS

Up until now, I have glossed over the question of how one goes about obtaining one enantiomer of a substance separate from the other. Except in very rare circumstances, every method relies on the fact that Mother Nature has provided us with a treasure chest of compounds that are already enantiomerically pure. Virtually any chiral substance you find in nature is there as one and only one enantiomer. We use these preseparated compounds to separate the enantiomers of synthetic racemic mixtures.

Resolution

Spontaneous Resolution of Crystals

For historical reasons, it is worth mentioning one exception. Chirality was discovered by Louis Pasteur in 1848, who was examining crystals of sodium ammonium tartrate under a microscope. He noticed that the crystals themselves (imagine doing this with grains of sugar) were not all the same, and in fact were of two types that were mirror images of each other. At this time the concept that *molecules* could involve mirror images had not been developed, because nobody knew the shapes of *any* molecules. Nevertheless, Pasteur used tweezers and painstakingly separated his crystals into two piles, each pile containing only one mirror-image form. He then did experiments on the two piles separately and discovered that the two piles had identical chemical properties in every respect, except that solutions of one of them rotated plane polarized light in one direction, and the other rotated the light in the opposite direction. Further, he discovered that if he mixed the crystals back together, solutions of the substance no longer rotated the polarized light. Pasteur concluded that the mirror image crystals must reflect some phenomenon at the *molecular* level, because the distinguishing property was still there when the crystals themselves had dissolved in a solvent. He did not know what the cause was, but this knowledge provided one of the key puzzle pieces that eventually led chemists in the 1870s to propose that carbon is tetrahedral, because if it is, isomeric mirror-image molecules become possible.

The process of taking a racemic mixture and separating it into its two enantiomers is called **resolution**. Pasteur was able to do his resolution by hand, because each molecule spontaneously sought out its mates and crystallized into different crystals. This is extremely unusual. Most of the time racemic mixtures crystallize (if they form crystals at all) into racemic crystals, and there are only a few compounds known that can be separated the way Pasteur did it.

Biological Discrimination

So if the molecules do *not* separate themselves spontaneously, how do we separate them? In every case, we take advantage of already-existing chiral compounds available from nature. The easiest approach is to take advantage of the machinery supplied by nature. As I have mentioned before, organisms can easily discriminate between enantiomers, because their entire environment is chiral. If you have an intestinal tract which has lots of right hands extending into it, and you feed this thing pairs of gloves, the right gloves will get trapped and digested by the intestine but the left gloves will get past. Similarly, if you feed a bacterium a pair of enantiomers (a racemic mixture), it is very likely that the organism will have a preference for one form over the other, possibly a strong preference. With luck, the bacterium will digest and destroy one of your enantiomers and leave the other unharmed. With outstanding luck, the one remaining will be the one you want! This method suffers from the fact that half the material you make (the wrong enantiomer) disappears during the resolution. Nevertheless, if you can find an organism that will do this work for you, it is an excellent method of resolution, because it is relatively easy to carry out. There is a lot of trial and error involved in finding such an organism, however, and the approach is not very common.

Chiral Chromatography

A second approach is similar in principle. It too relies on preexisting chiral compounds to discriminate between the enantiomers of your racemic mixture. But instead of digesting and destroying one isomer, this technique merely delays its passage through the "intestine." In this case we build ourselves a column of some inert material that has chiral molecules (read, right hands) glued onto it; the racemic mixture is percolated through this column, and the enantiomer that fits comfortably with the column material hangs around a bit, while the enantiomer that does not fit speeds its way on through. Eventually both come out, but one comes out before the other. This is called chiral chromatography, and it is an excellent way to separate a mixture into component enantiomers. However, this technique has its limitations also, because a chiral column is quite expensive, and there is a limit to how much material you can put through it at any one time. Thus, this technique is used most often for analysis rather than for bulk separation. In other words, it is a great way to find out what you have in your mixture (is it 50:50 or some other mixture of the two enantiomers?) but less useful if you want significant quantities of the two isomers separated from each other.

Resolution by Diastereomeric Salts

The most common method of separating enantiomers on a preparative scale is to convert them into stereoisomers that are *not* enantiomers; that is, diastereomers. Since diastereomers are not mirror images of each other, they do not have identical properties, and usually they can be separated from each other by some means. The trick here, then, is to find a method of converting enantiomers into diastereomers, perform the separation, and then reverse the original reaction to get back the enantiomers. Imagine, for example, that you had a box full of mannequin hands that you wanted to separate into a box of right hands and a box of left hands. For you this would be no problem, you would just look at them. But the worker to whom you have assigned this task cannot tell right from left. No problem: give him a box full of left gloves. He puts a glove on each hand. The hand–glove combinations that fit well he puts into one box, and the ones that don't he puts into a different box. Then he removes all the gloves in the separate boxes, and he has accomplished a resolution. He has separated right hands from left, even though he could not tell them apart, by combining them with something already chiral (left gloves, no right ones) so that the hand–glove combinations were *not* mirror images, and therefore he *could* tell them apart. Finally, he reversed the combining step to reveal the hands, now separated.

Chemically this is done in exactly the same way. You start with a racemic mixture, call it И and N, and combine it with a compound that is chiral and present as a single enantiomer, call it B. Where do you get this? Most likely from nature. You choose a reaction that is very efficient. You get two products: NB and ИB. These are *not* enantiomers, they are diastereomers. Since they are diastereomers you manage to separate them somehow by ordinary chemical means (perhaps by solubility differences). Put one in one bottle and the other in a different bottle. Then you undo the original reaction, removing the B. *Voila!* N and И are now in separate bottles!

Well, that is nice in theory, but how do we really pull this off? The key is to find a reaction that goes extremely well both forward and backward. The usual solution is to use acid–base chemistry. Imagine you had a racemic substance to resolve, and that it was a carboxylic acid, such as the one shown below. I have shown both enantiomers.

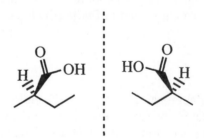

What would happen if we treated this with a base? You may recall from Chapter 8 that the proton of a carboxylic acid is relatively acidic (that's why it's called an acid) and can easily be removed. You may further recall that an amine is sufficiently basic to remove such a proton. If the amine we use happens to be part of a chiral molecule, such as quinine, and if we use it from a natural source so that we have only one enantiomer of it, we will get salts that are no longer mirror images. Further, we can expect this reaction to be extremely easy to do and presume that it will go pretty much to completion.

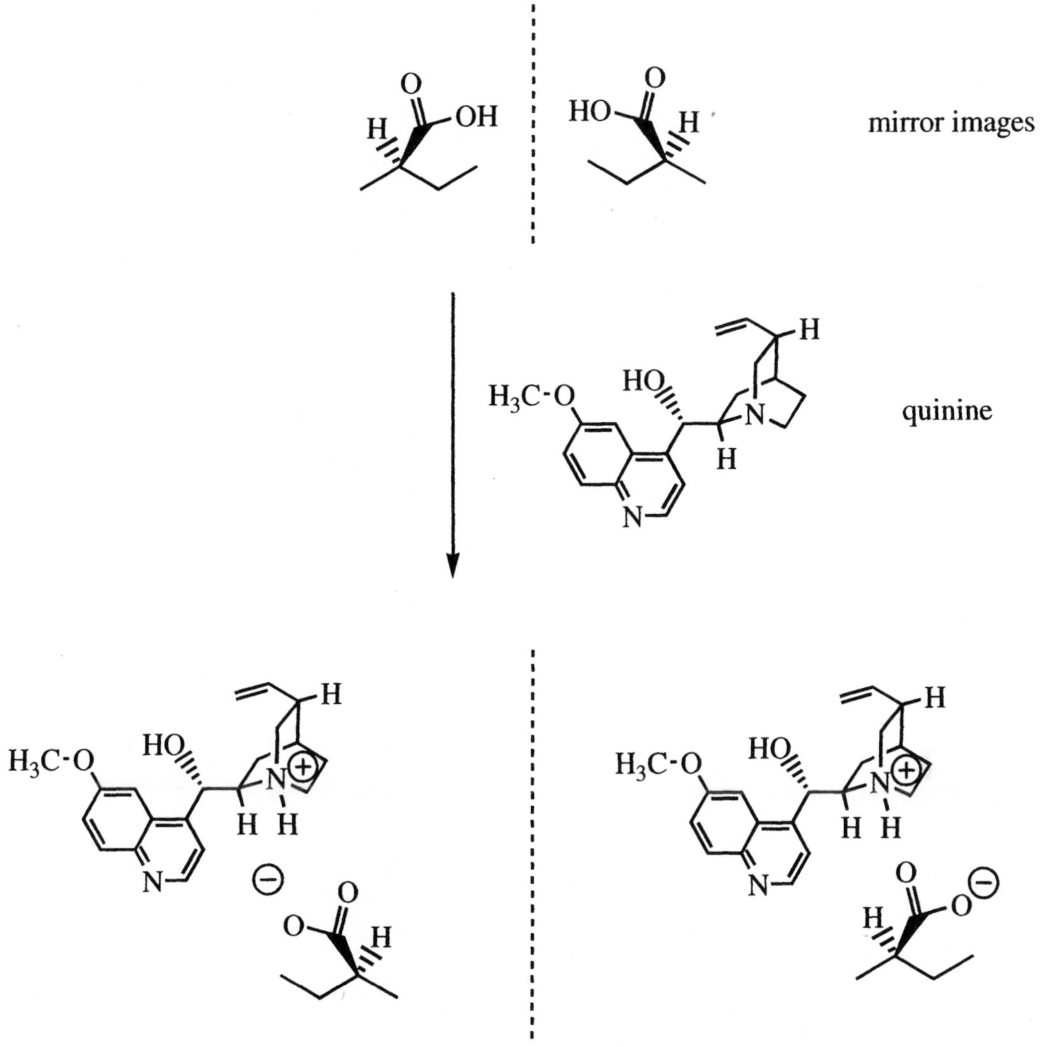

mirror images

quinine

no longer mirror images: diastereomeric salts

When these salts crystallize, they often (but not always) are sufficiently discriminating that one pair crystallizes in a different form from the other, and (of particular use to us) one set of crystals often forms *more easily* than the other. Thus, when crystals begin to appear, they are likely all to be of one diastereomeric salt and not the other. If you adjust your conditions just right, it is often possible to collect crystals that represent just the left-hand (or right-hand) side of the dotted line in the picture. If you then dissolve these crystals in water (they ought to dissolve, since everything is charged), and add acid (say, HCl), you will put the proton back on your carboxylic acid. This is now no longer charged and should separate from the water solution, so you can skim it off. Thus, you have been able to collect one of your enantiomers in pure form. What follows is a schematic representation of this process, where the letters *d* and *l* refer to the optical rotation of the substances, and the + and – signs in circles refer to charges. Which salt crystallizes first is a random matter—you do not know in advance which will come out for you.

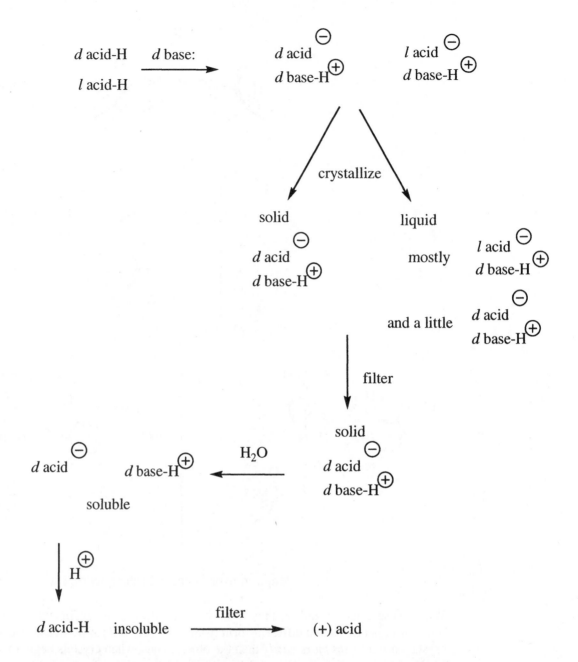

Of course, no crystallization is perfect. Although the crystals you collect have a decent chance of being pretty well confined to one enantiomer or the other, there is very little chance that *all* of that enantiomer crystallized. In other words, the material left over in the liquid is composed mostly of the enantiomer that did *not* crystallize, but there is some of the other enantiomer present as well. Shortly we will learn how to analyze mixtures like this to determine their composition.

Chiral Synthesis

All methods of resolution suffer from one enormous drawback: you expend tremendous effort to synthesize something, only to throw half of it away. It is occasionally possible to isolate the enantiomer you want, and then take the enantiomer you don't want and racemize it (make it a 50:50 mixture again) and do another resolution, but this can get tedious really fast. By far the best solution to the problem of needing a single enantiomer is to *make* only that enantiomer. That is, after all, the way Mother Nature does it. But Mother Nature has an advantage over us: her lab is chiral. All reactions in nature are carried out in a chiral environment, namely, on the surface of some enzyme. How can we, in a real lab, do the same thing? By using the same trick: make the reactions happen in a chiral environment. Chemists have been hard at work on this problem for many years, and there are now several reactions that can be carried out reliably and predictably with nearly perfect "enantioselectivity." In other words, for certain reactions you can start with an achiral starting material, run the reaction, and obtain only the desired enantiomer in high yield. This way you do not need to throw away any of your material.

Enzymes are the world's experts at this, and they do everything in a chiral environment. To accomplish the same thing in our lab, we need to create our own chiral environment. The best procedures so far have involved reagents attached in some way to preexisting chiral molecules. That way the reagent drags its own chiral environment around with it. When the reagent approaches the substrate, it has a preferred direction of approach that makes one enantiomer form more easily than the other. The best chiral reagents are catalysts, so that only tiny amounts are needed; as soon as a molecule of catalyst is used up, a reagent comes along to regenerate it. Students who want or need to know more about chiral synthesis can take additional chemistry courses; for our purposes, it is important only to realize that this is being done and to understand the principle behind it.

ANALYZING MIXTURES OF ENANTIOMERS: ENANTIOMERIC EXCESS

Whether by chiral synthesis or by resolution, one often acquires a mixture that contains both enantiomers of a substance, but not in equal amounts. The sample is not a single enantiomer, but it is not a racemic mixture either. We need a way to determine what exactly we *do* have in such a mixture, and a vocabulary with which to express it.

Chiral Chromatography

I mentioned above that enantiomers can be separated by chiral chromatography, and that although this is not especially useful for bulk separation it is quite useful for analysis. Analysis is precisely what we are looking for in this situation: given some mixture, what is it composed of? If we have a chiral column and some method of detecting what is coming off the column, we should see two peaks. The two substances are enantiomers, so they are identical in every respect except optical rotation and interaction with the column. Therefore, whatever method we use for detection will be equally sensitive to both enantiomers, and the sizes of the two peaks observed will be a direct measure of the amounts of the enantiomers. So a chart record from a chiral column that looks like the following, with peak sizes in the ratio of 2:1, could be interpreted as evidence that there is twice as much of one enantiomer as the other. The actual makeup of the solution would be $1/(2+1) = 33.3\%$ of one enantiomer and $2/(2+1) = 66.7\%$ of the other. Of course, a good integrating recorder could determine these numbers more precisely. Solutions like these are traditionally described by how much more of one enantiomer is present than the other, on a percentage basis. The term we use is **enantiomeric excess**, or "*ee*").

$$ee = \% \text{ major isomer} - \% \text{ minor isomer}$$

This particular mixture has an enantiomeric excess of 33.3%, which is obtained by

$$ee = 66.7\% - 33.3\%$$

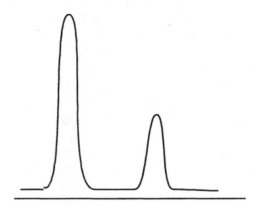

Optical Rotation

A chiral column is an excellent tool for determining the composition of an unknown mixture of enantiomers, but not everyone has ready access to one (they are pretty expensive). A lower tech solution to the problem is to go back to our old friend the polarimeter. You already know that if you have a pure enantiomer of some substance, it will have a characteristic optical rotation in some direction, which you can measure. You also know that if you have a 50:50 mixture of the two enantiomers, there will be no rotation at all, because the clockwise rotation of one cancels the counterclockwise rotation of the other. It should not surprise you that a solution that is not pure, but also not racemic has an optical rotation somewhere in between, and, indeed, a predictable and calculable rotation.

Consider the above mixture, with 33% of one isomer and 67% of the other. For the sake of argument, let's imagine that the 67% represents the (+) isomer, the one that rotates the light clockwise. (We could not tell this from the chart above, unless we knew ahead of time which peak came from which enantiomer.) Also, randomly, let's imagine that the (+) isomer has a specific rotation of +56° (so obviously the (–) enantiomer has a specific rotation of –56°). What will we see with a polarimeter for this solution? There are some (–) molecules and some (+) molecules in this solution. To the extent that they match up, they will cancel each other. That is, the 33.3% of the solution that is (–) will cancel 33.3% of the solution that is (+), leaving 33.3% of the solution able to rotate light. The light will rotate, in a clockwise direction (because (+) molecules are left over), but only 33.3% as far as it would have if the solution had been pure (+) enantiomer. We therefore expect a specific rotation of 33.3% of +56°, or +19°.

In other words, the part of the solution that is affecting the light represents only the *excess* of one enantiomer over the other—the enantiomeric excess once again. The amount the light rotates compared to what one ought to find for a pure enantiomer is often called the optical purity of a sample, and it is the same thing as the enantiomeric excess. Thus, you can calculate the enantiomeric excess of any sample by measuring its specific rotation and comparing that to the specific rotation of a pure enantiomer of that substance.

$$ee = [\alpha]_{\text{sample}} / [\alpha]_{\text{pure isomer}}$$

Test Yourself 4

As the campus drug dealer, imagine you have received a shipment of cocaine, but it came in racemic form. Since you are a respectable drug dealer, you do not want to sell racemic cocaine to your customers, because you are afraid that the wrong enantiomer [the (+) one] will have bad side effects on them. (Apparently you are quite unconcerned about the bad effects of the *right* enantiomer!) Anyway, you undertake a resolution of the cocaine by the method of diastereomeric salts. Since cocaine is a base, to do this you need an acid in a single enantiomeric form, and you are fortunate enough to find on your shelf a sample of lysergic acid to use for this purpose. You mix the racemic cocaine with the chiral lysergic acid and get crystals. Upon separation of the crystals and regeneration of the cocaine, you measure a specific rotation of +35°. (This can be assumed to be pure enantiomer.) When you regenerate the cocaine from the material that did not crystallize, you find a specific rotation of –28°. What is the enantiomeric composition of this second mixture [% (+), % (–)]?

STEREOCHEMISTRY OF SUBSTITUTION REACTIONS

Now that we have a better feel for stereochemistry, we can add somewhat to our understanding of substitution reactions. You will recall that we know about two distinct kinds of nucleophilic substitution (Chapter 13).

- S_N2 reactions have two species involved in the rate-determining step: the nucleophile and the substrate. There is only one step in these reactions. The nucleophile enters and the leaving group leaves simultaneously. The reaction happens best at primary centers and worst at tertiary centers. It requires a good nucleophile, because the nucleophile is what pushes out the leaving group, but the solvent does not have to be exceptionally polar, as long as it dissolves everything.

- S_N1 reactions have only one species involved in the slow step. The leaving group leaves, and in a second, faster step, the nucleophile enters. These reactions happen best at centers that can easily support carbocations, namely, tertiary and resonance-stabilized centers, and worst at primary centers (unless they are resonance-stabilized). Further, because a cation is involved, these reactions are most common in polar solvents like water and alcohols, and frequently the solvent itself is the nucleophile; that is, the nucleophile does not have to be very strong, because it does not do any pushing.

There is one other aspect of substitution reactions that we did not consider in **Chapter 13**. *Where* does the nucleophile enter, relative to where the leaving group departs? **In other** words, which of the following occurs?

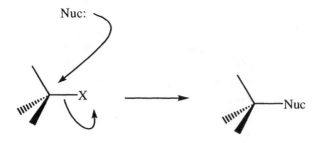

In the case of an S$_N$1 reaction, this is pretty easy to answer just by contemplating the mechanism of the reaction as we understand it. The first thing that happens is that the leaving group leaves, creating a carbocation on the carbon that used to be attached to the leaving group. Carbocations, as we know, are sp^2-hybridized, and they are flat, trigonal things. The empty p orbital with the positive charge is equally accessible from either direction, so we should expect (and largely do obtain) a random mixture of product molecules formed from either side. This is shown below:

How can we know this? If we were to start with, say, *tert*-butyl chloride, as in the above picture, the two products shown are not the slightest bit different. Who is to say we got the product in the way we claim rather than some other way? No one can. In this particular instance, we simply cannot tell what occurred. But there are other cases where we can. If the carbon where the leaving group is attached is a stereocenter, that is, if the three lines I have shown do not all represent methyl groups but three *different* groups, then the two products shown are not exactly the same, they are mirror images.

432   Chapter 15   Chirality

So if you start with a halide that is optically active (meaning not just chiral but also with an excess of one enantiomer over the other; it is easiest to think about if you imagine a single enantiomer), and you run this reaction, your product will *not* be a single enantiomer but a racemic mixture. The reaction is referred to as proceeding with **racemization**. Another way to describe this is that it proceeds with 50% **retention** (the product on the right, where the nucleophile took over precisely the place of the leaving group) and 50% **inversion** (the product on the left, where the nucleophile is attached to the same carbon, but the carbon "turned inside out"). Many S$_N$1 reactions have been tested, and although there are some complicated details we will not discuss, the general finding is that they do indeed proceed with racemization. This phenomenon can also be observed in cyclic cases, such as the following:

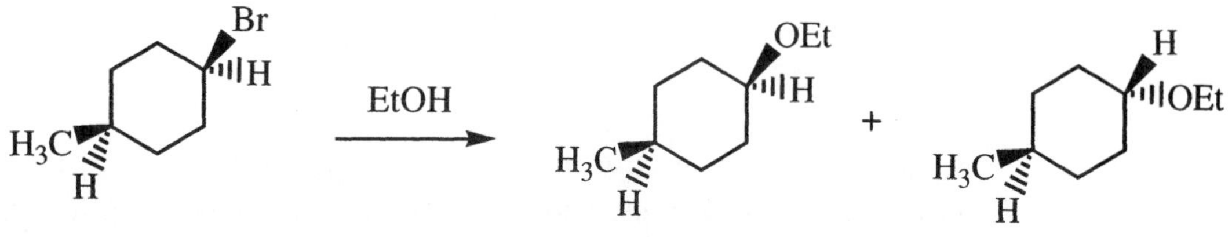

Here there is no chirality, but the two products are still distinguishable by being either *cis* or *trans* with respect to the methyl group on the other side of the ring, which obviously stays put during this reaction. So again the reaction goes half by retention, half by inversion, as expected.

Test Yourself 5: Show the mechanism of the reaction just discussed.

ORM: There is a nice animation of the S$_N$1 reaction and a summary of the supportive stereochemical evidence in Organic Reaction Mechanisms. Look under "Substitution Reactions" and choose "unimolecular, S$_N$1."

The S$_N$2 reaction is different. If you run the reaction shown below starting with enantiomerically pure chloride, you find that the product you get this time is enantiomerically pure. Further, if you determine the absolute stereochemistry of the product, you will find that the reaction goes with 100% inversion, as shown.

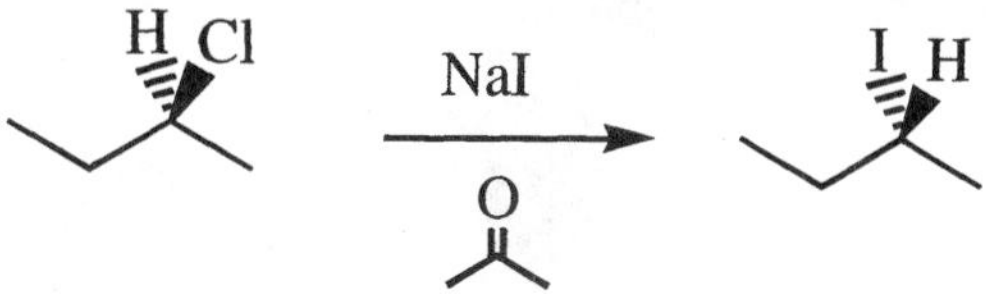

How do we account for this? The answer is shown below. We believe that the nucleophile *always* enters from behind the C–X bond. This is referred to as backside attack. Remember in Chapter 4 I told you that hybrid orbitals have little tails, and that those tails would become important to us later on? Well, later on is now. The nucleophile goes after the little tail, causing it to grow bigger. As the tail grows bigger, leading to better bonding with the nucleophile, the big end of the sp^3 orbital grows smaller, gradually breaking the bond to the leaving group. At the halfway point the orbital is a p orbital (and the other three orbitals are sp^2), bonded partially to the incoming nucleophile, which has lost part of its charge, and partially to the outgoing leaving group, which has gained part of a charge. The geometry around the atom at this point is trigonal planar, just like any other atom with a p orbital and three σ bonds. As the reaction continues, the end of the orbital toward the nucleophile grows even bigger, and the bond grows stronger, while the part toward the leaving group shrivels into a tail as the leaving group collects its charge and goes away. In the overall process, the carbon atom has turned inside out—inverted, much like an umbrella in the wind.

Transition State

How do we know all this? By experiments like the one described above, in which clean inversion can be demonstrated. Similar experiments with cyclic molecules show similar results: S_N2 reactions always go with inversion:

NaI

acetone

The requirement of backside attack makes it even more understandable that primary substrates react faster than secondary, and tertiary substrates not at all by this mechanism. The "tail," illustrated previously and again below, is well hidden inside the molecule, behind a bunch of other stuff sticking out to the left in this picture. The carbon in this molecule is secondary; you can imagine that if it were tertiary there would be even more trash getting in the way. There would be so much stuff, in fact, that the nucleophile would have an impossible task brushing all that aside to get to the tail. At a primary site, on the other hand, two of the three groups are tiny hydrogens, and the tail is pretty well exposed to any approaching nucleophile.

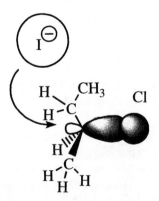

ORM: Under Substitution Reactions, choose "classic reaction." You can watch this with or without solvent. Click on the "Supportive evidence" box for a summary of everything we have learned about the S_N2 reaction (and a few things we have not).

One of the classic experiments that demonstrated the stereochemistry of the S_N2 reaction is described below.

(S)-(+)

The researchers began with optically pure (S)-(+)-2-iodooctane, as shown, and treated it with radioactive iodide. The labeled iodide displaces unlabeled iodide in a substitution reaction. At various stages they interrupted the reaction, collected the iodooctane present, and made two measurements on it: how much of it contained radioactive iodine atoms, and the extent of its optical rotation. The percentage of radioactive molecules can be taken as a measure of how fast substitution is occurring, because, at least at first, every substitution will change a non-radioactive ("cold") molecule into a radioactive ("hot") one. At the same time, one can measure the rate of loss of optical activity: at the beginning there will be some value obtained for the optical rotation, and at the end of the reaction it will have dropped to zero, since the product will be racemic. If you measure the two rates, substitution vs. loss of optical activity, you find that, at the beginning of the reaction, the compound loses optical activity at exactly twice the rate that it undergoes substitution. What does this experiment say about the stereochemistry of the reaction? Why is it important to compare the rates only during the early stages of the reaction?

Hint: this can be understood by using percents, but I think it is easier to understand by imagining that the flask contains some small, easily understood number of molecules; say, 100. The best approach, I think, is to consider the two substitution mechanisms we have studied, S_N1 and S_N2, and figure out what results you would expect *if* the reaction followed each path. If one of these is the result you see (and the other mechanism would predict different results), you have pretty good evidence.

Do not turn the page. Spend a minimum of a day on this question (unless you get an answer you are confident of sooner). Discuss it with your classmates and argue back and forth. Do not look at the answer until you have an answer of your own that you are willing and able to defend.

OK, I hope it is tomorrow.

Start with 100 molecules, all (S)-(+). Allow 2 of them to react with the radioactive iodide. We now have 98 molecules of "cold" 2-iodooctane, and 2 of "hot." Two percent of the molecules have been substituted.

If the reaction goes by the S_N1 mechanism, of the two new ones, one will be (S)-(+), and the other will be (R)-(−). Of the 100 molecules, we will now have 99 (+) and 1 (−). The one (−) will cancel one (+) and there will be 98 (+)'s causing the rotation; in other words, the optical rotation now will be 98% of its original value. So after 2% of the molecules substituted, 2% of the optical activity is lost. The rate of substitution and the rate of loss of optical activity should be the same.

If the reaction goes by the S_N2 mechanism, both of the new molecules will be (R)-(−). The two (−) molecules will cancel two (+) molecules, and there will be 96 (+) molecules remaining to rotate the light; in other words, the optical rotation will be 96% of the original value. So after 2% of the molecules have been substituted, 4% of the optical activity will be lost. The rate of loss of optical activity is double the rate of substitution.

Why does this apply only at the beginning of the reaction? Let's allow it to go for a while, say 20%. What is in the pot now? 2-Iodooctane, some of which is cold and (S) and some of which is hot and (R). Also, a bunch of I⁻, some that is hot and some that is cold (the leaving groups from the original molecules). I⁻ attacks a 2-iodooctane. Which kind of I⁻ will it be? At the beginning of the reaction we can rely on it being radioactive, but by now there is a reasonable chance that it is not. What will it attack? At the beginning of the reaction we could rely on the attacked molecule being (S)-(+) non-radioactive substrate, but by now there is a reasonable chance that it is not. We can have hot iodide attacking hot (R) substrate, changing the stereochemistry but not registering as a substitution, so we will begin to gain a population of hot (S) molecules; or we can have cold iodide attacking cold (S) substrate, generating a new population of cold (R) molecules. Or we could have cold iodide attacking hot (R) molecules leading back to cold (S) ones that have actually been through two substitutions. And so on. Keeping track of all this is nearly impossible. The common ploy is to look only at the first few percent of the reaction, when all these complications are very unlikely to crop up. At the beginning, for an S_N2 reaction, the rate of loss of optical activity is double the rate of substitution, leading to the conclusion that every substitution involves an inversion.

Problems

(Answers are provided at the end of the chapter for italicized problems.)

1. Build a model of the following compound and its mirror image. Is it chiral? Are you surprised by your answer?

$$BrHC=C=CHBr$$

2. Build a model of the following compound and its mirror image. Is it chiral? Are you surprised by your answer? Name it.

3. *For each of the following questions, assume that measurements are being made in a 10-cm (1-dm) tube.*

    a. *A solution of 0.4 g of optically active 2-butanol in 10 mL of water displays an optical rotation of –0.56°. What is its specific rotation?*

    b. *The specific rotation of sucrose is +66.4°. What would be the observed optical rotation of a solution containing 3.0 g of sucrose in 10 mL of water?*

    c. *A solution of pure (S)-2-bromobutane in ethanol is found to have an observed $\alpha = 57.3°$. If [α] for (S)-2-bromobutane is 23.1°, what is the concentration of the solution?*

    d. *Think carefully about that last answer you got. What is fishy about it? (That doesn't make it wrong, I just made up unusual numbers for the problem.)*

4. The following compound was reported recently in the *Journal of Organic Chemistry*. If the chemists who did this work dissolved 3.0 g of (+) **1** in enough ethanol to make 10 mL of solution and placed it in a 1-dm cell, what reading (between +90° and –90°) would they get on a polarimeter? How would they determine that this was not a correct reading?

(+)-**1**

[α] = +536

5. A compound is measured to have the following optical rotations in a polarimeter. What is its specific rotation? Assume a 1-dm cell length.

5g/10 mL	+43°
4 g/10 mL	+70°

6. Draw structural representations of each of the following molecules. Be sure that your structure clearly shows the configuration at each stereocenter.

a. (*R*)-3-bromo-3-methylhexane

b. (3*R*,5*S*)-3,5-dimethylheptane

c. (*S*)-1,1,2-trimethylcyclopropane

d. *(1R,2R,3S)-1,2-dichloro-3-ethylcyclohexane*

7. One of the enantiomers of chlorophenylacetic acid has a specific rotation of +192° (based on a Na lamp at 20 °C).

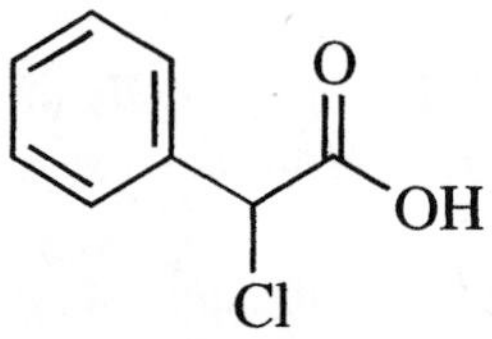

a. What is the observed rotation for a 10-mL benzene solution containing 6.0 grams of the solute? Assume a path length of 1 dm.

b. What value will the polarimeter indicate as the observed rotation?

c. Draw the (*S*) enantiomer of this compound. Is this the one that gives the rotation in part a?

8. Draw as many isomers (of all types!) as you can of methylchlorocyclobutane. In each case involving stereocenters, label (*R* or *S*) each stereocenter.

9. *Suggest a synthesis of (3R,4S)-3,4-dibromohexane starting with anything you like containing three or fewer carbons. Repeat for a diastereomer of this substance.*

10. *How many stereoisomers of 2,3,4-trichloropentane are there? Draw them, label each stereocenter (except on the middle carbon), indicate the relationship each has to the others (enantiomer, disatereomer, etc.), and identify any meso compounds.*

11. Thalidomide has the following structure. For reasons we have not discussed (take my word for it), the nitrogens are sp^2-hybridized and therefore flat.

a. How many stereocenters are there in thalidomide?

b. How many possible stereoisomers are there for this structure?

c. Draw all the possible isomers, and label each stereocenter as *R* or *S*.

d. Which of the above isomers is (+)-thalidomide, and which is (−)-thalidomide? (Trick question!)

e. You are a chemist working for Celgene, the company now trying to bring thalidomide back to market. [This is true. See *New York Sunday Times Magazine*, Jan 25, 1998.] One of your first tasks is to determine which isomer of thalidomide is the one that cures leprosy, and which is the one that causes birth defects, so of course you have to separate them. You start with a sample of thalidomide that is racemic. How much does this solution rotate plane-polarized light?

f. What is the ratio of (+)-thalidomide to (−)-thalidomide in this sample?

g. By doing a resolution, you are able to produce a sample of thalidomide that rotates light in a polarimeter. You dissolve 12 g of this sample in enough ethanol to make 20 mL of solution and observe a rotation of +48° in a 1-dm cell. You then add 20 mL of ethanol to your solution and observe a rotation of +24°. What is the specific rotation of your sample? If you cannot get an answer, explain why not and what you would have to do to get one.

h. Using a chiral chromatography column, you are able to determine that the sample you created in part g above had an enantiomeric excess (of the (+) form) of 64%. What is the ratio of (+)-thalidomide to (−)-thalidomide in this sample g?

i. Using a reasonable value for the specific rotation of your part g sample, what is the specific rotation of pure (+)-thalidomide?

j. What is the specific rotation of pure (−)-thalidomide?

k. Which of the structures you drew in part c is the isomer that causes birth defects?

l. Which of the structures you drew in part c is (+)-thalidomide?

12. Testosterone, which about half of you are suffering from an excess of, has the following structure:

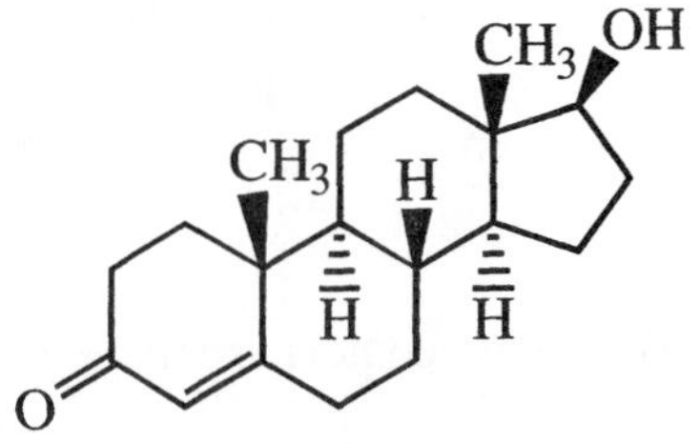

a. How many possible stereoisomers are there of this compound?

b. *Label each stereocenter.*

c. Draw a diastereomer of testosterone.

d. *When 9.0 g of testosterone is dissolved in enough ethanol to make 10 mL of solution, a rotation of –82° is observed in a 1-dm cell. When 7 mL of this solution is diluted with enough additional ethanol to make 10 mL of new solution, a rotation of +69° is observed. What is the specific rotation of testosterone?*

13. You have run a stereospecific reaction, one that is supposed to give more of one enantiomer than another for a particular product. According to the literature, your product has a specific rotation $[\alpha]$ of –375°. The compound you obtained was purified to exclude any diastereomers or other impurities, and then 4 g was dissolved in 10 mL of ethanol and a rotation of +80° was observed. When 6 g was dissolved in 10 mL of ethanol, a rotation of +30° was observed. How good was your reaction? (What was your *ee*, and what is the actual composition of your product?) Did you get more *R* or *S* product?

14. A man has died after eating a homemade mushroom soup. As the forensic chemist in the case, you have isolated from the soup a poison that comes from a mushroom that looks almost identical to the mushroom the man *wanted* to put in his soup. Thus, he must have mistaken the poisonous mushroom for the delicious one, cooked it, ate it, and died. Case closed. However, the man's son appears and insists that his late father was very well aware of this look-alike poisonous mushroom and would never have made such a mistake. The son claims that someone must have made some of the mushroom toxin in a laboratory and added some of this synthetic toxin to the soup. If this were true, you would be looking at a murder case. How would you go about checking whether the son's story is true? [For a full-length description of this scenario, see *The Documents in the Case* by Dorothy Sayers.]

15. *Compound A has a specific rotation, when pure, of +244°. You have created a sample of this with an ee of 75% favoring the (–) enantiomer. You make a solution of 3.8 g of this sample dissolved in 7 mL of solution. What reading will you get on a polarimeter?*

16. a. You have made a chiral compound, call it **1**, under special conditions that allowed you to make more of one enantiomer than the other. You take 5 g of your product, dissolve it in enough ethanol to make 10 mL of solution, and observe a polarimeter reading of +65° in a 1-dm cell. You take this solution and add 15 mL of ethanol and now observe a polarimeter reading of –46° (again in a 1-dm cell). Given that the specific rotation of pure (+) **1** is +347°, what *ee* did you obtain in your reaction?

b. Based on your answer to part a, what percent of your sample is (+) **1** and what percent is (−) **1**?

17. S_N2 reactions always proceed with inversion. Nevertheless, when the following compound, (*S*)-1-iodo-1-thiomethylpropane, is treated with $NaOCH_3$, the product still has the (*S*) configuration. Why?

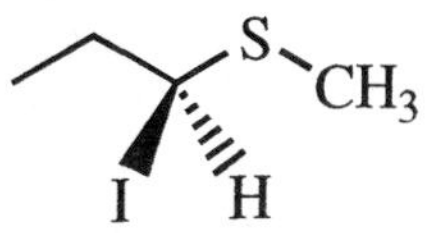

Selected Answers

Internal Problems

Test Yourself 1. -247°

Test Yourself 2.

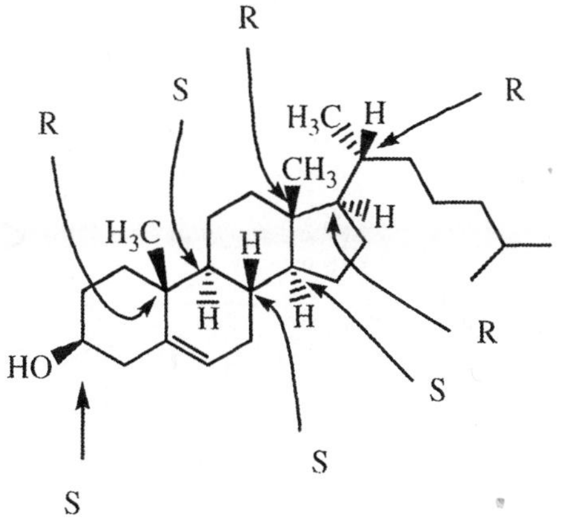

Test Yourself 3.

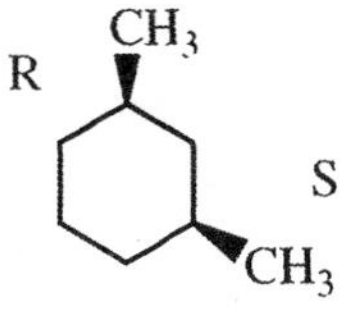

A

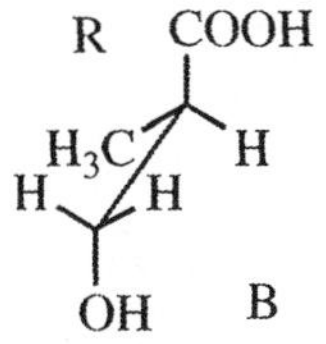

R COOH
B
(+)

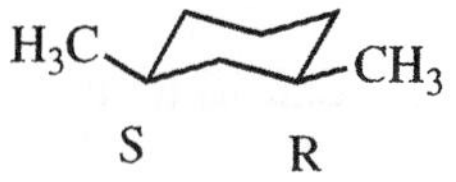

C

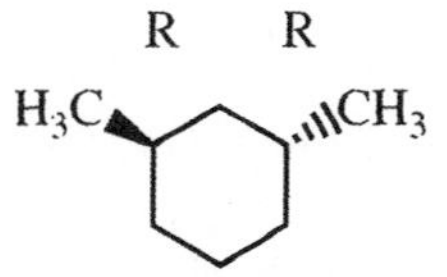

D

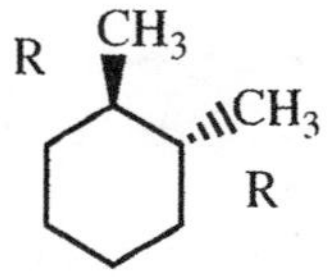

E
(+)

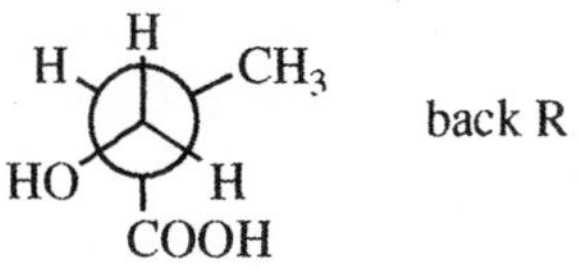

F
back R

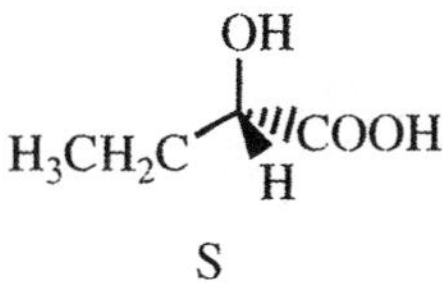

G

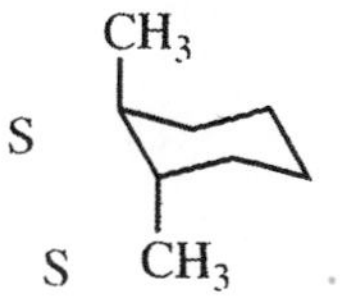

H

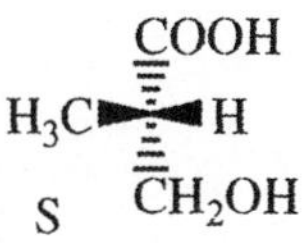

I

A and C are identical and *meso* therefore rotation is 0.

A (C) and D are diastereomers. D's rotation is unknown.

E and H are enantiomers. Therefore H's rotation is (–).

B and F are identical, therefore F's rotation is (+).

I is enantiomer of B and F, therefore rotation is (–).

G is a structural isomer of A, F, and I, rotation unknown.

Test Yourself 4. The mixture has an 80% excess of (+) enantiomer. Actual composition, 90% (+), 10% (–).

Test Yourself 5.

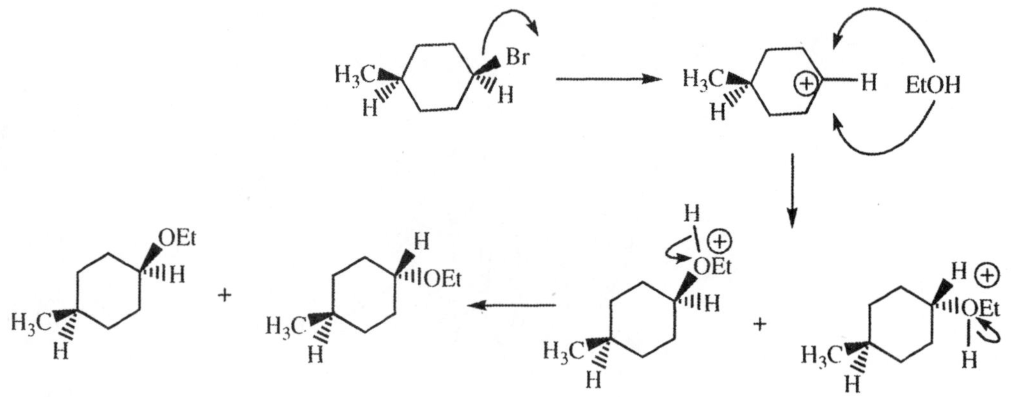

End of Chapter Problems

3. a. –14°

    b. +19.9°

    c. 2.48 g/mL

    d. 2.48 g/mL???? Few substances are that dense even when pure!!! And this one is not, so it is impossible to prepare such a sample!

5. The +43 could be +43, +223, +403, +583, –137, –317, –497, etc.
    The +70 could be +70, +250, +430, +610, –110, –290, –470, –650, etc.

    The second reading should be 4/5 of the first. The only pair that fits is –137 → –110.
    So when c = 5g/10 mL = 0.5 g/mL, α = –137. Thus, [α] = –274°

6. d.

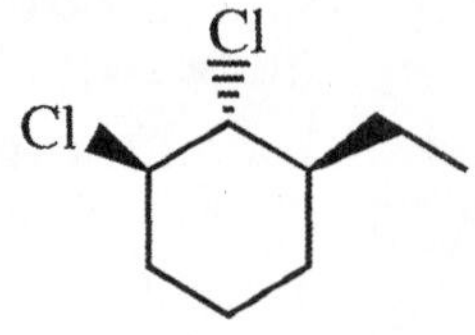

9.

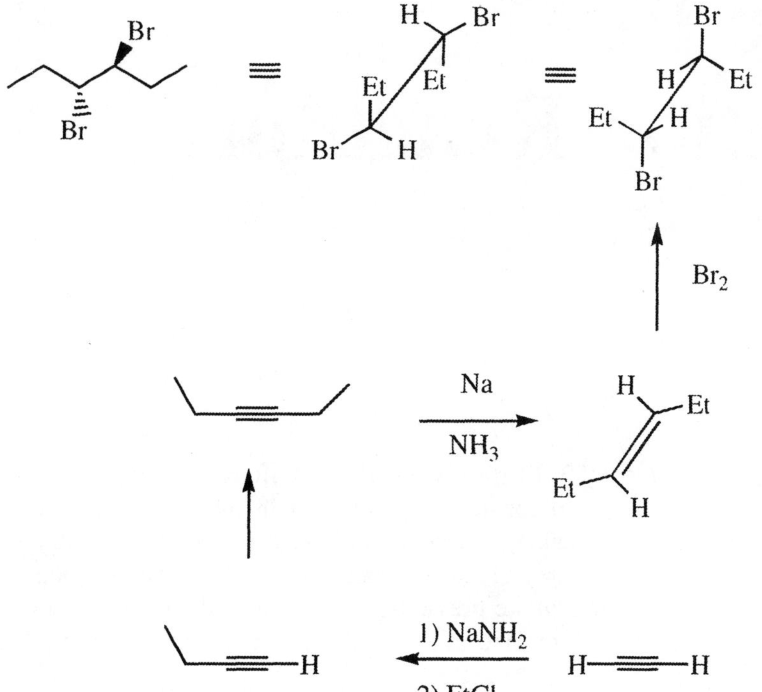

To make a diastereomer reduce with H_2/ Lindlar.

10. 4:

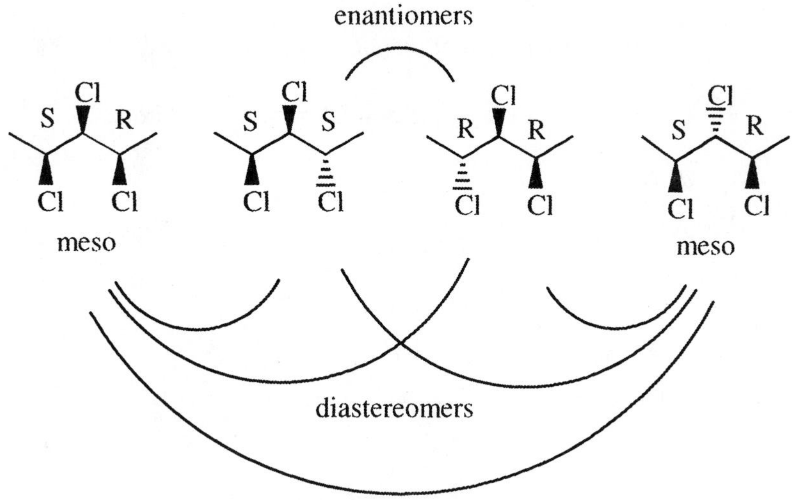

12. b.

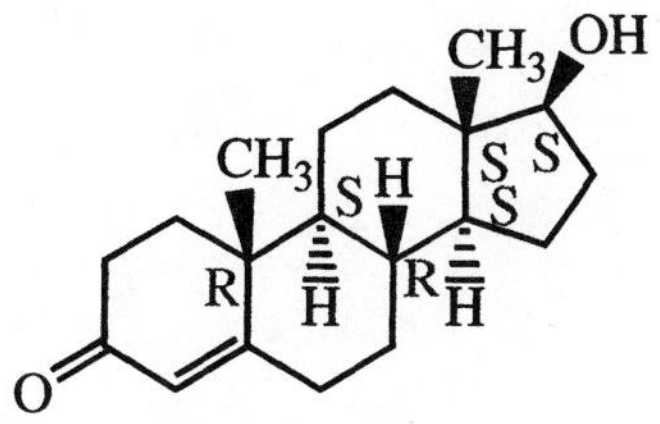

d. +109°

15. +81°

Chapter 16
Elimination Reactions

We spent Chapters 10–12 discussing the reactions of double and triple bonds. It seems sensible at this point to consider how one might make such things. Of course we already know one method of making double bonds: we can reduce triple bonds to double bonds. But that is not quite the same as *creating* unsaturation. We want to focus in this chapter on how to put a multiple bond where previously there was only a single bond. It turns out that most of the methods are the reverse of the addition reactions we recently learned. In this chapter we will

- Learn about elimination reactions of alkyl halides, both E1 and E2

- Discuss the competition between elimination and substitution reactions

- Discuss dehydration of alcohols as a second way of making double bonds

- Learn how to make triple bonds from double bonds.

All the acid-catalyzed addition reactions we learned in earlier chapters share the following characteristics:

$$C=C \xrightarrow{\text{HQ}} \underset{C-C}{\overset{H \qquad Q}{}}$$

A substance HQ, where Q is an atom that is more electronegative than H, adds to a carbon–carbon double bond, with the H atom going on one carbon, and the Q atom going on the other. In this chapter we will study elimination reactions, which involve the *removal* of H from one carbon and Q from a neighboring carbon. Further, with addition reactions, we found that the mechanism varies depending on whether Q^- is compatible with acid solution. If Q^- can exist in acid solution (e.g., halides), then it comes on as an anion, and the reaction stops. If Q^- cannot exist in acid solution (OH^-, OR^-), then it comes on as a neutral species, carrying an extra hydrogen, which is dropped in a subsequent step.

$$C=C \xrightarrow{\text{H}^+} \underset{C-C}{\overset{H \quad \oplus}{}} \begin{array}{c} \nearrow X^{\ominus} \\ \searrow H_2O \end{array} \begin{array}{c} \underset{C-C}{\overset{H \qquad X}{}} \\[2em] \underset{C-C}{\overset{H \qquad \overset{\oplus}{OH_2}}{}} \xrightarrow{} \underset{C-C}{\overset{H \qquad OH}{}} \end{array}$$

Similarly, we will find that the mechanism of the elimination reaction varies with the nature of Q^- for similar but not quite identical reasons.

Let's consider a generic elimination reaction and contemplate likely mechanisms.

$$\underset{C-C}{\overset{H \qquad Q}{}} \xrightarrow{} C=C$$

First we have to do some bookkeeping. We notice immediately that H and Q have disappeared in the product, so our mechanism has to account for that. What is less obvious, but equally important, is that two electrons are also missing: in the starting material there are two electrons between C and H, and another two between C and Q; in the product only two of these four electrons are accounted for in the alkene, so the other two must have departed. There are three obvious ways those electrons could have left: both with the H, both with the Q, or one with each. Except for the sodium/ammonia reduction we discussed in Chapter 12, we will not be dealing with radicals, so let us ignore the one-with-each possibility. Of the other two, we have a strong prejudice that if two electrons are going to leave with either H or Q, it ought to be with Q, because Q is more electronegative than H. Furthermore, two electrons with H would be H^-, which ought to make you blanch. (There is such a thing, but it is very nasty stuff, and we will not see it except under special circumstances.) So if Q is taking two electrons with it, H must be taking none and is therefore leaving as H^+. So already we can guess some pretty interesting things about this reaction:

$$\underset{C-C}{\overset{H \qquad Q}{}} \xrightarrow{} \overset{\overset{\oplus}{H} \qquad \cdot\cdot Q}{} \quad C=C$$

Another question has to do with timing. If H⁺ and Q: are leaving the molecule, there are three ways it might happen:

1. H⁺ could leave first, and Q: second;

2. Q: could leave first, and H⁺ second;

3. Both could leave simultaneously.

It turns out that under various circumstances all three of these possibilities occur; for now, we will worry only about 2 and 3. With both of these mechanistic routes, one of the things that happens right away is that Q leaves with its electrons; it acts as a leaving group. We discussed leaving groups at length in Chapter 13. If you don't remember that stuff, now is a good time to go back and reread it. In a nutshell, good leaving groups are weak bases.

E1 REACTIONS

Let's imagine some molecule with a good leaving group attached—say, 2-bromopropane— and think about making propene from it:

The leaving group is a good one and is ready to go. One option is simply to wait around for the bromine to take off on its own, maybe even heat the solution up a little to encourage the process. Although such a process does occur, there are several problems with this approach. There is more to breaking a bond than simply having a good leaving group around. When the leaving group leaves, the carbon it was attached to becomes a carbocation.

If the result is not a very good carbocation, the process will not happen readily. Thus, you should predict (correctly) that halides attached to primary carbons would undergo this kind of reaction only very rarely, if at all, and that tertiary halides should be more willing than secondary halides to see such a departure happen. Sound familiar?

There is another problem with this idea. We have taken a perfectly good, neutral molecule and suggested that it simply disintegrates, all by itself, into a cation and an anion. If the cation is decent and the anion is decent, this has a chance, but there is one more important criterion to be met: the molecule must be sitting in a medium that can support charges, that is, a polar solvent. Otherwise, creating charge separation will be prohibitively energetic. You need a polar solvent to lower the energy of the ions. Again, sound familiar?

Let's assume all these factors are in place, and ionization occurs. We now have a carbocation in a polar solvent, let's say methanol (CH_3OH). As we learned way back in Chapter 10, a carbocation's main goal is to stop being a carbocation, and one of the options it has for accomplishing this is to kick off a proton:

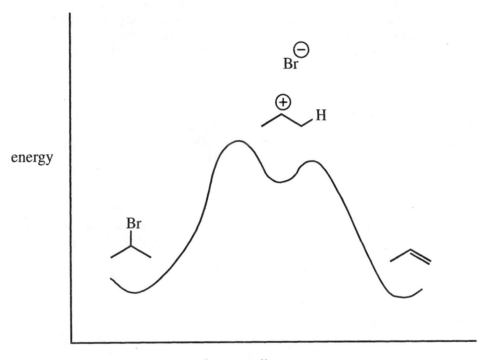

Voila! We have witnessed an elimination reaction. This process has a name, and like the names of the substitution reactions, it relates to the number of molecules involved in the slowest step of the reaction. Which step is that? It should not be much of a stretch to figure out that it will be a lot harder to make a carbocation from some neutral compound than it is to make something neutral from a carbocation. Indeed, once a carbocation is formed, the second step is very rapid. So even though another species shows up in the second step, the first step (the slow one) involves only the alkyl halide: one molecule. We have here a unimolecular reaction, and since the reaction is an elimination, unimolecular, we call it **E1** for short. An energy diagram for an E1 reaction is shown below.

Directional Preferences

The example I used to illustrate this reaction suffers from one shortcoming (as an example, not as a reaction): no matter what proton is lost, you get the same product. What if this is not the case? What if we started with 2-bromobutane? What would we expect to get? Try it now!

In this case three different products are possible. If this does not surprise you, you are either a very good or a very naughty student. If you first wrote down the structure of 2-bromobutane and determined for yourself that there are three different products, you are a very good student. Congratulations! If you did *not* write down the 2-bromobutane structure and didn't even *try* to figure out how many products there would be, shame on you! By now you should realize that the only way to learn this stuff is to try to answer questions on your own before letting me do it for you. If you are not doing this, you are throwing away your tuition money.

Most of you fell in the middle ground, I suspect: you did try the problem, but decided that there were only two possible products, 1-butene and 2-butene. That is a standard (and pretty good) answer; you simply forgot that there are two different versions of 2-butene—it can be either *cis* or *trans*.

The question before us is, will we get all of these, and if so, in what proportions; i.e., will there be any preference with respect to the direction of elimination? Simply from a probability standpoint, you might argue that you should get more 1-butene, since there are three protons that could be lost to give that product, while only two would lead to one of the 2-butenes. Nevertheless, we find the opposite to be the case. You get more 2-butene molecules, and of those, most are *trans*. Why? Remember, double bonds are better off when they have more carbons around them (beginning of Chapter 10). 2-Butene has a disubstituted double bond, and the one in 1-butene is only monosubstituted. Further, *trans*-2-butene is more stable than *cis*-2-butene. In elimination reactions, you usually get the more stable of the various possible alkene products. This statement is called the Saytzeff (or Zaitsev; same guy, different spelling—he spelled it in Russian) rule.

Why do we get more of the more stable product? The answer comes from a close inspection of the energy diagram for this reaction, and a good guess about the transition state for proton loss. Let's look at a close-up of the second step of this E1 reaction.

What does the mountaintop look like? It must be something between a carbocation and an alkene, a transition state where the proton has begun to leave but is not yet gone. The bond between carbon and hydrogen is partially broken, and the carbon–carbon double bond is already partially formed. This could occur on either side of the positively charged carbon, leading to either one of the following transition states:

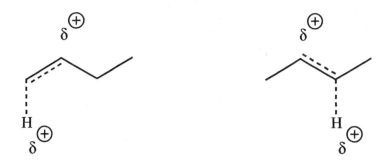

But look! The partial double bond in the transition state on the right is disubstituted, while the one on the left is monosubstituted. Since disubstituted double bonds are lower in energy than monosubstituted ones, it stands to reason that disubstituted *partial* double bonds are lower in energy than monosubstituted *partial* ones, at least by a little bit. This is illustrated in the following energy diagram.

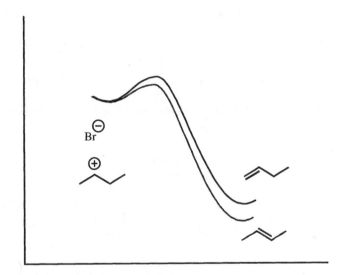

So you see, the molecule has an easier time (i.e., has a lower energy barrier to go over) making the disubstituted vs. the monosubstituted double bond. The same reasoning applies to the *cis vs. trans* situation.

ORM: In "Organic Reaction Mechanisms," under "Elimination Reactions," look at "E1 unimolecular reaction."

Test Yourself 1

Which of the following would react faster in an E1 reaction? Why? Show the product(s) for each. If more than one product is possible, would any of them be preferred?

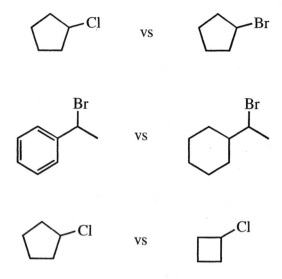

Now I must confess to you, having spent all this time describing an E1 reaction, hardly anyone actually carries out a reaction this way. Why not? Too many things can go wrong. Recall, in Chapters 10–13 we generated lots of carbocations in polar solvents, like methanol and water, and found that very often the carbocation chose to solve its problem *not* by dropping a proton but by capturing a molecule of the solvent surrounding it, that is, by undergoing an S_N1 reaction. As I suggested in several hints earlier, the conditions necessary for an E1 reaction are exactly the same ones you would choose for an S_N1 reaction, so you should expect both to occur simultaneously. Indeed, it is almost impossible to do one without doing the other.

Carbocations are also subject to rearrangements (although this particular one is not). So if an elimination is what you want, it is usually best *not* to allow nature to take its course, as in an E1 reaction, but instead try to make things happen your way. How?

E2 REACTIONS

Remember, in an elimination reaction there are two things that need to happen: the leaving group has to leave (with its electrons, by definition), and the H has to leave, without its electrons and therefore as H^+. Once the leaving group is primed, what might you do to encourage a reaction to occur? Clearly, you should go to work on the hydrogen. You want to pluck off the hydrogen without its electrons, as H^+. What could attract H^+? A base, that's what. Indeed, we defined a base as a proton acceptor, something that is attracted to H^+, and the more attracted it is, the better it is as a base. So we should apparently tempt our molecule with a base to encourage reaction.

In the presence of a base, here is what happens:

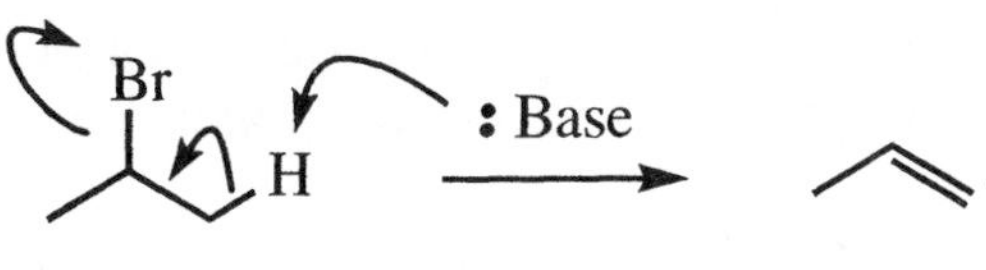

I'm suggesting that this is happening all at once, not in separate steps. We have here a one-step reaction, involving the collision of two molecules. Since the slow step (the only step) involves two molecules, the process is called elimination, bimolecular, or **E2**.

As we discussed with S_N2 reactions, it is important how you draw the three arrows in the above picture, even though, once drawn, no one can tell in what order they appeared. But *you* know. And you know that this reaction does *not* start with the leaving group leaving, because if it did it would be E1. Furthermore, if the bond to the bromine begins to break first, why do we bother adding any base? The base was put in for a reason: to accelerate the process. So it is logical to assign it some important role in initiating the reaction. The proper way to show an E2 mechanism is first to draw the arrow from the base to the proton, then the arrow from the proton to the carbon (the middle one), and finally the arrow showing the leaving group leaving, so that in your head you are thinking, "The base is what makes this happen."

Let's do an example, to illustrate many of the points that can come up in writing elimination reactions. Show a mechanism for the following reaction. Do it now, without looking at the answer.

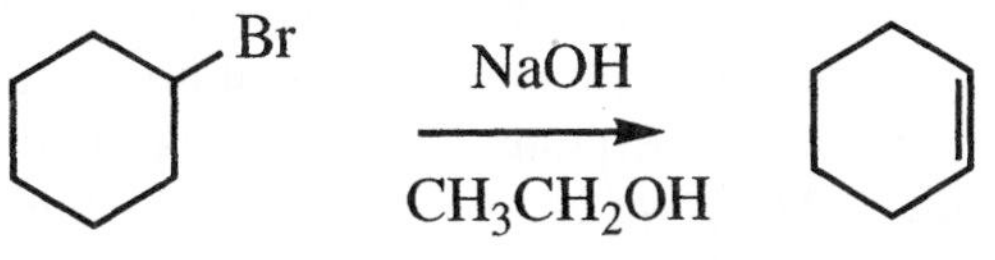

Answer:

In solving a problem like this the first order of business is to look at where we start and where we are going:

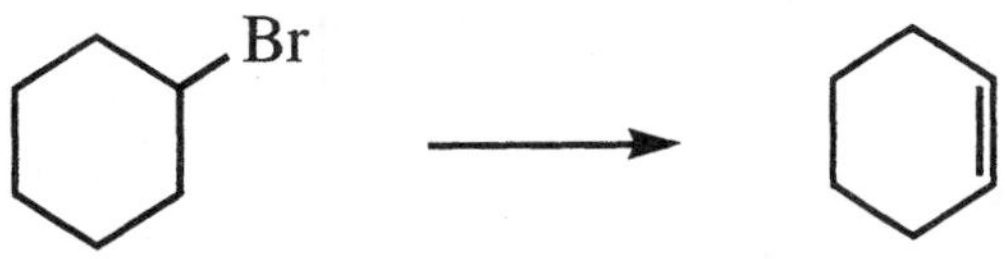

What has changed? Clearly, the Br is gone and a double bond has appeared. But what you might not have seen, because it is not written explicitly, is that there is something else missing. On the left there are two H's on the lower-right carbon; on the right there is only one. This is a good time to remind you that it never hurts, and often helps, to write in all the assumed hydrogens in the vicinity of a reaction.

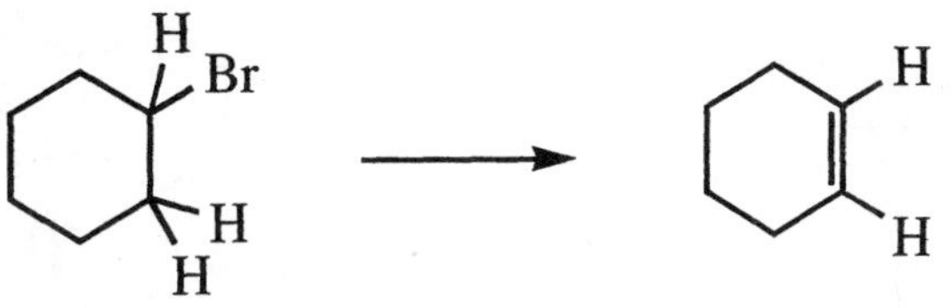

Any correct mechanism must show not only how the Br goes away, but also how the H goes away. The reaction involves departure of a Br from one carbon *and of an H from a different carbon*, and the mechanism must show both things. There are three questions remaining: in what order do the two things leave, which electrons go where, and does this happen by itself, or do we need some external agent to make it happen?

Which electrons go where is answered by a consideration of electronegativity and leaving group ability, which we have discussed already. Clearly, the Br likes to hang out with extra electrons, and the H does not, so we should favor a mechanism that allows the Br to depart *with* its electrons, and the H without. Which happens first? This is actually answered by considering the third question: is there something we added that makes this happen, or does it occur on its own? If we need to add something to make it happen, then we should not draw a mechanism that shows the molecule spontaneously doing something. We added sodium hydroxide to make this reaction occur. The implication is that if we did not add sodium hydroxide, it would not occur. Therefore, the first instruction we send in should involve the sodium hydroxide in some way. In other words, we should not expect the Br to leave first, on its own, without some external push.

It is also important to recognize something about sodium hydroxide, which applies equally to sodium chloride or any other "salt." Ionic substances tend to come apart into their constituent ions, in this case Na^+ and OH^-, provided there is some polar solvent around. Further, Na^+, for our purposes, never does anything. It has no electrons to share, and it wants none, being happy with its filled octet. Na^+ is what is called a "spectator ion": it sits on the sidelines and watches the action, but virtually never gets involved. Its only role is to hang around in the vicinity of negative charges to maintain electrical neutrality, but we rarely need to show this.

Thus, we should focus all our attention on its partner, OH^-. The oxygen here also has an octet, but it has a minus charge it would prefer not to have, and we recognize OH^- as a reasonable base (we can verify this by looking on our pK_a chart). It is looking for a partner with which to share its electrons. One possible partner is H^+, a "proton." Sure enough, there is a proton on our molecule that we are actually trying to dispose of. Perhaps this would be a good place to start.

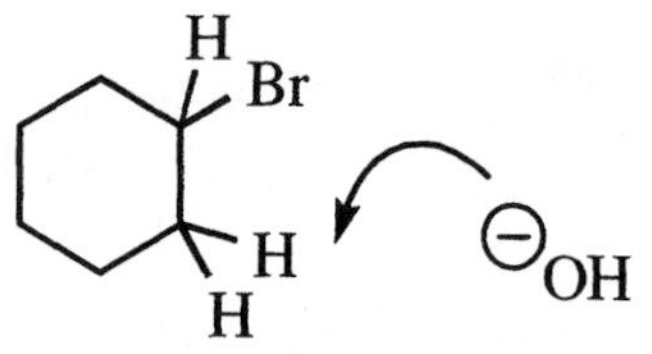

Notice the arrow goes *from* the OH^- *toward* the H, because an arrow always shows the movement of electrons, not atoms. However, this is insufficient, because the instruction we have given requires creation of a new bond between the O and the H, but the H already has a bond and it cannot have two. The instructions above would lead to the following molecule, which is chemical nonsense:

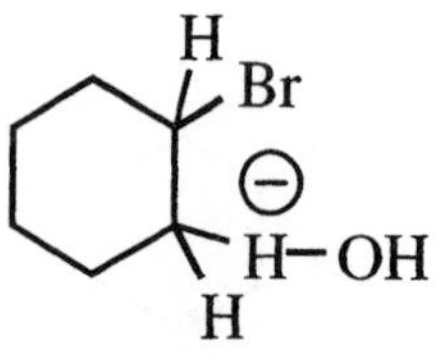

Therefore *as these electrons come in, some others must leave simultaneously from the H!* This is shown in the next picture.

Now this process, at least, makes chemical sense, but is it reasonable? It is, in fact, an acid–base reaction, and we can evaluate the likelihood of its occurrence by looking at the pK_a's of the species involved. The acid on the right (water) has a pK_a of about 16; the acid on the left (a normal C–H bond) has a pK_a near 50. In less numerical terms, we have traded a minus charge on an O (uncomfortable, but not unheard of) for a minus charge on a C (extremely unhappy; C is much less electronegative than O), a very unlikely process. Conclusion: the reaction shown above will not occur; the electrons do not want to camp out on a carbon atom. The carbanion above is not a viable *intermediate* in this reaction.

So, what to do? The OH⁻ comes and snares the H, but the electrons do not want to retreat to the carbon. Perhaps we can get them to do something else rather than just sit there. Maybe *at the same time* that the OH⁻ is taking the H, the electrons are doing their own thing, which is to move into the space between the two carbons:

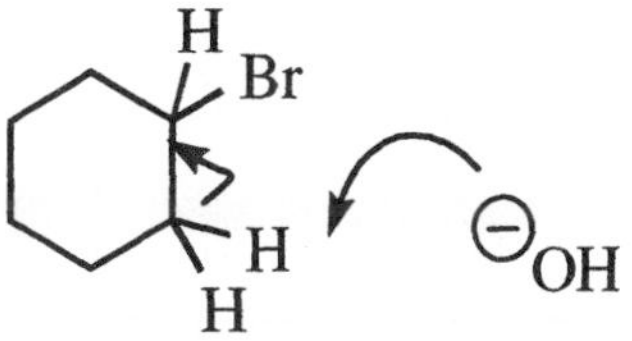

But this, too, is chemical nonsense, because the top carbon already has 4 bonds and cannot have 5. The instructions written above would create the following molecule, which does not make sense:

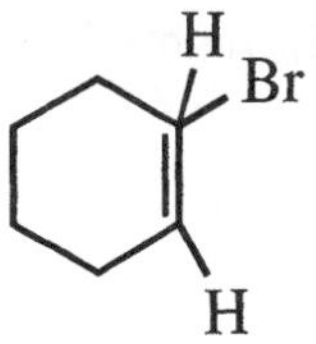

Thus, as electrons approach this carbon, other electrons *must* again leave. But this is actually fine, because we have already identified some electrons in the carbon–bromine bond that we want to have leave. So let them:

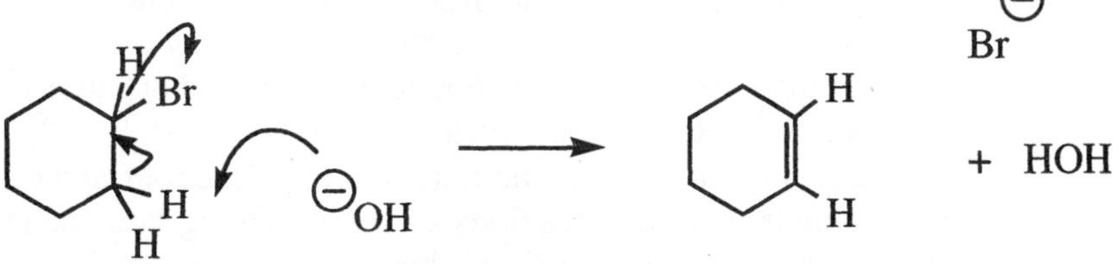

Thus, we can write this entire process in a single step; everybody has done something desirable. Notice that the Na⁺ is nowhere to be seen. It is hanging around near the OH⁻ at the beginning and near the Br⁻ at the end, but it never *does* anything. Never use Na⁺ as the cause for any process.

You should notice that normally OH⁻ cannot remove a proton from an ordinary carbon atom. The only reason it occurs here is that there is a leaving group on the carbon next door, allowing the electrons to go somewhere other than directly onto the carbon losing the proton.

We can also draw an energy diagram for an E2 process, but it is not very interesting, because everything happens at once.

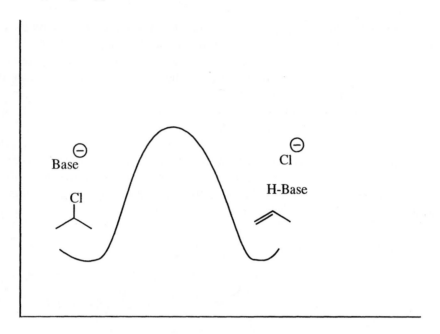

As with E1 reactions, and for the same reasons, E2 reactions generally follow Saytzeff's rule and lead to the best possible double bond whenever there is a choice.

ORM: In "Organic Reaction Mechanisms," under "Elimination Reactions," look at "E2 bimolecular reaction."

Kinetics

In naming the elimination reactions we have considered, we focused on the slow step as the one that defines the mechanism, just as we did with substitution reactions. We can measure the rates of these reactions, and see how the rates change when we change the amounts of various reagents. Indeed, experiment shows that E1 reactions depend only on the amount of substrate, while rates of E2 reactions also depend on the amount of base present, suggesting that a collision is necessary to make the reaction occur.

There is more we can learn with kinetics. Remember I told you that for E1 reactions, tertiary halides react faster than secondary, and primary not at all? We know this through kinetics. Similarly, we can make predictions about E2 reactions and test them with kinetic techniques. For example, consider reactions of the following compounds with NaOH. Would any of them react faster or slower than others in an E2 reaction?

You should write the expected products for each of these three starting materials (one, two, and two products, respectively), and the mechanism for each reaction. All of them will be E2 reactions. In an E2 reaction everything happens at once, so there is no intermediate, and the carbon holding the bromine does not become positive or even slightly positive during the reaction. Therefore, we should not expect molecules to care whether the bromine is attached to a primary, secondary, or tertiary position, and to a first approximation all three of the reactions should go at about the same rate (and they do). This is an important point to keep in mind, and we will come back to it later: S_N1, E1, and S_N2 reactions all care what kind of carbon is holding the leaving group. In the former two cases, it is because a cation has to be made; in the latter, it is because the nucleophile has to reach that carbon. But in an E2 reaction the aggressor is attacking a different place altogether, the hydrogen next door, and does not particularly care what kind of position the leaving group is on.

If you look in more detail, you might be able to suggest some small differences. In the first compound, there is only one hydrogen that can be involved in this reaction. The base must come to this specific hydrogen in order to make a reaction begin. In the second and third compounds there seem to be many more eligible hydrogens, so the chances of the base finding one that will work is somewhat larger, so you might predict that these two compounds would react faster than the first. This thinking is good and accurate to a first approximation, but as we will soon see there are not as many eligible hydrogens as you think.

Stereochemistry of E2 Reactions

We described the E2 reaction as happening in one step. The hydrogen and the leaving group depart simultaneously, leaving a double bond in their place. We now have kinetic evidence that this is correct, since the reaction does indeed involve a collision between the substrate and the base. But the process is also accompanied by some stereochemical requirements that are not obvious until you examine the substrate and the product closely.

A double bond consists of a σ bond and a π bond. The σ bond has been there all along, so the π bond is the one we are forming in this reaction. A π bond, you recall, is a side-to-side overlap of two p orbitals. Where did those orbitals come from? One came from the σ bond between the carbon and the leaving group, and the other came from the σ bond between the carbon and the hydrogen:

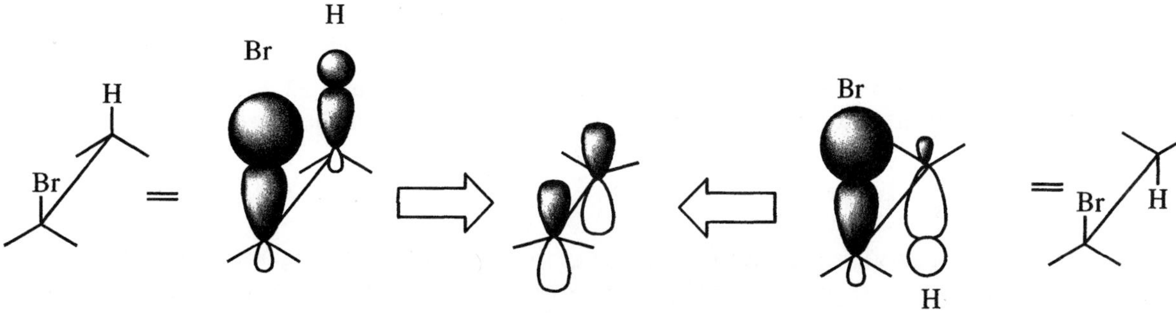

If you think about this for a moment, you should realize that the only way two σ bonds can morph into a π bond like this is for them to be parallel to each other *before the reaction occurs*. In other words, the substrate must be sitting there with the carbon–bromine and carbon–hydrogen bonds aligned when the base shows up to perform its role. If the bonds are not lined up properly, reaction cannot occur:

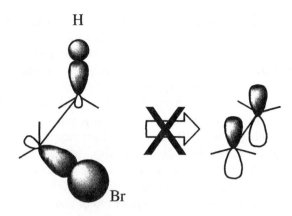

The interesting point here is that there are really only two ways the appropriate bonds can be lined up, and these are shown on the previous page: either the two departing atoms are on the *same* side of the molecule, as in the left picture, or they are on opposite sides, as in the right picture. Further, we learned way back in Chapter 7 that molecules don't like to sit around in the eclipsed form, which is what the left picture requires. Thus, the chances of the molecule looking like the left picture at the same time that the base shows up are pretty small. E2 eliminations, therefore, almost always proceed from a conformation like the right one, where the hydrogen and the leaving group are parallel to each other but on opposite sides of the molecule, in a so-called "anti" relationship (technically, "antiperiplanar," but most people leave off the long word).

So what? For most molecules this has relatively little consequence. If they are free to rotate, they simply rotate until the requirements are met and then undergo reaction. But there are certain cases in which the requirement leads to noticeable consequences. Consider, for example, the following reaction. What should be the product here of an E2 reaction? Show the mechanism. Do it now! In case you don't remember, "Et" is the chemists' shorthand for "ethyl," CH_3CH_2-.

$$\xrightarrow[\text{EtOH}]{\text{KOH}}$$

Answer: This is an E2 reaction. The requisite base is KOH, which dissociates in a polar solvent to K^+ and OH^-. Ethanol is the polar solvent and has no other role.

COMMON MISTAKES: DON'T LET THIS HAPPEN TO YOU!

Many students show alcohols like EtOH breaking apart into Et+ and OH-. NEVER do this. We will learn some of the chemistry of alcohols shortly, and more in a later chapter, but at no time will we see them simply dissociating on their own. As we have learned, an alcohol group, –OH, can be called a "hydroxy" group, but it is not related to hydrox*ide*, OH-. Do not dissociate alcohols!

As you know, elimination requires the loss of a leaving group from one carbon and a hydrogen from an adjacent carbon. The leaving group here is obviously chloride. The carbon to the right of the one with chlorine has no hydrogens, so the hydrogen to be lost is clearly the one (the only one) written to the left. The mechanism is that of our standard one-step reaction:

The result shown is consistent with the Saytzeff Rule since we have formed the best possible double bond, the one with the big groups *trans*. If your mechanism showed all this you get partial credit, but you missed the most important point. This elimination *cannot* occur, at least not as an E2, unless the departing groups are parallel to each other *before* reaction. Further, the molecule should be staggered, with its groups *anti* to each other. If we draw some conformation of this molecule and rotate it until the alignment is right, we discover that the only possible product of the reaction is one with the two phenyl groups *cis* to each other, even though they would rather be *trans*. Tough! This reaction makes a *cis* product.

Of course, one *could* make the *trans* product from this starting material, but in order to do so the hydrogen and the leaving group would have to be lined up in an eclipsed form. This would add several kcal/mol to the energy barrier of the E2 pathway, making it prohibitively high. The reaction gives the *cis* product because that is what arises in the pathway that is easiest to follow, and once the reaction gets to its destination, it cannot go back. The reaction is subject to kinetic control. The following energy diagram illustrates this situation.

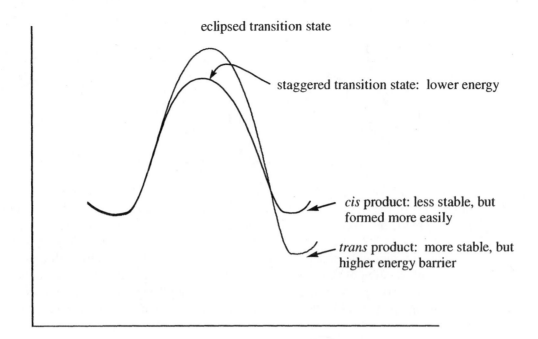

eclipsed transition state

staggered transition state: lower energy

cis product: less stable, but formed more easily

trans product: more stable, but higher energy barrier

Bear in mind, the restriction we're talking about applies only to an E2 reaction. If the same compound were simply dissolved in ethanol and heated, the leaving group could leave on its own, creating a carbocation. This has a finite lifetime and the ability to rotate, so it could lead to either the *cis* or the *trans* product, and it would naturally choose the *trans*.

rotate the back carbon

A context in which the geometric constraints of an E2 reaction are particularly interesting involves cyclohexane compounds. Consider bromocyclohexane. As you know, bromine has a strong preference for being in an equatorial position. However, if it is in an equatorial position, there is no hydrogen *anywhere* that is properly lined up to eliminate. The only way bromocyclohexane can undergo E2 elimination is for the leaving group to be in an axial position. Granted, this does not occur very often, and when it does occur the situation does not last very long, but it does happen often enough and for long enough for reaction to take place, and when the leaving group is axial there will usually be two (but only two) hydrogens in a position to eliminate. These are the two shown below. Both are *trans* to the bromine. In all cyclohexanes, only *trans* eliminations are possible, and even then only when the leaving group and the appropriate hydrogen are both axial. This is referred to as a *trans*-diaxial relationship.

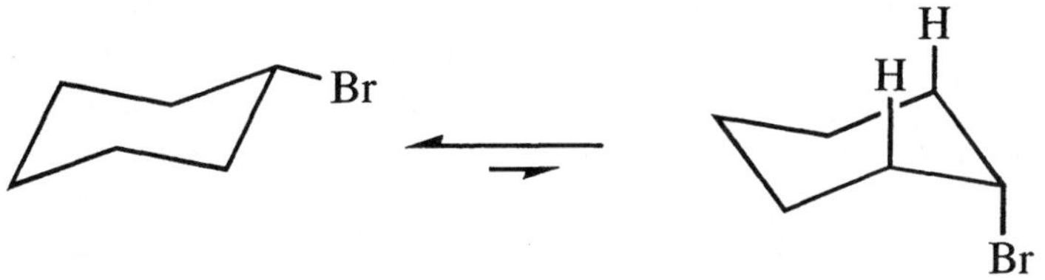

So consider the possibilities for the following elimination reaction:

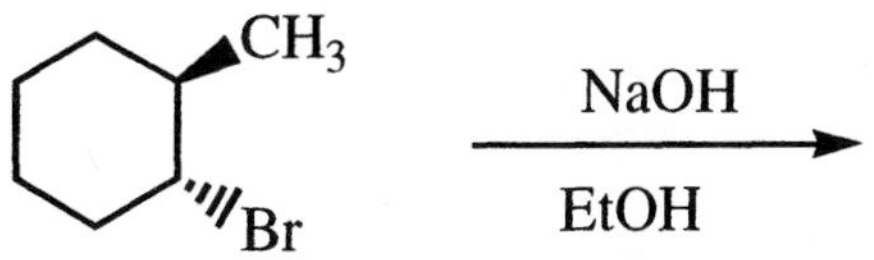

What will you get? Again, Saytzeff's rule suggests you should get 1-methylcyclohexene, but the stereochemistry of this reaction demands that you instead get 3-methylcyclohexene. Write out the structures yourself to make sure you understand why that is true. Build models if necessary.

Test Yourself 2

Menthyl chloride and neomenthyl chloride differ from each other in the stereochemistry of the carbon to which the chlorine is attached. In both compounds a methyl group is *trans* to an isopropyl group. [Notice the difference between the words "methyl" and "menthyl."]

a. Given the following information, determine the relative stereochemistries of the two compounds.

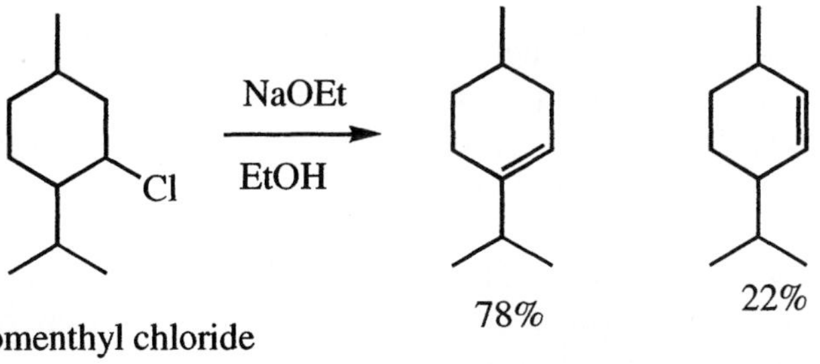

neomenthyl chloride

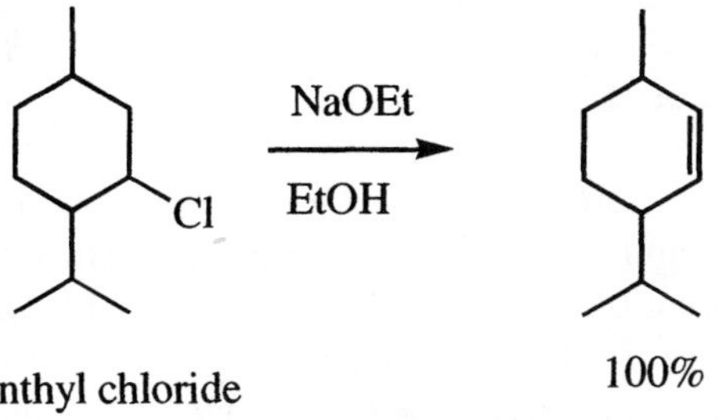

menthyl chloride

b. When the reaction with menthyl chloride is carried out in 80% aqueous ethanol with no added base, both of the possible elimination products are obtained. How do you explain this result?

COMPETITION WITH SUBSTITUTION REACTIONS

I told you earlier that E1 reactions are not carried out all that often (on purpose), because the conditions that lead to an E1 reaction [good leaving group, 3° (or 2°) position for the leaving group, polar solvent] are precisely the same conditions that lead to S_N1 reactions, so it is very difficult to control how much of the compound will eliminate and how much will substitute. There are a few tricks you can play in this game, but they are limited. For an S_N1 reaction to occur (in the absence of other nucleophiles), the solvent must act as the nucleophile; for an E1, the solvent must act as a base. If you want an E1 reaction, you should choose a solvent that is polar and can serve as a base but has trouble being a nucleophile. These are not common: most polar solvents have OH groups in them, like water and alcohols, and OH groups can act as nucleophiles and then lose an H to make neutral products. Solvents like this are called "protic" solvents because they have protons attached to electronegative atoms like oxygen or nitrogen. Solvents without OH groups are often not polar enough to support ionization reactions. The most common exception to this generalization is dimethyl sulfoxide (DMSO), shown below. This is a polar, aprotic solvent. It can dissolve many polar compounds, and support ions, but it is a lousy nucleophile, because if it attacked a carbocation it would gain a charge but would have no easy way to lose that charge.

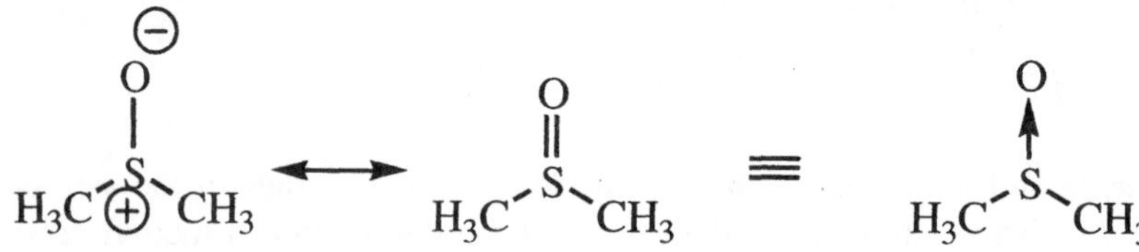

A second trick you can play is the reverse: if you want substitution by S_N1, you should deliberately choose as a solvent something that is polar but a lousy base. These, too, are not common, since it does not take much of a base to accept a proton from a carbocation; the most common examples are liquid carboxylic acids like acetic acid or trifluoroacetic acid.

The bottom line is that when you want a "one" reaction (E1 or S_N1) you need to generate a carbocation in a polar medium, and you have relatively little control over what that carbocation chooses to do. For that reason, such reactions are relatively rarely desired, with a single exception about to be covered in the next section.

What about "two" reactions? All nucleophiles have lone pairs, and are therefore bases, and *vice versa*. Any nucleophile you feed to a substrate in an effort to accomplish an S_N2 reaction has the potential to effect an E2 instead; any base you use in an effort to cause an E2 has the potential to substitute instead. Can you control these divergent pathways at all? Yes, there *are* some ways to steer a "two" reaction in the direction you want, but there is always a risk that the wrong thing will still occur.

You may recall from Chapter 13, where you learned about nucleophiles, that nucleophilicity and basicity do not always go hand in hand. Although *most* strong bases are also good nucleophiles, there are a few exceptions to this. One has to do with "polarizability": as you go lower on the periodic table, you find weaker bases but stronger nucleophiles. Thus, the halide ions (excepting fluoride) are pretty good nucleophiles but very weak bases;

nucleophiles based on sulfur and phosphorus also tend to favor substitution over elimination. Another species that is a pretty good nucleophile, but not much of a base (for reasons we have not discussed), is the cyanide anion, CN⁻. All of this does you little good, however, if you are trying to make a particular compound. Consider the following target:

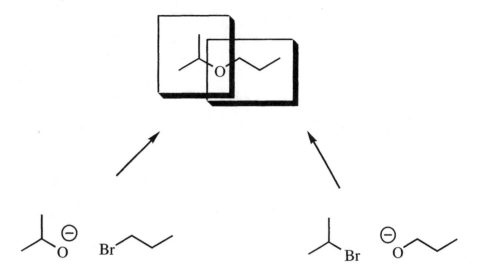

As we practiced in Chapter 13, there are two ways to dissect this, as shown. Both require an alkoxide (RO⁻) as a nucleophile, and alkoxides, as you know, are also good bases. The fact that a sulfur-based nucleophile would favor substitution over elimination does us no good here—we don't want a sulfur in our molecule! We have no choice but to use an alkoxide nucleophile and accept whatever losses we might suffer due to competing elimination. Our two choices are not equal, however. Remember that S_N2 reactions happen best on 1° centers, then 2°, then 3°. In the left pathway above, we are showing an S_N2 reaction on a 1° carbon. The right pathway shows an S_N2 on a 2° center. The left substitution should be more favorable than the right. That in itself may not be good enough: what if the elimination is *also* better for the left path? If it were, frankly, we would be up a creek. But remember what we decided about the E2 reaction: unlike all the other reaction types, this one does not care what kind of carbon is holding the leaving group, because the instigator, the base, is attacking somewhere else entirely, at a proton next door. So while substitution will be better in the left route than in the right, elimination will *not* be much different. So we conclude that the preferred way to carry out the above reaction is the left route, because choosing the right-hand route would expose us to a higher risk of elimination.

What if elimination is what we want? Consider the following transformation:

The leaving group is on a primary carbon, so E1 is out of the question. We have to do this by E2. But since the leaving group is on a primary carbon, and any base will also be a nucleophile, we run significant risk of a competing S_N2 reaction. Here, too, though, we can influence the partition between the two pathways. Recall from Chapter 13 that there is a second category of deviations from the general rule that better bases are better nucleophiles. This is the group of bases with such steric bulk that they have difficulty functioning as nucleophiles, the so-called "non-nucleophilic bases." The examples we considered at the time were potassium *tert*-butoxide and lithium diisopropylamide (LDA), both shown below.

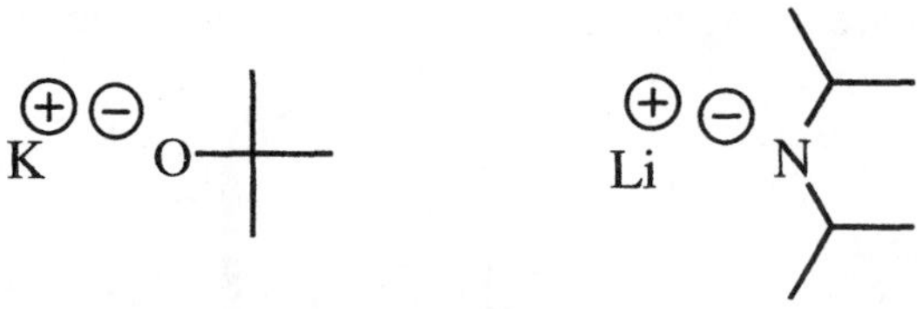

Using non-nucleophilic bases like these, one can often get elimination to occur with relatively little competition from S_N2 reactions. The most common of the reagents is potassium *tert*-butoxide, since LDA is really overkill for the current purposes—it is more basic than we need.

There is still a logistical problem we need to consider: what solvent shall we use for this? We need something that will dissolve the potassium *tert*-butoxide, therefore something polar, like water or an alcohol; but if we use water or one of our normal alcohols, like ethanol, the following equilibria will be set up immediately:

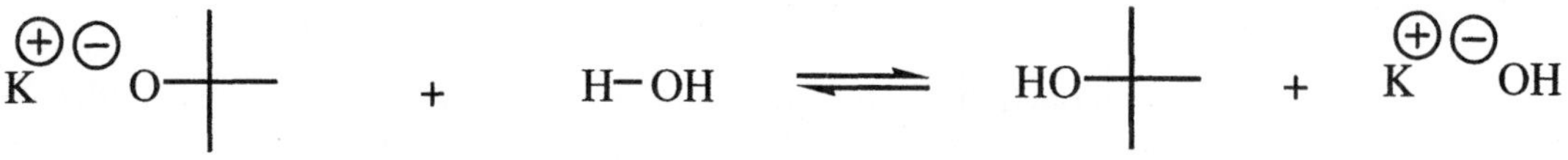

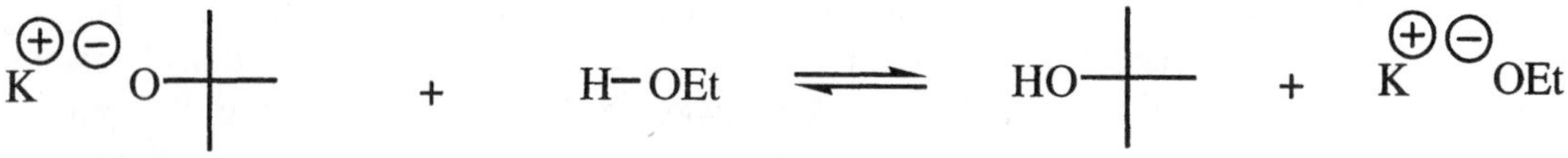

Use your pK_a table to determine whether these equilibria will favor the left- or right-hand sides, and by approximately how much. Now!

You should have discovered that both of these favor the right by a small amount. So even though we go to the effort of putting a hindered base (potassium *tert*-butoxide) into our reaction, what is actually there, at least most of the time, is the perfectly ordinary (and nucleophilic) hydroxide or ethoxide. Thus, water or ethanol will not do as a choice for the solvent. What will? Think about this a bit before you continue.

There are two choices that are logical, and you should have thought of one of them if not both. We need a polar solvent, but most of our polar solvents have an OH group that interferes with our plans by dumping its proton onto our reagent. We can solve this in two ways. One is to use a polar solvent that does *not* have an OH group. We know of one such solvent, which was mentioned earlier in this chapter: DMSO. DMSO is, in fact, a common solvent for E2 eliminations involving potassium *tert*-butoxide. The second trick is to use a solvent that does have an OH group, but one which, if it dumped its proton, would not change the makeup of the mixture. That solvent is *tert*-butyl alcohol, because the equilibrium below would not bother us at all!

tert-Butyl alcohol is also a common solvent for such eliminations, although DMSO is more common.

One other potential problem arises when you use potassium *tert*-butoxide for an E2 reaction. The reagent is so bulky that it sometimes goes after the protons that are most accessible, namely, the ones on the least hindered carbon adjacent to the leaving group, leading to the less substituted double bond, whereas Saytzeff's rule says that E2's should lead to the *more* substituted double bond. There is disagreement about this in textbooks: some say you get the Saytzeff product, some say you don't, and both groups support their positions with data from the literature. For our purposes, I suggest we ignore this problem and pretend that potassium *tert*-butoxide in DMSO will provide us with the expected (Saytzeff) elimination product with little complication either from substitution or the wrong elimination product.

In summary, E1/S_N1 reactions are difficult to predict. You sort of take whatever you get. There is no good recipe for deciding whether the product will be that from E1 or S_N1, unless one or the other of these is unlikely due to the nature of the solvent or the substrate. The same applies to many E2/S_N2 reactions, but in some instances it is possible to be pretty confident about what you will get. Sodium iodide in acetone is a dead giveaway that you are doing an S_N2 reaction; potassium *tert*-butoxide in DMSO means a clean E2 reaction.

Here is a chart that might help you keep track of the various factors involved in these reactions.

Reaction	Conditions	Consequences
S_N1 } E1	Good Leaving Group (LG) Polar Solvent (H_2O, alcohol) Carbon with LG must lead to good carbocation Do NOT need good nucleophile or base Common conditions: plain solvent (alcohol or water), often warmed	3° > 2° > 1° (except allyl and benzyl)
S_N2	Good Leaving Group (LG) Good Nucleophile (which may or may not be good base) Access to carbon with leaving group Do not need polar solvent (but polar solvent is OK) Common conditions: NaI/acetone	1° > 2° > 3°

E2	Good Leaving Group (LG) Good Base (which may or may not be good nucleophile) H on adjacent C Do NOT need access to carbon with leaving group Do not need polar solvent (but polar solvent is OK) C-H bond and C-X bond must line up Common conditions: KOH/EtOH; KO✝, DMSO	1° , 2° , 3° all good

DEHYDRATION REACTIONS

In all the eliminations we have considered so far, we have had a molecule with a leaving group that was ready to leave—a halogen in each case. What if we don't have that? Consider the case of isopropyl alcohol. How can we accomplish an elimination here? Notice that we want to remove H from one carbon and OH from its neighbor; together they make H_2O, so we are trying to remove water, a reaction called **dehydration**.

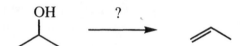

The problem is that the leaving group is lousy (OH⁻); further, if we were to treat this with a decent base, it is almost certain that the base would remove the proton from the OH group rather than one from the carbon next door. So for two reasons, we cannot do an E2 reaction, which requires a pushy base as the thing that starts the whole process. To overcome this obstacle, we use the same trick we used in Chapter 13, when we learned how to *substitute* alcohols: we turn the OH group, currently a bad leaving group, into a good one by protonating it. To choose an appropriate acid, you find one higher than the protonated alcohol on the pK_a chart, obviously higher on the acid side. The most common choice is sulfuric acid, H_2SO_4; another common one, not on your chart, is phosphoric acid, H_3PO_4. Both of these have the advantage of having very weak nucleophiles as conjugate bases, so you are not likely to suffer an unwanted substitution reaction, and they are both liquids.

Why does it matter that the acid be a liquid? Most acids, if they are solids or gases (like HCl), are used in a laboratory setting as solutions in water. When your object is to *remove* water from a molecule, it makes little sense to surround it with water. We will return to this point soon.

As we discovered in Chapter 13, when an alcohol is in the presence of a strong acid, it is reasonable to propose that the OH group gets protonated. This is an acid–base reaction, and the pK_a values tell us that the equilibrium favors the protonated alcohol:

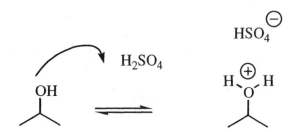

Now we have turned the bad leaving group OH⁻ into a good one, H_2O, and we are ready to do an elimination reaction. Will it be E1 or E2? As you know, E2 reactions are bimolecular, meaning that the reaction requires a collision between the substrate (here the protonated alcohol) and a base, and it is the base that initiates the process by grabbing a proton. There are two reasons you cannot get an aggressive base involved in this reaction. First, the reaction is taking place in sulfuric acid, silly! If you put a base into sulfuric acid, it will get protonated by the acid and no longer be a base! Second, if by chance the base managed to encounter a substrate molecule before being protonated by the solvent, the base would surely steal a proton from the oxygen rather than from the adjacent carbon. So an E2 reaction is out of the question. The only thing we can do here is wait for an E1 to occur. Sitting in sulfuric (or phosphoric) acid, the leaving group will eventually leave on its own, creating a carbocation:

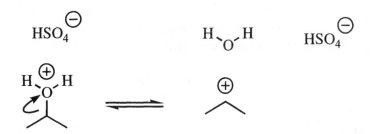

Now what? As you know, carbocations have three choices. One is to rearrange into a better carbocation. This particular carbocation has nothing better that it could become, but if it did, now would be a good time for it to rearrange. The second choice is to pick up a nucleophile from the surroundings. There are only two here: HSO_4^-, which is one of the worst nucleophiles in the world, and water, which just came off. Notice the equilibrium arrows in the equation, which remind us that this is a possibility. However (and this is why we used sulfuric acid), the molecule of water we just generated is the only one present in the solution (except for a few others that came from other substrate molecules by the same process), so the chances of the carbocation finding any water are slim. Further, in sulfuric acid, water is rapidly protonated to hydronium ion, which is not nucleophilic. Finally, if water did succeed in attaching to the carbocation, we would be back where we started, not someplace new. So even though competition with S_N1 reactions is *normally* a problem with E1 reactions, it won't be in this case.

The third option is to drop a proton from a carbon adjacent to the carbocation, generating a double bond. What will take this proton? This is a bit unclear. It could be HSO_4^-, although this is a pretty lousy base. It could be water, although there is little water around, and most of it is already protonated and therefore a *very* lousy base. It could be another molecule of the starting alcohol, although many of these have already been protonated by the sulfuric acid. The fact is, no one really knows precisely where the proton goes. For that reason, we often show for such reactions an unspecified base, labeled B:, as the agent removing the proton:

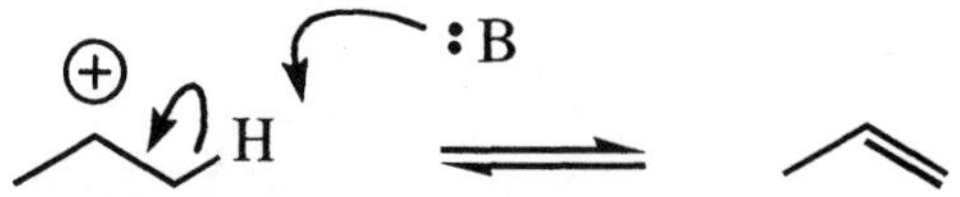

Overall, the reaction looks like this.

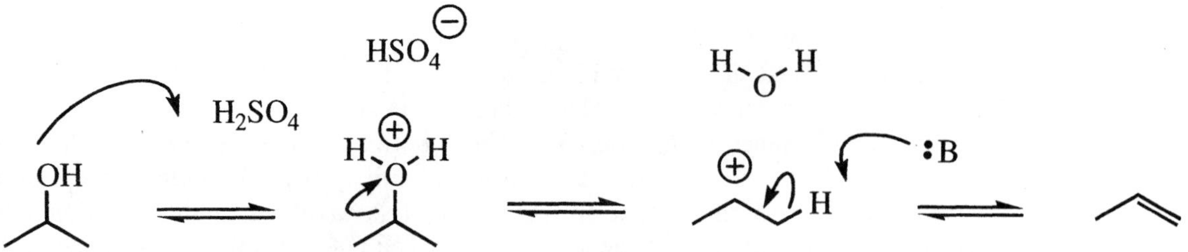

Whether the alcohol is tertiary, secondary, or even primary, the reaction *must* go by an E1 mechanism. This is the exception I referred to above, when I said that chemists normally do not conduct E1 reactions on purpose. Obviously, this reaction goes better with tertiary and secondary cases than with primary, but if you wait long enough even primary alcohols will eliminate by this mechanism. Thus, this reaction is also an exception to our general claim that primary substrates will never undergo "one-type" reactions. The key here, for both exceptions, is that there is really nothing else the molecule can do. Be careful, though: if you do try dehydrating a primary alcohol, there will very likely be rearrangements if any favorable ones can occur.

ORM: In "Organic Reaction Mechanisms," under "Elimination Reactions," look at "dehydration of n-butyl alcohol."

Test Yourself 3

Suggest a mechanism for the following reaction:

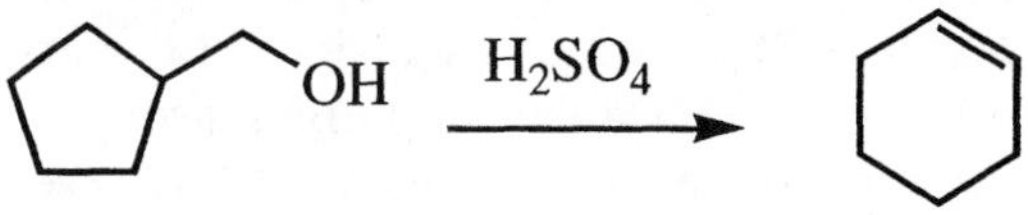

You may have noticed that the mechanism we have been discussing is precisely the reverse of a mechanism we learned earlier, that of addition of water to an alkene:

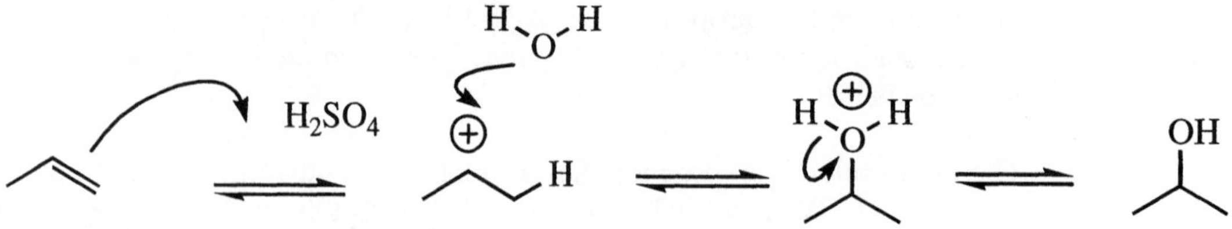

Why do we say in one case that

and in the other that

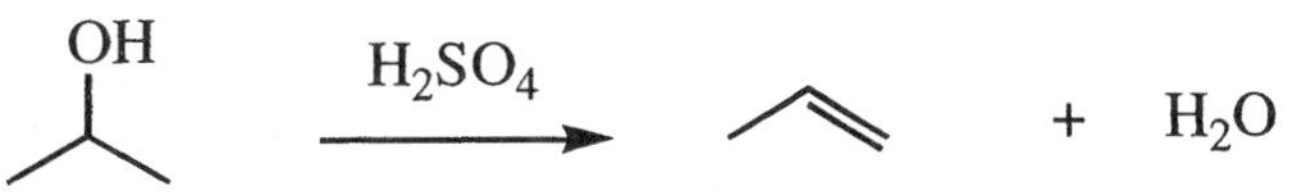

$$\underset{\text{OH}}{\bigwedge} \quad \xrightarrow{\text{H}_2\text{SO}_4} \quad \bigwedge \quad + \quad \text{H}_2\text{O}$$

How can we know what will occur?

Le Châtelier's Principle

The key lies in our understanding of equilibrium. This reaction does in fact go in both directions, and we chemists can control which direction predominates. The reaction is properly written like this:

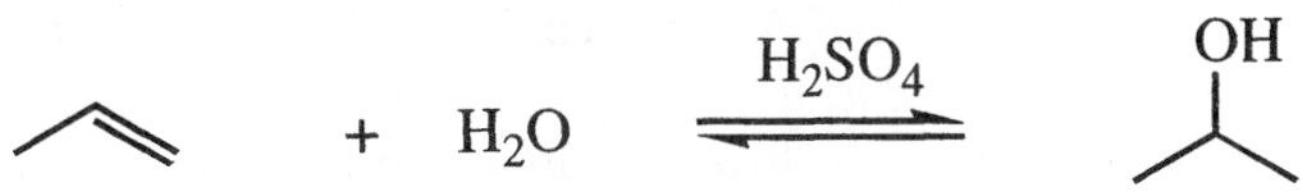

$$\bigwedge \quad + \quad \text{H}_2\text{O} \quad \underset{}{\overset{\text{H}_2\text{SO}_4}{\rightleftharpoons}} \quad \underset{\text{OH}}{\bigwedge}$$

and it has associated with it an equilibrium constant

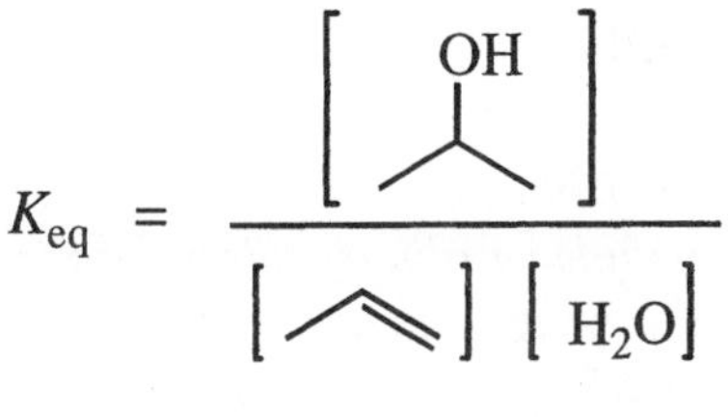

$$K_{eq} = \frac{\left[\ \underset{\text{OH}}{\bigwedge}\ \right]}{\left[\ \bigwedge\ \right]\left[\ \text{H}_2\text{O}\right]}$$

which can be rearranged to

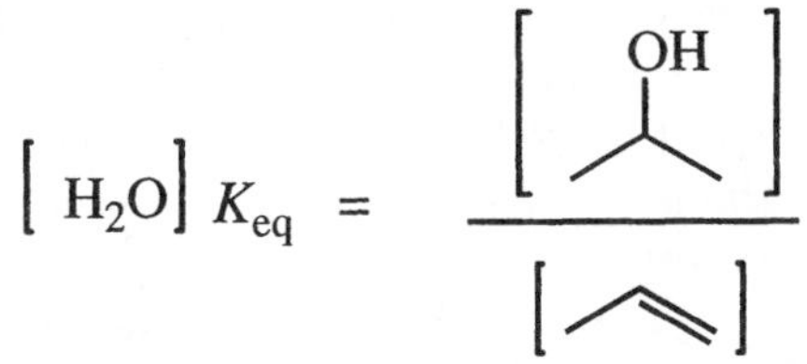

$$\left[\ \text{H}_2\text{O}\right]K_{eq} = \frac{\left[\ \underset{\text{OH}}{\bigwedge}\ \right]}{\left[\ \bigwedge\ \right]}$$

Remember, K_{eq} does not change. But the product of K_{eq} and $[\text{H}_2\text{O}]$ *can* change, since $[\text{H}_2\text{O}]$ is *not* a constant. (In Chapter 8 we treated $[\text{H}_2\text{O}]$ as a constant, because we used water as our solvent, so the concentration of water did not significantly change during the reaction. The above reaction is not done in water solvent, unless we choose it to be.) If you're trying to convert an alkene into an alcohol, you want the above numerator to be larger than the denominator; that is, you want the ratio of alcohol to alkene to be a large number. You can make this be the case by forcing $[\text{H}_2\text{O}]$ to be large, so you run the reaction in water, with a little sulfuric acid added. On the other hand, if you want to convert an alcohol into an alkene, you want the numerator to be small and the denominator to be large; that is, you want the ratio to be a *small* number. You can make it so by causing $[\text{H}_2\text{O}]$ to be very small, so you run the reaction in the *absence* of water, using concentrated sulfuric acid. This is another way of saying what we said in an earlier paragraph: "When your object is to remove water from a molecule, it makes little sense to surround it with water."

The previous paragraph is couched in mathematical terms, with ratios and numbers and everything. It is not difficult math, and you should be able to handle it without difficulty, but there is another, more qualitative approach to the same problem. It is called "Le Châtelier's principle" (biologists refer to this as the "law of mass action"). It states that if you have an equilibrium such as we are considering,

$$\wedge \quad + \quad H_2O \quad \underset{\longleftarrow}{\overset{H_2SO_4}{\longrightarrow}} \quad \overset{OH}{\underset{}{\searrow}}$$

if you add or remove participants in the equilibrium, the reaction will shift to partially undo what you just did: that is, if you add something on the left, the reaction will shift to the right, and *vice versa*; if you remove something on the right, the reaction will shift to the right, and *vice versa*. Thus, to make the reaction shift to the right, we should add water, and to make it shift to the left, we should remove water. This is the same conclusion we came to above.

Incidentally, another common technique for driving dehydration reactions is to distill off the alkene product during the reaction. This is often possible, because alcohols have high boiling points due to hydrogen bonding. The corresponding alkenes, unable to hydrogen bond, usually have lower boiling points, so it is possible to heat the reaction mixture and remove the alkene product, leaving the alcohol in the reaction mixture to react further.

WHO CARES? YOU DO!!

Many of the reactions occurring right now in your body are ones that have unfavorable equilibrium constants; that is, under ordinary circumstances they would favor starting materials over products. How do living cells arrange for these reactions to go in the proper direction? By using Le Châtelier's principle! Given the concentrations of the various components present in the cell, and more importantly at the active site of the enzyme, the reactions are forced to go in the direction that the body requires. This is often accomplished by what is called "coupling," in which the product of one reaction serves as the starting material for a second. The second reaction removes that substance, lowering its concentration and causing the first reaction to make more. Indeed, one can imagine that as the body runs short of any particular substance, equilibrium constants ensure that more of that substance is made, and vice versa.

CREATION OF TRIPLE BONDS

So far we have discussed how to make double bonds from halides and alcohols. Can the same reactions apply to making triple bonds? Yes and no. You cannot make a triple bond by E1 elimination of an alcohol, because the alcohol would have to be a vinyl alcohol, also known as an "enol," and we learned in Chapter 12 that enols do not sit around as enols, but instead tautomerize into ketones or aldehydes. So the necessary starting material is hardly ever available. Even if it were, in order to become a triple bond an enol would have to proceed through a vinyl carbocation, which is pretty bad. So for both reasons, triple bonds cannot be made from alcohols in sulfuric acid.

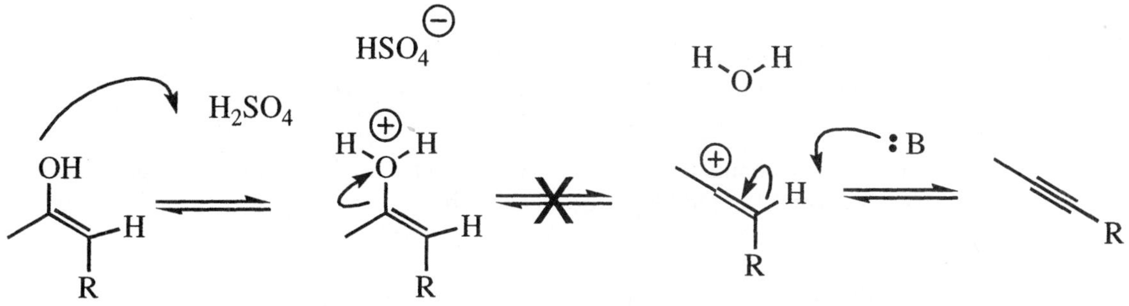

On the other hand, it is possible to make triple bonds by E2 elimination of vinyl halides. Elimination of a vinyl halide is more difficult than elimination of a saturated halide, and one typically needs a stronger base than we have so far been using. Sodium amide (sodamide), which we first saw in Chapter 12 as a base for removing hydrogen from the end of a terminal alkyne, is the most common base used to *make* alkynes also. Since vinyl halides can be made by elimination of vicinal (adjacent) dihalides, the most common approach to triple bonds is the following:

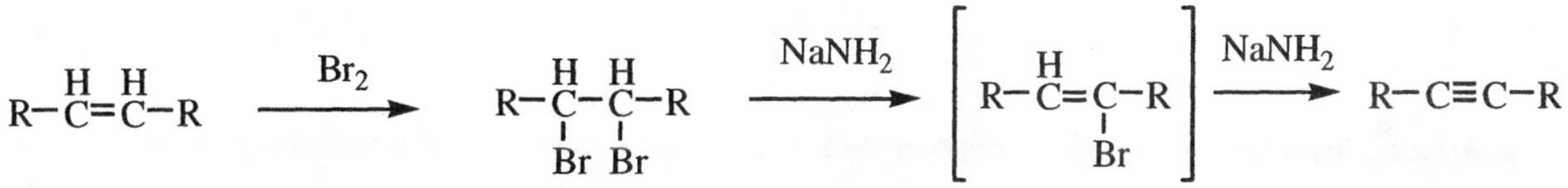

This represents a nice way to turn a double bond into a triple bond, and it is the only way you know (now) to make a triple bond where there was none before. But it remains true that most alkynes are not made this way: they are made the way we learned in Chapter 13, by alkylation of previously existing alkynes.

Problems

(Answers are provided at the end of the chapter for italicized problems.)

1. Show the mechanism of the following reaction. Show how both products are formed and indicate which should be preferred and why.

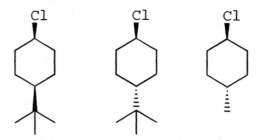

2. One of the following three compounds undergoes E2 reaction with potassium *tert*-butoxide much more slowly than the other two (essentially, not at all). Which is it, and explain why it is slower than *each* of the others. In other words, if your choice is **C**, why is it slower than **B**? Also, why is it slower than **A**? Which of the three should be the fastest, and why?

3. *Acyclic alkenes (such as 3-hexene) react with bromine, as you know, to give dibromides, which can react with a strong base like NaNH₂ to give alkynes. The latter reaction occurs in two steps: first an elimination to give a vinyl bromide (bromine attached to the carbon of a double bond) and then a second elimination to give an alkyne. However, cyclohexene reacts reliably by the same sequence to give 1,3-cyclohexadiene. The bromination works as expected, but the first elimination, rather than giving a vinyl bromide (1-bromocyclohexene), gives 3-bromocyclohexene, which then continues in a second elimination to give 1,3-cyclohexadiene. Explain why this sequence occurs as it does.*

4. When (2R,3R)-2-chloro-3-methylpentane is treated with potassium *tert*-butoxide in DMSO, an alkene is formed, but when a diastereomer of the starting material is used, a different (stereoisomeric) alkene is formed. Show which product comes from which starting material, and explain why there is a difference.

5. *When meso-2,3-dichlorobutane is treated with ethanol/KOH to eliminate one equivalent of HCl, all the product has the two methyl groups cis to each other, whereas the same reaction with (2S,3S)-2,3-dichlorobutane gives a product with the two methyl groups trans.*

   a. Why is this true?

   b. What product would be obtained from the (2R,3R) isomer?

6. There are several stereoisomers of 2-bromo-1,2,3-trimethylcyclohexane, but only one of them reacts with ethanol/KOH in such as way as to produce an alkene whose double bond is *not* in the ring. Which isomer is it, and why is its reactivity unique?

7. An alcohol A is treated with phosphoric acid to give a mixture of alkenes B and C, which are treated with ozone followed by zinc to give the following four compounds. Show structures for A, B, and C.

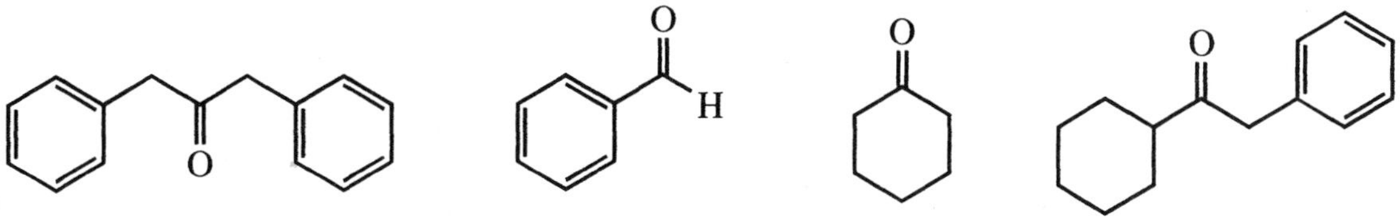

8. Show the mechanism of the following reaction. Show how both products are formed and indicate which should be preferred and why.

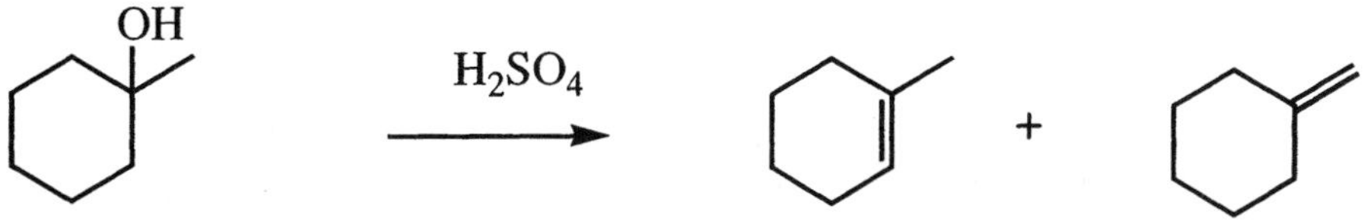

9. Imagine you were trying to make 3-heptene from an alcohol by treatment with sulfuric acid. Show two different alcohols you might start with, indicate which would be the better choice, and explain why.

10. Suggest a mechanism for the following reaction:

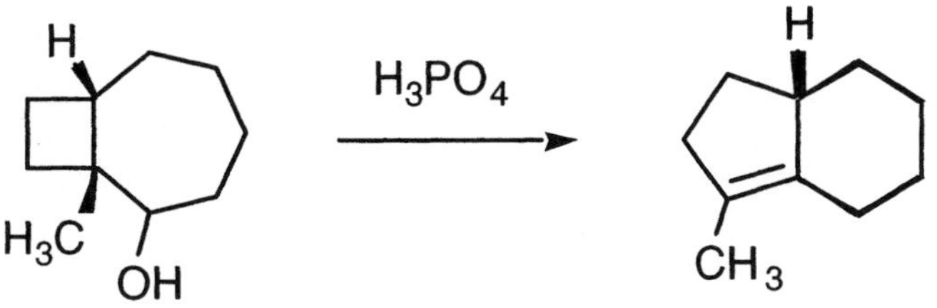

11. *Advanced question. Try to figure out the following rearrangement. There is only one rearrangement step. You will probably need a model to see what is going on.*

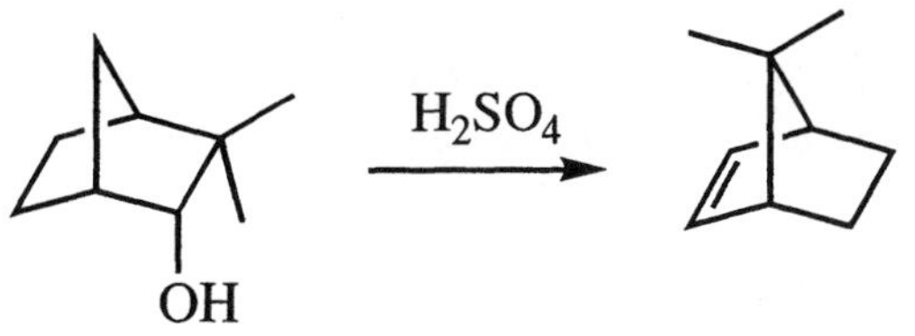

12. *Superadvanced question: Why does the above molecule violate Saytzeff's rule? (Not in what way, but for what reason.)*

13. The following substance could undergo several different reactions. Choose conditions that would encourage each. Identify the type of reaction in each case, and explain why the products are the ones shown.

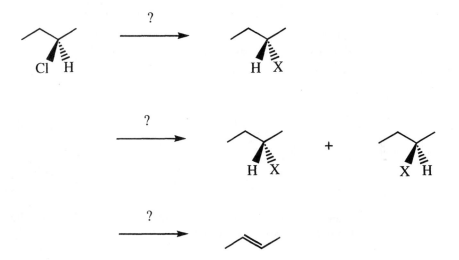

14. *Suggest a way of accomplishing the following conversion, using any reagents you are familiar with.*

(specifically *trans*)

15. Suggest multistep ways to accomplish the following conversions:

16. *In each group, circle the compound that will react fastest with NaI/acetone, put a box around the one that will react fastest in hot ethanol, and a triangle around the one that will react fastest with KO⁺ in DMSO. If you feel that none of the candidates will react at a measurable rate under the given conditions, put a circle, box, or triangle around the word "none."*

a.

NONE

b.

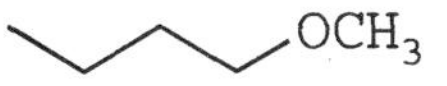

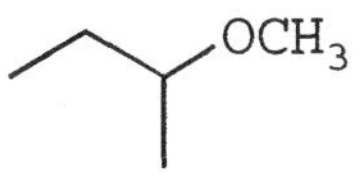

NONE

c.

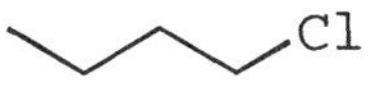

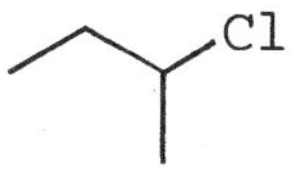

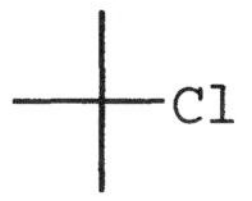

NONE

d

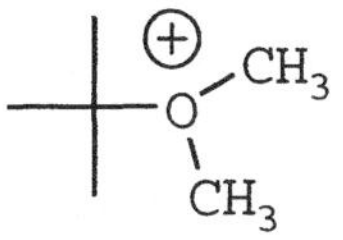

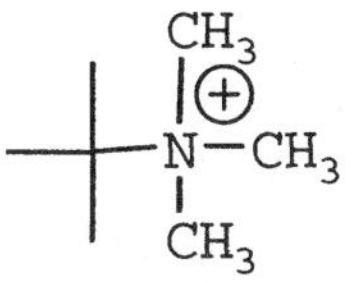

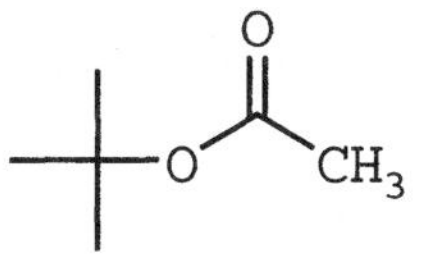

NONE

e.

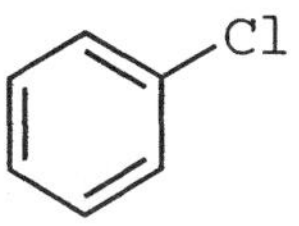

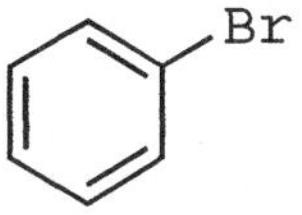

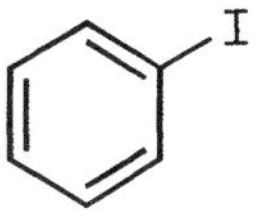

NONE

17. There are many ethers with the formula $C_6H_{13}IO$. Upon treatment with sodium methoxide in methanol, both S_N2 and E2 reactions could occur.

Find an isomer that is chiral and for which ...

a.   ... both the S_N2 and the E2 products are chiral.

b.   ... neither the S_N2 nor the E2 products are chiral.

c.   ... the S_N2 product is chiral but the E2 products are not.

d.   ... the E2 products are chiral but the S_N2 product is not.

e.   ... will undergo S_N2 reaction but not E2 reaction.

f.   ... will undergo E2 reaction but not S_N2 reaction.

18. *A contains only carbon and hydrogen. Upon treatment with acidic water in the presence of mercuric sulfate, a single product B is obtained. A reacts with hydrogen and Lindlar catalyst to give C, which reacts with acidic water to give D and essentially nothing else. D reacts with concentrated sulfuric acid to give almost entirely E and not C. E reacts with ozone followed by zinc to give ketone F and aldehyde G. E reacts with acidic water to give almost exclusively H and not D. H reacts with sulfuric acid to give a mixture of E and I. I reacts with ozone followed by zinc to give J, a single molecule that is both an aldehyde and a ketone.*

Identify A–J. There is a compound for each letter in that series. Note that there is no unique answer; there are many, but they all share some important features. Assume no rearrangements and normal reaction conditions (I haven't tried to trick you by leaving out an ingredient).

19. Compound **A** has the formula C_9H_{16}. When 124 mg of this substance is treated with hydrogen and a palladium catalyst until it refuses to react further, 2 mg of hydrogen has reacted. When **A** is treated with ozone, two different aldehydes are formed. When **A** is treated with HBr, two different compounds, **B** and **C**, are formed in similar amounts. When **B** is treated with potassium *tert*-butoxide in DMSO, **A** is formed almost exclusively, but when **C** is treated the same way, **D** is formed almost exclusively. **D** is an isomer of **A**. When **D** is treated with ozone, an aldehyde and a ketone are formed. When **D** is treated with HBr, a single product **E** is formed almost exclusively. **E** reacts with potassium *tert*-butoxide in DMSO to give **D** and **F** (a new isomer of **A**) in similar amounts. **F** reacts with ozone to give a single compound, **G**, which contains both an aldehyde and a ketone.

    Identify the structures of **A** through **G**. Remember, all are neutral, organic compounds.

    Hint: there are no rearrangements involved in this problem, and stereochemistry (*cis* vs. *trans*) has been ignored. Assume ozone is always followed by zinc.

20. a. Compound **A** is chiral, and contains only C, H, and Cl. When **A** is treated with sodium iodide in acetone, a rapid reaction occurs, and the product **B** is still chiral. When **A** (or **B**) is treated with hot ethanol, there is no reaction. **A** reacts with potassium *tert*-butoxide in DMSO to give **C**, which is also chiral, and **C** reacts with HCl to give almost entirely **D**, not **A**. **D** is also chiral. **D** reacts slowly with sodium iodide in acetone to give **E** (also chiral), and **D** also reacts slowly with hot ethanol. **D** reacts with potassium *tert*-butoxide in DMSO to give mostly **F**, which is not chiral, instead of **C**, which is. **F** reacts with HCl to give mostly **G** instead of **D**. **G** is not chiral and it does not react with sodium iodide in acetone, but it reacts rapidly with warm ethanol to give mostly **F**, which contains only carbon and hydrogen, and **H**, which also contains oxygen. Suggest structures for **A** through **H** that are the simplest ones that meet these descriptions.

    b. What does the information in this problem tell you about the stereochemistry of an S_N2 reaction?

Selected Answers

Internal Problems

Test Yourself 1.

a.

Bromide; better leaving group; product is 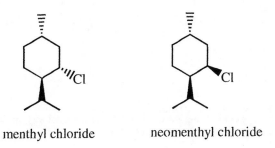 for both.

b.

faster, due to
resonance-stabilized
carbocation

preferred

c.

faster, because
carbocation is
less strained

Test Yourself 2.

These are E2 conditions. Menthyl chloride eliminates in only one direction, which suggests that in the other direction the hydrogen is not properly disposed for elimination. The hydrogen has to be *trans* to the chloride to eliminate, so in menthyl chloride it must be *cis*, making the methyl group *trans* to the chloride. In neomenthly chloride, both directions are OK so the hydrogen must be *trans* to the chloride making the methyl group *cis*.

menthyl chloride   neomenthyl chloride

In aqueous ethanol we are looking at an E1 reaction, where there is no stereochemical requirement for the elimination. Once the carbocation is formed, elimination can occur in either direction. It prefers one, but the other is still possible.

Test Yourself 3.

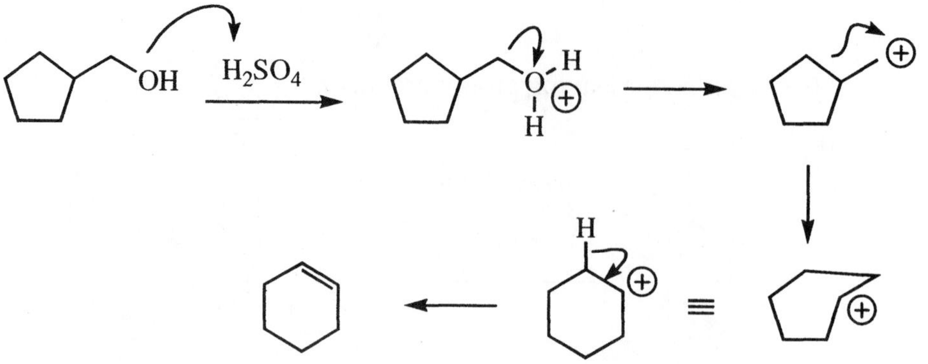

End of Chapter Problems

3.

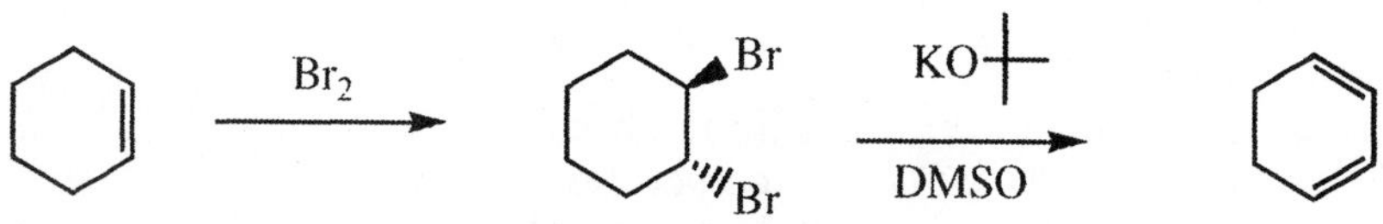

exclusively *trans*

E2 elimination requires the hydrogen to be *trans* to the leaving group. The only qualifying hydrogen, for either bromide, is one of the H's on the CH_2 group. Elimination *must* occur away from the other bromide, leading eventually to 1,3 cyclohexadiene.

5. a.

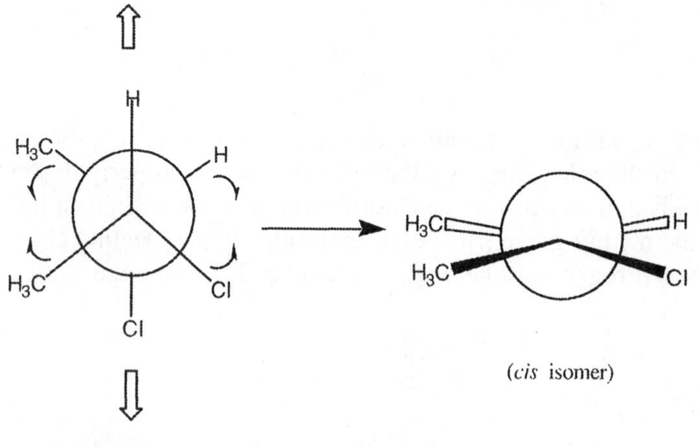

(*cis* isomer)

Reactive conformation of the
meso compound

Reversing the stereochemistry at one of the stereocenters gives a model that, in the reactive conformation, would be compressed to a *trans* alkene when H and Cl leave.

b. (2R,3R) and (2S,3S) give exactly the same product. Mirror-image starting materials give identical (achiral) products.

11.

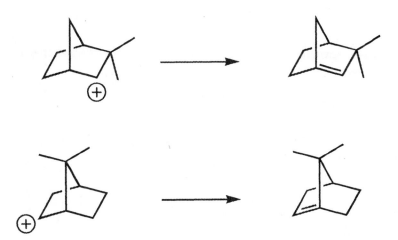

12.

Either before or after the rearrangement one might imagine an elimination to give a trisubstituted double bond rather than the disubstituted one we get:

However, both of these double bonds are impossible. Try building a model of either. You will find that the orbitals that are supposed to be making the π bond are in fact perpendicular to each other and unable to share their electrons.

Double bonds at "bridgehead" positions like these are violations of what is called "Bredt's Rule." Of course, there are many chemists who try to make such molecules deliberately, just to see how good Bredt's Rule is, and there has been some success, but it remains true that bridgehead double bonds are extremely difficult to make.

14.

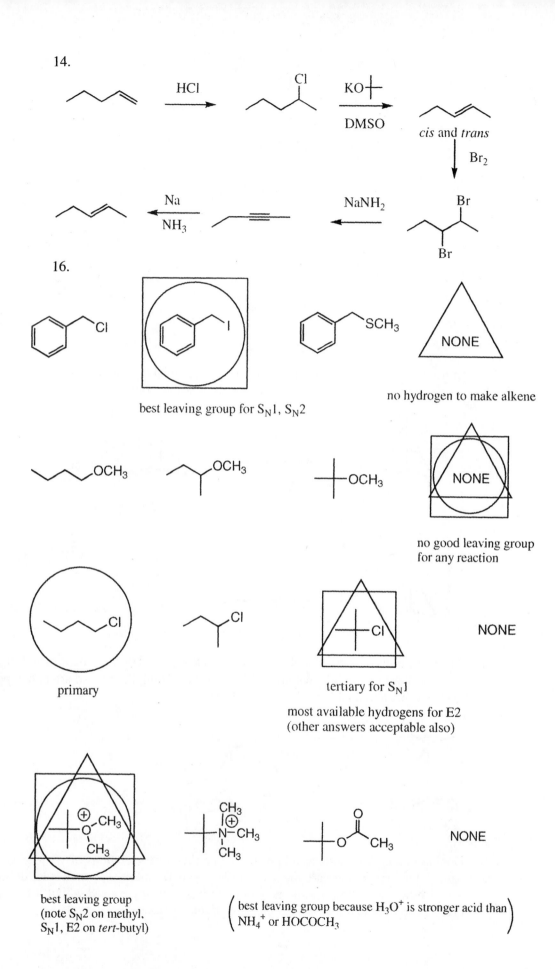

16.

best leaving group for S$_N$1, S$_N$2

NONE

no hydrogen to make alkene

NONE

no good leaving group
for any reaction

primary

tertiary for S$_N$1

most available hydrogens for E2
(other answers acceptable also)

NONE

best leaving group
(note S$_N$2 on methyl,
S$_N$1, E2 on *tert*-butyl)

(best leaving group because H$_3$O$^+$ is stronger acid than
NH$_4^+$ or HOCOCH$_3$)

NONE

none of these reactions
occur with phenyl halides

18.

This kind of problem is called a "road map," and it can be among the most fun of all problems if you let it be. The trick is to organize the information given, decide what it tells you about the unknowns, and then slowly build up what the structures must be. You *cannot* simply make wild guesses and write down trial structures from the beginning. Actually, solving this kind of problem is not unlike a doctor diagnosing a disease from an array of symptoms. You look at all the evidence and decide what fits best.

So what do we know? **A** reacts with acidic water in the presence of mercuric sulfate. There is only one functional group we know that does this: there must be an alkyne in our unknown. Further, since there is only one product, this alkyne must be either at the end of a molecule (a terminal alkyne) or it must be symmetric. Also, the product **B** is a ketone. It turns out that the problem could be answered using either guess, but certainly it will be simpler if we take the former route. So we can guess that **A** is a terminal alkyne, recognizing that we may at some point need to revise that guess, and **B** is therefore a methyl ketone.

A reacts with hydrogen and Lindlar catalyst. Certainly if we had not previously realized that **A** is an alkyne, we should know now. Further, **C** is an alkene. **D** must be an alcohol. Assuming that **A** is a terminal alkyne, **D** must be a secondary alcohol with a methyl group on one side and we don't know what on the other:

$$R\!-\!\!\!\equiv\!\!\!-H \longrightarrow \overset{O}{\underset{R}{\|}}\!\!\!\diagup$$

A   **B**

$$R\diagup\!\!\!\diagdown \longrightarrow R\diagup\!\!\!\overset{OH}{\diagup}$$

C   **D**

Let's continue. **D** reacts with sulfuric acid to give **E**, not **C**. This is clearly an elimination, and if we do not get **C**, the elimination must go to the left, which suggests that whatever is there is more substituted than what is on the right, so that following Saytzeff's rule creates the alkene on the left. Now we can guess something about the piece labeled R: there must be at least one H on the next carbon (so it can eliminate), and also at least one carbon (so it will *want* to eliminate in that direction). So **D** must look something like this

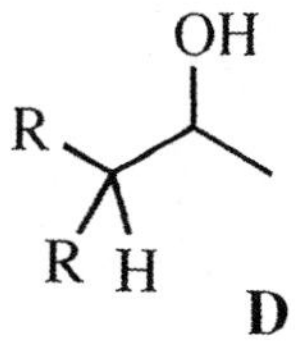

D and **E** must be

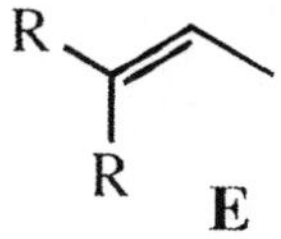

where one of the R's (but not both of them) might be an H. Of course, at this point we can go back and add to what we know about **A**, **B**, and **C**.

E reacts with ozone to give a ketone and an aldehyde. This requires that *both* R groups in the above picture be carbons and not hydrogens.

E reacts with acidic water to give **H** and not **C**. Of course it does—Markovnikov's rule. **H** must be the alcohol with OH on the tertiary center:

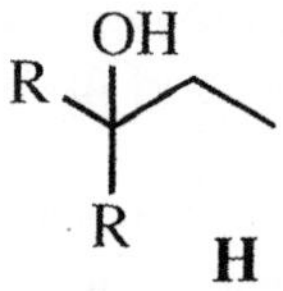

This then eliminates to give two different alkenes, in similar amounts. One of them is **E**, so elimination to the right must be just as good as elimination to the left. This means that the first carbon on each of the R groups is secondary, and elimination to the left produces **I**.

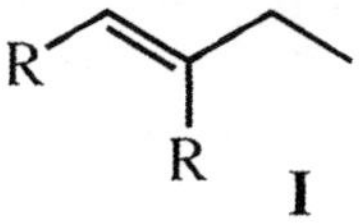

I reacts with ozone to give only one compound. The only way this can occur is if the two R groups are tied together; that is, the left side of this molecule is a ring. It does not matter how large a ring you propose, since we have no further information than this. Call it six:

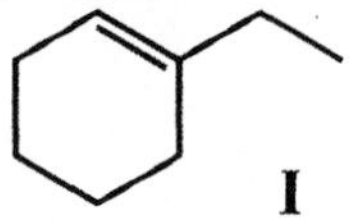

Now you can go back and adjust the structures of all the other lettered compounds.

Chapter 17
Alcohols, Thiols, and Ethers

We already have learned quite a bit about alcohols and ethers, but there are a few aspects of these interesting compounds that we have not yet covered. In this chapter we will

- Review the syntheses of alcohols that we have learned so far

- Review the reactions of alcohols that we have learned so far

- Learn new ways of turning alcohols into leaving groups

- Learn how to oxidize alcohols to carbonyl groups

- Review the syntheses of ethers we have learned so far

- See how thiols compare to alcohols

- Encounter our first protecting group

- Study the reactions of epoxides.

ALCOHOLS

We'll start with a review of what we know about alcohols.

Nomenclature

The naming of alcohols was covered at the beginning of Chapter 13. If you don't remember this, go back and review.

Properties

Physical

We learned a long time ago that alcohols are capable of forming hydrogen bonds. Since alcohols have both OH bonds and lone pairs on O, they can hydrogen bond with partners that have OH bonds (other alcohols and water), with partners that have NH bonds (amines), with partners that have oxygen lone pairs (water, alcohols, ethers), and with partners that have nitrogen lone pairs (amines). This makes alcohols good solvents for many organic compounds. Alcohols are also miscible with many other solvents. Alcohols tend to have higher boiling points than other compounds with similar molecular weights that cannot engage in hydrogen bonding.

Spectroscopic

As we have seen, the OH group of an alcohol produces one of the peaks that is easy to identify in an IR spectrum, appearing to the left of the CH bonds at about 3300–3600 cm^{-1}. This tends to be a strong, broad peak, and it is hard to miss. However, we have to be a little careful: water, after all, has OH bonds, so if there is a little water in a sample you will see a peak in this region even if there is no alcohol group in your compound itself. So be careful when interpreting OH peaks. Nevertheless, if your sample is reasonably pure, an OH peak is an easy thing to see.

In NMR spectroscopy, we have established that the chemical shift of a hydrogen atom in an OH group is difficult to predict. It usually comes somewhere between 1 and 6 ppm. You will also remember that such protons exchange rapidly with each other and with the medium (if the medium contains any similar protons), leading to two phenomena: first, in the presence of D_2O an OH peak disappears due to the H's being replaced by D's; second, the protons in OH groups do not usually participate in coupling with neighbors, because they do not stay put long enough to affect (or be affected by) their neighbors. The chemical shift of hydrogen on a carbon *bearing* an OH group is significantly downfield (near 4 ppm) simply because of the electron-withdrawing influence of the oxygen.

Acid–Base

Alcohols can act as either acids or bases. Look on your pK_a table now and figure out the pK_a of a typical alcohol viewed as an acid, and also the pK_a of the OH group when it acts as a base (technically, the pK_a of the conjugate acid of the alcohol). You should find that the OH protons of an alcohol are quite similar to the OH protons of water, with pK_a values in the high teens; the pK_a of the conjugate acid of an alcohol is not included in the table, but you could predict that it is very much like the conjugate acid of water, about −2. So alcohols have acid–base properties very much like those of water, which means that in acidic solution some alcohol molecules will always be protonated. We have seen this phenomenon already in several reactions. Also, in the presence of a pretty good base, alcohols can lose their OH protons to become alkoxide anions. We have also seen this already, in the synthesis of ethers; for that purpose we used $NaNH_2$ as the base.

$$ROH \quad + \quad \overset{\ominus}{NH_2} \quad \longrightarrow \quad \overset{\ominus}{RO} \quad + \quad NH_3$$

Determine from your pK_a table whether the reaction above will go far in the direction written. Go on, do it. You should find that this reaction (actually an equilibrium) does indeed lie well to the right, which means that $NaNH_2$ is a fine choice as a base for making an alkoxide, but in fact this is not how the reaction is usually carried out. Many of you may already have seen the reaction of sodium with water. This is a fun reaction to watch (if a chemist does it for you; *do not try it on your own!!*), because sodium reacts so violently with water that the system frequently catches on fire. What is happening in this reaction? I will not describe for you the mechanism, because it is pretty complicated, but the result is shown below:

$$HO-H \quad \xrightarrow{\text{Na}} \quad \overset{\ominus}{OH} \quad \overset{\oplus}{Na} \quad + \quad H_2$$

Hydrogen gas is given off (it's actually the hydrogen that is burning if the reaction catches on fire), and what is left in the reaction mixture is sodium hydroxide. You should notice that the equation is not balanced: you actually need two waters and two sodiums, giving two NaOH's and one H_2. It is not the case that the sodium rips off both H's from some water molecule; the sodium atoms rips off one hydrogen from each of two waters. The second hydrogen on each water molecule does nothing, it simply goes along for the ride. Indeed, if the second H were not H, the same type of reaction would still occur.

Of course, if the second H were not an H, then the starting material would not be water but some alcohol, and that is precisely the point. Alcohols undergo this same reaction,

$$RO-H \quad \xrightarrow{\text{Na}} \quad \overset{\ominus}{RO} \quad \overset{\oplus}{Na} \quad + \quad H_2$$

and that is how alkoxides are usually generated. We write it like this:

$$ROH \quad \xrightarrow{\text{Na}} \quad \overset{\ominus}{RO}$$

This should now become part of your working vocabulary.

Synthesis

We have learned several methods for the synthesis of alcohols. The first was the addition of water to alkenes, catalyzed by a bit of acid. The reaction follows the normal path of an addition reaction, where the first step is addition of a proton from the catalyst, creating the most favorable possible carbocation, followed by reaction of the carbocation with water to create a positively charged oxygen, which in turn loses a proton to make an alcohol. The reaction follows Markovnikov's rule and puts the OH group at the position that corresponds to the most stable carbocation, which is usually the more substituted position. Make sure you actually remember the details by drawing out the mechanism of the following reaction:

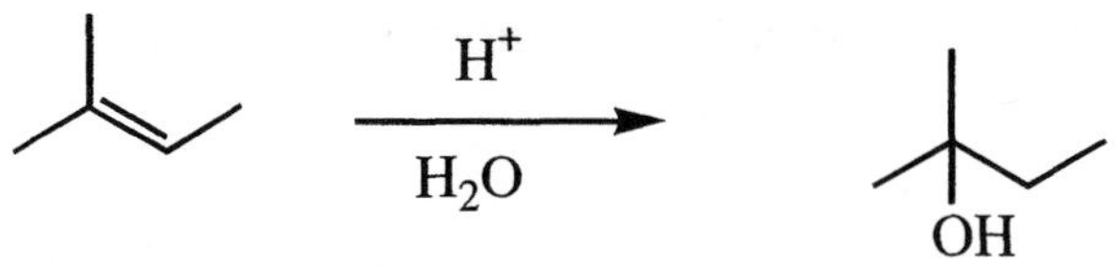

$$\text{(CH}_3)_2\text{C=CHCH}_3 \xrightarrow[\text{H}_2\text{O}]{\text{H}^+} \text{(CH}_3)_3\text{C-CH}_2\text{... OH}$$

Addition reactions like this are subject to rearrangements at the carbocation stage if a better carbocation is available.

The second method we learned was the reduction of carbon–oxygen double bonds (carbonyl groups). If the carbonyl group is from a ketone, you get a secondary alcohol; if from an aldehyde, a primary alcohol. For this purpose we used H_2 and a catalyst, usually Pd/C; I told you at the time that this is actually not the way the reduction is usually accomplished. In the next chapter we will learn a better way to do this.

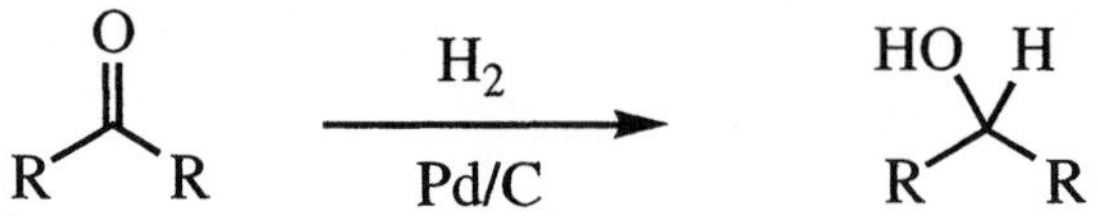

$$\underset{R \quad R}{\overset{O}{\parallel}}\text{C} \xrightarrow[\text{Pd/C}]{\text{H}_2} \underset{R \quad R}{\overset{\text{HO} \quad \text{H}}{\diagup}}\text{C}$$

The third method is replacement of a leaving group, say, halide, with OH. We discussed this in terms of the S_N2 reaction, but now we know that when you add OH$^-$ to a halide, it might substitute, but it might also cause elimination, and there is little we can do about influencing the outcome. The S_N2 reaction is actually not a very good way to convert halides into alcohols. We also discussed this type of substitution in terms of the S_N1 reaction. Again we are faced with competition between substitution and elimination, and again there is little we can do to affect the outcome. To tell you the truth, there is no good way we know of converting a halide into an alcohol. The good part of this is, you will hardly ever want to. As we will discover in the next chapter, alcohols of all types are easy to make in other ways, and as I will show in this chapter, most halides are actually made *from* the corresponding alcohols. So if you want any particular alcohol, you can usually make it without going through the corresponding halide.

Old Reactions

We have previously learned quite a few reactions of alcohols, and you should already have cards for each of them.

Alcohols can be used to make ethers in two ways. The first one we learned was the acid-catalyzed reaction of alkenes with alcohols:

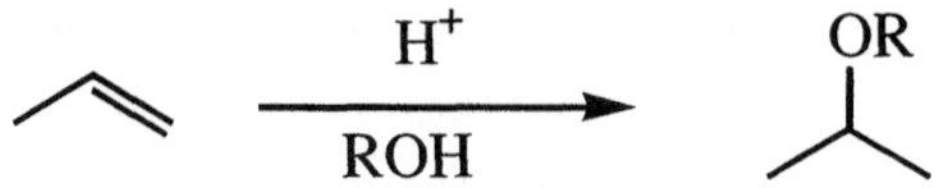

$$\text{CH}_2\text{=CHCH}_3 \xrightarrow[\text{ROH}]{\text{H}^+} \underset{}{\overset{\text{OR}}{\diagup}}\text{C}$$

For practice, write the mechanism of the above reaction. If you run into trouble, review Chapter 10.

The second approach we encountered for ether preparation was deprotonation of an alcohol to make an alkoxide, followed by treatment with an alkyl halide (the Williamson ether synthesis). This involves an S_N2 reaction on the halide, so it is subject to competition with elimination reactions, but if the halide is primary the method can work quite well. Further, we now have a better approach to the deprotonation process:

$$ ROH \xrightarrow{\quad Na \quad} RO^{\ominus} \xrightarrow{\quad R'X \quad} ROR' $$

In both of the above reactions, the carbon–oxygen bond of the alcohol remains untouched. But we also know two reactions that cleave this bond, too. Both occur under acidic conditions, which turn the OH into a leaving group.

$$ ROH \xrightarrow{\quad HX \quad} RX $$

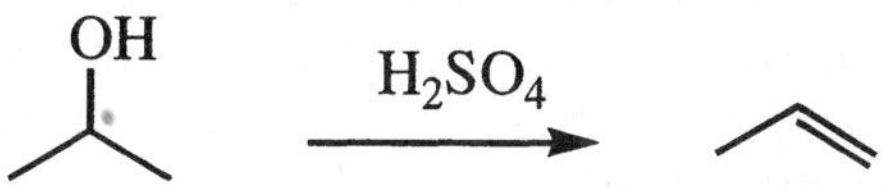

In the first case a nucleophile is provided, and a substitution reaction occurs, either S_N1 or S_N2 depending on the nature of the alcohol. In the second case a nucleophile is deliberately avoided, as is water, giving the molecule little choice but to undergo an E1 reaction leading to an alkene.

New Reactions

Conversion into Halides

There are really only three new reactions we need to discuss at this point. Two are related to substitution reactions. So far we have only one way of turning an alcohol into a halide, namely, treating the thing with strong HBr (or HCl or HI) solution. The reaction proceeds by protonation of the alcohol, making it a good leaving group, and replacement of the leaving group with the halide ion that is present. If the alcohol is primary, this proceeds by an S_N2 mechanism; if tertiary, by an S_N1 mechanism. Rearrangements are normally not a problem for tertiary alcohols, since the corresponding cations are already about as stable as they can get; primary alcohols substitute by S_N2 reactions, which do not involve carbocations, and therefore the reactions are not complicated by rearrangements. However, with a secondary alcohol, there is at least a strong possibility of S_N1 reaction, which brings with it the likelihood of rearrangement. Thus, if you want to change an alcohol into a halide without rearrangement, it would be useful to find alternative conditions that lean toward S_N2 rather than S_N1. Such conditions exist.

The standard reagent for converting primary and secondary alcohols into chlorides is thionyl chloride. To convert the same substrates into bromides, the reagent of choice is phosphorus tribromide.

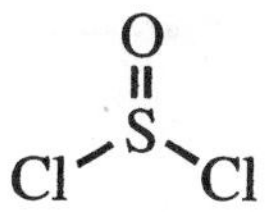

thionyl chloride 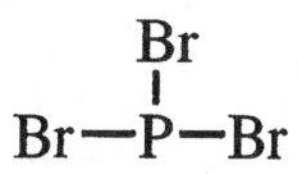   phosphorus tribromide

Both work essentially the same way. Below is the reaction with PBr_3.

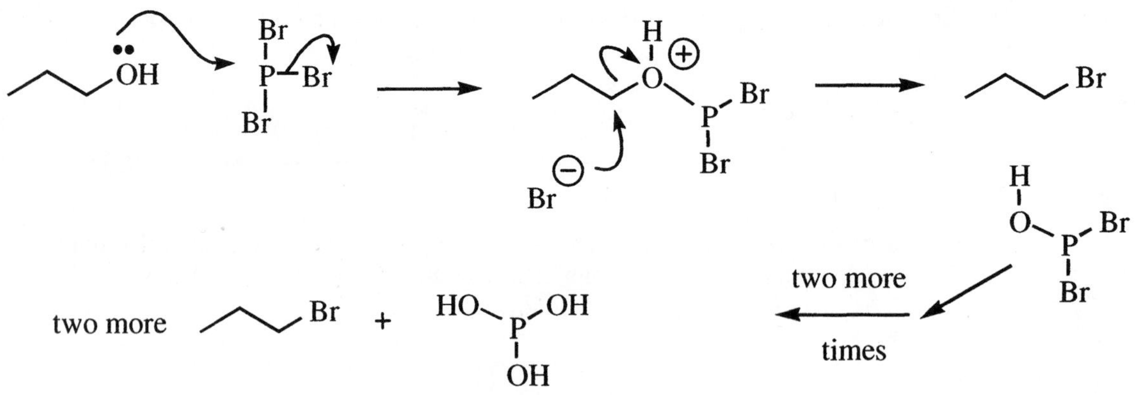

The first step of the reaction is analogous to the first step of the HBr reaction: we are transforming the bad leaving group OH into a good one by giving it a positive charge. The key point you should see in the mechanism is that the actual substitution step, the second step, is an S_N2 reaction. Mostly this is because PBr$_3$ reactions are carried out in non-polar solvents rather than water, so the reaction prefers an S_N2 mechanism. Therefore, secondary alcohols usually can be converted into the corresponding bromides *without* rearrangement. A corollary of this is that tertiary alcohols do *not* react well with PBr$_3$.

Each molecule of PBr$_3$ will change three alcohol molecules into the corresponding bromides. This is important to know when you are running the reaction, so that you will know how to measure out the appropriate amount of reagent; from the point of view of the organic chemist, however, we typically only want to know what the reagent *is*, so we would usually write the reaction this way in a synthesis:

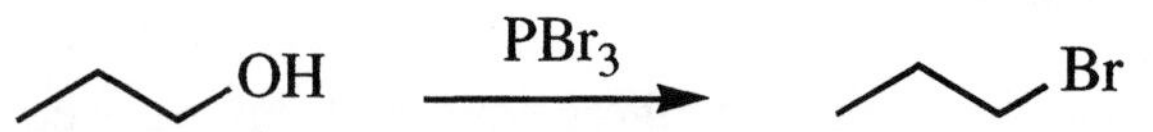

The mechanism for thionyl chloride is only slightly different: it, too, involves turning the lousy leaving group OH⁻ into a better one, and then doing an S_N2 reaction. However, each molecule of thionyl chloride can only convert one molecule of alcohol into chloride, because the reagent is lost in the process:

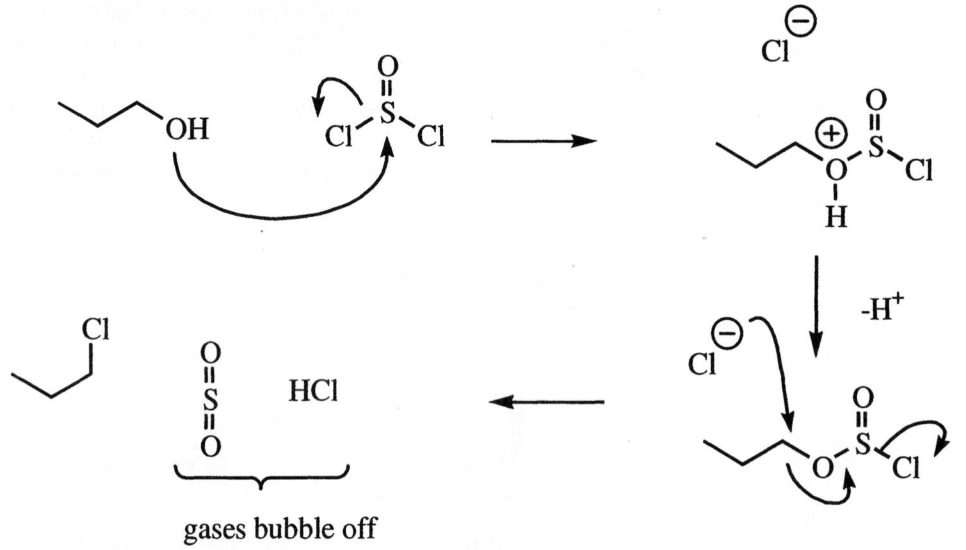

gases bubble off

Because these reagents prefer to substitute by an S_N2 mechanism, the reactions work very well with primary alcohols, and are usually preferred over using HBr or HCl. With secondary alcohols the reaction is not perfect but tends to be much more reliable than reaction with HBr or HCl, so PBr$_3$ and thionyl chloride are the preferred reagents for converting 2° alcohols into halides as well. However, PBr$_3$ and thionyl chloride are very poor choices for tertiary alcohols, and HBr and HCl work much better.

ORM: In "Organic Reaction Mechanisms," under "Alcohols and Ethers," look at "rx alcohol with thionyl chloride."

Tosylates and Related Species

In both of the above reactions, we have converted an alcohol into something else. But we are still quite limited as to what that something else can be. Indeed, in all the cases we have seen so far, the "something else" is a halide. What if we wanted our alcohol to be replaced by, say, a triple bond?

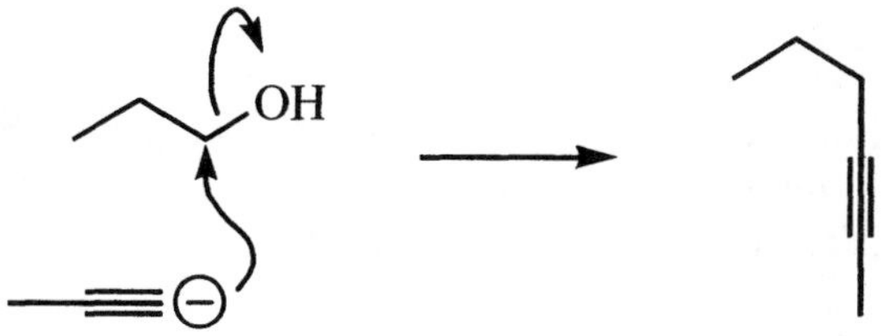

This clearly will not work, for several reasons:

1. OH$^-$ is a lousy leaving group.

2. The acetylide anion would steal the proton from the OH group before it even tried to cause a substitution reaction. The direction is favorable for proton transfer, as you should verify for yourself from your pK_a chart.

3. If you try to make the OH group into a better leaving group by adding acid, you will instead first protonate the acetylide anion and thereby lose your nucleophile.

However, we have an easy way around this problem: we *do* know how to turn an alcohol into a halide. So, if we wanted to accomplish the above transformation, we could simply do it in two steps. Clearly, in the second step, any nucleophile that can replace a halide could be used.

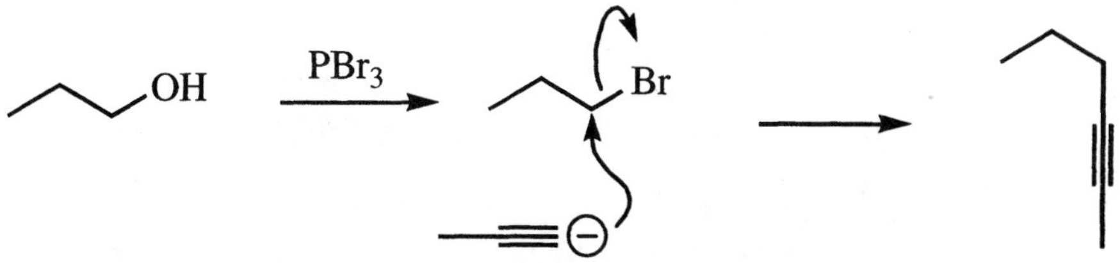

If the alcohol were secondary instead of primary, the same considerations would apply (the S$_N$2 reaction would be less favorable, and you would expect some competition from elimination, but the transformation could still be accomplished):

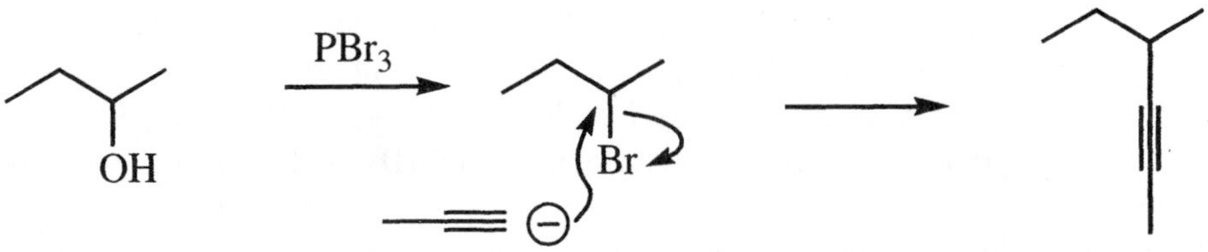

But the secondary case brings up a new consideration: what if you cared about the stereochemistry of the product? Imagine that you needed to achieve the following transformation:

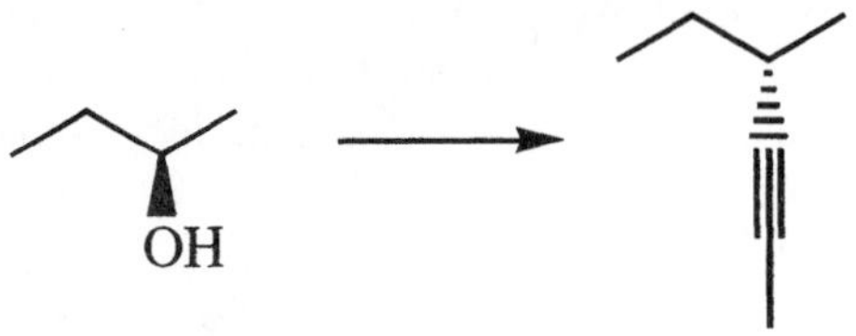

This would not work using the above sequence, because the PBr$_3$ reaction is an S$_N$2 process, resulting in inversion, and the next reaction would produce a *second* inversion, bringing you back to the original stereochemistry.

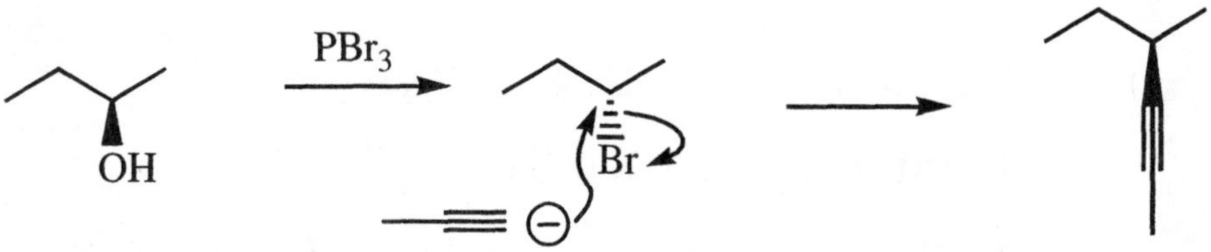

The point is that sometimes it is useful to be able to make the alcohol into a leaving group *without* changing its stereochemistry. We currently have no way to do that. So let's invent one.

Think back to Chapter 13, when we first discussed reactions of alcohols. We decided that we could not do a substitution reaction on an alcohol, because OH$^-$ is a lousy leaving group. But rather than give up, we looked under our list of leaving groups (weak bases on the pK_a chart, remember?) for an oxygen-based leaving group that *would* be good and came up with H$_2$O. If H$_2$O is to be the leaving group, then H$_2$O with a plus charge must be attached to the carbon, and that is how we discovered that acid-catalyzed substitutions of alcohols would work. Now we want to find an alternative, so we should go through the same process: haul out your pK_a table, and look for an oxygen-based leaving group that is good, but *not* positively charged when still attached. In other words, we need a good leaving group that will leave with a negative charge on oxygen.

This should be easy to find. Immediately above water on your pK_a chart you should find PhSO$_3^-$. What is this? Remember, "Ph" is an abbreviation for a benzene ring, so PhSO$_3^-$, should be the following:

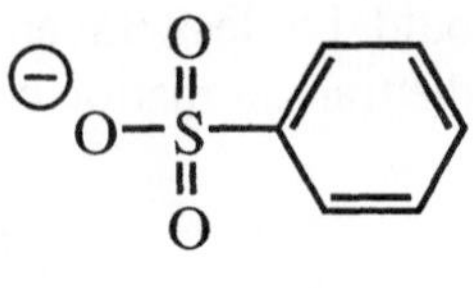

The official name for this is the "benzenesulfonate anion." It should not surprise you that it is a very stable anion and a good leaving group—after all, the anion part of it looks almost like the conjugate base of sulfuric acid, the strongest acid on the chart, and we previously discussed how that is strongly resonance stabilized. Can we make our alcohol into something that looks like this, so that we will have a good leaving group? To do that, we would have to convert the OH group into an OSO$_2$Ph group, as shown below. It turns out there is an easy way to do this.

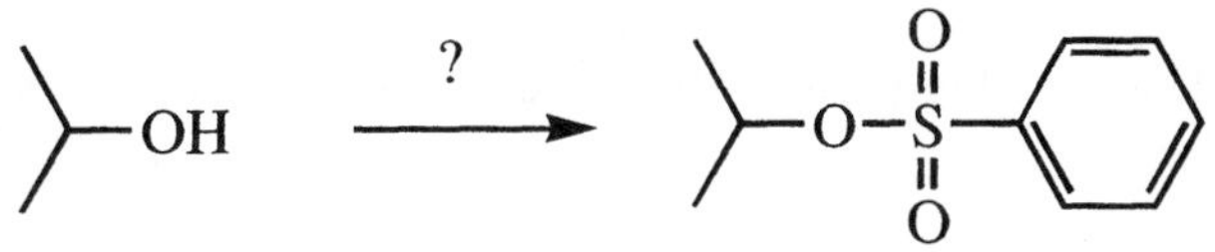

At this point, I am going to add an extra methyl group to the benzene ring we have been drawing, but only because the most commonly used example of this business happens to have one. The methyl group actually has no effect on the chemistry we are going to discuss.

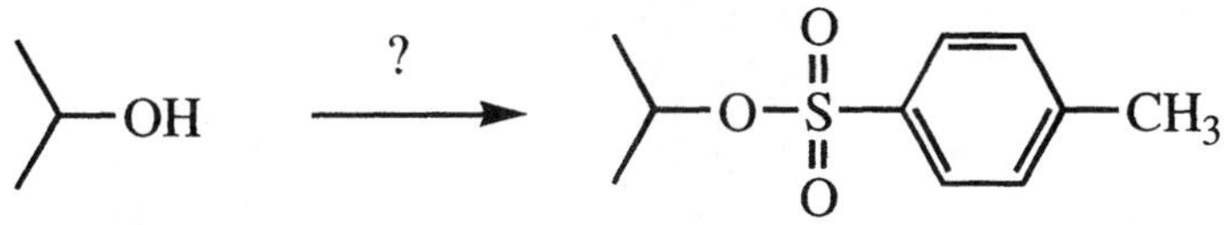

The reaction we use is analogous to an S$_N$2 reaction, but it occurs at a sulfur atom rather than at a carbon. The reagent is *para*-toluenesulfonyl chloride in the solvent pyridine. You do not need to worry about the details of the reaction, except for one very important point: the bond between the carbon and the oxygen of the alcohol is *not involved*, and it remains intact throughout the process. Therefore the stereochemistry of this bond remains intact as well.

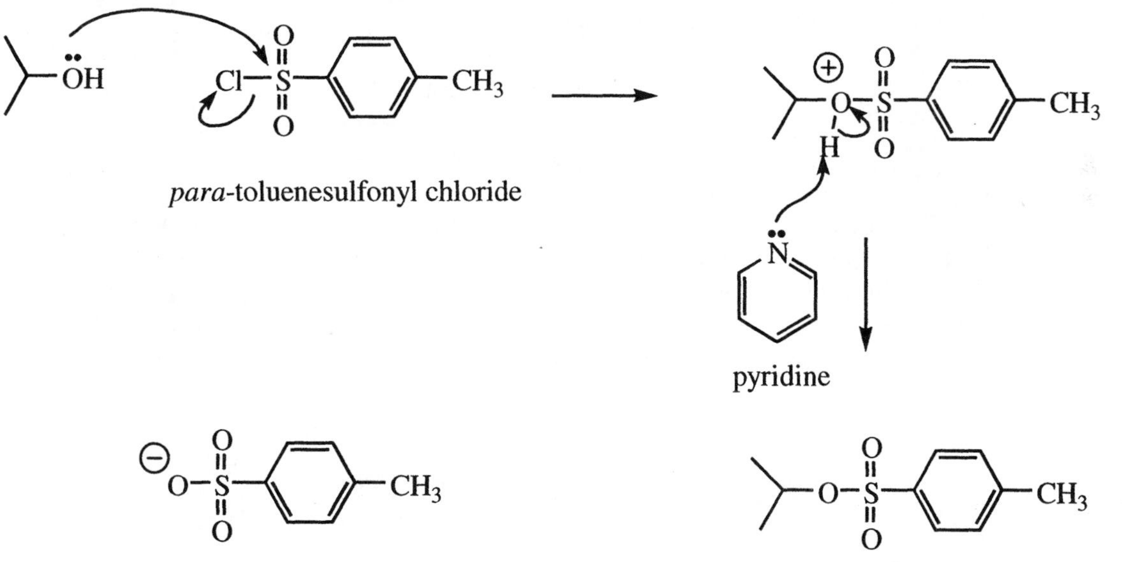

para-toluenesulfonyl chloride

pyridine

para-toluenesulfonate anion

a "*para*-toluenesulfonate ester"

So by treating an alcohol with *para*-toluenesulfonyl chloride in pyridine, we can change the OH group into a good leaving group, the *para*-toluenesulfonate anion, without altering the stereochemistry of the carbon–oxygen bond.

The names for (and structures of) these chemicals are very cumbersome, and they are used often enough that chemists have invented abbreviations for them. The root "*para*-toluenesulfonyl*" is shortened to "tosyl" in spoken language, and at the same time the picture

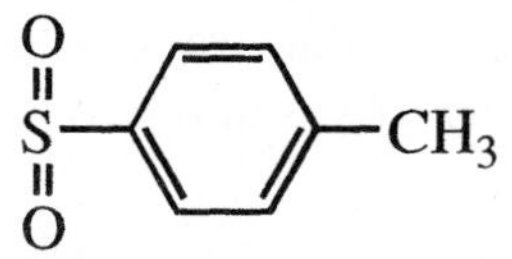

is shortened to Ts

Thus, the big picture above can be written as follows, and the meaning is exactly the same as before.

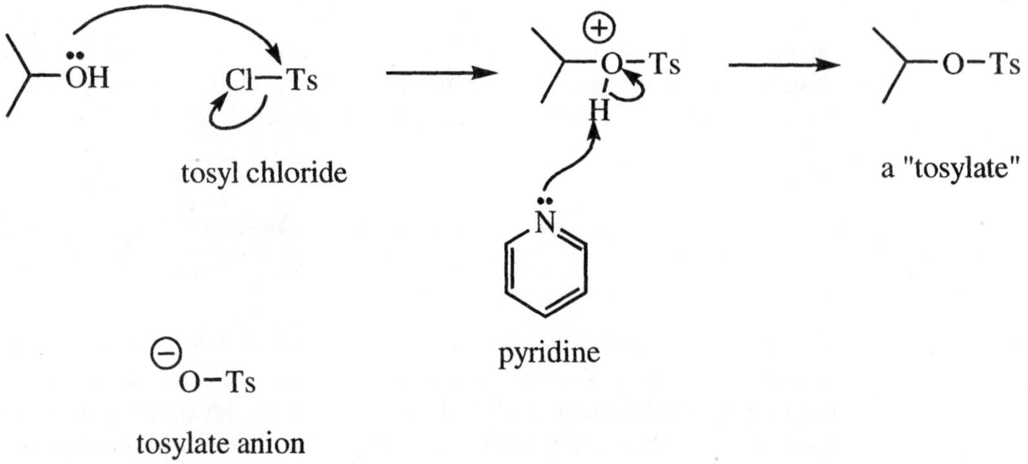

tosyl chloride

pyridine

a "tosylate"

tosylate anion

The process is often abbreviated still further by showing it in the following way, where "Pyr" is a standard abbreviation for "pyridine":

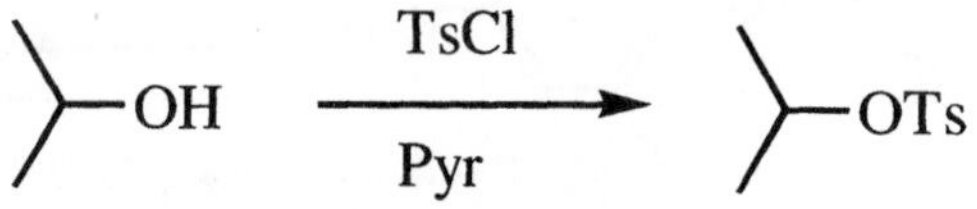

$$\text{-OH} \xrightarrow[\text{Pyr}]{\text{TsCl}} \text{-OTs}$$

The product is a tosylate ester, usually called just a tosylate. Tosylates behave exactly like bromides or chlorides: the tosylate anion is an excellent leaving group, and molecules that contain tosylate groups will undergo all the same substitution and elimination reactions that halides undergo. So let's return to the problem that brought up this subject: you want to accomplish the following transformation:

This is now trivial:

There are many variations on this theme, since the group attached to the sulfur in the tosylate has nothing to do with the reaction. Although tosylate is the most common version, there is also one with a bromine on the benzene ring, called "brosylate," –OBs; one with a nitro (NO_2) group, called "nosylate," ONs; one with only a methyl group attached to the sulfur (instead of a benzene ring), called a "mesylate," –OMs; and one with a trifluoromethyl group there, called "triflate," –OTf. Triflate is one of the world's best leaving groups.

Now, I do understand that most of you will never be doing syntheses in which you need to worry about stereochemistry (or syntheses of any other type, for that matter) in your future careers. So why do I bother you with tosylates and their relatives in this course? The main reason is that biological systems use an almost identical strategy to accomplish the same goals. If your body is processing something you ate, and it decides that it needs to turn some OH group into a leaving group, it can't simply add PBr_3 or TsCl to accomplish the job. Both are too toxic. But it can do essentially the same thing. The biological version of a really good leaving group is a phosphate group:

phosphate ester   phosphate anion

Like the tosylate anion, phosphate anion is strongly resonance stabilized and an excellent leaving group. (Actually, at physiological pH this trianion picks up one proton from its surroundings to become a dianion.) Let's say your body decides that it would be a good idea to convert a particular alcohol into a phosphate ester. How might it do this? Well, how did we do it in the tosyl case? We took the piece we wanted to add, tosyl, attached a leaving group, chloride, and allowed the alcohol to knock off the leaving group. So the body should try the same trick.

Therefore

However, we have just established that the body's version of a good leaving group is not chloride, but phosphate. So we should replace the chloride in the above equation with phosphate.

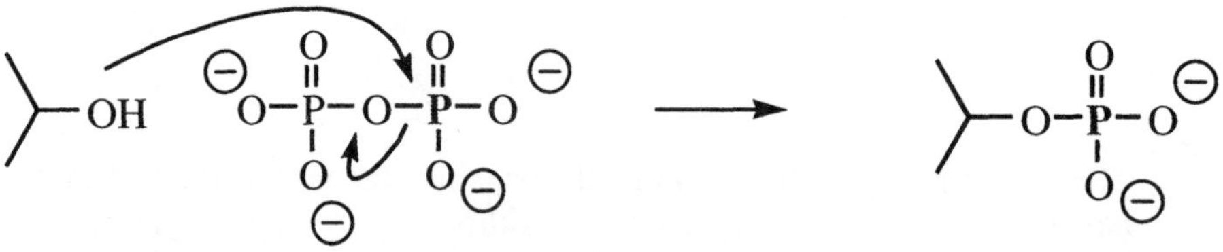

We have now come very close to what actually happens. The difference between this stripped-down version and real life, as is often the case, is a bunch of unimportant garbage attached. Of course, what is unimportant garbage to an organic chemist can be crucial to an organism; the attachments serve extremely important functions in identification and regulation processes. But the guts of the organic chemistry is exactly the same as what we have just invented logically. Below is the actual reagent organisms use to convert alcohols into phosphates.

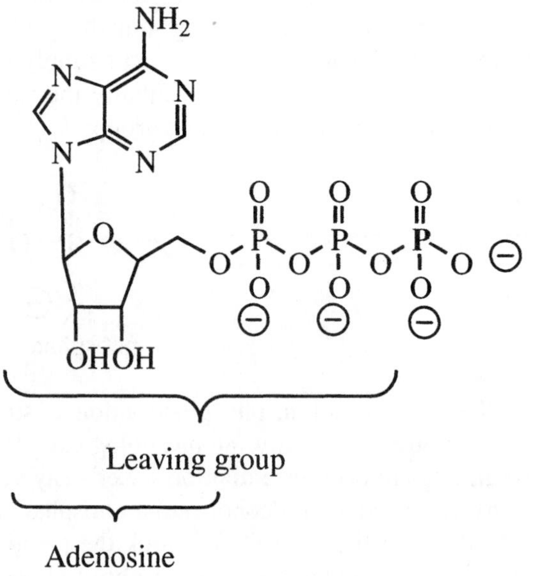

Leaving group

Adenosine

The phosphorus on the right is the one that will attach to the alcohol to turn OH into a leaving group. The entire rest of the molecule is itself a leaving group, serving the same role as the chloride in tosyl chloride. This leaving group has two phosphates—according to our analysis above, only one is necessary, but there is nothing wrong with two. The thing is still a good leaving group. The left-most part of the molecule is a group known as "adenosine," so the whole molecule is called "adenosine triphosphate," otherwise known as ATP. By now you have almost certainly heard of ATP in some biology course. It is usually discussed as the energy carrier within a cell. What they don't tell you is that it is also the tosyl chloride of the cell.

Why is ATP such a good energy carrier? Because during this reaction, and others like it, one of the phosphate groups comes off:

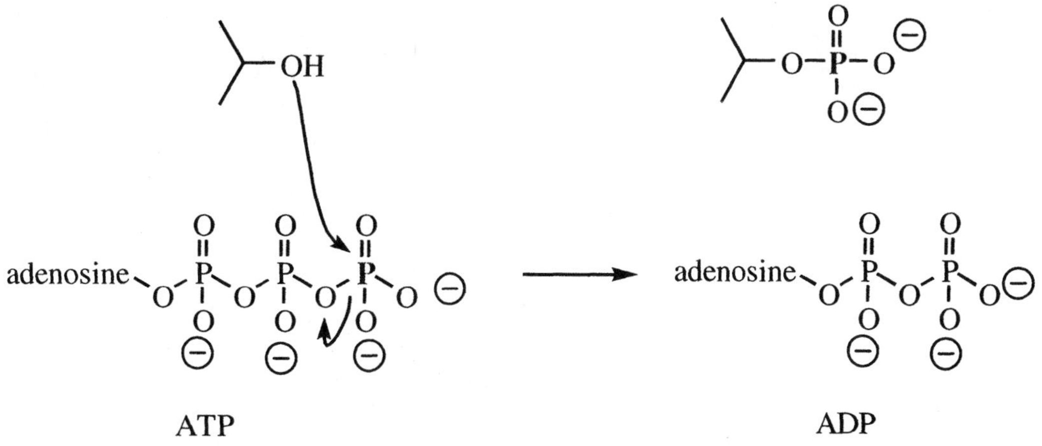

ATP   ADP

This is a great relief for ATP. You will notice that ATP carries four negative charges in proximity. These are all resonance stabilized, but still, the repulsion among them is real and some unhappiness (energy) builds up in this ion. The product ADP only has three negative charges near each other. Three is also not trivial, but it is better than four. Therefore, this reaction gives off energy (is exothermic). So converting ATP to ADP is a way of releasing energy for a cell to use; conversely, if there is extra energy around, converting ADP to ATP is a good way to sop it up. The feedback loops that keep all this in balance are fascinating topics for other courses to deal with.

Oxidation

There is one other extremely important reaction of alcohols, namely, oxidation. The general topic of oxidation–reduction chemistry is best left for another course, but we have already seen some reactions that fall in this category: hydrogenations of double and triple bonds, for example, are reductions, and ozonolysis of alkenes is an oxidation. There are rules that you may have learned in previous chemistry courses for figuring out oxidation states of atoms, and then determining whether oxidation or reduction has taken place. We will not deal with such rules in this course. A simple rule of thumb will get us through: when bonds to hydrogen are introduced at carbon, the carbon is being reduced. When bonds to oxygen are added to carbon, the carbon is being oxidized. This will cover 90% of the cases we will encounter.

The reaction we want to discuss right now is the following:

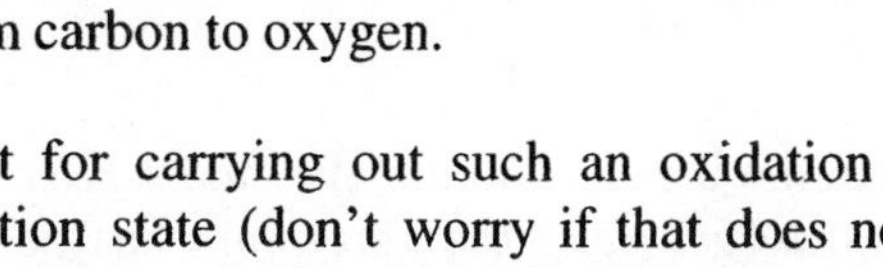

We have already learned how to carry out this reaction in the reverse direction, by using H_2 and a catalyst, and I have mentioned several times that we will soon learn a better way to do it. The right-to-left reaction is a reduction, since hydrogen gets added. The left-to-right reaction is an oxidation, both because it is the reverse of the reduction, and because we have introduced more bonds from carbon to oxygen.

The most common reagent for carrying out such an oxidation (an "oxidizing agent") is chromium in the +6 oxidation state (don't worry if that does not mean anything to you). Below are several common forms of this. Note that in each of them the chromium atoms have six bonds to oxygen, so all the species are equivalent in terms of oxidation state.

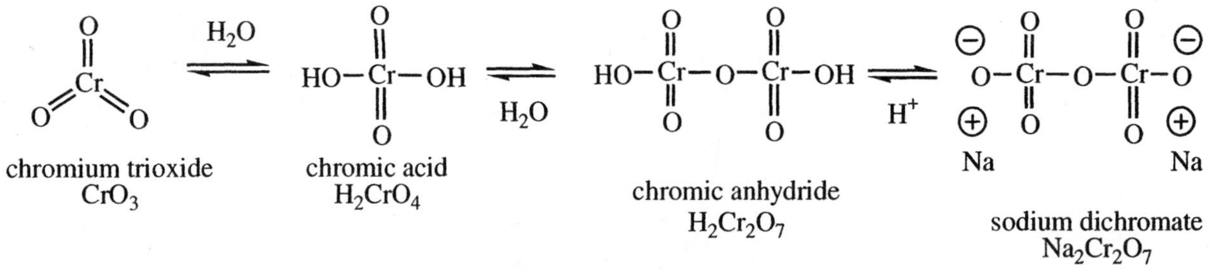

chromium trioxide
CrO_3

chromic acid
H_2CrO_4

chromic anhydride
$H_2Cr_2O_7$

sodium dichromate
$Na_2Cr_2O_7$

It is likely that the actual species causing the reaction is the same no matter which substance you add to the reaction mixture, since they can all interconvert. The most common additive is sodium (or potassium) dichromate, a bright orange crystalline substance, dissolved in water containing sulfuric acid. It should be clear that under these conditions there would certainly be at least some chromic acid present. For the sake of simplicity, I will show how this reaction would work if chromic acid were the true oxidizing agent.

Under acid conditions, the OH group of chromic acid can be protonated; the leaving group water can then fall off and be replaced by the OH group of our alcohol. This is nothing more than an S_N1 reaction taking place on chromium rather than carbon.

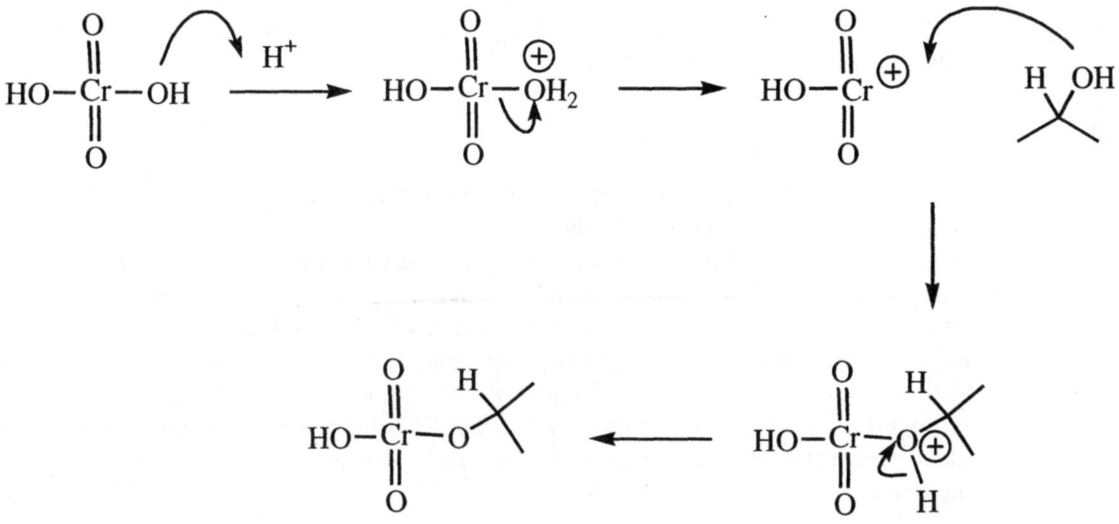

Notice that neither the carbon nor the chromium has changed its oxidation state in this process (each has the same number of bonds to oxygen that it started with). At this point the bonds reorganize themselves in a process you could not predict:

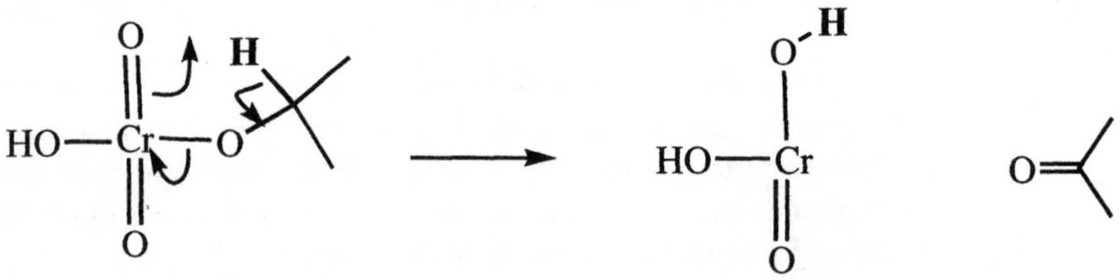

Now both the carbon and the chromium have changed. The carbon has been oxidized (more bonds to oxygen) and the chromium has been reduced (fewer bonds to oxygen). The net reaction is the following:

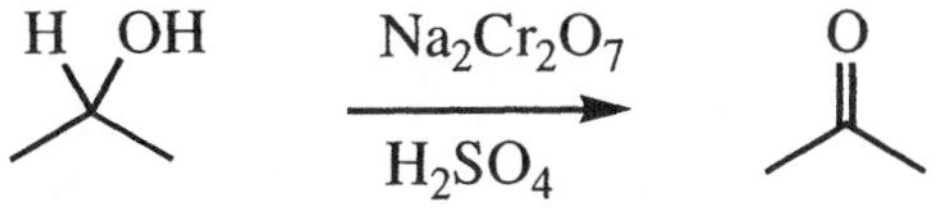

$$\text{H OH} \quad \xrightarrow[\text{H}_2\text{SO}_4]{\text{Na}_2\text{Cr}_2\text{O}_7} \quad \text{O}$$

This is a new kind of transformation for you, and it is one you will use often, so you should learn it well. You should also make up your normal three cards to catalog it.

Unfortunately, the actual reaction is a bit more complicated than I have let on so far, so we need to discuss it further. I illustrated the reaction above using a secondary alcohol. What I showed is correct. But what if the alcohol we started with were a tertiary alcohol? The first part of the reaction (the S_N1 reaction on chromium) could proceed as written, but at that point the reaction could go no further (and it would actually go back where it came from). Why? Because the second step of the oxidation reaction, the one in which the bonds reorganize, requires a hydrogen on the carbon, and in a tertiary alcohol there is none. Thus, tertiary alcohols do not undergo this reaction.

What about primary alcohols? Here is where the situation gets a little messy. The reaction begins exactly as predicted:

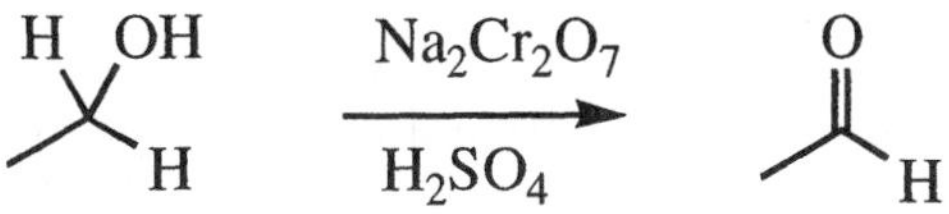

$$\text{H OH} \atop \text{H} \quad \xrightarrow[\text{H}_2\text{SO}_4]{\text{Na}_2\text{Cr}_2\text{O}_7} \quad \text{O} \atop \text{H}$$

However, as we mentioned in Chapter 12 (and will cover in more detail in Chapter 18), aldehydes and ketones participate in the following equilibrium:

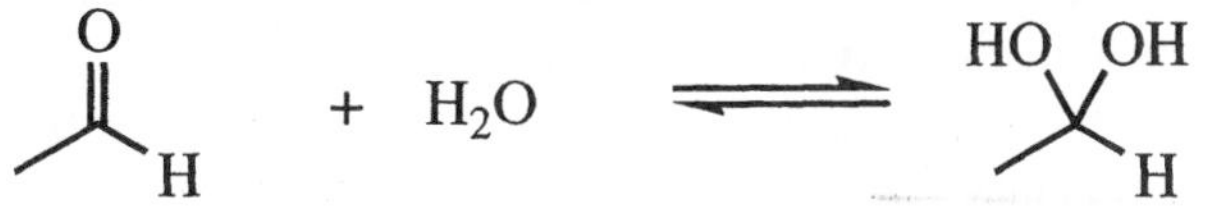

$$\text{O} \atop \text{H} \quad + \quad \text{H}_2\text{O} \quad \rightleftharpoons \quad \text{HO OH} \atop \text{H}$$

"hydrate"

If you start with a ketone (the H in the above picture would be a carbon), then the product can do nothing but go back where it came from. But if you start with an aldehyde, as shown above, the product hydrate has both an OH and an H on the same carbon, which makes it susceptible to further oxidation by chromic acid. Since all this takes place in the reaction mixture, where there is lots of chromic acid around, the reaction continues:

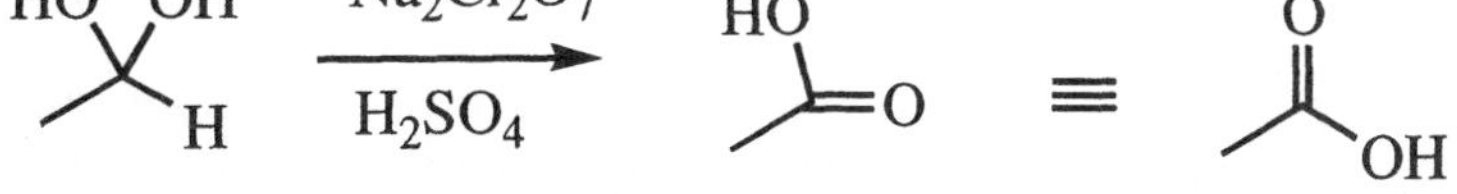

$$\text{HO OH} \atop \text{H} \quad \xrightarrow[\text{H}_2\text{SO}_4]{\text{Na}_2\text{Cr}_2\text{O}_7} \quad \text{HO} \atop =\text{O} \quad \equiv \quad \text{O} \atop \text{OH}$$

You should recognize the product of this step as a carboxylic acid. So treatment of a secondary alcohol with chromic acid produces a ketone, but treatment of a primary alcohol under the same conditions does not produce an aldehyde, but instead gives a carboxylic acid.

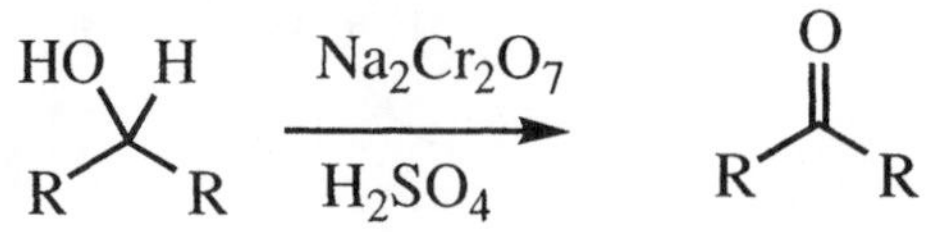

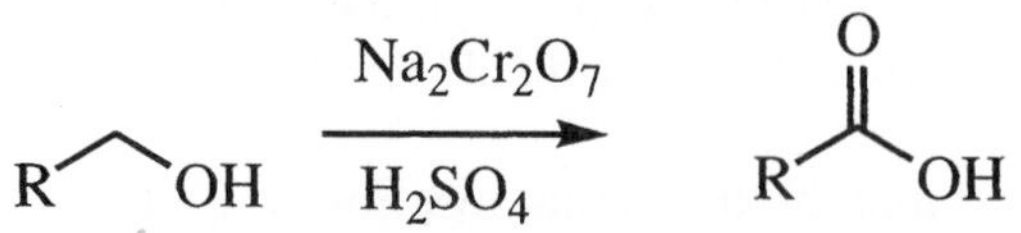

$$\underset{\substack{R \quad R}}{\overset{\substack{HO \quad H}}{\bigwedge}} \xrightarrow[\substack{H_2SO_4}]{\substack{Na_2Cr_2O_7}} \underset{\substack{R \quad R}}{\overset{\substack{O}}{\bigwedge}}$$

$$R\diagup\!\!\diagdown OH \xrightarrow[\substack{H_2SO_4}]{\substack{Na_2Cr_2O_7}} \underset{\substack{R \quad OH}}{\overset{\substack{O}}{\bigwedge}}$$

That is fine if you want to make a carboxylic acid out of your primary alcohol, and this is in fact a perfectly good way to make carboxylic acids (make cards!!!). But what if you *wanted* an aldehyde? Is there any way to oxidize a primary alcohol to an aldehyde *without* getting the carboxylic acid? The answer is yes, and one convenient method was derived from a knowledge of how the above reaction works. Below is a short review of how a primary alcohol turns into a carboxylic acid:

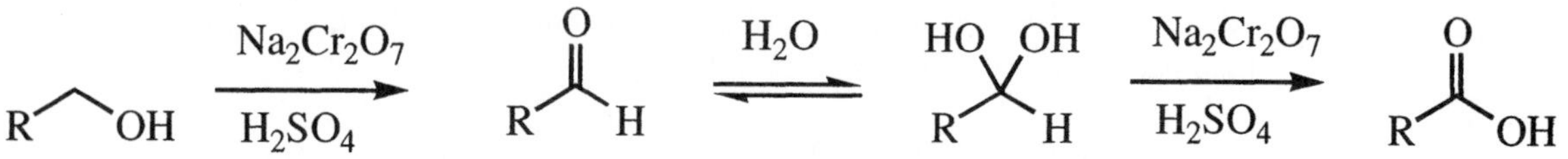

$$R\diagup\!\!\diagdown OH \xrightarrow[\substack{H_2SO_4}]{\substack{Na_2Cr_2O_7}} \underset{\substack{R \quad H}}{\overset{\substack{O}}{\bigwedge}} \xrightleftharpoons{\substack{H_2O}} \underset{\substack{R \quad H}}{\overset{\substack{HO \quad OH}}{\bigwedge}} \xrightarrow[\substack{H_2SO_4}]{\substack{Na_2Cr_2O_7}} \underset{\substack{R \quad OH}}{\overset{\substack{O}}{\bigwedge}}$$

The important point is that it is not the aldehyde itself that becomes the carboxylic acid, it is the aldehyde *hydrate*; the aldehyde cannot become a hydrate without water! So the key to keeping aldehydes from becoming carboxylic acids in this reaction is to avoid water. But therein lies a problem: sodium dichromate is a salt that is highly insoluble in everything other than water. Using sodium dichromate as the reagent, you cannot avoid water! So the challenge becomes, can we design a reagent that looks and acts like sodium dichromate (or chromic acid, or chromium trioxide...) but does not require water to dissolve it? The answer is yes: there are now many ways to do this, but E. J. Corey, a Nobel Prize-winning chemist from Harvard, devised one of the first and most popular versions, shown below:

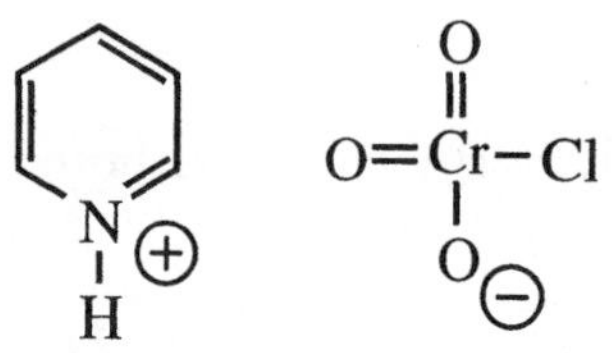

pyridinium chlorochromate
PCC

Although PCC is a salt, consisting of a minus piece and a plus piece, and most salts are soluble in water and not much else, *this* salt happens to be soluble in organic solvents, so it can be used in the absence of water. It is the absence of water (that is, the solubility properties of the reagent) and not any magic reactivity that allows PCC to be used to make aldehydes from primary alcohols. The reaction can be shown like this:

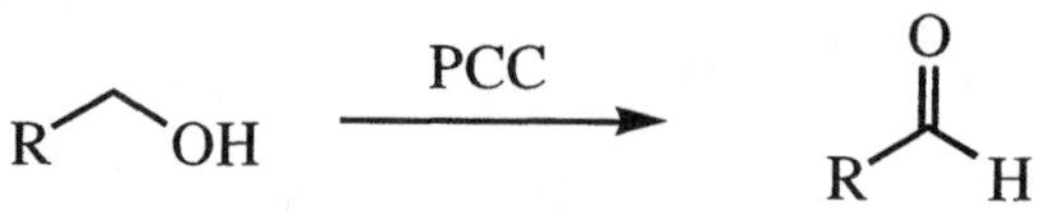

$$R\diagup\!\!\diagdown OH \xrightarrow{\substack{PCC}} \underset{\substack{R \quad H}}{\overset{\substack{O}}{\bigwedge}}$$

Add this one, too, to your card file and your memory banks.

It should be noted that PCC works equally well to make ketones from secondary alcohols, but so does sodium dichromate in sulfuric acid, and the latter is a lot cheaper, so most of the time PCC is used only for oxidizing primary alcohols to aldehydes.

COMMON MISTAKES: DON'T LET THIS HAPPEN TO YOU!

Many students, far too often, propose converting alcohols into carbonyl groups by treatment with base. And why not? After all, the only difference between an alcohol and a carbonyl group is that pesky hydrogen on the oxygen, right? Let's just yank it off!

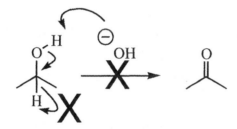

What these students are forgetting is that there is *another* hydrogen, on the carbon, one that is not shown but is assumed in the picture of the alcohol. In order to carry out the above reaction, there would have to be a third arrow, as shown in the scheme below, which makes it clear that the proposed reaction involves H⁻ as a leaving group. This is not going to happen.

Be sure you always use an oxidizing agent for an oxidation!

I should mention that the inorganic chemistry of this oxidation reaction is also more complicated than I have let on. I showed chromium +6 being reduced to chromium +4, which does occur, but the reaction does not stop here: the ultimate product in terms of chromium has chromium in the +3 state. This has no effect on the organic chemistry, so I will not discuss it further. Perhaps when you get to inorganic chemistry you will learn more about the subject.

THIOLS

Thiols, also called "mercaptans," are the sulfur equivalent of alcohols, thus, RSH. (In general, the prefix "thio" points to "sulfur.") Thiols behave very much like alcohols in many respects, and most of the differences are predictable. One immediate difference that you should recognize is that while alcohols are capable of hydrogen bonding, thiols are not. Thus, thiols should be less soluble in water than the corresponding alcohols, and they should have lower boiling points. This tends to be true.

Another difference you should be able to predict, based on our previous discussions of acidity and nucleophilicity, is that thiols should be more acidic than alcohols, so the conjugate bases, the thiolate anions, should be weaker bases than alkoxides, but the same thiolate anions should be *stronger* nucleophiles than alkoxides. This is consistent with trends we have observed previously with the halides—for example, where we found that I⁻ is the weakest base but the strongest nucleophile among the halides. One thing this says is that making thioethers should be particularly easy. You will recall that making ethers using alkoxides as nucleophiles in S_N2 reactions is complicated because of elimination reactions:

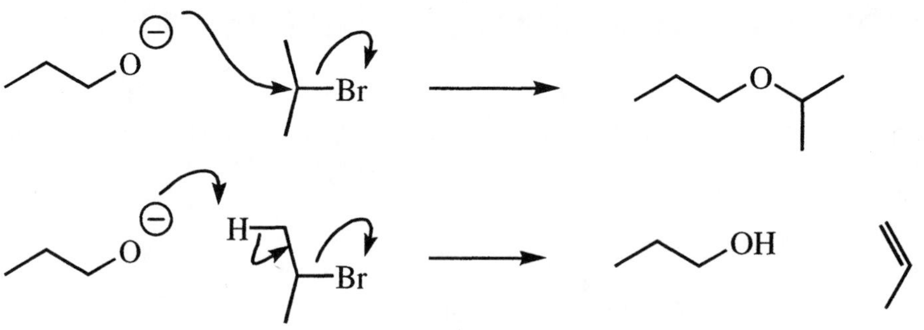

Based on the above considerations (S^- is a weaker base but a better nucleophile than O^-), we should expect—and we actually find—that the top reaction below goes with very little competition from the bottom one.

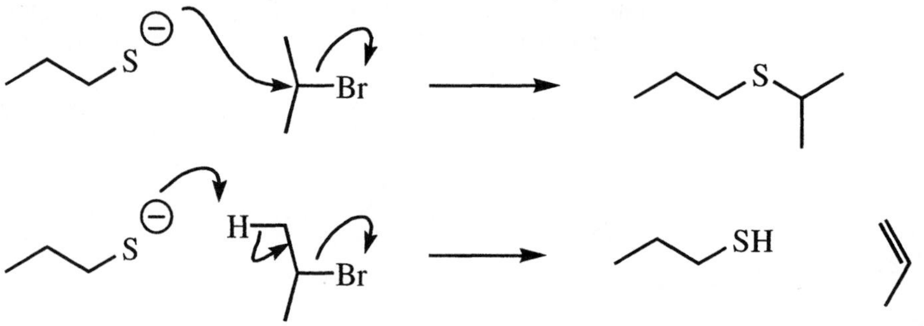

There are two aspects of thiol chemistry that you could not predict by applying the knowledge you so far have. One is that they STINK! Thiols are the active component in skunk smell, fart smell, rotten egg smell, the smell in your urine after you eat asparagus, and almost every other nasty smell you can think of. (For everything you never wanted to know about farts, check out www.heptune.com/farts.html). A trace of a thiol is intentionally added to natural gas so that you can easily tell when there is a gas leak in your house, since the hydrocarbons (methane, ethane, etc.) that serve as fuels have no smell themselves.

The second important extra difference between thiols and alcohols is that the following reaction involving sulfur compounds is pretty easy to accomplish, in both directions, and this is not true with oxygen. Indeed, in the oxygen case, the reaction goes very rapidly, sometimes explosively, but only to the right, because the oxygen–oxygen bond is very weak, while sulfur is quite comfortable in either form.

$$\text{RS}-\text{SR} \rightleftharpoons \text{RSH} \quad \text{HSR}$$

disulfide $\qquad$ thiols

Of course this does not happen all by itself: there are hydrogens added in the forward direction ("reduction") and removed in the backward direction ("oxidation"), and these have to come from and go to somewhere, but the important point is that appropriate sources and destinations for hydrogen are not hard to come by, and such a reaction is easily accomplished, in either direction, with the right reagents.

So what? Recall that enzymes are the workers of the cell, each specializing in a particular transformation and then passing the modified substrate on to the next enzyme, like an assembly line. Every enzyme is perfectly suited for its special job by virtue of its shape. But most of an enzyme's structure is actually nothing more than a huge long piece of spaghetti, all coiled and tangled in a ball; it is "clefts" (gashes, depressions) in this ball that allow a substrate to come in and go out, orienting it just so, and the necessary tools also project into the cleft's "active site" in just the right places to do their jobs. Thus, an enzyme's function is critically dependent on shape.

Unlike a piece of spaghetti, an enzyme has the remarkable ability to land on its plate and curl up in exactly the same way every time you drop it. We have already discussed some of the reasons for this, including hydrogen bonds that stabilize certain pieces and encourage certain parts to be near water, and hydrophobic interactions that encourage certain other parts to bury themselves on the inside and stay away from water. There is a third aspect to enzyme folding that we have not mentioned yet, one that is not so much involved with making the spaghetti fold up right, but which is very important in keeping it that way once it has found its preferred shape.

One of the natural amino acids is cysteine:

$$\text{H}_2\text{N} - \overset{\displaystyle \overset{\text{SH}}{|}}{\text{C}} - \text{COOH}$$

When this piece is incorporated into a long chain made up of other amino acids, the CH_2SH group sticks out from the chain. Normal enzymes (and other proteins) have SH groups like this scattered in various places along their structure. After the spaghetti has folded up into its preferred structure, it often happens that two such CH_2SH groups, from distant parts of the chain, happen to be near each other. The molecule (the protein) then often arranges for these SH groups to be oxidized and joined in a disulfide bond (called a "cystine" bond) as shown in the cartoon below. This in turn makes it easier for the protein to maintain its shape.

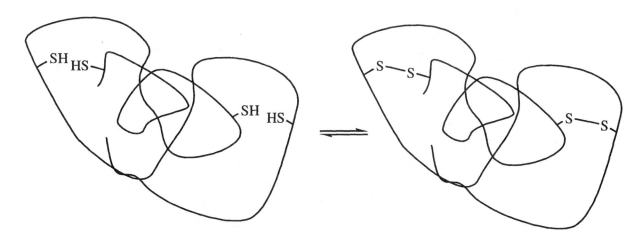

Another special application of this chemistry occurs in hair salons. As it happens, hair is a protein that has a lot of cysteine residues in it, and strands of hair have a lot of disulfide bonds holding the protein in place. If you are unhappy with the current shape of your hair, you can change it: simply undo all the disulfide bonds, lay your hair out the way you want it to be, and in the new shape lots of cysteines will again turn out to be near each other. Then simply make new disulfide bonds and the hair will stay locked into its new shape. This is what a perm is!

ETHERS

Nomenclature

Ethers are named by naming the two groups attached to the oxygen, one after the other, followed by the word "ether," as in the following examples. Sometimes it is easier to name an ether as an alkoxy substituent on something else—this is OK also. And there are a few ethers with special names that appear to have nothing to do with the ether functional group. These are exceptions, and you probably don't need to memorize many such names, although THF shows up often enough that it certainly would not hurt to know what it is.

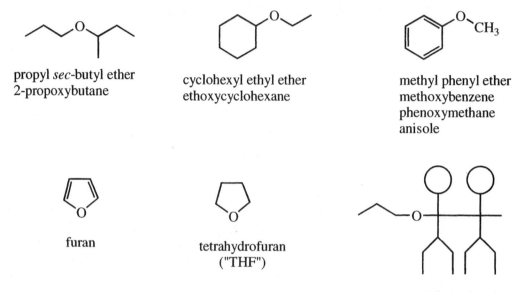

propyl *sec*-butyl ether
2-propoxybutane

cyclohexyl ethyl ether
ethoxycyclohexane

methyl phenyl ether
methoxybenzene
phenoxymethane
anisole

furan

tetrahydrofuran
("THF")

propyl people ether
(get it? Purple people eater?)

There are two special cases of the use of the word "ether" that you probably should remember. One particular ether is almost always referred to simply as "ether," rather than "something something ether." This is diethyl ether. When you read that a substance is dissolved in ether, or extracted with ether, it is diethyl ether that is being used. Diethyl ether is by far the most common ether, due to its low cost and low boiling point.

The second special use of the word "ether" is a holdover from ancient history. Before the discovery of various gases, the term "ether" was used to describe a supposedly invisible substance all around us through which light and sound were thought to be propagated. Later the term became used for "vapors" in general. The portion of petroleum that vaporizes easily, and then is condensed and used as a solvent, is often referred to as "petroleum ether," or "pet ether" for short. It is not an ether at all, in the chemical sense; indeed, it is nothing but alkanes, mostly pentanes, hexanes, heptanes, and octanes, depending on what boiling range you select. Pet ether is a common solvent in organic labs, but do not be deceived into thinking that it is chemically an ether.

Properties

Ethers, as might be expected, are more polar than hydrocarbons but less polar than alcohols. As a result, they tend to be pretty good solvents for a wide range of organic compounds. "Ether" (diethyl ether) and THF are very common solvents in organic labs. Besides being a good solvent for organic compounds, ether is quite insoluble in water and has a low boiling point (it cannot hydrogen bond to itself), which makes it ideal for separating organic materials from inorganic materials in a process called "extraction." You will probably learn about extraction in some lab course, if you have not encountered it already. It is interesting to note that THF, which is simply diethyl ether with its ends tied together, *is* soluble in water. It is not clear (even to me!) why these two substances should be so different.

Ether is also an anaesthetic, and was once commonly used by doctors to put patients to sleep. An organic lab with lots of ether in the air can put students to sleep (as opposed to a normal organic lab, which is so exciting the students are alert all the time, right?). Ether is also highly flammable. For both of the latter reasons, one should be very careful in handling ether.

There is a third caution that organic students should be aware of: on standing in the air, ethers tend to form peroxides. The reaction is shown below for diethyl ether:

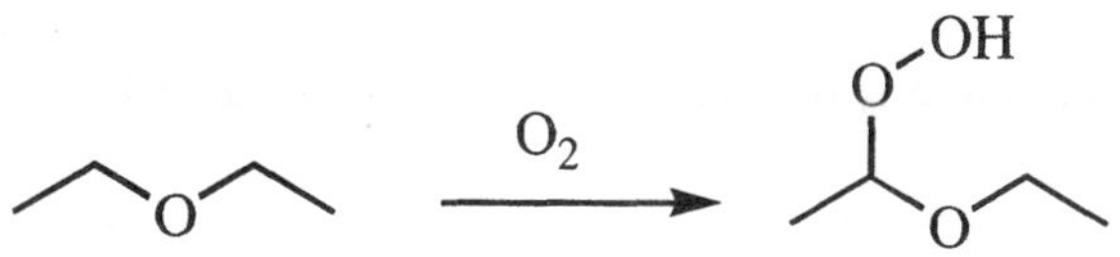

I will not discuss the mechanism of this, and I do not intend for you to learn it like the other reactions we have covered (no need for cards, for example). I just want you to be aware that a bottle of ether that has been open for a while could have some of this peroxide in it. If that ether is then used as a solvent, and later evaporated (as often occurs), the small amount of peroxide will not evaporate as fast as the ether itself (it is bigger, and can hydrogen-bond), so it will become more and more concentrated until it becomes the major substance left in your flask. So what? Peroxides are explosive when they are concentrated! Many reactions (and chemists) have been ruined by violent explosions happening during the evaporation of ether.

Preparation

Simple ethers like diethyl ether and THF are cheap and easily purchased, but many ethers must be custom made. We have already encountered two ways of making ethers. One of our first reactions, way back in Chapter 10, was the reaction of an alkene and an alcohol in the presence of acid. This is a way of making ethers from alkenes, or a way of making ethers from alcohols, depending on your point of view. In real life, your "point of view" in this sense will depend on what is your most valuable starting material. For example, the following two reactions amount to exactly the same chemical process, but the first would almost certainly be described as the preparation of an ether from an alcohol, while the second would be thought of as the preparation of an ether from an alkene:

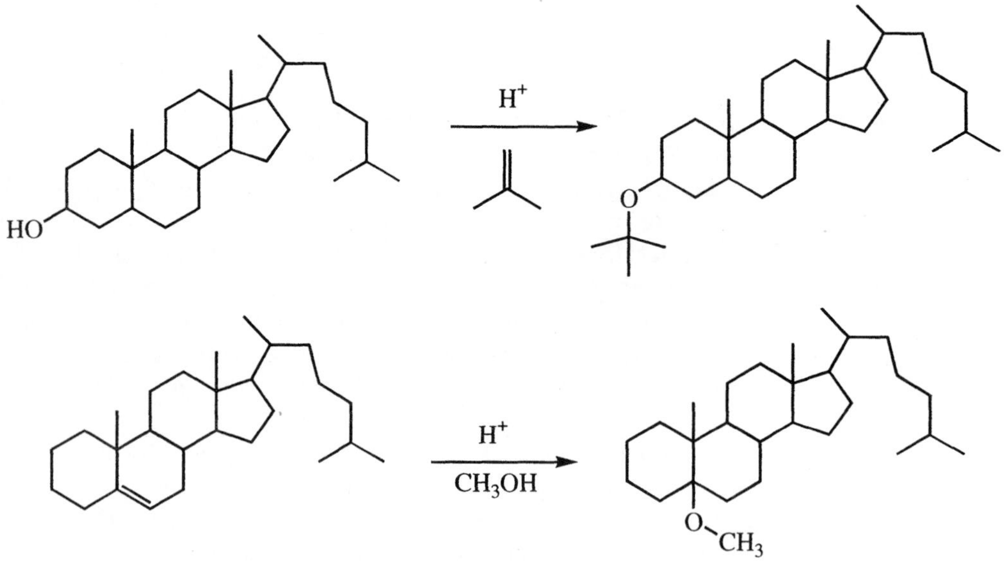

Be sure you can show the mechanisms of both of these reactions.

The second synthesis of ethers was covered first in Chapter 13, then again in Chapter 16, and again earlier in this chapter. This is the Williamson ether synthesis, an S_N2 reaction between an alkoxide and an alkyl halide. The reaction works best if the alkyl halide is primary, acceptable if it is secondary, and miserable if it is tertiary. You should also remember that tertiary alkoxides, like potassium *tert*-butoxide, are pretty lousy nucleophiles due to their steric bulk—recall that this is one of the classic non-nucleophilic bases—and certainly other tertiary alkoxides will be at least as reluctant to react as nucleophiles. So making an ether that is tertiary on either side of the oxygen is quite difficult by the Williamson route:

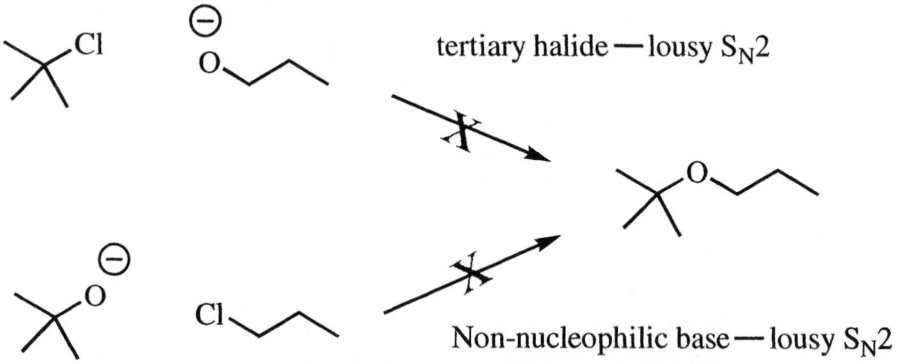

tertiary halide — lousy S_N2

Non-nucleophilic base — lousy S_N2

Such ethers are best made the first way you learned, which works quite well:

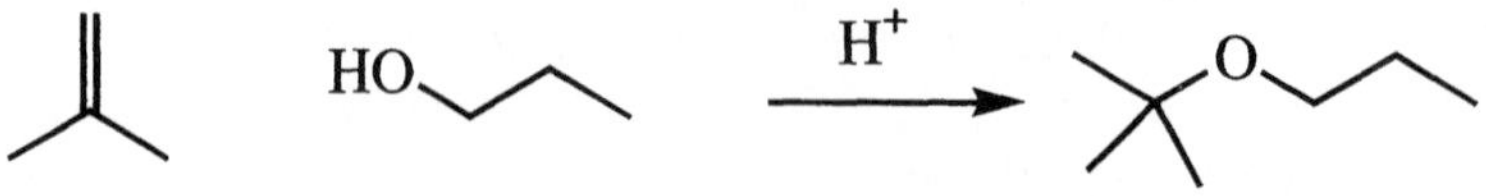

Reactions

As a general rule ethers tend to be fairly inert. (That is why they make such good solvents: they dissolve many compounds quite easily, but they do not actually participate in many reactions.) Ethers lack a good leaving group, so they do not react with nucleophiles. They contain no acidic protons, so they do not react with bases. But they *do* contain an atom with a lone pair, so under acidic conditions they can be protonated, which suggests that they could be made to react with HX to give halides. This is true; however, bear in mind that most ethers are made from alcohols, which themselves react with HX to make halides, so why would anyone want to make a halide from an ether? It is, in fact, fairly rare that anyone intentionally does anything at all to an ether, with one important exception.

Protecting Groups

The reaction of an ether with acidic water to make two alcohols is, in most cases, pretty difficult:

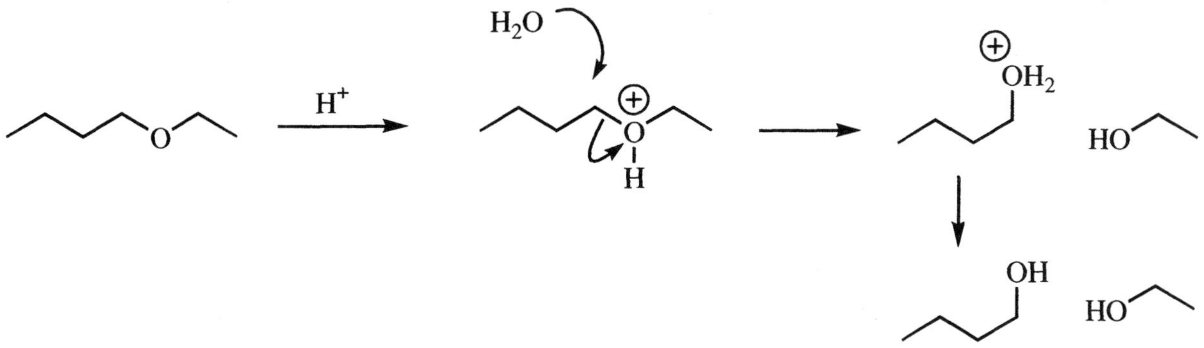

Water is not a particularly good nucleophile, generally not good enough to be the aggressor in an S_N2 reaction, so the reaction shown above is unlikely. And of course, it is certainly not possible to imagine that OH^- would be available to accomplish this S_N2 reaction. So, if you want to hydrolyze an ether into its component alcohols, it will have to be by an S_N1 mechanism, and S_N1's work best when a good carbocation can be formed, which most often means that one side of the ether is tertiary. Therefore the following reaction is one of the few cases in which hydrolysis of an ether is easy:

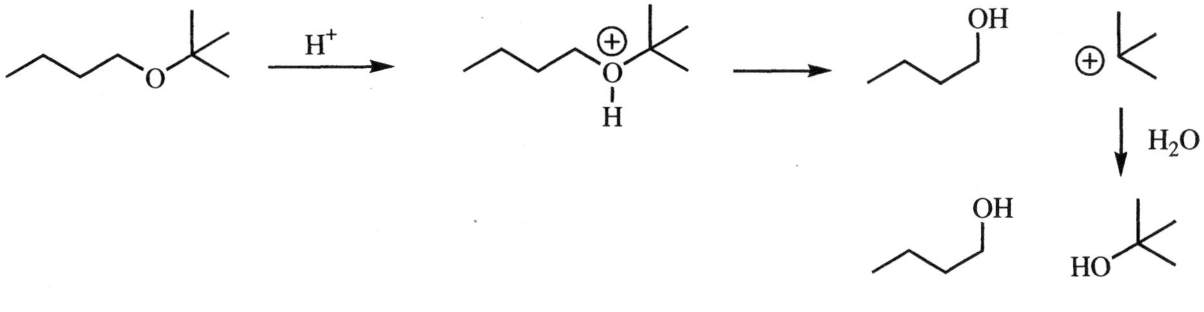

Hmm, now here is something interesting: at the end of the previous section, I said that the acid-catalyzed reaction to *make tert*-butyl ethers is pretty easy. Now I am telling you that the reaction to *unmake tert*-butyl ethers is easy. In other words, the following reversible reaction is easy to set up, and further, by controlling the conditions, we can easily dictate the direction in which it will go:

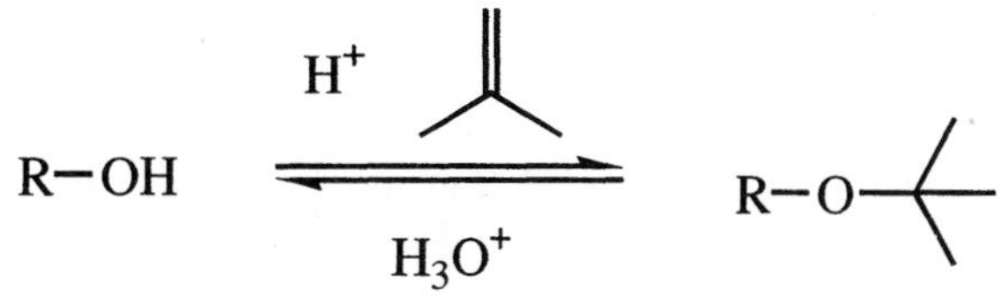

So what? Let's imagine that you went into the lab to attempt the following reaction:

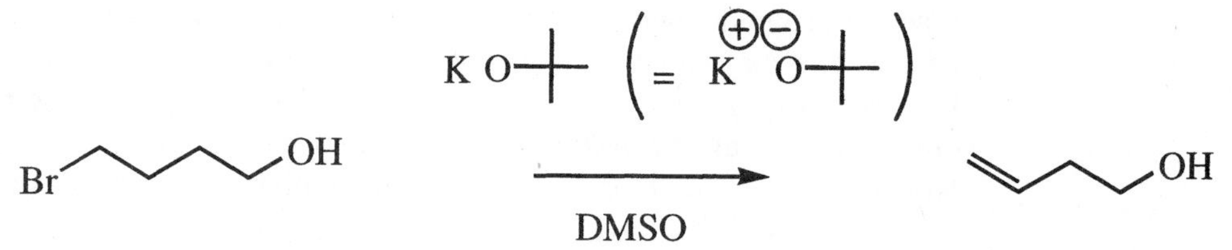

This ought to be pretty trivial. You are trying to cause an elimination reaction, and the leaving group is on a primary position, so you will obviously choose E2 conditions. You have cleverly selected a non-nucleophilic base, which you know will prefer to remove a proton rather than itself undergoing an S_N2 reaction with your primary leaving group. Nevertheless, in spite of your careful planning, this reaction will not work. Instead of the desired product, you will get mostly tetrahydrofuran (THF).

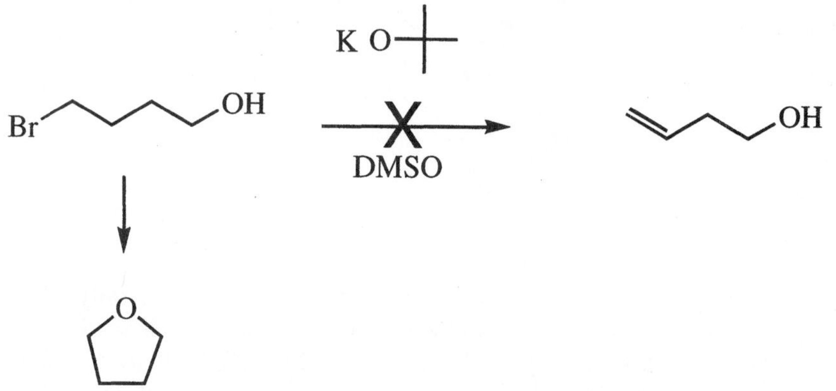

Why, and how, does this occur? You should take some time to try to figure out the answer now, before you read my explanation.

We actually discussed this before, in what appears to be a different context. In Chapter 16, in the section on "Competition with Substitution Reactions," we worried about what solvent to use for an E2 elimination. We decided that water or ethanol would be a bad choice, because the *tert*-butoxide anion could remove a proton from the solvent, so that we actually would have hydroxide or ethoxide present instead of *tert*-butoxide, and therefore nucleophilic reactions could occur. The same is true here: there is an OH group in our substrate, and the *tert*-butoxide anion can remove a proton from *it*, creating a nice nucleophile that is directly attached to (and therefore near) the halide. No special collision needs to occur; no choice of solvent can prevent the proton exchange. An intramolecular S_N2 reaction is inevitable:

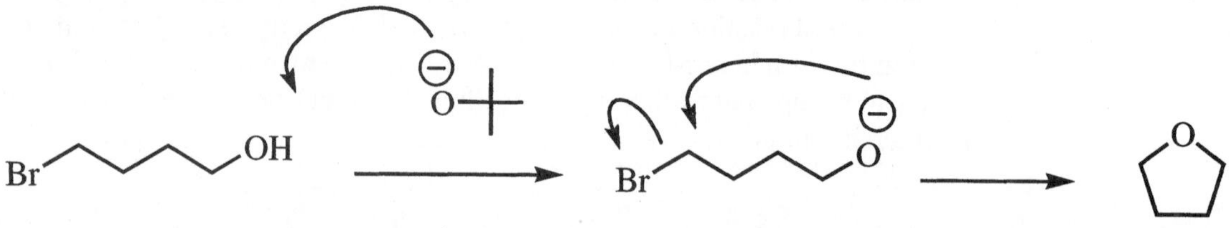

There is one (and only one) way to get around this. The thing that is preventing us from accomplishing the desired reaction is the OH group on our substrate. If it were not there, the elimination we wanted would work just fine. However, we cannot simply remove the OH group, because we need to have it in the final product. What we really need to do is *hide* the OH group—figure out some way to remove it for a brief period of time, and then bring it back. And we want both the hiding operation and the revealing operation to be fairly simple and easy to carry out.

We suddenly have a use for a reaction that easily and reversibly converts alcohols into ethers. We *can* achieve the desired transformation, in the following way:

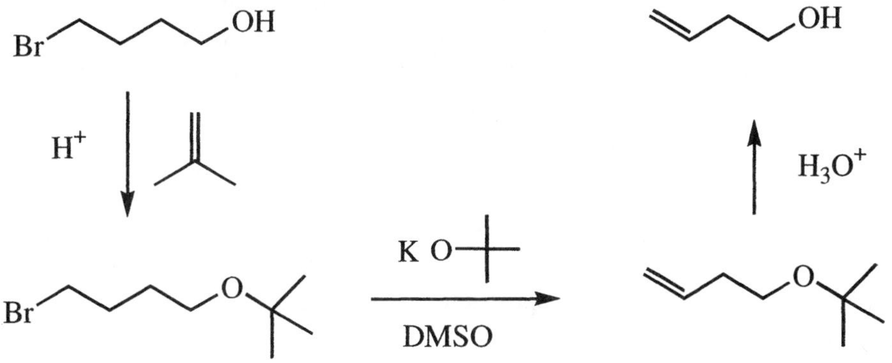

The net transformation is precisely what we originally set out to do. True, it has taken us three reactions rather than one to accomplish it, but on the other hand each step now proceeds as desired with very few complications. It is sometimes worth the extra time to run a couple of extra reactions in order to ensure a clean transformation.

The official term for the type of process we have developed is "protection"; we have protected the alcohol in the form of an ether, and then deprotected to reveal the alcohol again. The *tert*-butyl group here is the protecting group.

There are many protecting groups in organic chemistry, and as you learn more reactions you will more often encounter situations in which protection is necessary. Special protecting groups have been developed for almost every functional group, but the most commonly protected functionalities are the alcohol and carbonyl groups. Whole books have been written on the process of protection in organic chemistry, and there are over a dozen pretty good protecting groups for alcohols alone. We will not be covering this wide range of protecting groups for alcohols. For now what is important is that you understand the concept of protection and the criteria for a good protecting group, and also that you are able to make use of protecting groups when necessary.

Test Yourself 1

Suggest a method for accomplishing the following transformation. It may take you several steps.

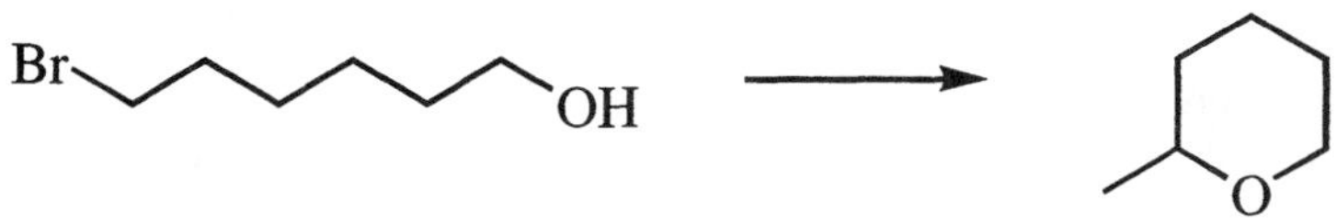

EPOXIDES

Nomenclature

Epoxides (also known as "oxiranes"), you will recall, are characterized by three-membered rings containing one oxygen atom. They are ethers, in that they have a C–O–C bond, but unlike other ethers they are reactive due to strain in the three-membered ring. We will discuss the reactivity of epoxides shortly.

Although there is an official IUPAC method for naming epoxides, almost no one uses it. Instead, epoxides are named in a very strange fashion, but one that works quite well. An epoxide is construed as a double bond with an oxygen planted on top of it (consistent with the most common method of making one), and it is named as such. The common name is the name of the alkene the epoxide is *derived* from, followed by "oxide." See the following examples.

$$\text{△ with O} \quad = \quad \text{O} + \text{=} \quad = \text{ ethylene oxide}$$

= propylene oxide

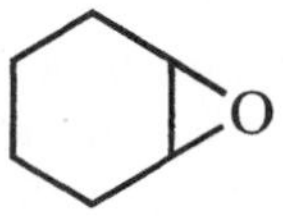

= cyclohexene oxide

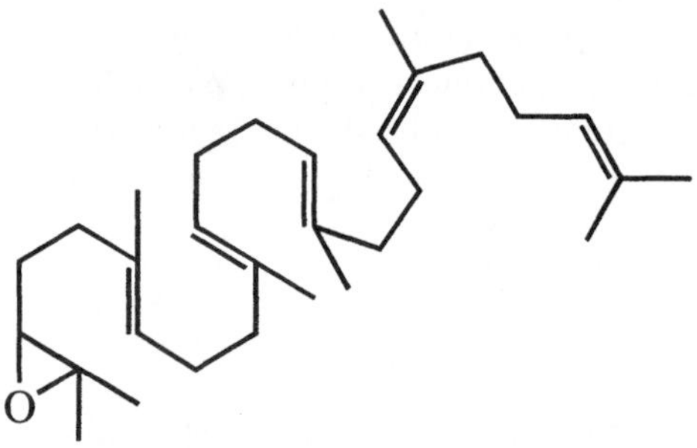

= squalene oxide

Preparation

You already know the most common way of making epoxides, literally plopping an oxygen onto a double bond. You know the reagent used most often to do it: a peracid, most often *meta*-chloroperbenzoic acid. This was discussed in Chapter 11, which you may want to review.

Epoxides can also be made from halohydrins, which are substances with an OH on one carbon and a halide on the next carbon. The reaction is basically an intramolecular Williamson ether synthesis. The fact that it works is a bit surprising in that you are forming a strained three-membered ring; however, the constant, unavoidable proximity of the alkoxide to its target carbon allows such a reaction to proceed at a reasonable rate, and as long as the conditions do not allow the ring to open again, you can isolate the strained product.

We will not use this particular method much, because we do not know how to make halohydrins in the first place, but in fact they are not difficult to make.

Reactions

Most ethers are relatively inert, but epoxides are quite reactive. Why? Epoxides have a lot of energy—strain energy—pent up in them by virtue of being three-membered rings. During a reaction that breaks this ring, that energy is released. I like to think of epoxides as being spring-loaded, like a cocked gun. Touch the trigger, and they'll pop open pretty easily. But you do have to at least touch the trigger. There are two general approaches to setting off an epoxide: pushing and pulling.

In Base

Up until now, we have insisted that O⁻ is a lousy leaving group, and it will not leave in an S_N2 reaction. Epoxides are the exception to this: the release of strain that accompanies an epoxide reaction does allow this particular O⁻ to serve as a leaving group. Of course, in one sense it does not "leave" at all, since it is still part of the molecule, but it has left the carbon that was substituted, so it is still considered a leaving group. The other point is that after leaving, the oxygen has a minus charge and is therefore a strong base, and if the solvent has available protons the O⁻ will grab one. Below are a couple of examples of this. Notice that the *result* of each reaction shown is a compound with an alcohol on one carbon and an alcohol or ether on the adjacent (vicinal) carbon.

Thus, one of the ways to trigger an epoxide opening is to provide a reasonable nucleophile to *push* it open in an S$_N$2 reaction. There are a couple of consequences to this approach that are not obvious from the above pictures.

Stereochemistry

As you know, all S$_N$2 reactions go with inversion. That means that the nucleophile comes in from one side while the leaving group leaves from the opposite side. But the leaving group in this case is the oxygen of the epoxide, which is directly attached to the carbon next door. Therefore, at the end of the reaction, the oxygen that was the leaving group (now an alcohol) and the newly introduced nucleophile must be on opposite sides of the molecule. This is illustrated in a cyclic case below. The two substituents end up necessarily *trans* to each other.

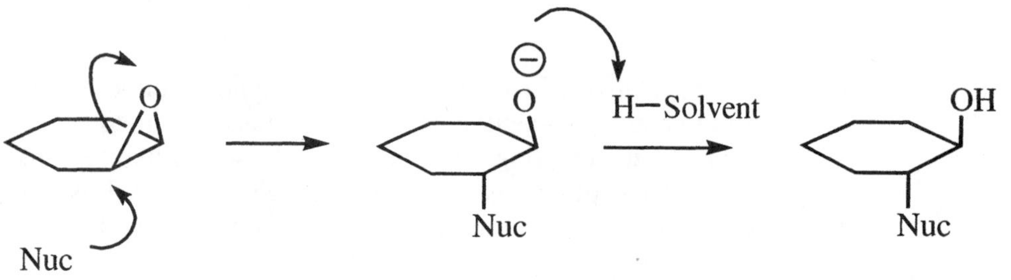

Of course, in an acyclic case there could also be stereochemical consequences, leading perhaps to the formation of only a single diastereomer. This will be illustrated in a problem soon.

Regiochemistry

"Regiochemistry" is a word we use in conjunction with specifying *where* in a molecule a reaction will occur. If the epoxide-opening reaction occurs by an S$_N$2 mechanism, then like all other S$_N$2 reactions it should occur better at less substituted positions. None of our examples so far have addressed this point, but consider an epoxide that is different at its two ends, like the following:

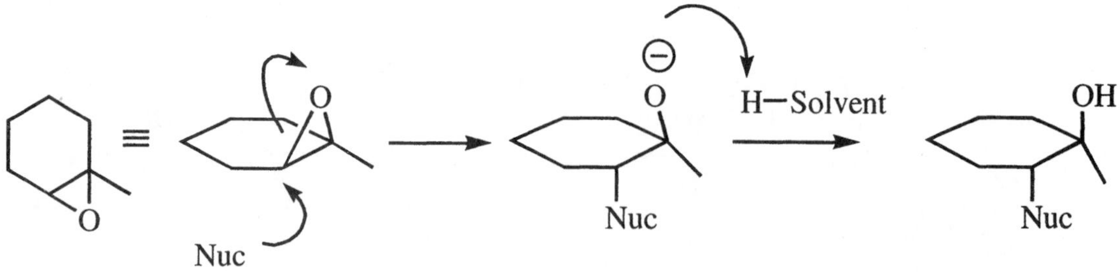

In this case, or any other that is not symmetrically substituted, the nucleophile will attack at the less substituted, more accessible end (provided the nucleophile really is pushing the ring open). This means that the OH group will end up on the more substituted end of what used to be the epoxide.

ORM: In "Organic Reaction Mechanisms," under "Alcohols and Ethers," look at "rxs epoxides" and choose "ring opening, basic."

The second way to trigger epoxides is to add acid. The only possible thing acid can do to an epoxide is protonate the oxygen. This makes the oxygen a good leaving group, and one could imagine that the epoxide would spring open immediately in an S_N1 reaction, creating a carbocation. There is good evidence (that we will see soon) that this does not occur; nevertheless, the carbon–oxygen bonds loosen somewhat, and *some* positive charge develops on the carbons. This positive charge is sufficient to attract nearby species with lone pairs— that is, nucleophiles. Notice that the nucleophile does not need to be very aggressive in this case, because it is not the nucleophile that is initiating the reaction. The (partial) positive charge is sucking the nucleophile in. This is why I refer to this approach as "pulling," as opposed to the S_N2 approach, which I called "pushing." The acid reaction is not quite S_N1, but it is S_N1-like.

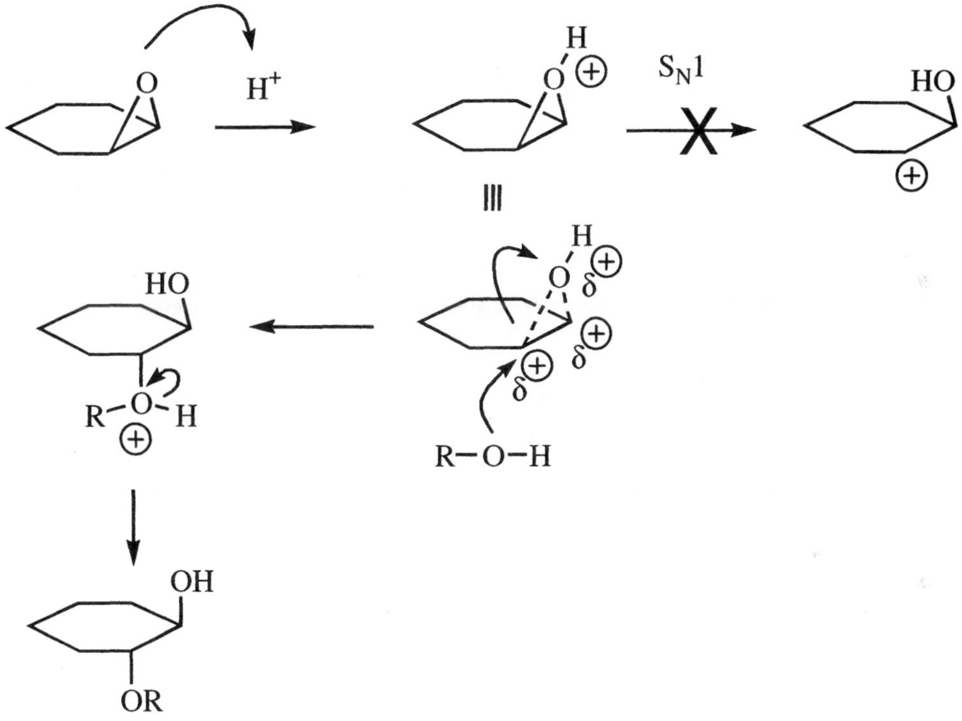

H^+ S_N1 $R-O-H$

III

HO HO OH OR

Stereochemistry

How do we know the reaction is not following a plain old S_N1 path? The evidence is in the picture above: just as in the S_N2 case, this reaction also leads exclusively to *trans* products. If the reaction proceeded by way of an open cation, the nucleophile would be able to approach from either face of the molecule, and while it might prefer one face over the other, it is unlikely that its preference would be perfect. But epoxide openings, even in acid, usually give only the *trans* product. This evidence has forced chemists to propose that the intermediate is not completely open, but a three-membered ring with shared charges. This intermediate strongly resembles the bromonium ion we encountered back in Chapter 11; like the bromonium ion, this cation can be shown as resonance structures that violate our rules of resonance in that there are single bonds made and broken as we go from one resonance structure to another. I told you back in Chapter 11 that there would be a second example of such weird resonance. Here it is.

What happens if the epoxide is not symmetric? You could make a reasonable guess, based on the structure of the cation involved. This cation is sharing charge among three places, which suggests that it is a resonance hybrid. We should look at the three resonance structures that comprise it:

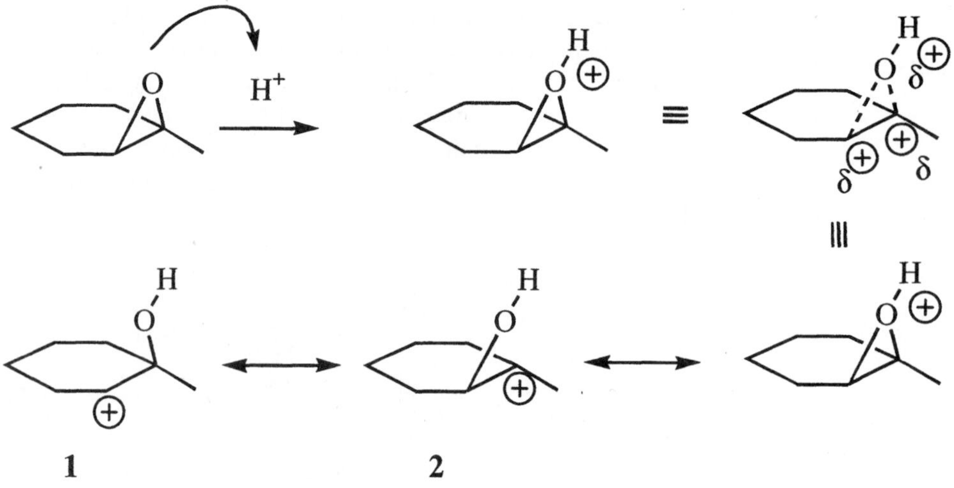

1   **2**

Of the two structures with positive charge on carbon, it should be apparent that **2** is better than **1**, because **2** is tertiary and **1** is secondary. Therefore, this resonance hybrid, the combination of the three structures, will be more like **2** than **1**; i.e., there will be more positive charge on the more substituted carbon. When a nucleophile drifts by with a lone pair, it will be more attracted to the position with the greater charge. Thus, we can predict (correctly) that the nucleophile will usually end up attached to the *more* substituted of the two possible positions, which leaves the OH group on the *less* substituted position. This is precisely the opposite of what we observed in the case of openings induced by bases.

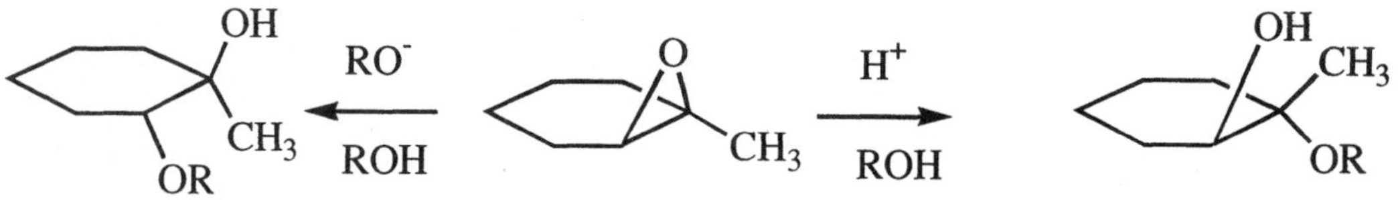

ORM: In "Organic Reaction Mechanisms," under "Alcohols and Ethers," look at "rxs epoxides" and choose "ring opening, acidic."

Test Yourself 2

Show how you could make the following compound starting with any hydrocarbons you like as your only sources of carbon atoms. As usual, you should try to solve the puzzle yourself before you look at my answer.

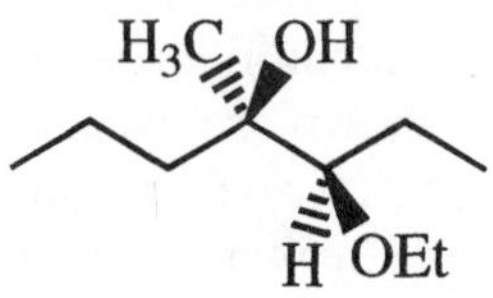

Biological Epoxidations

Epoxidations are used by biological systems in many ways. One of these accomplishes the removal of hydrocarbons from the body. Hydrocarbons tend to be very insoluble in water, the biological solvent, so they are difficult to move around and process. One of the first things that happens to hydrocarbons biologically (if they are unsaturated) is an epoxidation, followed by opening of the epoxide with water. The result is an easy approach to a diol, which is much more soluble in water.

Occasionally this strategy backfires. Benzo[a]pyrene, which was mentioned in Chapter 11, is epoxidized in this way, and a diol is formed, but then a second epoxidation occurs, and the new epoxide ends up reacting with DNA, causing genetic damage and eventually cancer.

How do biological systems epoxidize molecules? Not with peracids. One of the standard methods involves an enzyme called cytochrome P-450. The active site of this enzyme contains an iron atom, which picks up an oxygen atom from O_2 and ultimately delivers it to the alkene. The mechanism of the process is very complex and far beyond the scope of this text.

Problems

(Answers are provided at the end of the chapter for italicized problems.)

1. Show the organic products of the following reactions. Write NR if there is no reaction.

    a.

    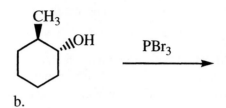

    CH₃ ···OH $\xrightarrow{\text{PBr}_3}$

    b.

    CH₃ ···OH $\xrightarrow[\text{pyridine}]{\text{TsCl}}$

    c.

    CH₃ ···OH $\xrightarrow[\text{2) CH}_3\text{CH}_2\text{Br}]{\text{1) Na}}$

    d.

    CH₃ $\xrightarrow{\text{mCPBA}}$

    e.

    H₂C–OH $\xrightarrow{\text{PCC}}$

    f.

    H₂C–OH $\xrightarrow[\substack{\text{H}_2\text{SO}_4 \\ \text{H}_2\text{O}}]{\text{Na}_2\text{Cr}_2\text{O}_7}$

2. When cyclohexylmethanol is treated with sulfuric acid, three different alkenes are formed, as shown below. When the same compound is treated with tosyl chloride in pyridine followed by potassium *tert*-butoxide in DMSO, only one of these products is formed. Show the mechanism for formation of the three products in sulfuric acid, identify which is formed in the second process, and explain why there is a difference.

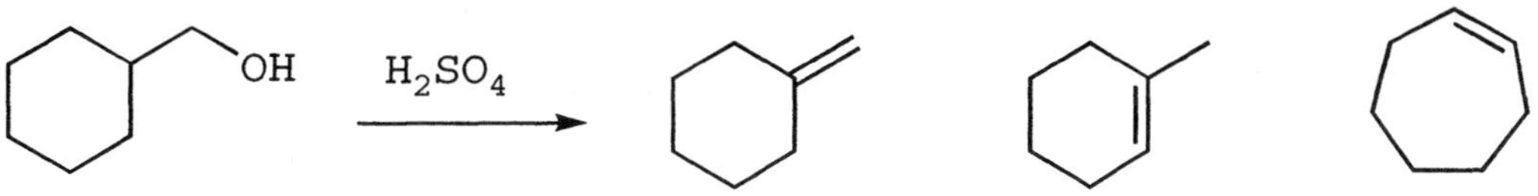

3. *a. When 1-cyclohexylethanol is treated with HBr, three different bromides are obtained (ignoring stereoisomers). Identify them, and propose a mechanism that accounts for their formation.*

   b. When the same starting material is instead treated with PBr₃, only one product is obtained. What is it, and why is the result different from that in part a?

   c. When cyclohexylmethanol is treated with HBr, only one product is obtained. What is it? Again, why is the result different from that in part a?

4. Suggest an efficient synthesis of 4-methoxyoctane starting with organic compounds containing three or fewer carbons.

5. Starting from anything you like containing three carbons or less, suggest syntheses of the following molecules.

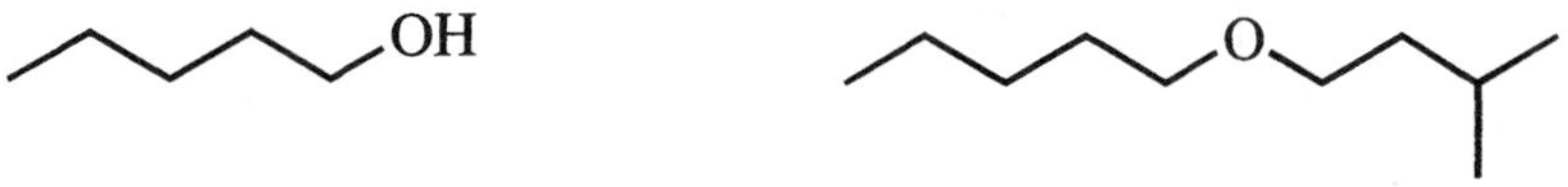

6. Show the mechanism of the following reaction. Explain why only one of the two ethers was cleaved, and why it was the one shown.

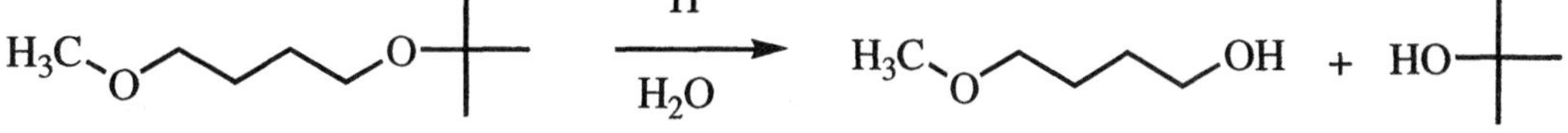

7. Most ethers are relatively inert, but, as shown in problem 5, *tert*-butyl ethers can be cleaved rather easily in aqueous acid. Another type of ether that is easily cleaved is a diether such as 1,1-dimethoxycyclohexane. Explain why ethers of this type are so easily cleaved by aqueous acid. Imagine the product to be 1-methoxycyclohexanol.

8. *The product of the reaction described in problem 7 is actually not 1-methoxycyclohexanol. What you instead get is cyclohexanone and two molecules of methanol. See if you can suggest a mechanism for this that does not require you to postulate anything you have not seen before.*

9. We have seen the use of *tert*-butyl ethers to protect alcohols during reactions. The *tert*-butyl groups are easy to put on and easy to take off. An even more useful protecting group is the tetrahydropyranyl (THP) ether shown below. Explain why this group is so easy to put on and take off.

dihydropyran   tetrahydropyranyl ether

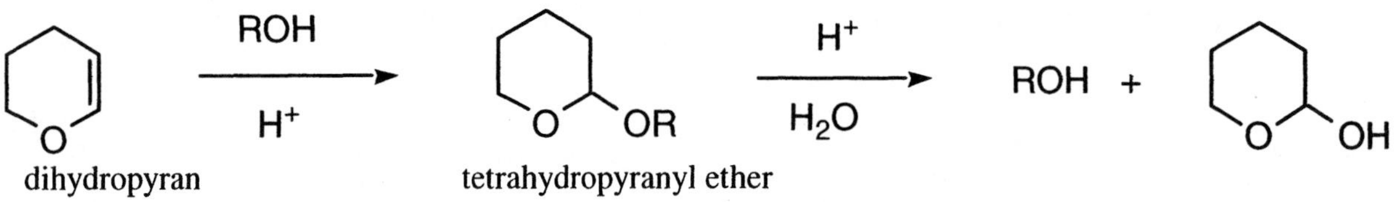

10. Another useful protecting group is the benzyl ether shown below. Explain why this group is also easy to put on and take off. Show the mechanisms involved and illustrate your explanation as appropriate.

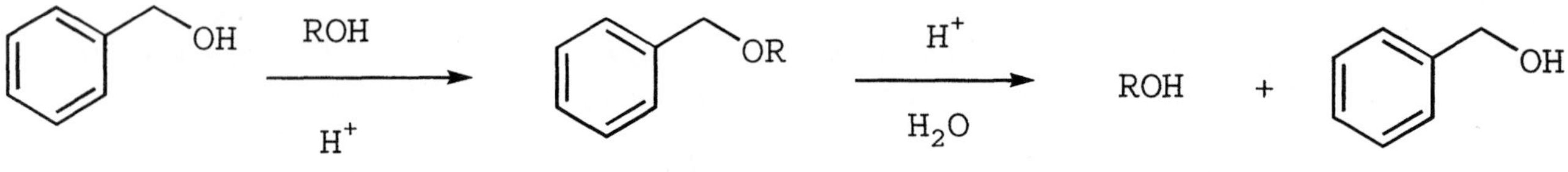

11. *Chemistry student Oscar Erlenmeyer attempted the following reaction. He was dismayed to discover that his product was a diketone rather than the product shown. What did he get? Show a mechanism that accounts for what went wrong. How could Oscar accomplish what he wanted to do?*

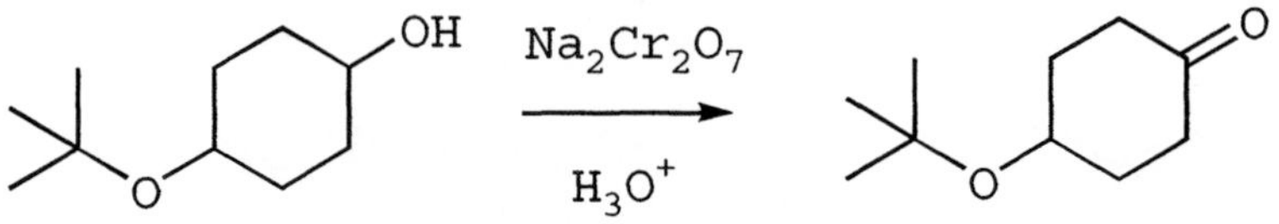

12. Starting with *R*-2-butanol, show the structure of the ethers you would get by:

a. treatment with PBr₃ followed by sodium methoxide, and

b. treatment with tosyl chloride followed by sodium methoxide.

13. How might you obtain the following product, starting with 1-bromo-1-methylcyclohexane?

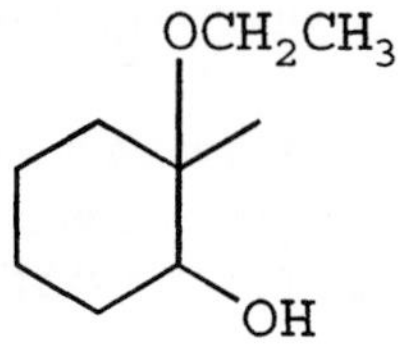

14. *Suggest a synthesis of the following compound starting with anything you like containing three or fewer carbons. By this I mean that all the carbons in your product must originate in compounds containing three or fewer carbons. Other reagents that come and go may be larger.*

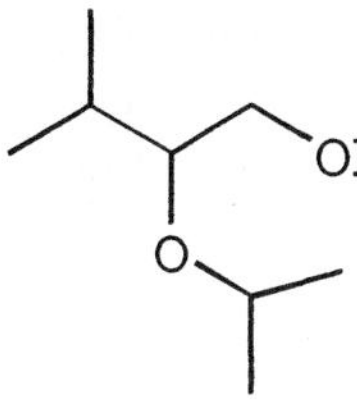

15. One of the following reactions can be accomplished with acidic methanol, the other with basic methanol (NaOCH₃ in methanol). Which is which? Show both mechanisms and explain why the reactions proceed differently.

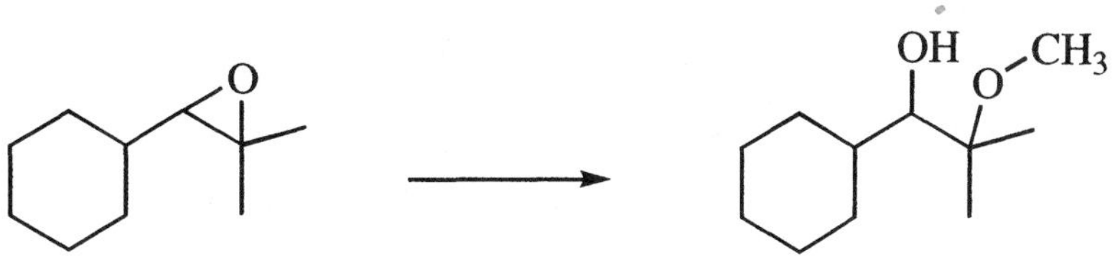

16. *Recall that epoxides can be opened by nucleophiles. One such nucleophile is the anion derived from a terminal alkyne. Use this reaction to make the first compound in problem 5 in fewer steps than you used before. Would the same strategy work for the right half of the second compound in problem 5?*

17. When epoxides are opened by alkoxides, the corresponding alcohol is usually used as solvent (e.g., NaOEt in EtOH). Could you use an alcohol solvent with an alkyne nucleophile? Why or why not?

18. *Suggest methods for accomplishing the following conversions, using any reagents you need containing four carbons or less.*

a.

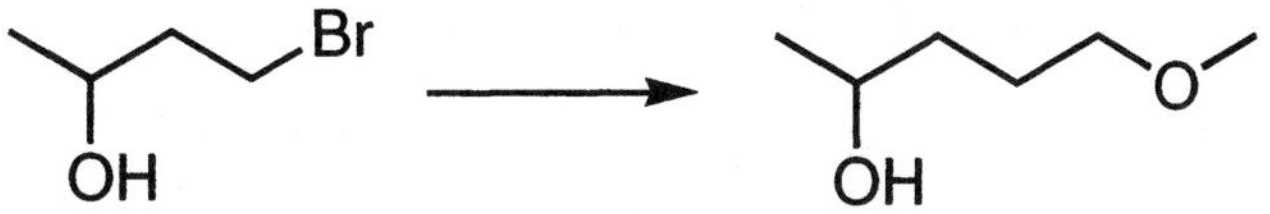

b.

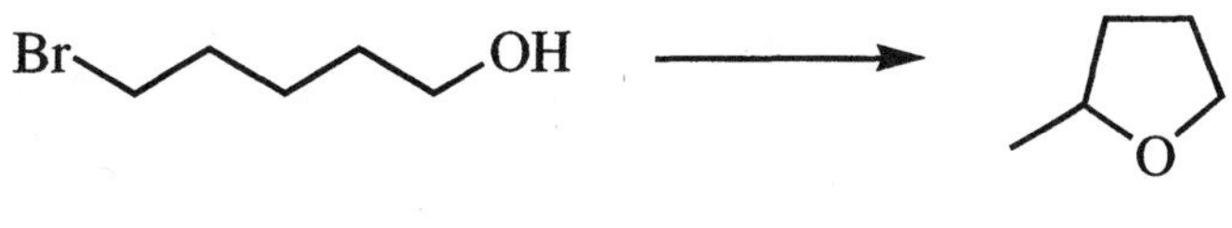

19. **A** is an organic liquid. Upon treatment with PCC **A** is converted into **B**, but with chromic acid one obtains **C** instead. **A** reacts with PBr_3 to give **D**, which reacts with sodium acetylide ($NaC \equiv CH$) to give **E**. **E** reacts with hydrogen and Lindlar catalyst to make **F**, which reacts with acidic water to give **G**. Alternatively, **E** reacts with acidic water in the presence of mercuric sulfate to give **H**. **H** can be converted into **G** with hydrogen and palladium catalyst; in turn, **G** can be converted into **H** with either PCC or chromic acid. Propose structures for **A** through **H** that are consistent with this information and also with the following spectra. Each lettered substance is a neutral organic compound.

A:

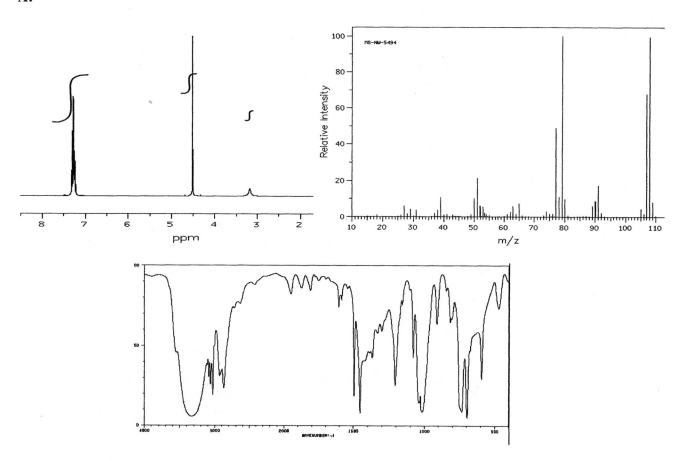

B:

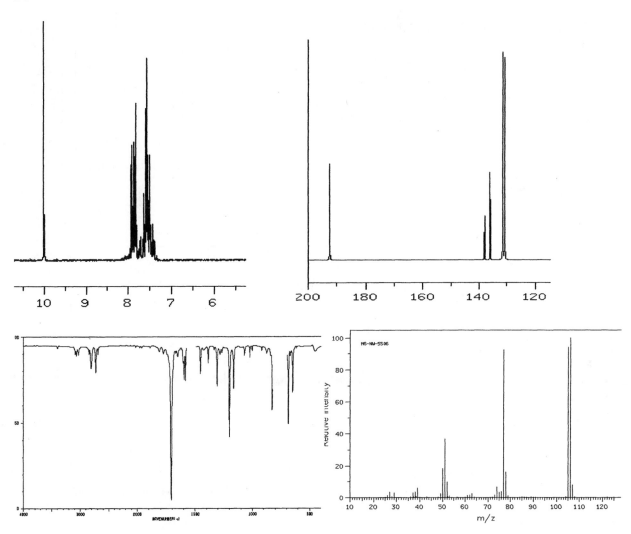

C:

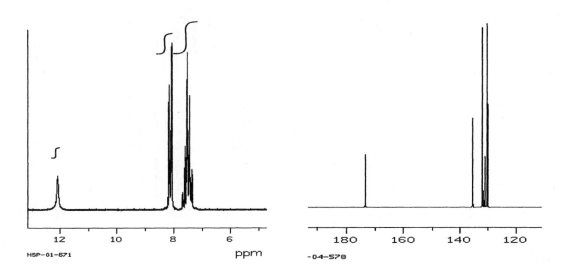

HSP-01-671

-04-578

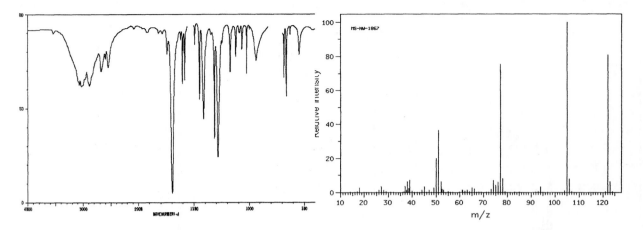

D:

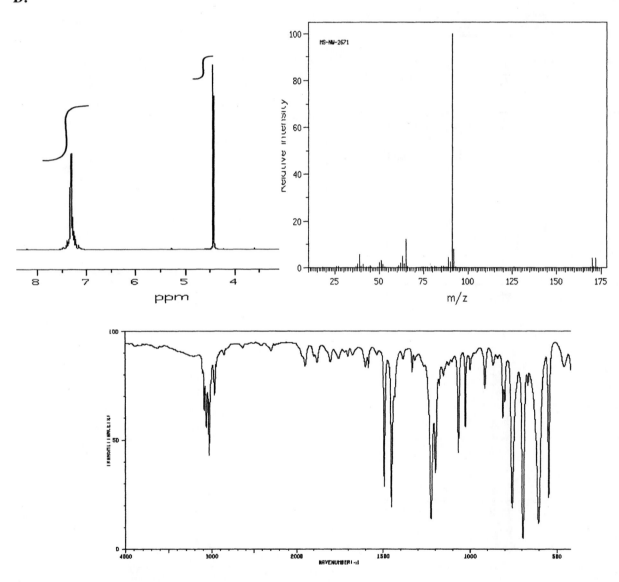

E: No spectral data available

F:

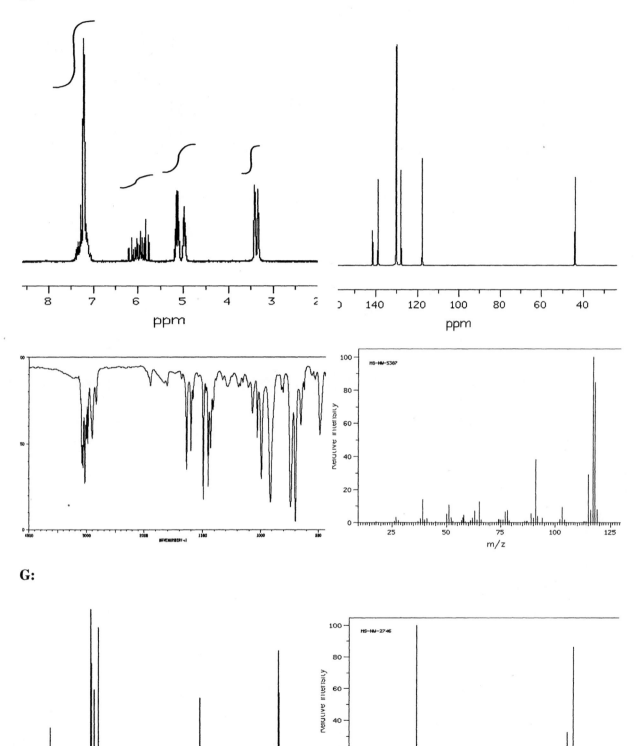

G:

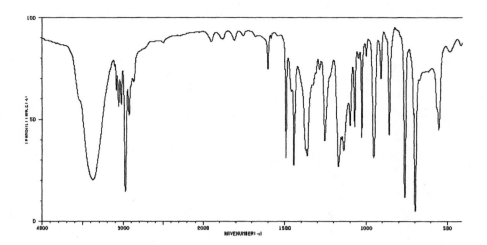

H:

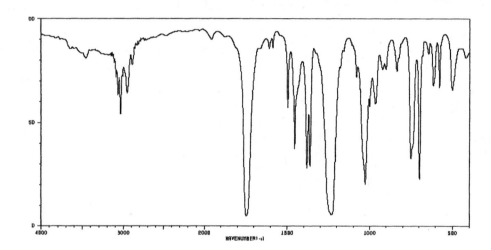

Selected Answers

Internal Problems

Test Yourself 1.

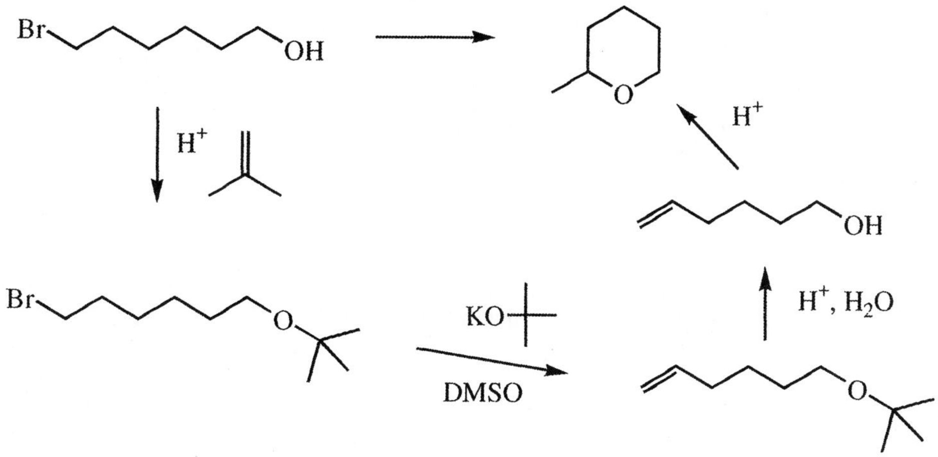

Alternatively,

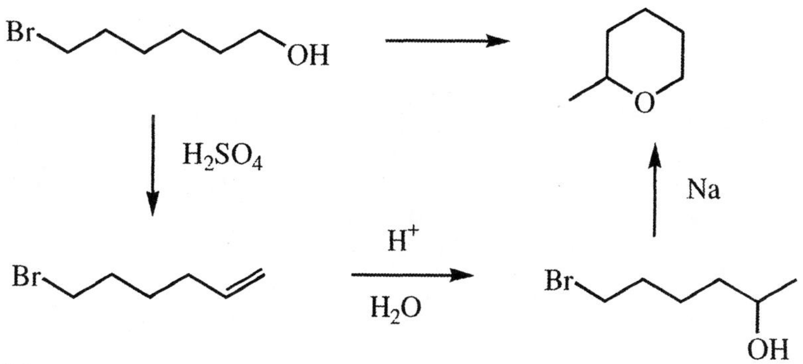

Test Yourself 2.

Since this is a synthesis problem, we will again start at the end. We see two functional groups in this molecule, an alcohol and an ether, and they are on adjacent carbons. It ought to be possible to form them both simultaneously by the opening of an epoxide. Which epoxide? Clearly, the one that spans the two carbons in question:

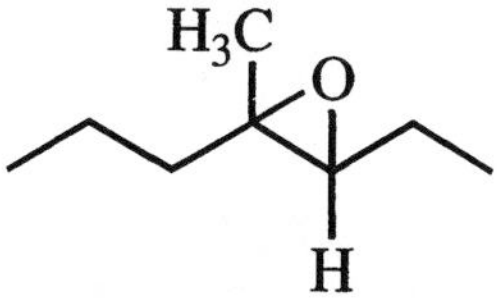

Further, the OH group in our product is on the *more* substituted position. This means that the nucleophile, the OEt, must come at the *less* substituted position, which means that it will need to push its way in, in an S$_N$2 reaction. So we should carry out the following reaction:

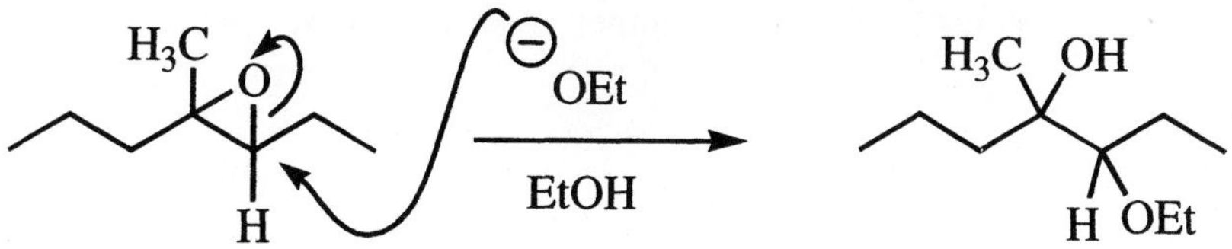

That takes care of the regiochemistry; now we need to worry about stereochemistry. When an epoxide opens, the two groups (OH and OEt, in this case) end up opposite each other. So let's take our desired compound and look at it in the conformation in which this is the case:

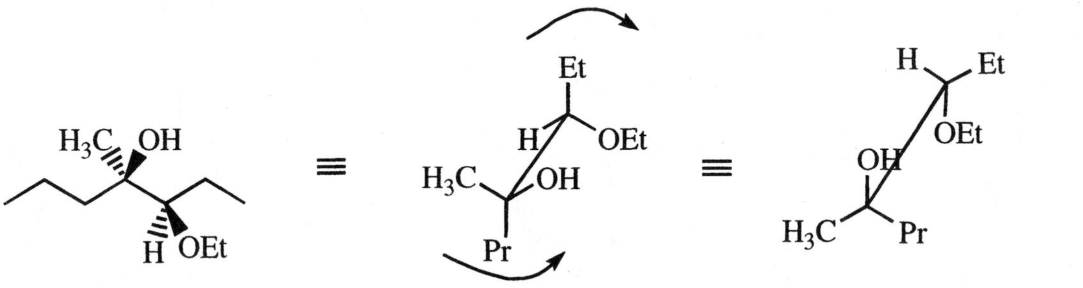

The oxygen of the OH group used to be the oxygen of the epoxide, and it must have been attached to the back carbon (as pictured here) from the top:

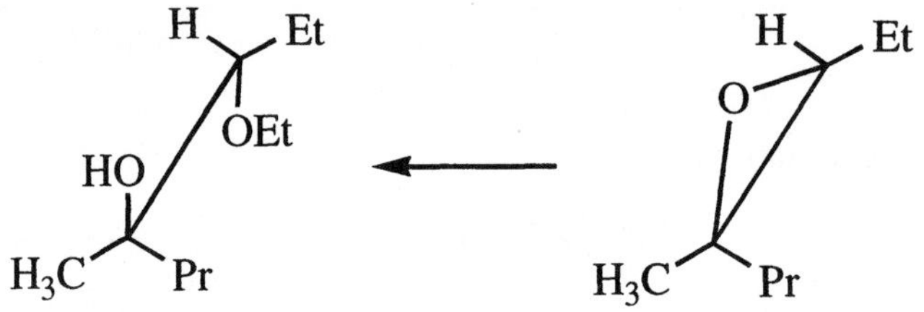

This tells us that the epoxide we need is actually not the first one we wrote when we started trying to solve this problem: it has different stereochemistry. Furthermore, we can see that the required epoxide must itself have come from the alkene with a corresponding shape: that is, it must have been the Z-alkene.

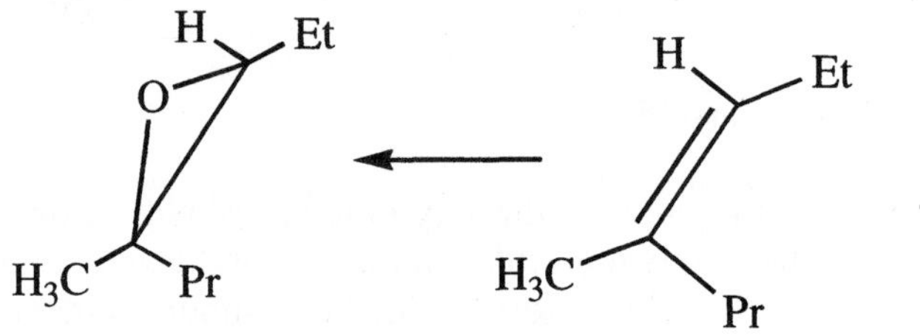

So now we can show our overall synthesis to this point as follows:

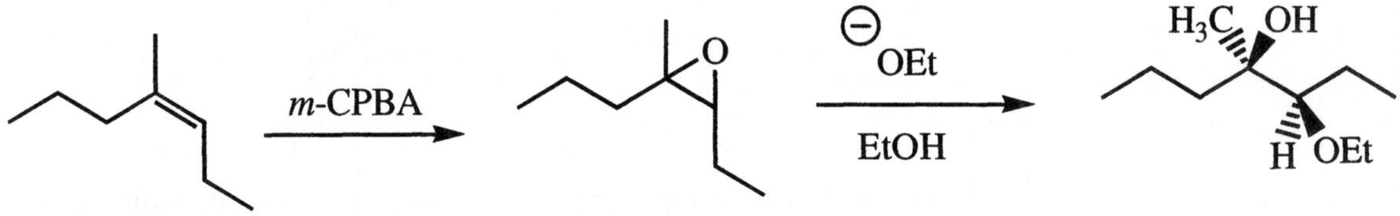

To provide a complete answer to the question, we would also need to explain how we propose to make the ethanol from some hydrocarbon, but I know you were able do that in Chapter 11, so I won't bother you with it now.

End of Chapter Problems

3. a.

b. First one. PBr₃ reacts by an S$_N$2 mechanism, which avoids rearrangements.

c.

Primary position avoids S$_N$1, so reaction is S$_N$2, which prevents rearrangements.

8.

Resonance-stabilized carbocation is easy to make!!

11.

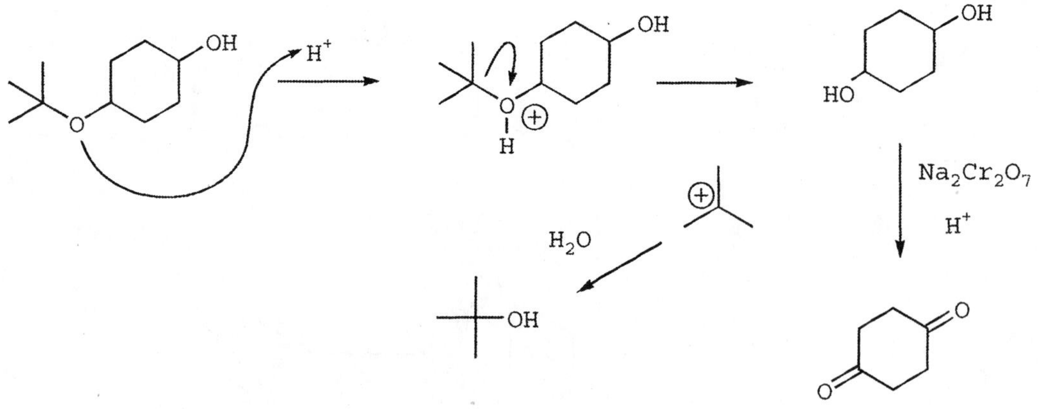

To prevent this, avoid the use of aqueous acid. Try PCC!

14.

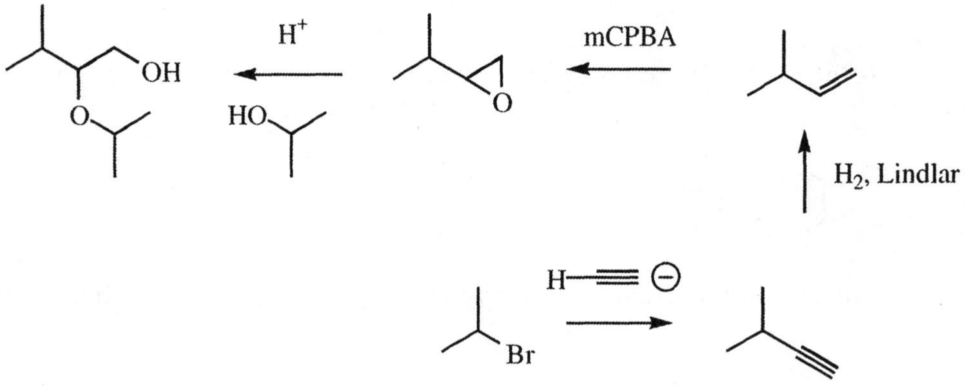

16.

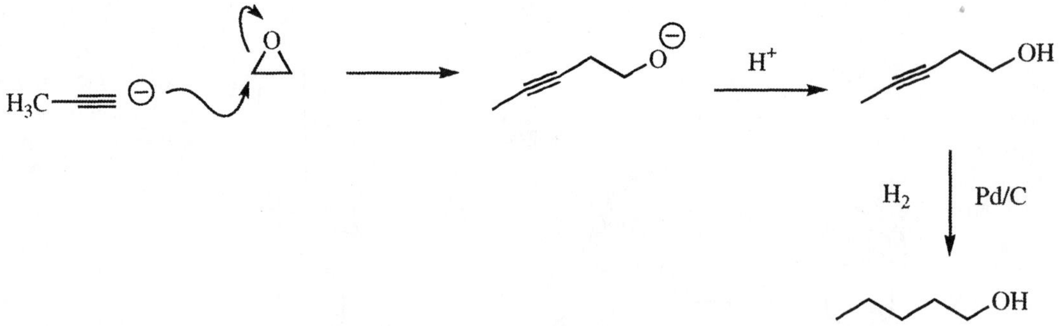

No, because in order to make the right half by this strategy, you would need to use an isopropyl anion, which we do not (yet) know how to make.

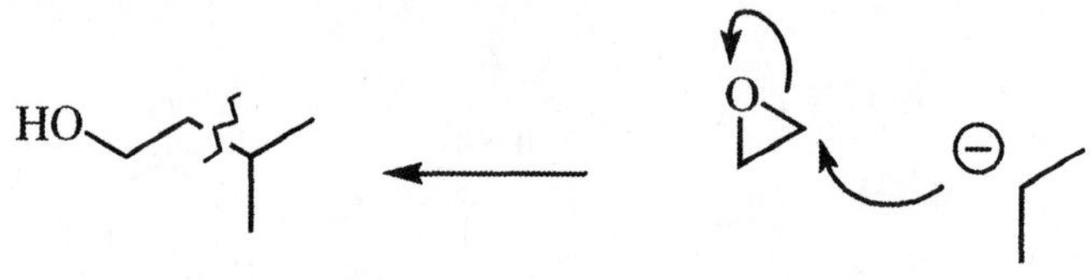

18. a.

b.
Reasonable strategy:

However, this is likely not to work because

So protect the alcohol first:

(Actually the last two steps would probably happen together.)

Chapter 18
Carbonyl Chemistry I: Aldehydes and Ketones; Carbohydrates

The most important functional group in organic chemistry is without a doubt the carbonyl (that's "car-bon-EEL," not "car-BON-ill") group. Before I became an author, I evaluated the books on my shelf in part by how quickly they got to carbonyl chemistry, because I wanted to cover it as soon as possible. So here I am, just barely getting to carbonyls in Chapter 18. And yet, I can see nothing I would sacrifice of what we have already covered! So be it. In this chapter we will

- Learn the proper way to reduce carbonyls

- Discover how to do synthesis with Grignard reagents

- Consider the reactions of aldehydes and ketones with oxygen nucleophiles

- Learn how to protect aldehydes and ketones

- Discuss carbohydrate structure and reactivity.

NOMENCLATURE

As with almost everything else we have seen, there are two ways to name aldehydes and ketones—the official way, and the way everyone uses. The official way to name a ketone is to name the entire chain it is on, dropping the "-e" of the alkane name and adding "-one." Thus, the ketone with a carbonyl group at the 2 position of pentane would be called 2-pentanone. If there are attachments on the chain you name them the usual way. The other way of naming ketones is to name the two attachments to the carbonyl group, followed by "ketone." Thus, 2-pentanone might be called methyl propyl ketone. Finally, there is one ketone that has a special name that every chemistry student should know: 2-propanone (actually, the "2" is not necessary here—why not?) or dimethyl ketone is hardly ever called either of these—everyone knows it as acetone. Below are some examples.

propanone
dimethyl ketone
acetone

3-methylcyclohexanone

3-buten-2-one
methyl vinyl ketone

methyl phenyl ketone
acetophenone

COMMON MISTAKES: DON'T LET THIS HAPPEN TO YOU!

Ketone is spelled k-e-t-o-n-e. It does not have a "y" in it. Don't EVER write keytone!!!!

Aldehydes have only one substituent, because on one side of the carbonyl there is guaranteed to be a hydrogen. Aldehydes are named according to the name of the group that includes the carbonyl carbon, officially by replacing the "-e" of the alkane with "-al," but sometimes by adding the word "aldehyde." Here are some examples:

methanal
formaldehyde

ethanal
acetaldehyde

butanal
butyraldehyde

cyclohexanecarboxaldehyde

trans-3-phenyl-2-propenal
trans-**cinnamaldehyde**
(smells like cinnamon)

benzaldehyde
(smells like almonds)

The generic way to write an aldehyde is RCHO. You should recognize "–CHO" as signaling an aldehyde, and understand that its structure is a carbonyl group attached to a hydrogen (and why this must be so!).

PROPERTIES

Polarity

The chief feature of aldehydes and ketones is the presence of the carbonyl group. Since the oxygen is more electronegative than the carbon, this double bond is polar, with the oxygen being partially negative and the carbon partially positive. This is the only fact we need in order to understand most of carbonyl chemistry.

The mild polarity of the carbonyl group (and the ability of the lone pairs on oxygen to engage in hydrogen bonding with O–H bonds) suggests that the carbonyl part of a molecule should be moderately soluble in water and, indeed, if the carbonyl group is the dominant feature of a molecule, the compound usually is water soluble. Thus, small aldehydes and ketones tend to be water soluble: acetone can mix with water in any proportions. On the other hand, acetone is also sufficiently non-polar to dissolve most organic substances. This makes acetone an excellent solvent. Unfortunately, the carbonyl group is too reactive to allow acetone to be used as the solvent for most types of reactions (it is generally useful only for S_N2 reactions), but you will find acetone in frequent use in many organic labs simply as a wonderful solvent to clean glassware, since most crud dissolves in it, and it evaporates quickly. Larger aldehydes and ketones, in which the carbonyl group constitutes a smaller fraction of the molecule, are normally insoluble in water.

Acid–Base

The oxygen of a carbonyl group has lone pairs, so it can be protonated under acidic conditions, as shown in the top equation below. A carbonyl group also has a double bond, which, as we know, can also be protonated, as shown in the bottom equation below. So, when an aldehyde or ketone is presented with acid, which process occurs, and which cation do you get? Think about this before you read on. *Hint*: it is a trick question.

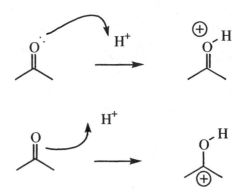

Let's take the second question first: which cation do you get? The trick is that neither cation represents an actual species, because this is another case of a resonance-stabilized system. The actual structure of the cation corresponds to neither the top picture nor the bottom one, but something in between. In other words, the question is meaningless because both cations are lousy pictures of the same thing: namely, the resonance hybrid of the two (and there should be a resonance arrow between them, but I left it off to see if you would realize that it belongs there). Since both pictures of the cation represent the same species, it follows that it does not matter which process arrow we draw, since both lead to the same thing. You can choose to illustrate the process as either the protonation of a double bond, or the protonation of a lone pair, as long as you recognize that what you get is the same either way. As we will see, this protonation is the first step in most acid-catalyzed carbonyl reactions.

Carbonyl compounds can also lose protons. Of course, the carbonyl group itself cannot—it consists of nothing but a carbon and an oxygen—but if the carbon next door has any protons, these are somewhat acidic. Again, you should be able to explain why before I tell you:

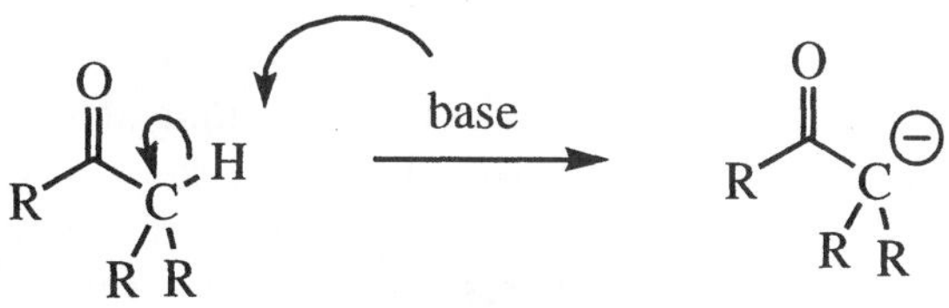

Again, the answer is resonance. The anion that is produced (called an "enolate anion") is resonance stabilized:

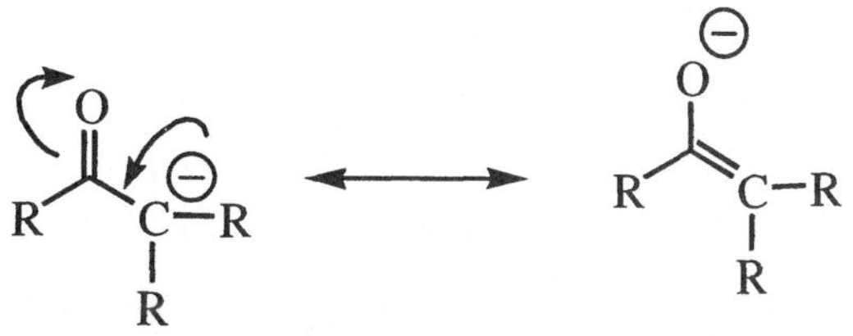

Enolate anions are therefore more stable than ordinary anions on carbon and can be made under much milder conditions. A typical aldehyde or ketone has a pK_a of about 19, which is not that different from the pK_a of an alcohol. In other words, removing a proton from the α carbon of (i.e., the carbon next to) a carbonyl group is about as easy as removing a proton from an alcohol. As we will see, deprotonation of this sort is often the first step in base-catalyzed carbonyl reactions.

Spectroscopy

Nearly everything here should be review. Carbonyl groups are among the few groups we can detect unambiguously in the infrared spectrum. They show up between 1650 and 1800 cm^{-1} as strong, sharp peaks. Ordinary aldehydes and ketones are usually about in the middle of this range, say, 1700–1730. In the next chapter, we will discuss why certain carbonyl peaks appear exactly where they do within the standard range. Aldehydes also have a hydrogen bonded to the carbonyl group. This C–H bond is a little different from other C–H bonds, and it appears near 2850 cm^{-1}, which is on the right-hand edge of the normal C–H range. Sometimes you can pick this out and use it as a tip-off that you have an aldehyde.

In the ^{13}C-NMR spectrum, the carbonyl carbon appears at the very left part of the normal range, usually close to 200 ppm, and it is usually a pretty small peak. This is due to the fact that it has no hydrogens on it, and (for reasons I have not explained) carbons with no H's produce peaks that are smaller than normal in a ^{13}C spectrum.

In the proton NMR, of course, the carbonyl group itself is invisible, but protons on the α carbon are shifted downfield by about 1 ppm relative to their normal position. Thus, acetone and other methyl ketones produce sharp peaks near 2 ppm. If you record any NMR spectra on a real instrument with real samples, beware of acetone: many students rinse out their NMR tubes with acetone and then wonder what the humongous peak at 2.1 ppm is!

Aldehydes have a unique type of hydrogen, and it appears in a unique region of the ^{1}H-NMR spectrum: almost nothing else shows up between 9.5 and 10 ppm. If you see something there, it almost certainly indicates an aldehyde, and if you think you have an aldehyde, there had better be a peak near 10.

PREPARATION

You already know several ways of making aldehydes and ketones. The first one you learned, in Chapter 11, is the ozonolysis of alkenes. Recall an ozonolysis reaction requires two operations: treatment first with ozone, and second with zinc. The purpose of the zinc will be explained shortly. The result is a ketone if the alkene had two alkyl groups at one end, and an aldehyde if either of the R groups was a hydrogen. For example,

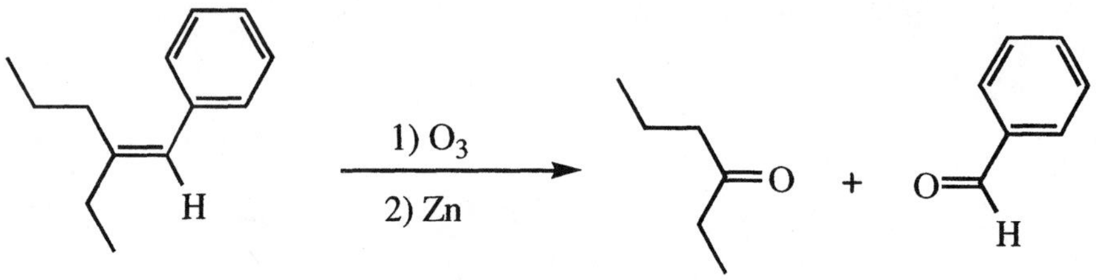

The second reaction you learned is the hydration of alkynes. Recall that this reaction occurs through the addition of water to an alkyne to give an enol, which promptly tautomerizes to a ketone. You cannot make aldehydes this way, unless the starting alkyne is acetylene itself (why not?). Remember that this reaction is pretty good for making methyl ketones from terminal alkynes, and ketones from symmetric internal alkynes, but internal alkynes that do not have identical substituents usually give close to 50:50 mixtures:

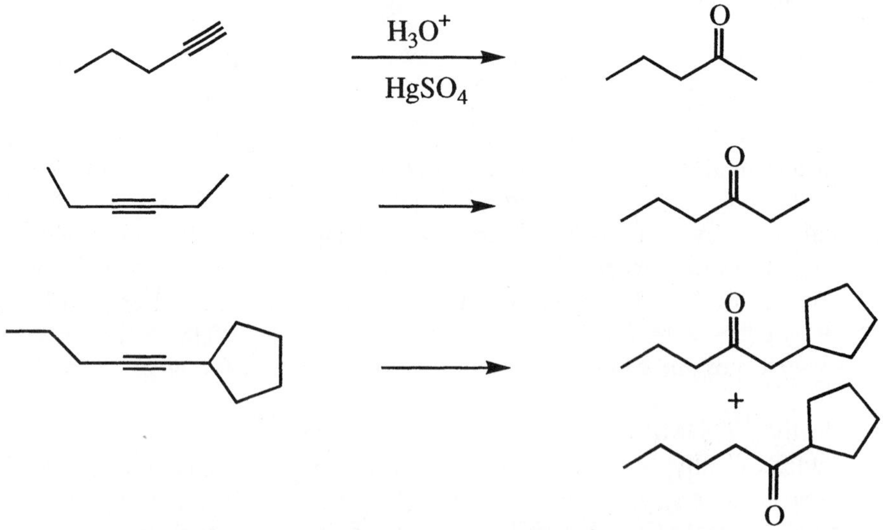

The third method you learned, in Chapter 17, is the oxidation of alcohols. For this we use chromic acid, usually in the form of sodium dichromate in sulfuric acid solution; tertiary alcohols do not react, secondary alcohols become ketones, and primary alcohols become carboxylic acids. Primary alcohols ought to give aldehydes, but aldehydes (which *are* formed) are converted into the corresponding hydrates with the water that is present, and these hydrates are rapidly further oxidized to carboxylic acids. We did learn a trick for making an aldehyde from a primary alcohol, however: use PCC as our reagent. This causes the reaction to stop at the aldehyde stage because PCC can be used in the absence of water.

REACTIONS

General Considerations

As I mentioned earlier, the reactivity of a carbonyl group derives from the polarity of the carbon–oxygen bond, which puts a partial positive charge on carbon. Positive charges, as you well know, are stabilized by substitution: tertiary carbocations are more stable than secondary, which in turn are more stable than primary. Replacing a hydrogen with an alkyl group on a positive site makes the positive charge more comfortable; conversely, replacing an alkyl group with a hydrogen makes a positive charge more *un*comfortable. The same logic applies to *partial* positive charges: replacing an alkyl group with a hydrogen makes the system less stable. If you compare an aldehyde to a ketone, the difference is that an alkyl group in the ketone has been replaced with a hydrogen. It stands to reason, then, that aldehydes should be more uncomfortable, and therefore more reactive, than ketones.

There is also another factor at work here: in order for a carbonyl carbon atom to react, some nucleophile has to approach it and become attached. As you learned back in Chapter 13, nucleophiles have an easier time approaching less substituted positions. So for both steric and electronic reasons we should expect aldehydes to be more reactive than ketones. Sure enough, in general, they are. This will become important later on, so keep it in mind.

Test Yourself 1

The following ketone is unusual in that it is more reactive than most aldehydes. Explain why.

Oxidations

As you have seen, aldehydes can be oxidized to carboxylic acids. This often happens through the intermediacy of the hydrate; the hydrate of an aldehyde (but not that of a ketone) has both an H and an OH on the same carbon and is therefore susceptible to the same sort of oxidation as an alcohol. (There are other mechanisms for aldehyde oxidation as well, but we will not go into them.)

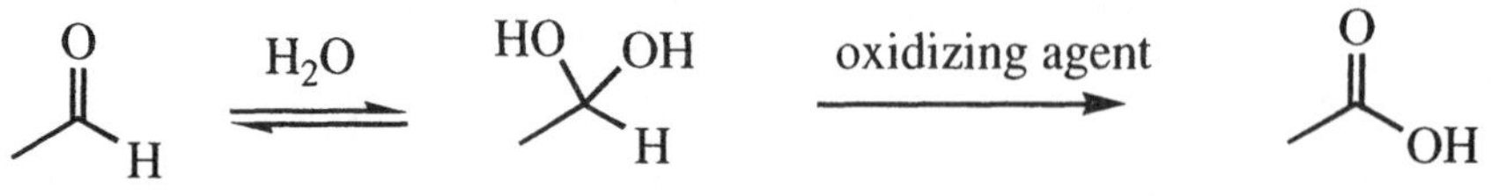

hydrate

Almost any oxidizing agent can accomplish this transformation; that is, the oxidation of an aldehyde to a carboxylic acid is a particularly easy one. An oxidation of this sort that has visual consequences involves silver ion, Ag^+. (The mechanism by which silver ion oxidizes aldehydes is probably not the one shown above.) When Ag^+ oxidizes an aldehyde, the silver ion is reduced to Ag^0; that is, to metallic silver. If the reaction is carried out in a clean test tube, silver coats the inside of the tube and turns it into a mirror. In fact, very few other functional groups react in the same way with Ag^+. You can therefore test an unknown sample to see if it is an aldehyde by combining it with a solution containing Ag^+; if you get a mirror, it is most likely an aldehyde. The most common version of this experiment is called the "Tollens test."

Another oxidizing agent that can convert an aldehyde into a carboxylic acid is oxygen, that is, air. A bottle of aldehyde that has been sitting around for a while is very likely to contain some of the corresponding carboxylic acid simply because of contact with the air. Thus, chemists using aldehydes obtained from stockroom bottles are well advised to purify them before use. (If you are given an old bottle of benzaldehyde, a liquid, from the stockroom, you will probably see solid in it, because the benzoic acid it is contaminated with is a solid.)

Peroxides, compounds that contain O–O bonds, are also good oxidizing agents. If an aldehyde were treated with a peroxide, a carboxylic acid would result. This is why, in the ozonolysis of alkenes, we add zinc in the second step. The first step does not actually produce the products we expect, but instead a strange, complicated intermediate called an "ozonide." Later, when water is added, aldehydes and ketones are indeed formed, along with hydrogen peroxide. If one of the products is an aldehyde, this would normally be oxidized immediately by the H_2O_2 to a carboxylic acid. Zinc metal kills any hydrogen peroxide present to prevent this further transformation. Of course, if you *want* a carboxylic acid, you can just leave out the zinc! Note that you just learned a new way of making carboxylic acids.

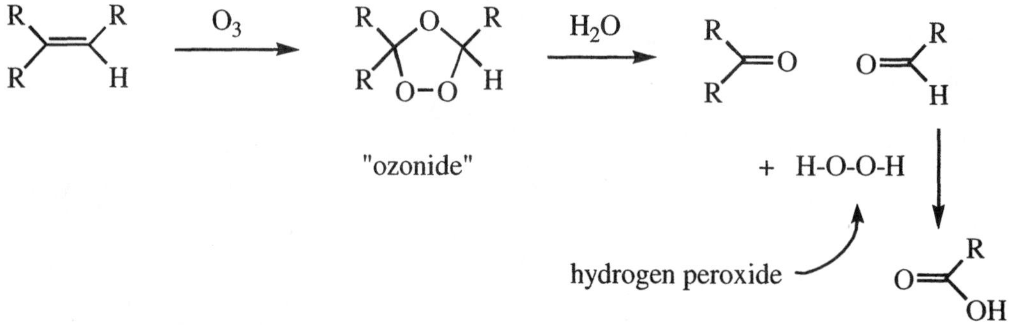

"ozonide"

When you want to convert an aldehyde into a carboxylic acid on purpose in the lab, chromic acid is probably the best choice of a reagent. However, it is quite rare that one actually wants to do this.

Since aldehydes are easily oxidized, they are themselves good reducing agents, in the sense that they reduce the other reagent you added, such as silver ion. In the Tollens test, silver ion is reduced from Ag^+ to Ag^0 by the aldehyde. Later on we will be discussing carbohydrates, or sugars, which are naturally occurring compounds that often contain aldehyde groups. Biochemists and biologists frequently refer to sugars that are aldehydes as "reducing sugars." Now you know why.

Reductions

Ever since Chapter 11, we have been reducing carbonyl groups, especially aldehydes, by catalytic hydrogenation, as if they were just like any other double bond. But also ever since Chapter 11, I have been telling you there is a better way. At last it is time to learn the better way. There are actually two of them we will learn now, and for the time being we will regard them as interchangeable, but in the next chapter we will discover some differences between them.

It is true that aldehydes and ketones can be reduced by catalytic hydrogenation, because carbonyl groups contain double bonds, and catalytic hydrogenation operates on most kinds of double bonds. But carbonyl groups are *not* just like ordinary carbon–carbon double bonds, because they are polar. This makes them *less* reactive in the catalytic hydrogenation reaction (don't ask me why), but it also opens up new possibilities that do not exist for ordinary carbon–carbon double bonds.

The carbonyl group has a partial positive charge on carbon and a partial negative charge on oxygen. This suggests that whatever reacts with the carbon ought to be a nucleophile, and what reacts with the oxygen ought to be an electrophile. This will turn out to be true.

To make a new carbon–hydrogen bond at the carbon of a carbonyl group, hydrogen is required to act as a nucleophile; that is, hydrogen with a pair of electrons, or H^-, which is called "hydride ion." We have not seen this yet, and although ionic compounds such as NaH do exist and behave as if they contain H^-, free hydride like this turns out to be a terrible nucleophile. (Indeed, NaH is another entry in our catalog of strong, non-nucleophilic bases, although these days most chemists use LDA for such purposes.) However, there exist compounds in which hydrogen is *bonded* to an electropositive partner, which allows it to *act* as a nucleophile. The most common partners of this type for hydrogen are aluminum and boron, and the most common such compounds are lithium aluminum hydride (LAH) and sodium borohydride. Note that LAH is pronounced "el-ay-aitch," not "lah."

lithium aluminum hydride   sodium borohydride

Both of these act as sources of nucleophilic H^-, and for mechanistic purposes we can imagine the aggressive species as being H^-, while recognizing that the aluminum (or boron) is necessary to give the hydride its nucleophilic properties. Both of them work, more or less, like this:

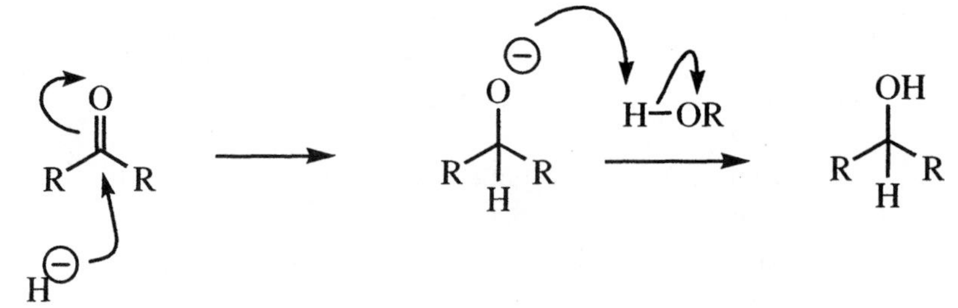

As I mentioned above, there are some significant differences between the two, and you ought to keep them in mind. Sodium borohydride is a relatively mild reducing agent; indeed, we will discover in the next chapter that it does little else but reduce aldehydes and ketones. It is so stable that it can be dissolved in alcohols, and the reaction shown above is usually carried out in ethanol solvent, so the "ROH" shown in the second step would be the ethanol solvent:

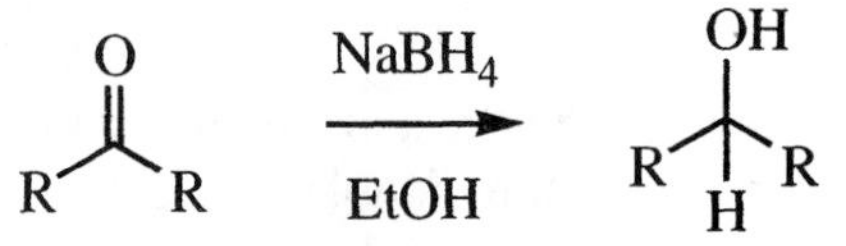

Lithium aluminum hydride, on the other hand, is extremely reactive—so reactive that it will catch fire (because of the formation of H_2) if it runs into water or anything else with O–H bonds, including alcohols. It must be kept strictly away from such solvents and is typically used in ether (or THF, another ether). Therefore, the species from which the O⁻ steals a proton (H^+) in the second step above cannot be present from the beginning of the reaction, or the LAH would react with it instead. LAH reactions are typically run in two steps: first, the LAH is added to the carbonyl compound in ether; later, after the reaction has stalled at the halfway point, with a negative charge stuck on the oxygen, water is added (*very carefully!!!*) in order to destroy any extra LAH remaining and to provide a source for the hydrogen that will appear on the oxygen. LAH reactions have to be written indicating that there are two steps:

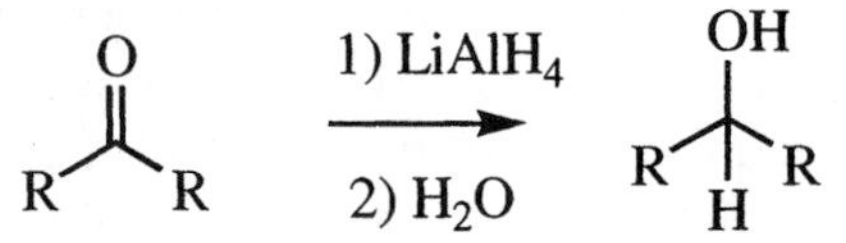

If you write this reaction without the 1) and 2), you are indicating that everything is thrown together at once, and anyone following your instructions had better duck.

It can be important, under certain circumstances, to keep track of the hydrogens that are involved in this reaction. First, from a stoichiometric point of view, you should recognize that both of our reducing agents, $NaBH_4$ and $LiAlH_4$, have four hydrogens capable of reacting, and all four of them do. Thus, each molecule of LAH can reduce four molecules of a ketone or aldehyde; each mole of ketone or aldehyde requires 1/4 mole of LAH or $NaBH_4$. Second, from a bookkeeping point of view, you should remember that the two new hydrogens on the new alcohol (one on the carbon, the other on the oxygen) did *not* come from the same source: the one on the carbon came from the reducing agent while the one on oxygen came from the solvent (water or ethanol). Who cares? Sometimes it is useful to incorporate *labeled* hydrogens, ^2H (deuterium, or D) or ^3H (tritium, or T) into molecules, so that you can find out what happens to these hydrogens during some subsequent reaction. In order to do that, you need to know where the particular hydrogens in question come from.

Test Yourself 2

Suggest a synthesis of the following molecule starting from any unlabeled organic starting materials you like. Assume that any reagent containing hydrogen is available with deuterium instead. As usual, you should try this yourself before you look at my answer.

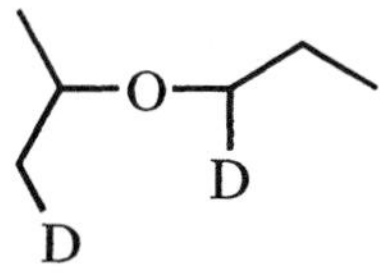

Biological Reductions and Oxidations

When organisms need to accomplish reduction reactions, they cannot pull LAH or sodium borohydride off the shelf. Instead, nature has devised its own approach to providing a source of H⁻ for reductions. Although there are a few variations on the theme, the standard biological reducing agent is something called NADH for short:

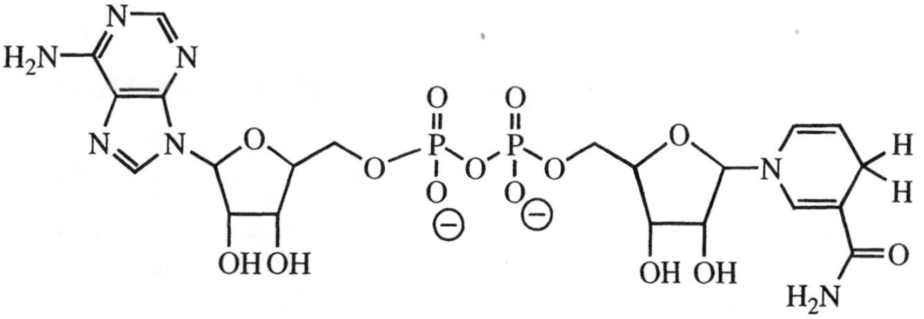

Most of this molecule is totally irrelevant to the chemistry it engages in (though extremely relevant to how it knows where to be and what to do), so we will abbreviate the structure as follows:

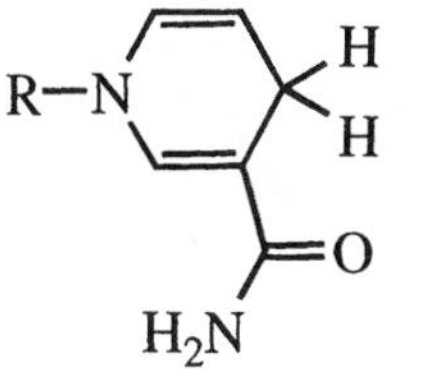

where "R" represents all the stuff on the left.

In the presence of a carbonyl group, this molecule can deliver hydride, H⁻, to its target:

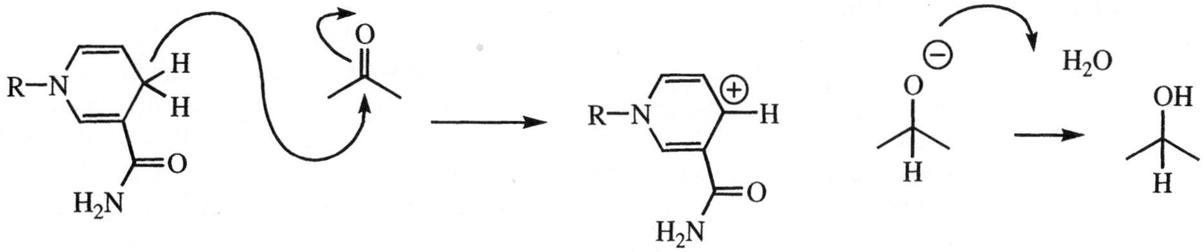

Why can it do this? Because the cation that is generated is especially stable due to resonance:

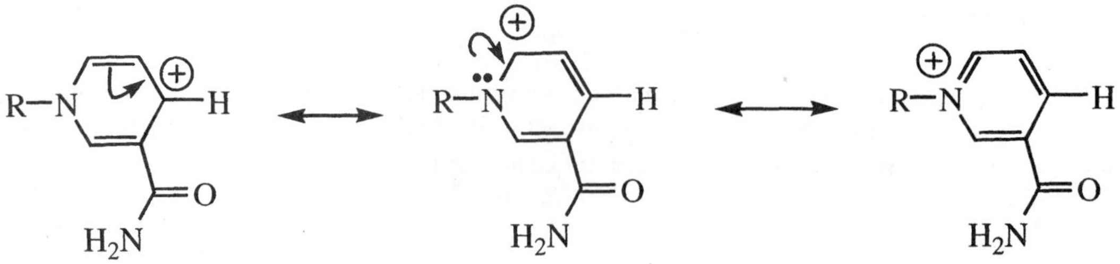

Indeed, this cation is even more special than a normal, resonance-stabilized cation: you will notice that in the last resonance structure I drew, there is a completely conjugated six-membered ring, just like the one in benzene, and like benzene, this thing has exceptional stability called "aromaticity" (we will discuss this topic in Chapter 23). The new molecule, called NAD$^+$, just loves being aromatic, but on the other hand, it is not so thrilled about that positive charge. How can it get rid of the positive charge? One way is to accept a hydride from some other molecule:

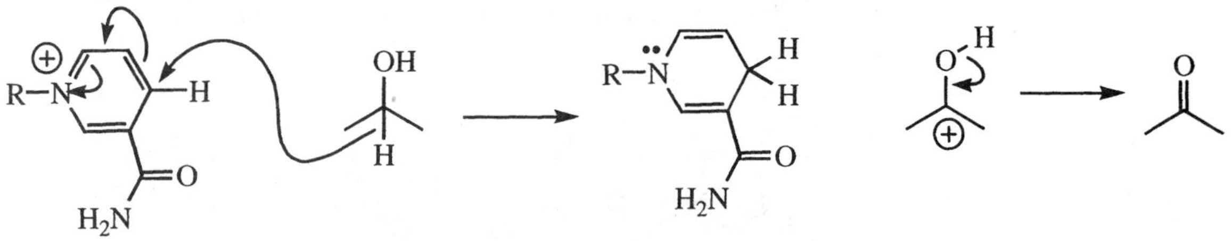

Now the positive charge is gone, but so is the aromaticity! This is one seriously conflicted molecule! It wants to give away hydride, to become aromatic, but then it wants it back again to get rid of the positive charge! NADH wants to serve as a reducing agent (donating hydride), but the product of the reaction wants to act like an oxidizing agent (accepting hydride). It is, in other words, perfectly designed to be a biological redox reagent. When oxidation is called for, NAD$^+$ is sent in to become NADH. When reduction is called for, NADH goes in and becomes NAD$^+$. How this process is controlled is a subject for another course.

Some of you may have heard of another biological redox reagent called NADP$^+$, which reversibly turns into NADPH. This pair is identical to the NAD$^+$ and NADH combination except that an extra phosphate group has been stuck on one of the OH groups of NAD. Functionally, the system is identical. There is yet another pair of this sort, with a molecule called FAD that becomes FADH$_2$ and *vice versa*. The structures and mechanisms in this case are somewhat different and we will not discuss them.

Addition Reactions

Irreversible: Grignard Reaction

In discussing reductions, we discovered an important trick in the chemists repertoire: while hydrogen normally is bonded to carbon or something more electronegative than carbon, and therefore carries a partial positive charge, we can also create molecules in which hydrogen is bonded to something *less* electronegative than itself, thereby creating hydrogen with a partial *negative* charge, and this can react with other molecules as if it had a real negative charge.

Thus, NaBH$_4$ and LiAlH$_4$ *behave* as if they were sources of H$^-$ and can be used to carry out reduction reactions. It turns out that we can play the same trick with carbons; indeed, we already introduced this possibility way back in Chapter 9! You may recall from the very end of that chapter that we met things called Grignard reagents:

$$R-Br \xrightarrow{\text{Mg}} R-Mg-Br$$

We did not talk at the time about how this reaction occurs, and we will not do so now either. Consider it magic. But the results are extremely interesting: carbon (the "R" group) is suddenly attached to magnesium instead of bromine. Instead of having a partial positive charge, it now has a partial negative charge. Although the reactions Grignard reagents undergo are in truth very complicated, we will pretend they are very simple: pretend that this is a source of R$^-$, and you will be able to predict a great deal about its chemistry.

For example, what would happen if we made a Grignard reagent and added to it some water or an alcohol (something with an O–H bond)? One possibility would be an acid–base reaction:

$$\overset{\ominus}{R} \;+\; H_2O \;\rightleftharpoons\; RH \;+\; \overset{\ominus}{}OH$$

In which direction would this reaction go? One could find out by consulting a pK_a chart, but you should also be able to answer the question without even looking. What this reaction proposes to do is trade a minus charge on carbon for one on oxygen. Chapter 2 should have let you predict that this would be a very favorable thing to do! Oxygen is far more electronegative than carbon, so a minus charge will of course be better on O than on C! (Using your pK_a table, you will find that H$_2$O is about 30 orders of magnitude more acidic than RH.)

So what does this tell us? Grignard reagents must be carefully shielded from exposure to water or alcohols or any other "loose" proton sources or they will destroy themselves. In this respect they are like LAH: duck when you add water. Grignard reagents are almost always used in ether (indeed, for esoteric reasons, the ether seems necessary for their formation in the first place). We have also learned something else: if we want to change a halide into a hydrogen in a molecule, we can. In other words, the reaction of a Grignard reagent with water is a reaction to remember:

$$R-Br \xrightarrow[\text{2) } H_2O]{\text{1) Mg}} R-H$$

Note that this now gives us a fourth way of incorporating hydrogen into a molecule, and that if here you used D$_2$O instead of H$_2$O, there would be a deuterium in your product at precisely the position that the halogen previously occupied.

But of course, this transformation is not why Victor Grignard won the Nobel Prize. Grignard reagents are far more useful than simply a vehicle for putting hydrogen into a molecule. The most important use of Grignard reagents is reaction with carbonyl groups, in what is called the "Grignard reaction." In this respect, Grignard reagents parallel exactly what we learned about LAH, except that the attacking nucleophile, instead of H$^-$, is C$^-$:

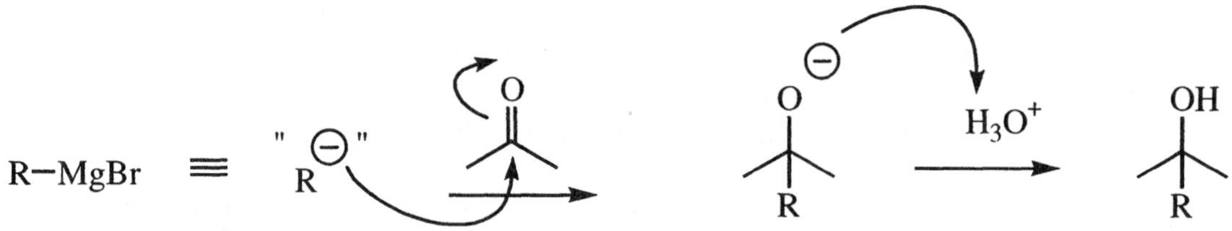

Focus your attention on the first step, where the Grignard reagent, behaving as R⁻, attacks the carbonyl group of an aldehyde or ketone. Is this a favorable process? From a stability point of view, the most important thing we have done is exchange C⁻ for O⁻, which is, as we just determined, an extremely favorable thing to do. Thus, you should expect such a reaction to be very rapid and irreversible. It is.

Two important points you should note: first, the reaction takes place in two steps, just like the LAH reaction. Your equations should always indicate that:

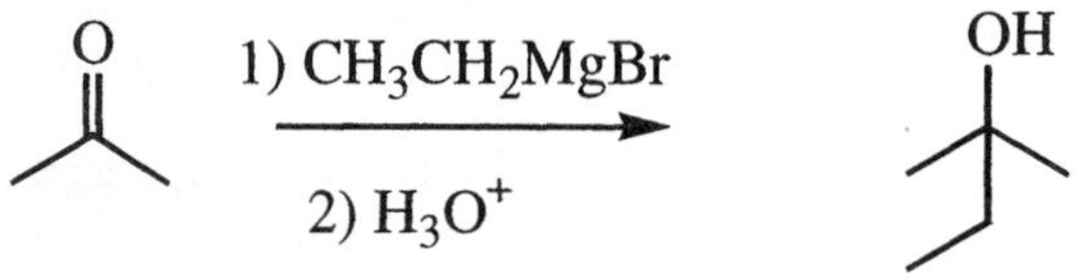

Second, and most important, *this is a new way of making carbon–carbon bonds!* Indeed, until now, the only way you have seen to stitch carbon chains together required using an acetylide anion as a nucleophile for attacking a carbon (preferably primary) bearing a leaving group. There are intrinsic limits to the structures you can make using that technology, and although you should not forget the method, since it is sometimes the best way to make a particular target, it is about to be placed on the back burner. The Grignard reaction opens up our horizons tremendously. To see how the Grignard method can be used synthetically, we need to analyze the reaction carefully.

Consider the example I just showed you. I've redrawn it below to emphasize some points. First, we need to keep track of the carbonyl carbon. In the product, that carbon is the one with the OH group directly attached to it. That means if we want to work backwards (i.e., given some alcohol, where did it come from?), we need to focus our attention on the carbon bearing the OH. It used to be a carbonyl group, meaning that there was a double bond between that carbon and what is now the oxygen of the alcohol; the hydrogen of the alcohol was *not* present in the starting material, and one of the three other bonds to this carbon (apart from the bond to the oxygen) was also missing in the starting material. That bond is now to the group that forced its way in as a Grignard reagent, making a new carbon–carbon bond.

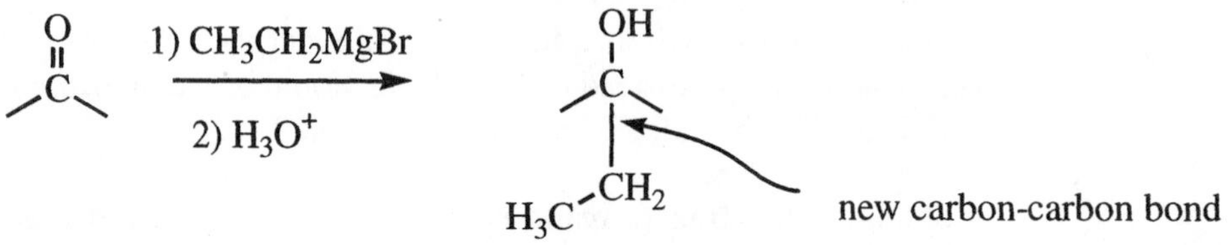

Thus, we can look at our product alcohol and dissect it into some component parts, as shown below. The funny hollow arrow means "came from."

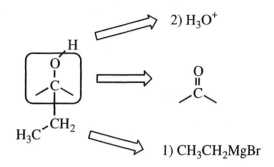

At this point, we need to recognize that some of the things we did in this exercise were necessary, and some were choices. The hydrogen of the alcohol came from the water we added after the reaction was over. No choice about that. The carbon–oxygen single bond must have begun life as a carbon–oxygen double bond if we are to use a Grignard reaction to make this compound. No choice there either. But in our alcohol there are *three* carbons attached to the carbon *bearing* the OH group. It is not written in stone that the new group had to be the ethyl group that was not there in the ketone we started with—it could have been one of the *other* groups, *if* we had started with a different carbonyl compound. In other words, we *might* have dissected the alcohol this way:

So there are two different ways we might have considered for making this particular alcohol by a Grignard reaction. (Is there a third? Why or why not?)

Virtually any alcohol you can write down can be made using Grignard reactions. Indeed, there may be multiple approaches to any given alcohol. Let's take an example. Show how you could make the following alcohol starting from stable organic molecules containing four or fewer carbons. Do it now!

First, to help you keep track of everything, let's write in the invisible hydrogen on the carbon bearing the OH group:

Since this is an alcohol, we can contemplate making it from smaller pieces by means of a Grignard reaction. There are two different Grignard syntheses possible here, shown below:

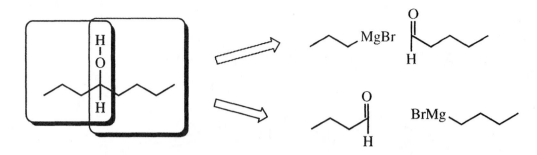

Clearly, if we have to make this from pieces containing four or fewer carbons, the second (blue) approach is more efficient. We then need to complete our plan of action by indicating where the Grignard reagent will come from, since organomagnesium compounds are not stable organic molecules:

COMMON MISTAKES: DON'T LET THIS HAPPEN TO YOU!

In writing down this process, many students tend to gain or lose carbons. This is easy to do, but also easy to avoid. The reason it happens is that students often take, for example, the piece circled below, and then try to write a corresponding Grignard reagent as shown:

What this misses is that there are four carbons in the piece we are trying to attach; the person who wrote the above process only indicated *three*. Writing down the corresponding Grignard reagent takes four lines, not three, because you have to introduce the bond between the last carbon and the magnesium as an extra line.

A similar common mistake involves predicting the products of a Grignard reaction. For example, I have often seen students write one of the top answers to the following problem rather than the (correct!) bottom one:

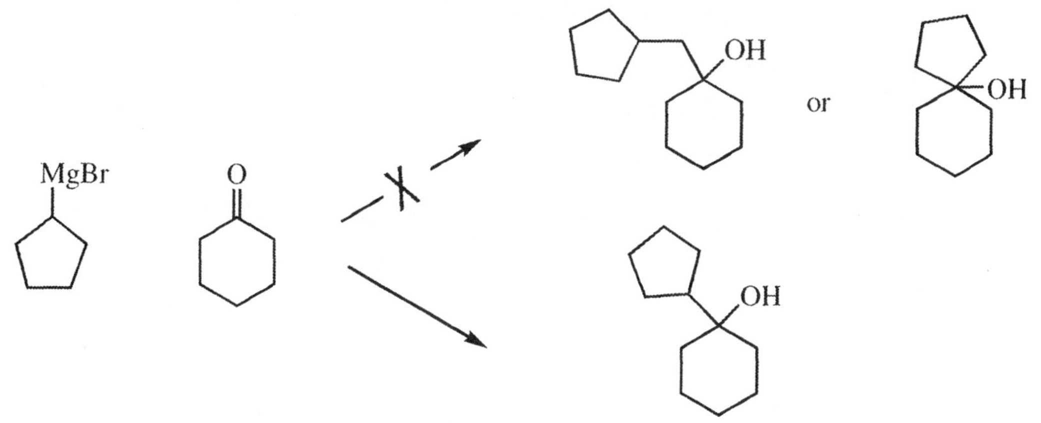

The easy way to avoid this kind of problem, especially once you recognize that you are susceptible to it, is to count your carbons at each step along the way. Gaining and losing carbons like this is a sure (but avoidable) way to lose points on an exam.

What if I gave you the same target compound we just worked on, but now changed the ground rules to require starting with compounds containing *three* or fewer carbons? Work on this for a while. Remember, synthesis problems are rarely solved all at once. You do what you can do, and then work backward from there.

The first part is easy:

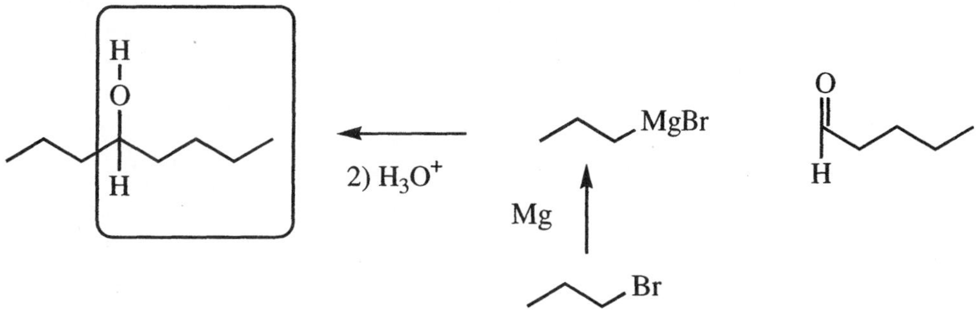

We have already established one three-carbon starting material. But the other piece has five carbons. What shall we do? The key here, and in any Grignard synthesis, is the fact that you can make almost any alcohol you want with a Grignard reaction, but an alcohol is the only thing you can make. So when you have to make substances that are not alcohols, see if you can make them *from* alcohols. So let's show how we could make the required aldehyde from an alcohol:

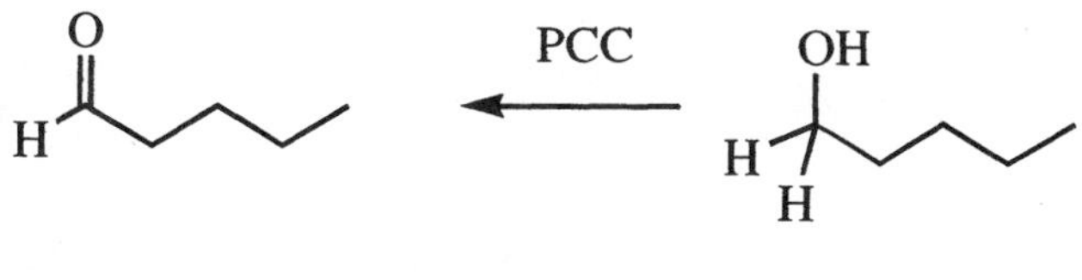

Now *this* we can make by a Grignard reaction:

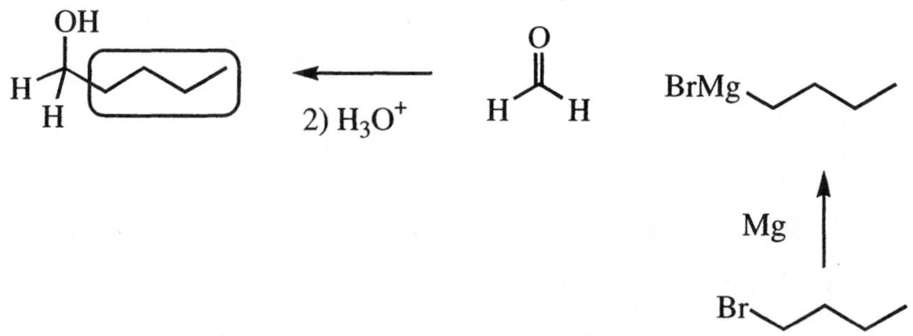

So far we have gotten down to four carbons; again, the thing we need is not an alcohol, but again, we can make it *from* an alcohol:

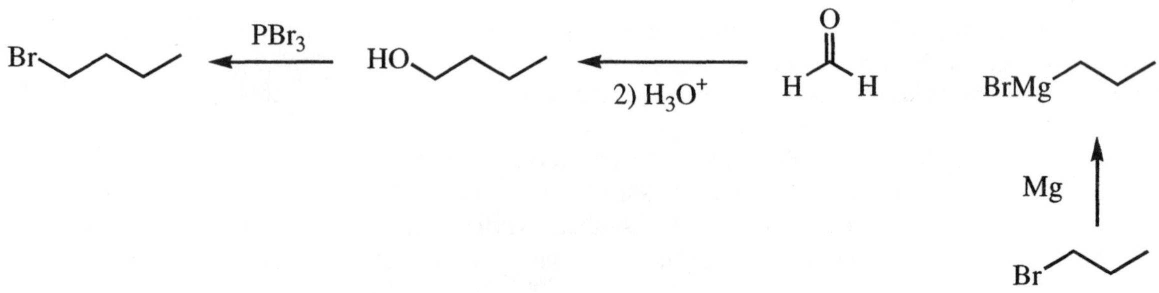

And at last, we are working with a stable, three-carbon unit. So you see, a molecule, almost *any* molecule, can be made by sequential Grignard reactions together with appropriate functional group manipulation. The whole sequence in the present case, put together, looks like this:

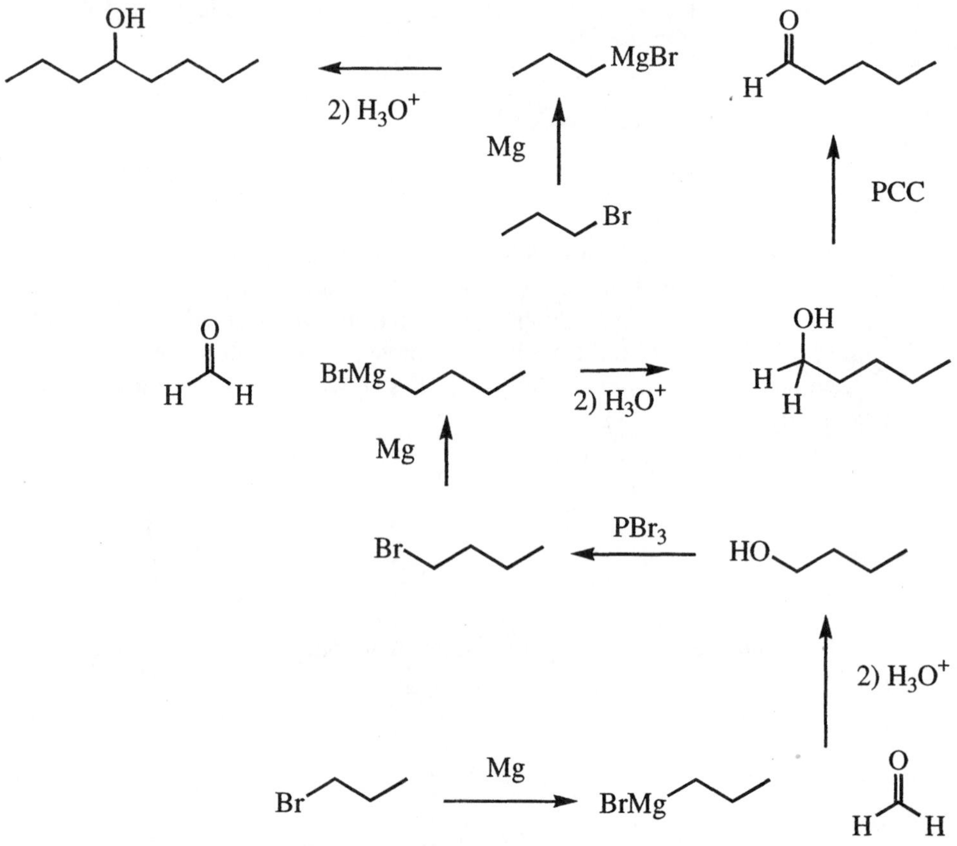

The grind-it-out strategy exemplified by the above sequence works, and there is nothing wrong with it, but some of you may be itching for a shortcut. Do you really have to build up these chains one carbon at a time? No, you don't, and now we'll examine a shortcut that may occasionally save you some time. Be sure you understand it before you use it, however.

You might recall that epoxides are susceptible to attack by nucleophiles, and that the nucleophile always attacks at the less hindered position of the epoxide. Also, the product of nucleophilic attack on an epoxide is an alkoxide anion (just like the product of Grignard attack on a carbonyl), which, if there were nothing to quench it in the reaction mixture, would sit around and wait for you to add something. Using this approach, you can add two carbons at a time to make longer chains, or create branches. However, it is important that you keep careful track of where your substituents end up:

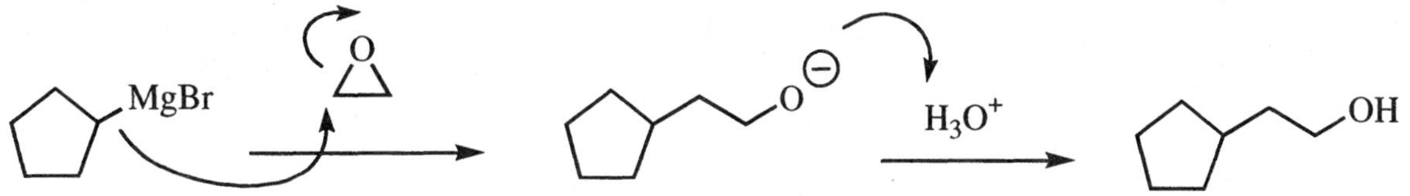

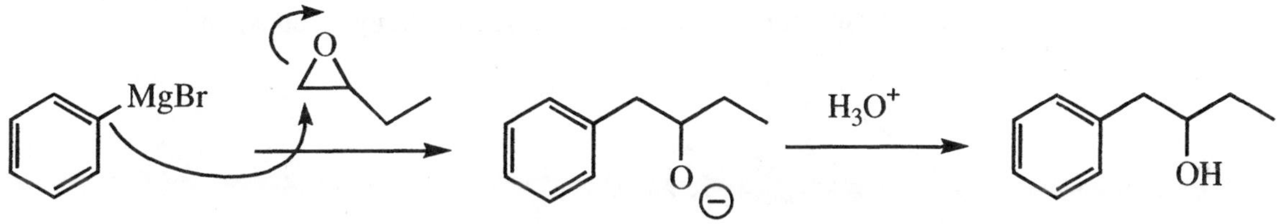

Notice from the above example that this reaction, unlike all the others we have learned that apply to alkyl halides, does work with halobenzenes. In other words, you *can* take bromobenzene and treat it with magnesium to get the Grignard reagent phenyl magnesium bromide, which reacts just like a typical Grignard reagent. Every other halide reaction we have learned (substitution, elimination) is *inapplicable* to halobenzenes. Benzene has special chemistry that we will learn about in Chapter 23.

Using the epoxide trick, you could shorten the synthesis we worked out before:

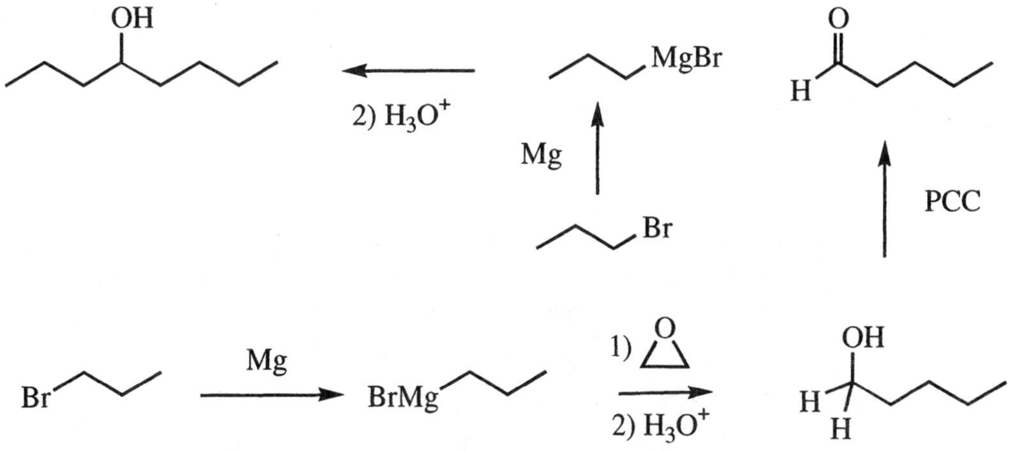

By the way, while we are on the subject of shortening syntheses, don't forget the old-fashioned reactions you learned in previous chapters. The following synthesis is also a very good solution to the problem:

H—≡—H $\xrightarrow[\text{2)}\ \ \diagdown\diagup\text{Cl}]{\text{1) NaNH}_2}$ H—≡—$\diagup\diagdown$ $\xrightarrow[\text{2)}\ \ \diagdown\diagup\text{Cl}]{\text{1) NaNH}_2}$

$\Big\downarrow \begin{array}{l}\text{H}_2\\ \text{Lindlar}\end{array}$

$\xleftarrow[\text{H}_2\text{O}]{\text{H}^+}$

OH

There are two further points I need to make about the Grignard reaction. First, with the exception of the epoxide reaction I just discussed, Grignard reagents are very poor nucleophiles for S_N2 reactions. It is not clear why this should be the case, but it is. Do not try to displace leaving groups like halides with Grignard reagents as a way of making carbon–carbon bonds. It rarely works. Stick to reactions involving carbonyl groups or epoxides.

Second, there is one type of carbonyl substrate you probably have not yet considered: carbon dioxide is a molecule containing two equivalent carbonyl groups, and one of them *can* react with a Grignard reagent:

MgCl $O=C=O$ $\longrightarrow$ $\xrightarrow{\text{H}_3\text{O}^+}$ OH

Voila! A new way of making a carboxylic acid! The carbon dioxide for this reaction is usually used in the form of dry ice, which is solid carbon dioxide.

Your ability to make a huge variety of molecules from smaller pieces has just multiplied by several-fold in this section on Grignard reactions. You will need a lot of practice stringing together these reactions and the relevant functional group transformations. Do not skimp on the process or you will regret it! There are plenty of practice problems at the end of the chapter.

Reversible Additions: Oxygen Nucleophiles

Hydrates

As you have now seen, a good nucleophile can force its way to the carbon of a carbonyl group, pushing electrons out onto the oxygen. One good nucleophile you are familiar with is negatively charged oxygen. Consider, for example, the reaction of a ketone (or aldehyde) with OH^-:

OH $\rightleftharpoons$ O OH $\xrightarrow{\text{H}_2\text{O}}$ HO OH + $^{\ominus}$OH

a "hydrate"

There is an important difference between this reaction and the ones we just finished discussing: whereas the reaction of hydride or a Grignard reagent with a carbonyl group results in an anion (on oxygen) that is much more stable than the one we started with (on hydrogen or carbon), in this case the product anion is really very similar to the attacking anion. The first step is not dramatically downhill in energy, and it is bad from an entropy point of view (converting two species into one), so we might predict that this reaction would be unfavorable. That prediction is (in most cases) correct. Such a reaction is reversible, and the equilibrium favors the starting materials over the product. The starting material is so preferred that in most cases you cannot detect the presence of *any* of the hydrate by the usual methods of measurement.

The above paragraph should cause you concern for several reasons. First, when the reaction goes from right to left, OH⁻ is the leaving group, and I have told you repeatedly that OH⁻ is a terrible leaving group. This is true for substitution reactions, and also for the elimination reactions we have studied. But here we have a special kind of elimination, one in which the species undergoing elimination is already an anion (the technical term for this type of reaction is E1cB, which stands for **E1** from the **c**onjugate **B**ase of the starting material). In reactions creating a double bond from a pre-existing anion, OH⁻ is *not* such a terrible leaving group, so this elimination is allowed.

The second reason you should be concerned about the hydrate reaction, as I have described it, is that I said the reaction favors the starting materials so much that you cannot even detect any hydrate at all. If we can't see it, how do we know such a thing really happens? The bottom line answer is, we don't exactly *know* it. But there is evidence that we can't explain any other way. The most convincing evidence, perhaps, is that you can exchange the oxygen in an aldehyde or ketone by allowing the substance to sit in basic water for awhile. If the water initially contains a labeled oxygen (^{18}O, usually symbolized by an asterisk on the O), and the carbonyl group does not (or *vice versa*), after a while exchange will become apparent:

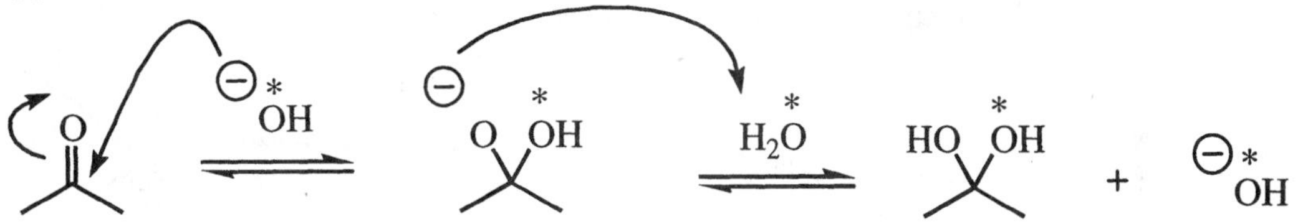

As the hydrate heads back into the starting material, either OH group can lose its proton, and if it happens to be the one with the label, then the ketone will suddenly contain a labeled oxygen (write out this process and see for yourself!). Over time virtually all the ketone molecules will acquire ^{18}O. Chemists have been able to come up with no logical explanation for this that does not involve a hydrate.

The third reason you should be concerned about the hydrate reaction as I described it is that I claimed that equilibrium favors the starting material, and yet we have already studied a reaction (chromic acid oxidation) in which an intermediate (aldehyde) is converted to the product (carboxylic acid) *via* just such a hydrate. If there is so little hydrate present that we cannot even detect it, how is this possible? You are actually already familiar with a similar phenomenon. Consider a helium-filled balloon. The helium is trapped in the balloon—the number of atoms getting through the balloon and escaping into the surroundings is negligible compared to the number of atoms trapped in the balloon. The percentage of atoms escaping would be an extremely small number. And yet, as you know, over time the balloon will deflate. Why? Because once a helium atom has escaped it never goes back. One escapes, then another, then another. Every time one escapes, it disappears, and other helium atoms keep pushing at the balloon. The same is true with hydrates: the percentage of molecules in the hydrate form is very small, but if every time one shows up, it is removed from the equilibrium (by being converted to a carboxylic acid, for example), pretty soon another molecule will make the same trip, and over time they will all be converted. Also, you must remember that the number of molecules we are dealing with in any measurable sample is immense: 10^{20} or more. Even a very small percentage of such a huge number represents a lot of actual molecules.

There is one further piece of evidence supporting the existence of hydrates. Certain molecules that are particularly uncomfortable when they contain carbonyl groups actually prefer to be hydrates. This includes carbonyl groups that are especially reactive by virtue of having electron-withdrawing groups attached to the carbon, such as hexafluoroacetone and trichloroacetaldehyde (chloral):

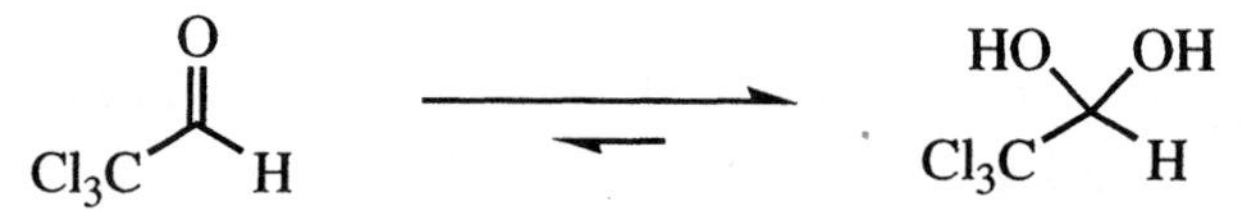

"chloral"   "chloral hydrate"

Chloral hydrate is the substance that villains always put in drinks to knock out the good guy in detective stories (a "Mickey Finn").

So far we have seen how an aldehyde or ketone can change into its hydrate and back in base. The same process can also occur in acid:

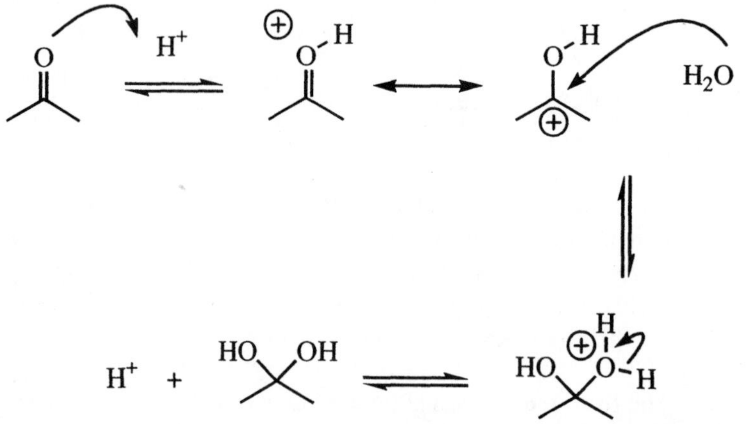

Like the base process, equilibrium normally greatly favors the starting materials, so much so that the hydrate is normally undetectable. (This, of course, must still be the case, since an equilibrium constant depends only on the energy difference between the starting material and the products, *not on the pathway* by which the molecules get from one to the other: $\Delta G^{\circ} = -RT\ln K_{eq}$.) Like the base process, you can detect that this is happening by using labeled water and watching the label appear in the carbonyl group (show a mechanism for this!). And like the base process, at the end you get back the H$^+$ (or OH$^-$) you started with, so the reaction is catalytic. Therefore you should remember that wherever there is aldehyde (or ketone) and water together, there *will* be a small amount of hydrate as well. It is unavoidable, because it is virtually impossible to create an environment that does not have at least a trace of acid or base present. Even the walls of your container or reaction vessel are usually slightly acidic or basic.

COMMON MISTAKES: DON'T LET THIS HAPPEN TO YOU!

Now, pay attention here, because what I say will apply to almost everything we do in the next three chapters, and many students get it wrong. As a general rule, a weak nucleophile and a weak electrophile will not react with each other. One or the other has to be pumped up before a reaction will occur. A neutral oxygen is a weak nucleophile. A carbonyl group is a weak electrophile. These will not react with each other until one or the other is pumped up. A neutral oxygen can be pumped into a strong nucleophile by removing a hydrogen and making it an anion. *Then* it can attack a carbonyl group, a weak electrophile. On the other hand, a neutral carbonyl group can be pumped up by *gaining* a proton, giving it a positive charge and making it a strong electrophile. Then it can draw in a neutral oxygen, a weak nucleophile. Once the carbonyl is juiced by having a positive charge, it does not *need* a strong nucleophile to attack it. Indeed, if there were one around, it would most likely simply steal the proton from the carbonyl group (or, more accurately, the carbonyl group could not have been protonated in the first place!). The take-home message here is this: some carbonyl reactions involve a succession of anions reacting with neutral compounds, and others involve a succession of cations reacting with neutral compounds, but you almost never find cations reacting with anions.

Look at the hydrate-forming reactions shown above. One shows only negative charges; the other only positive charges. Don't mix them!

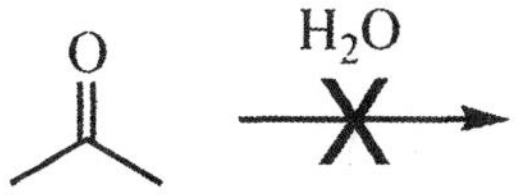

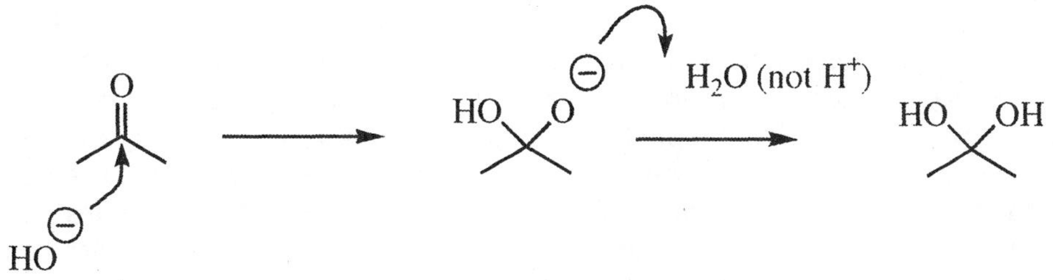

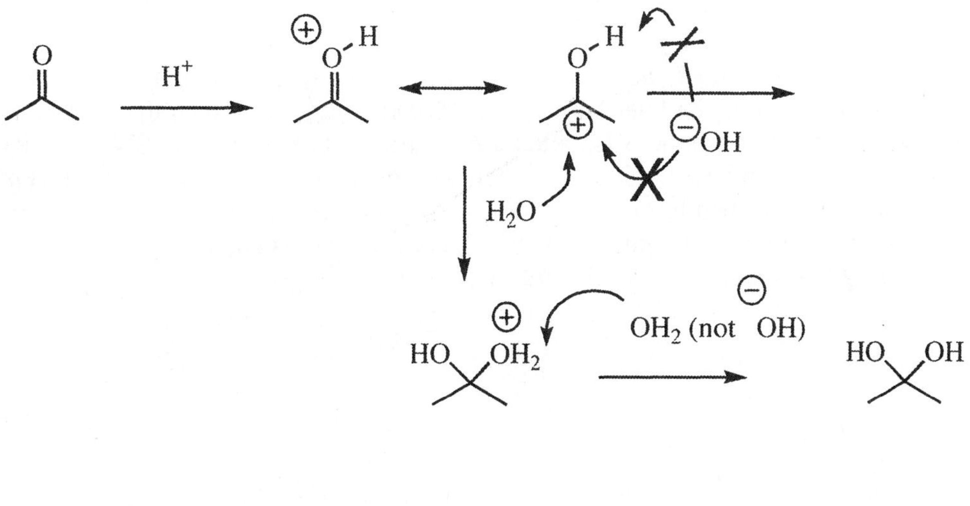

You may recall, from way back in Chapter 10 when we first started looking at reactions, that we discovered alcohols do most of the same things water does. In the above reaction, for example, one of the protons of water takes off, but the other one remains stuck to the oxygen. Would the reaction be any different if that second proton were, instead, an alkyl group? Of course not! The two reactions we have just studied (formation of a hydrate in either base or acid) work exactly the same way if the solvent is an alcohol rather than water:

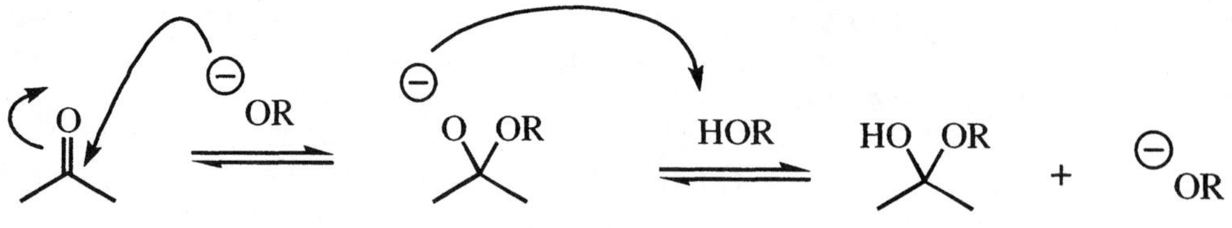

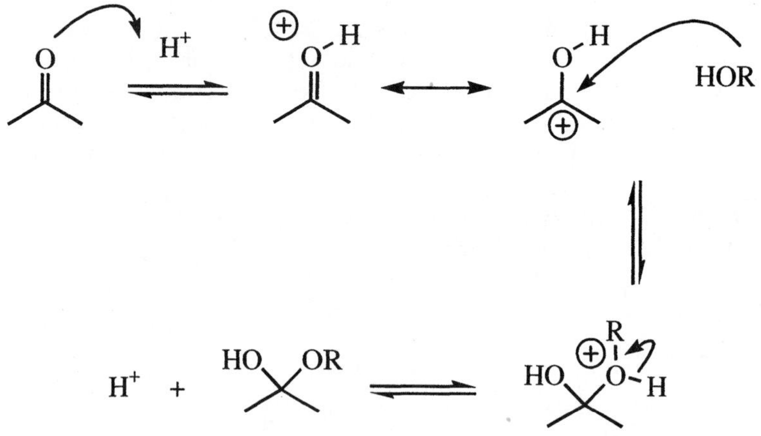

If you recognize these reactions as being nothing new, you are on the right track to learning this material. Notice that one mechanism uses only negative charges and the other only positive charges. This reaction, like the hydrate-forming reaction, can happen any time an aldehyde (or ketone) is brought together with an alcohol, because it needs only a trace of acid or base to get the process going. The only difference between this reaction and the hydrate reaction is that there is only one way to get back where you came from, so ^{18}O-labeled alcohol could not introduce ^{18}O into your starting ketone.

The product, in this case, has a different name, and, I'm sorry to say, the exact name depends on whether you start with an aldehyde or ketone. If it is an aldehyde (that is, if there were a hydrogen at the end of one of the lines coming off the carbonyl group), then the product is called a *hemiacetal*; if it is a ketone it is called a *hemiketal*. (That's "hemi-ASS-it-al," not "hemi-uh-SEE-til"; and "hemi-KEE-tal.") A carbon with *both* an OH and an OR attached is either a hemiacetal or a hemiketal, depending on what the other two groups are. (Technically, according to IUPAC, this distinction has been eliminated, and both types are called hemiacetals, but everyone ignores this rule!)

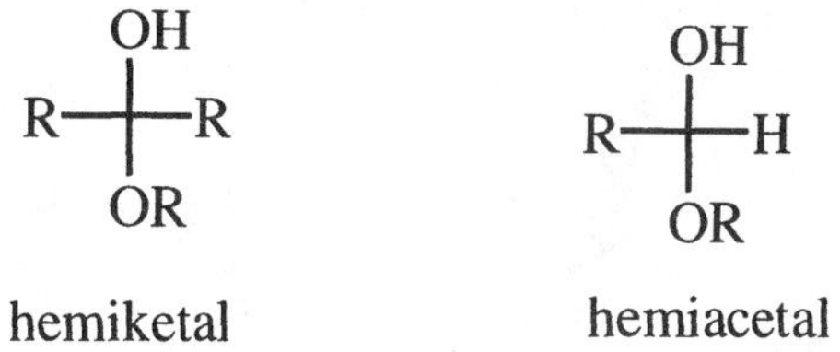

hemiketal   hemiacetal

These names of course apply to the species themselves, even if they are not made by the above process, so you should train yourself to recognize the corresponding functional groups.

Like hydrates, hemiacetals and hemiketals are in equilibrium with the corresponding aldehydes and ketones; as with the hydrates, these equilibria prefer the starting materials, so hemiacetals and hemiketals are rarely seen. The reason, as with the hydrates, is that there is little thermodynamic difference between the starting materials and the product (they have about the same energy), but entropy much prefers the situation with two separate molecules compared to one in which they are tied together:

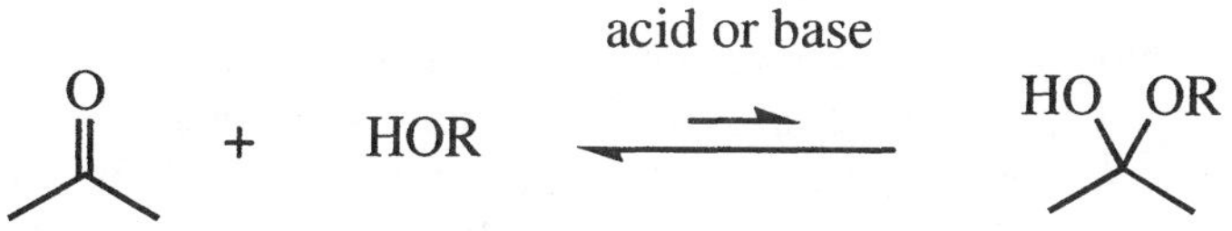

This is especially important, because hemiacetals and hemiketals will enter into such an equilibrium no matter where they come from, but in most cases what actually predominates is the carbonyl compound and the alcohol.

Test Yourself 3

Show the mechanism of the following reaction. *Hint*: all carbocations involved are resonance stabilized, and you should show that resonance.

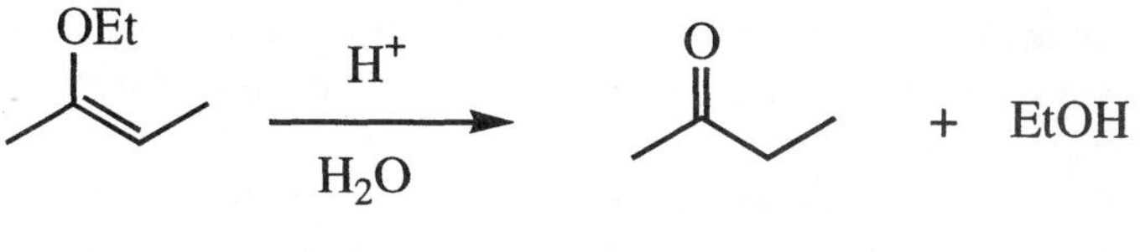

Cyclic Hemiacetals and Hemiketals

The fact that the position of equilibrium is determined by entropy brings up an interesting question: suppose there were no entropic advantage to the carbonyl compound over the hemiacetal or hemiketal. Can we imagine a situation in which that would be the case?

Let's go after the second part first. The reason for the usual entropic preference is that in the equation as written there are two molecules on the left and only one on the right. One way to even the score would be to make the two molecules on the left become one, by connecting them. That is, if the alcohol that participates in this process is physically attached to the carbonyl part of the molecule in the first place, then we would be converting one molecule into a different molecule. The only difference would be that on the right there would be a ring. Here is a conceptualization of this idea, followed by a real example.

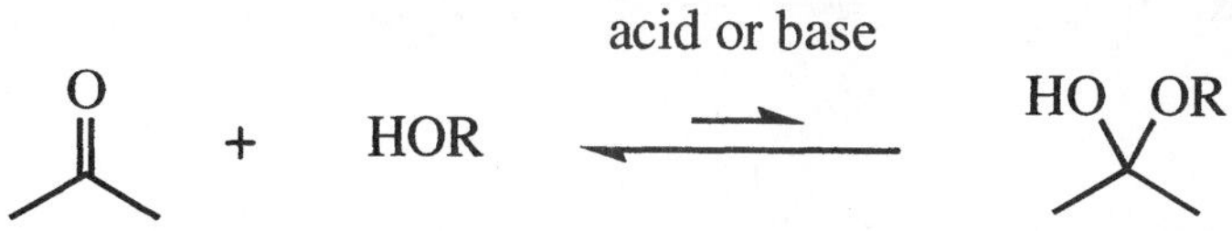

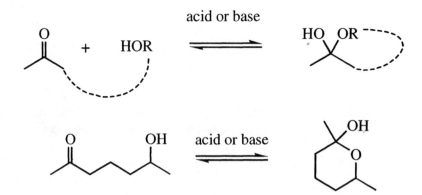

Take time out now to show the mechanism for this last reaction, in both acid and base, and both forward and backward. It is exactly the same as what we have done before, although students are often thrown off for some reason by the ring.

Now for the issue we started with: in cases like this, where does equilibrium lie? The answer is implied in the arrows above: there is no longer a strong preference for the ketone–alcohol side. Indeed, if the ring is of a favorable size (5 or 6 members, like the above example), the hemiketal (hemiacetal) is often the preferred form. We will return to this point at the end of the chapter when we discuss carbohydrates.

Test Yourself 4

Show the structure of some hemiacetal you would predict to be stable (i.e., preferred over its aldehyde–alcohol form).

Ketals and Acetals

I am getting tired of writing (and you are probably getting tired of reading) "hemiketal (or hemiacetal)," so I am going to invent the word "hemiXtal" to stand for that phrase. OK?

When a carbonyl group becomes a hemiXtal in base, can anything more occur? About the only imaginable thing is that the base (still present) might remove a proton from the OH of the hemiXtal, and the molecule could go back where it came from, as shown below. Thus, a hemiXtal is the end of the line under basic conditions.

However, this is not the case under acidic conditions. By way of review:

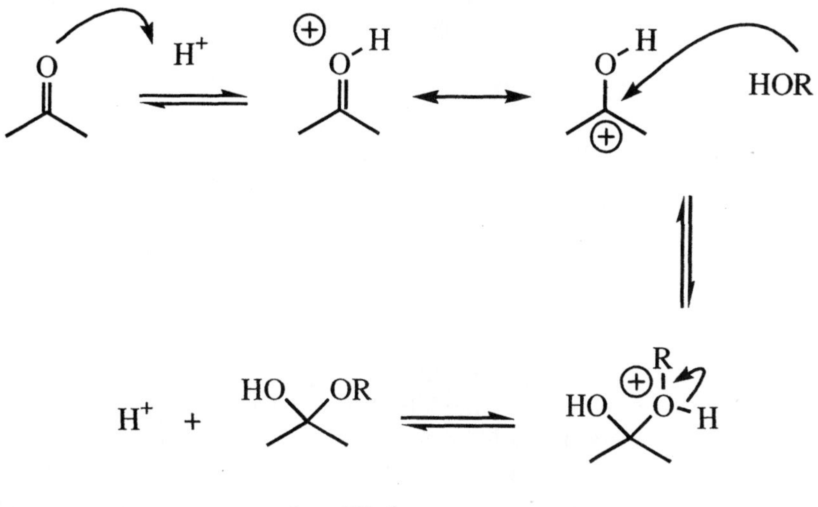

hemiXtal

If you now start at the hemiXtal and work your way backwards, you'll discover you have two choices for the first step: you can of course protonate the OR group, which sends you back where you came from, but the lone pairs of the OH group are no less available, and no less attractive to the proton. You could equally well protonate the OH group. What would that lead to?

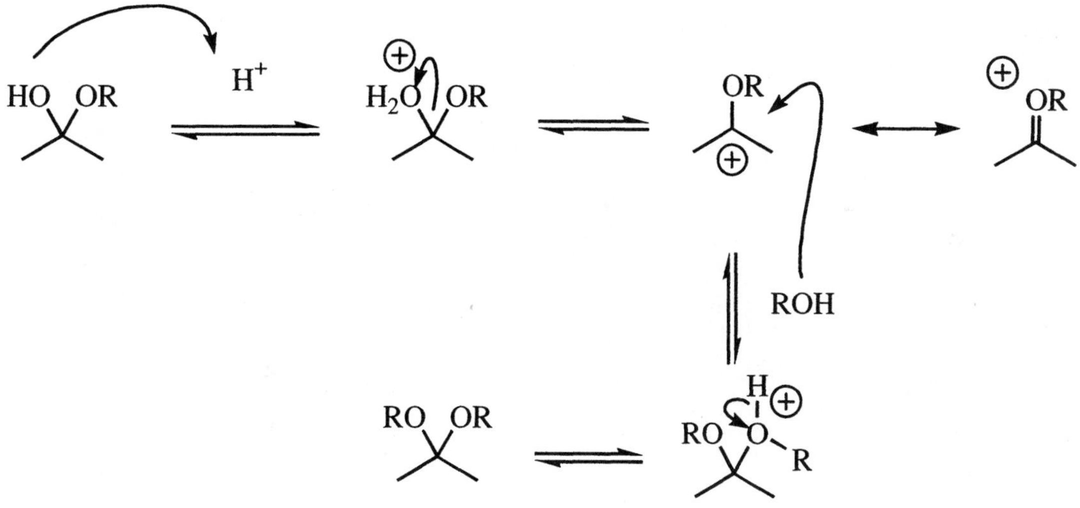

ketal or acetal

Now we have something new, a species with a carbon bearing *two* OR groups. This is called a ketal, if related to a ketone (i.e., if we started with a hemiketal), or an acetal, if related to an aldehyde (i.e., if we started with a hemiacetal). There are several points that need to be made about Xtals.

1. Xtal is a word I made up so I wouldn't have to keep saying "ketal or acetal." Do not use it outside the context of this class or no one will know what you are talking about! Any specific example is either a ketal or an acetal, and you should refer to it as such. It is only when referring to the family in general that things get wordy.

2. You can only make an Xtal in acid. Further, you can only get back to the aldehyde or ketone again in acid (can you think of anything reasonable that a base could do to an Xtal?). Xtals are stable in basic solution!

3. The overall process of going from a carbonyl compound to an ordinary Xtal is still unfavorable from an entropy point of view. You are trying to take three separate molecules and make them into two.

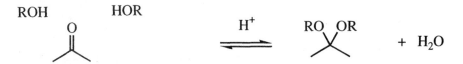

4. Therefore, you should not be able to make these things at all; that is, equilibrium should normally favor the left side of the above equation, and the amount of Xtal present should be undetectable. And indeed, this is usually true.

5. However, we can get around the above problem. There are actually several ways to do it. One is to play the game we tried before with hemiXtals: tie some of the starting materials together. Below are examples of ways to do this, where the dotted lines represent carbon chains of unspecified lengths.

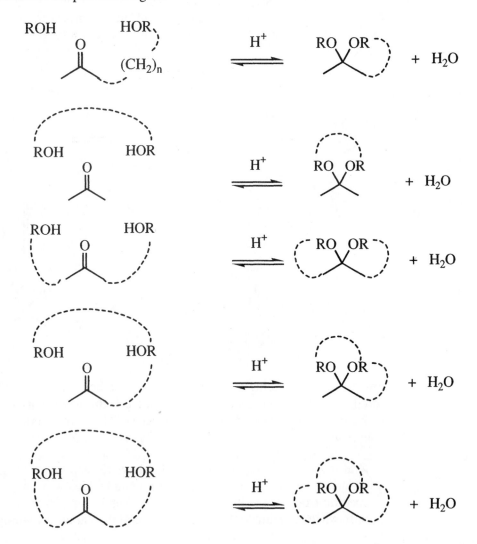

The last three should be particularly good, since they have entropy working in our favor. Take some time now to devise real examples representing all the above possibilities. Here's one that corresponds to the second reaction on the list:

(ethylene glycol)

Show a mechanism for this process.

Even the original equilibrium, however, can be manipulated in our favor if the goal is to make an Xtal from three separate pieces. Recall equilibrium constants:

$$K_{eq} = \frac{\left[\underset{RO \quad OR}{\diagup\!\!\diagdown} \right] \left[H_2O \right]}{\left[\underset{O}{\diagup\!\!\diagdown} \right] \left[ROH \right] \left[ROH \right]}$$

If you want lots of Xtal and not much carbonyl compound (i.e., you want the reaction to go from left to right), you just have to make [ROH] large and/or [H₂O] small. So carry out the reaction in alcohol solvent with a little acid added, or figure out a way to remove water as soon as it is formed. If you want the reaction to go from right to left, you reverse these considerations, and make [H₂O] large by running the reaction in acidic water. This is another simple application of Le Châtelier's principle.

6. You need to be able to recognize Xtals immediately when you see them, and also to visualize easily the carbonyl and alcohol components they relate to. (The same is true of hemiXtals.) Here is an easy reminder, but beware of memorization as a substitute for understanding what really happens!

A carbon bonded to two different oxygens by single bonds is related to the same carbon bonded twice to a single oxygen (that is, a carbonyl compound). The two OR groups attached to that single carbon are related to alcohols that must have reacted with the carbonyl group.

ORM: In Organic Reaction Mechanisms, under "Carbonyl Compounds," check out "reaction with alcohols, acetals."

Why have we spent so much time on this subject, when one has to go to such lengths just to see if anything occurs? It turns out that ketals and acetals are exceedingly useful to organic chemists, and also exceedingly common in biological systems. I'll cover their chemical usefulness first.

Imagine you wanted to accomplish the following transformation:

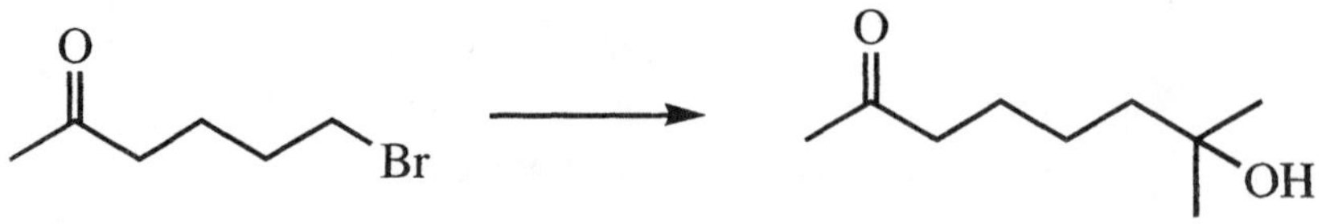

Based on what we have learned earlier in this chapter, it should be easy: you just run a Grignard reaction:

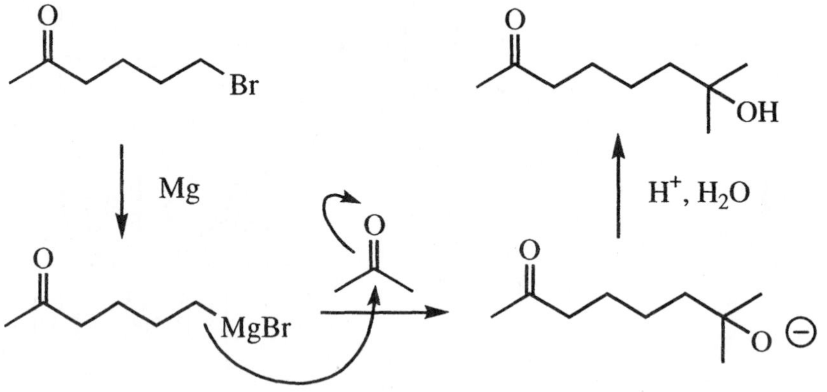

But there is a big problem with this synthesis. Can you see it? Take a minute and look for it.

The problem is in the first step. You have made a Grignard reagent. What do Grignards love? Carbonyl groups. What is hanging at the other end of your molecule? A carbonyl group. Do you really think this Grignard reagent will stick around and wait for you to add some acetone to it? No way! It will bite its own tail the moment it is formed:

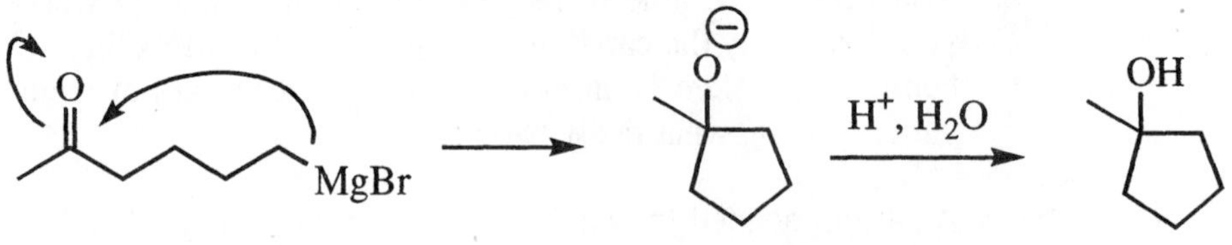

The desired transformation cannot be accomplished so long as there is a carbonyl group at the other end of the molecule. But wait! We've encountered similar situations before! The solution is to *hide* the carbonyl group while we run our Grignard reaction, and then bring it back after we are done. To accomplish this wizardry, we need a reaction that temporarily turns a carbonyl group into something that is not sensitive to base (i.e., Grignard conditions), and this reaction must be easily reversible, so we can release the carbonyl group again at will. Now do you see why we spent so much time on ketals and acetals? They fit this description exactly! To an organic chemist, the main use for ketals and acetals is as protection for ketones and aldehydes.

So here's how we solve our problem:

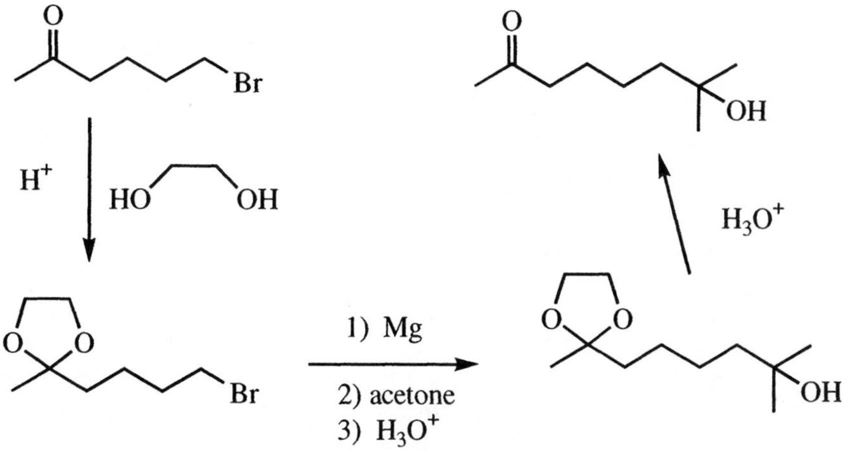

H⁺ | HO⌒OH

1) Mg
2) acetone
3) H_3O^+

H_3O^+

This process is actually even shorter than it looks, because the last reaction would probably occur by itself during the last part of the previous reaction.

The need for protecting one functional group while you work on another is extremely common in organic synthesis, and protection is well worth practicing.

Test Yourself 5

Show methods for achieving the following transformations:

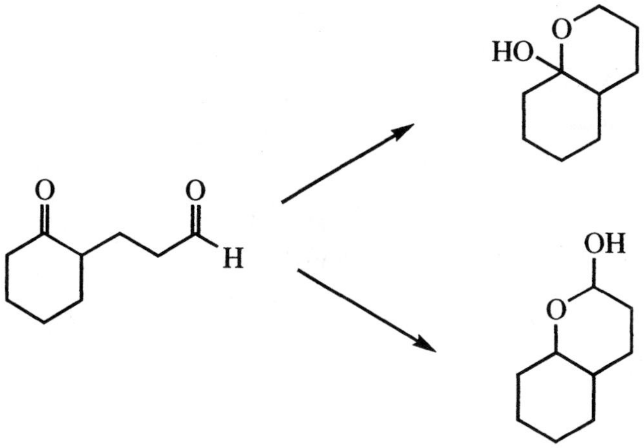

Follow a Synthesis

You have now studied enough reactions so that you can follow along with the experts as they synthesize some compound of real interest. The following compound is called brevicomin. It is a pheromone (chemical signal) emitted by the Western pine beetle to attract other insects to an infested tree. Brevicomin is an important tool in efforts to defend trees against the beetle.

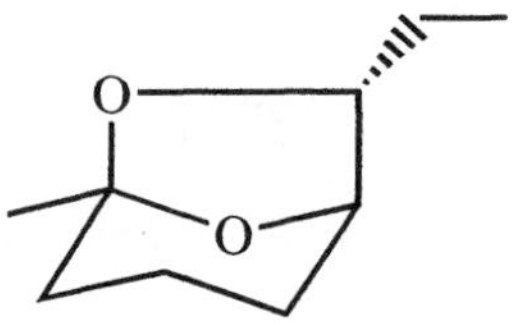

brevicomin

Now, you may be wondering how one can defend trees by preparing a compound that attracts their enemies. There are several ways this can work. One involves setting up traps in the woods to draw target insects to them. The insects can then be killed or sexually neutralized in the traps, or simply monitored for population. (One needs to be careful with such traps, though, because insects are not only sensitive to the particular compounds they are attracted to, but also sensitive to concentration. The amount of pheromone emitted by a female insect is very small, but it can attract a male from miles away. However, the male knows that when the amount of attractant gets to a certain level, he must be in the vicinity of a female, and so he starts looking around for her. There is at least one instance reported in which insect traps were set up and laced with sex attractant at too high a level. Males were attracted toward the trap, but they stopped about a mile away, where the level seemed just right.) A second way to use sex attractants to control an insect population is simply to bathe the target area with the attractant. Everywhere a male goes there is attractant, so he has no idea where the females really are. Lack of breeding reduces a population in a hurry.

Anyway, let's look at brevicomin and see if we can figure out how to synthesize it. What you should see immediately is a ketal; that is, the compound contains a carbon with two different OR groups attached to it. This should tip you off that brevicomin is actually equivalent to the following open-chain compound, and if you made *it*, you would have succeeded in making brevicomin:

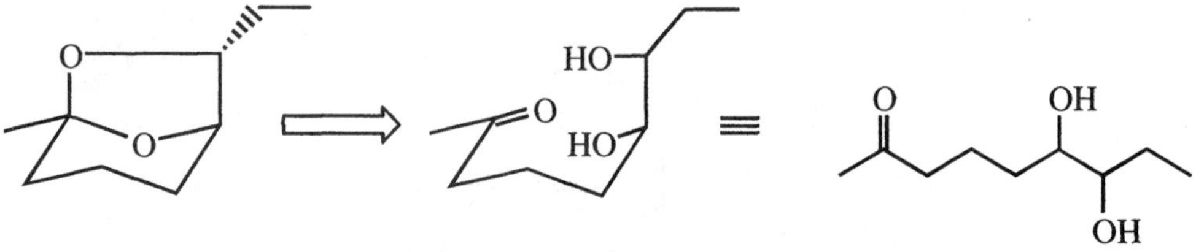

Further, if you keep track of stereochemistry (building a model will help here), you will discover that the compound you need has the following configuration:

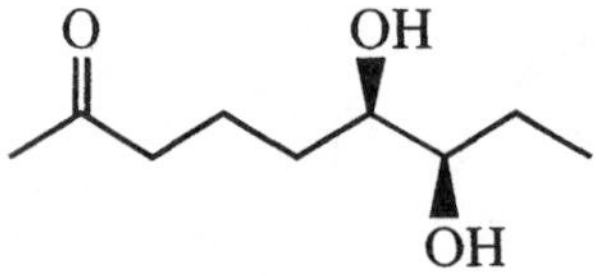

There is only one method we know (though others exist) for creating a vicinal diol, and that is the opening of an epoxide; further, we know that when we do this, either in acid or in base, the two OH groups will be opposite each other in the compound as it forms. In other words, if we used this approach, we would make the open form of brevicomin in the following conformation (again, your models will help here), and that tells us which epoxide and therefore which alkene is the necessary starting material, as shown below:

Since we need a *cis* double bond, we know we should make it from a triple bond by reduction with hydrogen and Lindlar catalyst. So here we have a reasonable synthesis of brevicomin starting from the commercially available ketal shown. Every reaction is one you have studied and can understand.

As Dave Barry would say, "I am not making this up." Well, only a little bit. The original research article I got this from is by Johnston and Oehlschlager and appeared in *J. Org. Chem.,* **1982**, *47*, 5384. What they did differs from the above in a couple of key ways. First, they wanted to make the product optically active. The reaction that introduces chirality is the epoxidation (the material has no stereocenters until then), and there is a method you do not yet know for making epoxides enantiospecifically. Thus, these authors did not actually use mCPBA for their epoxidation. Furthermore, the epoxidation method they used requires that the double bond have an OH group one carbon atom away (at the allylic position). Johnston and Oehlschlager had to go to some extra effort to place an alcohol near their double bond and then later remove it. The overall strategy, however, is essentially that presented above.

CARBOHYDRATES

Among the most common organic molecules in the world are carbohydrates, also called sugars. These are molecules with the general formula $C_nH_{2n}O_n$, or $C_n(H_2O)_n$. Strictly on a formula basis, these would appear to be combinations of carbon and water, each taken n times (hence the name, carbo-hydrate). This provides almost no insight at all into the chemistry of carbohydrates, but there is a really neat demonstration that I hope your professor will do for you in which you take sugar (table sugar) and remove all the water, leaving just carbon. The neat part of the demonstration is that during the reaction the water turns into steam, and the formation of carbon accompanied by steam generation creates a carbon foam that is really cool. Even my teenage daughter says it's cool, so it must be.

There are many different carbohydrates, but they all share certain common features. For example, they all have straight chains, anywhere from three to seven carbons long, in which every carbon carries an oxygen. Most of the oxygens are present as alcohols, but each sugar also has one carbonyl group, either at an end (an aldehyde—a reducing sugar, remember?) or (usually) one carbon in from an end (a ketone). Most of the carbons bearing OH groups are stereocenters, so there is a whole family of six-carbon aldehydes whose members differ only in their stereochemistries, then another whole family of six-carbon ketones, and another of five-carbon aldehydes, and so on. The sugars have names like glucose, galactose, mannose, fructose, ribose, etc. It is not necessary that you remember all these names (unless your professor disagrees!) and the structures that accompany them. It is, however, necessary that you learn how to read the drawings that people make of these things, and to recognize them both when they are open and when they are closed as hemi- and full acetals and ketals.

Below is one such sugar, as an organic chemist might write it and as biochemists do. Perhaps you can see immediately why organic chemists have bought into the biochemists' approach. What remains now is for you to learn how to make the translation.

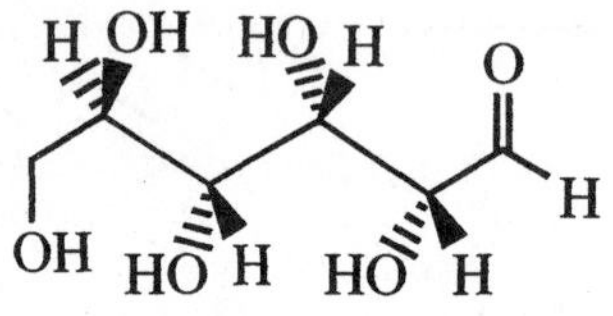

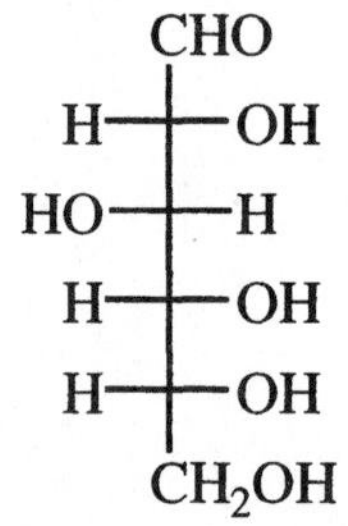

(2R,3S,4R,5R)-2,3,4,5,6-pentahydroxyhexanal

D-glucose

I do not intend to teach you the nomenclature system for sugars, so I will not discuss how this thing comes to be named D-glucose. I will simply mention that its enantiomer would be named L-glucose, and that these D's and L's do *not* refer to optical rotations. There is a complicated relationship among molecules that leads to the labels, such that the capital D and capital L actually refer to absolute stereochemistry in an unambiguous way, but we'll ignore it. Nevertheless, I do need to describe the system of drawing utilized in the right-hand structure above.

What you see there is a shorthand way of depicting stereochemistry. It is called a "Fischer projection," after the famous biochemist who invented it, Emil Fischer. When you are in this field, you quickly tire of drawing wedges and dotted lines, and a new system that avoids them is welcome. The new system depicts each stereocenter as a horizontal cross—a big plus sign—that has certain rules associated with it. The rules are these: The center of the cross is the carbon that is the stereocenter. The vertical bonds are (by definition) going back away from you; the horizontal bonds are (by definition) coming out toward you.

$$ + \quad \equiv \quad \blacktriangleright \overset{\vdots}{\underset{\vdots}{C}} \blacktriangleleft $$

This definition has the immediate consequence that Fischer projections (unlike real molecules or our usual pictures of them) cannot be rotated at will in the plane of the paper. For example,

the molecule $Z \blacktriangleright \overset{W}{\underset{Y}{C}} \blacktriangleleft X$ can be represented by the Fischer projection $Z {-}\!\!\!\underset{Y}{\overset{W}{|}}\!\!\!{-} X$

$Z \blacktriangleright \overset{W}{\underset{Y}{C}} \blacktriangleleft X$ can be rotated by 90°, and $Y|\!\!\cdot\!\!\cdot C\!\!\cdot\!\!\cdot|W$ with $\overset{Z}{\underset{X}{\big\updownarrow}}$ is the same stuff. But the Fischer

projection $Y{-}\!\!\!\underset{X}{\overset{Z}{|}}\!\!\!{-}W$ does not refer to this $\curvearrowleft$, it refers to this: $Y \blacktriangleright \overset{Z}{\underset{X}{C}} \blacktriangleleft W$, which

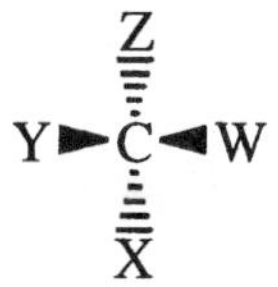

is *not* the same molecule: it is the enantiomer of the original. Get out your models (again!) and confirm this for yourself. Also confirm that a 180° rotation *does* lead to the same molecule.

Writing a single stereocenter as a Fischer projection generally is not a problem. The difficulty comes in writing a chain, like glucose, because a Fischer projection is not a realistic picture of a molecule in any sense. Look again at the pictures of glucose, repeated below. If you grab the left-hand picture by the aldehyde end and let it dangle vertically, so that it resembles the vertical picture on the right, you should realize that there is no angle from which you can look at this and see anything like the Fischer projection picture. Build a model and confirm this for yourself.

CHO
H—OH
HO—H
H—OH
H—OH
CH$_2$OH

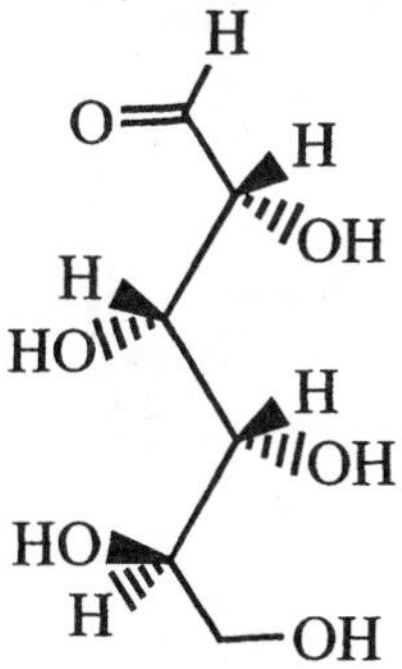

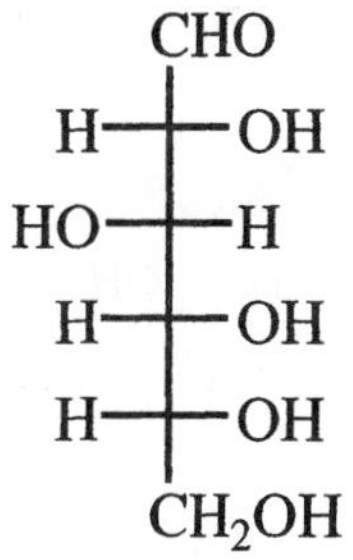

So what is the Fischer projection a picture of??? It represents the individual stereocenters of glucose, taken *one at a time*, each one looked at according to the definition. To construct (and interpret) such a picture, it is necessary to change your point of view for each stereocenter. In order to take a real picture of a molecule, like the left one, and convert it into a Fischer projection (or *vice versa*), you need to move around. To decide how to show (or interpret) the Fischer projection at carbon 2, you need to be where eyeball #2 is in the picture below—the OH is on your right; but to show carbon 3, you need to go to eyeball #3, where the OH is on your left. This is correct, even though, in the extended actual molecule, both OH groups are on the same side! Again, use your model to confirm all these statements.

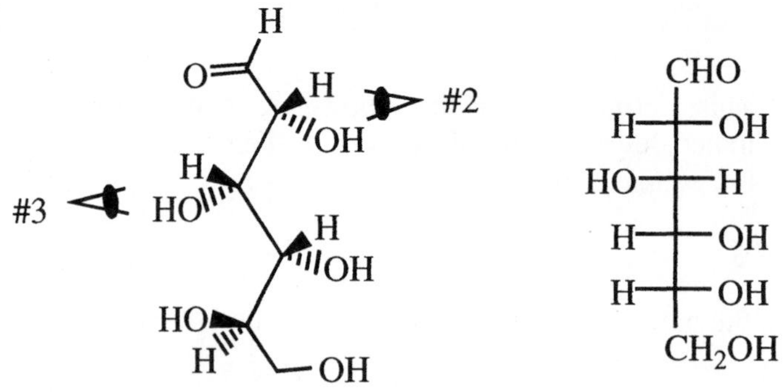

There is one way you can hold the model in which the Fischer projection bears some resemblance to the actual molecule: if you wrap the structure around in a circle, as shown below, then you can perhaps see some relationship between the Fischer projection and the model, but of course the relationship breaks down in the sense that the Fischer projection is drawn as a line and not as a circle.

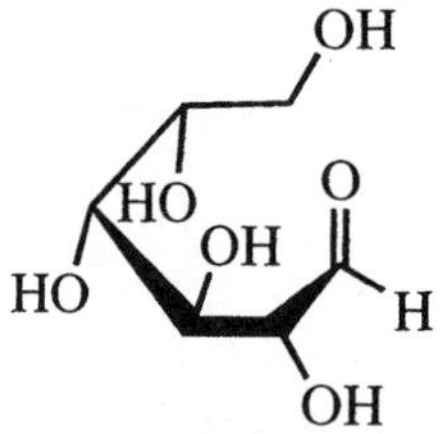

There is a good reason why the most commonly drawn picture of glucose (and other sugars) resembles the molecule when it is wrapped up in a circle. This is because, in nature, glucose most often *is* wrapped up in a circle, though not quite the one drawn above. Recall our discussion of hemiacetals: a hemiacetal can be the favored side of an equilibrium if it forms as a five- or six-membered ring. Glucose, with an OH on every carbon, could form any size hemiacetal ring it wants, but it most often chooses a six-membered ring, giving a hemiacetal based on the OH group on the #5 carbon. I show below the mechanism for creating this in acid; you should be able to show the comparable process in base. Either will do.

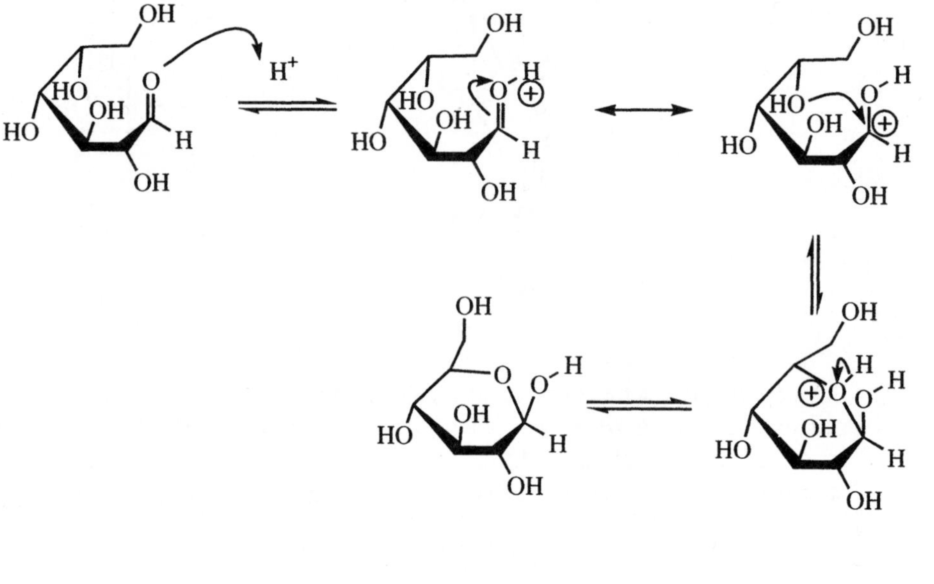

There are several things I need to mention about this:

1. I started with the aldehyde pointing up in my picture, but this was random. If it had been pointing down (by rotating the 1–2 bond), we would have gotten a different product, with the hemiacetal OH pointing down. Since glucose can go back and forth between the aldehyde form and the hemiacetal form, and since it can rotate around the 1–2 bond while it is an aldehyde, it can also go back and forth between the two stereoisomeric hemiacetals. These are called the "anomers" of glucose. The up one is called the β-anomer, or β-D-glucose, and the down one is the α-anomer. The process of going from one to the other is called "mutarotation."

ORM: Under "Carbonyl Compounds," watch "anomers of glucose." This is a good movie of the process of opening and closing the glucose ring, so you can see how mutarotation occurs. However, I have one quibble with the movie: on two occasions it shows your compound being protonated by water, creating a cation on the molecule together with OH⁻. I wish they had shown this protonation being caused by some accidental acid in the medium, because I have tried to train you NOT to use neutral water to protonate an organic molecule. I cannot claim that it NEVER happens, but it is certainly a very rare occurrence.

2. The cyclic forms are most often depicted as a slight variation on the above pictures, in which the up and down bonds are shown as truly vertical lines, the way they are depicted below. This is called a Haworth projection. Sometimes, however, someone tries to show the same thing with a Fischer projection. To do this requires a ridiculously long bond that goes around corners, a truly unsatisfactory way to show any molecule, but you might still see it occasionally.

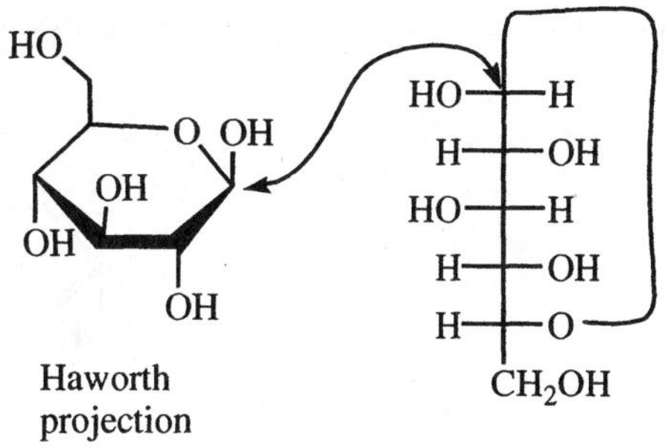

Haworth
projection

3. If you draw these structures out as chairs, you will discover that the β-anomer has its new OH group in an equatorial position, and the α-anomer has it in an axial position:

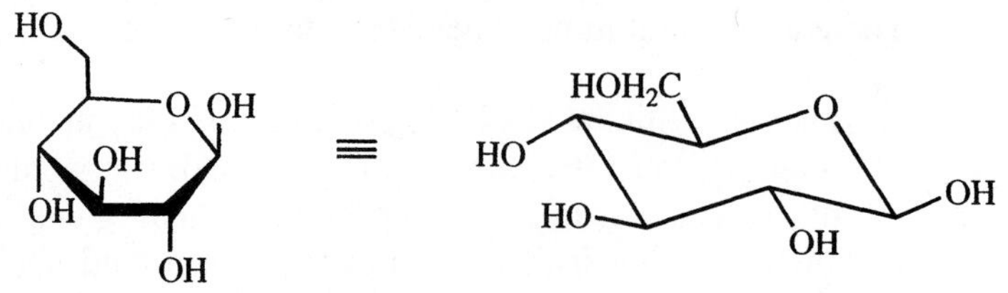

Haworth
projection

β-D-glucose

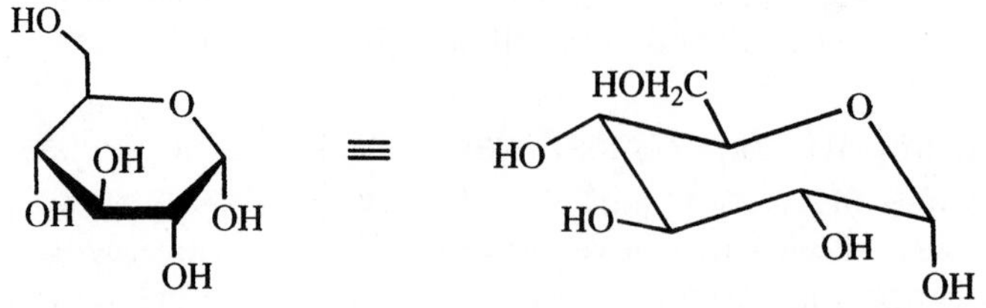

Haworth
projection

α-D-glucose

4. You should notice that in the β-anomer, not only the new OH group, but also all the *other* projecting groups are in equatorial positions. In other words, this should be the most stable sugar molecule there is. Indeed, there is probably more β-D-glucose in the world than any other organic substance. Free glucose, in nature, is about 64% in the β form, 34% in the α form, and 0.02% in the aldehyde form.

5. When you want to indicate the cyclic form of a sugar, but do not wish to specify which anomer you are referring to, you can use a squiggly line for the unspecified bond. In general, a squiggly line like this means either that you do not know, or you do not care, which isomer is being referred to, or that a mixture is present.

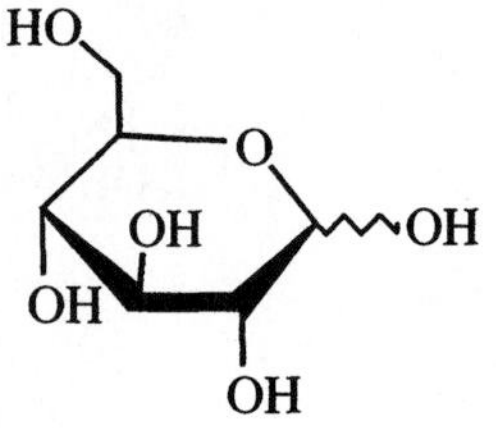

6. The five-carbon sugars, such as ribose, could choose to make six-membered rings also, but instead they most often choose to close as five-membered rings. Ribose in this form is an important part of the backbone of RNA (*ribo*nucleic acid). DNA is very similar to RNA, but its backbone pieces are missing the OH group at carbon 2 (*deoxy*ribonucleic acid).

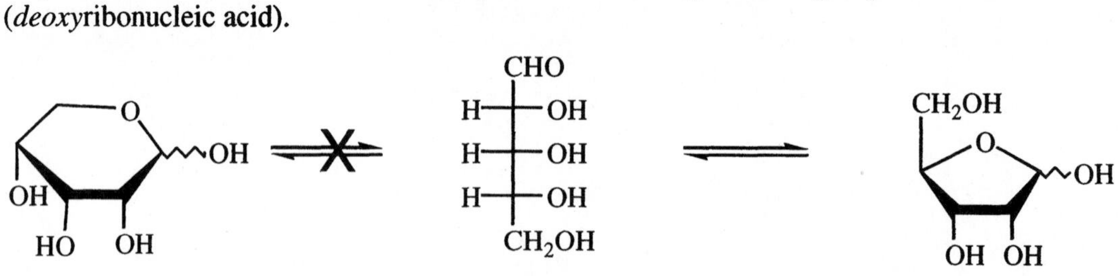

ribose

7. All the examples shown so far have been hemiacetals. As you know, hemiacetals, in the presence of more alcohol and acid catalyst, can be converted into acetals. The sugar environment is not lacking in alcohols—nearly every carbon has an OH group!

So it is not surprising that under appropriate conditions, the acetal OH group can be replaced with a new alcohol group. Notice the resonance possible in the carbocation intermediate. The full acetal (or ketal) of a sugar is referred to as a glycoside.

acetal   hemiacetal

8. When an alcohol approaches the carbocation, it can approach from either the top or the bottom, as shown in the mechanism above. If it comes from the top, you get a β-glycoside, which is the product shown. If it comes from the bottom, you get an α-glycoside, which is not shown. Both processes occur.

9. The product contains an acetal, as indicated, but it is also still a hemiacetal. The latter function can have the same thing happen to it, and the process can continue, allowing for the formation of long chains of glucose molecules. A long chain of glucose joined in the β fashion at the 1 and 4 carbons is called cellulose. This is the structural backbone of plants, and is the most common molecular species on our planet. A long chain of glucose joined in α fashion is called amylose, which is part of the structure of starch.

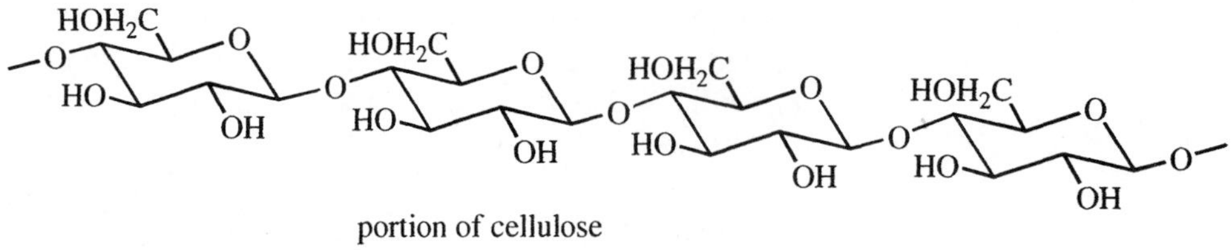

portion of cellulose

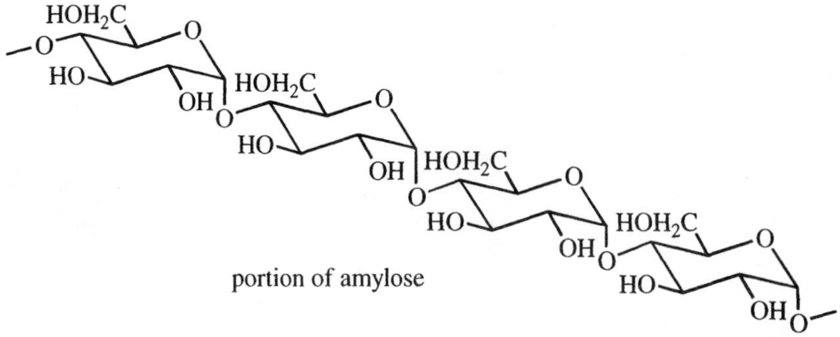

portion of amylose

10. Mammals have an enzyme that can hydrolyze α-glycoside linkages, so we can digest starch. They do not have an enzyme that can handle β–glycoside linkages, so no mammal can digest cellulose. Mammals (like cows) that live on cellulose do so by hosting in their intestines special bacteria that *can* hydrolyze β-glycosides. So again, the shape of a molecule can have profound biological significance.

11. All these structures still have one end that is a hemiacetal, which of course is in equilibrium with an aldehyde and the compound is therefore a reducing sugar. Some sugars have *all* their carbonyl groups tied up as full acetals or ketals. These are *not* in easy equilibrium with carbonyl forms, and can therefore be called non-reducing sugars. One such non-reducing sugar is sucrose, ordinary table sugar, a dimer consisting of a combination of an acetal of glucose (an aldehyde sugar) and a ketal of fructose (a ketone sugar).

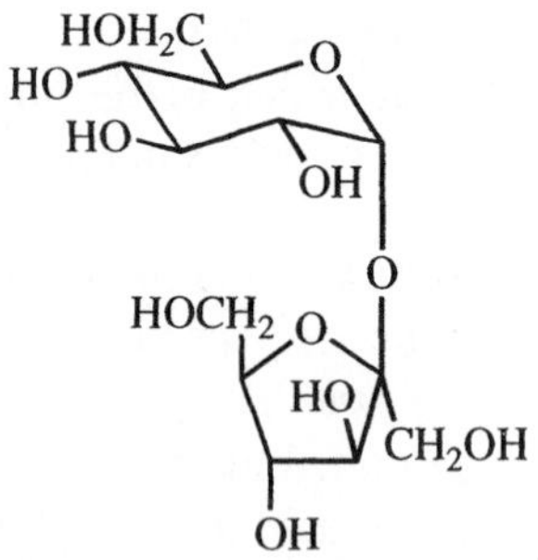

sucrose

12. Sugars are often referred to as "saccharides." Two sugars glued together, like sucrose, would be called a disaccharide. A long chain of sugars glued together, like cellulose or starch, would be called a polysaccharide.

In summary, sugars behave exactly as you would expect based on their structures, which normally are combinations of alcohols and a carbonyl group. They can and usually do form cyclic hemiacetals and hemiketals, which are in equilibrium with the open forms unless tied up as full acetals or ketals. Their carbonyl groups, when exposed, behave like ordinary carbonyl groups, and the alcohols behave like ordinary alcohols.

Problems

(Answers are provided at the end of the chapter for italicized problems.)

1. Show the organic products of the following reactions. Write NR if there is no reaction.

a.

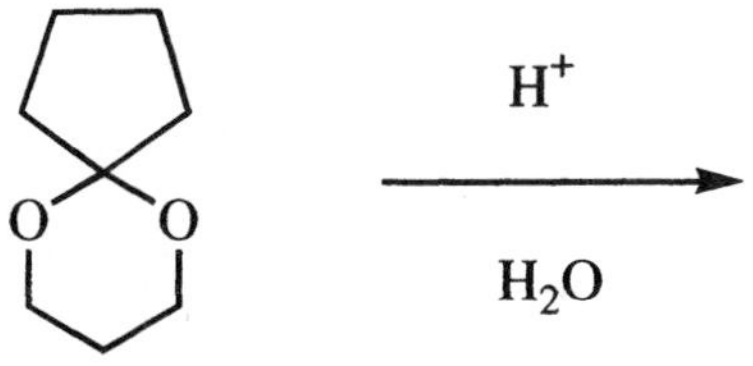

H^+

H_2O

two products

b.

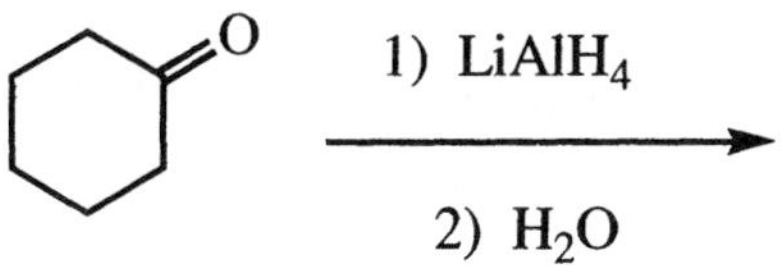

1) $LiAlH_4$

2) H_2O

c.

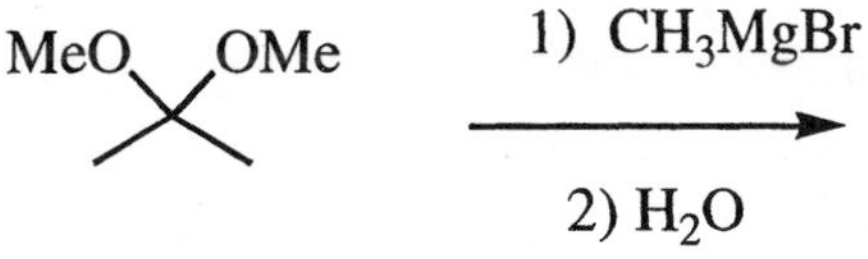

MeO   OMe

1) CH_3MgBr

2) H_2O

d.

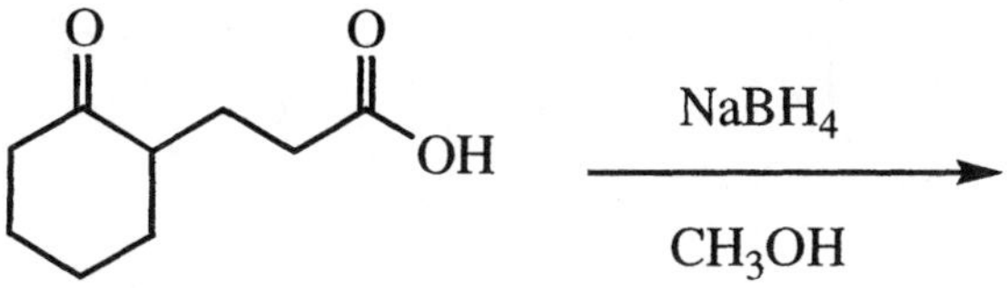

OH

$NaBH_4$

CH_3OH

e.

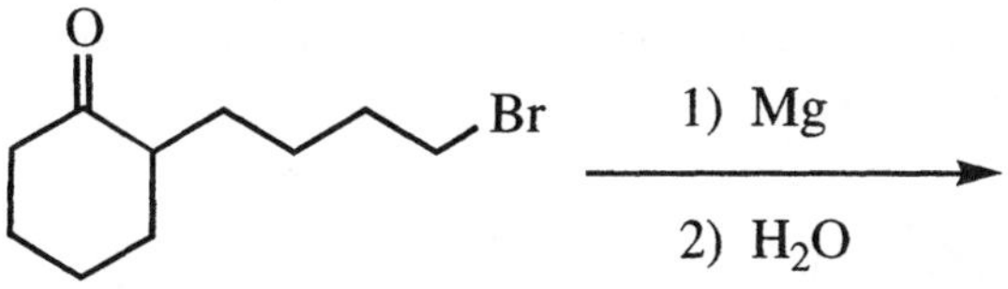

Br

1) Mg

2) H_2O

f.

OH

OH

H^+

remove H_2O

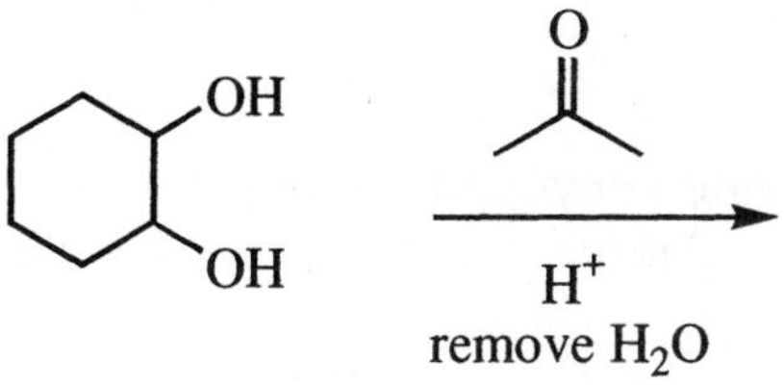

2. Show the mechanism for the conversion of an enol into a ketone in both aqueous acid and aqueous base. Also, do the same for the reverse reaction.

3. *2,4-Pentanedione, with a pK$_a$ of 9, is dramatically more acidic than acetone, pK$_a$ 19. Explain why this is the case.*

4. Suggest a mechanism for the following reaction:

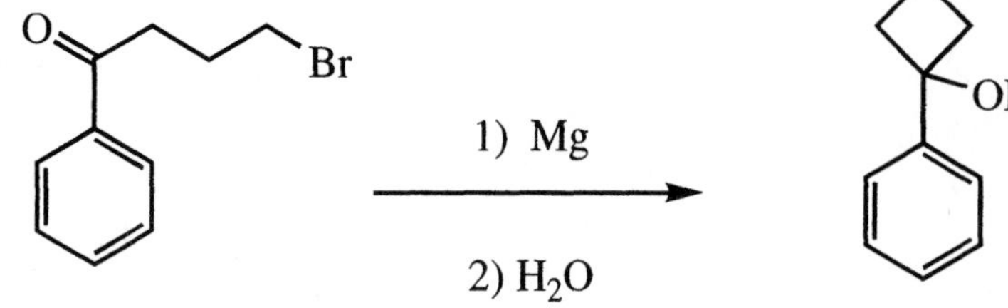

1) Mg

2) H$_2$O

5. Starting with anything you like containing three carbons or less, suggest syntheses of the following molecules:

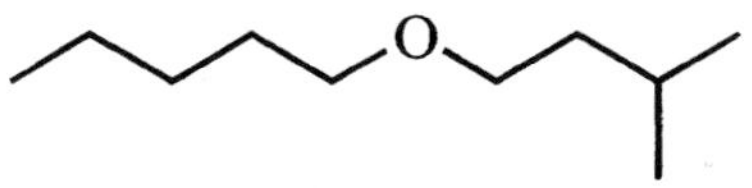

6. *Suggest a way of accomplishing the following conversion. Recall that D is ^{2}H, an isotope of H that behaves chemically exactly the way H does. Any reagent that delivers hydrogen can be obtained in deuterated form.*

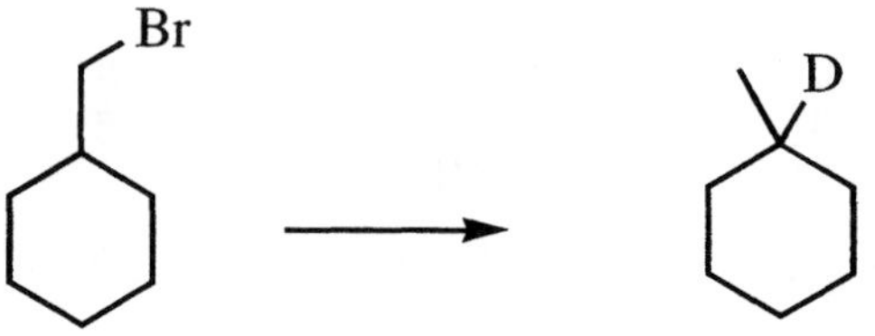

7. Starting from isopropyl alcohol as your primary source of carbon, incidental organic reagents such as bases and solvents, as well as inorganic reagents of your choice (including labeled materials as needed), devise syntheses of the following labeled compounds. It is not necessary to write mechanisms.

8. *Propose a synthesis of the following compound starting with pieces containing three or fewer carbons and no deuterium, and using any of the standard reagents you have learned about:*

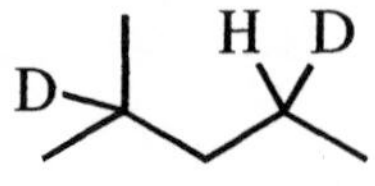

9. *Suggest reasonable syntheses for each of the following compounds using only ethanol and bromocyclohexane as sources of carbon atoms. Use the Grignard reaction as your method of creating new carbon–carbon bonds.*

    a. *Propanal*

    b. *Methyl cyclohexyl ketone*

    c. *2,7-Dimethyl-2,7-octanediol*

    d. *2,3-Dimethyl-3-chloropentane*

10. Suggest a synthesis of the following compound, starting with anything you like containing three carbons or less:

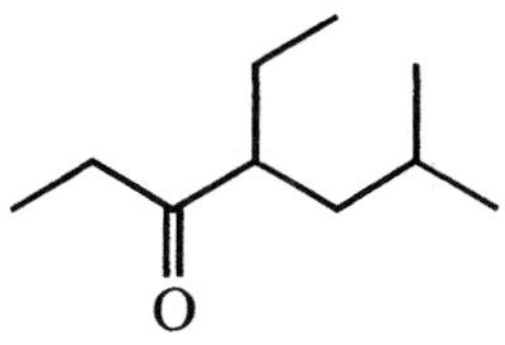

11. Suggest a synthesis of the following molecule, using anything you like containing three carbons or less as your only sources of carbon:

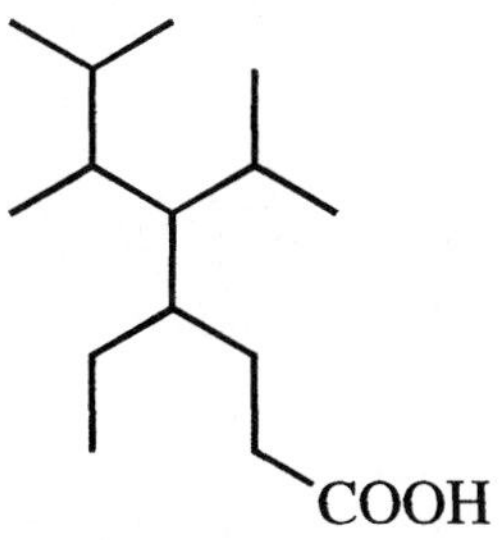

12. *Suggest two syntheses of 1-cyclohexylcyclopentanol, one by way of an intermolecular Grignard reaction and one that involves an intramolecular Grignard reaction.*

13. *Suggest a synthesis of the following molecule beginning with hydrocarbons as your only sources of carbon, and using any other reagents you are familiar with:*

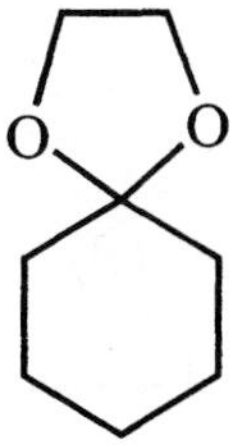

14. Suggest a synthesis of the following compound, starting with bromocyclohexane and anything else you wish containing three carbons or less:

15. *Show how the use of propylene oxide (shown below) could make the synthesis in problem 14 a lot easier.*

16. Beginning with bromobenzene, show a synthesis whereby 3-phenylpropanoic acid is produced by the use of exactly **two Grignard Reactions** and any other relevant non-Grignard reactions.

17. *Suggest a method of achieving the following conversion, using any reagents you need containing four carbons or less:*

18. When cyclohexanone is allowed to sit in ^{18}O-labeled water, in the presence of a trace of HCl, the cyclohexanone slowly becomes labeled with ^{18}O. Show a mechanism that explains this observation.

19. Suggest mechanisms for the following reactions:

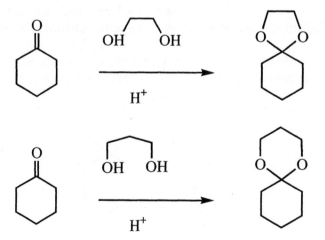

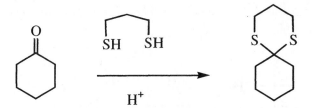

20. *Show mechanisms for the reverse of the reactions from problem 19, in aqueous acid.*

21. Suggest a mechanism for the following reaction:

22. When the following compound is reduced with one equivalent of $NaBH_4$ or $LiAlH_4$, an unusual product is obtained. Suggest a mechanism for this process. Assume water or alcohol is present at the end of the reaction.

23. Starting from alcohols containing three carbons or less as your only sources of carbon, as well as any reagents you are familiar with, suggest syntheses of the following molecules:

24. *Ribose and galactose are two monosaccharides that are important in nutrition. The D- forms have the following structures:*

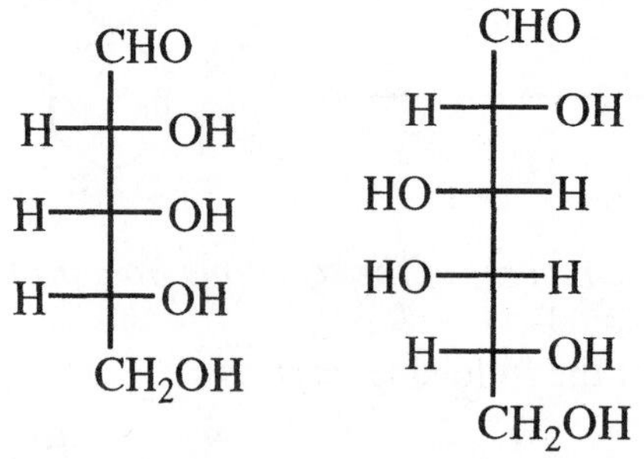

D-ribose   D-galactose

Write two anomeric cyclic hemiacetal forms for each.

25. *Cotton is largely cellulose, a polymer with the following structure. Explain why chemists often have acid holes (H^+/H_2O) in their blue jeans. Show a mechanism for the process. Would you expect blue jeans to be as sensitive to aqueous base?*

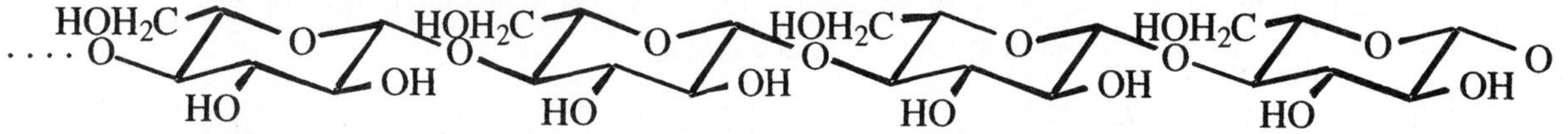

26. Suggest a synthesis of the following molecule starting from 1-bromo-1-methylcyclohexane and anything else you wish containing three carbons or less:

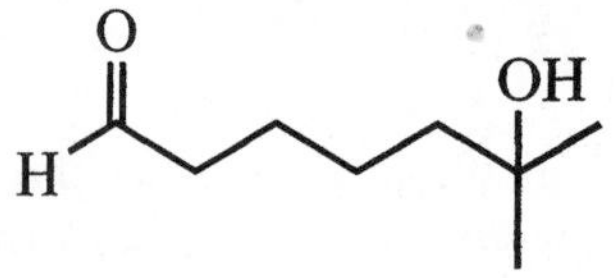

27. Compound A, $C_6H_{10}O_2$, is optically active. B is also optically active. All the other compounds involved are optically inactive. E is identical to A except for its optical activity. Suggest structures for A–J. Each letter stands for a different (neutral) organic compound. There are several possible answers. A hint is provided in the answer section.

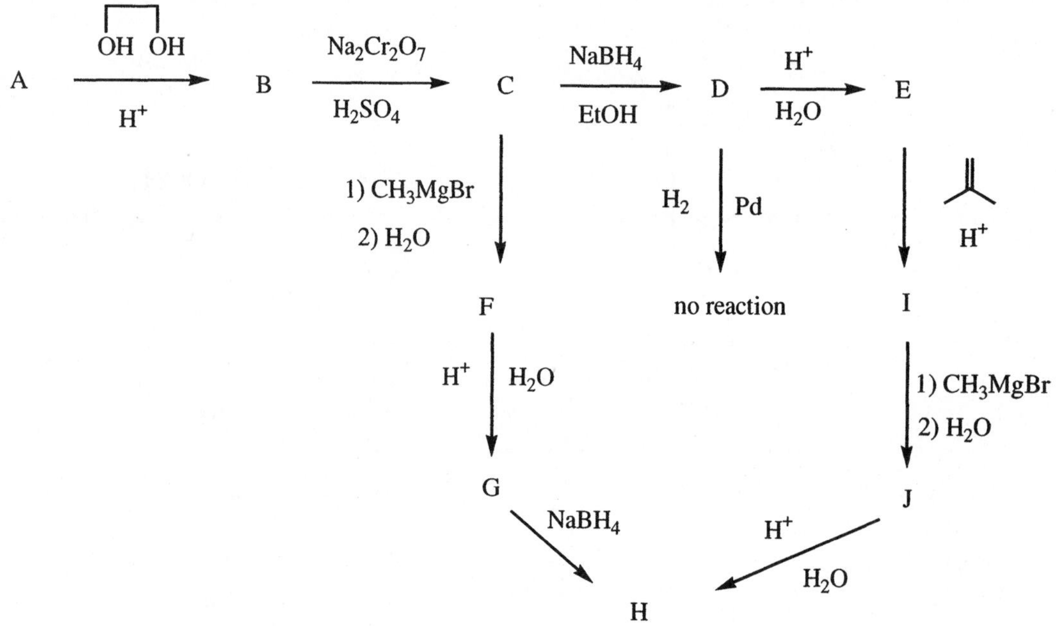

Selected Answers

Internal Problems

Test Yourself 1.

Usually ketones are *less* reactive than aldehydes because the alkyl groups are electron donating, therefore making the partial positive charge on the carbon of the carbonyl group less severe. However, *these* alkyl groups are actually electron withdrawing, due to the fluorines, so they make the positive charge *more* severe and therefore the carbonyl group is more reactive.

Test Yourself 2.

Like all other synthesis problems, we do this one backwards. This is an ether, so we dig out our methods for making ethers. We could do this in acid, starting with an alcohol and an alkene; or in base, starting with an alkoxide and a halide. In acid the alkene would have to be the left piece, and the alcohol would be the right piece; in base either half could play either role, but it would be better for the left piece to be derived from the alcohol and the right piece from the halide to minimize competition with elimination reactions. (If you did not understand the previous sentence then you should go back and reread the parts of Chapter 16 that deal with competition between substitution and elimination.) So we have two good options for this synthesis:

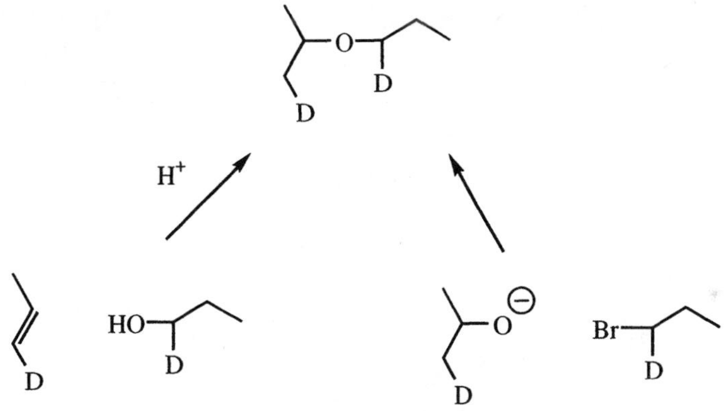

Both approaches would work, but I'm going to follow through on the left one. When I get done, you should try to pursue the right one.

Let's review the ways we know of attaching hydrogen atoms to molecules. There are two that should jump into your head immediately, and a third that you will probably kick yourself for forgetting. The first is catalytic hydrogenation of an alkene or alkyne. This method always puts two hydrogens on at once. If we use D_2 instead of H_2, we will necessarily get two deuteriums, but on each piece we want only one, so this is not an appropriate method. (Note that you cannot use HD for catalytic hydrogenation. Let me rephrase that: you could, but it would be pointless, because there is no way to control which end of the double bond will get the H and which will get the D. They would go on randomly, and the product would be a mixture.) The second method of putting hydrogen in a molecule is reduction of a carbonyl group with $LiAlH_4$ or $NaBH_4$, as we just learned. If we use $LiAlD_4$ or $NaBD_4$, we can put a single D in a specified place. Indeed, this will be a good way to make the primary alcohol we need:

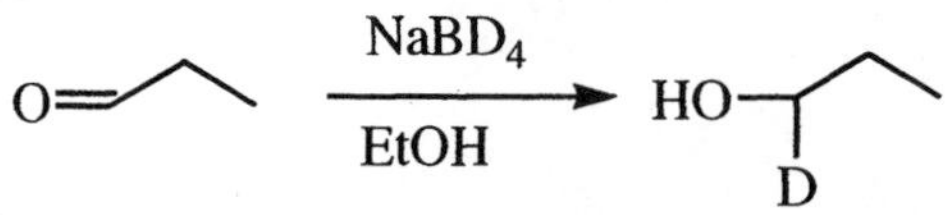

The alkene we need is a little tougher. One obvious approach, now, is to make it from the alcohol we just prepared:

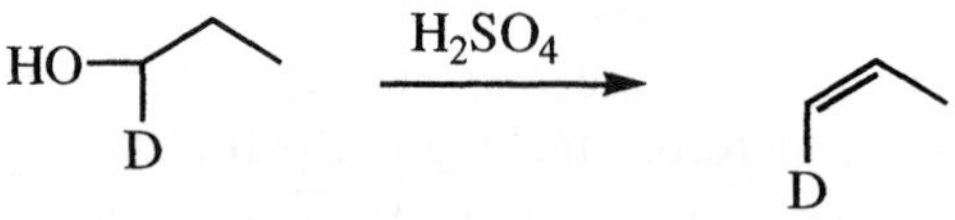

These two pieces can then be combined as originally planned:

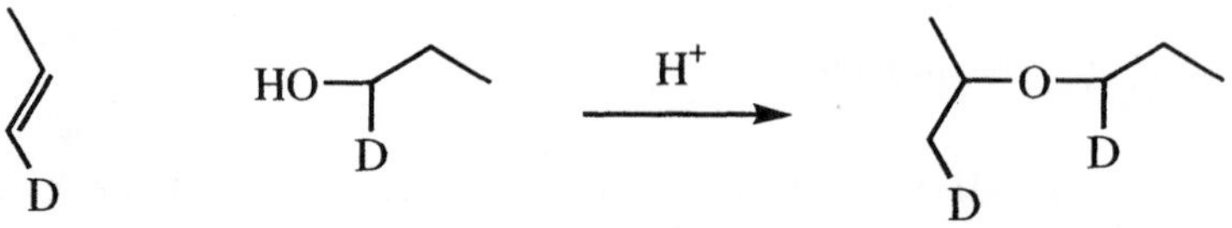

There is a shortcut, however, that you might find interesting. Recall the mechanism of the above reaction: H⁺ adds to the double bond, on the end carbon, creating a secondary carbocation that is then trapped by the alcohol. The H⁺ over the arrow actually ends up as part of the final product. This is the third way we have of incorporating hydrogen into molecules, and like the other methods, it works just as well with deuterium. So instead of starting with the two pieces shown above, using H⁺ as the catalyst, we could have done the same thing as shown below:

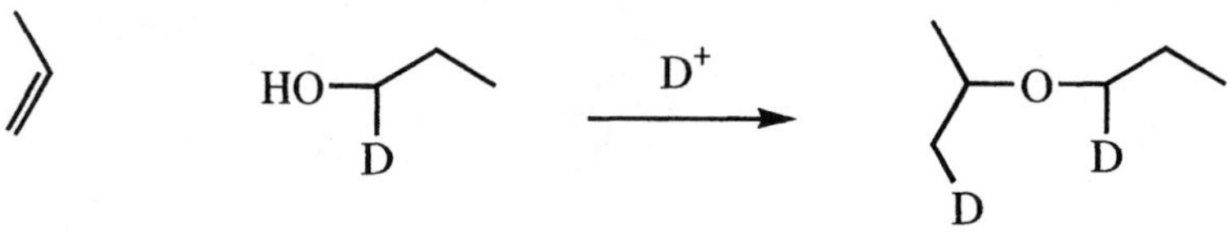

There is a logistical problem with this approach, however: recalling the mechanism again, you will remember that at the end of this reaction, in the last step, a proton is lost from the oxygen atom. If we use the above compounds, we will gradually be increasing the concentration of H⁺ in our reaction mixture at a time when we want our alkene to react only with D⁺. A chemist actually running this reaction would be forced to deal with this problem.

Test Yourself 3.

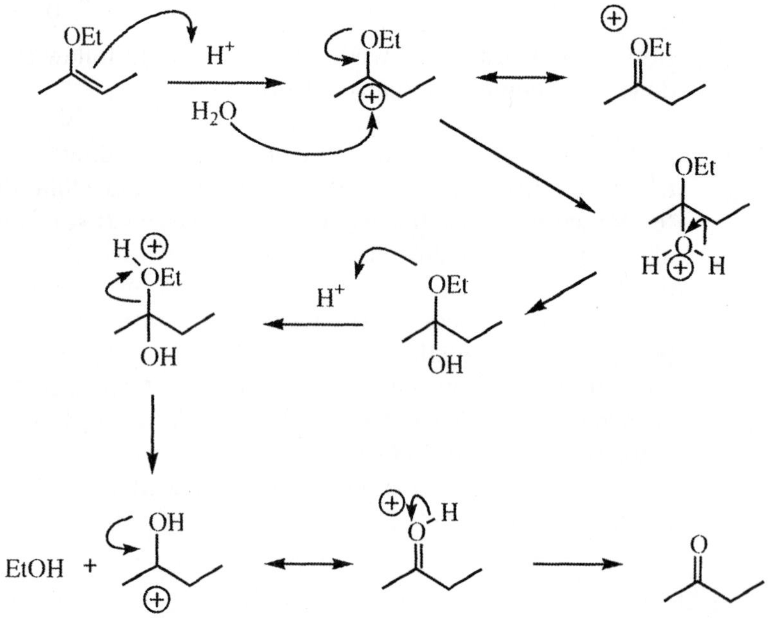

Test Yourself 4

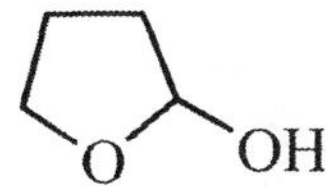

Test Yourself 5

Let's look at the top one first. This product you should recognize as being a hemiketal. Further, you should be able to recognize it as a hemiketal made from the following ketone:

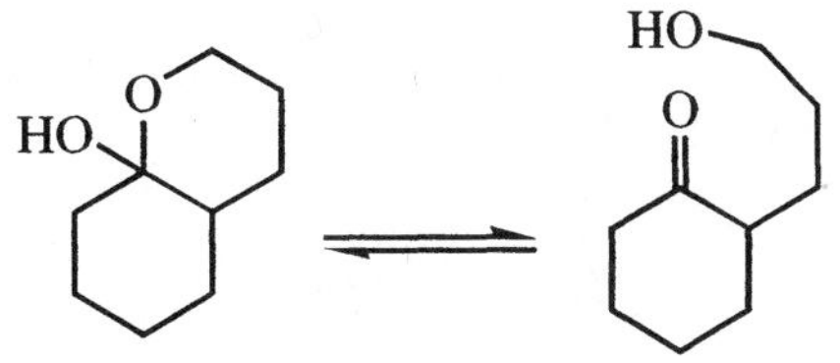

But that starting material can be made simply by reducing the aldehyde in the original compound. So all we have to do is figure out how to reduce an aldehyde without reducing its accompanying ketone. This is easy, because we already know that aldehydes are more reactive than ketones. So, we simply add an appropriate mild reducing agent, say, NaBH$_4$, but only enough to reduce one of the two groups (the official term is "one equivalent"). Done.

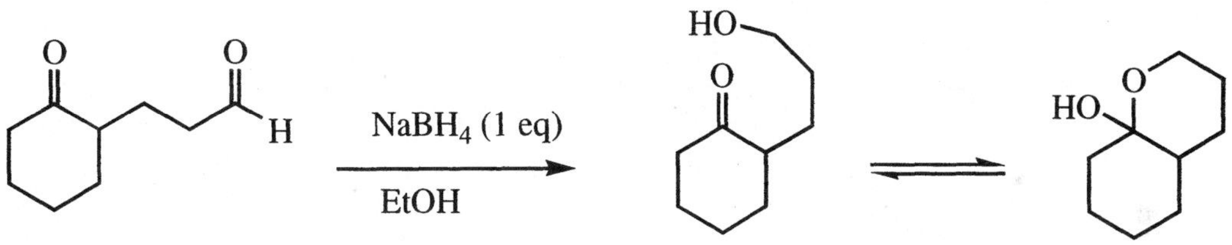

On to the bottom process. This you should recognize as requiring reduction of the ketone and not the aldehyde. How could we do that? Again, we can take advantage of the fact that an aldehyde is more reactive than a ketone. If we put in one equivalent of ethylene glycol and a bit of acid, we can legitimately expect to protect the aldehyde as an acetal while leaving the ketone alone. We can then reduce the ketone, release the aldehyde, and allow the newly revealed alcohol–aldehyde to fold into its cyclic (hemiacetal) form. See if you can write equations to illustrate the prose in this paragraph.

End of Chapter Problems

3.

The second anion is far more stable than the first.

6.

8.

9. a.

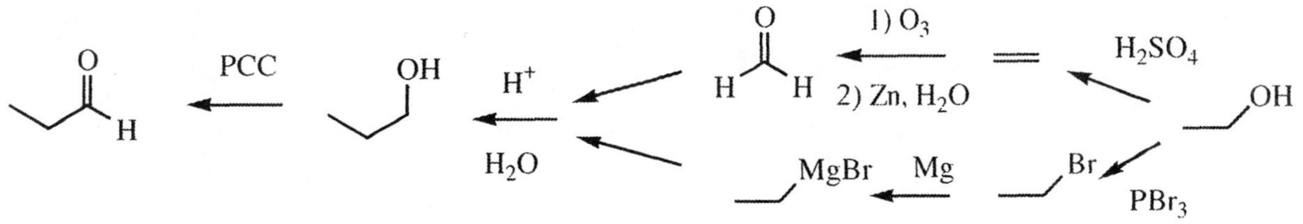

1) O_3

2) Zn, H_2O

H_2SO_4

PCC

H^+

H_2O

MgBr   Mg

Br

PBr_3

b.

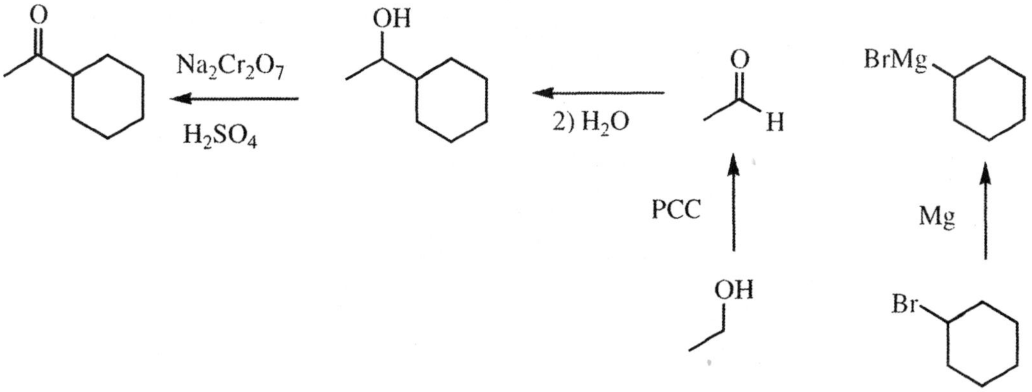

$Na_2Cr_2O_7$

H_2SO_4

2) H_2O

PCC

BrMg

Mg

Br

c.

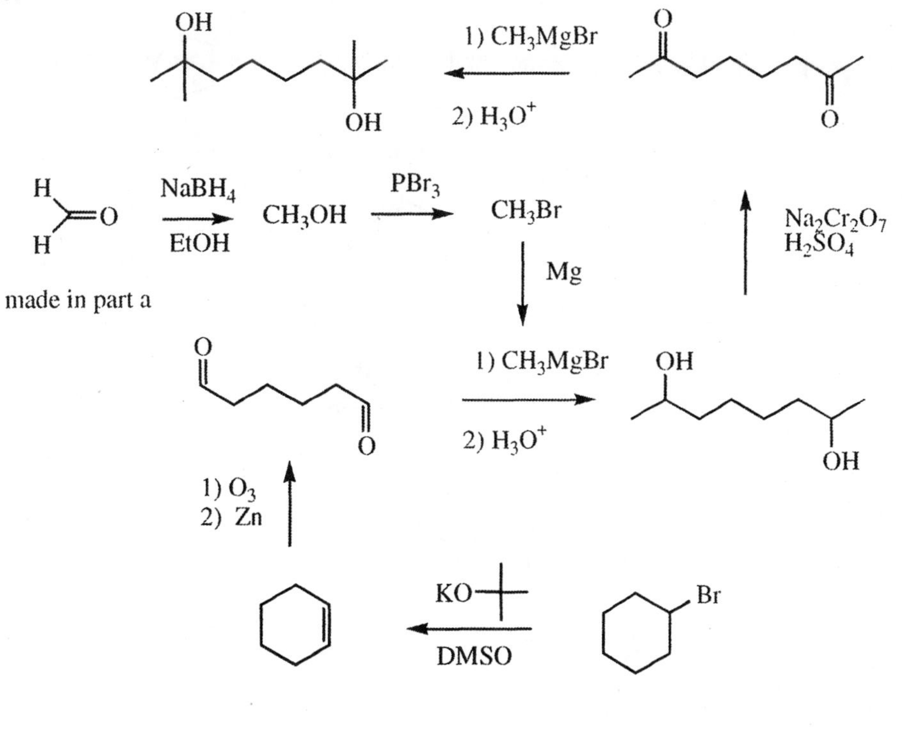

1) CH_3MgBr

2) H_3O^+

NaBH$_4$

EtOH

CH_3OH

PBr$_3$

CH_3Br

made in part a

Mg

$Na_2Cr_2O_7$
H_2SO_4

1) CH_3MgBr

2) H_3O^+

1) O_3
2) Zn

KO

DMSO

Br

d.

12.

13.

15.

17.

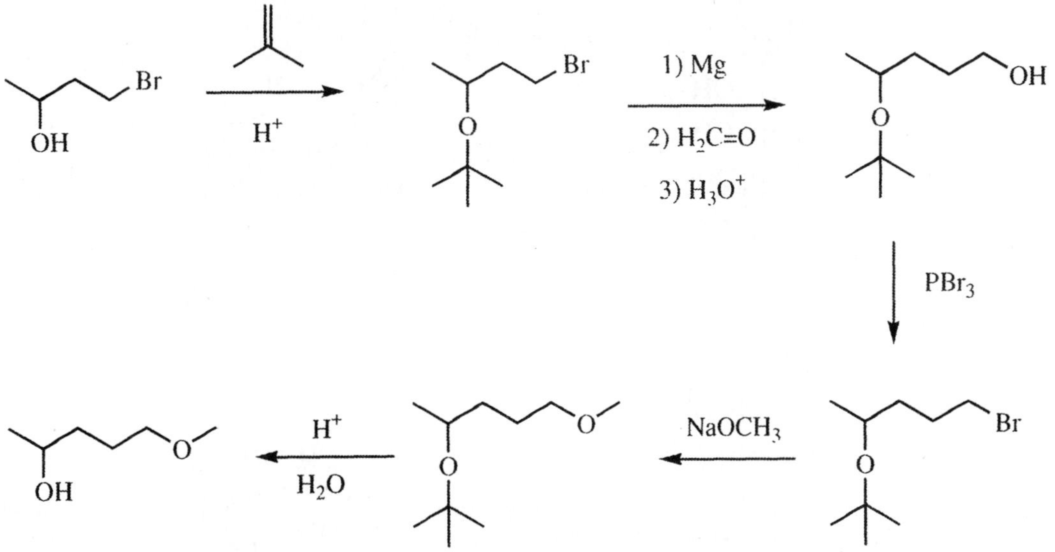

20.

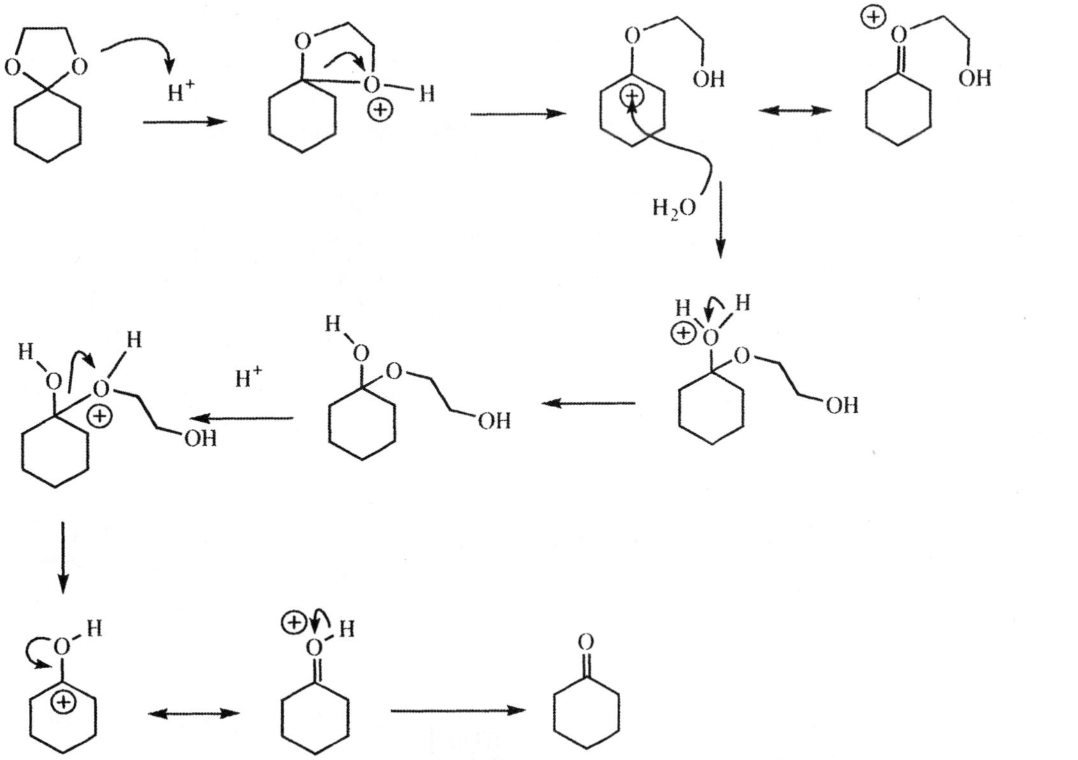

Notice all carbocations are resonance stabilized. Mechanisms exactly the same for the other two.

24.

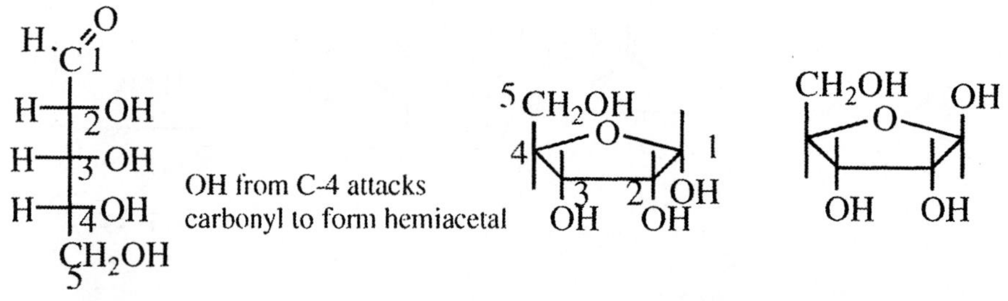

OH from C-4 attacks carbonyl to form hemiacetal

D-Ribose

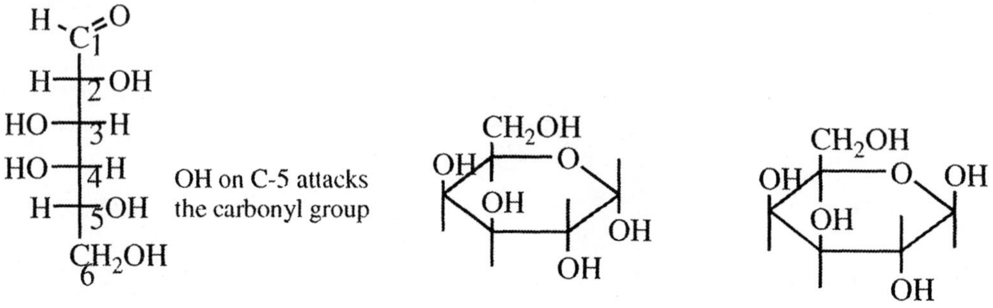

OH on C-5 attacks the carbonyl group

D-Galactose

25.

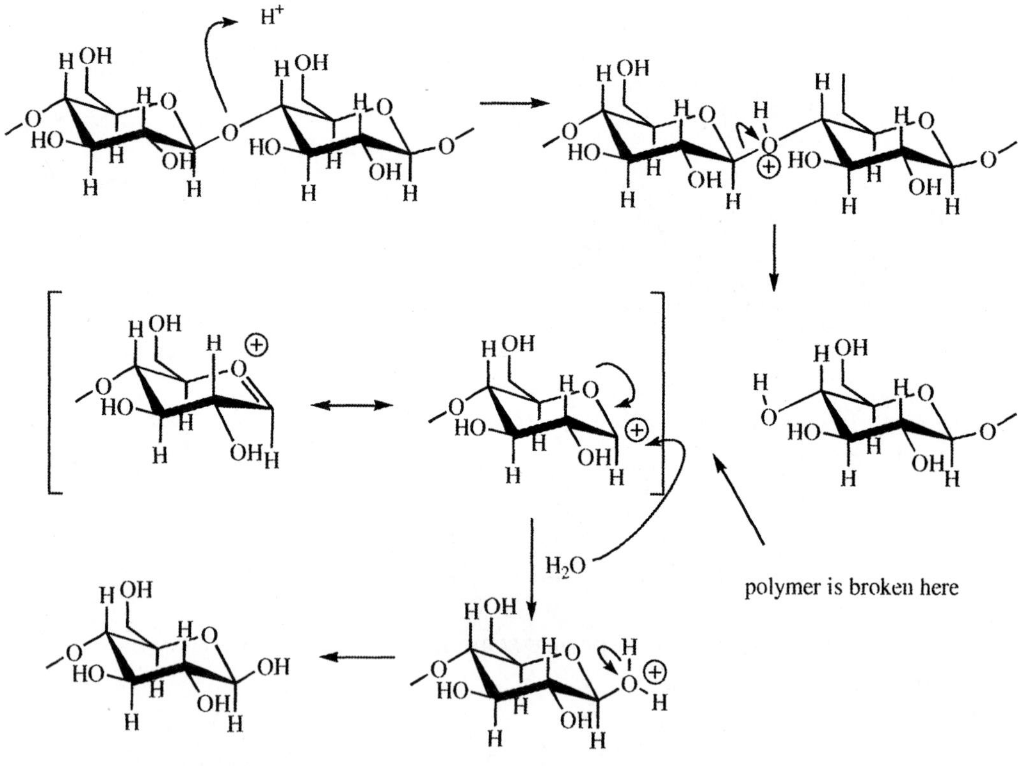

polymer is broken here

No, acetals are stable to base.

27. *Hint*: recall that a racemic mixture is not optically active even though it contains chiral molecules.

Chapter 19
Carbonyl Chemistry II: Carboxylic Acids, Acid Chlorides, Anhydrides, and Esters

Those of you who learned the previous chapter well will find this chapter boring, since almost everything we will do is a straightforward extension of what we already know about aldehydes and ketones. In this chapter we will learn about

- Esterification of acids

- Hydrolysis and transesterification of esters

- Reductions of acids and esters

- Grignard reactions of esters.

CARBOXYLIC ACIDS

Be sure you recognize that a carboxylic acid, $R\overset{\overset{\displaystyle O}{\|}}{C}OH$, is frequently represented as RCOOH or RCO_2H, and when you see either of the latter you should immediately conjure up a mental picture of the former.

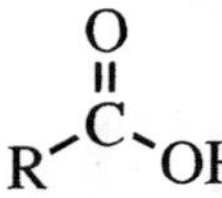

Nomenclature

Carboxylic acids are named based on the length of their carbon chains, including the carbon of the functional group. The "-e" of the alkane name is replaced with "-oic acid." Thus, CH_3CH_2COOH is propanoic acid, and CH_3COOH is technically ethanoic acid, although no one actually calls it that, the common name being "acetic acid," which you should remember. There is an important abbreviation I should introduce here: the *acetyl* group, carbonyl-with-a-methyl-attached, is abbreviated Ac. Therefore, acetic acid can be written as HOAc.

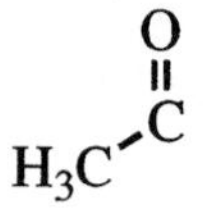

$$H_3C\overset{\overset{\displaystyle O}{\|}}{C} = Ac \qquad\qquad H_3C\overset{\overset{\displaystyle O}{\|}}{C}OH = AcOH = HOAc$$

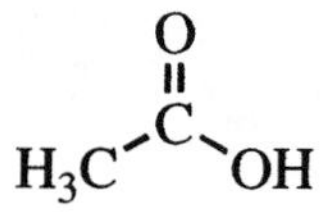

You should also know the structure of "benzoic acid":

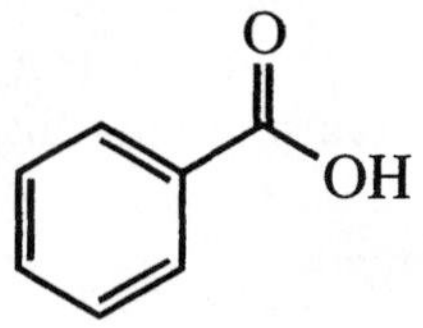

The group COOH is called a "carboxyl" group. If you need to name a carboxyl group as a substituent on something else, you use the word "carboxy," as in the following example:

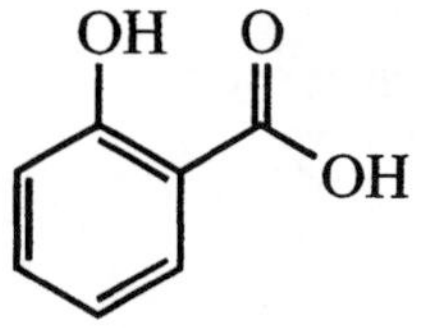

2-carboxyphenol
2-hydroxybenzoic acid
salicylic acid

Names exist, too, for the conjugate bases of carboxylic acids: you change the ending "-ic acid" to "-ate." Thus, CH_3COO^- is the acetate anion, or just "acetate," as in sodium acetate, and it is often written OAc^-. The generic expression for the anion of a carboxylic acid is "carboxylate anion."

Properties

Acidity

The single most important property of carboxylic acids is inherent in their name: they are carboxylic *acids*! As you know, carboxylic acids lose their protons quite readily because the resulting anion (the conjugate base) is resonance stabilized, sharing the negative charge equally between two oxygens. The pK_a of a typical carboxylic acid is about 5, substantially lower than the pK_a of an alcohol (about 18).

An important thing to notice about the resonance of a carboxylate anion is that these two resonance structures are the only ones you can draw. No matter what the R group, there can be no resonance that moves the charge onto it, so an R group can stabilize or destabilize the charge only by an inductive effect. Therefore changing the R group has only a small effect on the acidity of a carboxylic acid. Below are some numbers that show the inductive effect at work.

Acid	pK_a Value
CH_3COOH	4.75
$ClCH_2COOH$	2.85
$Cl_2CHCOOH$	1.48
Cl_3CCOOH	0.7

Solubility

The carboxylic acid group is fairly polar, and therefore small carboxylic acids are soluble in water. (Vinegar is largely a solution of acetic acid in water.) Of course, as with all the other groups we have seen, as the rest of the molecule gets bigger, the polarity of the acid functionality becomes less and less relevant, and large carboxylic acids are insoluble in water. Most carboxylic acids are soluble in organic solvents, but they often dissolve with a twist. The carboxylic acid part of the molecule is still polar, and thus looks for a polar environment. It can find some satisfaction by aggregating with others like itself, which permits the formation of a nice, cyclic, double-hydrogen-bond system. This is sort of the reverse of the micelles we discussed in Chapter 5.

The above discussion applies only to carboxylic acids themselves. In the presence of a base, most carboxylic acids look rather different. They become the conjugate bases of carboxylic acids, namely, carboxylate anions, with a full negative charge on the carboxylate group. The result is a salt, and almost all salts are soluble in water but not in organic solvents. This transformation gives us an ability to manipulate the solubility properties of carboxylic acids, and that can be very useful. Imagine, for example, that we had a mixture of the following two compounds, and we wished to separate them.

There are many ways this might be done. If you have had some organic lab experience by now, you might consider recrystallization, distillation, chromatography, etc. (if you don't know what these terms mean, don't worry about it). But those methods are somewhat tedious, and they all rely to some extent on luck. In this case, as in few others, a method exists that is nearly guaranteed to work and is, at the same time, easier than the others. Both of the compounds shown are neutral, organic systems, so we can be pretty confident both will dissolve in a standard organic solvent like ether (remember, "ether" means "diethyl ether"), but not in water. Let's dissolve this mixture in ether. Now we just learned that in the presence of base, the carboxylic acid (but *not* the alcohol) can lose a proton to become an anion, and therefore will cease to be soluble in ether but become soluble in water. So if we add to our solution of these two compounds in ether a solution of sodium hydroxide in water, and shake it up, here is what will happen: Water and ether do not mix, so we will get two layers. In the top layer will be the alcohol dissolved in ether. The bottom layer will contain carboxylate anion dissolved in water, along with excess sodium hydroxide. The two species we're interested in are in physically separate places, with a nice visual line separating them, so it becomes a simple matter to get them into separate containers.

Now one container houses the alcohol dissolved in ether. If we evaporate the ether, what's left is our alcohol, and we've finished half of the job. What about the other container? This contains the carboxylate anion dissolved in water. If we next evaporate the water, we will *not* be left with a carboxylic acid, but rather its anion (actually, a salt). What to do? The answer should be obvious: put the missing proton back on. This will require adding enough acid (HCl, H_2SO_4, etc.) to make the solution acidic. Suddenly we no longer have a carboxylate anion, but a carboxylic acid, and this is no longer soluble in water. If the carboxylic acid (like this one) happens to be a solid, we could simply filter it away from the water; more often it would be dissolved in fresh ether. Once the new ether solution has been separated from the water, the ether can be evaporated. Overall, with a few simple manipulations, we can effect an almost perfect separation of these two organic compounds. This process is known as "extraction."

Test Yourself 1

Phenol has a pK_a of about 10. Explain why the pK_a of phenol is so low compared to other alcohols, and then suggest a way to separate benzoic acid from phenol. Do it now!

phenol

Spectroscopy

You already know that in the IR, a carboxylic acid shows both a carbonyl stretch at about 1700 cm^{-1} and an OH stretch that is very broad, usually running from 3400 or so all the way to 2600 or so, smearing right through the CH region near 3000 cm^{-1}. In the NMR there are two kinds of hydrogens we need to worry about: those adjacent to the carbonyl group, which, like their counterparts in ketones, for example, appear near 2 ppm, and the hydrogen that is part of the carboxylic acid function itself, which usually appears between 11 and 13 ppm. In the ^{13}C-NMR spectrum you should expect to see the carbonyl carbon on the far left side, as usual, above 160 ppm.

Preparation

We have already learned several methods of making carboxylic acids. For example, we have seen that ozonolysis of alkynes leads to acids:

$$R\!\!-\!\!\equiv\!\!-\!\!R' \quad \xrightarrow[\text{2) } H_2O]{\text{1) } O_3} \quad R\overset{O}{\underset{OH}{\diagdown}} \quad \overset{O}{\underset{HO}{\diagup}}\!\!R'$$

We learned earlier that ozonolysis of alkenes bearing hydrogens usually is used to make aldehydes, but if we leave out the zinc we can get carboxylic acids from these also, because the aldehydes get oxidized to acids by hydrogen peroxide present in the reaction mixture:

$$\overset{R}{\underset{HO}{\diagup}}\!\!=\!\!O \quad O\!\!=\!\!\overset{R'}{\underset{OH}{\diagdown}} \quad \xleftarrow[\text{2) } H_2O]{\text{1) } O_3} \quad \overset{R\;\;\;\;\;H}{\underset{H\;\;\;\;\;R'}{>\!\!=\!\!<}} \quad \xrightarrow[\text{2) Zn, } H_2O]{\text{1) } O_3} \quad \overset{R}{\underset{H}{\diagup}}\!\!=\!\!O \quad O\!\!=\!\!\overset{R'}{\underset{H}{\diagdown}}$$

Primary alcohols can be oxidized to carboxylic acids using chromic acid, again because the intermediate aldehydes are oxidized under the reaction conditions:

$$R\!\!-\!\!\overset{H_2}{\underset{}{C}}\!\!-\!\!OH \quad \xrightarrow[H_2SO_4]{Na_2Cr_2O_7} \quad \left[\overset{O}{\underset{R\;\;\;\;H}{\diagdown\!\!\diagup}} \right] \quad \longrightarrow \quad \overset{O}{\underset{R\;\;\;\;OH}{\diagdown\!\!\diagup}}$$

Aldehydes can actually be oxidized to carboxylic acids by many reagents, including chromic acid, hydrogen peroxide, silver ions, or just air, so aldehydes themselves would make good starting materials for carboxylic acids.

Finally, we learned that the Grignard reaction produces carboxylic acids if carbon dioxide is used as the carbonyl partner:

$$R\!\!-\!\!Cl \quad \xrightarrow{Mg} \quad R\!\!-\!\!Mg{\cdot}Cl \quad \xrightarrow[\text{2) } H_3O^+]{\text{1) } CO_2} \quad R\!\!-\!\!\overset{O}{\underset{OH}{\overset{\parallel}{C}}}$$

At this point we'll refrain from learning any *new* syntheses of carboxylic acids.

Reactions

There are really only two reactions of carboxylic acids that we need to deal with. One involves the OH group of the acid, and one involves the carbonyl part.

Acid Chloride Formation

In general, the OH group of a carboxylic acid does not undergo the same reactions that alcohols do, but it *does* react with PBr_3 and $SOCl_2$ to make bromides and chlorides. The resulting halides are pretty reactive compounds, and the bromides are even more reactive than the chlorides; most people make the chloride when they want such a thing, so I will only discuss the chlorides.

You may recall that alcohols react with thionyl chloride (SOCl$_2$) to make alkyl chlorides. This was the first new reaction discussed in Chapter 17, a section you may want to review. By what is apparently the same mechanism, thionyl chloride also converts carboxylic acids into the corresponding "acid chlorides":

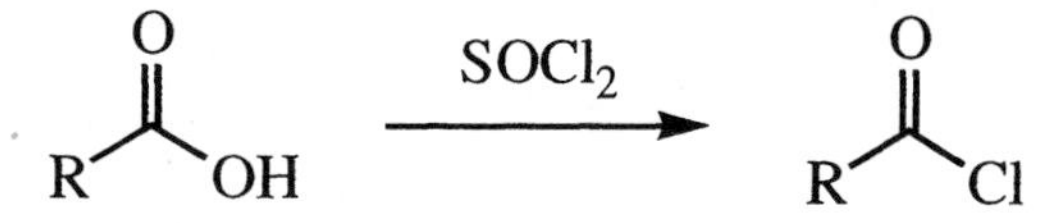

At the moment you do not know what to do with these creatures, but this will change shortly!

Esterification

The second important reaction of carboxylic acids is called the Fischer esterification. It is a reaction between a carboxylic acid and an alcohol. Let me remind you here of something I tried to establish in the last chapter: "As a general rule, a weak nucleophile and a weak electrophile will not react with each other. One or the other has to be pumped up before a reaction will occur. A neutral oxygen is a weak nucleophile. A carbonyl group is a weak electrophile. These will not react with each other until one or the other is pumped up."

An alcohol should not react with the carbonyl group of a carboxylic acid unless either the alcohol is made more nucleophilic (by becoming an anion), or the carbonyl group is made more electrophilic (by becoming a cation). Are both of these viable options for us? What would happen if we presented a carboxylic acid with the anion of an alcohol? Would it attack the carbonyl group?

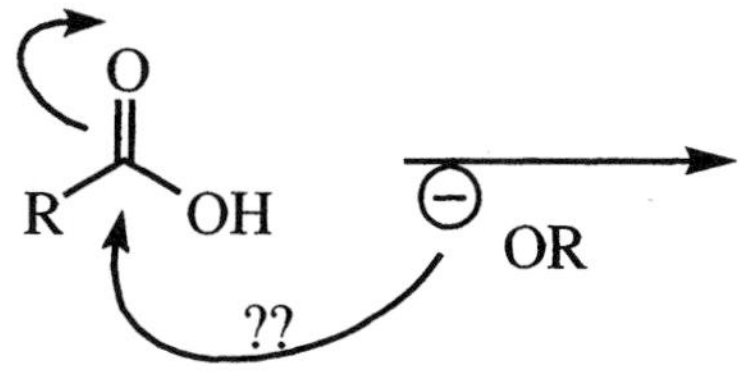

Remember a principle we have covered many times before: proton transfer is a very rapid reaction, and if such a transfer can occur, it will usually take precedence over any other process. A carboxylic acid is an acid; an alkoxide anion is a base. The reaction shown below is in fact very favorable.

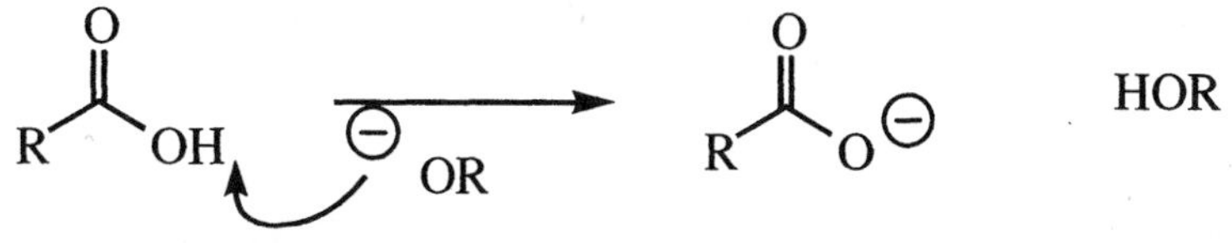

This reaction will occur as soon as you mix the two species, resulting in a carbonyl system that already has a negative charge and is not electrophilic, and an alcohol that no longer has a negative charge and is therefore not very nucleophilic. There is no obvious way around this problem, so encouraging reaction between an alcohol and an acid by improving the nucleophile is doomed to failure.

On the other hand, it is not unreasonable to expect that, in acid solution, at least some of the time the carboxylic acid will be protonated. When the carbonyl group of an acid is protonated, the resulting cation has three resonance structures, so it should be easier to make than the cation of a ketone, for example, which has only two resonance forms. (It is also easier to protonate the carbonyl oxygen of a carboxylic acid as opposed to the OH oxygen, since the result of the latter protonation would have no resonance structures at all. Prove this to yourself!)

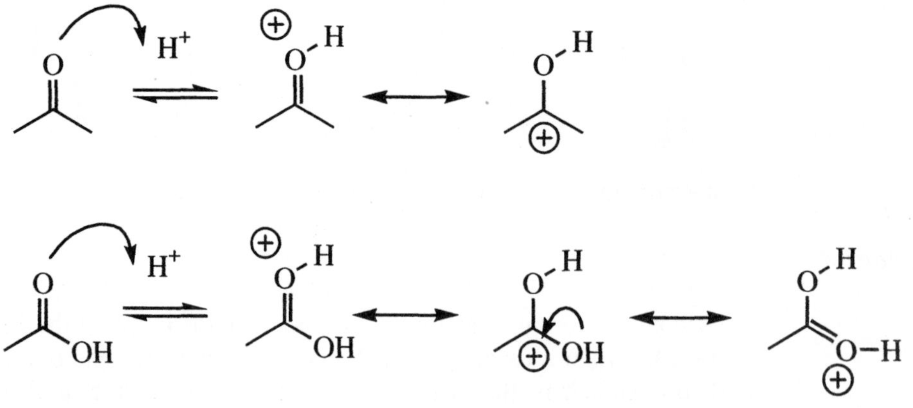

Once the carbonyl group is protonated, of course, it becomes a strong electrophile and can draw in a nearby alcohol (*in its neutral form, not as an anion!*) to help take care of the positive charge:

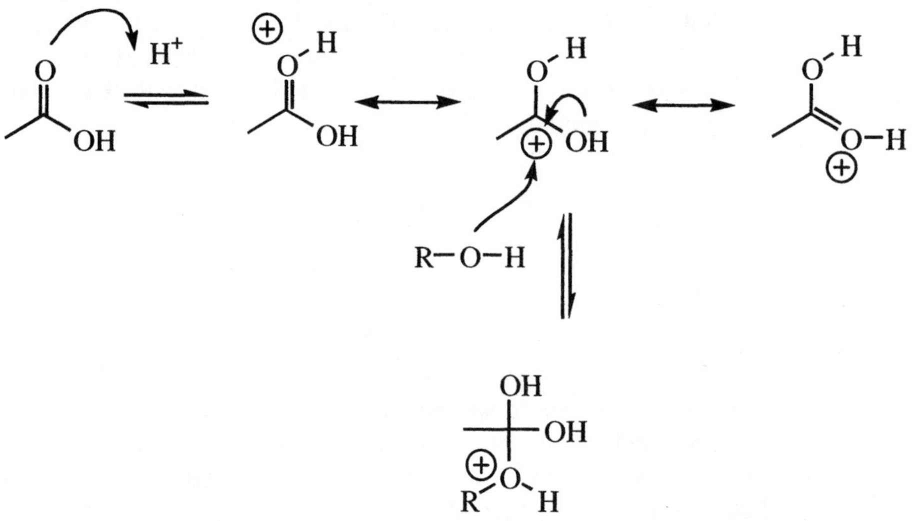

This new cation is admittedly not resonance stabilized, but it can easily solve its charge problem by dropping the extra proton:

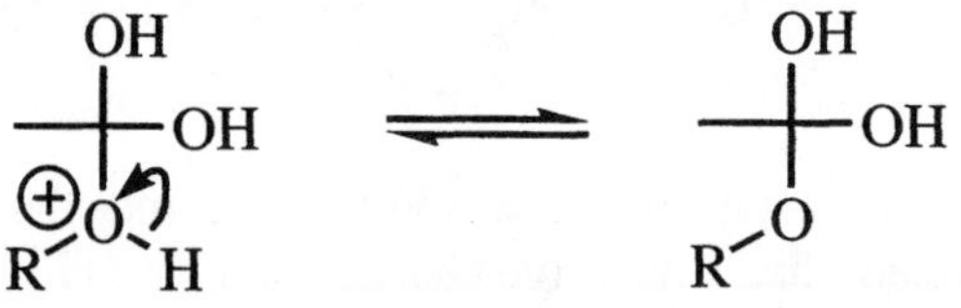

Now we have managed to create a new species, one with three C–O single bonds. As we learned in the last chapter (in the context of hydrates, hemi and full ketals and acetals), molecules will usually trade two C–O single bonds for one C–O double bond if they can find a way to do it. Is there such a way here? Our new molecule is sitting in acidic solution. Clearly, at random times, any of the three oxygen atoms could become protonated. If the OR group becomes protonated, the molecule might undo what it just did, returning all the way back to the carboxylic acid. This process does occur, but (at the moment) it does not interest us. On the other hand, if one of the OH groups becomes protonated, the system can head off in a new direction:

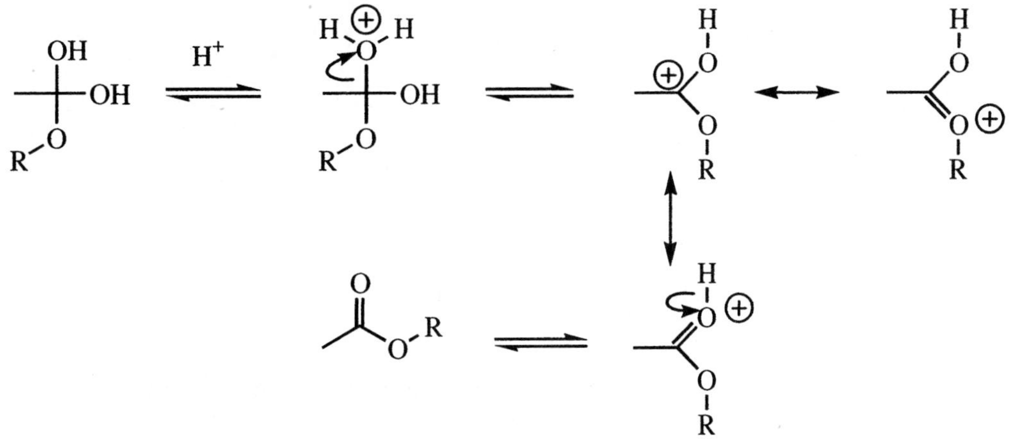

The result of this new direction is a new type of *compound,* an ester, and you have just seen one of the important ways of making esters. As I hinted before, it is called Fischer esterification, and the net reaction is shown below:

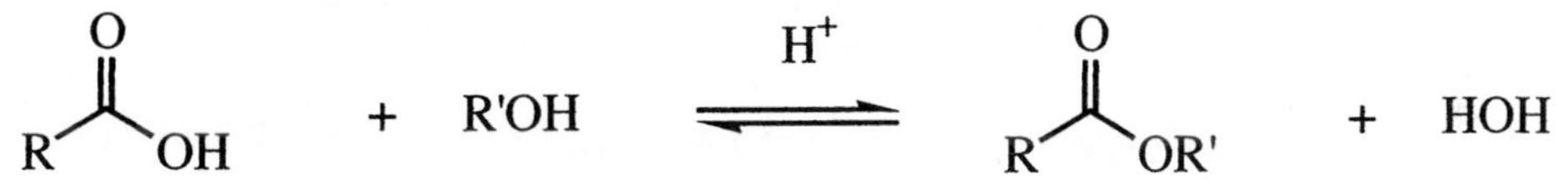

Notice that in this summary equation, as well as in the mechanism we went through, every step (and therefore the whole process) is shown as being reversible. If it's reversible, how can you force the process to go either to the left or the right? By now you should know the answer to this one, because we have faced the same problem several times, and always solved it the same way: you use the equilibrium constant to manipulate things in the direction you want, applying Le Châtelier's principle. To make things go from left to right, use a lot of alcohol (or a lot of carboxylic acid); to go from right to left, use a lot of water. We will come back to this later.

I should mention that this esterification reaction (or its reverse) is a relatively slow process, and to make it occur in a reasonable length of time the mixture is usually heated, often boiled overnight. Thus, you'll often see "Δ," the heat symbol, written over or under the arrow on these reactions.

Below are some examples of Fischer esterifications. They all represent exactly the same reaction, proceeding by exactly the same mechanism as that discussed above, even though some are written in slightly different form. You should take the time *now* to write out a mechanism for each one of them. You will notice, of course, that I have deliberately slipped in some of the abbreviations we have learned, in an effort to get you to see how they are used, and to get you to start seeing structures rather than letters when abbreviations appear.

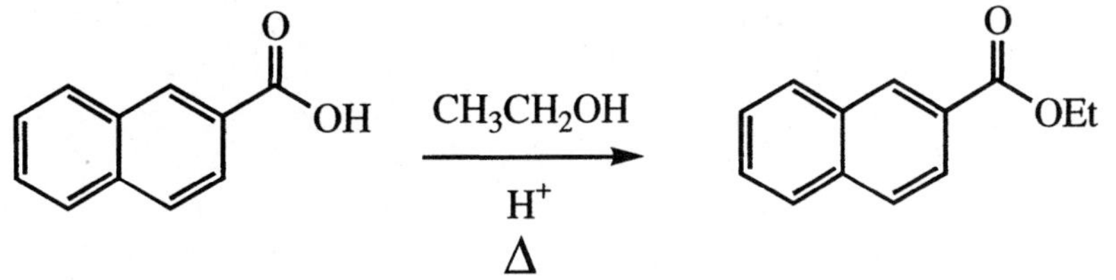

By the way: a cyclic ester like the last one above is called a "lactone," but it is still an ester in every sense.

CARBOXYLIC ACID DERIVATIVES

Compounds that are usually made from carboxylic acids, and which, upon treatment with water (hydrolysis), revert to carboxylic acids, are called "derivatives" of acids. Among these are acid chlorides, acid anhydrides, and esters.

acid (acyl) chloride   anyhdride   ester

Nomenclature

A carboxylic acid's name always ends in "ic acid," as in benzoic acid or acetic acid or, generically, alkanoic acid. The portion of the structure that comprises everything except the OH of the acid is called an acyl group, and it is named by replacing the "-ic acid" with "-yl," as in the benzoyl group or acetyl group (which you already knew). Acid chlorides are named by taking this acyl term and adding "chloride." The general expression should be (and is) acyl chloride, but more often these compounds are referred to as "acid chlorides."

alkanoic acid   acyl group   acyl chloride

benzoic acid   benzoyl group   benzoyl chloride

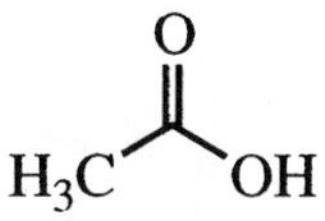

O
‖
H₃C   OH

acetic acid

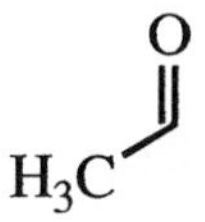

O
‖
H₃C

acetyl group

O
‖
H₃C   Cl

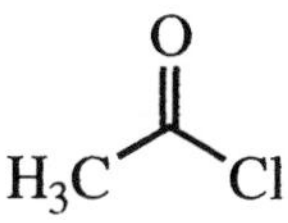

acetyl chloride

Anhydrides consist formally of two molecules of carboxylic acid from which a molecule of water has been removed; hence the term "anhydride," meaning "without water."

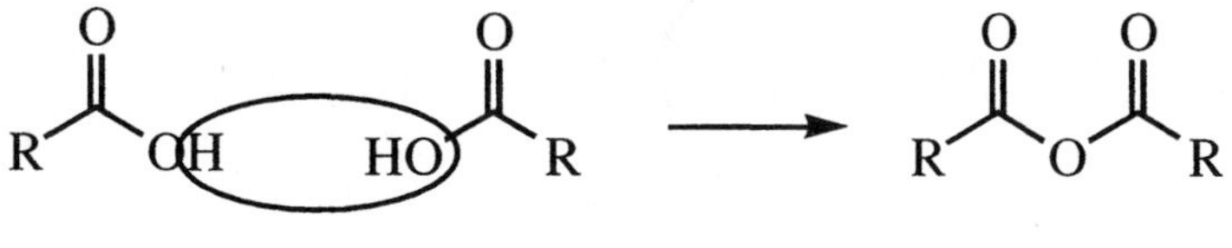

O   O   O   O
‖   ‖   ‖   ‖
R   C⟨OH   HO⟩   R   ⟶   R   C   O   C   R

Note that the above equation is not intended to represent a real chemical reaction, simply a relationship. Anhydrides are named by listing the names of the two acids involved, followed by the word "anhydride." Most of the time, however, the two acid molecules involved are both the same, in which case you only need to name the acid once.

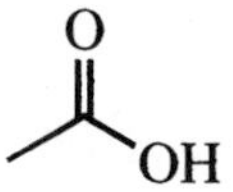

O
‖
OH

acetic acid

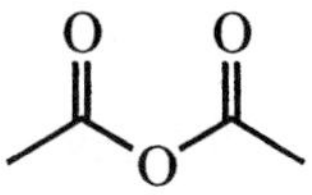

O   O
‖   ‖
O

acetic anhydride

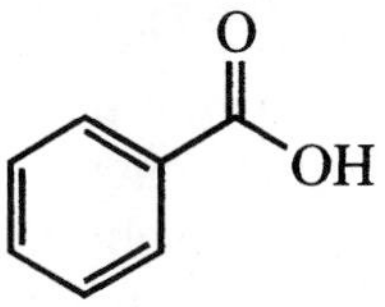

O
‖
OH

benzoic acid

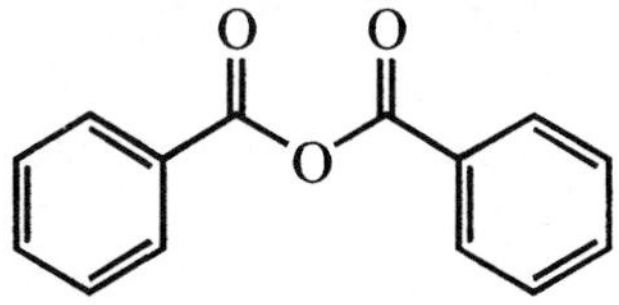

O   O
‖   ‖
O

benzoic anhydride

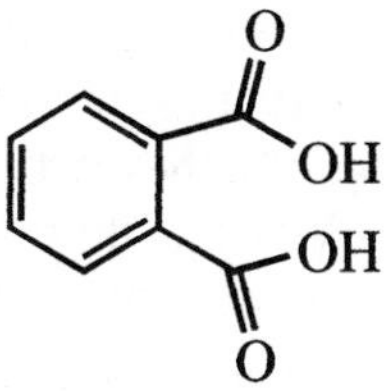

O
‖
OH
OH
‖
O

phthalic acid

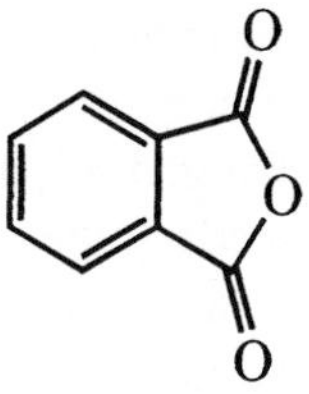

O
‖
O
‖
O

phthalic anhydride

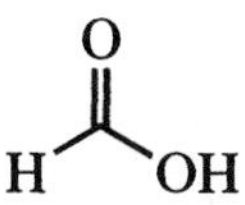

O
‖
H   OH

formic acid

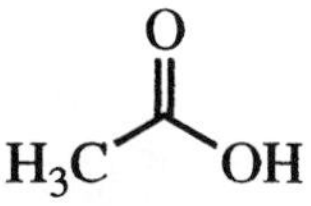

O
‖
H₃C   OH

acetic acid

O   O
‖   ‖
H   O   CH₃

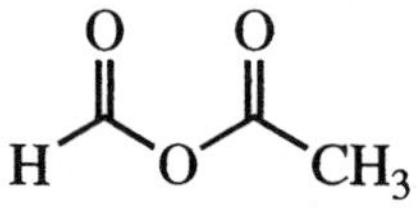

formic acetic anhydride

An ester, as you now know, is made from a combination of a carboxylic acid and an alcohol. To name an ester, you first name the alcohol part, and then, in a new word, name the acid in the form of its conjugate base.

2-naphthoic acid   CH₃CH₂OH / H⁺   ethyl alcohol   ethyl 2-naphthoate

cholesterol CH_3COOH / H^+ acetic acid   cholesteryl acetate

Properties

Reactivity

We have already determined that, in general, aldehydes are more reactive than ketones. Where do acids, acid chlorides, anhydrides, and esters fall along this scale?

Carbonyl groups undergo reaction when a nucleophile attacks at the carbon atom of the carbonyl group, which always carries a partial positive charge. It stands to reason that the reactivity of any particular carbonyl group should depend on the intensity of that partial charge; stated another way, groups that help make the partial positive charge more comfortable by distributing the charge over a greater area should make the carbonyl group *less* reactive.

Acid chlorides and anhydrides contain carbonyl groups that have, directly attached to them, good leaving groups (chloride or carboxylate anion). Good leaving groups are electron withdrawing; the partial positive charge on the carbon of the carbonyl group is therefore accentuated by the presence of these electron-withdrawing groups. That in turn makes acid chlorides and anhydrides *more* reactive than other carbonyl compounds. Indeed, these functional groups are so reactive that in most cases you do *not* need to further juice up either the nucleophile or the electrophile to make a reaction occur.

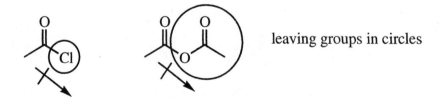

leaving groups in circles

On the other hand, carboxylic acids and esters have an OH or OR group attached to the carbonyl carbon. These are not good leaving groups, and even though the oxygen atom is electronegative, we also know from experience that oxygen atoms are pretty efficient at donating their lone pairs to help stabilize adjacent carbocations by resonance. (Recall the formation and hydrolysis of ketals and acetals.) The same is true here, so acids and esters, in general, are *less* reactive than aldehydes and ketones.

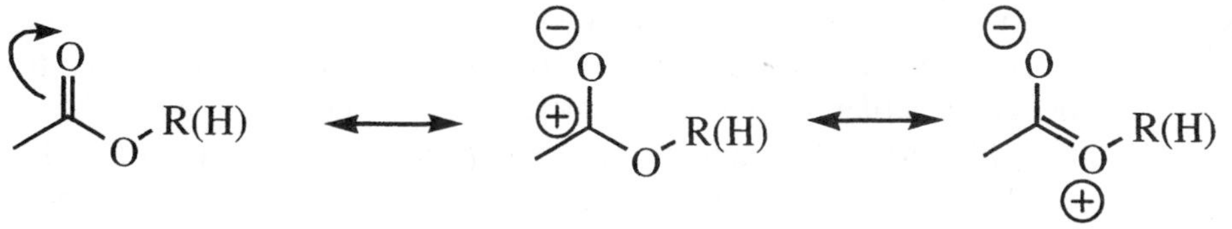

Alert students will wonder why the stabilizing effect of an oxygen atom referred to above does not apply to anhydrides, which also have an oxygen atom in the same place. This is an excellent question and I may not be able to satisfy you with my answer, but I'll try. The problem is that the oxygen atom of an anhydride has a carbonyl group on *both* sides of it. To the extent that it is helping the positive charge on one carbonyl group, it is unavailable to help the other. So one or the other of the two (usually equivalent) carbonyl groups is feeling pretty good, but the other is feeling pretty bad. On average, then, a carbonyl group in an anhydride is not as well stabilized by electron donation from oxygen as is the carbonyl of an ester or an acid.

So the order of reactivity of carbonyl groups is as follows:

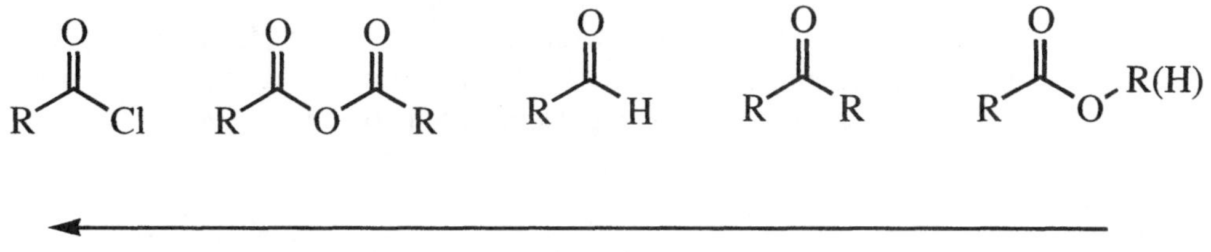

more reactive

This relationship is one you should keep in mind from now on.

Test Yourself 2

Figure out where you would expect amides to fall on this reactivity scale.

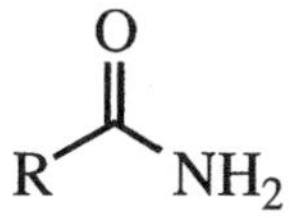

an amide

Spectroscopy

All the compounds we have considered in this chapter have carbonyl groups; all therefore show a peak in the IR spectrum between 1650 and 1800 cm^{-1}. It is instructive to look at precisely where in this region each type of compound appears. Recall that the location of such a peak is a measure of how hard it is to stretch the corresponding C–O double bond. In a very real sense this is also a direct measure of bond strength. One can describe a carbonyl group as a cross (resonance hybrid) between double-bonded and single-bonded forms:

Clearly the second structure should be pretty poor, and would not contribute very much to the overall structure of a carbonyl group, which is pretty much a C–O double bond. Nevertheless, one might argue that substituents that make the second structure—the one with a single bond—better would have the effect of weakening the overall C–O bond, while substituents that make that structure worse would strengthen the C–O bond. This is in fact what we see: acid chlorides and anhydrides, with their electron-withdrawing groups, make the single-bonded structure contribute even less than usual, leading to a stronger C–O bond, one that is harder to stretch. The corresponding peaks show up at the high end of our region, near 1800 cm^{-1}. Oxygen substituents tend to cancel themselves: the electron-withdrawing nature of oxygen makes the single-bonded structure worse, but the electron-donating lone pair allows for a third resonance structure that makes it better:

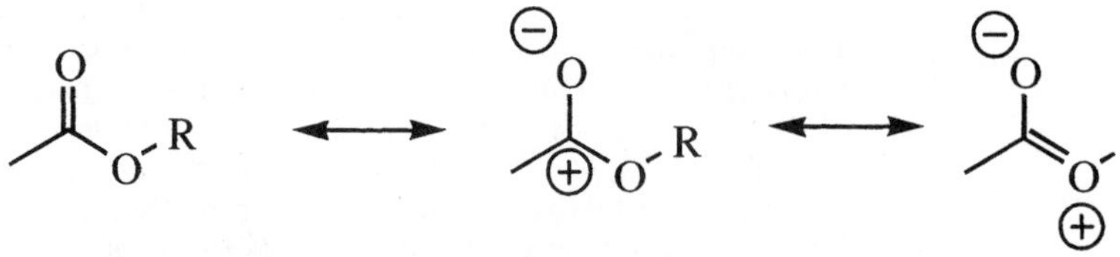

Acids and esters show up in the middle of the region, near 1700–1735 cm^{-1}, about the same place as aldehydes and ketones. Amides, which we will cover in Chapter 22, differ from acids and esters in two respects: nitrogen is less electronegative than oxygen, so it is simultaneously less electron withdrawing and more electron donating. This accentuates the C–O single bond structure relative to other carbonyl compounds, so amides have the weakest C–O bond of the lot, giving a peak between 1650 and 1690 cm^{-1}.

Test Yourself 3

Explain why enones show a lower stretching frequency than ordinary ketones:

1718 cm^{-1} 1691 cm^{-1}

In the NMR, all of these groups are similar, at least to a first approximation. If there are hydrogens on the α carbon, these should appear between 2 and 2.5 ppm. Hydrogens on a carbon next to the oxygen of an ester appear near 4 ppm, sometimes a little downfield of 4 (left) due to the electron-withdrawing nature of the carbonyl group. In the ^{13}C-NMR, the carbonyl carbon itself appears, as always, between 160 and 220 ppm.

Preparation

Acid chlorides are almost always made according to the method described earlier, using a carboxylic acid and thionyl chloride. Here's an example.

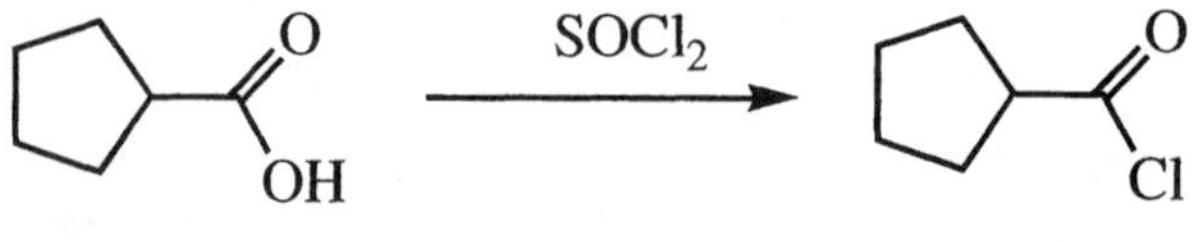

There is one case in which this procedure does not work: you cannot make formyl chloride by this (or any other) method, because the product is unstable and falls apart. So do not try. All other acid chlorides are OK.

Anhydrides are rarely made in labs. Most are symmetric and can be purchased. Manufacturers prepare them by heating the hell out of a carboxylic acid, thereby driving off water in a reaction that, on paper, really ought not to take place. We won't worry about it. Anhydrides can also be made, if necessary, by combining an acid chloride with an acid:

Notice that this procedure would allow you to make an unsymmetric acid anhydride, if you ever wanted to.

Esters can be made by Fischer esterification, using the appropriate carboxylic acid and alcohol. The most common application of this is in making a methyl or ethyl ester, or some other ester involving an alcohol that is easy to use in large excess because it is cheap and easily evaporated (for removing the excess alcohol after the reaction is over).

If neither the acid nor the alcohol piece of the ester you want falls in the "cheap-and-volatile" category, Fischer esterification is often not the best choice, however. Another fairly common way to make esters is first to turn the acid into a more reactive derivative, most often the acid chloride, and then to react *that* with the appropriate alcohol. Since acid chlorides are so reactive, no acid or base catalyst is required to get this to take place:

Chapter 19 Carboxylic Acids and Derivatives 593

Notice that because an acid chloride is so reactive, it can be attacked by a neutral alcohol. This applies to anhydrides as well, although they usually react a bit more slowly. The intermediate (top right in the picture above) contains both a positive and a negative charge. Students often ask which to take care of first. My mechanism shows the negative charge disappearing before the positive charge does, but there is really no evidence that this is correct. Both events occur, and both happen so fast that the ester is formed rapidly. It would not be terrible if you wrote these steps in the opposite order.

ORM: Under "Acid Derivatives," check out "acid halide, rx with ROH"

An acid chloride can be made from any acid (except formic acid), so this procedure for making esters is pretty general. If that's the case, why would you ever use an anhydride for the same purpose? There is really no reason you would need to, but several of the small anhydrides are cheaper and easier to use than the corresponding acid chlorides. Thus, when making acetates, acetic anhydride (Ac_2O; you should recognize this abbreviation and be able to write out the full structure!) is more often used than acetyl chloride.

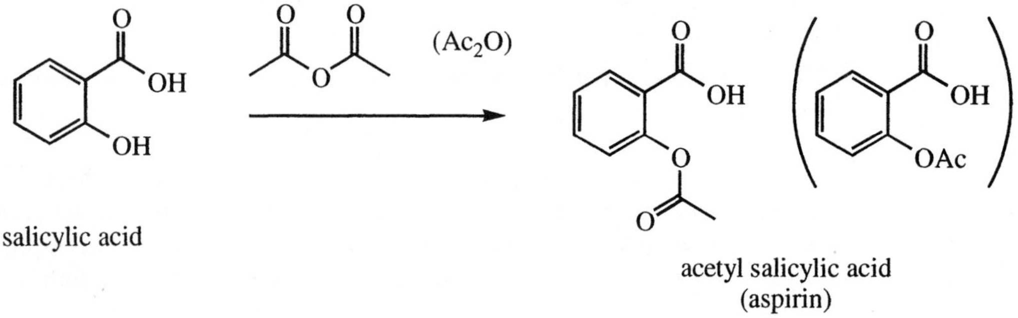

salicylic acid

acetyl salicylic acid
(aspirin)

Reactions

Based on the reactions you have already seen for aldehydes, ketones, and acids, and on what you know about the reactivity of these species, you should already be in a good position to predict what would happen to acid chlorides, anhydrides, and esters in the presence of water, alcohols, Grignard reagents, and reducing agents, including whether you would need to use acid or base catalyst and, if so, which would be better. Indeed, why don't you take an hour right now and try to work these things out for yourself before continuing your reading?

Reactions with Water (Hydrolysis)

As you know, **acid chlorides** and **anhydrides** are the most reactive carbonyl compounds— so reactive that, unlike other carbonyl compounds, they require neither acid nor base catalysis to get a reaction going. Thus, you should anticipate that acid chlorides and anhydrides would react with water all by themselves. Indeed, they do, leading to carboxylic acids. Since a bond in the starting material ends up breaking due to reaction with water, the process is referred to as a "hydrolysis."

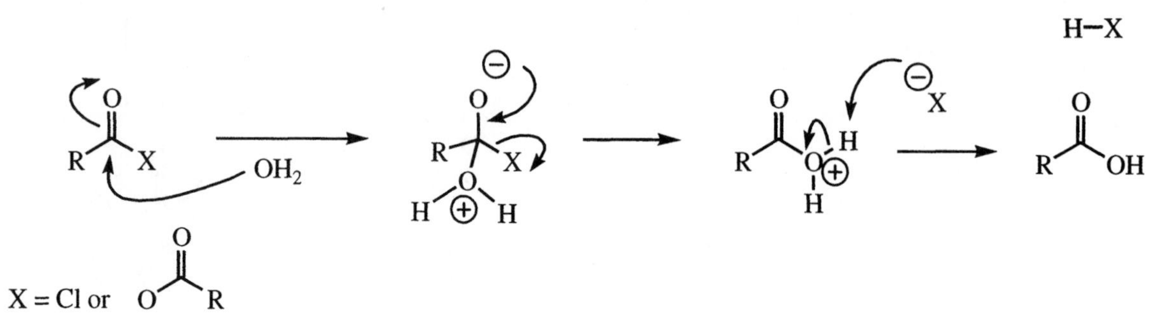

In other words, you can make carboxylic acids by hydrolysis of acid chlorides or anhydrides. (Of course, if the starting material were an anhydride, you would end up with *two* molecules of carboxylic acid, but they would both be the same material unless the anhydride was not symmetric. Draw some mechanisms to make sure this all makes sense to you.) Now most acid chlorides and anhydrides are made *from* the corresponding carboxylic acids. Why on earth would you want to make a carboxylic acid from an acid chloride? As a general rule, the answer is, you wouldn't *want* to, but that does not mean you wouldn't wind up doing it! Acid chlorides react so rapidly with water that every time you open the bottle, for example, some hydrolysis takes place due to moisture in the air. This is why most acid chlorides smell like HCl—it is not the acid chloride you smell, but actually HCl formed through hydrolysis! The hydrolysis of an acid chloride or anhydride is really something one would almost never want to do deliberately from an organic chemist's point of view, but we need to discuss it, because it illustrates the chemistry of these compounds, and because you need to be aware of the possibility of impurities being present in your bottles of these things.

There are two cases in which such a reaction might actually be done on purpose. One is quite common, the other is pretty rare. The first involves a change in our point of view. So far we have been looking at this as a reaction that changes an acid chloride (anhydride) into a carboxylic acid with the aid of water. But, it is also a reaction of water! Often chemists have to run reactions where water is a problem, so a process that removes water could be very useful. Acetic anhydride is sometimes added to reactions either to clear out water that might interfere with what's supposed to happen or to eat up water that is produced by the reaction itself. Anhydrides work efficiently for either purpose.

The second reason one might hydrolyze an acid chloride or anhydride on purpose is that you might want to prepare a carboxylic acid that has special characteristics. For example, I once wanted to run a reaction in trifluoroacetic acid solvent, and in order to see what was happening during the reaction, I wanted to follow the process by NMR. Using ordinary trifluoroacetic acid out of a reagent bottle would have given me an NMR sample that was almost entirely trifluoroacetic acid, and nothing else would have been visible in the NMR spectrum. The normal solution to a problem like this would be to use solvent in which all the H's had been replaced with D's: deuterotrifluoroacetic acid. Since this didn't happen to be available, I made some of it myself by adding D_2O to trifluoroacetic anhydride.

> Unfortunately, this story has a sad ending. When I added D_2O to my trifluoroacetic anhydride, I forgot that this particular anhydride is even more reactive than most anhydrides. The reaction occurred so rapidly that everything shot out of the bottle before I had finished adding the D_2O! Why is trifluoroacetic anhydride more reactive than other anhydrides? Don't wait for me—you answer it!

It is much more common to want to hydrolyze an **ester**. In this case we cannot expect a reaction to occur with just plain water. As with most carbonyl groups, either the nucleophile or the electrophile has to be juiced up. As you know, you can make a nucleophile hotter by giving it a negative charge; you make an electrophile hotter by giving it a positive charge. Let's consider the latter approach first:

You should recognize this reaction as being precisely the reverse of the Fischer esterification we discussed at length earlier in this chapter. The overall reaction is the following, and we have already discussed how one can use Le Châtelier's principle to control the direction of the equilibrium: to hydrolyze an ester, you use an excess of water; that is, you run the reaction in acidic water as solvent. To *make* an ester, you use an excess of one of the species on the right, most commonly the alcohol. And recall, we usually heat these reactions to make them go more rapidly.

The reason that a reaction occurred between this ester and water (compounds that normally ignore each other) is that an acid catalyst was present to protonate the carbonyl group and make it *want* to react with water. You should notice that, in the mechanism above, every time we use a proton in a step, we get one back in a later step, so the number of protons floating around (always attached to something, of course) does not change. The reaction does not consume H^+, so the acid is catalytic, and we do not need very much acid to make this reaction go.

You should also notice that in the entire mechanism for acid-catalyzed hydrolysis, all the species written are either neutral or positively charged. (Of course, you know there must also be negatively charged species running around somewhere in this solution, namely the counterions associated with the H^+, but you should also know that these anions do not get involved in the reaction, so there is usually no need to show them.)

The alternative way to make water react with an ester is to make the nucleophile, the water, more muscular. This means placing the ester in basic water, water that contains OH⁻. What happens then?

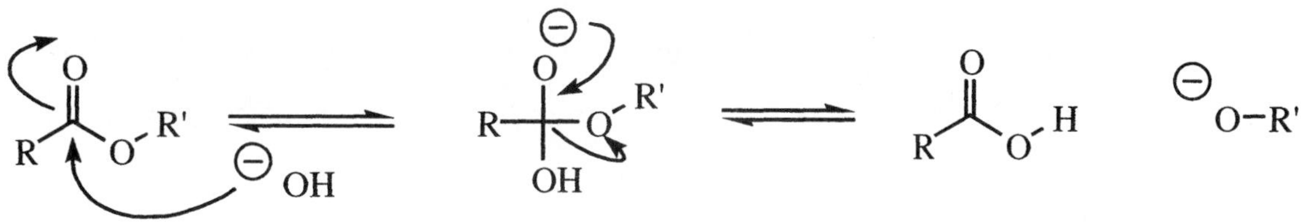

This time the nucleophile is able to force its way into the neutral carbonyl group. Electrons retreat onto the oxygen, regroup, and counterattack, driving out some leaving group. Recall that an O⁻ is perfectly capable of driving out another O⁻, but which O⁻ will it drive out? It could choose either the OH⁻ or the OR'⁻. Both options are reasonable and both occur, but if OH⁻ leaves we go back where we came from and cannot even tell that anything has occurred, so we tend not to worry about that possibility. If OR'⁻ leaves, of course, we head to the right in the mechanism above, and reach the products. But we are not done, and this is an important difference between basic hydrolysis and acidic hydrolysis of esters. We already know that an alkoxide anion and a carboxylic acid can react with each other rapidly in an acid–base reaction, so as to exchange a proton:

This equilibrium sits very far to the right, such that almost no molecules of neutral carboxylic acid remain. The carboxylate anion has no interest at all in a negatively charged nucleophile, so it just sits around. The point is that this last reaction, deprotonation of the carboxylic acid, effectively removes the carboxylic acid from equilibrium with its ester. In other words, the basic hydrolysis of an ester is always under pressure to go to the right, due to Le Châtelier's principle. For this reason, basic hydrolysis of an ester is often faster and more satisfactory than acidic hydrolysis. On the other hand, there is a price to pay: notice that the OH⁻ used in the first step is *not* regenerated at any later point in the reaction. Unlike acid-catalyzed hydrolysis, basic hydrolysis requires one mole of OH⁻ "booster" for every mole of ester, and the experimentalist would do well to measure reagents carefully to make sure there is enough base present.

There is also a practical problem associated with basic ester hydrolysis: the product is not the carboxylic acid we want, but rather the corresponding anion. How do we deal with that? Simple: put the proton back on. After you finish a basic hydrolysis, you add enough acid to make the solution acidic.

The basic hydrolysis of an ester is often referred to as "saponification," which is a term that means "soap-making." In the old days, soap was made by taking animal fat (long chain esters) and boiling it in water with fire ashes (also known as potash, which contains KOH). The long chain carboxylate anions that result from this process are soap, as described in Chapter 5.

ORM: Under "Acid Derivatives," check out "ester hydrolysis, basic"

Below are some examples of ester hydrolyses.

WHO CARES? YOU DO!!

Esters can be hydrolyzed in hot water under either acidic or basic conditions. You may recall from Chapter 5 that cell membranes are composed of phospholipids, which look like this:

Recall that these aggregate into lipid bilayers, which form a barrier between the inside and the outside of a cell. Under ordinary circumstances this poses no problem. However, recently a new class of organisms has been discovered that thrives in the extreme environments of hot springs and oceanic thermal vents, in temperatures near and sometimes above the boiling point of water. How do the phospholipids, which you should now recognize as esters, survive such conditions? They don't: these organisms have evolved in such a way that the links between the polar heads and the non-polar tails of the molecules are not esters at all, but ethers, which are not in danger of coming apart in hot water. This significant difference in the molecular makeup of their cells (among other factors) has forced biologists to create a new taxonomic division for them. There are now taxonomic divisions above "kingdoms," called "domains," and there are three of them: 1) Eukarya (eukaryotic organisms with membrane-bound organelles), 2) Eubacteria (the "true" bacteria), and 3) this new one, the Archaea.

Reactions with Alcohols

Reactions of **acid chlorides** and **anhydrides** with alcohols parallel in every way their reactions with water. The carbonyl group in this case is reactive enough to react without catalysis (although the reaction goes even faster if a catalyst is present). The oxygen of an alcohol attacks the carbonyl group, electrons retreat to the oxygen, then come back and eject the leaving group (chloride or carboxylate), and finally the oxygen loses its extra proton. Look back at the mechanism of the reaction of these compounds with water and follow exactly the same process this time using an alcohol instead of water. What you will discover is that the product is an ester. Indeed, this is the preferred way of making esters in which neither half is cheap and volatile (and therefore suitable for use in excess). Below are some examples. Again, be sure you can write mechanisms for them.

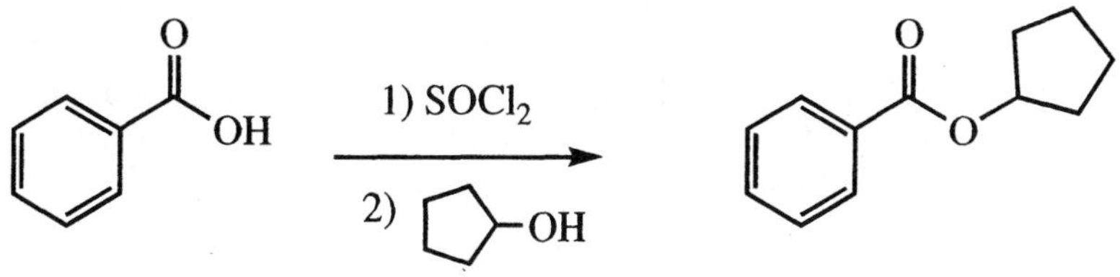

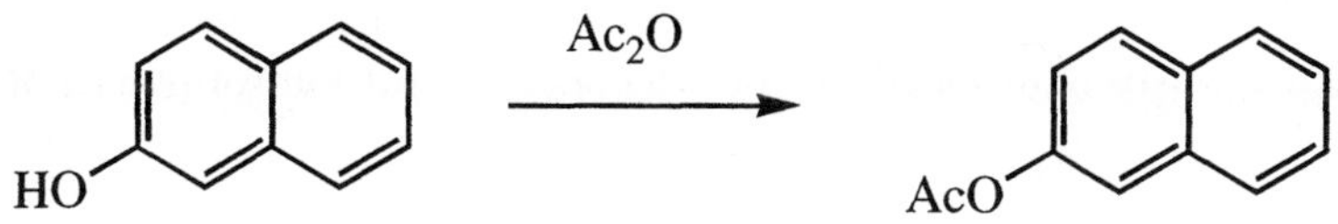

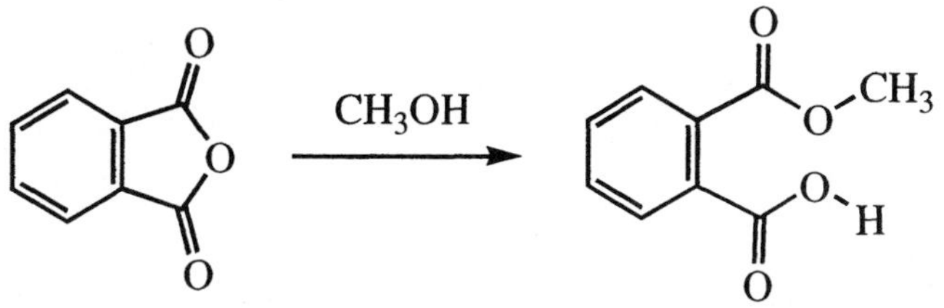

Esters can also react with alcohols, and again the reaction is an exact reflection of the reaction with water. Nothing at all happens in the absence of catalyst. If there is acid present, the carbonyl group can be protonated, making it a good enough electrophile to attract a neutral alcohol. If base is present, the alcohol can be deprotonated, making it a good enough nucleophile to attack a neutral carbonyl group. Again, you should look at these reactions, covered in detail above for the water case, and go through them yourself with alcohols. Now! What you will find is the following:

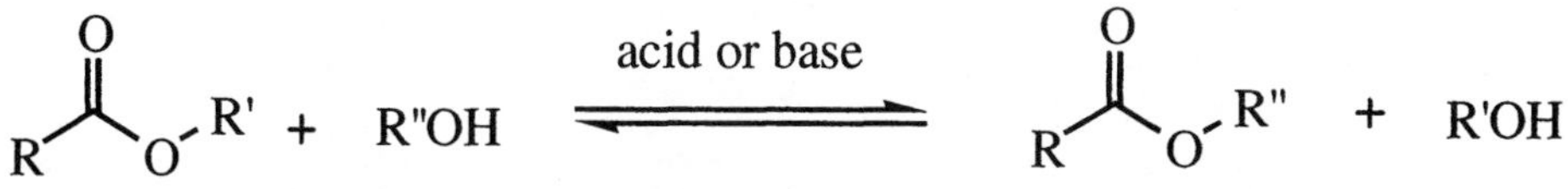

Just like the Fischer esterification or ester hydrolysis, this is an equilibrium that can be controlled by Le Châtelier's principle. By using a lot of R''OH, you can make the process go from left to right; in the presence of lots of R'OH it will go from right to left. The notion of trading one ester for another like this is called "transesterification."

Now, why would you ever want to do a transesterification? If you wanted the ethyl ester of some acid, you would just make it in the first place. No one in his right mind would first make the methyl ester and then change that into the ethyl ester. Indeed, this is true, and there are in fact very few examples of this reaction being done on purpose. But there are many examples of it happening to people by mistake. Consider, for example, the following reaction:

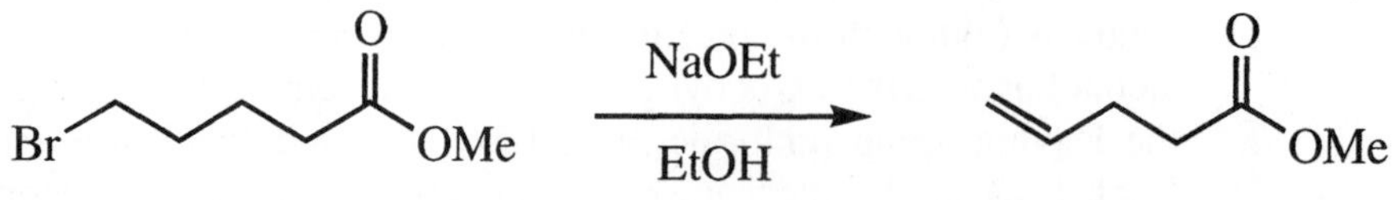

This is a standard elimination reaction, but it probably would not work especially well here. Why not? Your starting material is a methyl ester, and you have added ethoxide. What is to stop the ethoxide from attacking the carbonyl group and changing the methyl ester into an ethyl ester? Nothing! And it will happen, at least in some of the molecules. This leads to the possibility of ending up with two different alkene products, one a methyl ester and the other an ethyl ester. That complication could make isolation of your product difficult. You are better off making sure that the base you use for elimination matches the ester you started with: either use sodium methoxide for the reaction, or start with the ethyl ester. That way, if transesterification does occur, it will be invisible.

Test Yourself 4

Student Elmer Erlenmeyer attempted the following Williamson ether synthesis, and was surprised to obtain the wrong product. Show a mechanism for the formation of what he did get.

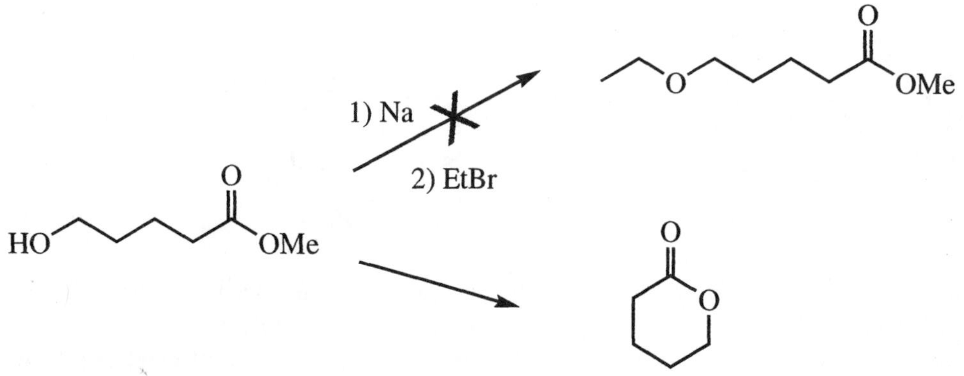

Grignard Reactions

Grignard reagents react with **acid chlorides** and **anhydrides** exactly as you might expect: nucleophile in, electrons out, electrons back in, leaving group out:

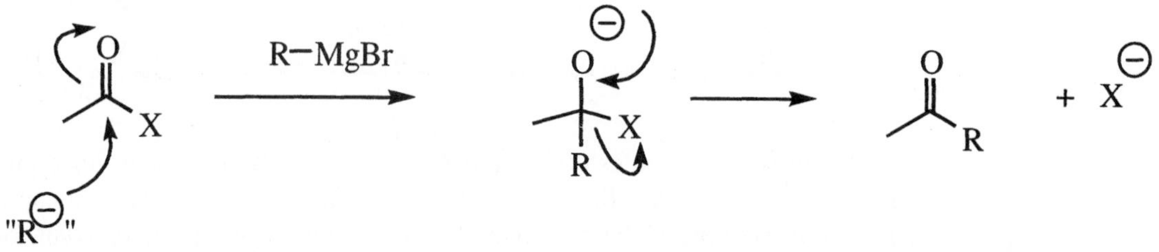

This brings up a problem, however: the product of this sequence is a ketone, and as we know, ketones also react with Grignard reagents—to make alcohols. So in the presence of excess

Grignard reagent, the product of this reaction will not be a ketone but in fact a tertiary alcohol, where two of the attachments are the same:

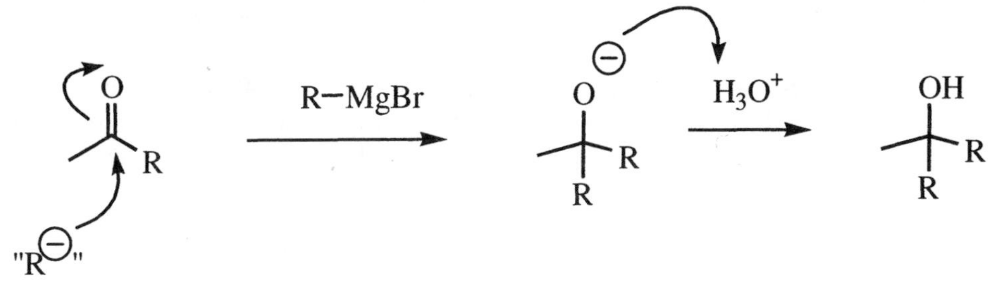

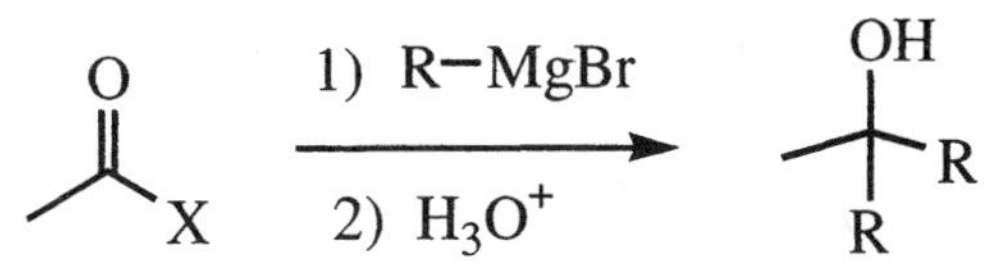

Net reaction:

So now we ask the standard question we always ask when we see two reactions happening in sequence: is it possible to stop at the halfway point? With careful measuring of reagents, could we make a ketone from an acid chloride?

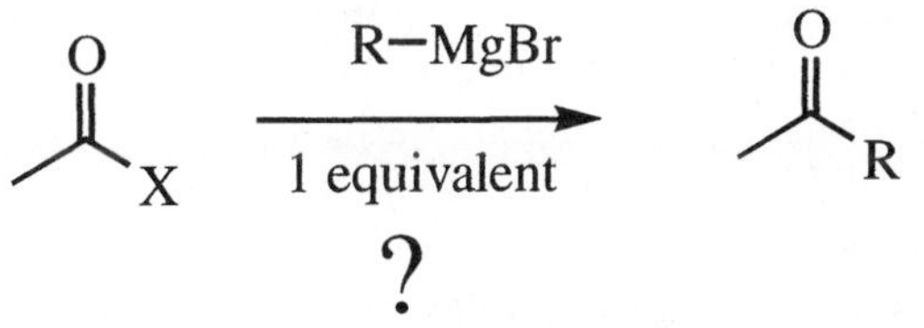

As usual, the answer to this can be deduced from consideration of the reactivity of the species involved (although we have seen some cases where our deduced conclusion gets contradicted by the facts). What do we know about acid chlorides as compared to ketones? Acid chlorides are more reactive, right? So after a few acid chloride molecules turn into ketones, and the next Grignard reagent is shopping for something to react with, it ought to prefer another acid chloride molecule to a ketone molecule. This suggests that it ought to be possible to convert a sample of acid chloride into the corresponding ketone with a limited amount of Grignard reagent, preventing much of the ketone from going on to alcohol.

Do the facts bear this out? Yes and no. Some ketones have been prepared this way, but the experimental logistics are somewhat cumbersome. The problem is this: although it is true that acid chlorides are more reactive than ketones, this is relevant only if the nucleophile is willing to wait for the best possible substrate to react with. It must try each, and then make a choice. But Grignard reagents are *so* reactive that they will react with the first carbonyl group they see, not waiting to determine whether there is an even better one nearby. In that situation you get a statistical mixture of unreacted acid chloride, ketone, and tertiary alcohol from this reaction. For that reason, this particular procedure for making ketones is not very common. From a practical point of view, you should write the tertiary alcohol as the sole product of a Grignard reagent reacting with an acid chloride or anhydride, always assuming an excess of the Grignard reagent.

Nevertheless, the logic by which we concluded that we *should* be able to make ketones from acid chlorides turns out to be correct. In order to make the possibility a reality, we have to modify the nucleophile so that it is pickier. Rather than one that reacts with the first carbonyl group it sees, we want to use a nucleophile that *will* try all the possibilities and wait for the best one. This has been accomplished by replacing magnesium in the Grignard reagent with copper or cadmium. I don't want to burden you with the details, so we will leave the subject at this point.

The probable reaction of Grignard reagents with **carboxylic acids** can also be deduced from what we already know. We have been treating Grignard reagents as if they contained carbons with negative charges (i.e., carbanions). As we know, carbanions are very strong nucleophiles, and also very strong bases. Carboxylic acids are among the most acidic of organic compounds. Without a doubt, the first thing that will happen when you mix a carboxylic acid with a Grignard reagent is proton transfer:

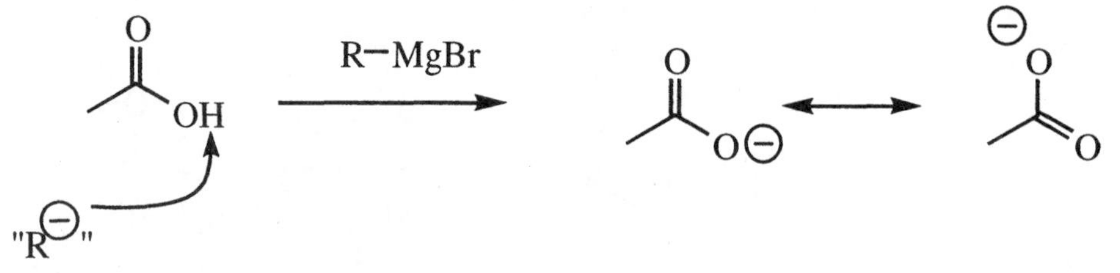

The carboxylate anion that results from this has a negative charge, so it is not likely to be very friendly toward a nucleophile that has its own negative charge. It would take a very strong nucleophile indeed to force its way into the carbonyl group of a carboxylate anion. (Such nucleophiles exist: organolithium compounds are capable of this reaction, but we will not cover them in this text.) Thus, we should expect the carboxylate anion to sit around in solution and wait for us to add acid at the end of the reaction, at which point we would recover the carboxylic acid we started with. Because the overall result of the reaction is the same carboxylic acid we started with, we usually say that carboxylic acids do not react with Grignard reagents, even though in fact they do donate a proton and then get one back upon acidification of the mixture.

Test Yourself 5

Show the mechanism of the following reaction and explain why it occurs the way it does.

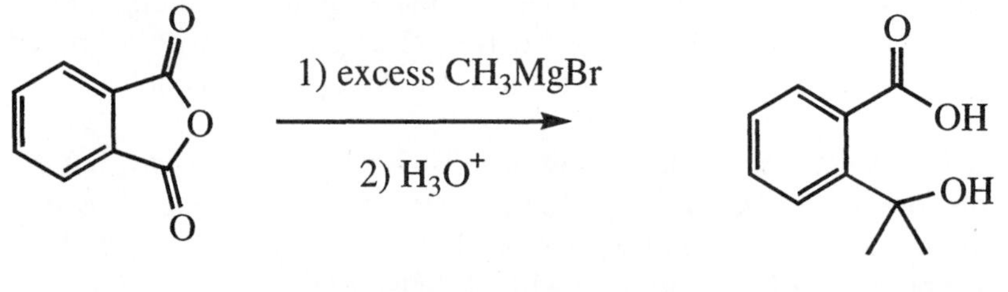

We can also predict the reaction of Grignard reagents with **esters**. The first part will be just like the reaction with acid chlorides: nucleophile in, electrons out; electrons back in, leaving group out, as shown below.

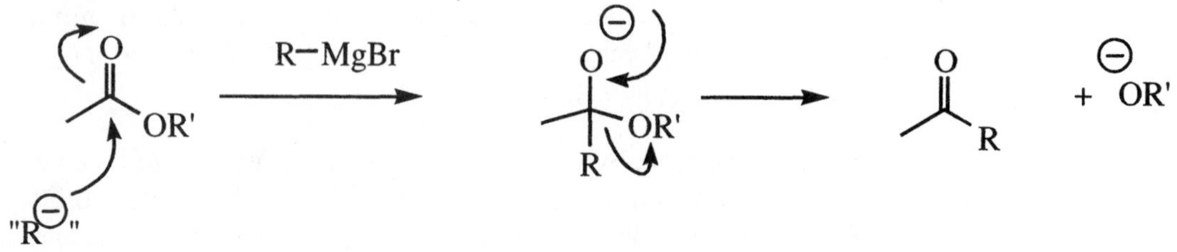

Once again we recognize that the product ketone can react with more Grignard reagent, resulting once more in a tertiary alcohol:

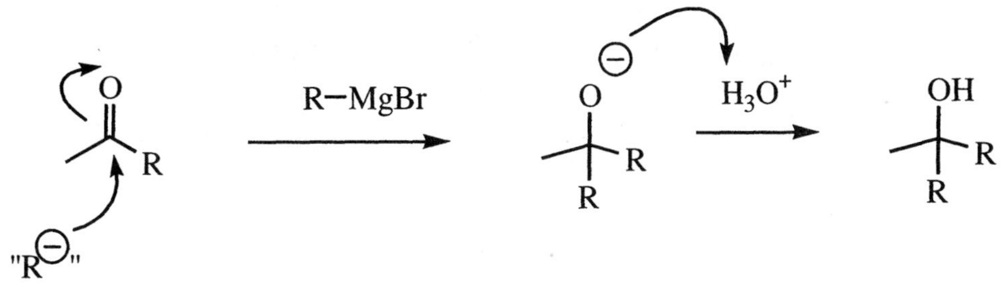

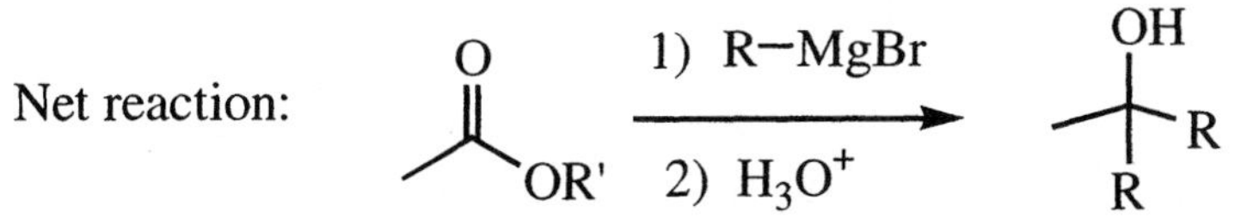

Net reaction:

And once again we ask if it might be possible to stop the reaction at the halfway point. This time we notice that ketones are *more* reactive than esters, so using a limited amount of Grignard reagent should *not* produce ketone; further, it should not even be possible to devise a pickier reagent that will allow us to isolate the ketone. (Students taking advanced chemistry courses may discover that such reagents have in fact been devised, but they rely on clever ways of bypassing the factors we have just considered.) Thus, we should always expect Grignard reactions of esters to consume an excess of Grignard reagent and result in tertiary alcohols.

Reductions

Hopefully you were able to predict at least some of the reactions we covered in the last several sections. Let's see how you make out with reductions.

We have studied two reducing agents: sodium borohydride ($NaBH_4$) and lithium aluminum hydride ($LiAlH_4$, or LAH). We know that LAH is by far the stronger of the two.

What will happen when one of these reagents reacts with an **acid chloride**? This is easily predictable and mirrors exactly the Grignard case just covered:

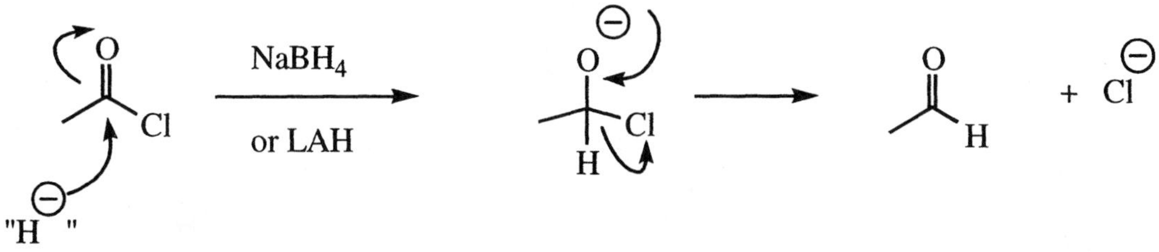

Again, we arrive at a product (an aldehyde) that can react with more of the reducing reagent, so we can anticipate that if excess reducing agent is present, the reaction will continue, creating eventually a primary alcohol:

NaBH₄ or LAH reaction scheme (aldehyde reduction):

$$\text{(acetaldehyde with "H}^{\ominus}\text{)} \xrightarrow[\text{or LAH}]{\text{NaBH}_4} \text{(alkoxide intermediate)} \xrightarrow{\text{H}_3\text{O}^+} \text{(alcohol)}$$

Net reaction:

$$\underset{\text{R}}{\overset{\text{O}}{\underset{\text{Cl}}{\|}}}\ \xrightarrow[\text{or, NaBH}_4,\ \text{EtOH}]{1)\ \text{LiAlH}_4\quad 2)\ \text{H}_3\text{O}^+}\ \text{RCH}_2\text{OH}$$

Again, we can ask the question, "Can we stop the process part way?", and again we can answer that it should be possible, because acid chlorides are more reactive than aldehydes. While one can, in theory, do this by careful measuring of reagents, it is more efficient to take advantage of a reagent that is pickier, one strong enough to react with an acid chloride but not with an aldehyde. Such reagents exist, but we will not cover them.

How will these reducing agents react with **esters**? First there is another question: *will* such reagents react with esters? Esters are less reactive than acid chlorides, aldehydes, and ketones. We know that LAH is ferocious and will react with anything it finds, but NaBH₄ is pretty mild. Indeed, in Chapter 18 we suggested that reaction with aldehydes and ketones is about as far as NaBH₄ can go. This is true: in general, NaBH₄ will not react with any carbonyl group less reactive than a ketone, so esters tend to ignore it. But LAH will work fine.

Now, *how* will LAH react with esters? Again we can figure this out. Reaction will occur like all the other reactions on a carbonyl group: nucleophile in, electrons out; electrons back in, leaving group out; another hydride in, wait for proton.

Ester reduction mechanism scheme:

$$\text{(ester with "H}^{\ominus}\text{)} \xrightarrow{\text{LAH}} \text{(tetrahedral alkoxide intermediate, OR)} \longrightarrow \text{(aldehyde + }^{\ominus}\text{OR)}$$

$$\text{RCH}_2\text{OH} \xleftarrow{\text{H}_3\text{O}^+} \text{RCH}_2\text{O}^{\ominus}$$

Can we stop this one part way? Presumably not: the intermediate aldehyde is *more* reactive than the ester we started with. (Again chemists have come up with clever ways of bypassing these considerations, but we will not cover them. The logic is still valid.)

Notice that the reduction of an ester results in *two* alcohols. One comes from the carbonyl side of the ester and is guaranteed to be a primary alcohol; the other comes from the "alcohol" side and can be anything—it is, indeed, the same alcohol you would get by hydrolyzing the ester.

$$\underset{R}{\overset{O}{\|}}\overset{}{C}\!-\!OR' \quad \xrightarrow[\text{2) } H_3O^+]{\text{1) LAH}} \quad RCH_2OH \qquad \textbf{HOR'}$$

How do **carboxylic acids** themselves react with reducing agents? The carbonyl group of an acid is less reactive than that of a ketone, so we should (and do) find that $NaBH_4$ is unreactive toward them. ($NaBH_4$ *does* remove the acidic proton, but after acidic workup there is no net reaction, so we generally describe $NaBH_4$ as not reacting with acids.) What about LAH? H^- is a strong nucleophile and also a strong base. As in the case of Grignard reagents, we should expect the first thing that will happen is an acid–base reaction:

$$\underset{R}{\overset{O}{\|}}\overset{}{C}\!-\!OH \quad + \quad "H^{\ominus}" \quad \longrightarrow \quad \underset{R}{\overset{O}{\|}}\overset{}{C}\!-\!O^{\ominus} \qquad H_2$$

This should be effectively irreversible, so we are stuck now with a carboxylate anion. As with Grignard reactions, we should anticipate that an anion like this will want nothing to do with a negatively charged nucleophile, and so will simply sit around until the reaction is over. This is good thinking, but it turns out to be wrong. For very complex reasons that I'd rather not describe, it turns out that LAH *can* react even with a carboxylate anion, so the reaction does *not* stop here. The reaction is complex, and no one seeing it for the first time would predict the actual result, but here it is:

$$\underset{R}{\overset{O}{\|}}\overset{}{C}\!-\!O^{\ominus} \quad \xrightarrow{\text{LAH}} \quad \underset{R}{\overset{H \; H}{\|}}\overset{}{C}\!-\!O^{\ominus} \quad \xrightarrow{H_3O^+} \quad RCH_2OH$$

I suspect it would do more harm than good to go through how this occurs, so this is one reaction you may as well just memorize. Notice that the reduction of a carboxylic acid to a primary alcohol requires *three* equivalents of H^-—that is, 3/4 of an equivalent of LAH: the first one removes the acidic hydrogen of the acid, and the next two attach to the carbon of the carbonyl group. The hydrogen on the oxygen of the alcohol did *not* come from LAH, but rather from water added during workup.

Test Yourself 6

Show the product of reaction of the following anhydride with either LAH or $NaBH_4$. Assume excess reducing agent is present. Do it now!

Test Yourself 7

Show the expected products of the above reaction when NaBD$_4$ and LiAlD$_4$ are used as the reducing agents.

In summary, with a few exceptions, the reactions of carboxylic acids and their derivatives are very predictable and sensible once you understand carbonyl groups and the factors that influence reactivity.

Problems
(Answers are provided at the end of the chapter for italicized problems.)

1. Show the organic products of the following reactions. Write NR if there is no reaction.

a.

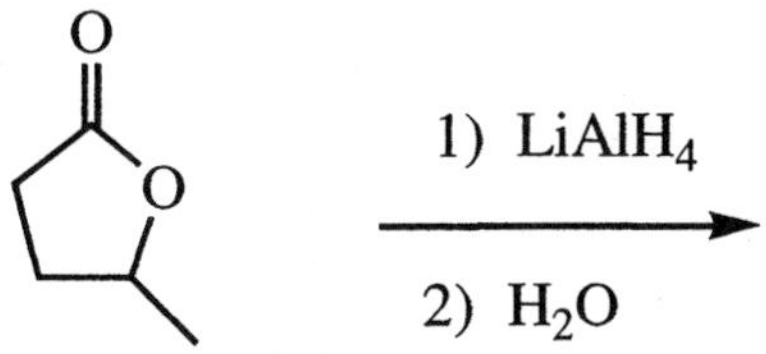

1) LiAlH$_4$

2) H$_2$O

b.

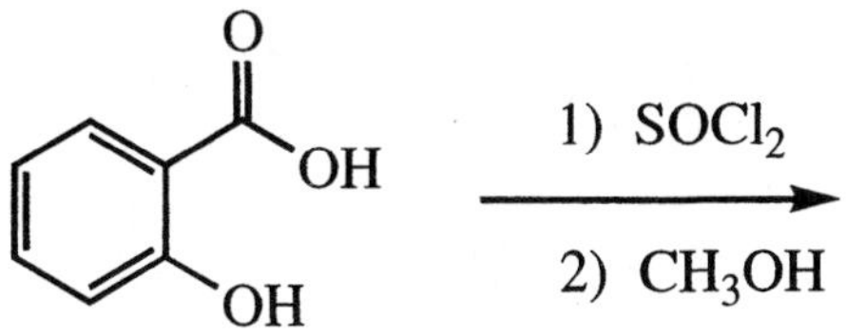

1) SOCl$_2$

2) CH$_3$OH

c.

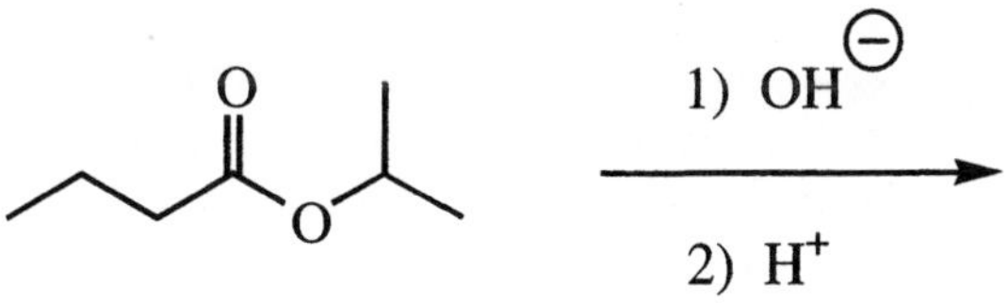

1) OH$^{\ominus}$

2) H$^+$

two products

d.

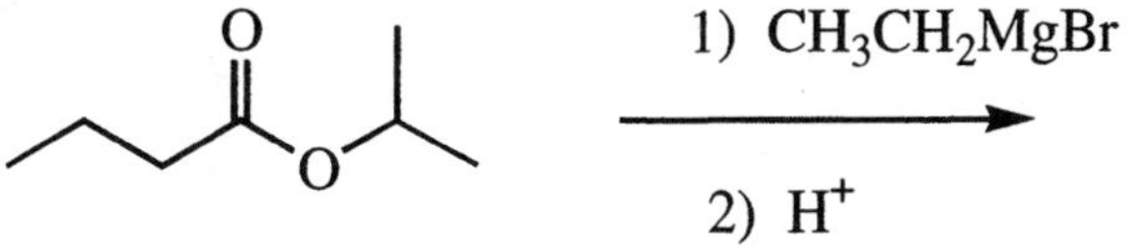

1) CH$_3$CH$_2$MgBr

2) H$^+$

two products

e.

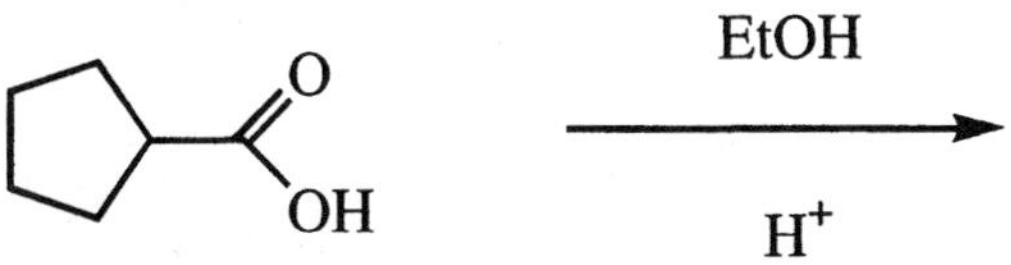

EtOH

H$^+$

f.

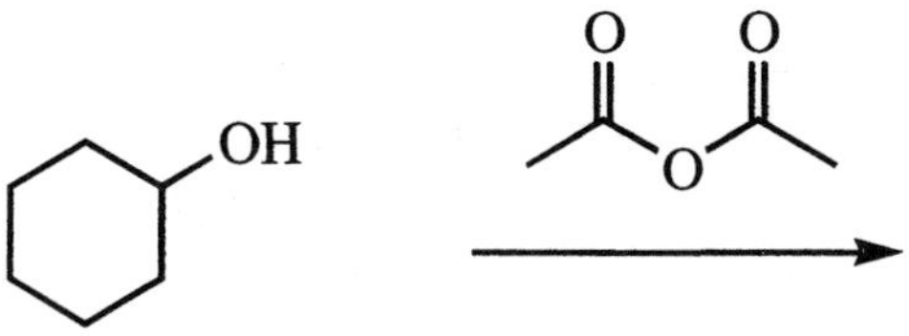

two products

2. What carboxylic acids and alcohols are the following esters derived from?

a.

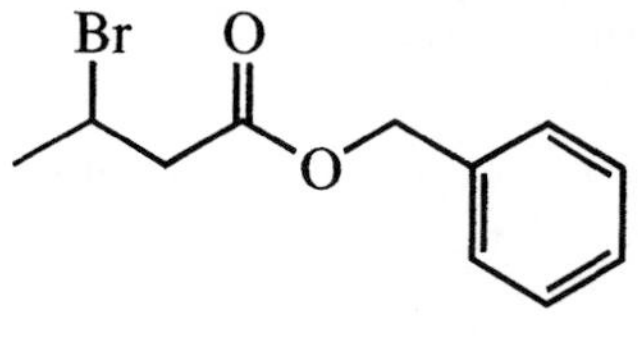

b.

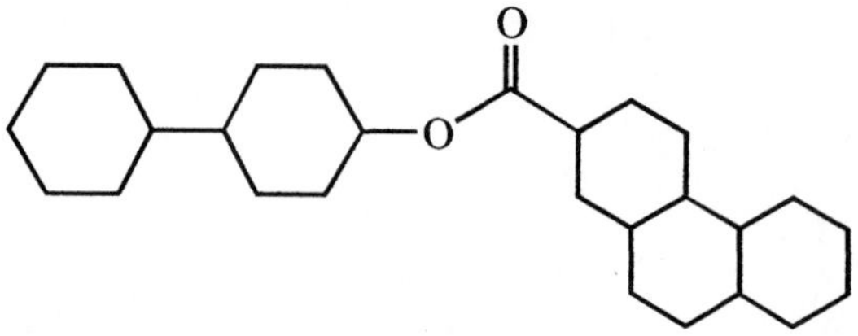

c.

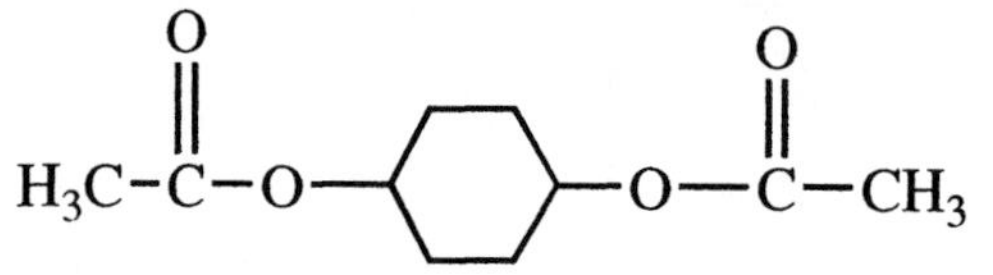

3. *The following compound cyclizes upon sitting around. Imagine there is a trace of acid catalyst present to help the process along, and show a reasonable mechanism for this reaction.*

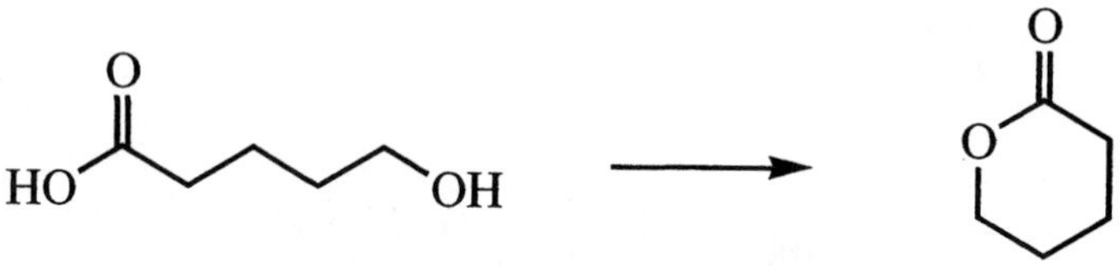

4. Show the expected mechanism for the following reaction:

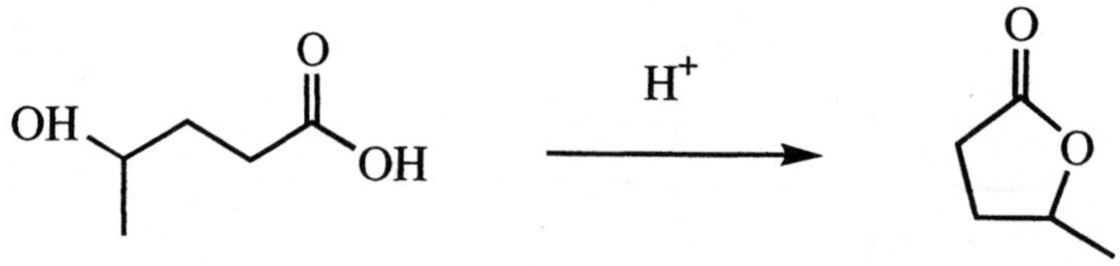

5. Using electronegativity values, explain why LiAlH$_4$ is a more potent hydride reducing agent than NaBH$_4$.

Na	0.9
Li	1.0
Al	1.5
B	2.0
H	2.1
C	2.5
O	3.5

6. Suggest a practical synthesis of the following molecule, using anything you wish containing three carbons or less as your only source of carbon.

7. *Suggest a synthesis of the following lactone starting with 1-methylcyclopentene and anything else you wish containing three carbons or less.*

8. Show the mechanism for base hydrolysis of the following molecules:

a.

b.

Do not forget to acidify the reaction mixtures in the end.

9. *Suggest a mechanism for the following transformation:*

10. Show what reagent(s) you would use to accomplish the following transformation in *one step*. Notice that one ester function is converted into an alcohol but the other is preserved. Hint: Do not try hydrolysis. Look carefully at how the ester groups are attached and consider an approach based on transesterification.

11. Suggest a mechanism for the following transformation:

12. *When methyl benzoate is treated with NaOH in water enriched with ^{18}O, the benzoic acid formed contains ^{18}O and the methanol does not. Under the same conditions methyl 2,4,6-trimethylbenzoate hydrolyzes to give unlabeled acid and ^{18}O-enriched methanol.*

a. *Show mechanisms that explain both reactions.*
b. *Which ester will hydrolyze more slowly?*
c. *What other types of esters are likely to hydrolyze by the second mechanism?*

13. As you know, formyl chloride cannot be made, so formate esters are harder to make than other esters. However, sometimes it is desirable to make them. One way to accomplish this is to treat acetyl chloride with formic acid to produce **A**, which then reacts with an alcohol to give a formate ester. Identify **A**, propose a mechanism for its formation, and suggest a

mechanism for its reaction with alcohols. Be sure you explain why the product you get is a formate ester and not some other ester.

14. Suggest a method of accomplishing the following transformation, using anything you wish containing three or fewer carbons. It can be achieved in two steps.

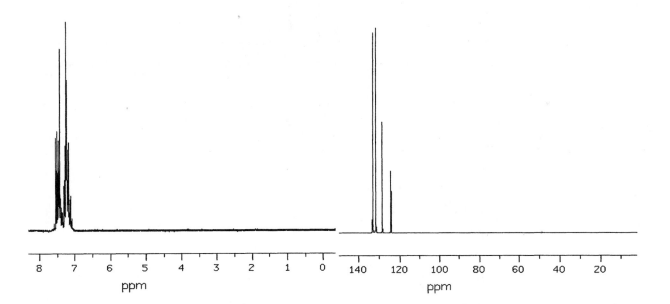

15. **A** is an organic liquid. When **A** is treated with magnesium followed by formaldehyde and then water, **B** is formed. **B** is converted into **C** by treatment with sodium dichromate in sulfuric acid solution. **C** can also be formed directly from **A** by treatment with magnesium followed by carbon dioxide and then aqueous acid. When **B** is treated with PCC, **D** is formed. **D** reacts with methyl magnesium iodide followed by water to give **E**, which reacts with either PCC or sodium dichromate in sulfuric acid solution to give **F**. **B** reacts with PBr₃ to give **G**, which reacts with magnesium followed by formaldehyde and then water to give **H**, an isomer of **E**. Finally, **B** and **C** react with each other, with a trace of acid catalyst, to produce **I**. **A** through **I** are all different, neutral, organic compounds. Identify them from the information above together with the following spectra.

A.

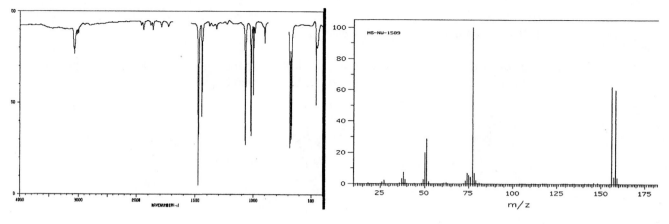

B.

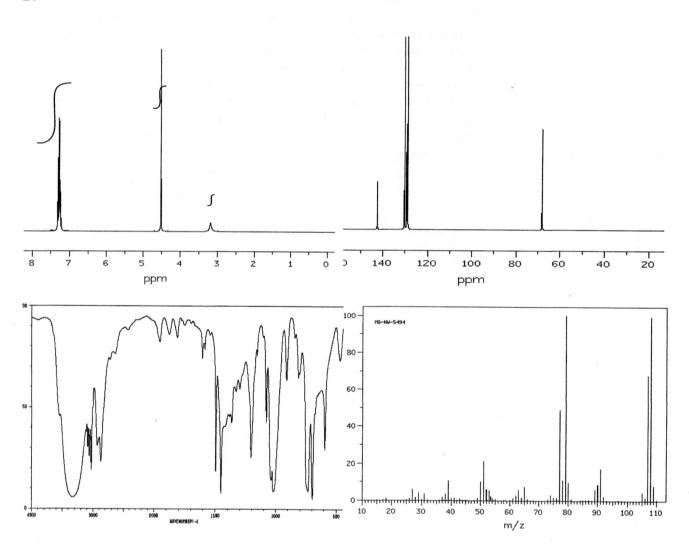

C.

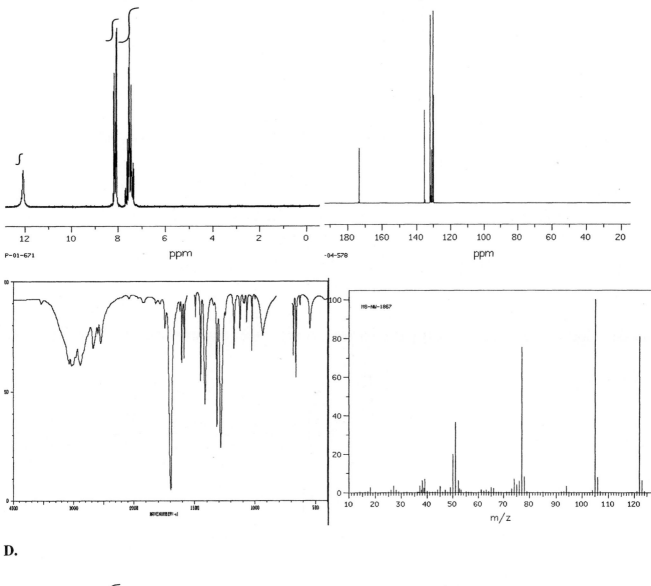

P-01-671

-04-578

HS-NW-1867

WAVENUMBER( )

m/z

D.

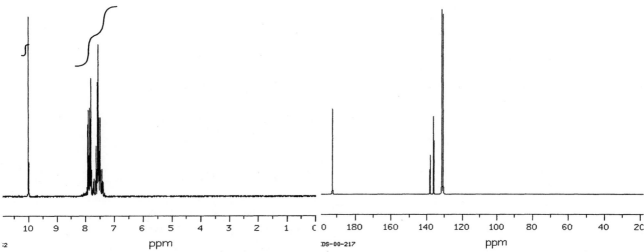

DS-00-217

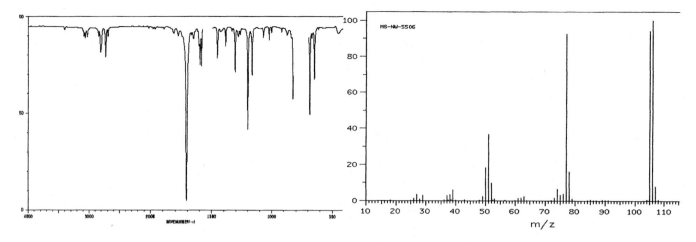

E. No NMR spectra available

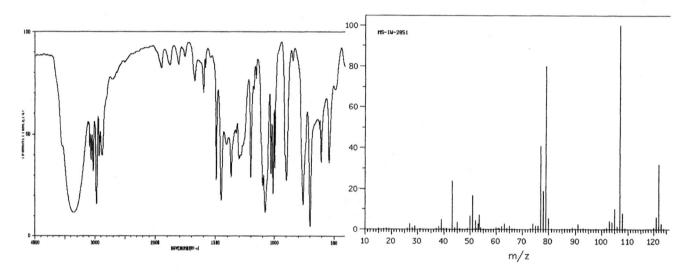

F.

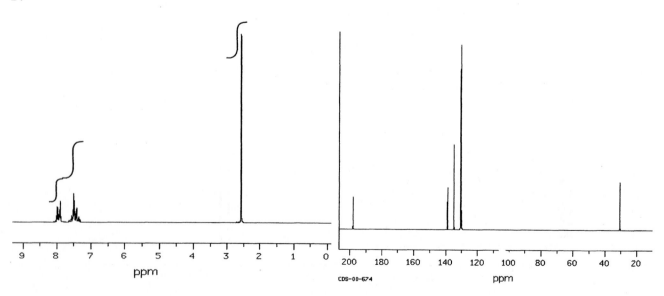

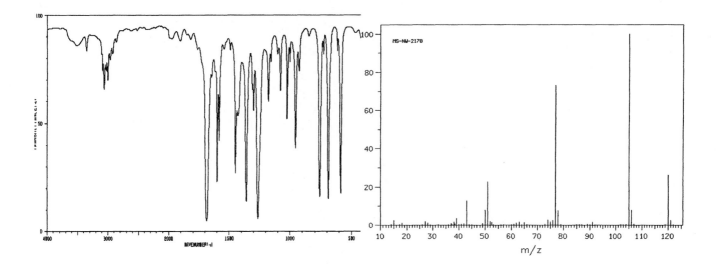

G.

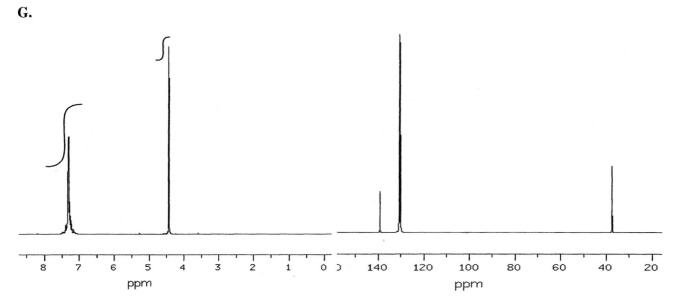

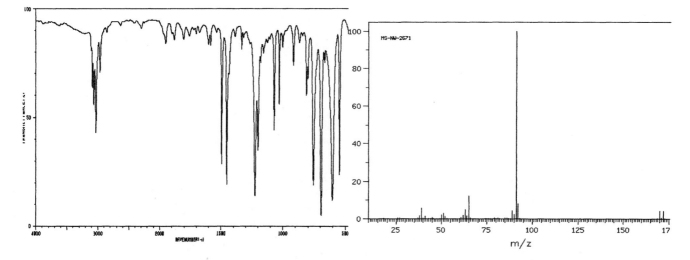

H.

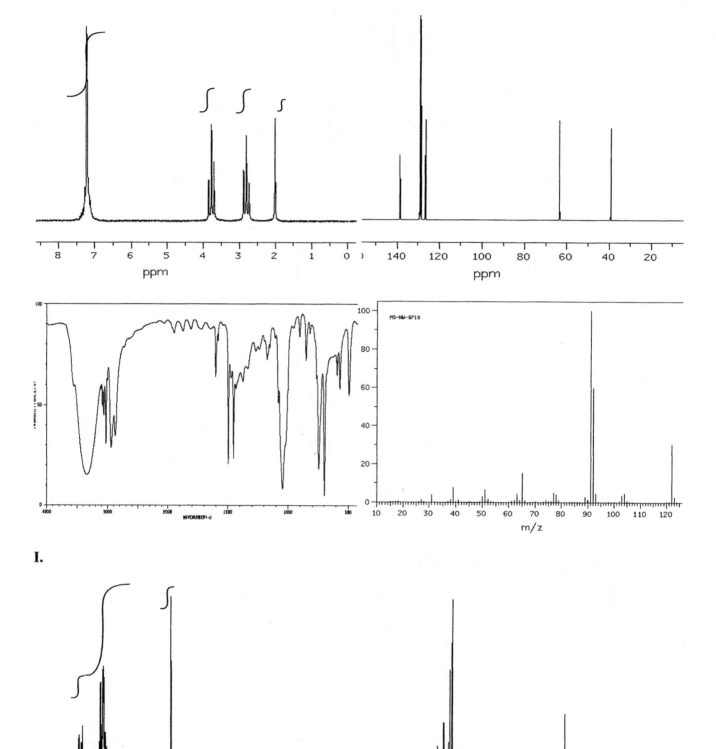

I.

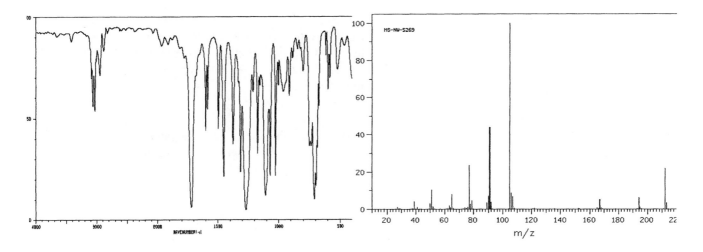

Selected Answers

Internal Problems

Test Yourself 1.

The first part you should have been able to do way back in Chapters 8 and 9, when we introduced resonance. The anion of phenol is resonance stabilized. You should be able to show four places that share this negative charge (although three of them are carbon atoms).

The rest of the problem is an exercise in using your pK_a chart. You should quickly discover that both of the following reactions will go well to the right, so using NaOH as in the above example will not accomplish a separation of these two compounds: if one loses its proton, so will the other!

Separating these two (at least by a technique similar to the one we just discussed) requires making one of the compounds into an ion while the other remains intact. Again, this should not be difficult: the carboxylic acid has a pK_a of 5, while phenol has a pK_a of 10. So any base with a pK_a somewhere between those values should effectively remove protons from the carboxylic acid while leaving the phenol alone. Therefore we look on the base side of the pK_a chart for something between 5 and 10. The most readily available species there is $HOCOO^-$, the bicarbonate ion, often written HCO_3^-. As its sodium salt, this is sold in every grocery store under the name "baking soda."

So here is what we do: we dissolve our two substances in ether, make a solution of sodium bicarbonate, combine these two solutions, and then shake the mixture. The phenol will stay in the ether, as itself, and the carboxylic acid will move into the water as its anion. We then separate the two layers, set the ether aside as "ether solution A," add acid to the water layer (careful, it will foam!) and extract the water with fresh ether to remove from the water anything that is now ether soluble. This becomes "ether solution B." The ether is now evaporated from ether solutions A and B and we have our two compounds separated (the phenol from solution A and the acid from solution B).

Test Yourself 2

Amides are the least reactive of all, since nitrogen is less electronegative than oxygen and therefore more willing to share its electrons with the neighboring carbonyl group.

Test Yourself 3

Enones have more single bond character than ordinary ketones due to the extra resonance structure.

Test Yourself 4

Test Yourself 5

carboxylate anion
will not react with Grignard

Test Yourself 6.

Let's start with NaBH$_4$. Will it react in this case at all? Anhydrides are *more* reactive than aldehydes, so we should expect that it will, and the reaction should proceed just like the reaction with an acid chloride:

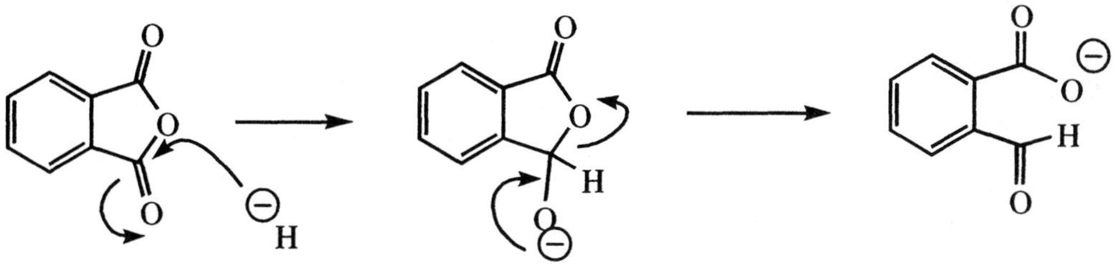

The product so far contains both an aldehyde group and a carboxylate anion. The aldehyde we know can react with NaBH$_4$, and it will. The carboxylate anion is *less* reactive than a ketone and will not react with any nucleophile we know except LAH, so it will sit around until the end of the reaction, at which point it can pick up a proton when we acidify the mixture.

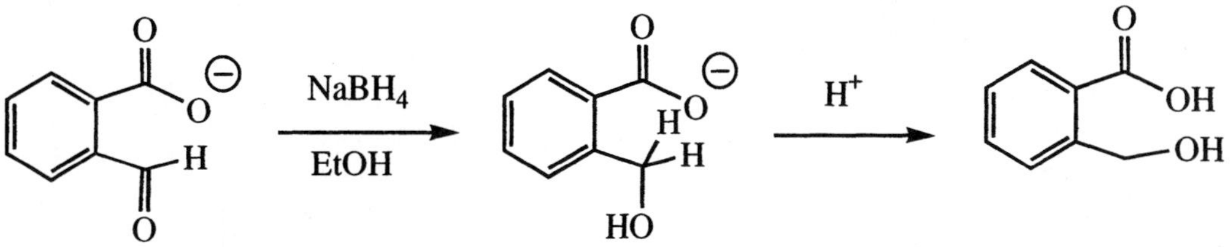

Net:

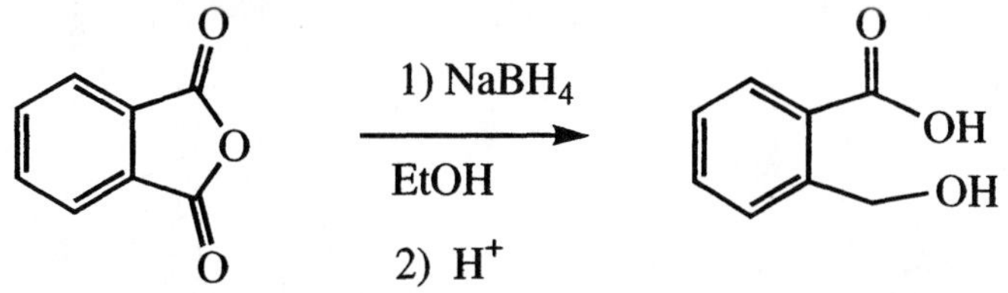

With LAH the reaction would begin in exactly the same way, but we now know that LAH *will* react with carboxylate anions. Thus, the final product with LAH will be the following:

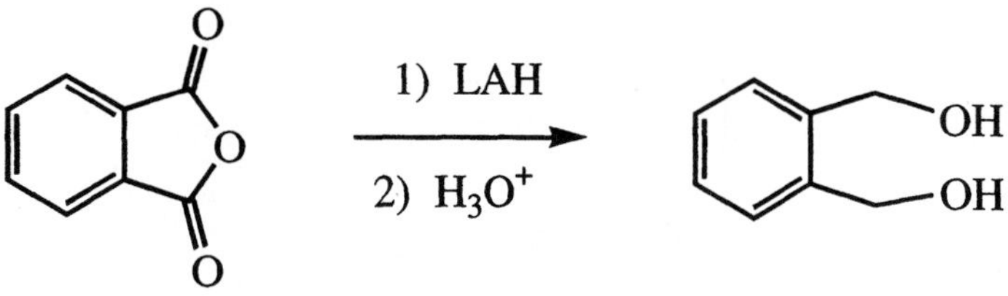

Test Yourself 7

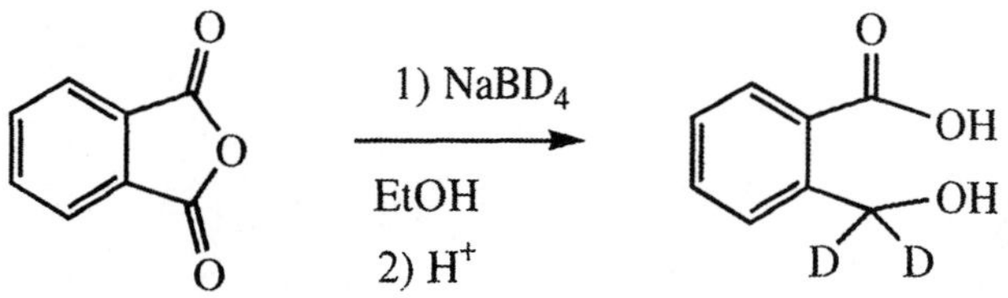

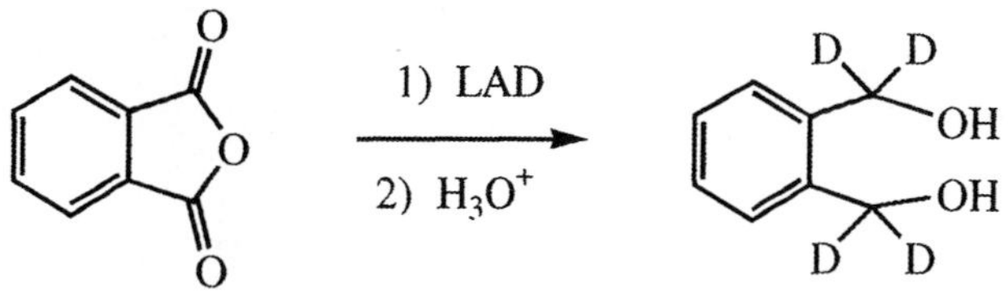

Chapter 19 Carboxylic Acids and Derivatives 619

End of Chapter Problems

3.

7.

Notice how the carboxylic acid protects *itself* from the Grignard reagent—IF you use two equivalents of the Grignard reagent. The first equivalent of Grignard reagent reacts (of course) with the carboxylic acid, destroying the Grignard reagent and making a carboxylate anion. The anion part of the molecule is now immune to further attack by a Grignard, and the remaining Grignard reagent can attack the ketone. This approach is OK, but it is wasteful in terms of Grignard reagent. If the Grignard had been one you worked hard to make, you would not want to use this approach, because it wastes half of it. In this particular case the Grignard is readily available, so it is not a problem to throw some away.

Of course, the aldehyde could have been protected as an acetal, but that would require two extra steps.

9.

12.

a.

Too hindered to do this. Only choice is

+ CH$_3$^{18}OH

b. This second process is surely slower, because if this S$_N$2 were faster than the usual reaction, it would always happen, but we know it does not.

c. We would expect the same thing whenever the ester group is particularly hindered.

Chapter 20
Carbonyl Chemistry III: Reactions Involving the α Carbon

For two chapters we have been concentrating on reactions that carbonyl compounds undergo at the carbonyl group. In all these cases the carbonyl compound has been the electrophile. While that theme will continue in this chapter, we want to expand our view to see how carbonyl compounds can also act as nucleophiles, either toward other nucleophiles we have encountered or toward other carbonyl groups. When we finish, we will have learned enough reactions to be able to understand the strategy of carbohydrate metabolism, which we will then look at in detail. In this chapter we will

- Expand on our previous discussions of enols

- Learn how to alkylate carbonyl compounds at the α position

- Discuss aldol and Claisen condensations

- Discover the Michael reaction

- Learn about active methylene compounds and their subsequent decarboxylation.

ENOLS

We have previously noted that carbonyl compounds are unusually acidic: a proton on the α carbon can be removed by a fairly weak base because the "enolate" anion generated is resonance stabilized:

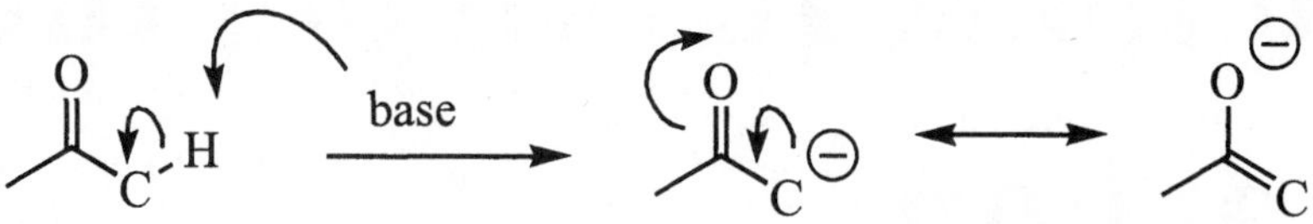

Protons α (meaning, on the carbon next door) to aldehydes and ketones have pK_a values of about 19, nearly the same as alcohols, so alkoxides and hydroxide are capable of removing these protons. Protons α to esters are less acidic, with pK_a values of about 25. The reason for the lower acidity of esters is that here the carbonyl group is already distracted by the lone pair on the oxygen of the ester and cannot give its full attention to stabilizing an anion on the other side of the carbonyl group.

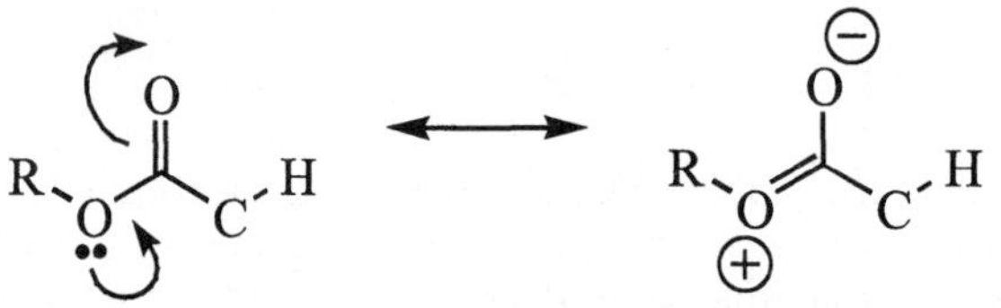

Alkoxides and hydroxide are still able to remove protons α to esters, but a much smaller percentage of the molecules is deprotonated at any given time; that is, the equilibrium lies farther in the direction of the anion in the case of aldehydes and ketones than it does in the case of esters.

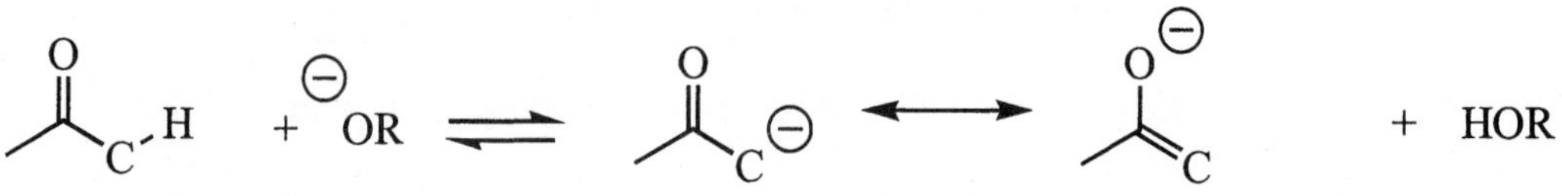

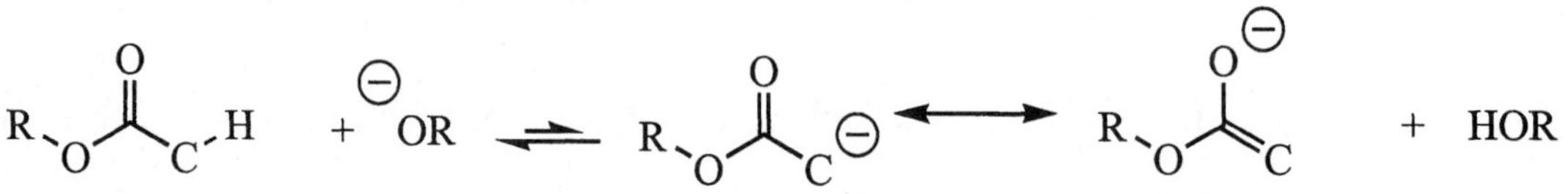

While we have already looked at this equilibrium from left to right, we have not yet considered the reverse reaction. When an enolate anion is protonated, the new proton could attach at either of the atoms that share the charge—either the carbon or the oxygen:

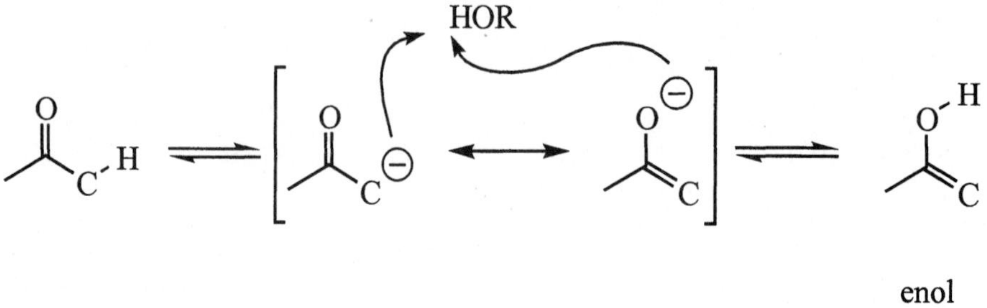

enol

If the carbon gets it, you are back to the ketone (aldehyde, ester) you started with; if the oxygen gets the proton, you have generated an enol. The point is that carbonyl compounds, in the presence of base, are in equilibrium with the corresponding enols; and this equilibrium almost always has a strong preference for the carbonyl compound over the enol.

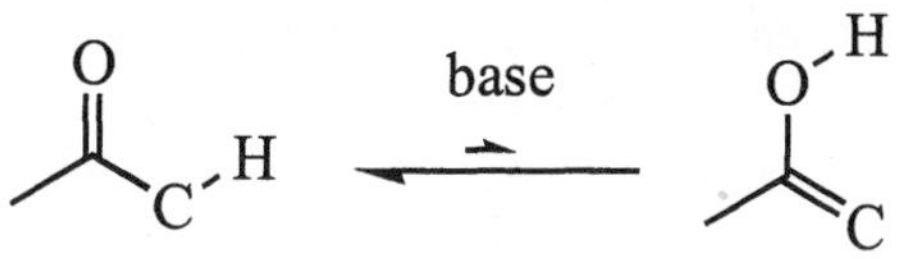

We have also seen that carbonyl compounds can be protonated in acid solution:

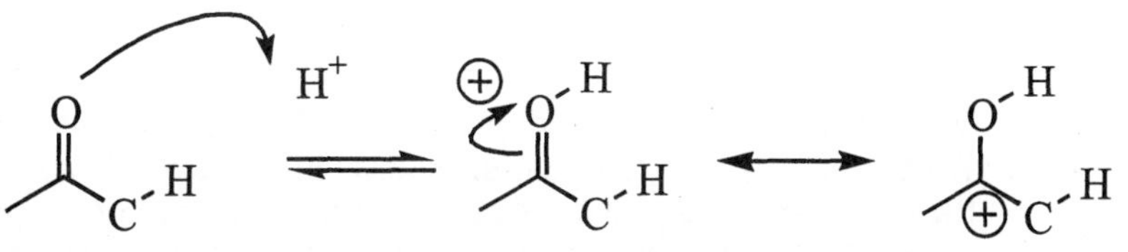

When a proton is removed from the resulting cation, it could be the proton on the oxygen, which gets you back where you started from, or the α proton, which makes the enol:

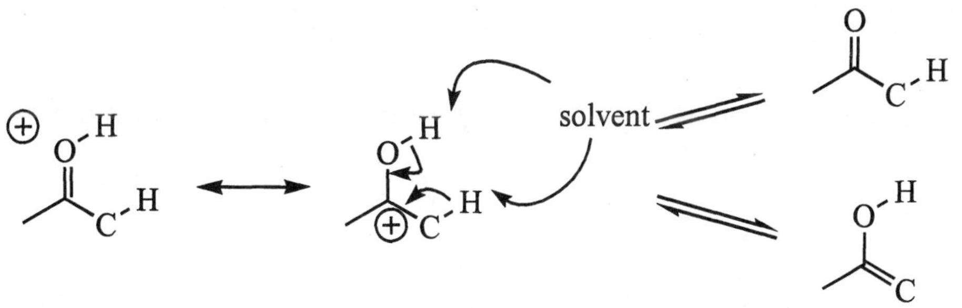

The conclusion is that in acid solution, as well as in base, carbonyl compounds are in equilibrium with their enols. In other words, this equilibrium can scarcely be avoided. Any time you run a reaction where it would appear that an enol is a product, you will in fact isolate the corresponding carbonyl compound (recall acid-catalyzed addition to alkynes in Chapter 12). More relevant to this chapter, almost any time you have a carbonyl compound, a tiny fraction of it will be in the enol form. Although the amount in the enol form at any particular moment is extremely small, usually considerably less than 1%, it is also true that, over time, every molecule will briefly become an enol and then go back again. You may recall that the same is true of hydrates. We conclude that carbonyl compounds are restless creatures, and even though we know they are best off as themselves, *they* do not always remember that, and periodically each one tries being a hydrate or an enol and then quickly learns its lesson and returns to its original state.

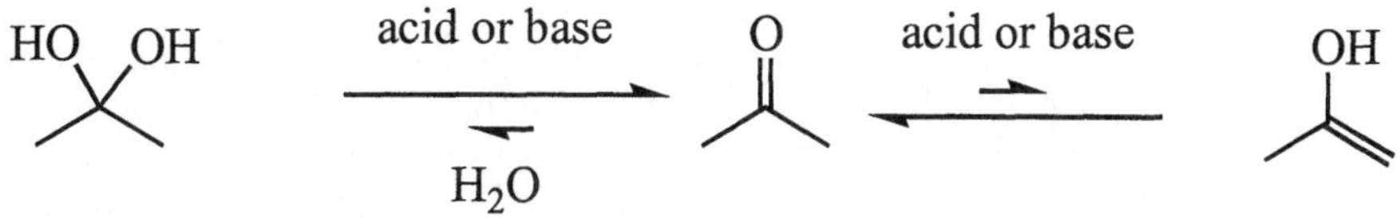

ORM: Under "Carbonyl Compounds," check out "enolization." You can watch it in both acid and base.

What are the consequences of this? We have already discovered some consequences of the hydrate equilibrium: it is through this process that aldehydes are oxidized to carboxylic acids, and if we want to avoid such an oxidation, we need to keep water away from the aldehyde (recall PCC). The consequences of the enol equilibrium are that hydrogens on the carbon atom α to a carbonyl group are insecure: they occasionally wander off and are replaced by new ones. This means that both the identity of the α hydrogens and their stereochemistry can change. For example, you can change all the α hydrogens into deuteriums by incubating a ketone in D_2O in the presence of either acid or base. One at a time, each α hydrogen comes off as an enol is made, and then a new one, almost always a D, replaces it when the enol returns to being a ketone.

You should be able to show the mechanism for this properly in both acid and base. Remember: in base, all your structures must be either negatively charged or neutral; in acid, they are either positive or neutral. In base, you remove a proton, then put one back. In acid, you first add a proton, then you remove one. Practice writing these mechanisms now!

Test Yourself 1

The following compound exchanges *five* protons with D_2O in acid or base. Explain why, and show mechanisms. Do it!

There is another consequence of the fact that carbonyl compounds are in equilibrium with their enols: enols are flat at the α carbon. If you have a compound that has a stereocenter adjacent to a carbonyl group, you should not be surprised when, over time, it racemizes. This is because, when the enol (or enolate) picks up a proton at the α carbon, the proton may come randomly from either the top or the bottom.

As a general rule, the ketone form is preferred over the enol to such an extent that you cannot really detect the presence of any enol; you infer that it must be there from the kinds of reactions I have described above. However, there are a few cases you should be aware of in which an enol is present in detectable amounts. All of these involve situations where the double bond of the enol finds itself in conjugation with some other double bond(s), and the system is therefore more stable than normal. Ketones with a carbonyl group in the β position, for example, usually have a significant amount of enol in their equilibrium mixture:

This particular type of enol benefits not only from conjugated double bonds but also from its ability to form an intramolecular hydrogen bond in a nice six-membered ring.

Another enol that is especially stable is phenol. In this case the enol double bond is part of a benzene ring; in the ketone form, the benzene-ring conjugation would no longer exist. Benzene rings are so special that this equilibrium barely deserves the name. There is effectively no ketone form to consider.

ENOLATE ANIONS

The term "enolate" refers to the anion you get upon removing a proton from an enol. This is, of course, the same anion you get by removing the α hydrogen from a carbonyl compound, so anions in which the negative charge is α to a carbonyl group are also called enolates.

As we have discussed before, resonance-stabilized anions of this type are similar in stability to alkoxide anions—ketones have pK_a values of about 19, alcohols about 17—so, not surprisingly, they do most of the things alkoxide anions do. Let's review some of the things alkoxides do, and see how enolates behave analogously.

To save space, I will draw the enolate anions in only one resonance form, but you should always recognize that ions like this are resonance stabilized, and a full description of them (on an exam, for example) would necessarily include both structures. This is important, because you need to remember this resonance stabilization, and you need to convince your professor that you are aware of it.

Nucleophiles in S_N2 Reactions: Alkylation

The first reaction of alkoxides that we studied was the S_N2 reaction of an alkoxide with a halide: the Williamson ether synthesis. Enolate anions undergo precisely the same reaction. It is referred to as "alkylation."

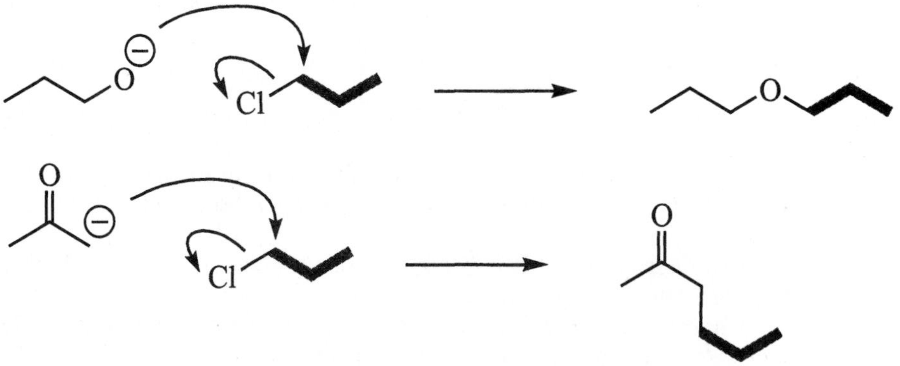

Notice that in this process, we have formed a new carbon–carbon bond. This represents our third way of making carbon–carbon bonds: first, we had alkylation of alkynes; second, Grignard reactions; now, alkylation of ketones.

It is worth exploring the logistics of this reaction: how should we make an enolate anion for the purpose of alkylation? We make an enolate anion by treating a ketone with a base. What base? One possibility is an alkoxide:

There are two problems with this. One is that the alkoxide might decide not to rip off a proton, but instead to attack the carbonyl. We know that alkoxide anions *do* attack carbonyl groups. But we also know that when they do, they tend to come right back off again, since the product (a hemiketal) is unstable with respect to the ketone. So this problem need not bother us.

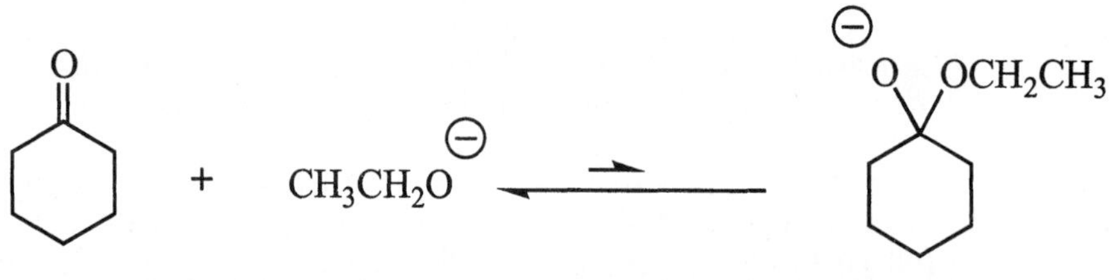

A second problem is that the acid–base equilibrium will lie slightly to the left, based on the pK_a values I just gave you. We will have some of each anion at equilibrium, but more alkoxide than enolate. This could still be a good way of generating enolates, and it is frequently used, but the alkoxide present must be one that will not interfere with what we are ultimately trying to do. What are we trying to do? We are going to add an alkyl halide in the hope that the enolate will react with it in an S$_N$2 reaction. Will a bunch of alkoxide floating around interfere with those plans? It sure will! The alkoxide will react with the alkyl halide in a Williamson ether synthesis!

There are two ways out of this dilemma: we could use as our alkoxide base one that will *not* react with an alkyl halide, or we could make sure that when we add the alkyl halide the enolate is the only nucleophile around. What alkoxide would not react with an alkyl halide? Well, we know that potassium *tert*-butoxide is a non-nucleophilic base, due to its steric bulk, so that would avoid the Williamson ether synthesis. But *tert*-butoxide is still a very good base, and most alkyl halides are capable of undergoing E2 reactions. So, although this lets us avoid S$_N$2 reactions, we probably cannot avoid having the base react in some other way with the alkyl halide.

The second approach is to make sure that at the crucial moment the enolate is the only nucleophile around. To do that we have to use just the right amount of base, and choose a base such that the following reaction goes essentially completely to the right:

This will require a much stronger base than alkoxide; i.e., something significantly farther down the pK_a chart, on the base side, than alkoxide. One base you will find down there, and that you have used before, is NH_2^-, which you may recognize more readily as the salt $NaNH_2$. We have used this to remove protons from terminal alkynes and also to make alkoxides. Would it work here? Certainly the acid–base equilibrium would turn out in our favor:

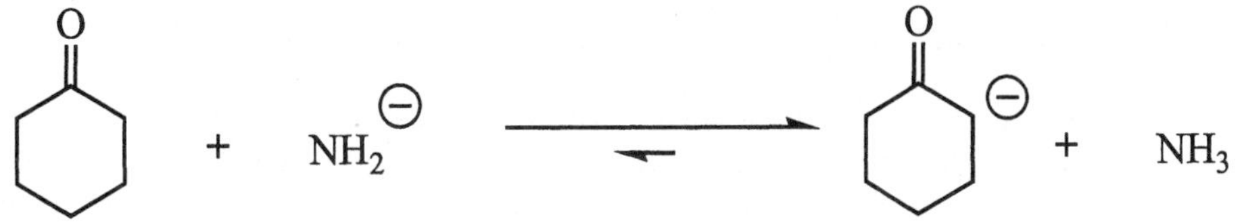

Is there anything else that could go wrong? What if the NH_2^- were to attack the carbonyl group instead of removing a proton?

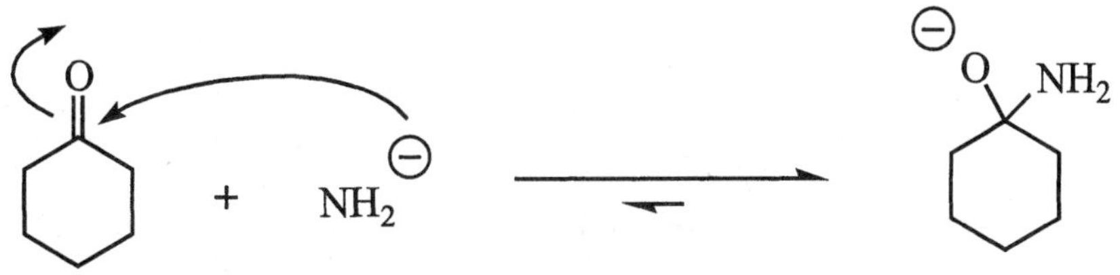

Hmm, this is an equilibrium that is also likely to sit far to the right, since O^- is so much more stable than N^-. So this is a potential problem. Any way to avoid it? Sure! Make the base large enough so that it will refuse to be a nucleophile. Indeed, that would suggest using the other non-nucleophilic base we encountered a long time ago, and this is our first occasion to take advantage of it: lithium diisopropylamide (LDA). Note that choosing LDA also frees us from the necessity of using exactly the right amount, since it will not act as a nucleophile toward the alkyl halide either.

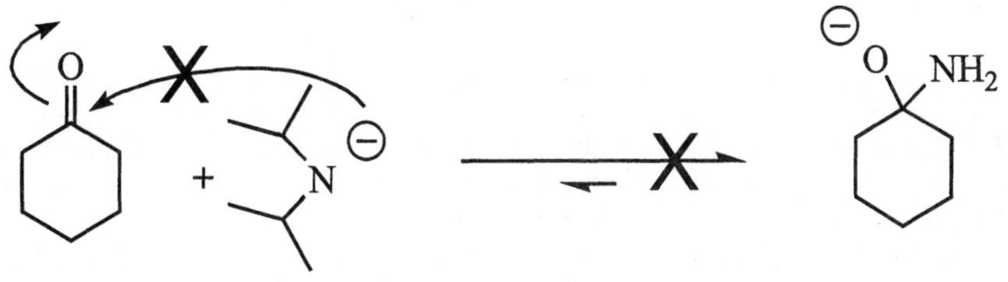

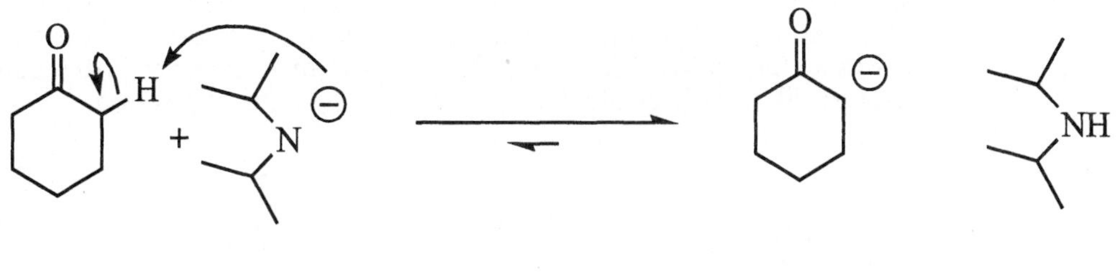

So the overall process involves treating a ketone with LDA, followed by an alkyl halide. Here are some examples:

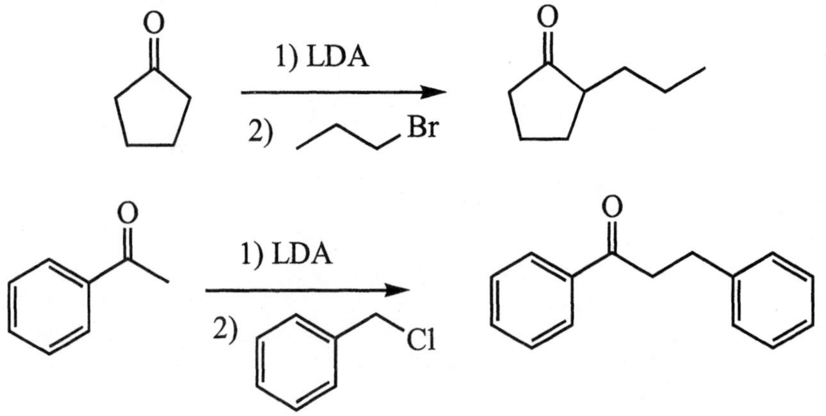

ORM: Under "Carbonyl Compounds," check out "alkylation of a ketone."

So far I have dodged the question of what happens if both sides of the ketone have hydrogens capable of being removed, but the two sides are not the same. For example, what would happen with 2-methylcyclohexanone, shown below? Which enolate anion is better? Try to reason this out for yourself.

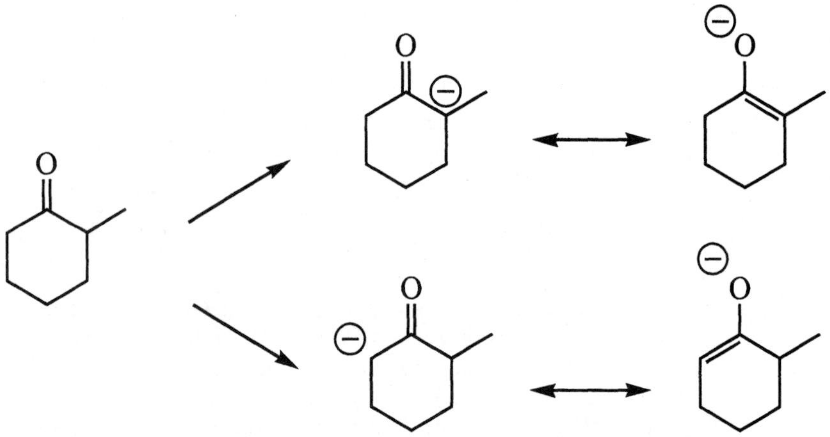

Most students conclude that the bottom anion is better. There is good logic to this: if alkyl substitution is good for cations (tertiary is better than secondary is better than primary), then the opposite should be true for anions; the bottom anion is secondary, and the top one is tertiary, so the bottom one should be better. Good thinking! But recall, these anions are resonance hybrids, and the negative charge is not stuck entirely on the carbon. It is shared with an oxygen, and since oxygen is more electronegative than carbon, we could guess that the hybrid will actually resemble the right picture more than the left in both of the hybrids. The essential difference between these sets of structures relates to the double bond: the C–C double bond in the top compound is tetrasubstituted, while in the bottom one it is trisubstituted. As you know, more substitution on a double bond leads to greater stability. So once again, we have one (valid) argument claiming that the bottom anion should be better, and another claiming that the top anion should be better. The only way to resolve such situations is by experiment. It turns out that, in most cases, the *more* substituted anion (in this case, the top one) is better. This means that, under ordinary conditions, unsymmetric ketones will usually be alkylated at the more substituted position:

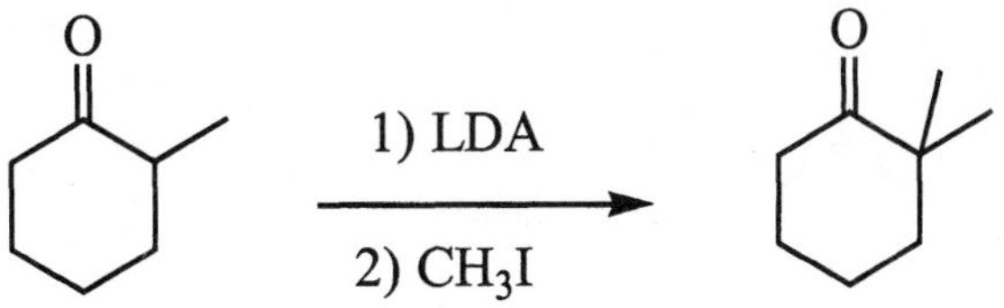

Later we will learn a clever way to steer this reaction to the other side.

In the preceding discussion I have swept one problem under the rug. I wonder if any of you noticed? I hope so. An enolate anion is a resonance hybrid, with the negative charge shared between the carbon and the oxygen. In all the reactions described above (and below), the enolate acts as a nucleophile by attacking some electrophile with its *carbon*. But at least half (likely, more: an anion is better on oxygen than carbon, so the resonance structure that has the charge on oxygen probably contributes more, as we just determined) of the charge is on oxygen. Shouldn't we see the oxygen acting as a nucleophile also, forming an ether as shown below?

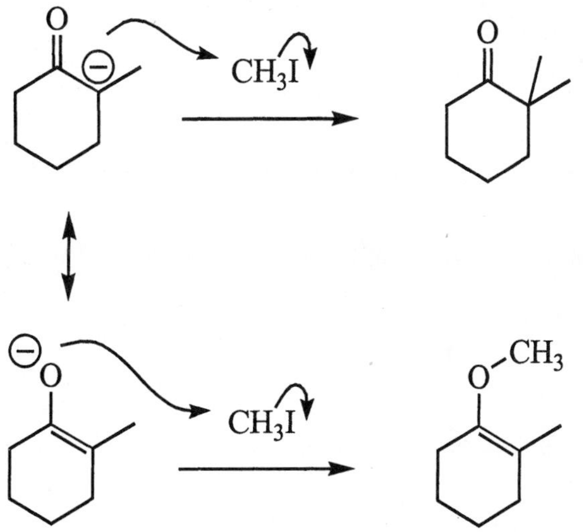

The fact is that reactions like this bottom one do happen sometimes; in certain cases (not this one) they are greatly preferred. But for reasons that I will not go into, we can assume for our purposes that enolate anions will always react at carbon rather than oxygen.

Bases: Proton Transfer

The second reaction of alkoxides you learned takes advantage of the fact that alkoxides are good bases and can remove protons from other molecules, resulting in elimination reactions. Enolates, too, are good bases, and can also grab protons. Of course, no one would ever make an enolate simply for the purpose of acquiring a base, since it takes a base to *make* an enolate—and you might as well use that base instead. But enolate anions that are formed during reactions frequently do grab protons from the solvent through an acid–base reaction. We will see some examples of this soon.

Nucleophiles with Respect to the Carbonyl Group: Aldol/Claisen Reactions

Alkoxides (and hydroxide) can attack carbonyl groups. If the carbonyl compound is an aldehyde or a ketone—that is, if it has no leaving groups attached to the carbonyl—the product will be a hydrate or a hemiXtal, depending on whether the nucleophile is OH⁻ or OR⁻. If the starting material is an ester (that is, if it *does* contain a leaving group) the product is either a carboxylic acid or a different ester, depending on whether the nucleophile is OH⁻ or OR⁻. These observations are summarized below.

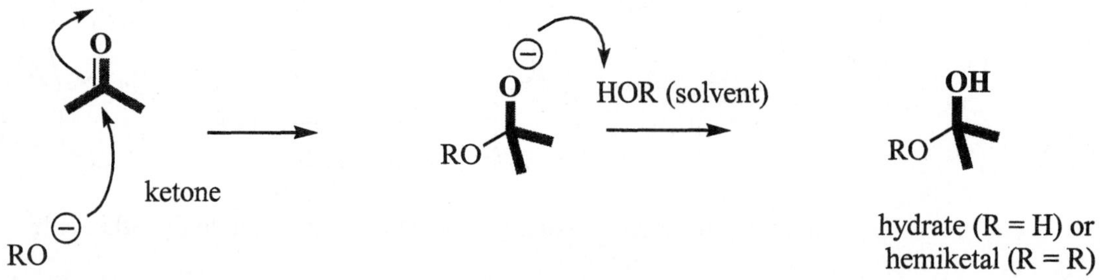

ketone

HOR (solvent)

hydrate (R = H) or
hemiketal (R = R)

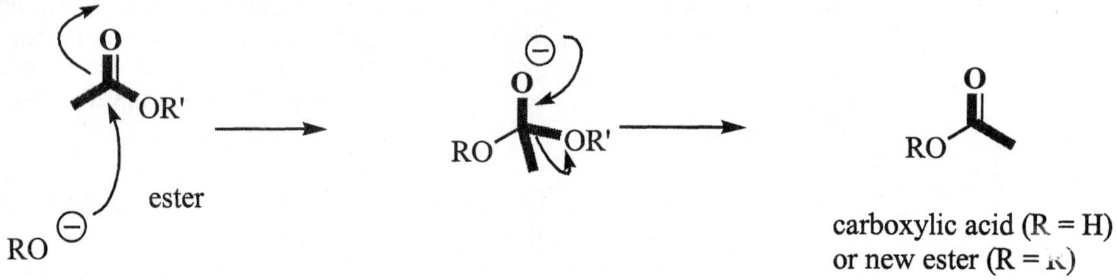

ester

carboxylic acid (R = H)
or new ester (R = R)

Enolate anions do exactly the same things:

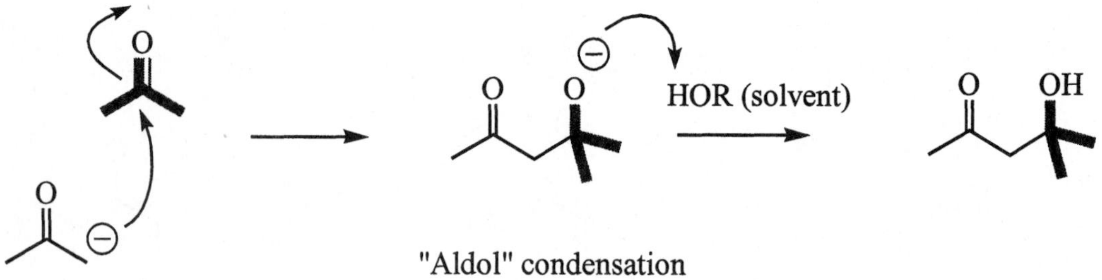

HOR (solvent)

"Aldol" condensation

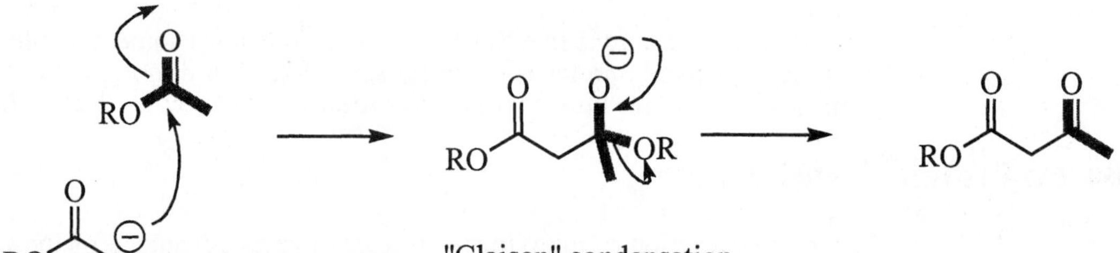

"Claisen" condensation

Once again it is worth considering the logistics involved in carrying out these reactions. How should we generate the enolate in this case, and what solvent should we use? Let's consider the first reaction, the aldol condensation, first. We start with some aldehyde or ketone and add a base. The base removes a proton from one molecule of the carbonyl compound, which then attacks a second molecule. The resulting anion picks up a proton from solvent:

I've shown the example above for the specific case of acetaldehyde. The product on the right is both an **ald**ehyde and an alco**hol**. This particular compound is called "aldol," and it is from this example that the family of reactions gets its name. Aldol condensations can occur as easily on aldehydes as on ketones, but for the rest of this section I will discuss just the ketone case. Aldehydes behave exactly the same way.

What properties do we need with respect to our base and our solvent? Clearly, the solvent needs to have some H's available for donation to the intermediate, which contains O⁻. A common solvent capable of exchanging protons with O⁻ would be an alcohol. A base strong enough to generate enolate anions from a small percentage of the ketone molecules is alkoxide. So one possibility here would be to use an alkoxide in some alcohol solvent: say, NaOEt in EtOH. Now we need to ask ourselves a few questions: An alkoxide base will deprotonate only a small fraction of the ketone molecules. Is that OK? Well, we *want* to have both enolate anion and neutral ketone present at the same time. Indeed, this should be an ideal base for accomplishing that. Anything much stronger would deprotonate *all* the ketone and leave no neutral molecules for the enolate anions to react with. So far NaOEt sounds OK. But if we use a base like that, there will also be some OEt⁻ floating around. Will that interfere with our reaction in any way? (Remember, that was a problem in the alkylation reaction.) Ethoxide ion might react with the ketone, not to make enolate, but rather to make a hemiketal. Still, this is reversible, and equilibrium favors the ketone, so this should not be a big problem. And there is really nothing else ethoxide can do. Finally, *how much* base do we need to run this reaction? You should notice that the reaction regenerates at the end of the reaction a basic anion just like the one that was used in the beginning. So after one cycle, there is the same amount of base present that we started with, and the process can occur again. In other words, the reaction is catalytic in base: a small amount of base can be used over and over to convert a large amount of ketone into product. This is illustrated as follows, this time with NaOEt.

Below are some examples of the aldol condensation. For those that are incomplete, you should fill in the blanks. You should also take the time *now* to show mechanisms for all of the reactions.

You notice, of course, that this is yet another new way of making carbon–carbon bonds! Furthermore, this is the first one we have learned that is capable of taking place in "protic" solvents, meaning solvents that contain –OH groups, such as water. It should not surprise you, then, to learn that this reaction (together with the next one we'll look at) is the approach used by biological systems for constructing carbon–carbon bonds. We will return to this topic in the next chapter.

ORM: Under "Carbonyl Compounds," check out "aldol condensation."

Now let's consider the Claisen condensation in the same way. A Claisen condensation (officially pronounced "CLEYE-sen" but many people say "CLAY-sen") is the same thing as an aldol condensation, except that the starting material is an ester instead of a ketone. I've reproduced it below, making a guess that NaOEt will be a good base for this case as well.

We should ask the same questions we asked for the aldol condensation: Is the base strong enough? Will it interfere in any way? How much of it will we need?

We know that esters are weaker acids than ketones (pK_a ~25 vs. ~19). So an alkoxide base will deprotonate a smaller fraction of ester molecules than of ketone molecules. But once again, we do not need lots of anions, and we *do* want plenty of neutral ester molecules present along with the anions. It turns out that alkoxide is indeed good enough for the purpose. So on to the second question: will alkoxide interfere? What can it do? Instead of deprotonating the ester, it might attack the carbonyl group. Is this a problem? Well, depending on what OR is, it might be. When OEt attacks the carbonyl, OR might leave, resulting in a transesterification. Of course, this is easy to correct for: just use as our base, not OEt specifically, but the OR that corresponds to the ester we have. That way, even if our base attacks the carbonyl group, no change will occur.

So far we have determined that an alkoxide base should be appropriate for a Claisen condensation. The next question is, how much of it do we need? For the aldol condensation, we only needed a little bit. Is the same true here? Let's look:

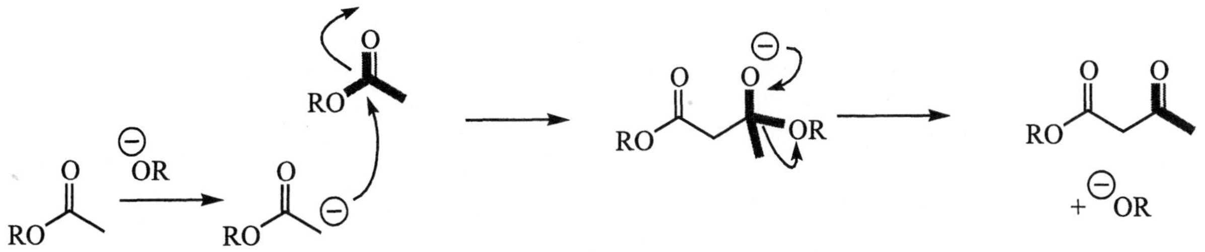

According to this description, we again get back at the end of the reaction the same kind of alkoxide anion we started with. But there is a problem here that you may not have noticed yet. The product we have created has a CH$_2$ group nestled between two different carbonyl groups. The protons on this CH$_2$ group should be considerably more acidic than normal α protons, because the resulting anion can be resonance stabilized in two different directions (show it!). According to our pK_a chart, compounds like these have a pK_a of about 11, making the new α protons by far the most acidic protons in the mixture, and as you know, proton transfer is one of the fastest reactions there is. Therefore the mixture of products will not sit around as shown previously, but the reaction will instead continue:

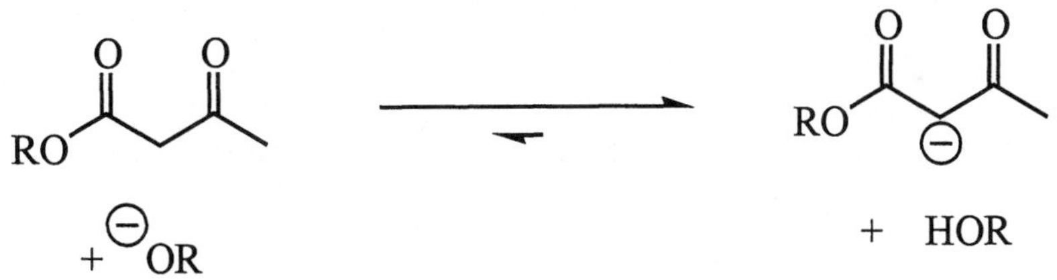

Now the reaction is over, but now we no longer are regenerating a base capable of deprotonating an ester. In order for the reaction to continue, a new molecule of RO$^-$ will need to become involved. In other words, we need one RO$^-$ ion for every pair of ester molecules if we want to accomplish a Claisen condensation in high yield. The reaction is *not* catalytic in base.

Of course, now the product is no longer a neutral β-ketoester, but rather its anion. In order to isolate the product at the end of the reaction, acid must be added to put a proton back on. This step is understood and is sometimes omitted in the description of the reaction.

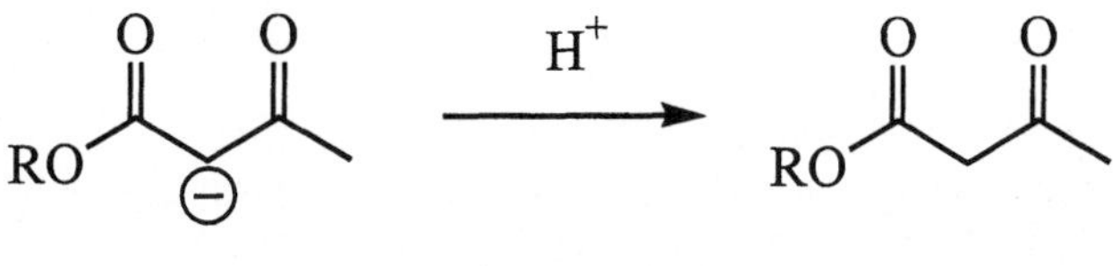

Below are some examples of the Claisen condensation. Again, for the ones that are incomplete, fill in the blanks, and show mechanisms for all the reactions.

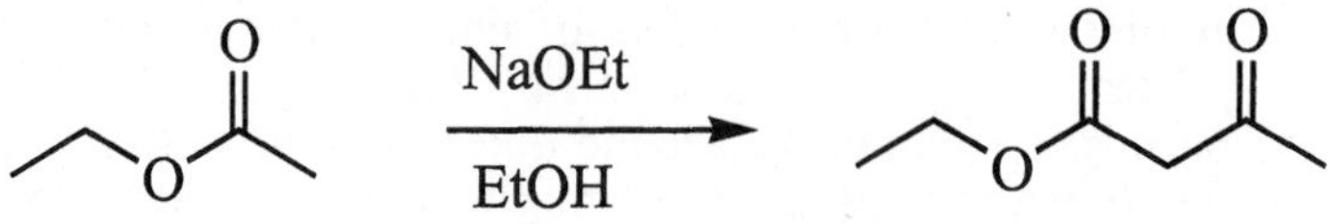

The next example is technically not a Claisen condensation, because one partner is a ketone instead of an ester, but we will treat it in the same category:

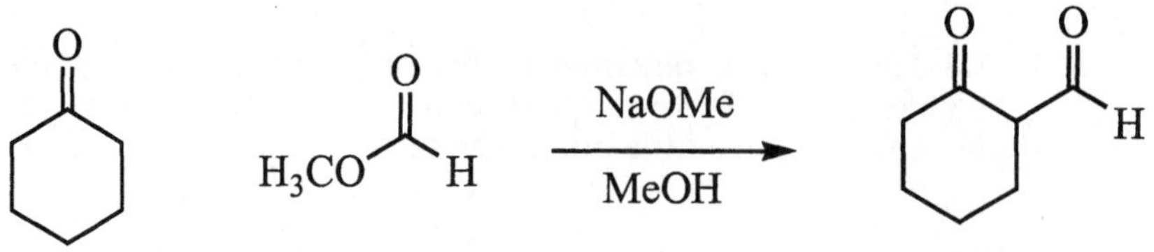

A Claisen condensation that makes a ring is officially called a "Dieckmann" condensation, but again we will include it in the present category. These reactions work especially well if a 5- or 6-membered ring is formed.

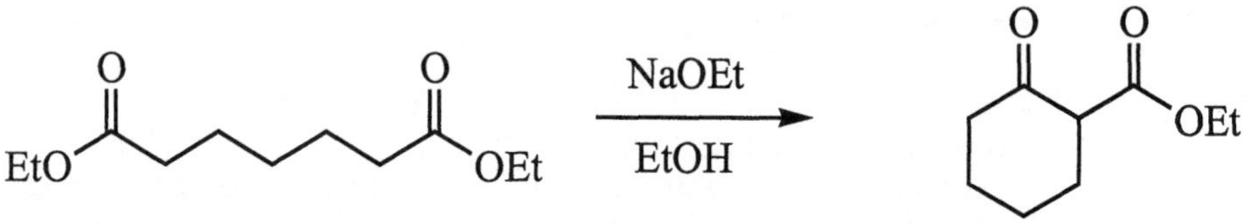

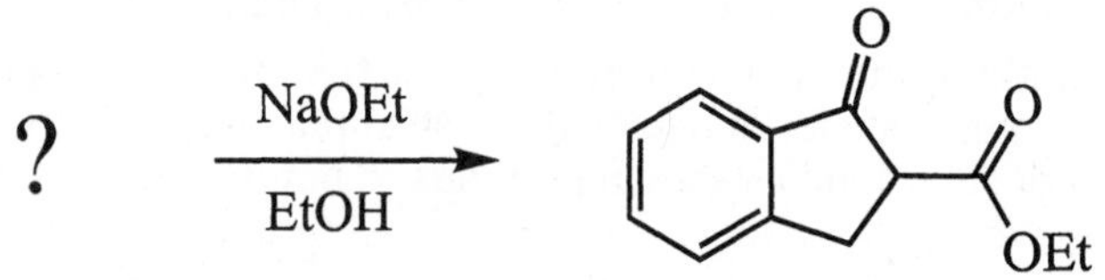

The next one is technically not a Dieckmann condensation, because one partner is a ketone instead of an ester, but we will again treat it as if it were. Of course, the ketone is the one that serves as the nucleophile here, since ketones are significantly more acidic than esters.

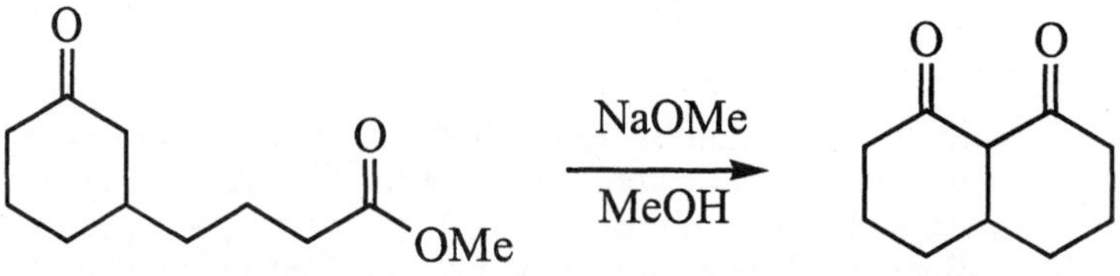

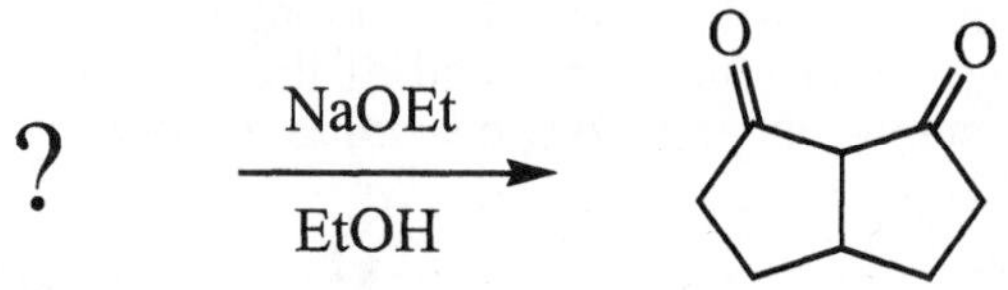

You should once more have noticed that we have here another new method for making carbon–carbon bonds, and one that is again compatible with protic solvents. We will return to this point when we discuss biological transformations in the next chapter.

Leaving Groups: Retro-Aldol/Claisen

You have also seen alkoxides act as leaving groups. Not in S_N2 reactions, of course: early on we identified O^- as a bad leaving group for substitution reactions. But when there is already an O^- in the molecule, and you can make a double bond by ejecting a different O^-, this ejection can occur. We have seen such a thing when a hydrate or a hemiketal reverts to a ketone, or when an ester reacts with an oxygen nucleophile:

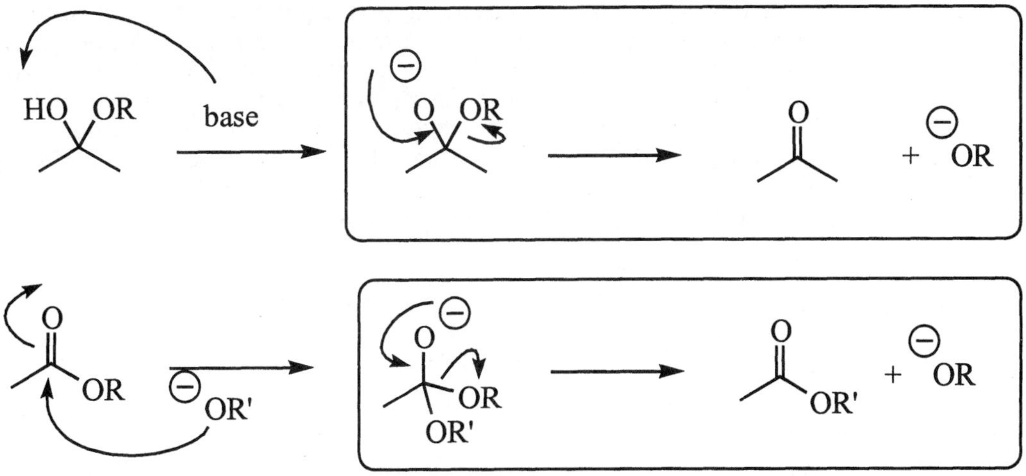

The reactions in the boxes are elimination reactions, in which the starting material is already an anion—that is, the conjugate base of some species. The technical term for such reactions is E1cB, meaning **E1** from the **c**onjugate **B**ase. So alkoxides can serve as leaving groups for E1cB reactions.

Enolates can also be leaving groups. The most obvious way to generate examples is to take the above equations and replace the OR leaving group with a potential enolate. Like this:

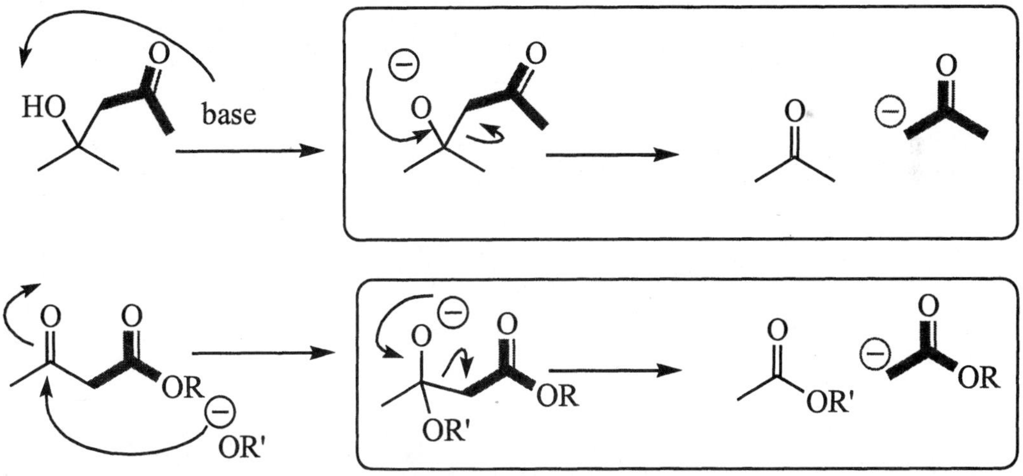

What you should see is that we have just written the reverse of both an aldol and a Claisen reaction. So these reactions must be reversible! Sure enough, they are, and we should have made that clear all along:

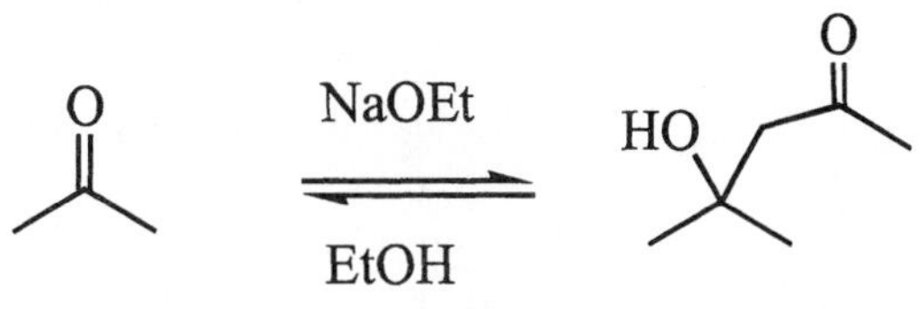

Notice that the equation above is a little fishy, because I have written the starting material only once when we all know that actually two identical molecules have to get together in this reaction. Remember, though: organic equations tend to be descriptive, not quantitative, and we don't usually bother to balance them. This is what we start with, this is what we get: the equation is accurate in that respect.

Since the reaction we're considering is an equilibrium, there must be an equilibrium constant associated with it; in other words, it must have a preference for being predominantly on the left or the right. Of course, there is no single "it" that is being referred to: there are many aldol condensations, and each has its own equilibrium constant. Nevertheless, if aldols are like hydrates or hemiketals, we might be able to say that equilibrium usually prefers the left side of the equation; that is, that the aldol reaction tends not to go forward. This is often true, although we will soon learn another factor that affects the equilibrium. Also, like hemiketals, aldols tend to be favorable if their formation involves making a five- or six-membered ring.

Test Yourself 2

Here's an example in which a reverse aldol process plays a significant role: suggest a mechanism for the following reaction. *Hint*: begin by deprotonating the OH, do a retro-aldol, then do a forward aldol.

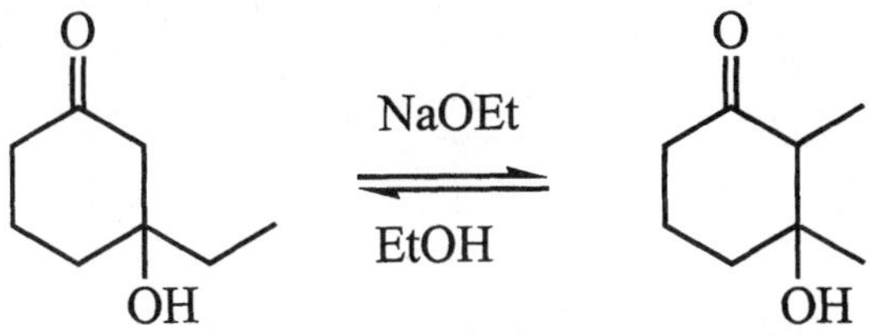

The Claisen condensation is also reversible, and it, too, has a preference for the starting material. However, the issue does not actually come up very often, because of a competing reaction we have already discussed. When a β-ketoester (can you figure out the derivation of that term?) reacts with an alkoxide, as shown in the equation below, it will much more often lose a proton from the position between the two carbonyl groups than undergo attack at carbonyl. This deprotonation removes β-ketoester from the equilibrium, pulling the reaction to the right, even though the first reaction by itself would rather go to the left. The Claisen condensation generally proceeds in the forward direction only because a very stable double-enolate anion can be formed.

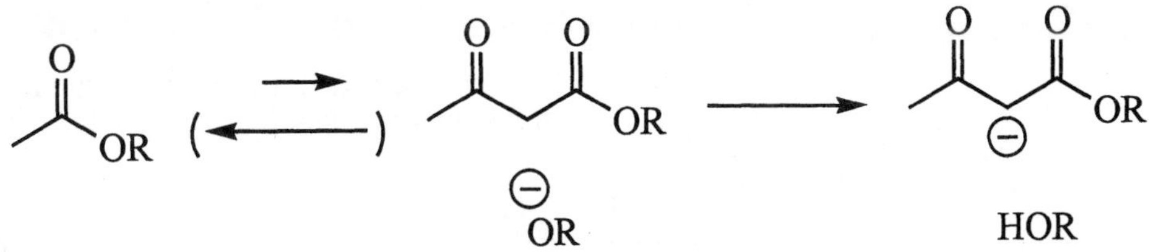

This brings up an interesting point: if the Claisen condensation, by itself, would rather not occur, and only does so because of the stable anion formed at the end, then it stands to reason that, if such an anion *cannot* form, the reaction should not proceed very well. What is necessary for the creation of a stable anion? There must be at least one proton on the carbon between the two carbonyl groups. If both of these positions are occupied by alkyl groups, no anion can be formed, so the reaction should not occur. This is generally true: the following Claisen condensation, for example, does not proceed; you simply get back the ester you started with.

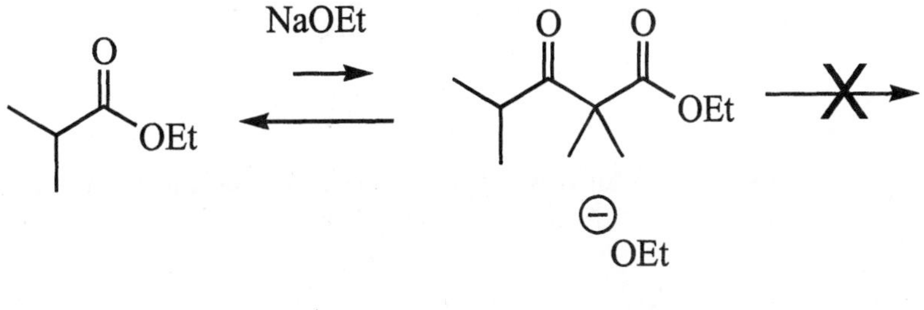

The fact that a Claisen condensation will not proceed when an α position on the ester bears only one hydrogen and the converse statement (that β-dicarbonyl compound with no protons between the two carbonyls will easily undergo a reverse Claisen reaction in base) can be used together in a very clever way to steer ketone alkylation. Consider 2-methylcyclohexanone undergoing a Claisen condensation with ethyl formate:

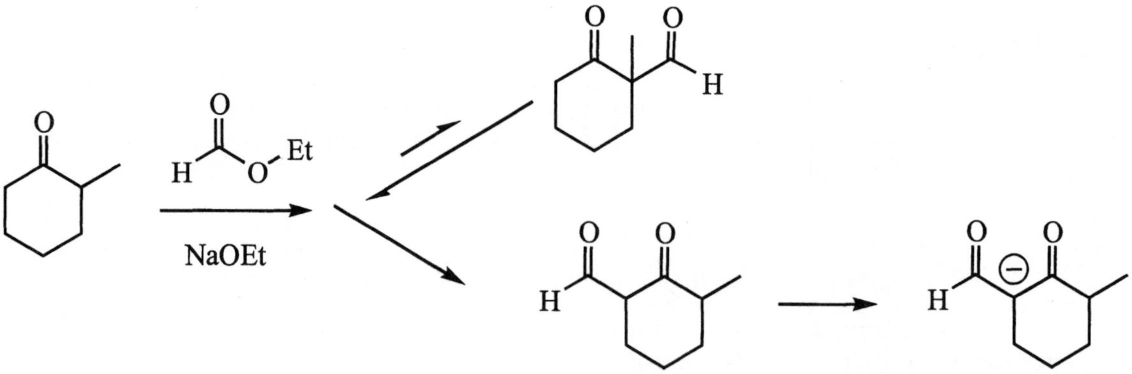

The reaction will proceed only on the less-substituted side of the ketone (the left side in the above picture), because there is only one proton on the right side, and after the (top) reaction occurs, there is no proton left between the two carbonyls. The top reaction would thus reverse itself and regenerate the starting materials. On the other hand, the bottom reaction *does* occur, because it can continue to make the stable anion shown. This anion is a good nucleophile, and if treated with an alkyl halide it can be alkylated:

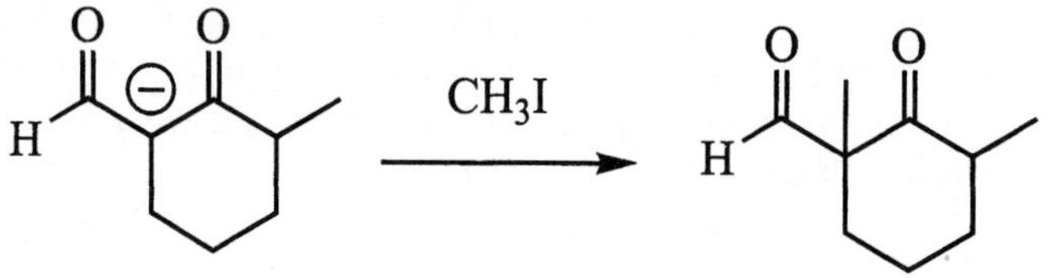

In this new compound, however, there is no longer a proton between the two carbonyl groups. If this substance is treated with ethoxide, a reverse ("retro") Claisen will occur:

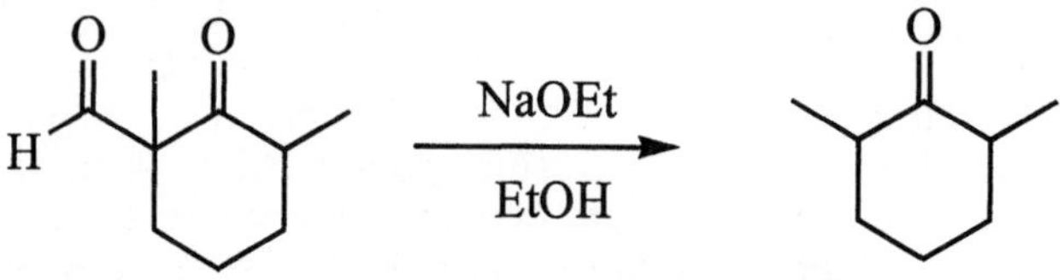

So an interesting overall process is the following:

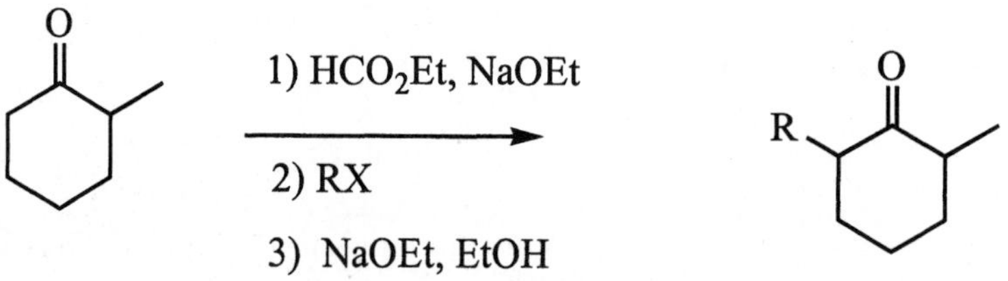

You should recall from an earlier section that the more straightforward method of alkylation usually leads to the alternative product:

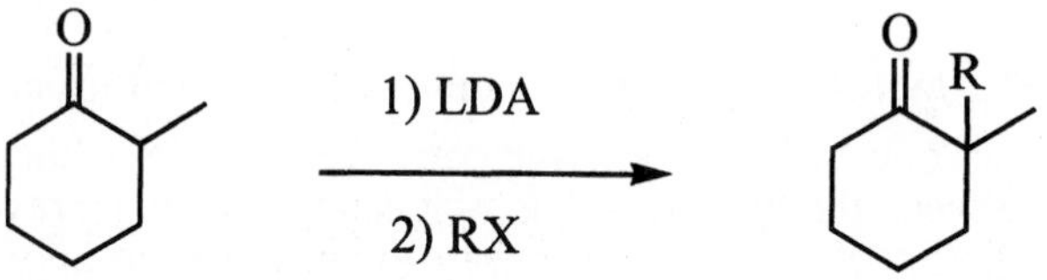

In other words, we can now steer alkylation to whichever side of a ketone we want.

Test Yourself 3

Show the mechanism for each step of the three-step alkylation process we just discussed. Explain why we use ethyl formate rather than ethyl acetate for this process. (*Hint*: recall that a β-diketone has two carbonyl groups that can be attacked by a nucleophile.)

Anions in E1cB Reactions: Enone Formation

The last section began with a consideration of E1cB reactions, as shown below:

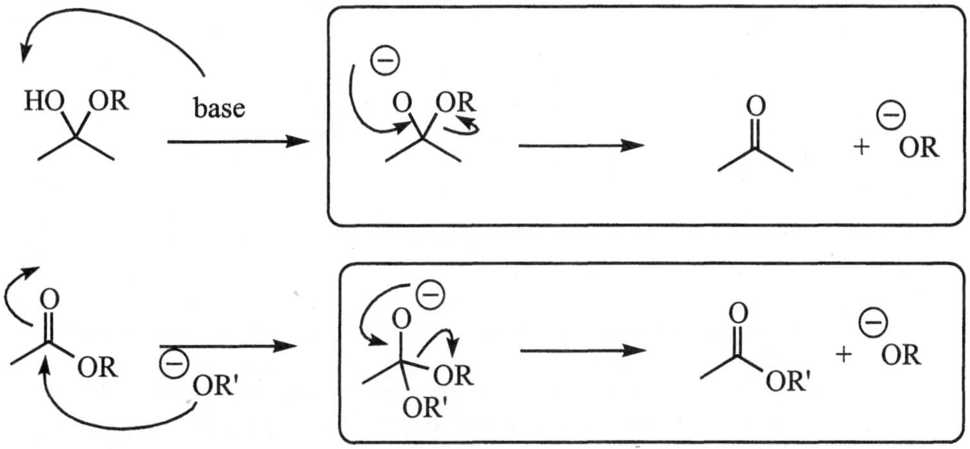

We focused our attention on the leaving groups in the boxed reactions, OR⁻, and reasoned that enolates should be able to perform the same function, and indeed, they could. Now let's change our focus: it is also O⁻ that is doing the pushing in these reactions, so enolates also ought to be able to perform *that* function, namely, throwing out O⁻ as a leaving group. And indeed, they can. Again we can easily illustrate this by taking the above equation (let's use the top one) and replacing the O⁻ in question with an enolate:

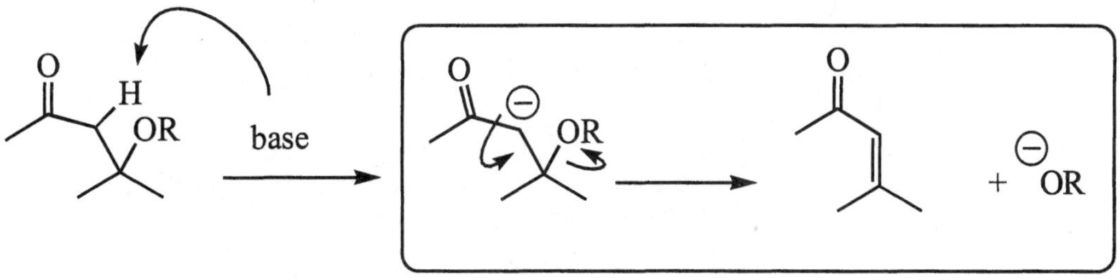

Remember, O⁻ is a lousy leaving group in general, but it can leave if the pusher is a preformed anion and if a double bond is being made—that is, in an E1cB reaction. That is precisely what we are looking at here. Under what circumstances are we likely to run into such a case? Look back at the aldol condensation.

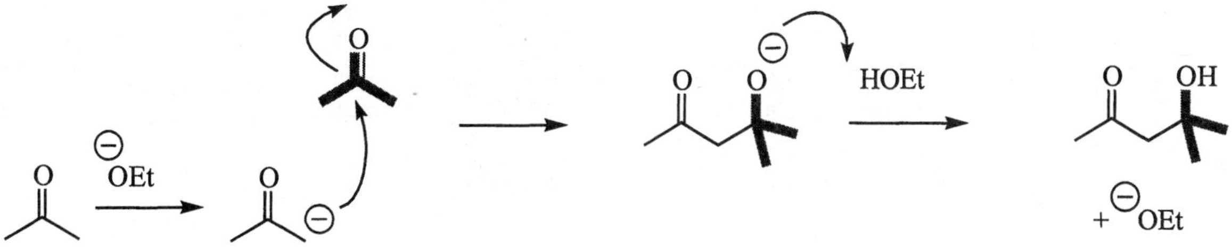

The product is a β-hydroxyketone, and it has been generated in basic solution. There are three protons in this molecule that are moderately acidic, and each is likely to be lost some of the time:

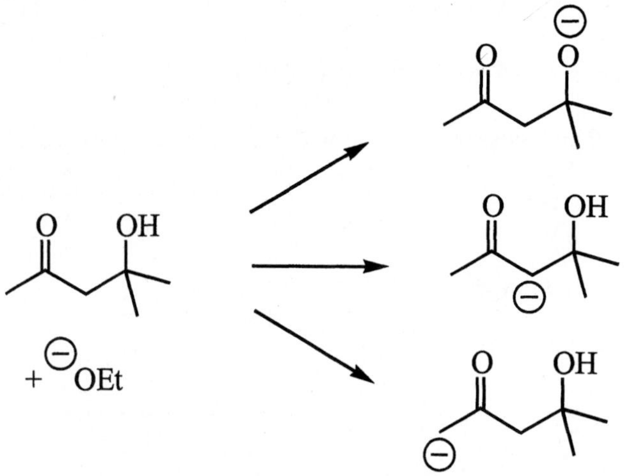

The top process, as we have already learned, is the first step in the reversal of the aldol condensation, and we know that does occur. The bottom reaction does not seem likely to lead anywhere useful. It is the middle one that is of interest to us now, because that represents the first step in the E1cB elimination of water:

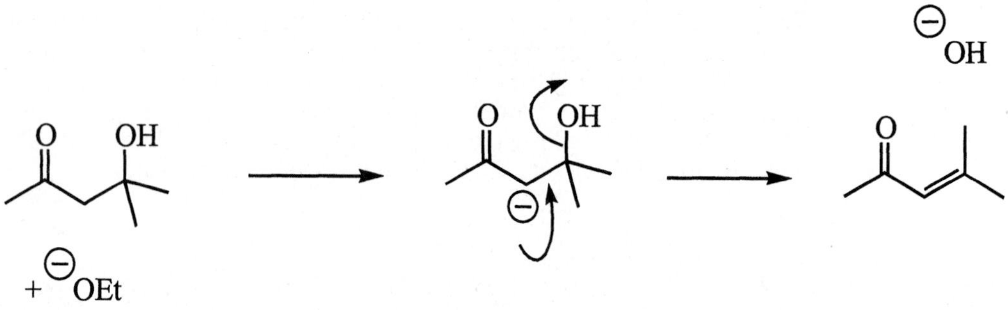

The final product is an α,β unsaturated ketone, or an "enone" for short. Technically the word enone could apply to any molecule that contains both a double bond and a keto group, but it is used almost exclusively for cases in which the double bond is conjugated with the ketone, as above.

Does such a thing really happen? Yes it does, and because it does, the product of an aldol condensation often is *not* a β-hydroxyketone but the corresponding enone. This is especially true when the new double bond has even more conjugation, as in the following example.

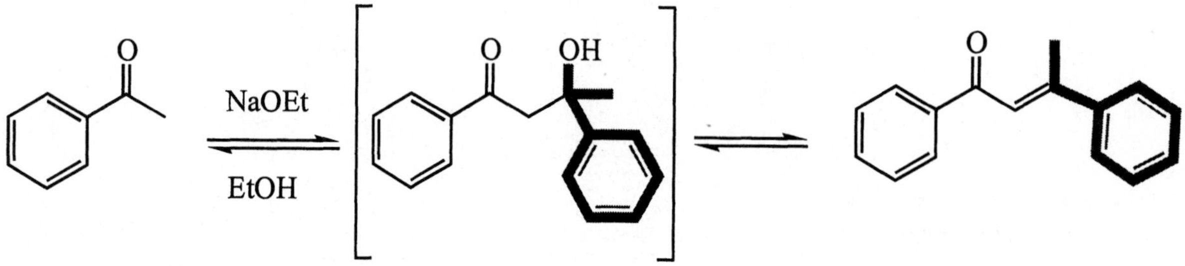

Sad to say, this makes your job harder, because you now need to be able to look at an enone and then figure out what pieces it might have come from in an aldol condensation. Like the following: what aldol condensation would lead to the following product? Try it before I show you how to solve this problem!

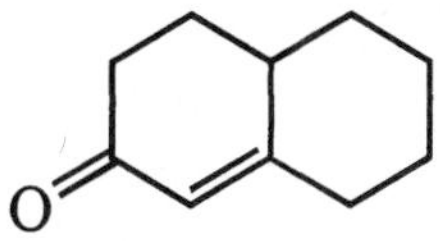

The first step is to find where there is an enone in the structure. This should not be difficult:

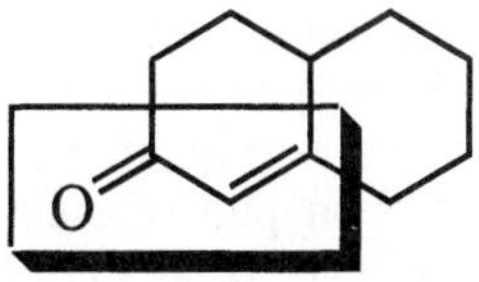

We know the enone must have come from an E1cB reaction, which means there was originally an OH group at the β position:

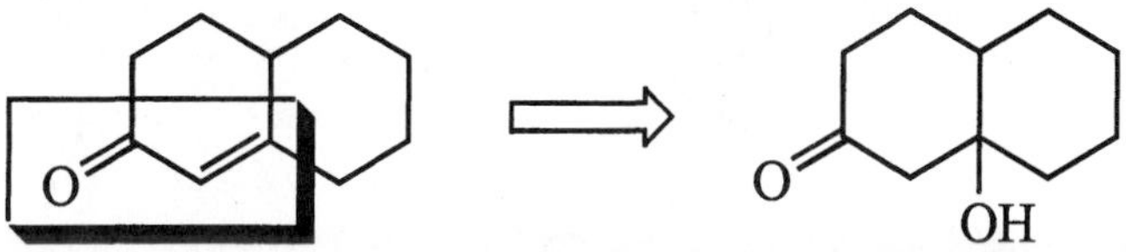

And we also know that this OH group was the carbonyl group of a ketone before it was attacked by an enolate. The attacking group must have been the carbon α to the remaining carbonyl group, which is still also attached to the carbon bearing the alcohol:

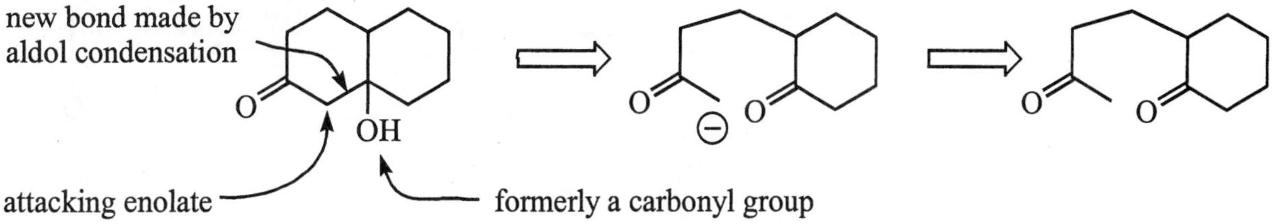

new bond made by aldol condensation

attacking enolate — formerly a carbonyl group

The bottom line for an aldol condensation, if it proceeds all the way to an enone, is that it eliminates the elements of water from the reacting ketones, creating in the process a double bond joining a carbon α to one ketone with the carbonyl carbon of the other ketone.

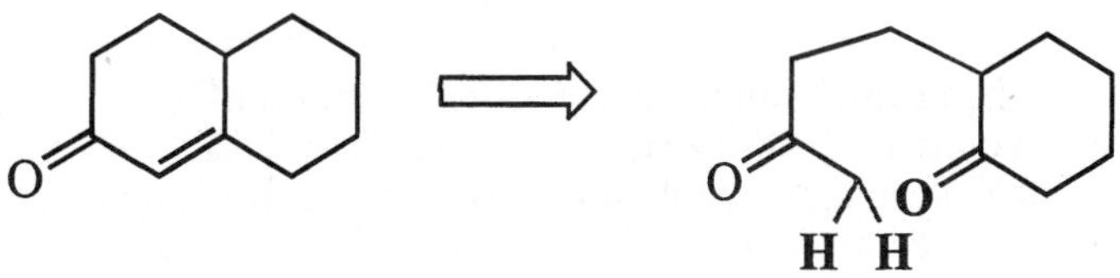

Test Yourself 4

What starting materials would be necessary to generate the following products through aldol condensations?

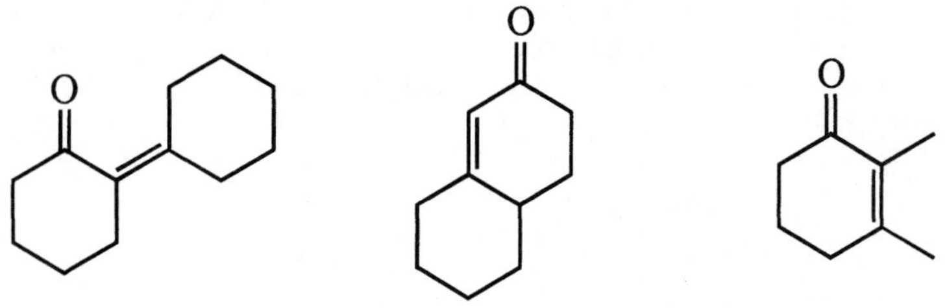

Result of Nucleophilic Attack: Michael Reaction

The final context in which we have seen alkoxide anions is when they result from nucleophilic attack on a carbonyl group. If a nucleophile is able to attack a double-bonded carbon (in a carbonyl) in such a way as to push the electrons out onto oxygen, making an alkoxide, it also ought to be able to attack a double-bonded carbon in the alkene of an enone so as to push a similar pair of electrons out onto the other carbon, making an enolate. Once again, the prediction is valid. And again, we can illustrate it by drawing an example of the first reaction and then replacing the O⁻ with an enolate anion:

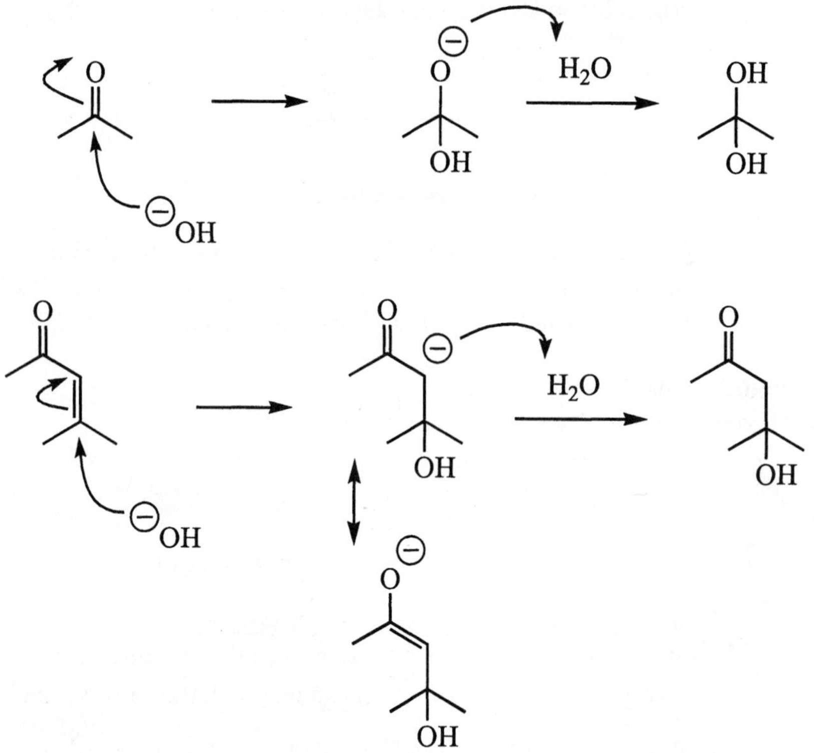

When a nucleophile adds to the β position of an α,β-unsaturated ketone, we call the process a "conjugate" addition, or "1,4 addition." You should notice that this is the first time you have seen a nucleophile add to a carbon–carbon double bond: all other alkene reactions have involved *electrophiles*. Of course, if a nucleophile were to add to a normal alkene double bond, the result would be one of the world's worst anions. The only reason we can get away with it here is that the resulting anion is resonance stabilized.

YUCK!

When the nucleophile adding to an enone is itself an enolate, there is a special name for the reaction: a "Michael" reaction. Strictly speaking, this name is supposed to be reserved for the narrow case of enolate nucleophiles, but it is often applied to any conjugate addition to an enone.

Michael reaction

When a nucleophile adds to an enone, an interesting problems arises: there are two sites on the molecule where that nucleophile could attack—the carbon atom in the β position or the carbon atom of the carbonyl group itself. Can you know ahead of time where the attack will occur? The ultimate answer to this is "no," but one can often make pretty good predictions. There are several approaches to this problem, of which I will describe only one here.

Regular (1,2) addition product

Conjugate (1,4) addition product

According to the model I will focus on, the nucleophile is more attracted to the carbonyl carbon: reaction happens fastest there. But it turns out that the product of a conjugate addition is generally more stable than the product of a regular carbonyl addition. *Given a choice*, the molecule would choose to become the conjugate addition product. So this is the key: if the molecule gets to choose—that is, if the reaction is under thermodynamic control, meaning that it is reversible and can sample all possibilities before selecting one—you'll obtain the conjugate addition product. On the other hand, if the reaction is such that once addition has occurred, it cannot reverse itself, you will be stuck with whatever forms first: i.e., the process will be subject to kinetic control. In practice, this means that nucleophiles that cannot themselves serve as leaving groups, such as Grignard reagents, hydride reducing agents, and so on, usually give 1,2 addition products, whereas nucleophiles that *can* leave, such as oxygen anions and enolate anions, tend to give conjugate addition products. Although there are some exceptions to these generalizations, most reactions can be correctly predicted using this approach.

Regular (1,2) addition product

Conjugate (1,4) addition product

Grignards
hydrides

alkoxides
hydroxide
enolates

The Michael reaction is yet another new way by which you can make new carbon–carbon bonds. It, too, is compatible with biological conditions; it turns out that organisms tend not to use conjugate addition as a way of making carbon–carbon bonds, but conjugate addition of other nucleophiles is a common biological reaction. We will return to this later.

Summary of Enolate Reactions

I have approached this topic in an unorthodox way, in order to emphasize the connections between enolate chemistry and the alkoxide chemistry you already know. The idea was to make all the reactions seem reasonable and predictable, rather than treating them as brand new reactions that I was suddenly springing on you. The downside is that some of the relevant reactions got discussed in many different places, which may have caused you to lose track of what was going on in each. So below I summarize what we have learned, organized by reaction type.

Alkylation

A strong base can remove a hydrogen α to a ketone to make an enolate anion. You generally get the more stable of the possible anions, which usually corresponds to the more substituted one. This anion can then be treated with an alkyl halide to effect an S_N2 reaction, resulting in a new carbon–carbon bond at the α position. Example:

1) LDA

2) CH_3CH_2Br

Aldol Condensation

An enolate anion can react with a neutral aldehyde or ketone in an aldol condensation. The reaction creates a β-hydroxyketone (or aldehyde), which might be the final product, but often the reaction continues, with elimination of H_2O and formation of an enone. The reaction is most commonly carried out using alkoxide in alcohol; only a catalytic amount of base is necessary, since base is regenerated at the end of the reaction. The reaction is reversible, and sometimes equilibrium favors starting material; product is most often favored when the reaction leads to rings or extra conjugation. Example (what other enone would you also expect to form?):

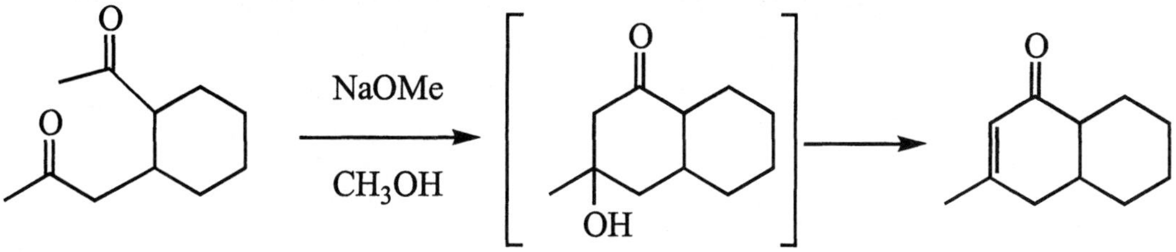

Claisen Condensation

Enolates react with esters in a reaction that results in the loss of an alkoxide leaving group to form a β-dicarbonyl compound. This product might be either a β-diketone or a β-ketoester, depending on whether the starting enolate was a ketone or an ester. The actual product of the reaction is usually a doubly stabilized anion, but a neutral molecule can be obtained by adding acid. The reaction is reversible: if the extra-stable anion cannot be formed (because there is not another α hydrogen to remove), the reaction has a preference for starting material. This reaction consumes base, so a stoichiometric amount of it must be present. Example:

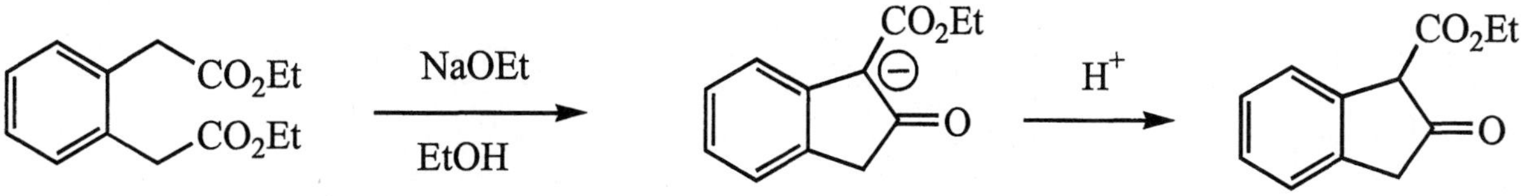

Nucleophiles can add at the β position of an enone, because the resulting anion is a resonance-stabilized enolate. Nucleophiles that themselves can serve as leaving groups tend to prefer this conjugate addition; strong nucleophiles like Grignard reagents and hydride reducing agents often go for the carbonyl group directly. Example:

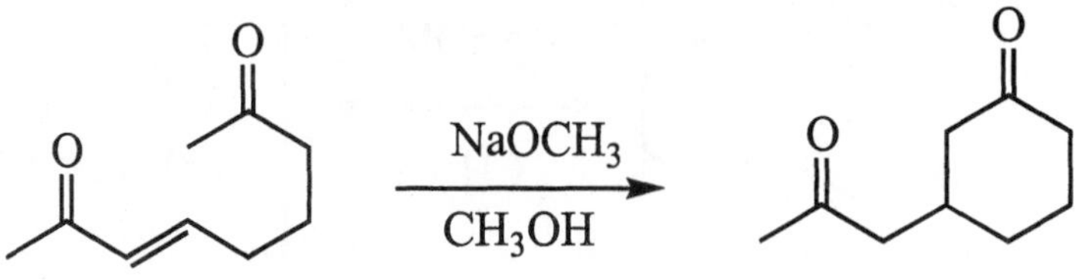

Test Yourself 5

The problem below puts together a lot of the chemistry discussed above. What follows is a well-known reaction called the "Robinson annulation," a really cool way of adding a second six-membered ring onto an already existing one. It was used extensively in steroid syntheses in the 1950s and 1960s. Propose a reasonable mechanism for this reaction. Go ahead, try it!

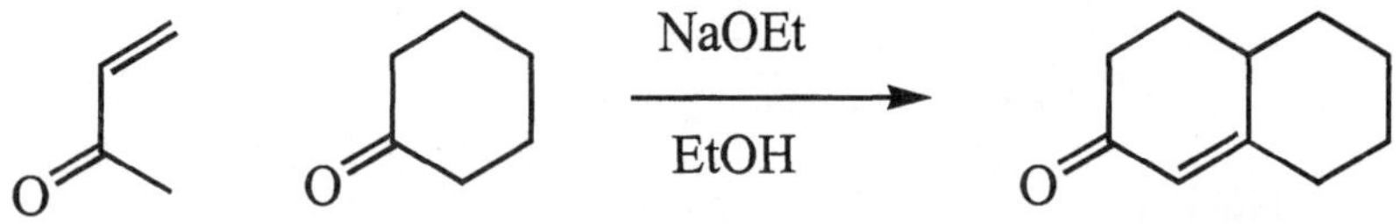

Test Yourself 6

Suggest a mechanism for the following reaction. *Hint*: start with a conjugate addition of hydroxide.

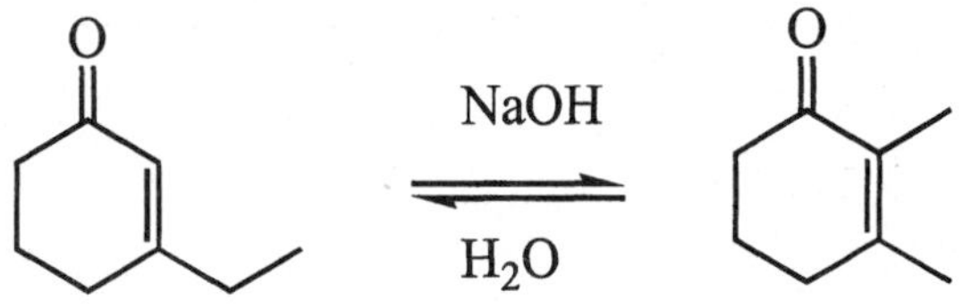

ALKYLATION OF ACIDS AND ESTERS
THE MALONIC ESTER SYNTHESIS; DECARBOXYLATION

We have seen above how we can add an alkyl group to the α position of a ketone. Could we do the same to an acid or an ester? Consider an acid:

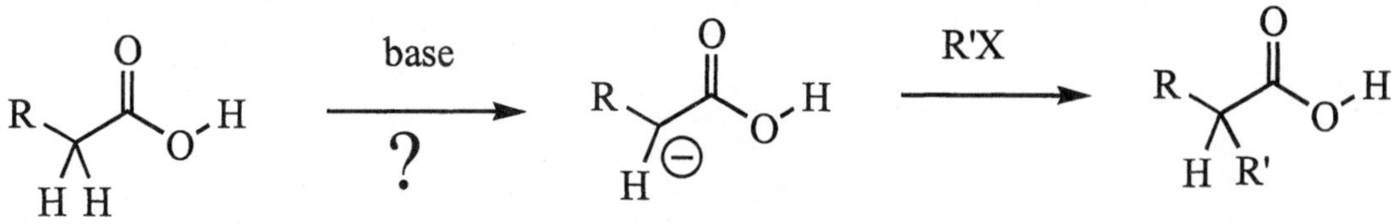

By now I hope that every one of you recognizes the above equation as silly. Clearly, a base picking a proton off this molecule would not choose *that* proton, but rather the one on the oxygen, leading to a resonance-stabilized carboxylate anion:

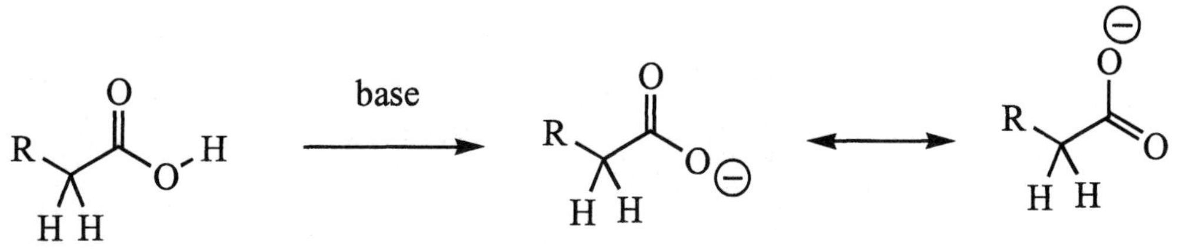

In order to pull off the α proton it would be necessary to make a *di*anion. In fact, this can be done, but it requires pretty drastic conditions and is not often attempted. So this is not a common way to alkylate the α position of an acid. Actually, there is no good way to alkylate an acid in the α position. So we are going to learn a roundabout way of making such compounds: acids that have anything we want in the α position. Eventually we will recognize that the method involves alkylation of a diester, hydrolysis, and subsequent "decarboxylation," but I will lead you through the process gradually.

First, you should remember that it is easy to make an acid from an ester, so if we were able to alkylate an *ester* in the α position, we would have accomplished essentially the same thing we set out to do.

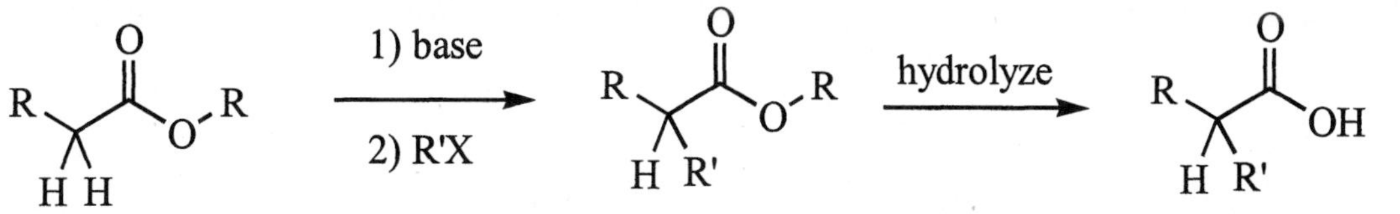

It is this new process we should analyze: the alkylation of an ester. Is it possible? There are two questions we need to worry about: can we remove the proton in question, and are there any side reactions that will interfere? We know that protons α to an ester group are less acidic than those α to a ketone, but we still have bases capable of doing the job. LDA, for example, is quite a bit stronger as a base than an ester enolate. So that part should not be a problem. However, as we begin to form the ester enolate, Claisen condensations are possible, and these lead to very stable anions that will not go back where they came from. So if we wanted to make the enolate of an ester we would have to do it in such a way that the enolate never sees a neutral ester molecule. This, too, is possible, and it is sometimes done, but it turns out there is an easier way.

We just established that we cannot easily alkylate an acid, but if we could alkylate something else (an ester) and then turn *that* into an acid, we would have accomplished functionally the same thing. That's the strategy we will apply now: working with something that is easy to alkylate and can later be turned into an acid. The alkylation part you can understand; the turning into an acid is a process we have yet to encounter, so I will have to introduce it when we get there.

We established a long time ago that we can easily make an anion α to a carbonyl group, because the anion is resonance hybridized. We also established that an anion α to *two* carbonyl groups is even better stabilized: on your pK_a chart you will find the following data:

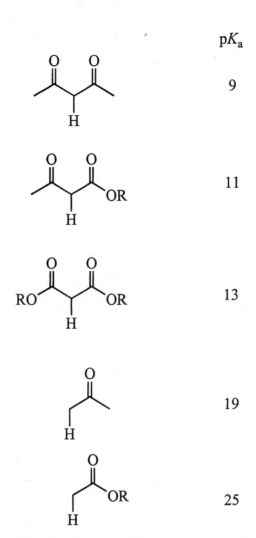

pK$_a$

9

11

13

19

25

This is entirely consistent with what we know about enolates: they are better with the charge next to a ketone rather than an ester. Now we find that being next to two ketones is better than next to one ketone and an ester, which in turn is better than next to two esters. But more important still is the actual values of the pK$_a$'s: these creatures are more acidic than water! It is relatively easy to make anions of this type: a base as mild as hydroxide or alkoxide will do a pretty efficient job of removing the proton, and the anion will sit around in water or alcohol without significantly stealing a proton back.

Compounds like this, with a methylene group (CH$_2$) nestled between two carbonyl groups, are called "active methylene compounds," because it is so easy to manipulate the corresponding methylene group in this way. The specific example shown above, with two ester functions, is called "malonic ester." Actually that is a generic term, since I have not specified *what* ester it is; the ethyl ester, for example, is officially "diethyl malonate," but as you will soon see, the exact nature of the ester turns out to be irrelevant for the current purposes.

These anions are very easy to work with. You can make them in alcohol solvent, using the proper amount of base, and there will be very little alkoxide left. You can then treat them with alkyl halide and not have to worry about the alkoxide interfering with your reaction. It turns out that these doubly stabilized anions are still pretty good nucleophiles in S_N2 reactions, so alkylations work pretty well:

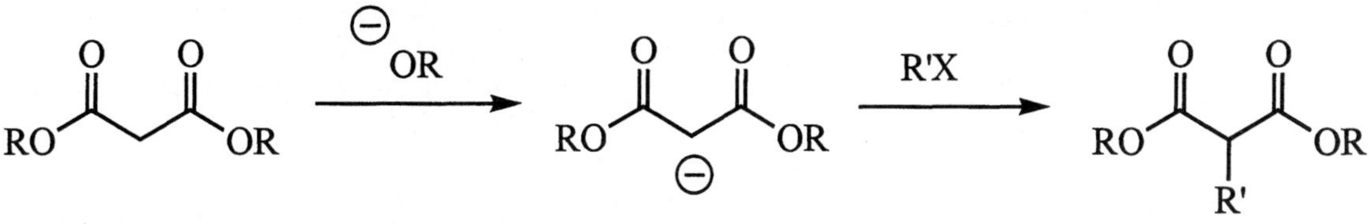

This process can be repeated, allowing a second alkyl group to be put on the central carbon (because there is still one hydrogen left, even though it isn't shown), and the new alkyl group could be the same as the first one or it could be different.

Now, where did this get us? We have a convenient and easy way of putting alkyl groups in the α position of something, but that "thing" is neither an acid nor a single ester. Is this progress?? It turns out that it is, because we are going to discover a way to get rid of one of the ester groups, and turn the other one into an acid. We actually do this in the opposite order. You know how to convert esters into acids. That part is easy: just hydrolyze, most conveniently in base, and then acidify:

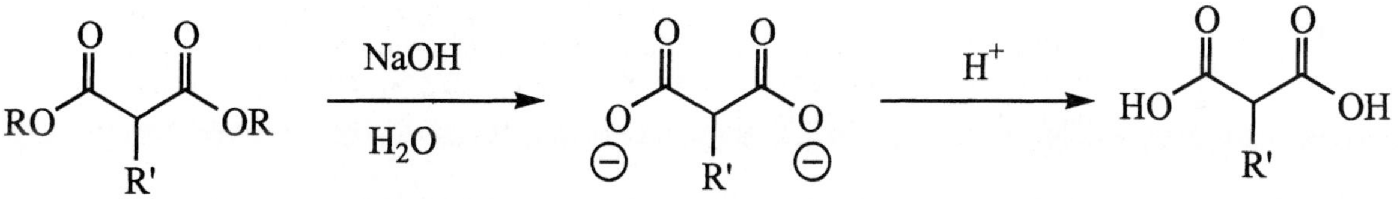

Now we are close to where we want to be. Where is that? It's been so long you may have lost track of the goal here. We are trying to figure out a way to synthesize a carboxylic acid with some random group (or groups) in the α position, placed there by alkylation of an enolate anion. What we have now is a *di*carboxylic acid, with one or more groups in the α position previously placed there by alkylation of an enolate. If only we could get rid of that pesky extra acid group!

As it turns out, we can. Certain carboxylic acids just lose carbon dioxide spontaneously, usually upon heating. The reaction, called "decarboxylation," formally involves the following:

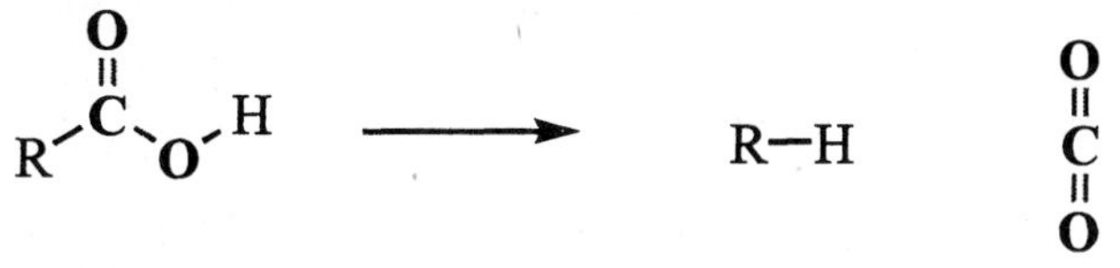

This will not happen to just any old carboxylic acid. There are ground rules. The main ground rule (which most books will not tell you) is that R must be capable of sustaining a negative charge. It is not obvious *why* this rule obtains. It is *as if* the reaction occurred the following way, although in most cases it does not:

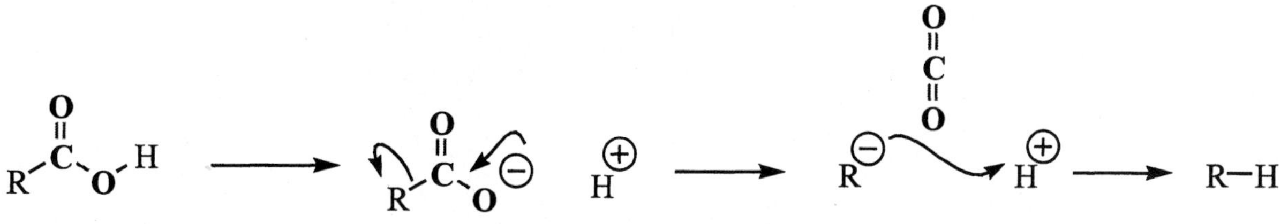

If the reaction worked this way, it would be clear why R would have to be able to sustain a negative charge: it is functioning as a leaving group in an E1cB reaction. As I said, in most cases it appears that the reaction does *not* actually occur this way; nevertheless, when R can serve as a decent leaving group in an E1cB reaction, decarboxylations happen relatively easily, and when it cannot, they don't. The first leaving group you saw in an E1cB reaction was OH⁻ (in the reverse of hydrate formation). If R in the above equation is OH, then we have carbonic acid, which is a compound that in fact spontaneously loses CO_2 to form water:

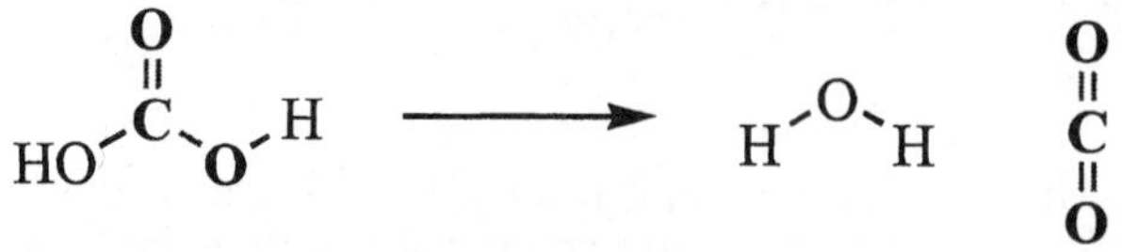

Carbonic acid decarboxylates very easily and quickly: you cannot really keep carbonic acid around. The anion of carbonic acid, however, sodium bicarbonate (more correct: sodium hydrogencarbonate), and also the dianion, sodium carbonate, are very common reagents. Sodium bicarbonate is baking soda, and as a child you probably mixed some with vinegar (acetic acid) to make a volcano using the above reaction. Baking soda works to make cakes rise because it mixes with acidic ingredients in the batter to form carbon dioxide, which in turn leads to little bubbles in the cake. Limestone is the dianion of carbonic acid, as the calcium salt, and geologists test for limestone by dripping HCl onto a rock and looking for bubbles. Acid rain causes buildings and statues to erode by the same process. You should note that both the mono- and dianion of carbonic acid are stable—it is not until both hydrogens are present that decarboxylation occurs. This suggests strongly that, as I indicated, the mechanism I wrote above is not correct. But carbonic acid still follows the basic pattern I described.

More to the point, any carboxylic acid with a carbonyl group in the β position also follows the rule:

I repeat again: this probably does *not* happen by loss of proton, loss of CO_2 to make an enolate anion, and recombination with the proton. I'll show you in a second what we think does happen. But the important point I want to make is that the reaction follows the ground rules: an enolate *can* serve as the leaving group in an E1cB reaction (reverse aldol, reverse Claisen) so we should anticipate that this reaction will also occur readily. It is not quite as easy as the decarboxylation of carbonic acid—but it does occur without too much trouble. Boiling water usually does the job.

How does it *really* happen? For compounds like the one shown, with a carbonyl group in the β position, it is almost certainly as shown below: there is a cyclic shift of electrons leading to CO_2 and the enol of the carbonyl group, which, like all enols, promptly reverts back to being a carbonyl system.

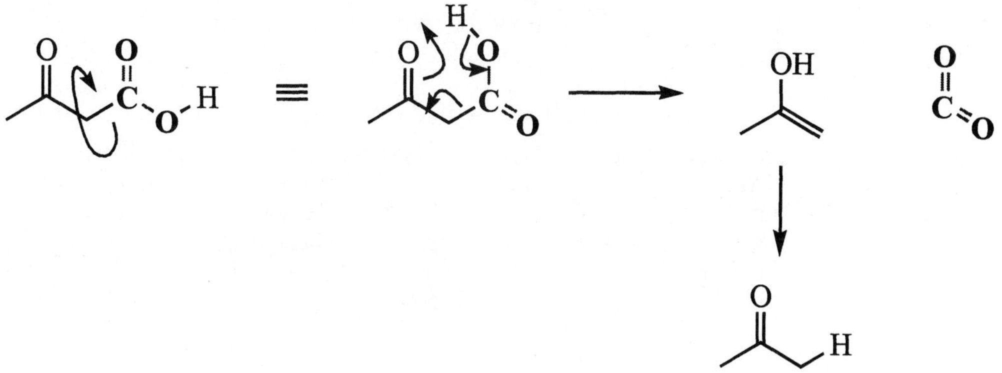

I hasten to point out that carbonic acid *cannot* decarboxylate by this mechanism, and there are other acids that have the same problem and nevertheless decarboxylate. So it is still useful to remember the ground rule. Nevertheless, *most* decarboxylations we will see probably occur by this cyclic mechanism.

You should notice that the diacid we made earlier fits into this category of acids that ought to decarboxylate easily: it has a carbonyl group in the β position. Indeed, when we heat a diacid like this, one (but not both) of the carboxylic acid groups disappears as CO_2.

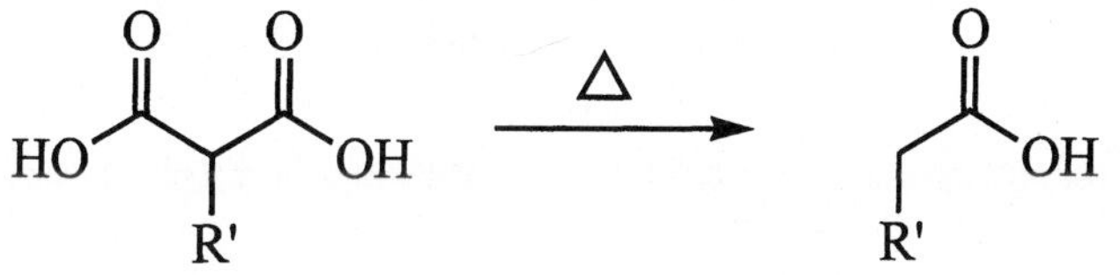

Test Yourself 7

Show the cyclic mechanism for the above reaction.

So we have finally figured out how to alkylate the α position of an acid: we don't do it directly, but we achieve the same result by starting with an ester of malonic acid, alkylating *that*, then hydrolyzing and decarboxylating:

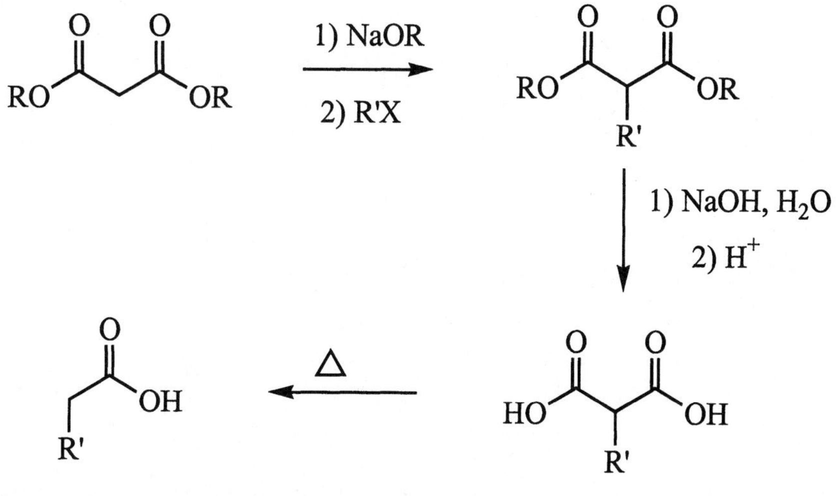

This sequence is called the "malonic ester synthesis," and although it is several steps long, all the reactions are pretty easy to carry out and they proceed in pretty good yields, so it is a common strategy for making substituted carboxylic acids. It should be clear to you that this strategy would also allow for insertion of a second alkyl group at the α position (potentially different from the first), but not a third.

The strategy exemplified by this reaction, that of using a second carbonyl group to enhance the acidity of an α proton for a reaction, and then disposing of the second carbonyl group, is also used by biological systems. More relevant to us, as we proceed into the next chapter, is that the ability of carboxylic acids to decarboxylate readily if (and only if) they are on a position capable of bearing a negative charge has important implications in the metabolism of the foods we eat. In the next chapter we will examine the strategy employed by living organisms as they deal with food and the problem of converting it into energy.

Problems

(Answers are provided at the end of the chapter for italicized problems.)

1. Show the organic products of the following reactions. Write NR if there is no reaction.

    a.

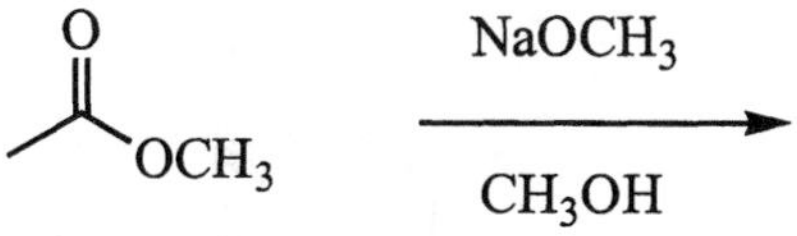

    b.

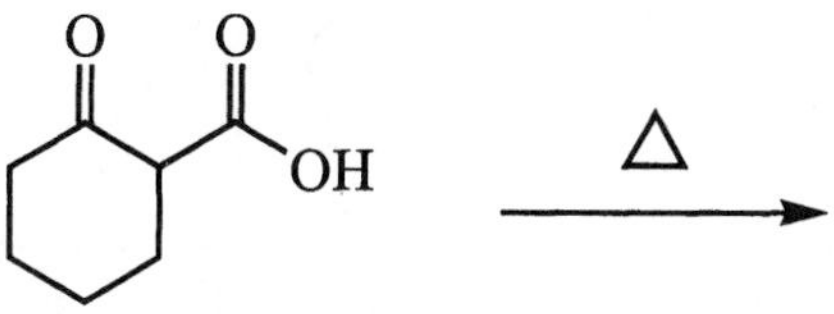

    c.

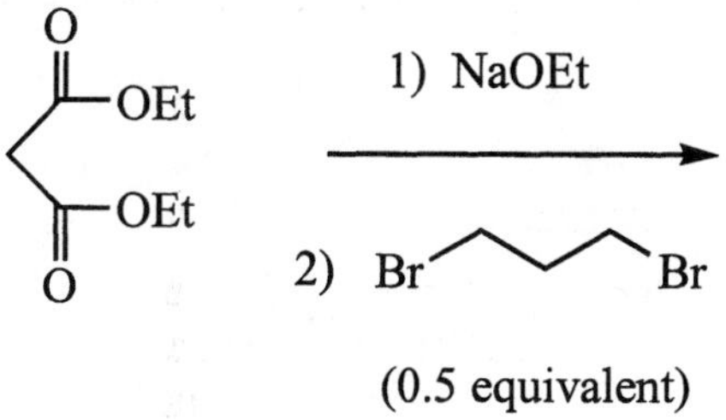

    d.

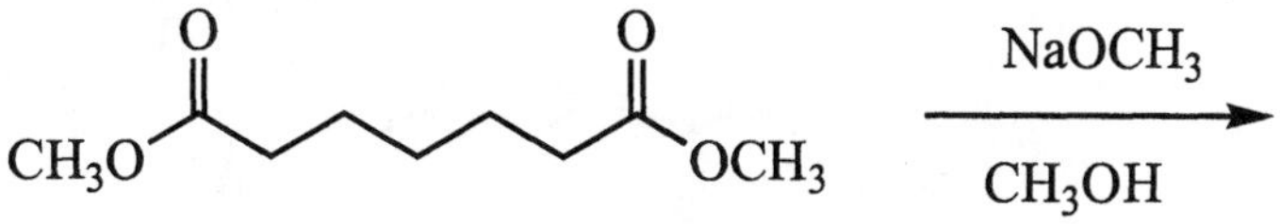

2. Suggest a mechanism for the following reaction. Be sure to show all relevant resonance structures.

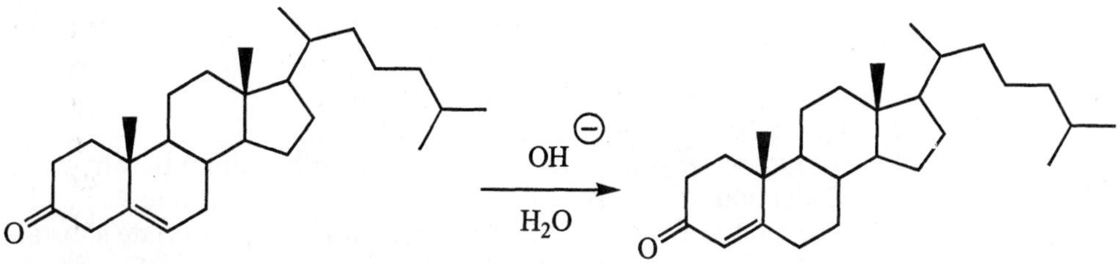

    You do not need to draw the whole structure for each step, just the relevant parts. Standard shorthand for the steroid structure is

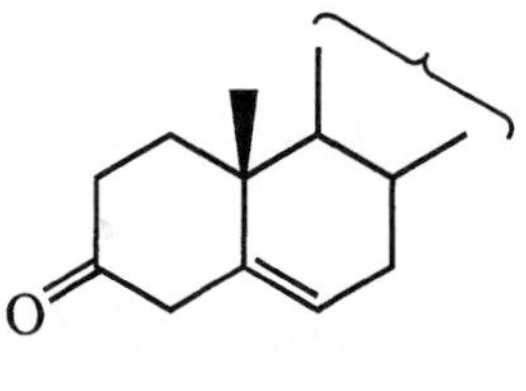

3. *Suggest a mechanism for the following reaction. Make sure all carbocations you show are resonance stabilized. Notice that the double bond has moved—this is not a typo!*

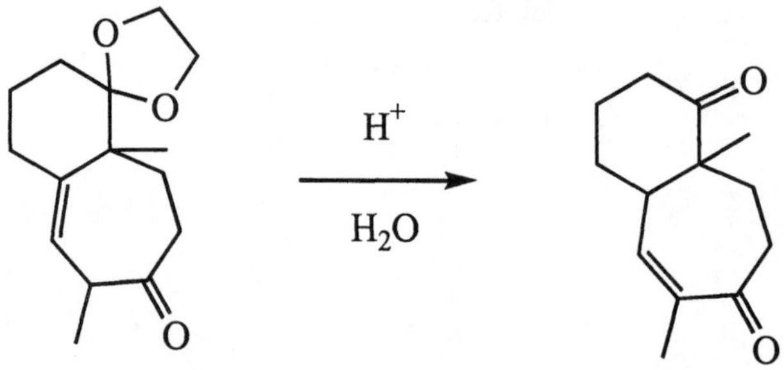

4. How can you account for the fact that *cis-* and *trans*-4-*tert*-butyl-2-methylcyclohexanone are interconverted by base treatment? Which of the two isomers do you think is more stable, and why?

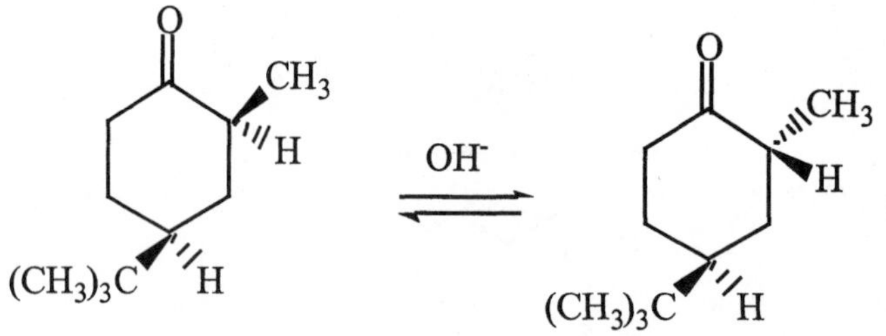

5. *Which hydrogens of the following compound should be readily exchangeable for deuterium?*

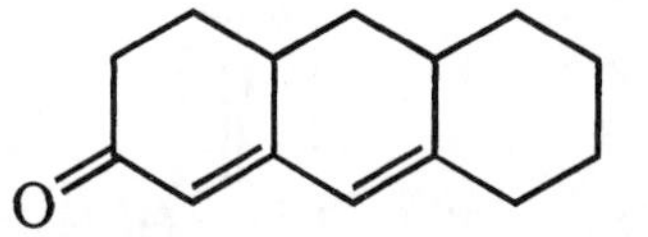

6. Explain why it is important to be careful when selecting a base for a Claisen condensation. Your answer should include a mechanism. Use the following starting material.

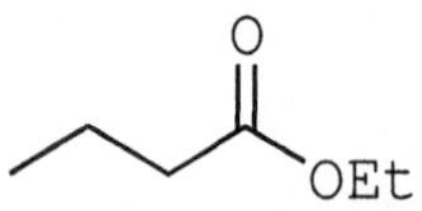

7. Write mechanisms for the following reactions:
   a. Claisen condensation of ethyl acetate with itself.
   b. Reduction of your product from part a with sodium borohydride.
   c. Reduction of your product from part a with LAH followed by acidic water.
   d. Reaction of the product from part a with the appropriate substances such that the final product obtained is

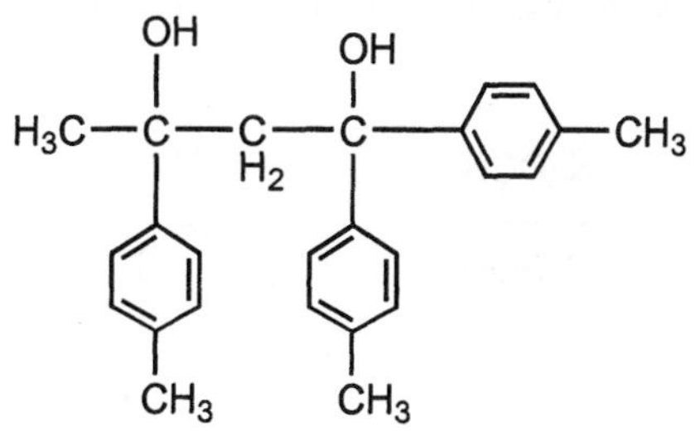

8. Remembering that bromine is an electrophile, show a plausible mechanism for the following reaction.

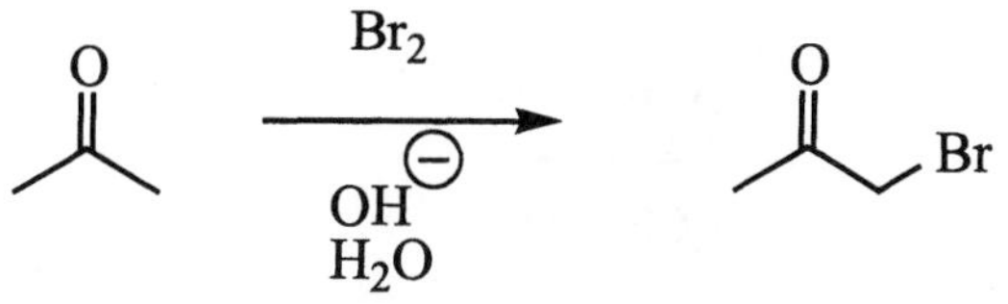

9. Remembering again that bromine is an electrophile, show a mechanism for the following reaction. This is much harder than the previous problem. Don't forget, there has to be some way of getting usable electrons on the α carbon to react with the bromine.

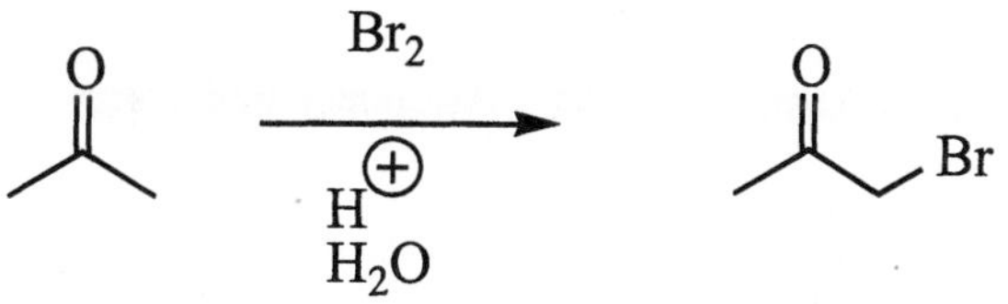

10. Given that bromine is electron withdrawing, which of the above two reactions should be difficult to keep from going farther?

11. Remembering that bromine is electron withdrawing, suggest a mechanism for the following reaction:

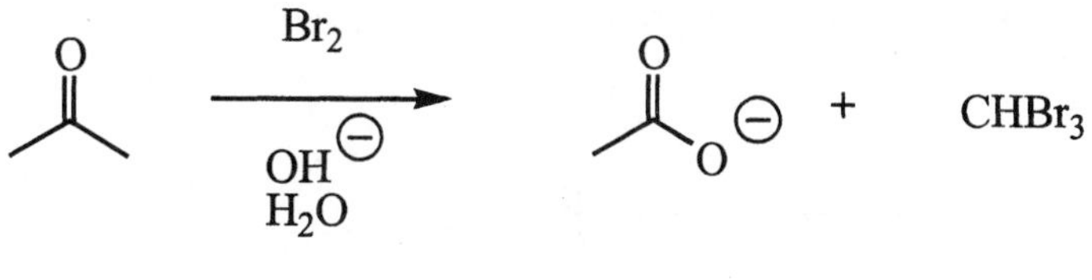

ORM: Under "Carbonyl Compounds," check out "halogenation of a ketone, acid catalysis" and "base catalysis."

12. *Show a mechanism that reasonably permits sodium ethoxide in ethanol to transform 3-methyl-3-hydroxycyclohexanone into heptane-2,6-dione.*

13. Fill in the blank structures for the following aldol condensations, as shown for the first example:

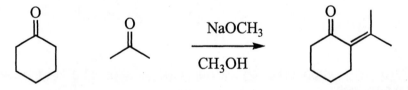

2 different products   4 different products

14. *A "crossed aldol condensation" such as the one below is not generally a useful process. Show the mechanism of the reaction illustrated. Explain why it would not be efficient. Can you suggest a pair of compounds for which this type of reaction would be efficient (not necessarily leading to this same product)?*

15. Do a mechanistic aldol **retrosynthesis** (that is, work backwards) to show how 3-phenyl-2-butenal can be formed. Choose any relevant reagents and identify the reactants.

16. *Show how the aldol condensation permits butanal (butyraldehyde) to serve as a convenient source for all the carbon atoms in 2-ethyl-1-hexene.*

17. Suggest a synthesis of the following compound, starting with bromocyclohexane and anything else you like containing three carbons or less:

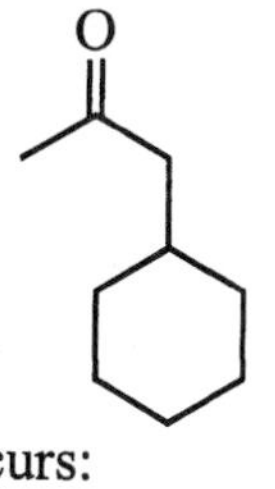

18. Explain why the following reaction occurs:

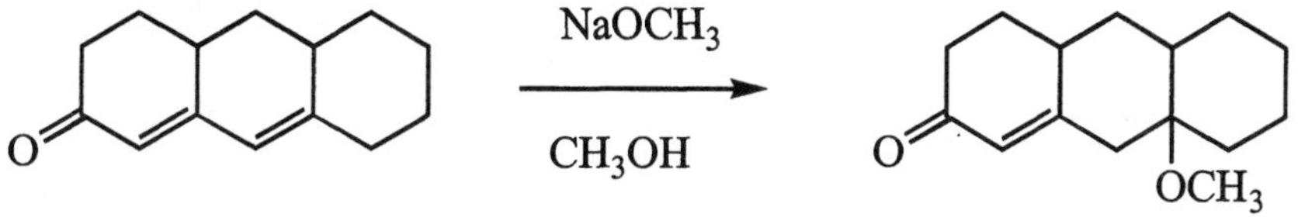

NaOCH₃

CH₃OH

19. Show a mechanism for the following reaction:

+

NaOCH₃

CH₃OH

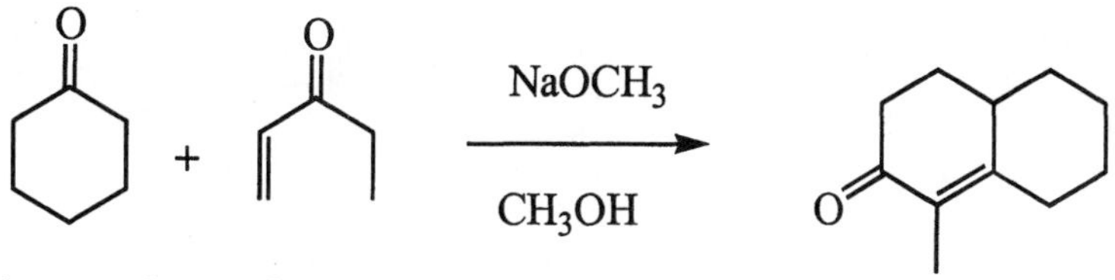

20. *Suggest a mechanism for the following reaction, showing all relevant resonance structures:*

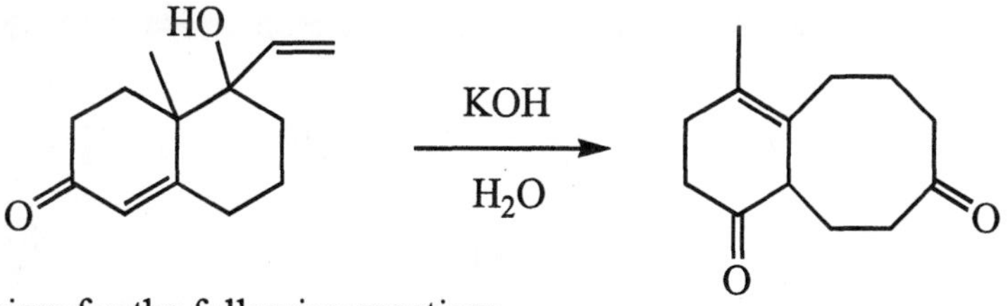

KOH

H₂O

21. Show a mechanism for the following reaction:

+

NaOEt

EtOH

H_3O^+

Δ

5,5-dimethylcyclo-
hexane-1,3-dione

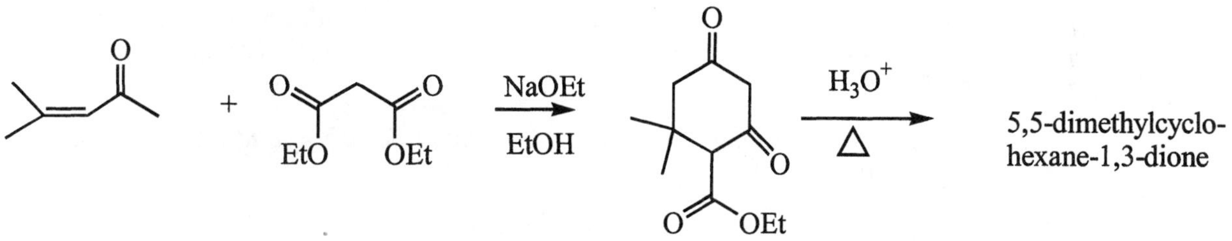

22. The malonic ester synthesis with dimethyl malonate, LDA, and 1,5-dichloropentane yields a diester with molecular formula $C_{10}H_{16}O_4$.

a. Write a mechanism for the synthesis including (obviously) the structure of the product.

b. Write a mechanism for saponification of the product from part a (that means, hydrolysis with sodium hydroxide followed by reaction with acidic water) and decarboxylation (removal of carbon dioxide). The molecular formula of the final product is $C_7H_{12}O_2$.

23. Supply the missing compounds or reagents:

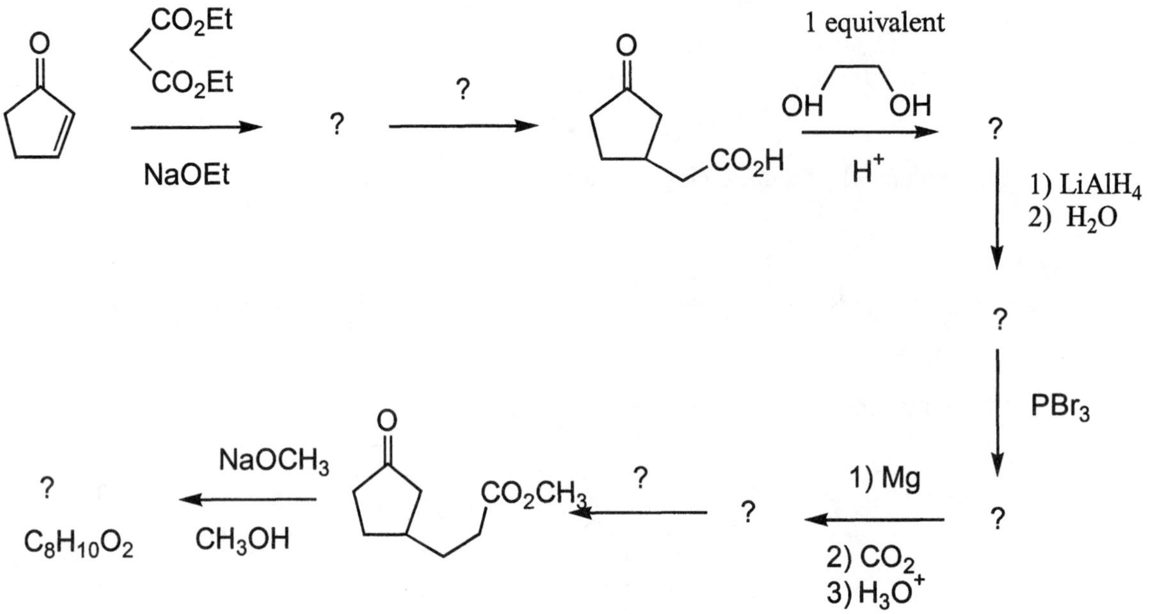

CO_2Et / CO_2Et

NaOEt

? → ?

1 equivalent

OH OH

H^+

?

1) $LiAlH_4$
2) H_2O

?

PBr_3

CO_2H

NaOCH$_3$

CH$_3$OH

? $C_8H_{10}O_2$

CO_2CH_3

?

?

1) Mg
2) CO_2
3) H_3O^+

?

24. a. Show how ethyl acetate can be used in a Claisen condensation to prepare ethyl acetoacetate (shown below; note how the name is derived!).

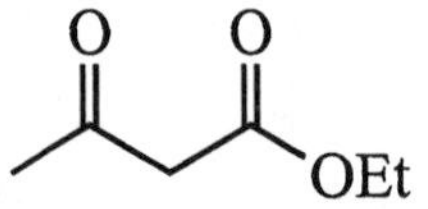

ethyl acetoacetate
(a derivative of acetoacetic acid)

b. Ethyl acetoacetate (also known as "acetoacetic ester") in turn serves as a convenient starting material for making 3-methyl-2-pentanone. Show how this multistep synthesis might be carried out.

25. *Which of the following carboxylic acids should decarboxylate readily? Careful—some of these are tricky!*

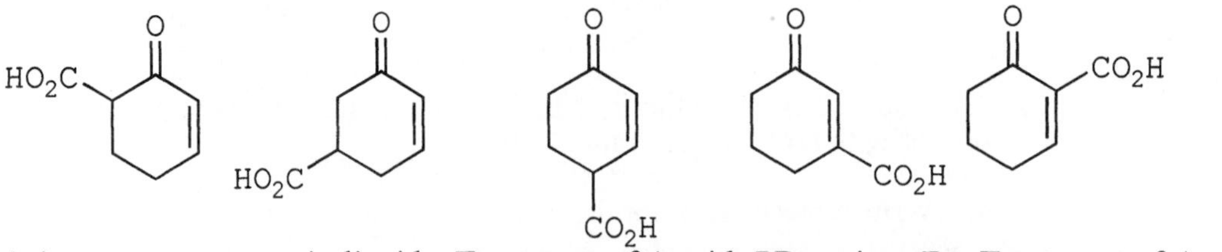

HO_2C

HO_2C

CO_2H

CO_2H

CO_2H

26. A is a common organic liquid. Treatment of A with PBr_3 gives B. Treatment of A with $Na_2Cr_2O_7$ in sulfuric acid gives C, which reacts with A and acid to give D. D reacts with sodium ethoxide to give E, which reacts with base followed by B to give F. F can be hydrolyzed in aqueous base followed by acid to give G, which on heating loses CO_2 to give H. H reacts with sodium borohydride to give I, which reacts with C and acid to give J.

^{1}H-NMR, ^{13}C-NMR, and mass spectra of many of these compounds are provided. Identify A through J, all of which are neutral organic compounds (i.e., they are not ions).

A: IR 3500 cm^{-1}

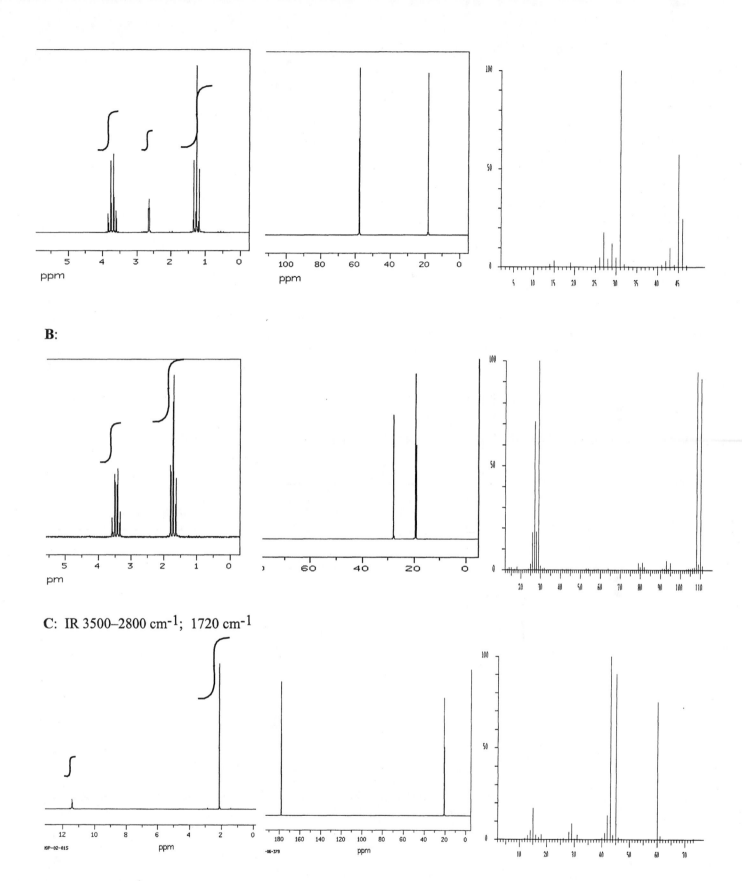

B:

C: IR 3500–2800 cm^{-1}; 1720 cm^{-1}

ISP-02-015

-06-379

D: IR 1725 cm^{-1}

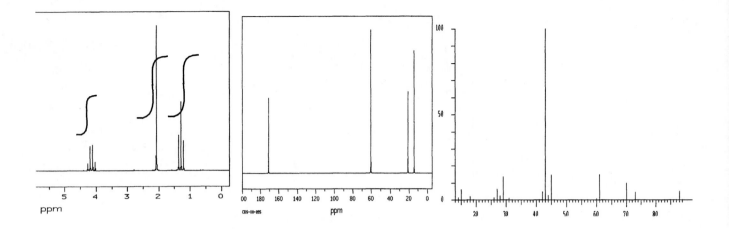

E: IR 1750 cm^{-1}

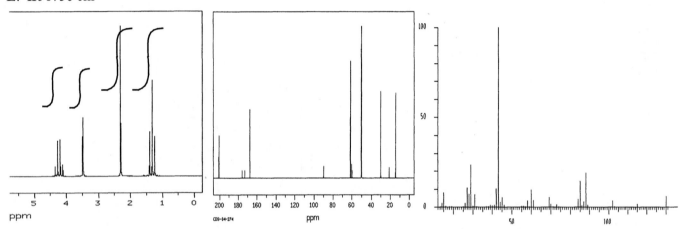

F: IR 1750 cm^{-1}

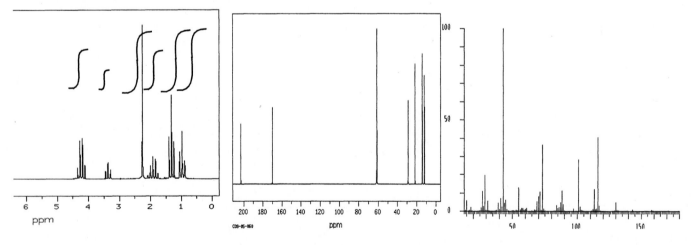

H: IR: 1735 cm^{-1}

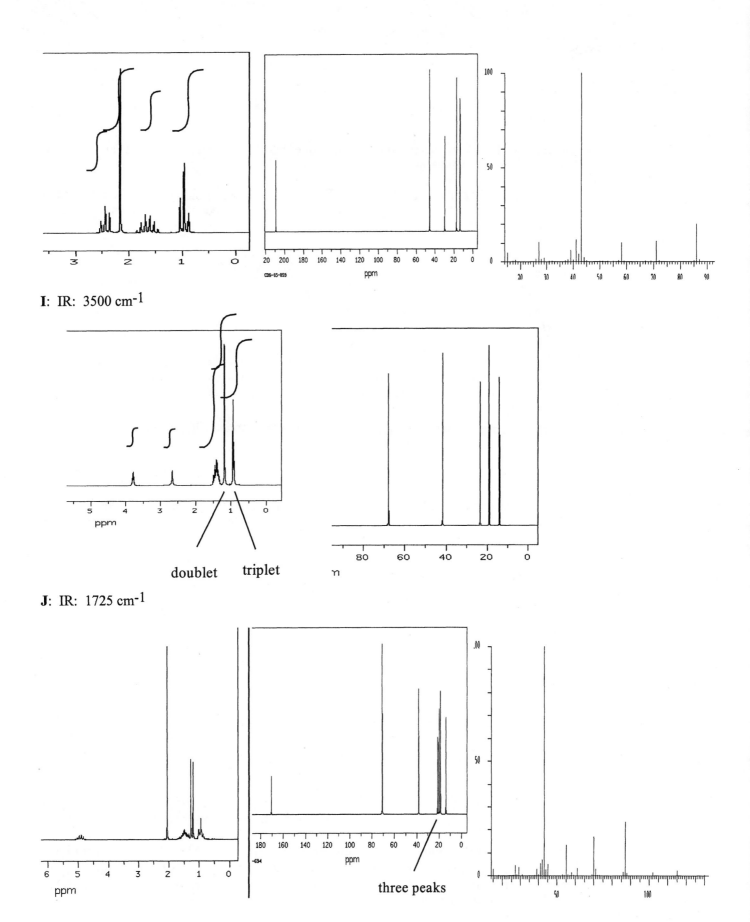

I: IR: 3500 cm^{-1}

doublet   triplet

J: IR: 1725 cm^{-1}

three peaks

Selected Answers

Internal Problems

Test Yourself 1.

The ones to the left of the carbonyl are just like the others we have done. They should be trivial, and I won't discuss them.

The ones on the bottom are a bit trickier. Let's try it in base: proton off, new one on. After we take the proton off, we need to justify our ability to remove it by showing resonance structures.

Then we show the new one coming on. To do this we use the most convenient resonance structure. Repeating the process will get rid of the other hydrogen at this position.

Can we do the same thing in acid? Sure: proton on, proton off; new one on, old one off.

Again we can repeat the process and replace the other hydrogen on the bottom.

The toughest proton to show exchanging is the one on the right. Most students just rip it off in base—after all, it is α to the carbonyl, right? Well, yes, but ... let's think about *why* α protons can be removed. It is because the resulting anion can be resonance stabilized, right? Could this be such an anion? Let's see:

Yuck! This resonance structure looks terrible! The three carbons would need to be *linear*, and they can't be. This anion, even though it shows a negative charge α to a carbonyl, is not stable like other enolate anions. So you cannot just rip this proton off. But you may recall that earlier in this answer, we *did* come across a minus charge on that carbon.

The middle resonance structure has a legitimate, stabilized negative charge on the target carbon. From this anion, we can add a deuterium, then remove the hydrogen to get back to the previous anion (now with deuterium instead of hydrogen) and finish off as we did before.

Now you should also be able to show how this particular proton can be exchanged in acid solution. Remember: protonate, then deprotonate. Protonate, then deprotonate. Make sure all your cations are resonance stabilized. No anions in acid. Go for it!

Aldol blanks, p 646:

Claisen Blanks, p 649:

Test Yourself 2

Test Yourself 3

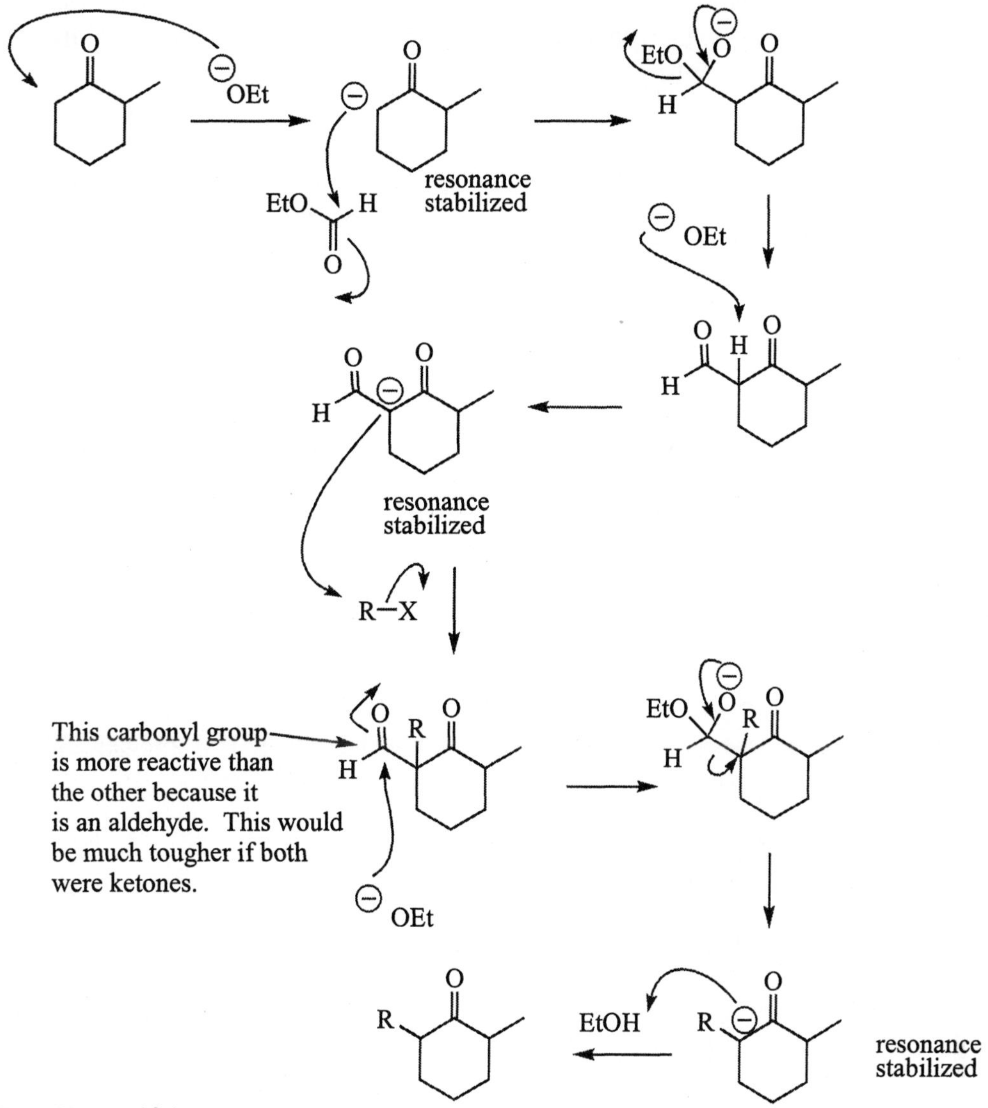

$\ominus$OEt

resonance stabilized

EtO H

O

$\ominus$ OEt

H

O   H   O

H

resonance stabilized

R–X

This carbonyl group is more reactive than the other because it is an aldehyde. This would be much tougher if both were ketones.

$\ominus$ OEt

EtO O R

H

R O

EtOH

R $\ominus$ O

resonance stabilized

R O

Test Yourself 4

O

O   O

O   O

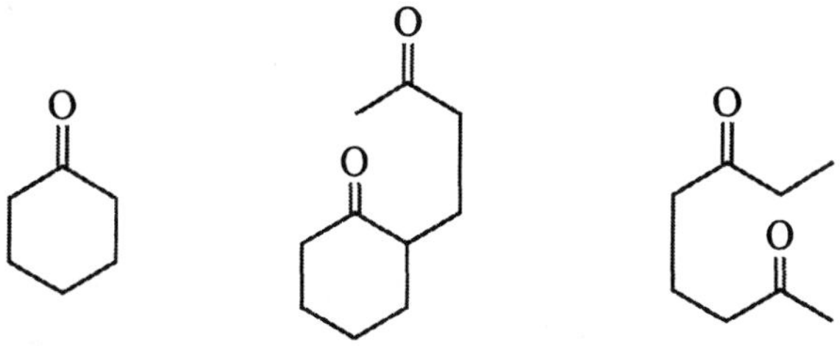

Test Yourself 5

There is actually no "correct" way to approach this problem. Here is one possibility, though: If you look at the product, you see an enone. You know that an enone can result from an aldol condensation, so draw the compound that would react to make this enone:

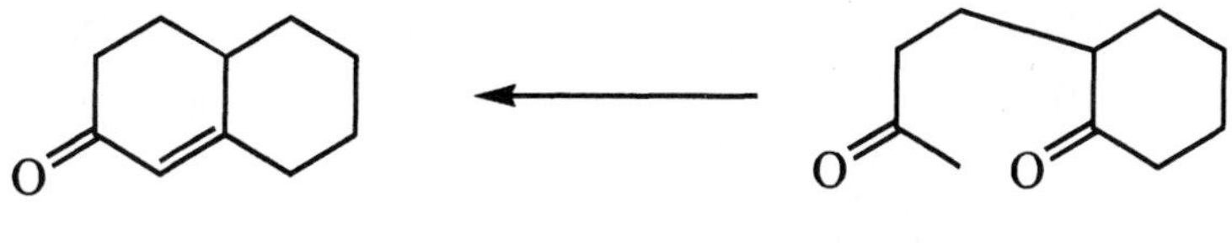

O ⟵ O O

Now concentrate on this new substance and see if it relates in any way to the starting materials we were given. Yes, there is in fact an obvious relationship: the target has acquired a bond between the α position of the ketone we were given and the β position of the enone we were given.

That sounds a lot like the consequences of a Michael reaction. So we should show a Michael reaction followed by an aldol. Here it is, put together:

Notice how I showed the resonance structures for all the enolate anions. I placed them in brackets to emphasize that each is a package deal. This is a common practice, but it isn't absolutely necessary.

Test Yourself 6

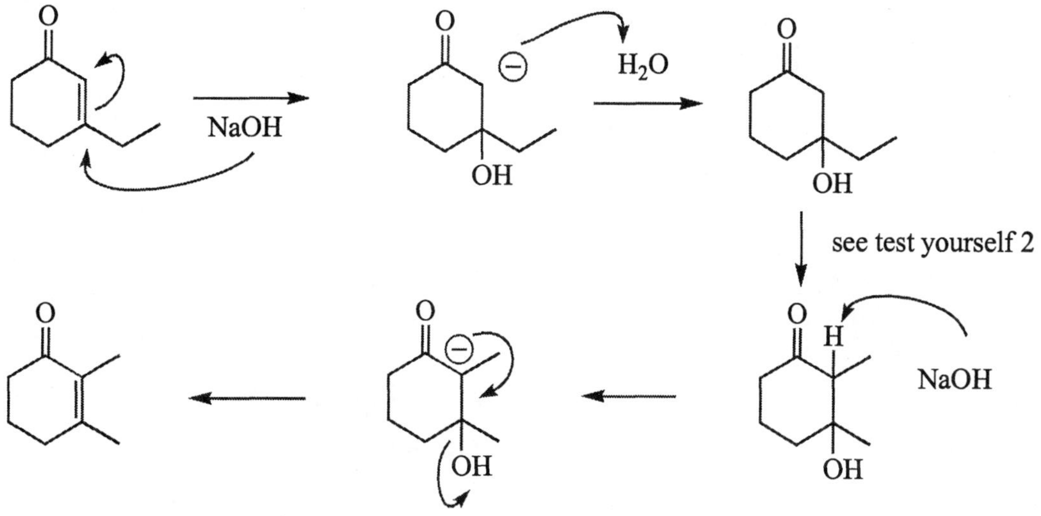

Test Yourself 7

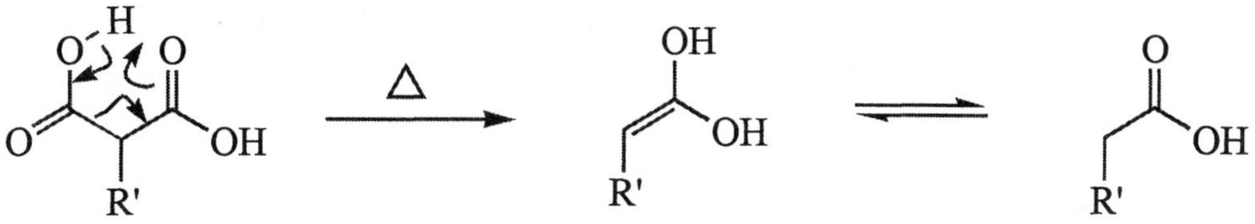

End of Chapter Problems

3.

5.

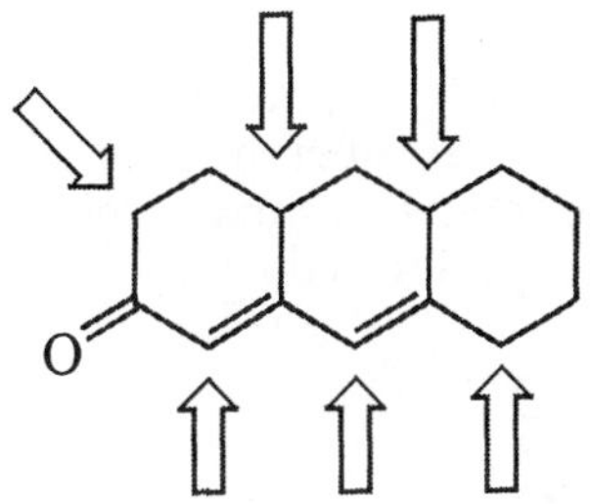

12.

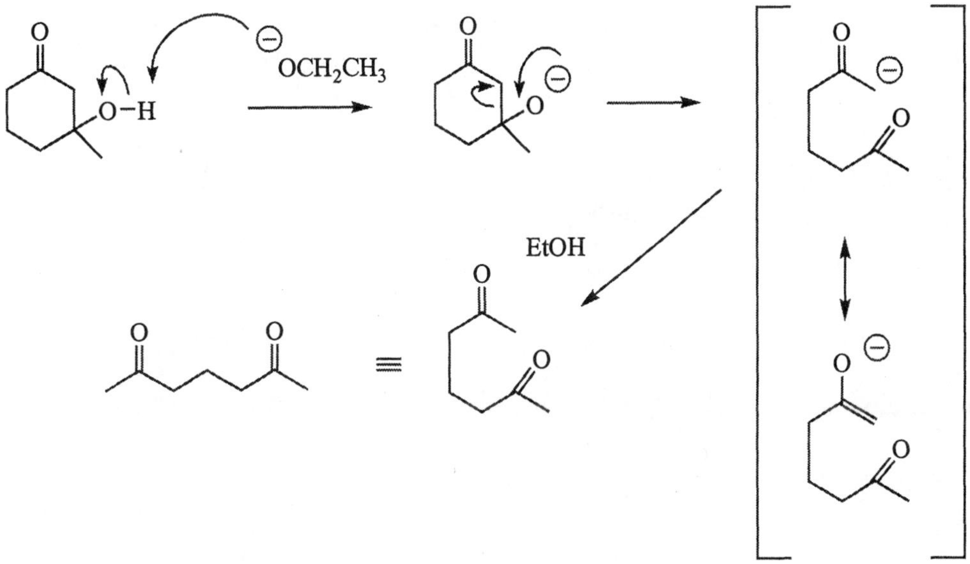

14.

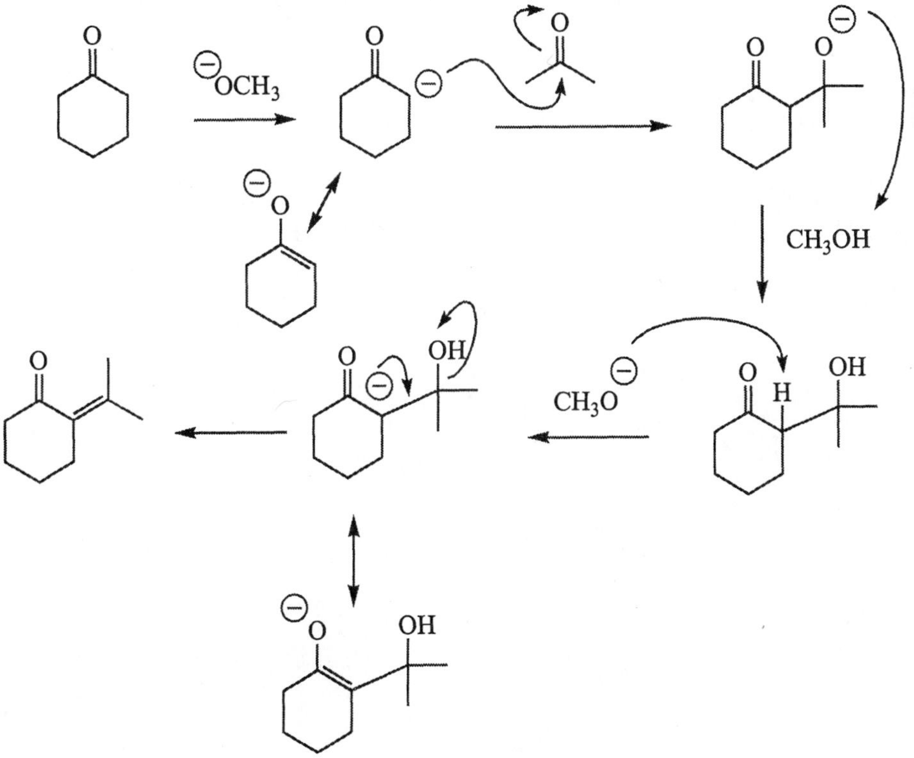

Both ketones can form enolate anions, and both ketones can be attacked by an enolate. Thus, this is only one of four possible products (show the others). You should expect at best a 25% yield, because all possibilities are equally likely.

To make this work, you would need to arrange for only one of the partners to be able to form an anion, and the other partner should be especially reactive toward nucleophiles. For example, the following mixed aldol works very well. Show the mechanism and explain why other possible products do not form.

16.

20.

HO
OH
O
O
O
O
O
O
O
O
O
O
O
H₂O
O
O

25.

HO2C
O
This one
O
HO2C
O
CO2H
This one
O
CO2H
O
CO2H

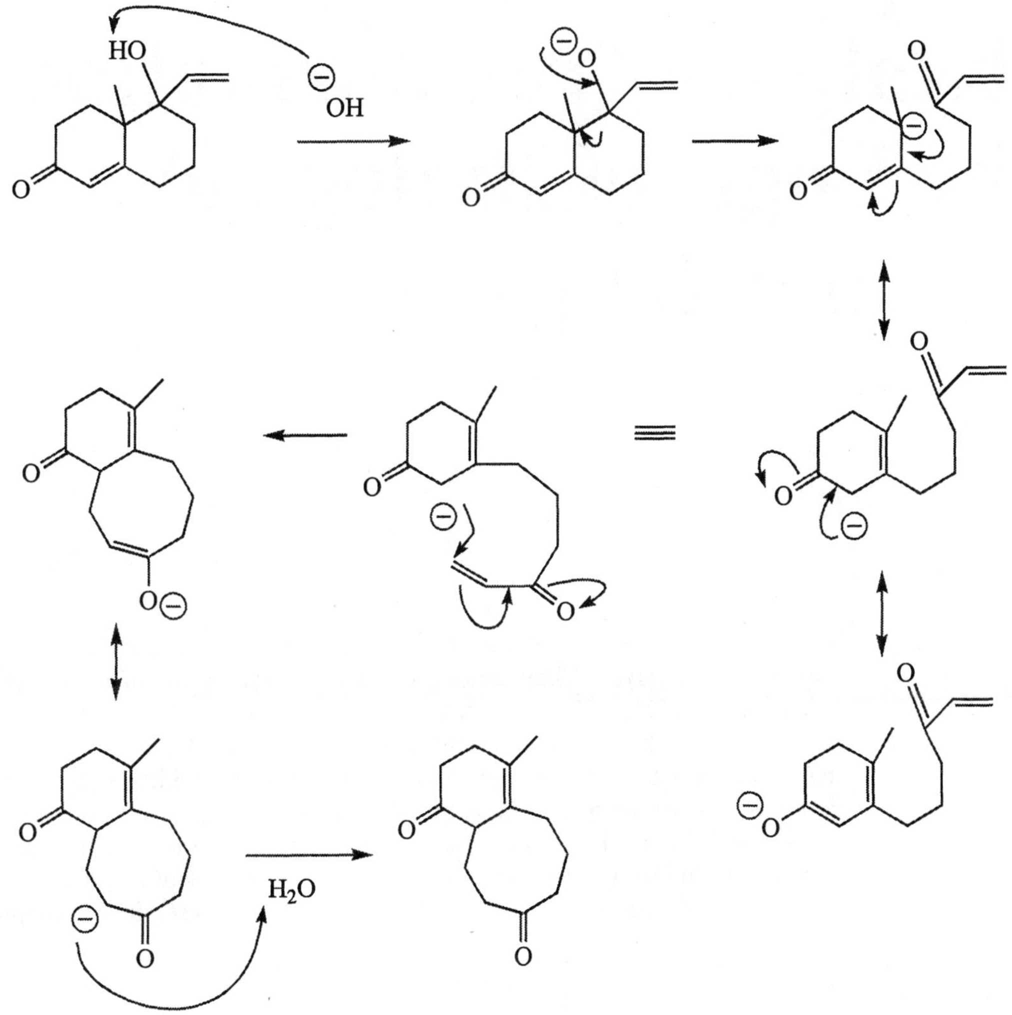

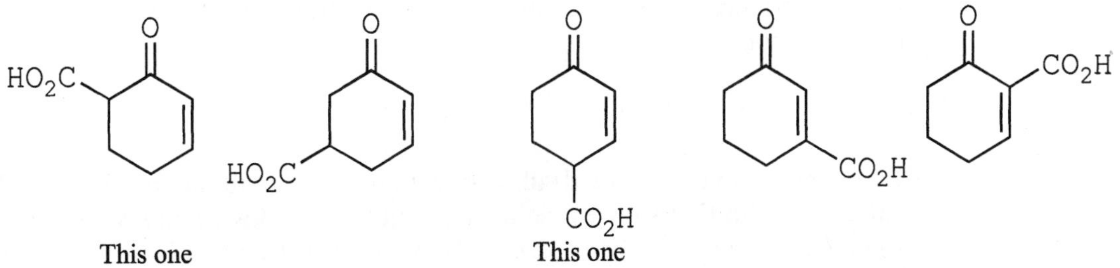

Chapter 21
Metabolic Transformations: The Logic of Biological Reaction Sequences

Many of you have encountered in biology courses a description of some of the standard metabolic pathways. Chief among these are "glycolysis" and the "TCA" (or "Krebs" or "citric acid") cycle. You may have learned the names of the compounds formed along the way (glucose, glucose-6-phosphate, fructose-6-phosphate, etc.); you may have learned the names of the enzymes that are involved (phosphofructokinase, etc.); but what you almost certainly did *not* learn is *why* nature has constructed the sequence in this way rather than some other. My goal in this chapter is to show you that there really is a logic to the pathway, and if you understand the thought process of Mother Nature, it becomes clear why these pathways (and many others) are organized the way they are. In this chapter we will

- Learn the five enzyme reaction types

- Discover how these five reaction types dictate the metabolic pathways

- See how Nature manages to bypass some restrictions in order to accomplish forbidden transformations

- Construct an energy budget for the metabolism of glucose.

Before proceeding, I would like to express my gratitude to Bob Abeles of Brandeis University, who first showed me the logic in these pathways in a visit to Haverford College, where I was teaching in 1978–1979. He was recovering from throat surgery at the time and could barely speak, but what he said spoke volumes.

THE FIVE ENZYME REACTIONS

Enzymes are usually presented as magic boxes. Substrate goes in, magic is done, product comes out. We describe them as masterful in what they do, and it often seems that for any desirable biological transformation, there exists a magic enzyme perfectly constructed to accomplish the task. There is a dirty little secret about enzymes that I will tell you now: the rumors are true, but only up to a point. Enzymes *are* perfectly constructed to do what they do, and they are supremely good at it, but they cannot do just anything. Indeed, to a first approximation (there are some exceptions, but this generalization covers a remarkable amount of biochemistry) there are only five things enzymes know how to do well. All of them are reactions we have already studied, reactions we understand pretty well. And here is the key: in order to get from point A to point B in any biological sequence, the organism has to use this particular limited tool kit. It must figure out how to string together these five reactions in such a way that it manages to get from A to B. In this sense, biological metabolism is not so different from synthesis problems you have been doing in this course, in which you have a limited set of tools with which to work and you have to string various steps together involving these tools in some sequence that gets you in a legitimate way from starting material to product. The cell's tools are even more limited than yours, but the "simple" cell has one big advantage over you: whatever it wants to do, within its vocabulary, it can do perfectly, and it can do all such things either forward or backward, at will.

So just what is the "vocabulary" of a typical cell? Here are the five allowed enzymic reactions, from which a cell will build all the observed sequences. Remember, all steps are equally good in either direction.

Oxidation–Reduction

As I discussed earlier, Mother Nature does not use chromic acid or PCC for oxidation; she does not use $NaBH_4$ or LAH for reduction. She has her own special reagents, which work very well for her purposes, and we discussed them previously in Chapter 18. Depending on what is to be oxidized/reduced, different redox agents will be used:

For carbon–oxygen bonds, NAD^+ is preferred:

Note that this reaction also covers the oxidation of an aldehyde to a carboxylic acid, since that reaction proceeds *via* a hydrate, which resembles the stuff above on the left.

For carbon–nitrogen bonds, NAD^+ is also the standard reagent:

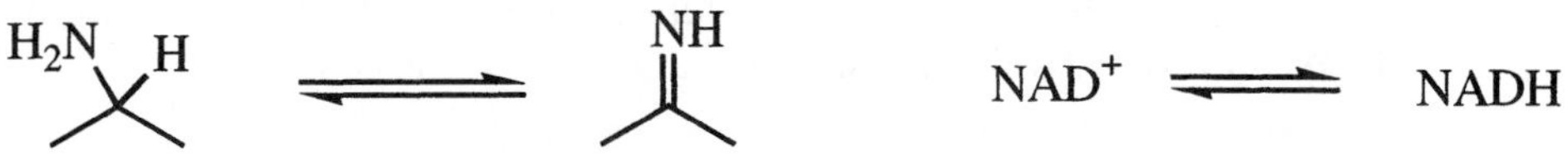

For carbon–carbon bonds, FAD is the more common choice:

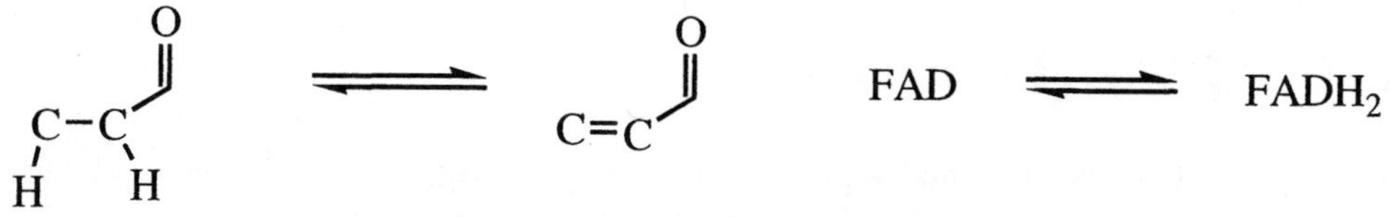

Note that this last reaction is both more restricted and less restricted than laboratory processes we have seen: although we know ways to hydrogenate any carbon–carbon double bond, Nature can hydrogenate an alkene only if it is conjugated with a carbonyl group. On the other hand, we can only carry out the above reaction from right to left, while Nature has no trouble with the reverse process—again, so long as the two carbons in question are α,β to a carbonyl group.

Elimination–Addition

Nature's vocabulary is somewhat limited in this instance, compared to ours: while we can add reagents to just about any double bond, Nature is largely restricted to working with double bonds conjugated to carbonyl groups; that is, Nature carries out Michael additions. Likewise, Nature accomplishes E1cB eliminations provided they lead to enones.

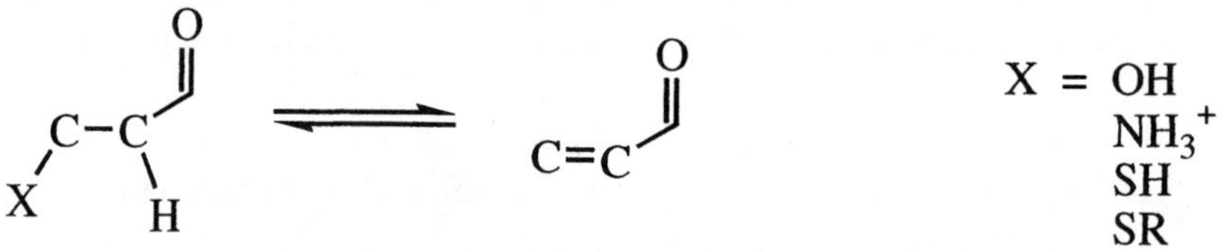

Notice that nucleophile/leaving group always enters/leaves at the position β to the carbonyl group.

Aldol (Claisen)

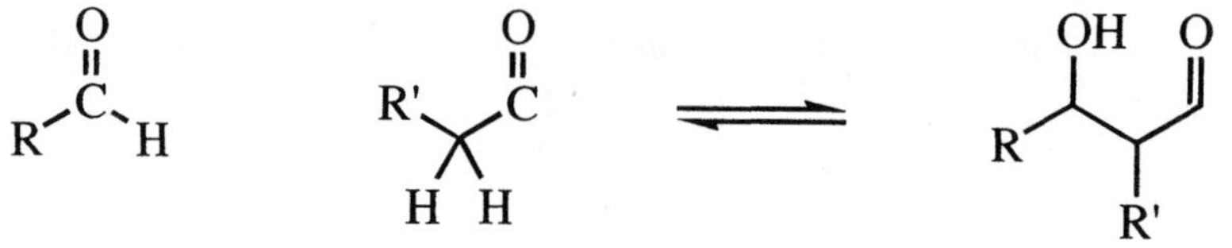

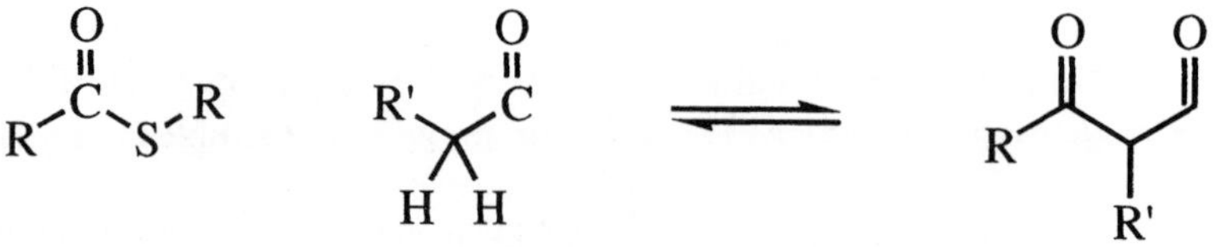

In biological metabolic reactions, esters usually take the form of thioesters, even though the chemistry is effectively the same as that for oxygen esters.

Decarboxylation

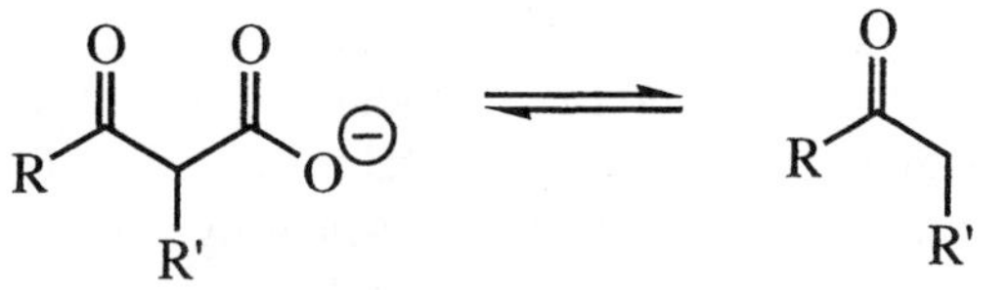

Recall that biological chemistry typically happens at pH ~7, which means that carboxylic acids will be deprotonated most of the time. You should also recall that this type of reaction is restricted to positions capable of sustaining a negative charge. Under normal circumstances, this means there has to be a carbonyl group β to the carboxylic acid undergoing decarboxylation, although we will later discover a sneaky way to avoid this restriction.

Acyl/Phosphoryl Transfer

When a carbonyl group and its associated garbage moves from one atom to another, we refer to the process as "acyl transfer." Hydrolysis of an ester is one simple example:

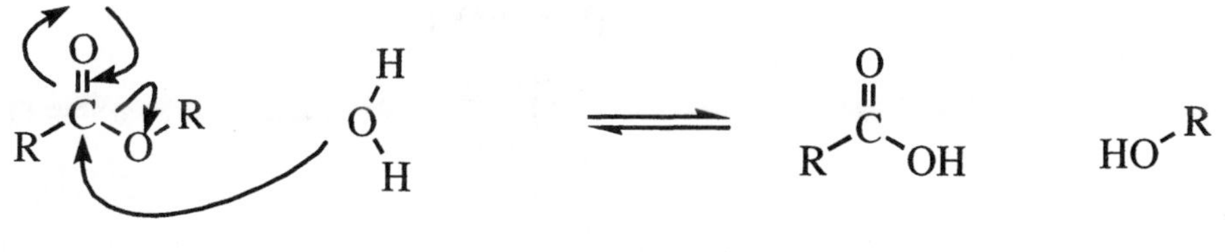

Although the actual mechanism is a little more complicated than depicted here (there are some proton transfers I've not shown), the net effect is that an acyl group hops from its original oxygen onto the oxygen of water. This could therefore be considered an acyl transfer reaction. (In another sense it is a substitution reaction, like an S_N1 or S_N2, although the mechanism is not quite the same.) Another example would be the hydrolysis (or formation) of an amide linkage, which is the structural basis for enzymes:

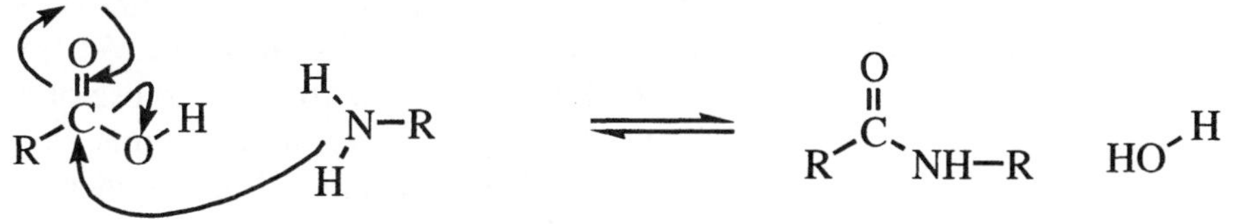

A phosphoryl transfer would be the analogous process with phosphorus, such as when ATP dumps a phosphoryl group onto an alcohol, or *vice versa*:

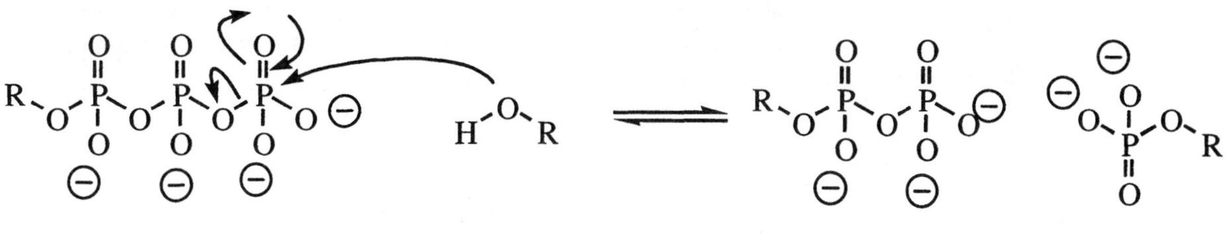

ATP   ADP

This, then is the chemical vocabulary of biology, and with relatively few exceptions, it is the complete vocabulary. Our job in the present chapter is to see how a knowledge of Nature's vocabulary leads to an understanding of the metabolism of carbohydrates. An important caveat, however: I will be discussing mainly the *strategy* of the reaction sequence. In most cases I will not be discussing details of *how* enzymes carry out these transformations, or how they know which direction to take a reaction, or when a reaction should be turned on or off, or any of a variety of other fascinating topics that are the subject matter for biochemistry courses. I hope many of you will be inspired to take such courses.

GLYCOLYSIS

The first order of business is to establish what Nature is trying to accomplish. Carbohydrates destined to be metabolized generally show up in the cell as glucose, so that is where we will start. The final destination of the first sequence, called glycolysis, is pyruvate, the anion of pyruvic acid. Again, since we are dealing with physiological conditions, pH 7 or so, carboxylic acids are almost always present in their ionized forms, so I will show them without H and with a charge. We are here trying to accomplish the following transformation, using reactions we know to be allowed.

glucose   pyruvate

I have shown glucose in its open, aldehyde form rather than the cyclic form in which it is usually found because it is easier to understand the chemistry from the open form. Since the two are in equilibrium, the cell can choose to operate on whichever form it likes. There are two other goals we ought to keep in mind: glucose has entered the cell through a (non-polar) cell membrane—and we ought to take some steps to keep it from drifting back out. Also, since the entire process of tearing up food is for the purpose of generating energy, we ought to capture energy wherever we can. Energy, in cell terms, is ATP, as you already know. What you may not know is that NADH is also a form of energy, in the sense that there is a pathway ("oxidative phosphorylation"; we will not discuss this in detail) that converts each NADH molecule into three ATP's while changing the NADH itself back into NAD^+.

OK, let's play God and design a strategic approach. Let's start by making sure our molecule does not escape. Glucose is neutral, and can pass through a cell membrane, but charged things cannot. An easy way to get a charge on a molecule is to add a phosphate group to an alcohol function in it, a phosphoryl transfer reaction:

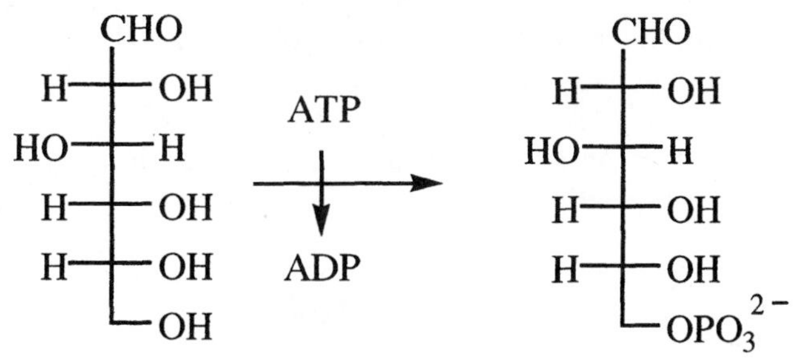

We have now trapped the glucose molecule in our factory. Unfortunately, this has cost us one ATP, the very thing we are trying to make. As in life, you sometimes need to invest some money to make a profit. As you will see, this particular investment will pay off later.

Glucose has six carbons. We are trying to make pyruvate, a three-carbon piece. Clearly the efficient way to do this would be somehow to cut glucose in half. The best carbon–carbon bond forming/breaking reaction we have is the aldol condensation, and in an aldol, the bond formed/broken has to be α,β to some carbonyl group. Our carbonyl group at the moment is on the top (#1) carbon; if we are to break the 3–4 bond, it needs to be shifted to the second carbon. How can we move it? This requires no magic at all; although an enzyme is called upon to steer the process, the reaction itself is trivial. Glucose is an aldehyde, and aldehydes with α hydrogens are in constant equilibrium with their enols:

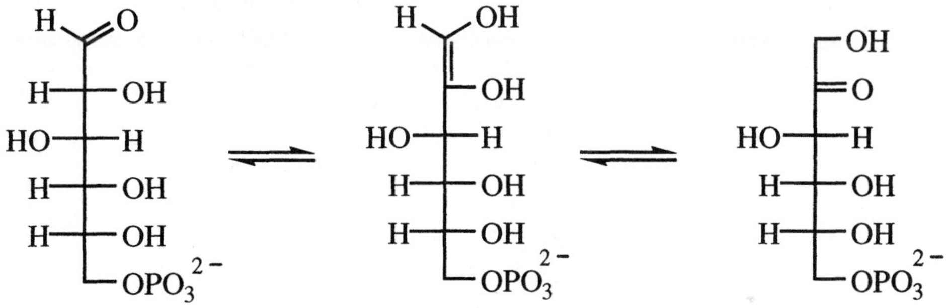

But this enol is special, in that it has OH groups at *both* ends of the double bond. So it is simultaneously the enol of two different carbonyl compounds, and when it stops being an enol it can do so in either of two directions: back to the aldehyde or to a ketone instead. So moving the carbonyl from carbon 1 to carbon 2 amounts to a trivial keto–enol tautomerism, doubled.

Now the carbonyl group is in the right place to contemplate breaking the sugar molecule in half. But wait! If we do that now, one half will break off without a charge, and we are in danger of losing it through the cell membrane. Before we bust this thing up, let's put a phosphate on the *other* end of the molecule too, using another phosphoryl transfer reaction.

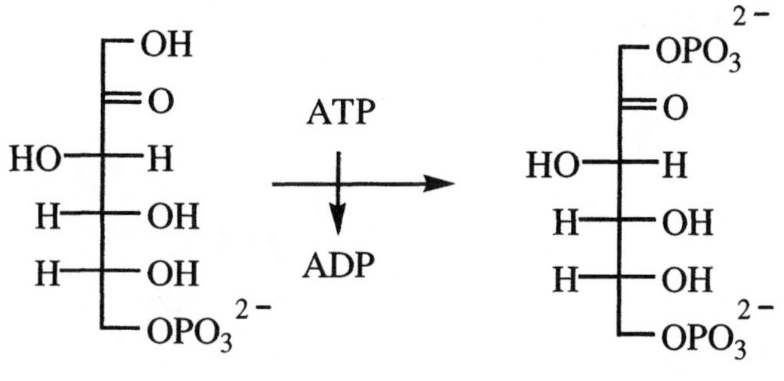

Hmm, this has cost us another ATP. Oh well, nothing ventured, nothing gained. So far we are down two ATPs.

Now we can break the molecule in half through a reverse-aldol condensation, one of our allowed reactions:

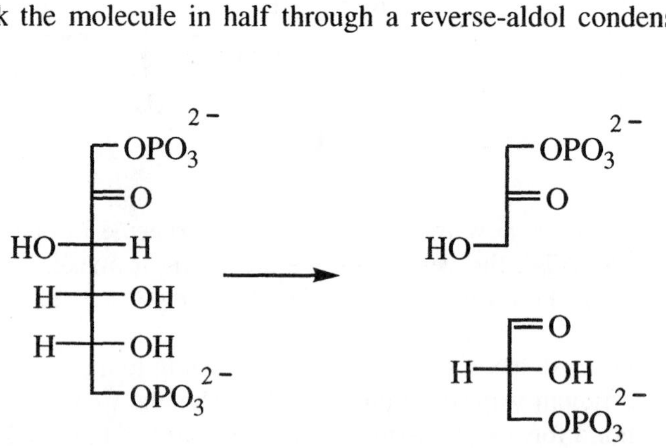

Write this process out to be sure you understand the sense in which it is a reverse aldol. Probably the easiest approach is first to write the mechanism of the reverse reaction, a forward aldol consensation.

Voila! We now have two three-carbon pieces. For efficiency, it would be useful if they were both the same, so that all future reactions could be accomplished on both fragments, rather than creating two different pathways for the two pieces. Currently they are not the same, but they differ only in whether the carbonyl group is on carbon 1 or carbon 2, and we have already seen that moving a carbonyl group in a sugar is a simple matter of keto–enol tautomerism. So we can easily make them alike:

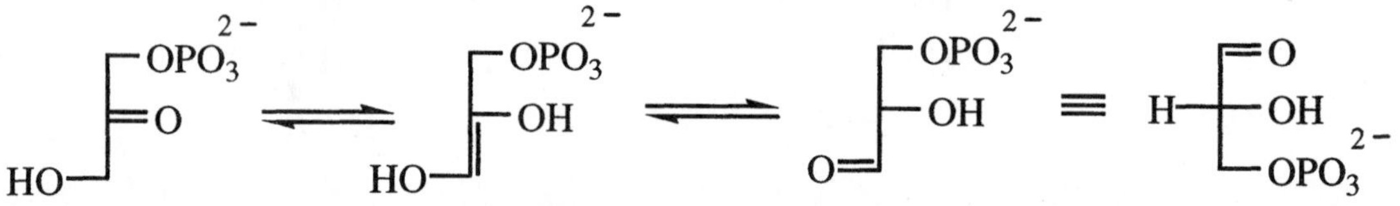

Provided the stereochemistry is handled properly (and enzymes are great at controlling stereochemistry), the two pieces are now identical. (Another good reason for adding the phosphoryl group in the previous step.) From here on we will show only one, but you should remember that every reaction we show will occur *twice* for each glucose molecule we started with.

Let's pause for a moment and see where we are, and how it relates to where we are going. We have glyceraldehyde-3-phosphate, a 3-carbon aldehyde with OH on carbon 2 and phosphate on carbon 3. We want pyruvate, a 3-carbon acid with a keto group on carbon 2. Clearly we need to oxidize our aldehyde to a carboxylic acid, then fiddle around with the substituents on carbons 2 and 3.

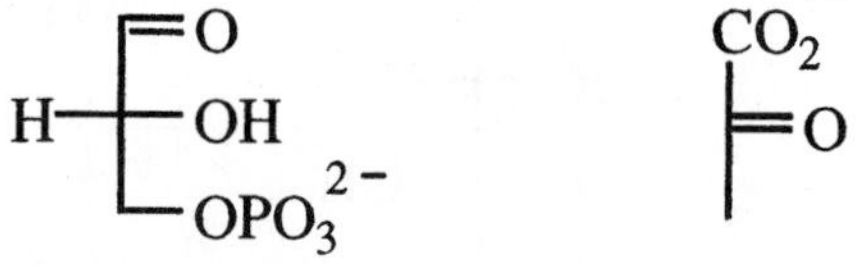

glyceraldehyde-
3-phosphate

pyruvate

The oxidation step should be easy: we simply make the hydrate of our aldehyde, taking advantage of all the water floating around, and oxidize the hydrate with NAD^+ to an acid, like so:

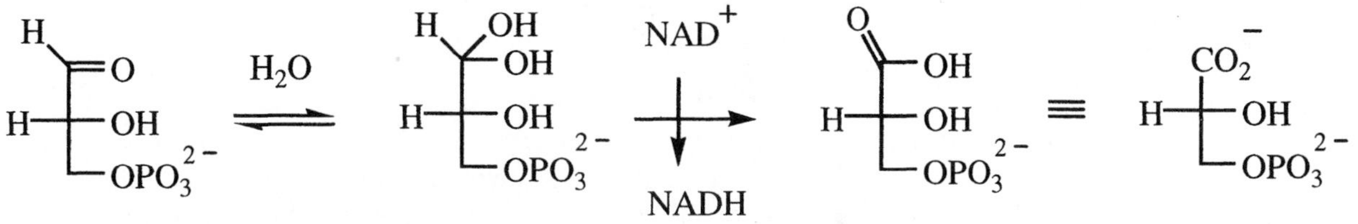

But this is not quite what happens. Remember, one of our tasks is to generate ATP whenever possible, and Mother Nature has devised a slight variation on this theme that allows us to capture some energy. Instead of water coming into the aldehyde, it is a phosphate ion, PO_4^{3-}, that comes in; the rest of the reactions are the same as those above, except that after NAD^+ oxidation, there is still a phosphate attached to the carbonyl group. This can be transferred to an ADP to make ATP:

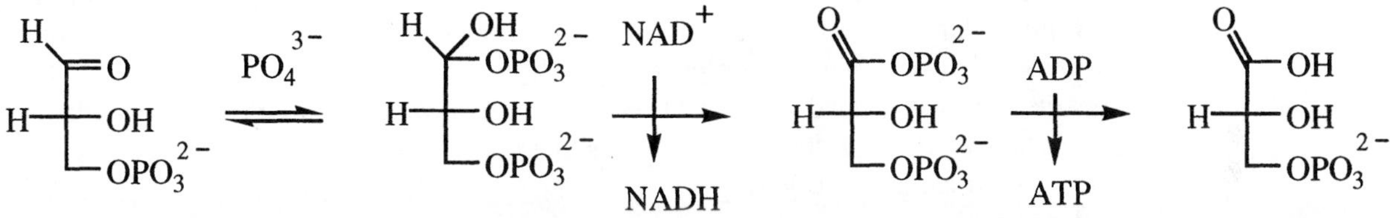

Notice that this little side excursion has netted us two ATPs (remember: each glucose provided two three-carbon units); since we used up two ATPs in the preliminary stages, we are now even.

We are also pretty close to our goal. Indeed, we could get there in one step, just by eliminating phosphoric acid. This is an allowed process, since the leaving group is β to a carbonyl group, and it would result in the enol of the ketone we are trying to make:

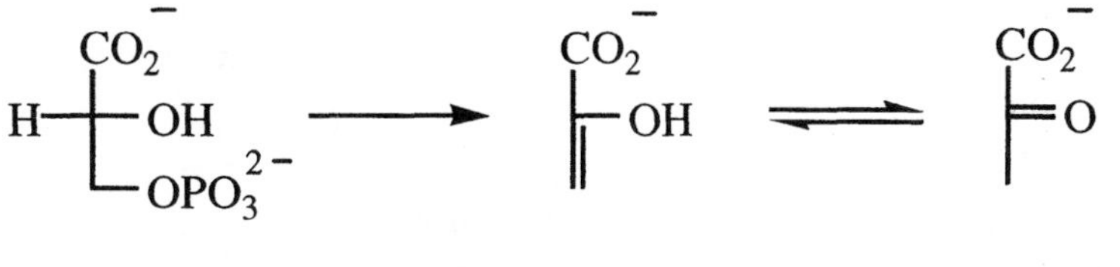

However, once again we will take a short detour in order to capture some energy. The strategy outlined above releases inorganic phosphate into the solvent (as a leaving group). If we can somehow hang onto the phosphate, we might be able to use it to make more ATP. Phosphate is certainly not necessary where it is, since OH can be eliminated just as well as phosphate, so let's move the phosphate from C-3 to C-2. This is an allowed phosphoryl transfer reaction. *Then* we can eliminate H_2O to form a double bond, making what is called "phosphoenolpyruvate." This can in turn transfer its phosphate group to ADP, at the same time tautomerizing to a ketone and generating the desired pyruvate:

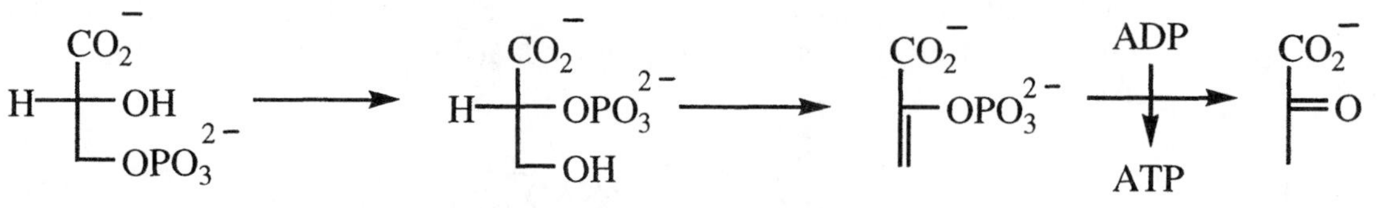

$$
\begin{array}{c}
\text{CO}_2^- \\
\mathrm{H-\!\!\!-OH} \\
\mathrm{OPO_3^{2-}}
\end{array}
\longrightarrow
\begin{array}{c}
\text{CO}_2^- \\
\mathrm{H-\!\!\!-OPO_3^{2-}} \\
\mathrm{OH}
\end{array}
\longrightarrow
\begin{array}{c}
\text{CO}_2^- \\
\mathrm{=\!\!-OPO_3^{2-}}
\end{array}
\xrightarrow[\text{ATP}]{\text{ADP}}
\begin{array}{c}
\text{CO}_2^- \\
\mathrm{=\!O}
\end{array}
$$

pyruvate

This detour nets us another two ATP's, so we are finally making a profit. In addition, we generated two NADH's in an earlier step, which, under normal conditions, would go off somewhere and create more ATP while reverting back to NAD^+. This process is important not only for the ATP it generates, but also because we need NAD^+ to keep glycloysis going. After all, we used up NAD^+ in our oxidation reaction. If there were no way to get more, the process would quickly grind to a halt. Actually, this brings up a problem with which you are already all familiar.

Under normal conditions, called aerobic conditions, the NADH goes off somewhere and creates more ATP while reverting back to NAD^+, using oxygen from the air. This is all well and good, but it takes some time. If time is not a problem, well and good. But occasionally you put your body in a situation where time is of the essence. You are exercising strenuously, using your muscles a lot, using up ATP (which is necessary to flex your muscles), and you need more ATP *now*. The fastest way for the body to generate ATP is from glucose, through the process we just discussed—glycolysis—and you get two ATPs for every glucose molecule you chew up. But you also use up NAD^+, and you can't continue with more glycolysis until the resulting NADH reverts to NAD^+. If the normal regeneration process is operating at capacity and still cannot keep up with the need for more NADH (this is often referred to as "anaerobic conditions," although there is really no lack of air), the body takes advantage of a quick fix: the pyruvate just made is reduced with NADH to lactate, in the process changing NADH back into NAD^+. This NAD^+ can then be used for the oxidation step of glycolysis, allowing the process to continue and more generation of ATP. But, the side effect is that more and more lactate is also made.

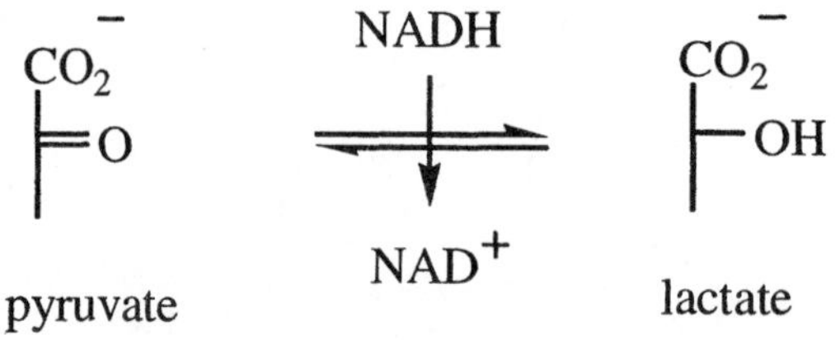

$$
\begin{array}{c}
\text{CO}_2^- \\
\mathrm{=\!O}
\end{array}
\quad
\underset{NAD^+}{\overset{NADH}{\rightleftharpoons}}
\quad
\begin{array}{c}
\text{CO}_2^- \\
\mathrm{-\!\!OH}
\end{array}
$$

pyruvate   lactate

Unlike pyruvate, lactate is a metabolic dead-end. Under normal conditions, pyruvate is made and immediately used up (as we'll see below), and the pyruvate level stays relatively low. But lactate has nowhere to go but back to pyruvate. So under conditions of strenuous exercise, lactate is being made, and it builds up in the muscles. Since lactate is equivalent to lactic acid, the muscle environment becomes more and more acidic. This leads to muscle fatigue, which you have all experienced at one time or another, I imagine. Once you stop exercising and start metabolizing normally again, the lactate is slowly converted back to pyruvate and used in various ways by the body. The conversion of pyruvate into lactate is merely a stopgap measure to generate energy under conditions where the complete aerobic pathway is too slow.

Here is an overview of the whole glycolysis process. We will come back to it later when we discuss the energy budget of glucose metabolism. I should mention that all the above reactions occur inside a cell, in what is known as the "cytosol," the solution within the cell.

Pyruvate is subsequently transferred into organelles (subcellular compartments) called mitochondria, where the rest of the metabolic reactions occur.

TRANSITION

The normal end point of glycolysis is pyruvate. The normal starting point of the next stage, the tricarboxylic acid (TCA) cycle, is acetyl CoA. The pathway that converts pyruvate into acetyl CoA has no official name—I call it "transition."

pyruvate   acetyl CoA

Conceptually, it involves three simple steps: decarboxylation, oxidation, and esterification (acyl transfer), all allowed reactions.

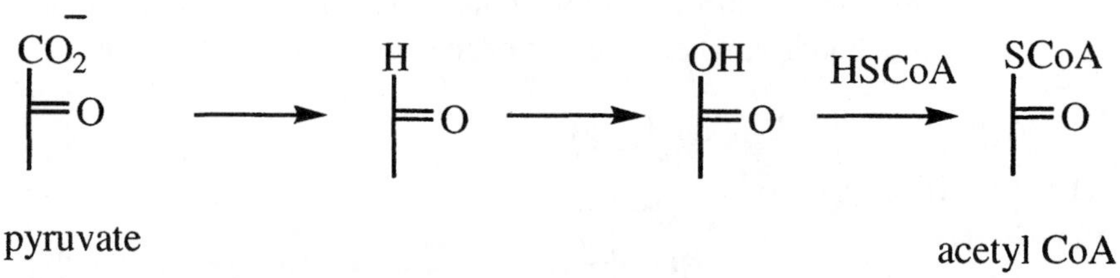

pyruvate   acetyl CoA

In practice, however, the process is considerably more complicated than this. One aspect of the complication I will discuss in detail; another I will sweep under the rug.

The first step of the process is a decarboxylation. Decarboxylation itself should not pose a problem, provided that the carboxylic acid group is attached to a position that can sustain a negative charge. This usually means the carboxyl group should be β to some carbonyl group—that is, the acid group should be attached to the α position of some carbonyl group. But this carboxylic acid is not attached to the α position of some carbonyl group—it is *directly* attached to a carbonyl group, and the carbon of a carbonyl group is not a place we have ever seen negative charge before. Something weird is going on here.

Thiamine Pyrophosphate

Something weird *is* going on here. In order to pull off this reaction, Mother Nature has to convince the molecule that it is OK to put a negative charge on the carbon of a carbonyl group, something that is normally not possible. The way she pulls this off is with a prosthetic group called thiamine pyrophosphate (TPP), also known as vitamin B_1. (A prosthetic group is a necessary adjunct to an enzyme but not part of the protein backbone of the enzyme; a vitamin is something the body needs but cannot make by itself. TPP is both of these.) TPP has the structure shown below, and it has the unusual property that the indicated hydrogen is acidic enough to be removed under physiological conditions. This is critical to its function, because we are going to use the resulting anion as a nucleophile.

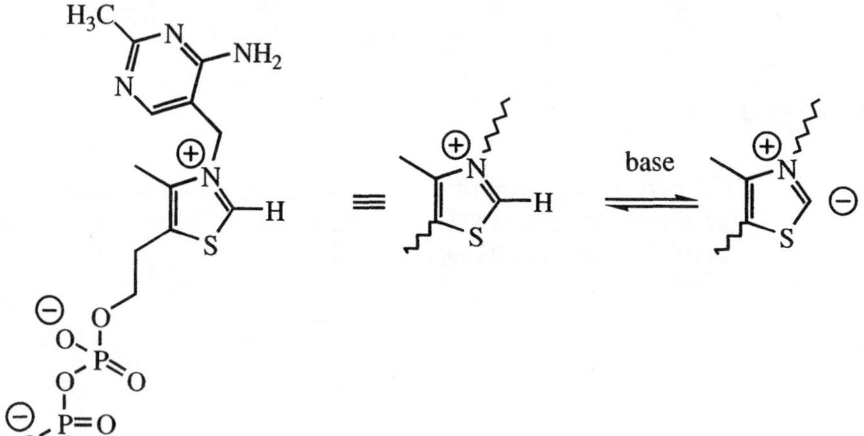

The anion we have generated is a good nucleophile, and it can attack the keto group of pyruvate:

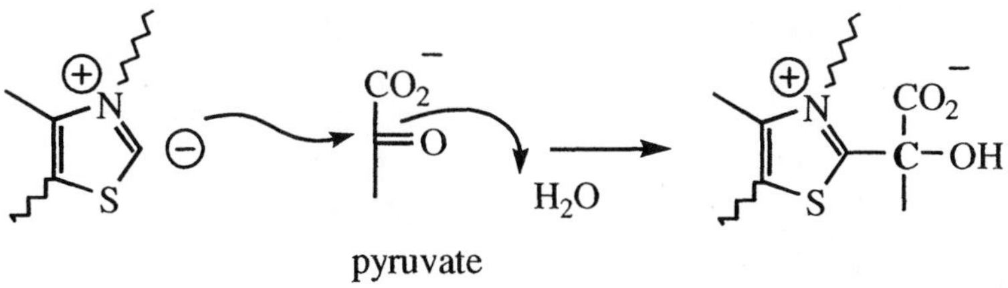

pyruvate

Now, if you inspect the carbon that holds the carboxylic acid group, you will discover that it suddenly is able to accommodate a negative charge. It has next to it a carbon–nitrogen double bond, something that resembles a carbonyl group; what is more, the nitrogen has a positive charge, so a negative charge on the bold carbon has a resonance structure that is neutral, as shown below. This is even *better* than a negative charge α to a carbonyl group, and decarboxylation happens readily:

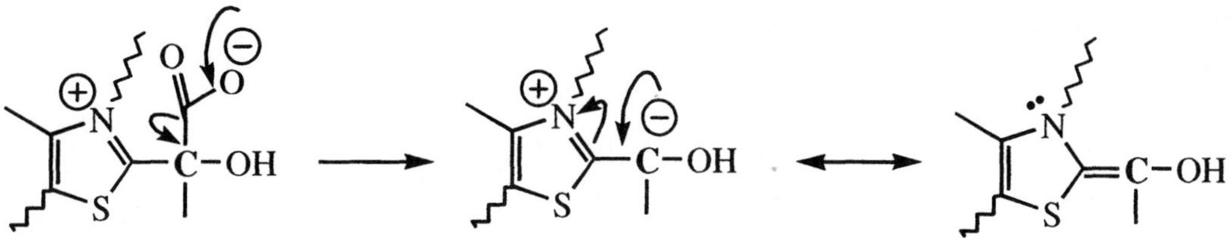

There are several other reactions in biological pathways that involve what appear to be anions directly on the carbon of a carbonyl group. When such a reaction is occurring, it's a good bet that the enzyme involved uses TPP as a prosthetic group as a way of helping that anion to exist.

What happens next is a complex sequence of reactions involving a variety of other cofactors (cofactors are all molecules besides the main enzyme that are necessary in a reaction, for example, NAD^+; prosthetic groups like TPP are a subcategory of cofactors). The exact nature of these reactions has no influence on our discussion of strategy, nor does it affect the energy budget we will construct, so I am going to simplify the process by pretending it goes a slightly different way. The net result will be the same.

Let's start by putting a proton back onto the position that lost the carbonyl group, and then let's lose the TPP. We know TPP can function as a leaving group, since it is stable enough to exist in water and so must be something like the extra OH in a hydrate.

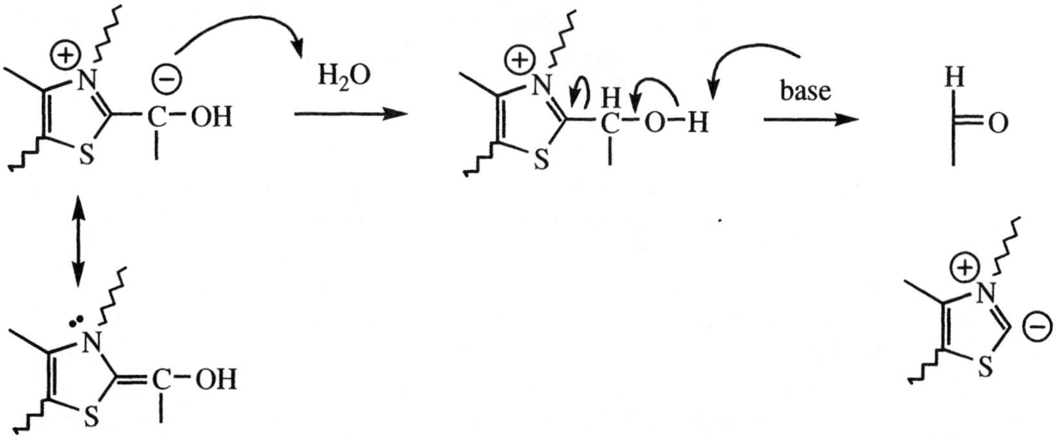

Now we have acetaldehyde. It should be simple, and allowed, to oxidize this to acetic acid with NAD^+—and just as simple and allowed to esterify the resulting carboxylic acid with some convenient thiol from the surroundings:

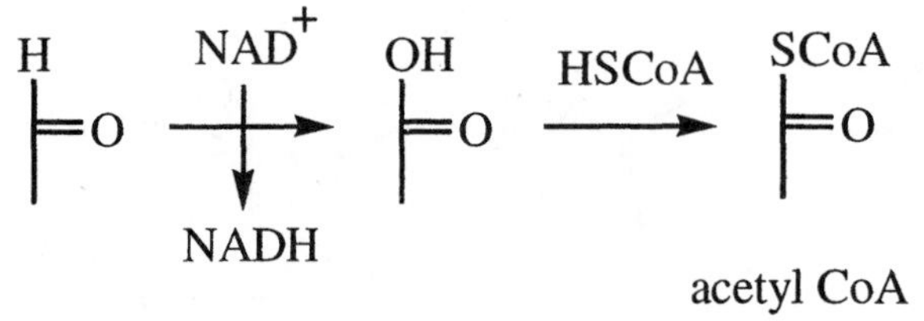

acetyl CoA

What is this CoA stuff? And why would we bother using it? The second question will become especially relevant soon, when we discover that in the very next step we are going to dump the CoA again.

Coenzyme A

CoA is an abbreviation for "Coenzyme A." It looks like this:

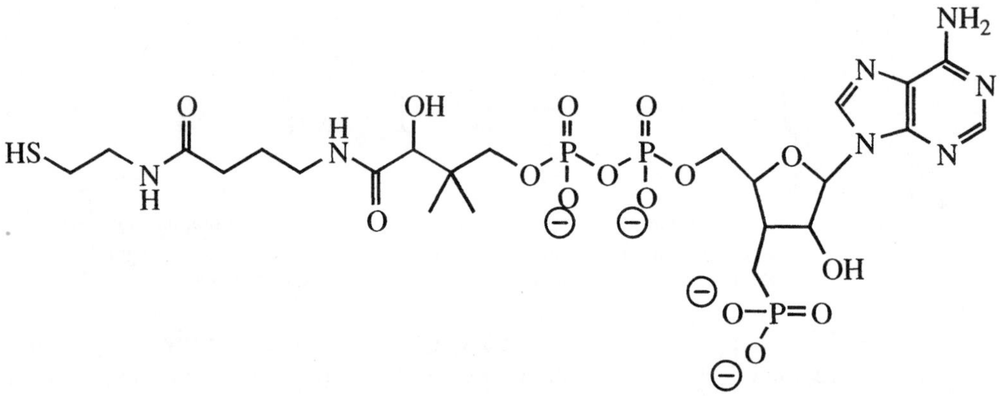

You can see why the compound is always abbreviated (as "CoA"). Actually, the CoA part is everything but the thiol group (SH) on the left end. We usually write out the SH, because that is the business end of the molecule, where all the important chemistry happens. The rest of the molecule is just the vehicle that carries it. The vehicle, as usual, is critically important in recognition and regulation, but not in the chemical reactions. So the whole molecule is generally written as CoASH or HSCoA.

Why is CoA involved? Precisely because it is an excellent vehicle. Just as ATP is the cell's currency for carrying phosphate groups (and therefore energy) around, CoA is the cell's standard carrier of acyl groups. When a surgeon calls for a scalpel, it is delivered by a nurse; whenever a reaction calls for an acyl group (in our case, the *acetyl* group from the acetic acid we just made), it is delivered by CoA.

There is another reason CoA is involved here, one that requires a preview of the subsequent reaction. The acetyl group in question is about to be involved in an aldol condensation, one that will require a negative charge to form at the methyl group of the acetyl group. If the acetyl group were part of acetic acid, there would already be a negative charge present from the carboxylic acid anion, and it would be difficult to put yet another negative charge on the methyl group. Thus, carrying the acetyl group around in the form of an ester (technically, a thioester) allows the next reaction to occur; after that, we no longer need an ester, so it can be hydrolyzed back to the acid. This will become more clear after we look at the next reaction.

Again I should remind you that I have made up some of this sequence. Everything from the decarboxylation step to here actually happens in a more complicated way in the cell. There are other cofactors involved, including lipoamide and FAD, and the sequence of events is different from what I have told you, but the net effect is the same: NAD^+ is reduced to NADH, and CoASH acquires an acetyl group. Of the three carbons we put into this pathway, one has gone away as CO_2, and the other two constitute the acetyl group of acetyl CoA.

TCA CYCLE

Let's step back a moment and consider again what we are trying to accomplish. We have set about digesting glucose, chopping it up into little pieces while harvesting energy from the chemical bonds we are breaking. The little pieces we end up making are, in every case, CO_2, which is why you exhale so much of that. Ultimately each glucose molecule will be converted into six CO_2 molecules. We have already accounted for two of them, which went away during the transition step (remember, each glucose molecule created *two* pyruvates); now we have the other four to worry about, and they currently exist as two acetyl groups tied up as acetyl CoA. Our general strategy will be to attach each acetyl group to an already existing molecule and use that framework to dispose of two CO_2 molecules, regenerating the "carrier" molecule we started with. You will discover that the carbon atoms we blow off are *not* the same ones we bring in, but the effect is the same: two in, two out. The "carrier" molecule is oxaloacetate.

The first thing that happens is that the acetyl group carried by CoA undergoes an allowed aldol condensation with the keto group of oxaloacetate. At the same time, CoA is hydrolyzed off in an allowed acetyl transfer reaction.

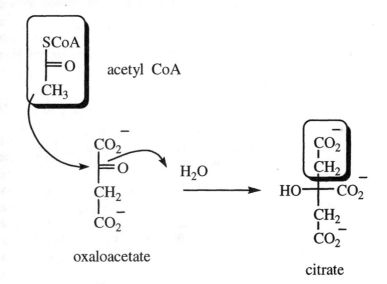

oxaloacetate

citrate

The citrate ion we have formed is the trianion of citric acid, a tricarboxylic acid. The name of the present pathway derives from citric acid: variously, "citric acid cycle," "tricarboxylic acid (TCA) cycle," or "Krebs cycle," after the person who discovered it.

Citrate contains six carbons. Our task now is to get rid of two of these and regenerate oxaloacetate so the process can happen again; at the same time, we want to harvest as much energy as possible, as ATP or NADH. Here goes.

None of the carboxyl groups in citrate are located in a position that would facilitate decarboxylation. They should be β to some carbonyl group. The only handle we have on this molecule is an OH group. This could be oxidized to a carbonyl, but not where it is now, because it is tertiary! If only the OH were on C-4 instead of C-3 (numbering from the top as drawn)! Hmm, let's see if we can get it there.

The OH group is now on C-3, which is β to the carbonyl group at C-5 (also the one at C-1, but the enzyme chooses to work in the direction away from the acetyl group that just entered). This means that water can be eliminated in an allowed reaction, creating a double bond between C-3 and C-4:

$$
\begin{array}{c}
CO_2^- \\
| \\
CH_2 \\
| \\
HO-C-CO_2^- \\
| \\
CH_2 \\
| \\
CO_2^-
\end{array}
\quad\longrightarrow\quad
\begin{array}{c}
CO_2^- \\
| \\
CH_2 \\
\| \\
C-CO_2^- \\
\| \\
CH \\
| \\
CO_2^-
\end{array}
$$

This new double bond is α,β to the carbonyl group on C-5, and also to the carbonyl group attached to C-3, so water can add back in either direction. If the oxygen of water adds to C-4, we get a new alcohol by another allowed reaction:

$$
\begin{array}{c}
CO_2^- \\
| \\
CH_2 \\
\| \\
C-CO_2^- \\
\| \\
CH \\
| \\
CO_2^-
\end{array}
\quad\longrightarrow\quad
\begin{array}{c}
CO_2^- \\
| \\
CH_2 \\
| \\
C-CO_2^- \\
| \\
HO-CH \\
| \\
CO_2^-
\end{array}
$$

This alcohol *can* be oxidized, and the product will have a carboxyl group β to the new carbonyl group, so it would be lost immediately. Indeed, CO_2 loss is so fast that most texts do not include the middle compound below as an intermediate along this pathway, but I do, because it lets you see how and why decarboxylation takes place.

$$
\begin{array}{c}
CO_2^- \\
| \\
CH_2 \\
| \\
C-CO_2^- \\
| \\
HO-CH \\
| \\
CO_2^-
\end{array}
\;
\begin{array}{c}
NAD^+ \\
\downarrow \\
NADH
\end{array}
\;\longrightarrow\;
\left[
\begin{array}{c}
CO_2^- \\
| \\
CH_2 \\
| \\
C-CO_2^- \\
| \\
O=C \\
| \\
CO_2^-
\end{array}
\right]
\;\longrightarrow\;
\begin{array}{c}
CO_2^- \\
| \\
CH_2 \\
| \\
CH_2 \\
| \\
O=C \\
| \\
CO_2^-
\end{array}
$$

We have now lost one of the two CO_2's we set out to lose. Furthermore, we have arrived at a compound that should seem familiar. Look back at the section titled "Transition." In that section we described a complicated process that converted pyruvate to acetyl CoA. The reaction required TPP, because we wanted to create an anion on the carbon of a carbonyl group, and we needed to fool the system into thinking this is OK, which is what TPP does. The reaction also involved a reduction of NAD^+ to NADH. And there were other complications involving lipoamide and FAD that I glossed over. The net result was that a carboxyl group directly attached to a carbonyl group was lost, and the product contained

SCoA attached to that same carbonyl group, in the form of a thioester. Precisely the same transformation takes place now, and the enzyme used is extremely similar to the one that does a similar job on pyruvate. (Of course they are not *precisely* the same, because part of the function of any enzyme is to provide a comfortable nest for its substrate, and the current substrate is almost twice the size of pyruvate.)

$$
\begin{array}{ccccc}
CO_2^- & TPP & & & CO_2^- \\
| & \overbrace{\quad\quad} & NAD^+ & HSCoA & | \\
CH_2 & \downarrow & \downarrow & & CH_2 \\
| & \longrightarrow & \longrightarrow & \longrightarrow & | \\
CH_2 & & & & CH_2 \\
| & CO_2 & NADH & & | \\
O{=}C & & & & O{=}C \\
| & & & & | \\
CO_2^- & & & & SCoA
\end{array}
$$

Mission accomplished! We have dumped two carbon atoms from citrate, both in the form of CO_2. All we have to do now is convert the resulting four-carbon unit back into oxaloacetate, so it can go around the cycle again. Here goes. The first order of business is to get rid of the SCoA unit hanging on there. This is a simple acyl transfer reaction. It turns out (you would have no way of predicting this) that the process is so exothermic that, at the same time, the same enzyme can grab a phosphate ion from solution and attach it to ADP to get ATP. (In fact, it actually converts GDP into GTP, but later the GTP converts an ADP into an ATP, so the net result is the same.)

$$
\begin{array}{ccc}
CO_2^- & & CO_2^- \\
| & GDP & | \\
CH_2 & \downarrow & CH_2 \\
| & \longrightarrow & | \\
CH_2 & & CH_2 \\
| & GTP & | \\
O{=}C & & CO_2^- \\
| & & \\
SCoA & &
\end{array}
$$

succinate

You may recall that among the allowed enzyme reactions was one you don't actually know how to do in the lab: one can strip two hydrogens off adjacent carbons, creating a double bond, provided the two carbons are α and β to a carbonyl group. Our two central carbon atoms are now α and β to two different carbonyl groups, so these qualify in spades. Oxidation to a double bond is therefore an allowed process. Recall that this reaction usually uses FAD as the oxidizing agent:

$$
\begin{array}{ccc}
CO_2^- & FAD & CO_2^- \\
| & \downarrow & | \\
CH_2 & \longrightarrow & CH \\
| & & \| \\
CH_2 & FADH_2 & CH \\
| & & | \\
CO_2^- & & CO_2^-
\end{array}
$$

fumarate

This new double bond is now susceptible to addition of water. Addition could occur from either end. In fact, the ends are identical, so the product is the one shown in either case.

$$
\begin{array}{c}
CO_2^- \\
| \\
CH \\
\| \\
CH \\
| \\
CO_2^-
\end{array}
\quad\longrightarrow\quad
\begin{array}{c}
CO_2^- \\
| \\
HO-C-H \\
| \\
CH_2 \\
| \\
CO_2^-
\end{array}
$$

Now all we need to do is oxidize the product and we are back to oxaloacetate:

$$
\begin{array}{c}
CO_2^- \\
| \\
HO-CH \\
| \\
CH_2 \\
| \\
CO_2^-
\end{array}
\quad\xrightarrow[\text{NADH}]{\text{NAD}^+}\quad
\begin{array}{c}
CO_2^- \\
| \\
O \\
| \\
CH_2 \\
| \\
CO_2^-
\end{array}
$$

oxaloacetate

What follows is a summary of the TCA cycle, all in one place. See if you can walk your way through it and remember the logic of each step.

The neat thing about this whole "logical" approach to metabolism is that everything makes sense. *If* you understand the organic chemistry behind it, *if* you know the five fundamental enzyme reactions, and *if* you remember a few tricks here and there, you can write down the whole process from beginning to end. There is no need to memorize a messy, confusing, apparently random sequence of reactions. I can do this without looking, and I swear I never sat down and memorized the pathways. I doubt if any of you could do so now, because your feeling for organic chemistry is not yet deeply entrenched, but pull out this chapter in a year or two, and you will find many aspects of the process that suddenly make sense.

ENERGY BUDGET

We had two goals when we set out on this mission: we wanted to chew up glucose into CO_2, and we wanted to harvest as much energy as possible. Clearly, we have succeeded in the first part of the mission: six CO_2's have been produced from the six carbon atoms of glucose. How did we do on the energy front? We have to count. But first, a couple of sneaky points I haven't told you about.

As you know, our currency is ATP. Each NADH gets converted to three ATPs by oxidative phosphorylation, which takes place in the mitochondria. Most of the NADHs are generated there, so this is no problem, but two molecules of NADH were made in the cytosol during glycolysis, and they can't undergo oxidative phosphorylation until they enter a mitochondrion. There are two different ways of getting in ("shuttles"); one costs an ATP, the other does not. So these NADHs might be worth either two or three ATPs depending on which shuttle is used. I'll call it three, and we will arrive at a maximum number of ATPs. But you will see some counts that end up two short, because they assume use of the other shuttle.

The second sneaky point is that $FADH_2$ does not accomplish oxidative phosphorylation in the same way NADH does, and it generates only two ATPs.

Don't forget that GTP is equivalent to ATP. Start at the beginning of glycolysis and count the ATPs you expect from the complete metabolism of one glucose molecule. Do it now!

If you did this right you should come up with 38 ATPs. Did you get that? If not, follow along with me.

During glycolysis, we started by using up two ATPs, but then we got four back (remember: everything after we get to three-carbon units is doubled). Further, we got 2 NADHs, but these were formed in the cytosol. We will still count them for three ATPs each, but some other bookkeepers count them for only two each. Total: 8 ATPs so far.

Once we get into the mitochondrion, we get 2 NADHs from the "transition," and 6 from the TCA cycle, for a total of 8. Each is worth three ATPs, so that's 24. In addition we get 2 ATPs from each of the $FADH_2$'s, so that's 4 more, and finally 2 ATPs from the GTP formed in the TCA cycle. The final sum is 38 ATPs.

KEEPING TRACK OF CARBON ATOMS

How do we know all these things I have told you? A great deal of work has gone into studying the pathways, and there is a tremendous amount I have not told you about. But one of the important techniques involves taking some glucose that has been labeled on a particular carbon atom and then seeing where the label appears in various molecules downstream. This technique helped chemists work out details for the pathways; now that we know the chemistry, we can deduce what they must have observed. For example, proceed through the glycolysis and TCA-cycle pathways starting with glucose labeled at carbon 2 with ^{14}C, which is radioactive. Where will this label appear in each of the successive compounds? Try to work it out now, before you look.

Here is the answer I get. I have labeled the special carbon in each intermediate with an asterisk.

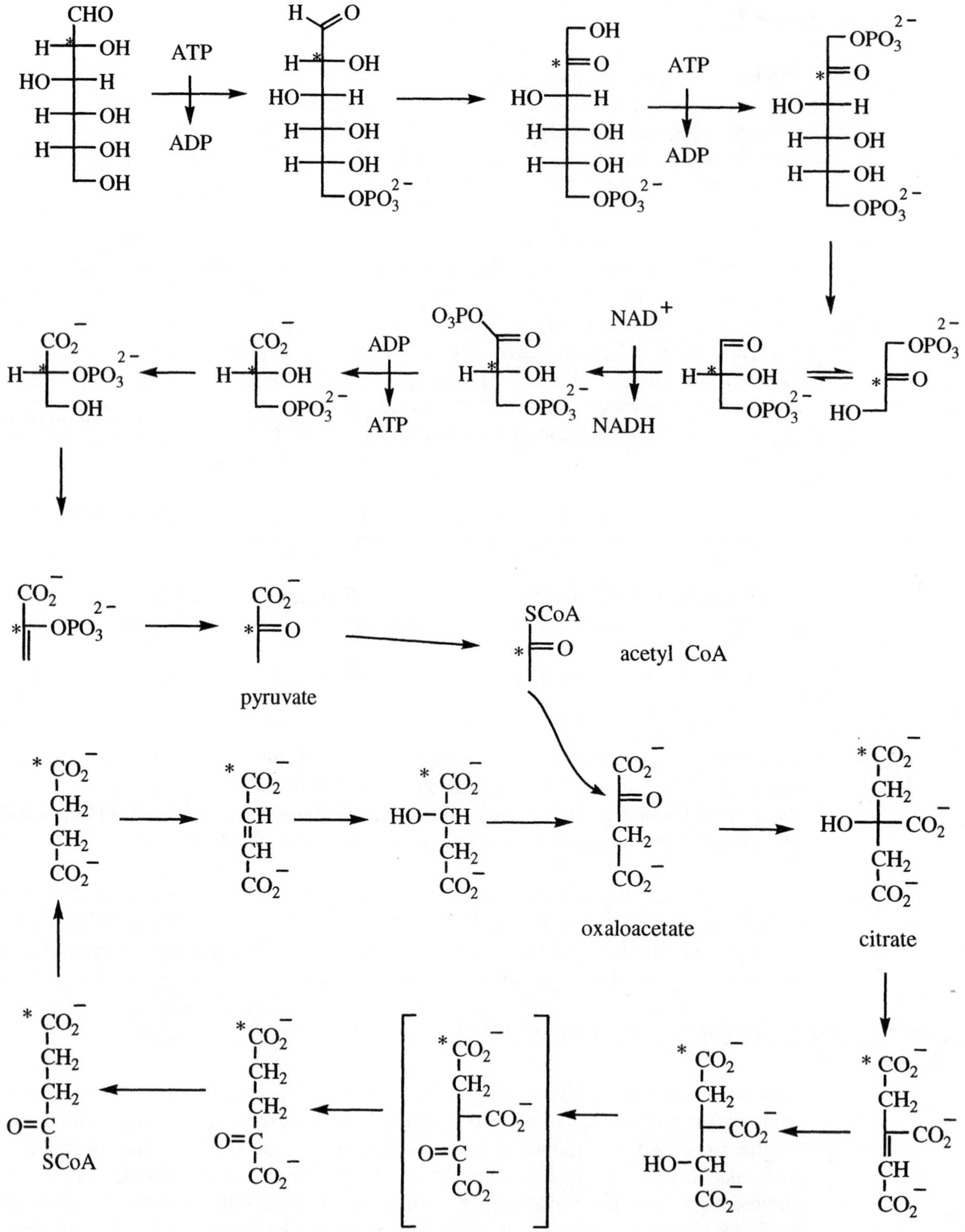

CHO
H—*—OH
HO——H
H——OH
H——OH
——OH

ATP → ADP

H C=O
H—*—OH
HO——H
H——OH
H——OH
——OPO$_3^{2-}$

——OH
*C=O
HO——H
H——OH
H——OH
——OPO$_3^{2-}$

ATP → ADP

——OPO$_3^{2-}$
*C=O
HO——H
H——OH
H——OH
——OPO$_3^{2-}$

C=O
H—*——OH
——OPO$_3^{2-}$
HO——C=O
——OPO$_3^{2-}$

NAD$^+$ → NADH

O$_3$PO C=O
H—*——OH
——OPO$_3^{2-}$

ADP → ATP

CO$_2^-$
H—*——OH
——OPO$_3^{2-}$

CO$_2^-$
H—*——OPO$_3^{2-}$
——OH

CO$_2^-$
*C——OPO$_3^{2-}$

CO$_2^-$
*C=O

pyruvate

SCoA
*C=O   acetyl CoA

CO$_2^-$
C=O
CH$_2$
CO$_2^-$

oxaloacetate

*CO$_2^-$
CH$_2$
HO——CO$_2^-$
CH$_2$
CO$_2^-$

citrate

*CO$_2^-$
CH$_2$
CH$_2$
CO$_2^-$

*CO$_2^-$
CH
‖
CH
CO$_2^-$

*CO$_2^-$
HO—CH
CH$_2$
CO$_2^-$

*CO$_2^-$
CH$_2$
CH$_2$
O=C
SCoA

*CO$_2^-$
CH$_2$
CH$_2$
O=C
CO$_2^-$

[*CO$_2^-$
CH$_2$
——CO$_2^-$
O=C
CO$_2^-$]

*CO$_2^-$
CH$_2$
——CO$_2^-$
HO—CH
CO$_2^-$

*CO$_2^-$
CH$_2$
——CO$_2^-$
CH
CO$_2^-$

What you find is that in fact this particular carbon never *is* released as CO_2—at least, not after one round of the TCA cycle. The carbons released as CO_2 during the cycle are not the same ones that came in with acetyl CoA. Here are some challenge questions for you:

Test Yourself

How many turns of the TCA cycle *does* it take to get rid of this carbon atom? What other carbon of glucose shares the same fate as this one? Which carbons of glucose get released the soonest? The latest? How many turns of the TCA cycle does it take to get rid of the carbon atom that leaves first?

SUMMARY

I hope many of you found the information in this chapter to be interesting. But what on earth should you do with it? Are you supposed to be able to reproduce the entire metabolic pathway from having read this a few times? Should you be able to describe the fate of some labeled atom, at any point in the process, without a crib sheet? The answers to all these questions ultimately depend on your professor, of course, but it seems to me that it would be a lot to ask for you to be able to reproduce the information in this chapter on an exam. What I would hope you will take away is an appreciation for what enzymes can and cannot do, and why pathways are what they are. Beyond that, you should know the five fundamental enzymic reactions, you should be able to predict simple new pathways based on pathways we have studied, and you should be able to predict when TPP is necessary in a reaction and why.

Problems

1. If you start with glucose labeled with ^{13}C at C-2, where will the label appear in oxaloacetate after the first turn of the TCA cycle? The second turn? The third? Repeat the analysis for C-4 of glucose.

2. Long-chain fatty acids, like many other natural molecules, are constructed biologically from acetyl CoA. Since the building block has two carbons, most fatty acids contain even numbers of carbons. Using any "allowed" reactions you wish, suggest how you think the body might add two carbons to a growing fatty acid chain:

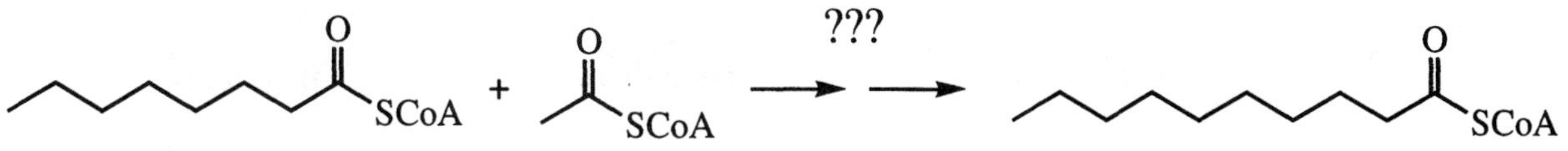

Selected Answers

Test Yourself. Carbons 2 and 5 are released on the second go-round of the TCA cycle. Carbons 3 and 4 are the first to go, before you even get to the TCA cycle. Carbons 1 and 6 last the longest.

Note that I have not specified precisely how long it *does* take for the final two carbons to be released. The reason for this is that things get complicated in the labeling scheme, because succinate is symmetric (previous diagram, next-to-last line, left structure). I have written the label on the top carbon, but in fact the enzyme cannot tell the difference between the top and bottom carbons, and so one can imagine this being labeled half on top and half on the bottom. Both of these carboxyl groups are lost in the next cycle, so this is not a problem; but if the succinate had been labeled in the middle rather than the end, as would be the case if the glucose were labeled on C1 or 6, then the label survives the second cycle and gets scrambled again when it gets to succinate for the second time. At this point, all four carbons of succinate are partially labeled, and after each round some of it will remain.

Chapter 22
Nitrogen Chemistry: Amines, Imines, and Amides; Protein Structure

By now you know a great deal about the chemistry of carbon, hydrogen, oxygen, and the halides, but relatively little about the chemistry of nitrogen. It is time to correct that deficiency. You will discover that the properties and reactions of nitrogen compounds are almost entirely predictable from the chemistry you have already learned. Some consequences of those properties and reactions may not be obvious until I point them out, but there should be no great surprises in this chapter. In this chapter we will

- Learn how to use acid–base chemistry to manipulate the solubility of amines

- Learn one way of making amines and the complications involved

- Explore the chemistry of imines, especially as regards biological systems

- Discuss the geometric features of amides, along with the chemistry of that functional group

- Revisit and expand upon our discussion of how proteins fold up reproducibly into functional blobs.

Compounds with R groups attached to nitrogen are called amines ("um-MEENS"). If there is a carbonyl group next to the nitrogen, you have an amide ("AM-mid," "am-MID," "am-MIDE," or "AY-mide"—chemists cannot agree on how to pronounce these things!). If the nitrogen is involved in a double bond with carbon, the compound is an imine ("IM-meen"). If there is a triple bond between carbon and nitrogen, it is a nitrile ("night-TRILE," "NIGHT-trile," or "night-TREEL").

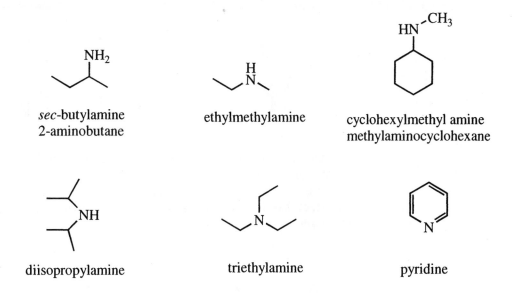

AMINES

Nomenclature

Amines are named by naming the attached group(s), followed by "amine." The amine group as a substituent is called "amino-." The following examples should illustrate the naming of amines.

Amines are derivatives of ammonia, NH_3. They can have one, two, or three carbons directly attached. These are called primary, secondary, and tertiary amines, respectively.

$$R-\overset{\bullet\bullet}{N}H_2 \qquad R-\overset{\bullet\bullet}{\underset{\underset{R}{|}}{N}H} \qquad R-\overset{\bullet\bullet}{\underset{\underset{R}{|}}{N}}-R$$

primary amine   secondary amine   tertiary amine

Note that the distinction here is *not* the same as that between primary, secondary, and tertiary halides and alcohols, where the names describe the *kind* of substituent attached to the group in question. With amines the primary/secondary/tertiary designation tells the number of carbons on the nitrogen itself. For example, *sec*-butyl amine is a primary amine, even though its amino group is attached to a secondary position. Pyridine is considered to be a tertiary amine.

An amine of special interest to biologists, because it is used in many experiments, is EDTA.

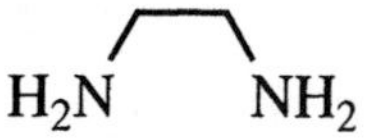

1,2-diaminoethane
"ethylene diamine"

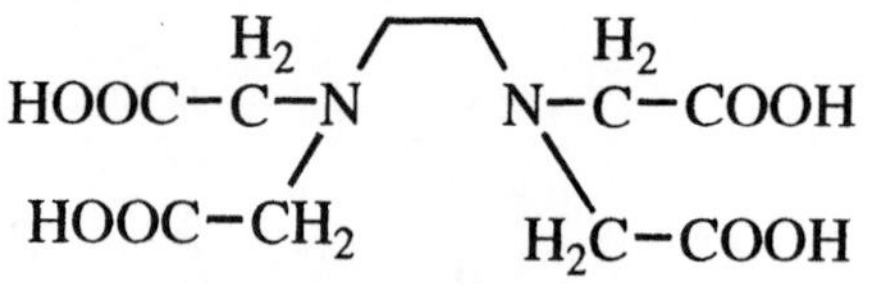

ethylene diamine tetraacetic acid
"EDTA"

EDTA is very useful because it beautifully encircles many inorganic ions to produce an octahedral complex, as shown below. You will learn more about octahedral bonding in inorganic chemistry.

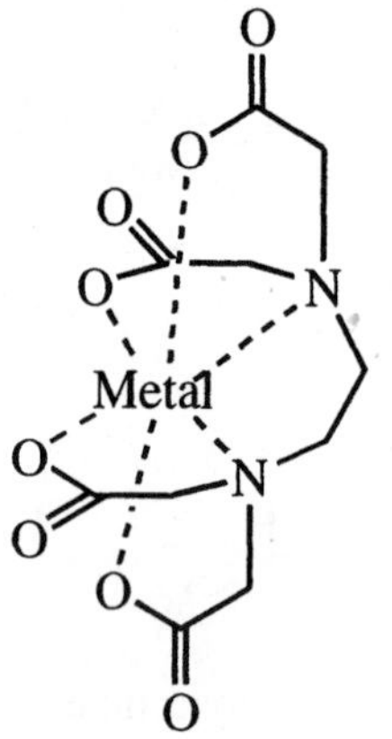

sometimes
written as

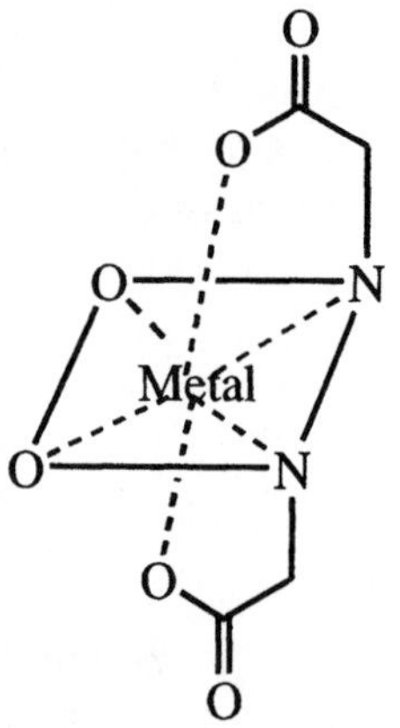

There are many natural products that contain the amine functional group. These mostly have names that have nothing to do with their chemical structure, and you need not memorize them. As a class, amine-containing natural products are called "alkaloids." The name derives from the fact that the compounds are basic, and "alkali" is an old term for a base. A few examples are shown below.

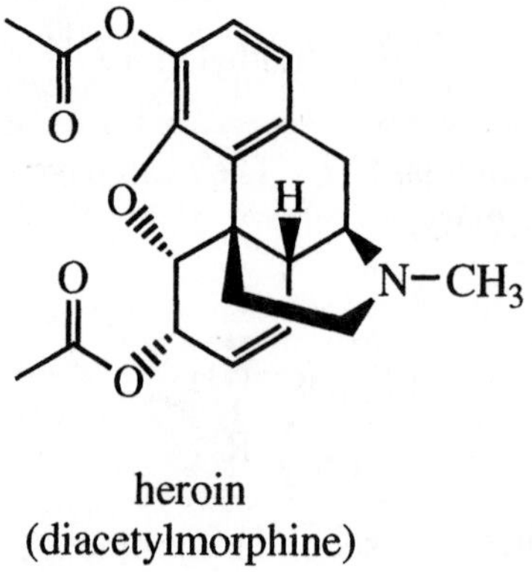

heroin
(diacetylmorphine)

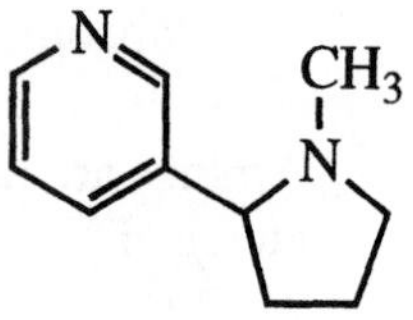

nicotine

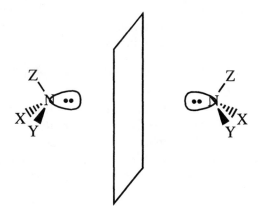

methamphetamine

strychnine

Properties

Geometry

Saturated amines have four regions of electron density around the nitrogen atom, so they should be sp³-hybridized and tetrahedral. That means that an amine with three different groups attached should be able to exist in two different forms, enantiomers of each other:

should be enantiomers

However, ordinary amines like this have never been found to come in two forms. Chemists have therefore concluded that the forms are able to interconvert. Because there is no partner on the end of the lone pair, it is not forced to remain pointing in any particular direction. Sitting around, it constantly flips back and forth between the two enantiomeric forms by enlargement of the tail and shrinkage of the head of the orbital containing the lone pair. This has the effect of inverting the molecule. Thus, our understanding is that there really *are* two enantiomers of amines like this, but it is impossible ever to keep around anything but a racemic mixture.

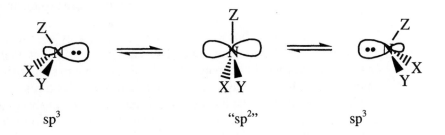

Of course, if the geometry of the molecule prevents inversion of this type, one *can* find enantiomeric amines:

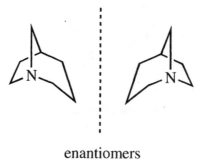

enantiomers

Acid–Base

The most important property of amines is their acid–base reactivity. You already know that primary and secondary amines can lose a proton from their nitrogen. Ammonia and its derivatives have pK_a values of about 33. This high value means that it is difficult to make the anions of amines, and once made they are eager to take a proton from something else. We have seen nitrogen anions used as bases on several occasions: $NaNH_2$ to deprotonate terminal acetylenes and alcohols, and LDA (which is effectively the same base with some steric baggage) to deprotonate ketones. The high pK_a value also means that in general, you will never encounter the anion of an amine unless you deliberately set out to make one or buy one.

On the other hand, every amine also has a lone pair of electrons. Since nitrogen is less electronegative than oxygen, it is more willing to share this lone pair than oxygen is. Amines are stronger bases and stronger nucleophiles than the oxygen compounds we have studied so far. In particular, amines are strong enough bases to deprotonate water to a noticeable extent, so solutions of amines in water are somewhat basic. More importantly, if you make available some acid, amines will be protonated in preference to most oxygen compounds, generating "ammonium salts."

HCl

triethylamine

triethylammonium chloride
Old name: triethyl amine hydrochloride

Most ammonium salts have pK_a values of about 9. This means that at pH 9, half the molecules are protonated as ammonium salts, and the other half are neutral, free amines. At pH 7, in biological systems, nearly all amine groups are protonated.

You will recall that carbocations are more stable if they are more highly substituted (tertiary carbocations are better than secondary, etc.). You might therefore expect that tertiary ammonium salts should be more stable than primary ones and that, as a consequence, tertiary amines should be more basic than primary amines. This turns out to be true in the gas phase and in non-polar solvents; however, in water, where pK_a values are measured, all amines are about the same. This is usually attributed to the fact that, while tertiary ammonium salts are intrinsically more stable than primary salts, as described above, they are less well solvated. Water has a tough time nestling up to a tertiary ammonium salt and can get much closer to a primary ammonium salt, making it more comfortable. These two effects appear to approximately cancel each other in aqueous solution.

Carbon–nitrogen bonds are polar, but not as polar as carbon–oxygen bonds. Like alcohols or ethers, amines are also capable of hydrogen bonding with water and other solvents. Therefore, small amines are soluble in water, but larger ones tend to be insoluble. Ammonium salts, however, are almost always very soluble in water. This means that we can manipulate amine solubility just as we did that of carboxylic acids. In aqueous acid, the amine will dissolve (as its ammonium salt); in base, it will not.

Manipulating the solubility of amines is a common laboratory operation. But this process has moved to the public arena in an unfortunate way. For example, cocaine is an amine with the following structure:

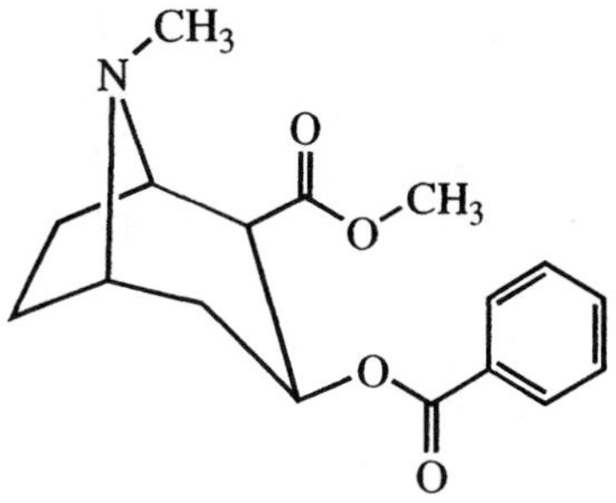

It is extracted from plants, crystallized, and shipped as the corresponding ammonium salt, with an extra H (and a positive charge) on the nitrogen and some anion, perhaps chloride (but perhaps something else) as the counterion. It turns out that cocaine is a more effective drug in its neutral form, as the "free base," and many users go to some lengths to convert the commercial salt into free base for their personal use. One approach is to dissolve the salt in water (it is soluble, of course), add some base like baking soda (sodium bicarbonate) to remove the proton, and then add an organic solvent to dissolve the cocaine. After removal of the water layer, evaporation of the organic solvent will leave free base. Unfortunately, most ordinary people do not have access to nice organic solvents like hexane, and they use whatever they can get their hands on, like gasoline. Gasoline actually works fine. The problem comes in evaporating the organic solvent: if you are in a hurry, you might be tempted to help it along with a little heat, perhaps placing it on the stove. Richard Pryor is one of many people who have been severely burned by fires caused by heating flammable solvents in a kitchen instead of a lab.

A second approach to making free base is even more dangerous. Again the commercial salt is dissolved in water and combined with sodium bicarbonate, but this time no organic solvent is added. The water is simply evaporated. The result of this is what is known as "crack" cocaine. The problem is that along with the free base, you also get the sodium from the sodium bicarbonate and the chloride (or whatever it was) from the cocaine salt, making sodium chloride or sodium something else that contaminates your cocaine. If this were nothing but sodium chloride it would not be so bad: you ingest sodium chloride all the time (though not often by smoking or inhaling it); but cocaine is rarely pure, and even worse, sodium bicarbonate often contains small amounts of aluminum and other impurities that are nasty. Plus you also wind up dealing with any extra sodium bicarbonate that may have been used. The net effect of this is that crack cocaine is more than just cocaine, and it does terrible things to the brain. This is aside from the effect of the cocaine itself, which is also not so wonderful.

So that's my moral essay for the day. Put slightly differently: don't do drugs!

Test Yourself

Suggest an easy chemical procedure for separating cyclohexylamine from cyclohexanol, obtaining both as pure, neutral compounds at the end of the process.

Spectroscopy

Like OH bonds, NH bonds produce peaks to the left of those from the CH bonds that are apparent in every IR spectrum near 3000 cm^{-1}. The peaks for NH bonds are usually much smaller than those for OH bonds, which, you will recall, are typically both broad and very strong. NH peaks are usually much weaker and in the 3300–3400 cm^{-1} range. Primary amines often show two peaks in this region, one for each of the NH bonds.

In NMR, nitrogen has the effect you might expect: it is less electronegative than oxygen, so protons on carbons next to nitrogen atoms are still pulled downfield, but not as far as those near oxygen atoms. (The same applies to ^{13}C–NMR.) Protons directly attached to nitrogen usually behave like OH protons in that they do not couple to their neighbors. They can appear anywhere between 0 and 6 ppm.

Preparation

There are many methods for preparing amines, but we will only consider one of them, the most straightforward possible. The lone pair of ammonia (or any amine) is a pretty good nucleophile. Therefore, ammonia and other amines are capable of performing S_N2 substitution on primary and secondary alkyl halides. Conceptually, this is a good way of making primary amines from ammonia, or secondary amines from primary amines, or tertiary amines from secondary ones. However, as we will see, there are significant problems with this approach, and it really is not used all that much.

You should notice several things here.

1. This is the first time you have seen a *neutral* nucleophile attacking an alkyl halide. All previous nucleophiles have been negatively charged, such as halide anions or oxygen anions. Neutral nitrogen is a better nucleophile than neutral oxygen. Other neutral nucleophiles capable of initiating S_N2 reactions are phosphorus and sulfur, which are better nucleophiles than either nitrogen and oxygen due to higher polarizability.

2. Ammonia is a base as well as a nucleophile. In theory, this reaction could lead to elimination as well as substitution. However, it usually takes a stronger base than an amine to induce an E2 elimination reaction. Elimination tends not to be an important complication in these reactions.

3. When an anion (like an alkoxide) attacks an alkyl halide, the result is a neutral compound (an ether), and the reaction is done. But when the nucleophile is a neutral compound like ammonia, the result is a cation. Something more needs to occur. The obvious thing is to drop a proton from the nitrogen. We have been fairly cavalier about

this, but I hope you have kept in mind that protons do not just float off into thin air; they are taken by something else in solution. What might that something else be in this case? By far the most basic thing in the solution is ammonia:

4. The above proton exchange accomplishes two things, both of them bad. It removes some of our nucleophile from the reaction, since NH_4^+ no longer has a lone pair and thus cannot react with alkyl halide; at the same time, it generates a new nucleophile, the alkyl amine. This new nucleophile should be approximately as reactive as the original one: the alkyl group is electron donating, so this lone pair should be even more nucleophilic, but on the other hand there is more steric bulk around the new nucleophile than the old one. The two factors approximately cancel. The net result is that there is a danger that the desired product of the reaction, an alkyl amine, will react with more alkyl halide to make dialkyl amine, and then tri- and even tetraalkyl compounds. (The tetraalkylated amine has a positive charge it cannot get rid of. Such systems are called "quaternary ammonium salts," and they are known.)

5. You should be able to devise a way around both of the problems mentioned in point 4. Think about it for a bit before you read on.

The problem of losing the starting nucleophile is easy to deal with: just use extra, so that you can lose some and still have enough left. The problem of a competing nucleophile is also easily solved: since the new nucleophile is about as good as the old one, the two will react in proportion to the amount of each that is present. If you want the original nucleophile to win this war, simply outnumber the other guys. Use enough extra of the original nucleophile so that there will always be a lot more of it than there is of the reactive product. In other words, *both* of the above problems can be solved fairly well by using an excess of the original nucleophile.

Below are some examples of this reaction, where "xs" means "excess" (get it?).

This direct approach to amine synthesis is OK if the amine you are starting with is cheap, easy to obtain, and also easy to remove from your product, but for many amines, such as expensive ones, you might not want to use such a big (costly!) excess. There are a number of other ways to make amines by indirect methods. Many involve the reduction of some other functional group. We will not take the time now to cover these alternate approaches.

Reactions

The most important reactions of amines we have already discussed. Amines are mild bases, with a pK_a of about 10 (for the conjugate acid). And they are mild nucleophiles, capable of undergoing S_N2 reactions even though neutral. They are also capable of attacking carbonyl groups in neutral form, something we have not seen before except in the case of acid chlorides and anhydrides, which are reactive enough to react with neutral oxygen nucleophiles.

Neutral nitrogen nucleophiles can also react with less activated carbonyl compounds. The results are predictable: if the carbonyl group has a leaving group, it will leave. If the carbonyl group does not have a leaving group, a proton will be dropped. The following examples should illustrate the principles.

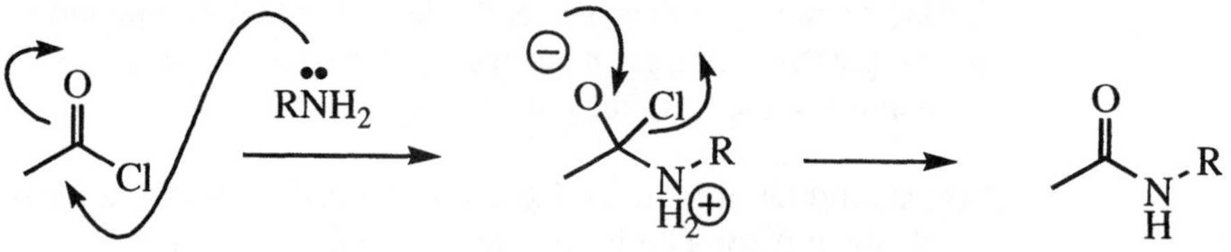

amide

similarly,

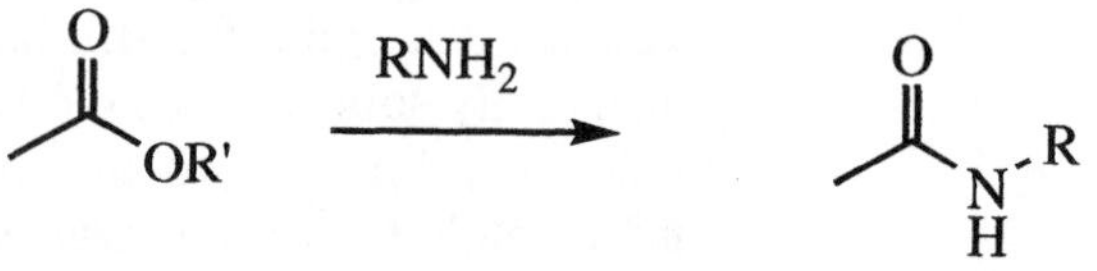

We will cover amides soon.

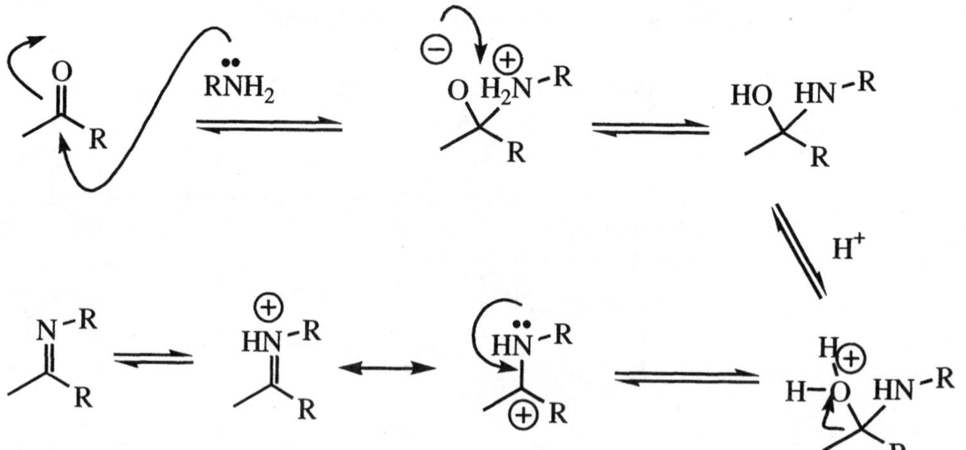

imine

Notice that in order to make an imine, *both* hydrogens from a primary amine are eventually lost. An amine must *be* a primary amine in order to be capable of making an imine. If the amine is secondary, a different hydrogen is lost, from the α carbon, and an enamine results (see below). Tertiary amines cannot get past the first step, so they just go back where they came from.

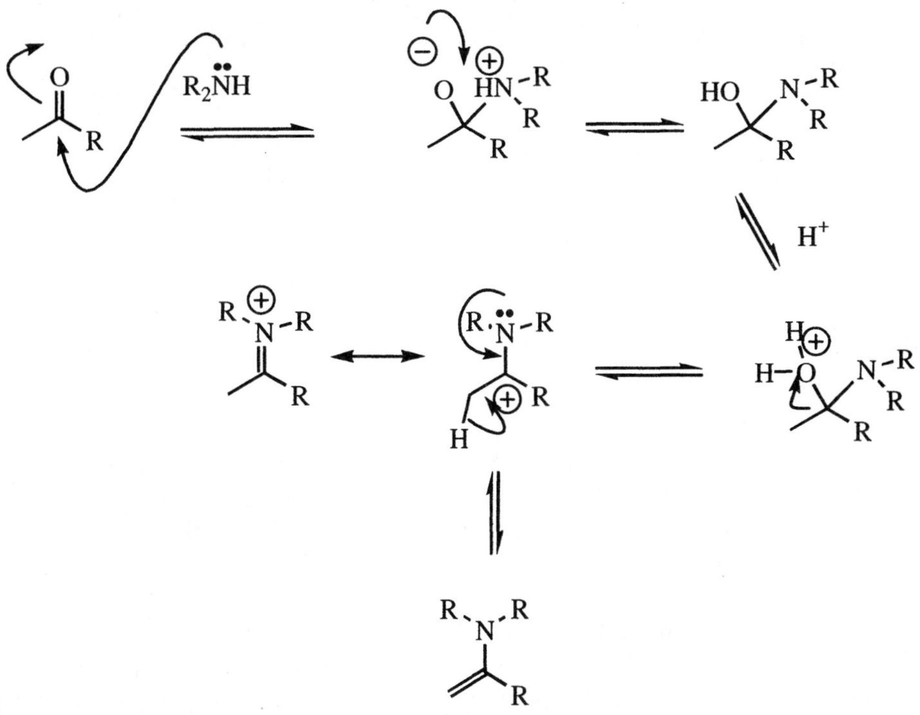

enamine

IMINES

Nomenclature

I have never heard an imine called by name, so I do not actually know how one names an imine. Usually you just name the ketone it comes from, and then say "imine with such-and-such amine." As you will discover shortly, with a few exceptions, imines are rarely isolated, so there is usually no need to name them. But you should know this: biologists and biochemists refer to imines as "Schiff bases."

Properties

Most imines are unstable and readily revert to the primary amine and carbonyl compound from which they came. However, there are a few imines that are not only stable, but are easily formed as crystalline solids. These imines all have immediately adjacent to the nitrogen atom of the imine some other heteroatom with a lone pair—the ability of the lone pair to conjugate with the new double bond apparently has a stabilizing effect on the imine. The following imines have special names and are stable.

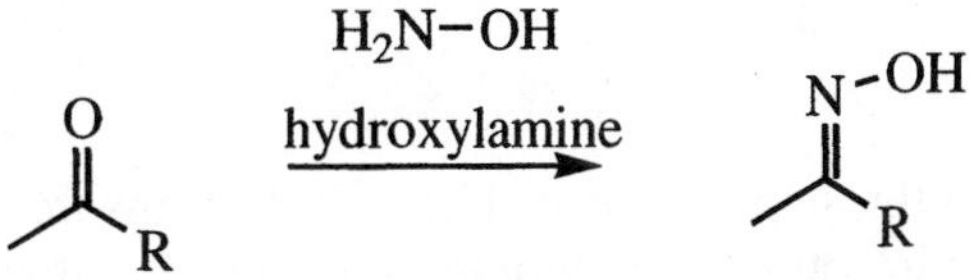

an oxime

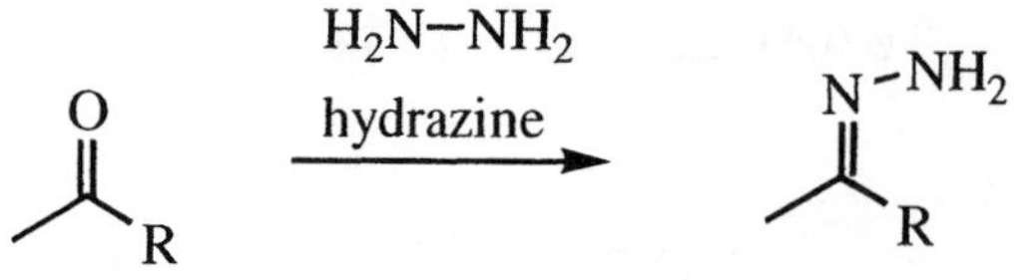

H₂N—NH₂
hydrazine

a hydrazone

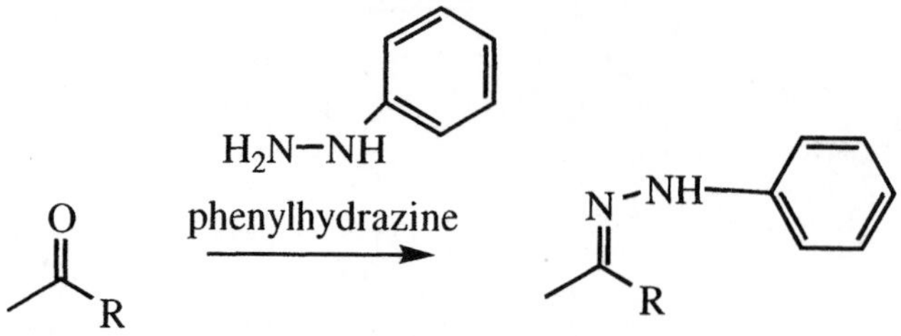

H₂N—NH
phenylhydrazine

a phenylhydrazone

These derivatives have historical significance, because they are stable, easily made, and have nice sharp melting points. Before the advent of spectroscopy, ketones and aldehydes were identified by making a series of derivatives like these, taking the melting points, and comparing the results with long, published tables. Every liquid ketone of course has some particular boiling point, but it could still be one of dozens of possibilities. If its oxime had a certain melting point, only a few of those dozens would share *both* characteristics. Add in a couple other derivatives and you could identify your compound with a fair amount of certainty. Since no one does this kind of chemistry any more, it is probably not necessary for you to remember the names of all the derivatives in the above equations.

Preparation

Here, once again, is the reaction for making imines. I have copied it here, because I want you to notice a few things about it.

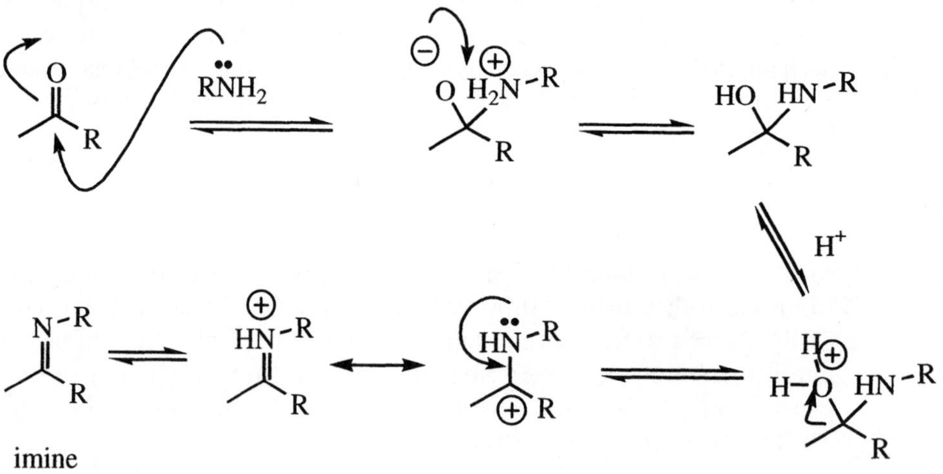

imine

Notice:

1. All the steps are reversible. This means that the starting materials, the ketone and amine, are in equilibrium with the products, the imine and water.

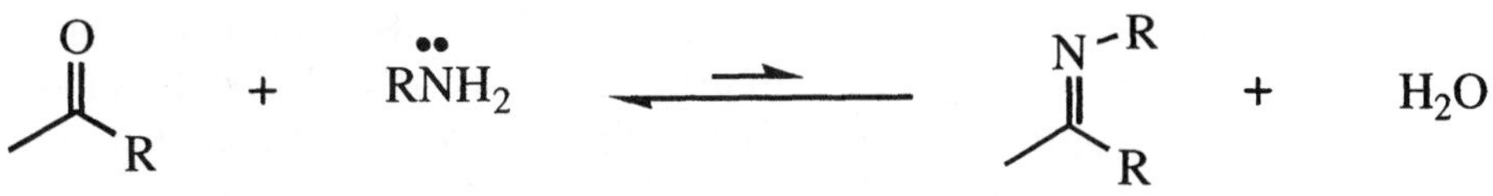

Like all equilibria, this one has a natural preference, and its preference is for the starting materials. Thus, imines have a natural tendency to come apart in aqueous medium. On the other hand, we have seen situations like this before (remember ketals and acetals), and we have learned how to manipulate conditions in such a way as to make the reaction go in an unnatural direction: by manipulating the water concentration, for example. By making the water concentration very low (usually by removing it as it is formed by some mechanical technique), a chemist could force this reaction to the right. An enzyme, of course, could do this quite easily, if it wanted to, by carrying out the reaction at an active site where there is no water.

2. In the first step of the reaction, a neutral amine attacks a neutral ketone. We have established that this is OK. The result has a negative charge on oxygen and a positive charge on nitrogen. I have shown the O^- reaching over to get an H^+ from the nitrogen. It is not clear that this is really what happens, since that would require a four-membered ring for a transition state. More likely, the O^- picks up a proton from the medium somewhere, and the N^+ loses one to the medium. It is also not clear (or important) in what order these two steps occur. The important point is that we eventually get the next structure shown, with neutral OH and neutral NHR attached to the same carbon.

3. The next step shows H^+ landing on the oxygen. This should really grate on your nerves. (If it didn't, you did not think about the process carefully enough. Also, if it didn't, figure out now why it *should*!)

There are two reasons why this should be a problem. The most obvious is that there are two sites on this molecule where a proton could land, the lone pair on the oxygen or the lone pair on the nitrogen. We know from our pK_a charts that neutral nitrogen is more basic than neutral oxygen by a factor of about 10 billion (H_3O^+ is more acidic than NH_4^+ by 10 pK_a units), so the chances of the proton landing on oxygen should be very small compared to the chances of it landing on nitrogen. Nevertheless, for this reaction to occur, the oxygen has to be protonated so that it can serve as a leaving group— remember, OH^- is a lousy leaving group except in E1cB reactions—and the nitrogen must be *not* protonated, because the lone pair needs to be available for resonance stabilization of the resulting cation. This combination of circumstances should be pretty rare, and it is no surprise that the reaction (in solution) is relatively sluggish. An enzyme, of course, can manipulate the placement of acidic groups in the active site so that the oxygen is near an acid and the nitrogen is not, making the desired combination quite probable.

The second thing that should bother you about this step is the fact that acid is shown at all. Where did this come from? Surely the reaction will not pause here and wait for us to add acid, so the acid must have been there from the beginning, but if it was there in the beginning, why didn't it protonate the amine in the first step? Once again we have to rely on an unfavorable equilibrium: although most of the amine molecules *will* be protonated under acidic conditions, there will always be a few that are not. It is these few that can react with the carbonyl group to get the reaction going. The percentage of amine molecules available for nucleophilic attack depends on how acidic the conditions are. If there is a lot of acid present, there will be almost no neutral amine molecules to get the reaction started, and the process will be extremely slow. But if there is not enough acid, then protonation of the OH group in the third step will never occur, and the reaction will be extremely slow. We should (and do) discover that such a reaction is pH-dependent: at high and low pH it is very slow, but there is an optimum pH somewhere in the middle. This turns out to be about pH 5 in this case.

Once again, all these problems pose difficulties for us chemists working in solution, but they are trivial for biological systems to solve simply because catalytic groups are available in the right place at the right time. Making and breaking imines are simple processes for enzymes.

Reactions

Imines are not commonly used in organic synthesis, so I will not discuss many of their reactions. The most important one, of course, is hydrolysis back to the corresponding carbonyl compound and amine. This is an extremely favorable process, both because the equilibrium favors the carbonyl–amine combination, and because the mechanism of acidic hydrolysis (unlike the mechanism of imine formation) has all its protonations happening on the *most* favorable positions rather than the least. Write this out and confirm it for yourself.

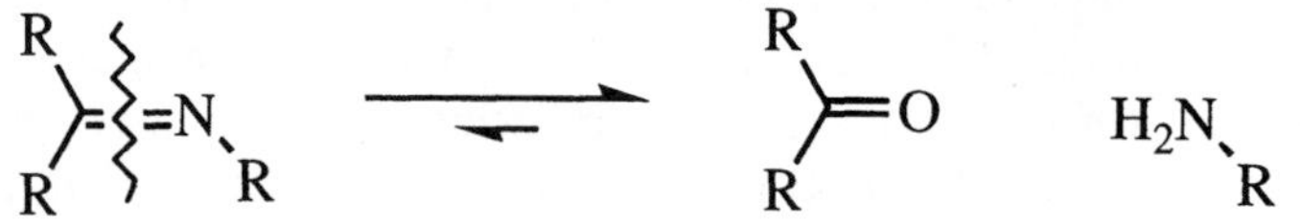

Imines ("Schiff bases," remember?) are very important biologically, because they are crucial intermediates in the metabolism of many amino acids, which, in turn, are the building blocks of proteins. The natural amino acids, as you know, are molecules with both an amino group and an acid group attached to the same carbon. In biological systems both groups are usually ionized, making what is known as a "zwitterion" but it is sometimes convenient to draw the corresponding molecules as if they were neutral:

There are some 20 naturally occurring amino acids, differing only in the R group in the picture above. As shown, the natural amino acids have the *S* configuration, which biologists always refer to for historic reasons as L. Some amino acids are acquired by ingestion (the "essential" ones, meaning that your body cannot make them and so must find an external source), but many are synthesized on site within your body. The immediate precursor of many amino acids is the corresponding α-ketoacid, the same compound but with a carbonyl group where the amine ultimately will be:

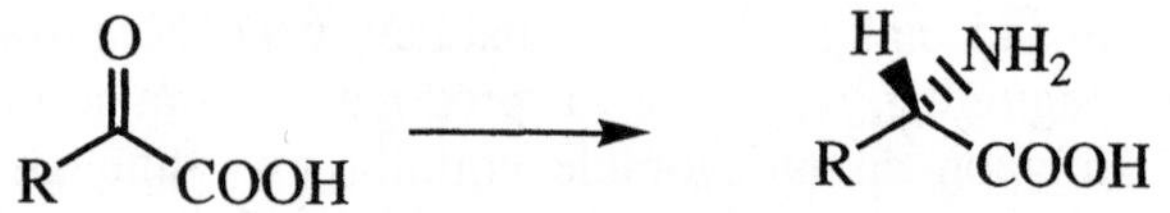

Nature has figured out a way to convert a carbonyl group into an amine and a hydrogen. How does it accomplish this magic? By using Schiff bases.

The starting point is the reaction of an α-ketoacid with an amine—specifically, an important "cofactor" named pyridoxamine:

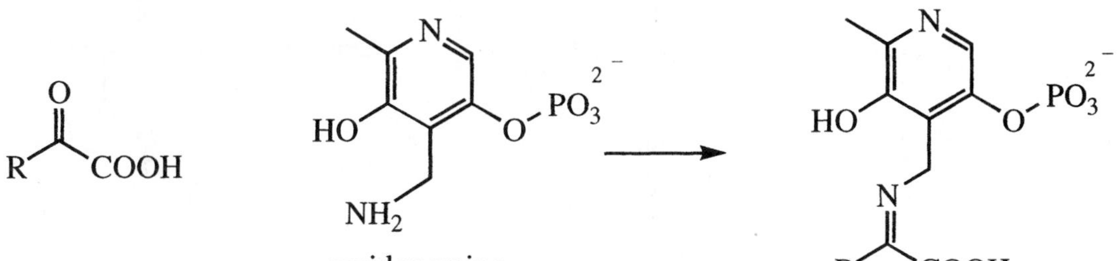

pyridoxamine

The two protons between the external nitrogen and the pyridine ring are particularly acidic. There are two reasons for this: one is that the resulting anion is extensively resonance stabilized (you should be able to show four other resonance structures for it), but even more important, the pyridine ring is usually protonated, with a positive charge on the nitrogen, so that one of the resonance structures cancels the negative charge on carbon, as shown below:

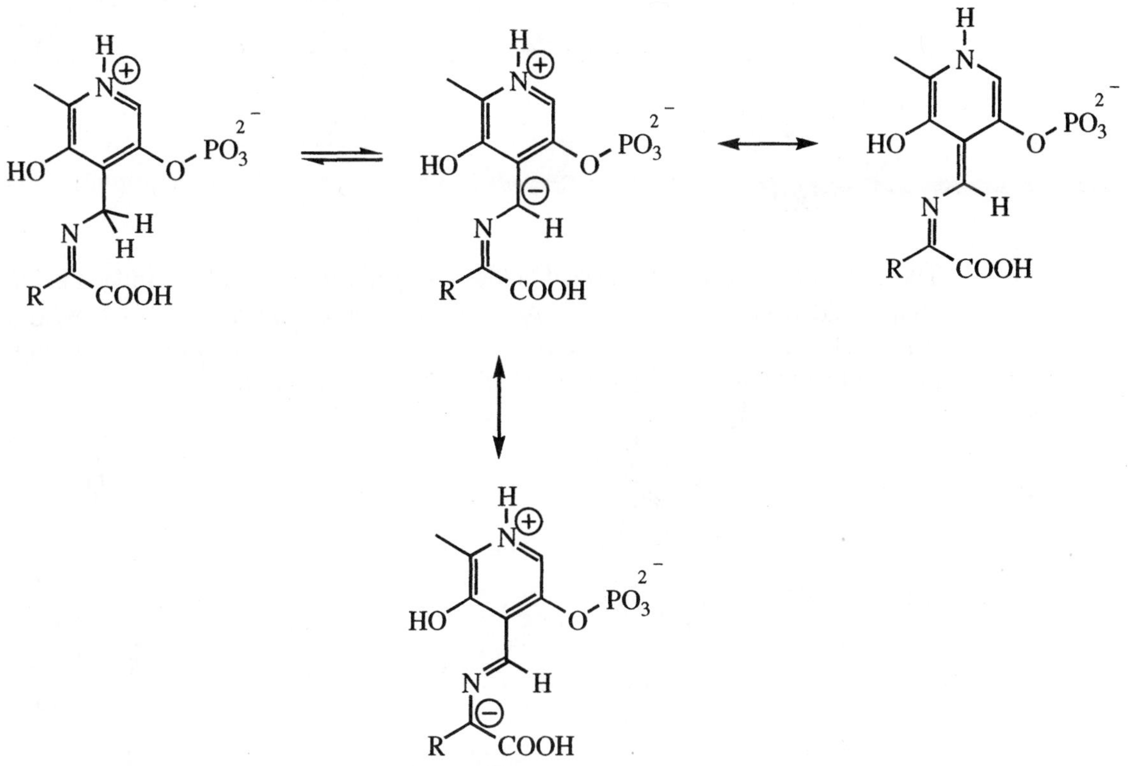

Another important resonance structure for this anion has a negative charge next to the carboxylic acid group, as also shown above. If the anion is reprotonated at this site, look what you get:

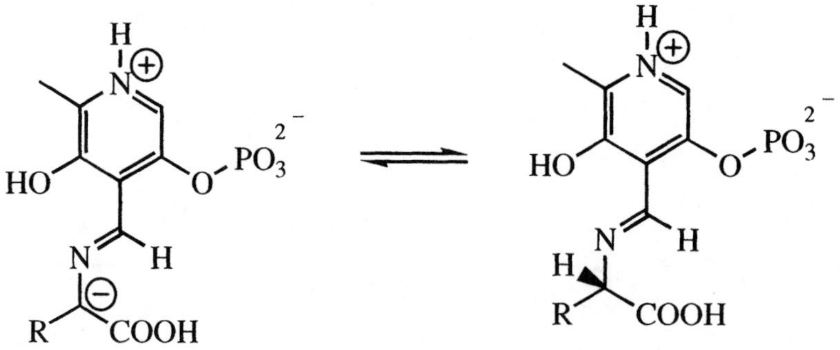

It should not surprise you to learn that enzymes can do this with great stereospecificity, creating the new stereocenter exclusively in the *S* configuration. The product we have made is once again a Schiff base, but this time it is a Schiff base derived from different partners, and hydrolysis leads to these partners:

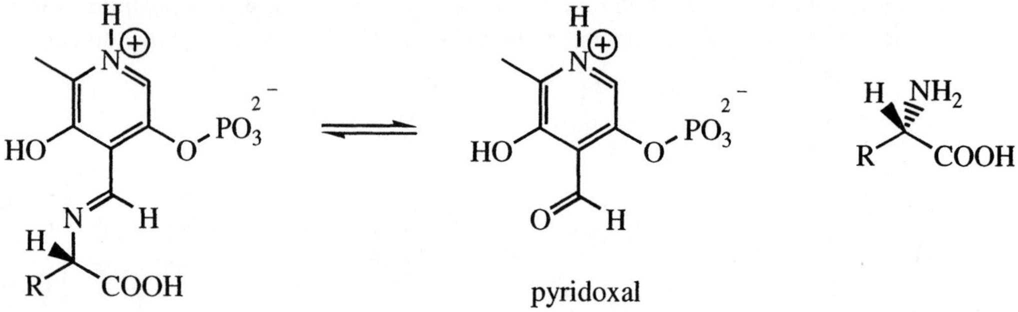

pyridoxal

The overall process is summarized below. As you can see, we have indeed made the desired amino acid from an α-ketoacid. At the same time we have converted pyridoxamine into pyridoxal. The key to the whole reaction is a proton transfer from one carbon to another, a transfer that you may recognize as actually a "tautomerization."

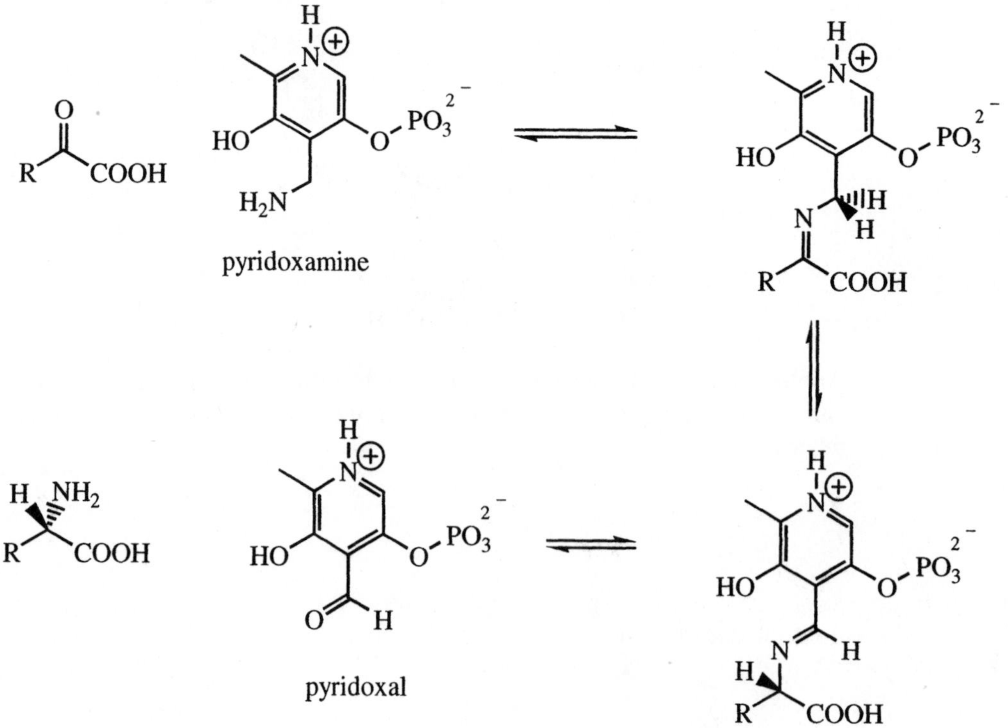

pyridoxamine

pyridoxal

In order for this to occur again (that is, in order to make another molecule of amino acid), the pyridoxal has to be returned to pyridoxamine. This is accomplished by a reverse of the above sequence, in which glutamate is converted into α-ketoglutarate. (It is probably more accurate to consider this the first reaction of the cycle, and the previously described amino acid synthesis is considered to be second; the resting enzyme contains pyridoxal, and it has to be converted into pyridoxamine before the amino acid synthesis can proceed.)

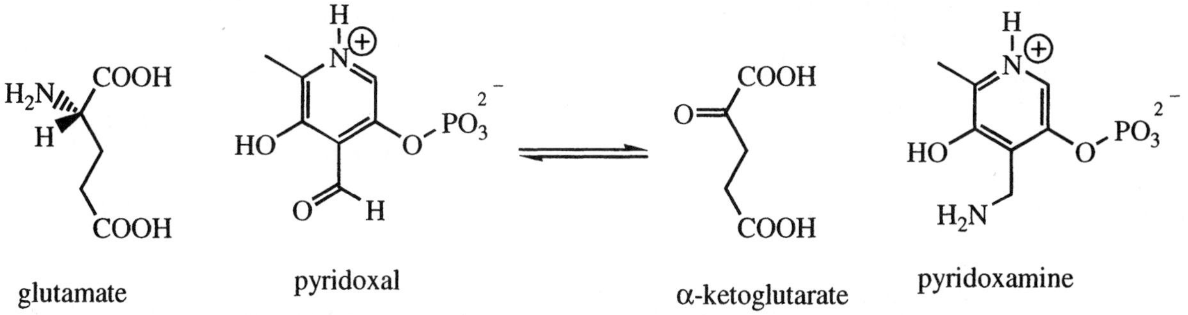

glutamate   pyridoxal   α-ketoglutarate   pyridoxamine

The pyridoxamine is now ready to transform another α-ketoacid molecule into the corresponding amino acid.

Obviously, pyridoxamine/pyridoxal must be critical for survival. Where does it come from? For the most part, you do not make it. One of the essential vitamins you have to ingest is vitamin B_6, shown below. This is converted by the body into the required form.

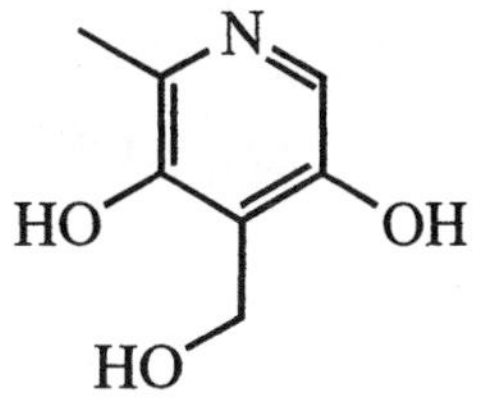

vitamin B_6
pyridoxine

These same reactions occur in reverse in the course of metabolism; that is, when it is time to break down an amino acid, one of the first steps is usually reaction with pyridoxal to make an α-keto acid.

AMIDES

Nomenclature

Amides are combinations of carboxylic acids and amines, and their names reflect the component parts.

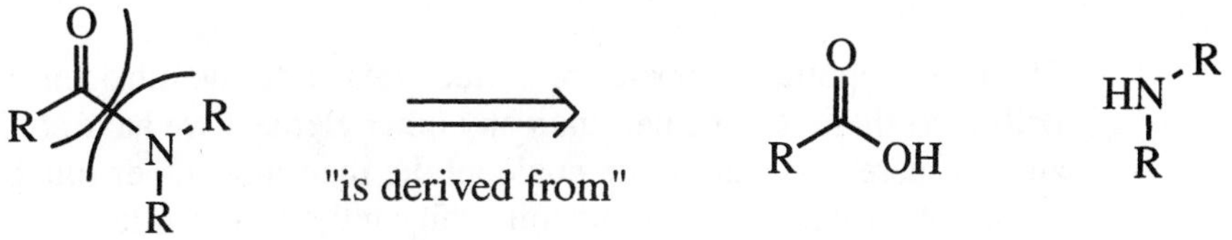

"is derived from"

You name an amide by starting to name the amine part of it, leaving off the word "amine"; then you name the carboxylic acid part, leaving off the "-oic acid"; then you add "-amide," and you are done. The following example should illustrate the process.

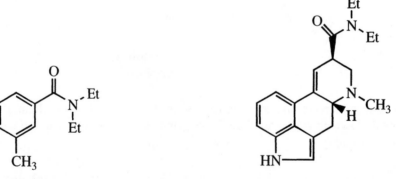

"is derived from"

benzoic acid   dimethylamine

"and is therefore called"

"dimethylbenzamide"

Here are a few other examples.

methylformamide   cyclohexylacetamide   2-methylpropanamide

diethyl 3-methylbenzamide
(insect repellant in OFF)

lysergic acid diethylamide
(LSD)

Properties

Resonance

The most important property of amides relates to the nitrogen, which, as you know, is more willing to share its lone pair than any other element we have studied, the property that makes amines more basic and more nucleophilic than most other functional groups. In amides, the ability of nitrogen to share the lone pair on the nitrogen leads to a resonance structure with a significant effect on the structure and reactivity of amides.

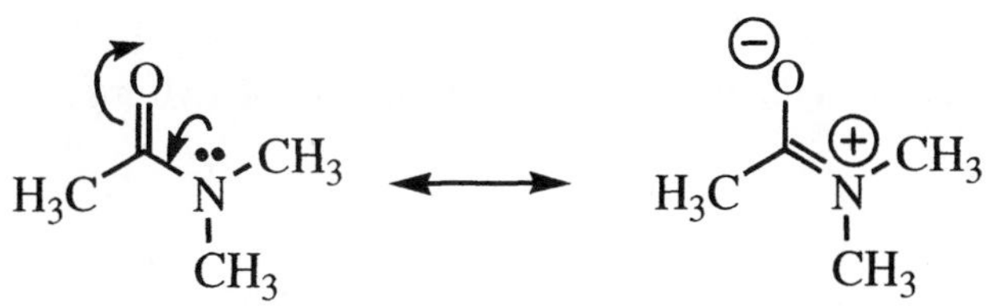

There are four pieces of evidence to suggest that this resonance structure is more than a figment of our imaginations. The first is a direct measurement. Using techniques we have not discussed in this book, one can measure bond lengths. The carbon–oxygen bond of an amide is longer than most carbon–oxygen double bonds, and the carbon–nitrogen bond is shorter than most carbon–nitrogen single bonds. This suggests that the second resonance structure contributes significantly to the resonance hybrid.

The second piece of evidence is a continuation of the reactivity series we discussed in Chapter 19. You remember:

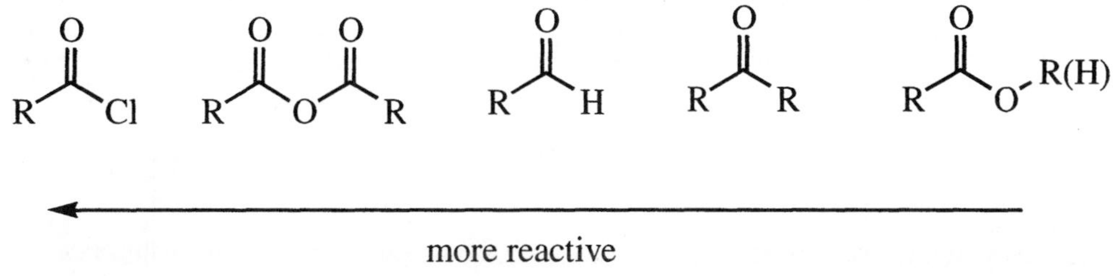

more reactive

We decided then that acids and esters have less reactive carbonyl groups, because the resonance structure with a positive charge on carbon is stabilized a bit by yet more resonance with the neighboring oxygen:

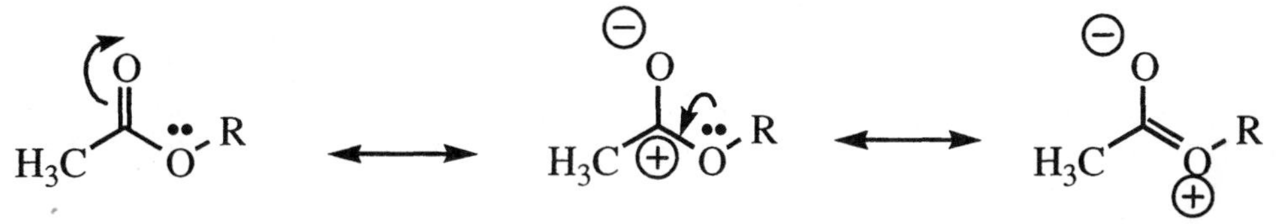

If the neighboring atom were instead nitrogen, this third resonance structure should be still better, so amides should be less reactive than acids and esters. The fact that amides really *are* less reactive than acids and esters suggests that our reasoning might be correct, and that the resonance structure with a carbon–nitrogen double bond is more important in amides than the analogous structure in other carbonyl compounds. By the way, it is also true that the nitrogen atom in an amide is a very poor base and a very weak nucleophile, another consequence of the lone pair being busy interacting with the carbonyl group in this resonance structure.

The other two pieces of relevant evidence involve spectroscopy. You should recall, also from Chapter 19, that we discussed IR stretching frequencies for the various types of carbonyl compounds. We found that those with weaker C–O bonds required less energy to stretch them; that is, their IR bands appeared at lower frequencies. The carbonyl groups of amides have about the lowest stretching frequency of any carbonyl group. This is evidence that the C–O double bond in an amide is more single-bond-like than that in any of the other carbonyl groups. In other words, the resonance structure with a carbon–nitrogen double bond (and therefore a carbon–oxygen single bond) contributes significantly to its structure.

The final piece of evidence for the importance of this resonance structure comes from NMR. Look at the resonance structure of dimethylformamide:

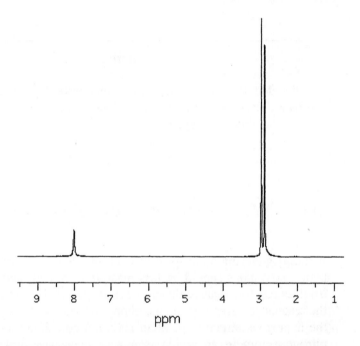

If there really is a carbon–nitrogen double bond (or even a significant partial one), it should be difficult for rotation to occur around the carbon–nitrogen bond. But if that bond cannot rotate, then the two methyl groups attached to it are not the same: one is *cis* to an oxygen, the other is *cis* to a hydrogen. The two methyl groups should therefore appear at two different places in the NMR spectrum. On the other hand, if rotation around the carbon–nitrogen bond is rapid, the two methyl groups should be identical, and they would appear at the same place in the NMR.

Below is the proton NMR spectrum of dimethylformamide at room temperature.

Clearly, there are two different methyl peaks. This means that rotation around the carbon–nitrogen bond is slow enough for the NMR instrument to distinguish between the two groups. Interestingly, if you heat up this sample, the two methyl peaks merge into one at about 110 °C. At this temperature, rotation around the carbon–nitrogen bond is fast enough so that the instrument can no longer tell the two methyl groups apart. If you know enough about the physics of the NMR process, you can actually calculate an energy barrier for this rotation, and it turns out to be about 21 kcal/mol, *much* more than the approximately 3 kcal/mol it takes to cause rotation around a typical single bond.

The bottom line here is that the carbon–nitrogen bond of an amide has many characteristics of a double bond, even though we usually draw it as a single bond. The most important property, as we will discover later (in the section on protein structure) is that, like with a double bond, the carbon, the nitrogen, and the four atoms attached to them all want to share the same plane:

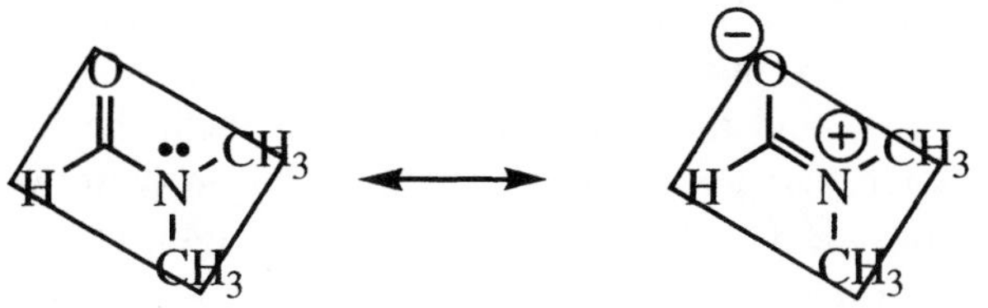

Preparation

Like esters, amides can be made most efficiently from acid chlorides or anhydrides by treatment with the appropriate nucleophile. Since the nucleophile in this case, an amine, is more nucleophilic than an alcohol, the reaction goes even better and faster than the corresponding reaction of an acid chloride with an alcohol (which leads to an ester).

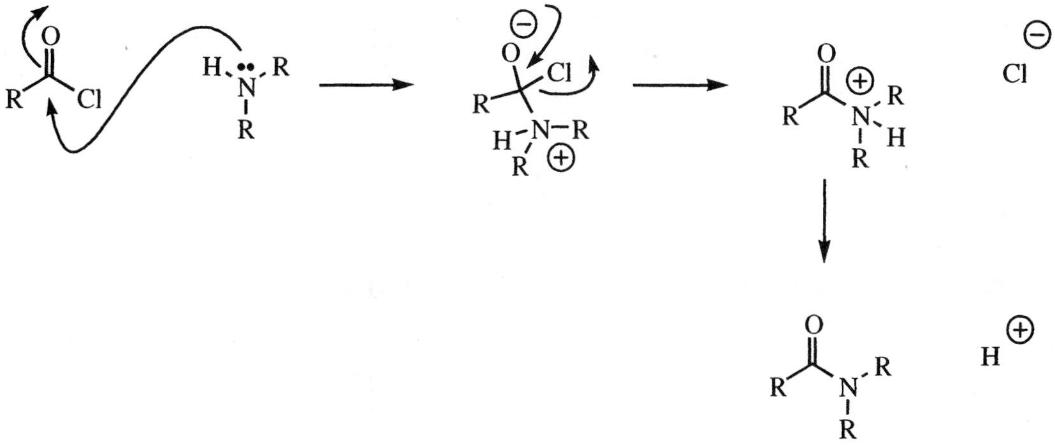

Based on our previous discussion of amine synthesis, you should be concerned about a potential problem here: the reaction generates HCl, and our starting material is an amine, which reacts readily with acid to accept a proton. A protonated amine no longer has a lone pair, and is thus not a nucleophile. This problem can be solved in one of two ways. One possibility is to use excess amine, so that you can afford to lose some and still have enough to react with all your acid chloride; this option is fine if your amine is cheap and easy to remove after the reaction. The second possibility is to run the reaction in the presence of some *other* base that will remove the newly released protons, thus leaving the amine free to react with the acid chloride. But what is to prevent that other base from reacting with the acid chloride itself? One clever choice is to use a tertiary amine, like pyridine, which cannot itself form an amide, because it has no proton to lose, but which can sop up the extra acid formed during the reaction. Below are some examples of amide formation.

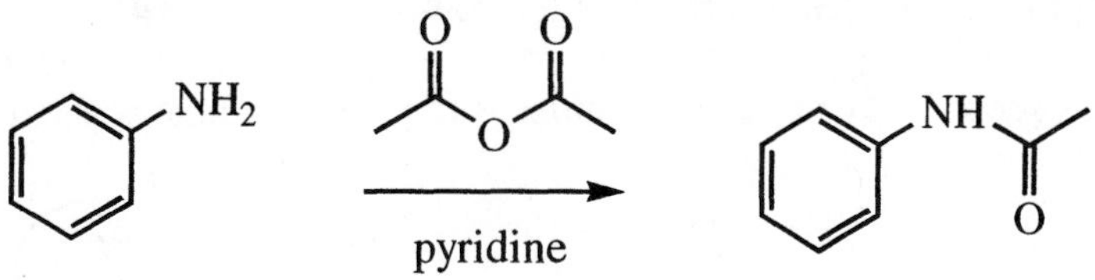

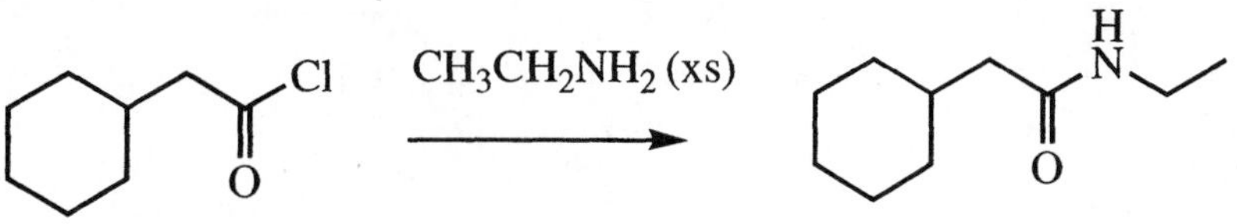

Sometimes sodium hydroxide is used for this purpose. Why can we get away with this? Hydroxide should be a *stronger* nucleophile than an amine, and the acid chloride should react with *it*. Nevertheless, there are many cases reported where an acid chloride (often dissolved in an organic solvent) is added to a solution of sodium hydroxide and an amine in water, and the acid chloride reacts with the amine. The explanation appears to involve solubility: an amine, being neutral, can dissolve in an organic solvent containing an acid chloride (or in droplets of the acid chloride itself), while the water solution of sodium hydroxide resists interacting with the acid chloride. A particularly interesting example of this is shown below.

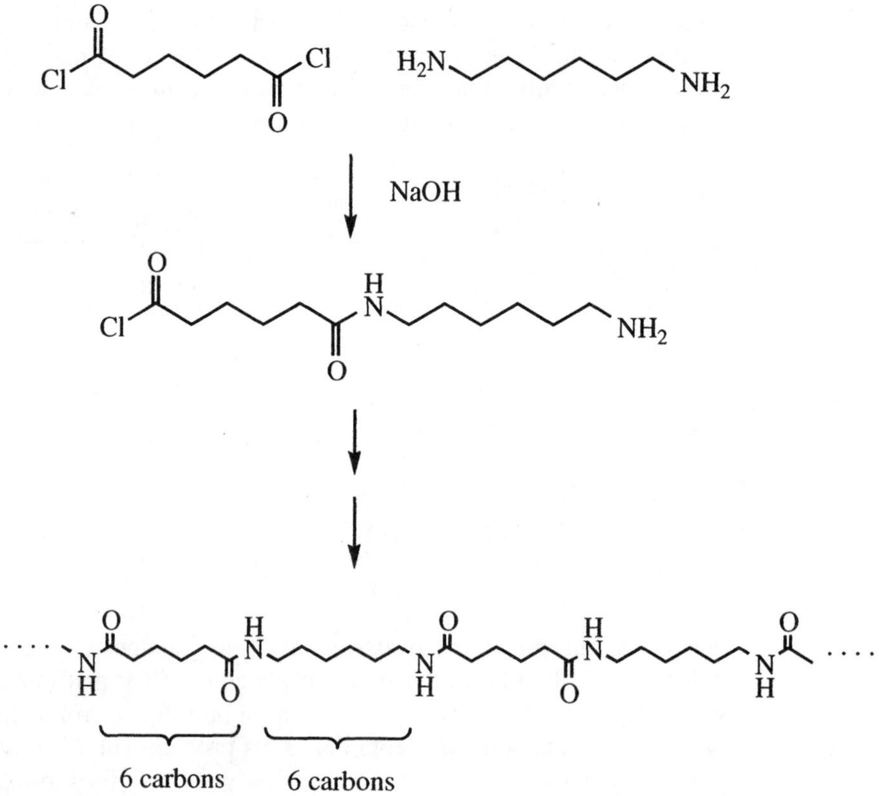

6 carbons   6 carbons

"Nylon 6-6"

The conditions here have been arranged such that reaction happens at the interface between two layers. The acid chloride is dissolved in an organic solvent like pentane, and the amine is dissolved in water containing some sodium hydroxide. Where the layers meet you get a scum of polymer. If you reach in with tweezers and pick this up, new polymer will form instantly, attached to the old, and you can draw out a continuous string of nylon. Ask your professor to demonstrate this for you—it's really cool!

A second way of making amides is to react an amine with an ester.

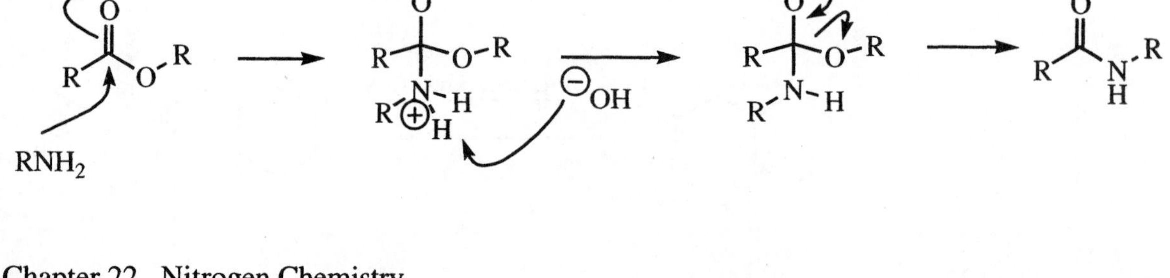

Esters are much less reactive than acid chlorides and anhydrides, so this is not a common approach in the lab. However, in the grand scheme of things it is certainly the way most amides in the world are made, since it is the biological approach and therefore accounts for all the protein in the world.

Proteins are made by stitching together amino acids one at a time to form amides, which in biological situations are called "peptides." The template that permits this to happen in a specific way is a string of RNA called "messenger RNA," which in turn got its instructions from DNA in the nucleus of a cell. Messenger RNA consists of a long string of chemical letters that are read in groups of three. Each three-letter sequence has a complementary sequence in a particular molecule of what is called "transfer RNA," and each type of transfer RNA has attached to it a specific amino acid. Transfer RNA–amino acid pairs of all types are floating around in the cytosol of the cell. Messenger RNA wanders out into the cytosol and a game much like Tetris begins. The correct t-RNA molecules drop into appropriate slots on the messenger RNA, bringing their attached amino acids into the necessary proximity for bonding to occur. The growing chain repeatedly shifts from one t-RNA to the next as the chain grows. This is shown schematically in the following crude picture of the RNA-Tetris game.

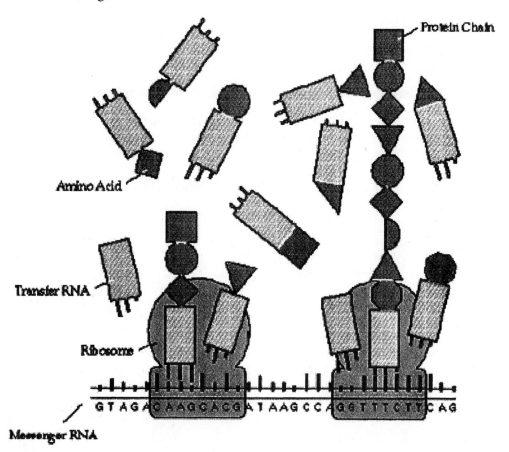

What this picture attempts to show is the way Mother Nature arranges to get amino acids strung together in the right order to produce a particular peptide, which will ultimately grow to be a protein. The actual stitching process is invisible in the picture above, so let's zoom in on an area near the top of the gray blob called the "ribosome," the organelle that catalyzes the stitching process. Here is what is going on there:

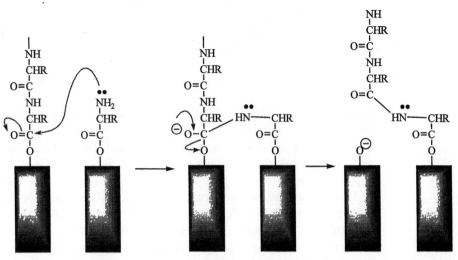

You see how a growing chain jumps from the left t-RNA onto the right t-RNA, making room for a new t-RNA to enter the picture still further to the right. At each step the chain grows by one amino acid residue, and the identity of that particular amino acid (i.e., the nature of the R group on it) is controlled by which t-RNA happens to fit into the next slot. The system is an extremely clever one! But what I want you to notice is that the chemical reaction taking place is the one we just discussed: formation of an amide from an amine and an ester.

Reactions

The only reaction of amides I want to discuss is hydrolysis back to carboxylic acids and amines. The reaction is simple, but it also involves some very interesting features that are worth discussing.

Amide hydrolysis can be catalyzed by either acid or base. Below is the acid-catalyzed process:

There are several things you should notice about this:

1. No negatively charged structures whatsoever are shown in the mechanism. In acid-catalyzed processes all structures are either neutral or positively charged.

2. When an amide is protonated, the proton lands on the carbonyl oxygen rather than on the lone pair on nitrogen. Why is this? Shouldn't nitrogen be more basic than the carbonyl oxygen? Certainly amines are much more basic than carbonyl groups, but this nitrogen is not part of an amine, it is part of an *amide*. The lone pair on this nitrogen is already pretty busy with the carbonyl group; indeed, protonating the nitrogen lone pair would prevent the amide resonance. The positive charge would be stuck on nitrogen with nowhere else to go. On the other hand, protonation on oxygen leads to a carbocation with three resonance structures, each with the positive charge in a different place. The charge is shared among three positions, and resonance interaction, which was limited in the neutral amide (because one resonance structure has two charges and the other has none), is much more effective in the cation (where each resonance structure has the same charge). Thus, if you think carefully about a protonated amide, it should not surprise you that the preferred site for protonation is oxygen. By the way, you should have drawn and examined several structures while you were reading this paragraph!

3. The next time a proton is added, you are in luck. There are three possible sites for protonation, either OH or the nitrogen. None of these involves resonance structures, either before or after protonation, so resonance need not enter into our considerations. Nitrogen is much more basic than oxygen, so we should expect that nearly every time it is the nitrogen that will be protonated. So the desired reaction—the one we've shown—should be the most likely outcome.

4. The products of the reaction are a carboxylic acid and an amine. Since the reaction is run in acidic medium, we should expect the product amine to become protonated, thus removing some acid from the mixture. The acid for this reaction gets consumed: you must be sure to take this into account so that you will use enough.

The corresponding base-catalyzed process is shown below:

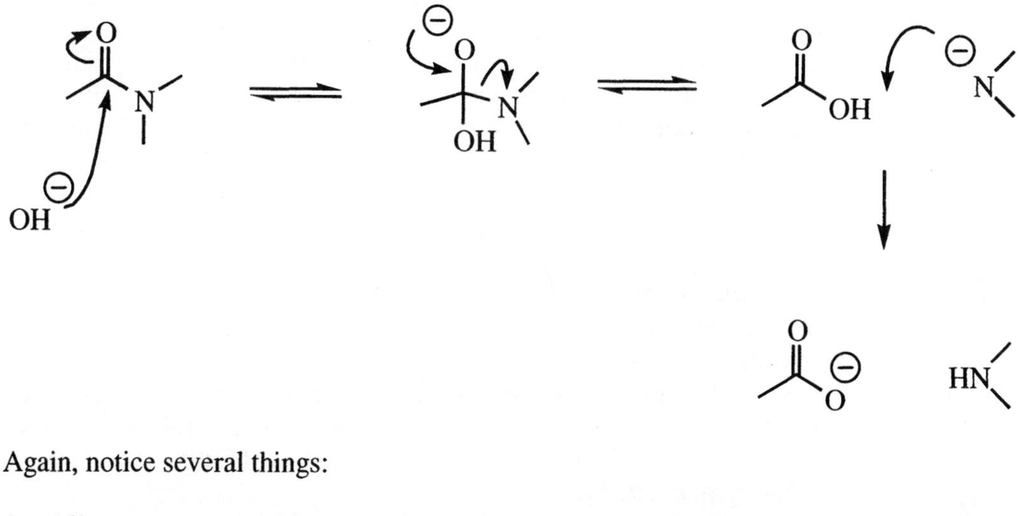

Again, notice several things:

1. All structures are either negative or neutral, as you should expect for a base-catalyzed process.

2. Hydroxide can act as either a nucleophile or a base. We have shown it acting as a nucleophile. Is there a competing base reaction we need to worry about? There could be protons on the α carbon, next to the carbonyl. However, α protons on an amide are not very acidic, with a pK_a near 30. Why are these so much less acidic than other α protons? The acidity of α protons depends on resonance stabilization of the enolate anion. But as we know, the carbonyl group of an amide is already quite busily engaged in resonance with the nitrogen next door, and has little interest in helping out an anion on the other side. Thus, this acid–base reaction is unlikely to cause a problem.

On the other hand, if there is a proton on the nitrogen, it *can* be removed by hydroxide. Protons on the nitrogen of an amide have a pK_a of about 17, very close to that of water, so hydroxide can remove these fairly well. Still, once a proton is removed there is not much the resulting anion can do, so there is little problem with this complication either.

3. After hydroxide attacks the carbonyl group, an O^- is generated on the oxygen of the carbonyl. As usual, this promptly goes back where it came from, kicking out some leaving group. What are its choices? It can kick out OH^-, returning to the amide, or it can kick out NR_2^-, to make carboxylic acid. Which is the better leaving group? By far, OH^-. Almost all the time, this intermediate will return to where it came from. We should anticipate that basic hydrolysis of an amide will be very slow and inefficient.

There is only one reason why we can get away with base-catalyzed hydrolysis at all: on those few occasions when nitrogen *is* ejected, the product immediately scarfs up a proton from the carboxylic acid, effectively removing it from the equilibrium (notice the arrow going down is not an equilibrium arrow, since this acid–base reaction is so favorable that virtually no neutral carboxylic acid will be present). Thus, the reaction very slowly drains to the right, even though the intermediate returns to starting material 99.999% of the time.

Protein Structure

As you know by now, enzymes are the assembly-line workers of the cell, taking in substrates, operating on them, and spitting out products. Enzymes do their jobs by having spaces in their structures that are just the right size and shape for the substrate to fit in, and once the substrate is there, it finds acid and base groups and polarity in just the right places to make the desired reaction proceed perfectly. The key to an enzyme's function, then, is its shape. But an enzyme is generally nothing more than a long piece of spaghetti, a polymer of amino acids stuck together by amide bonds, in this context called peptide bonds. How does an enzyme figure out the "right" way to fold up, and how does each copy of it manage to fold up in exactly the same way? Certainly if you took a piece of spaghetti 100 yards long and let it fall on a plate 10 times, it would not fall in exactly the same way each time!

The general term for long polymers of amino acids is "proteins." Not all proteins are enzymes, but most (though not all) enzymes are proteins. Proteins are not like pieces of spaghetti in a couple of crucial ways. A piece of spaghetti does not care how it lands: any pile of spaghetti is just as happy as the next. But proteins have certain constraints on the way they can fold up, which, taken together, manage to control the particular shape each protein molecule adopts. There are three levels of protein structure that we need to discuss: primary, secondary, and tertiary.

The key difference between one protein and the next is the make-up of the peptide chain. Proteins are composed of some 20 amino acids tacked together in long chains. Any of the 20 amino acids might occupy the first slot; any could occupy the second slot; and so on. For a protein that is n amino acids long, there are 20^n different ways of putting it together. Since most proteins are more than 100 amino acids long, a huge variety of possible structures could be made. One way of specifying a particular protein is to simply list off the amino acids that compose it, in the proper order. This information is called the "primary structure" of a protein. By convention, the constituent amino acids are normally listed starting from the end with a free NH_2 group (the "N terminus") and proceeding to the end with a free COOH group ("C terminus).

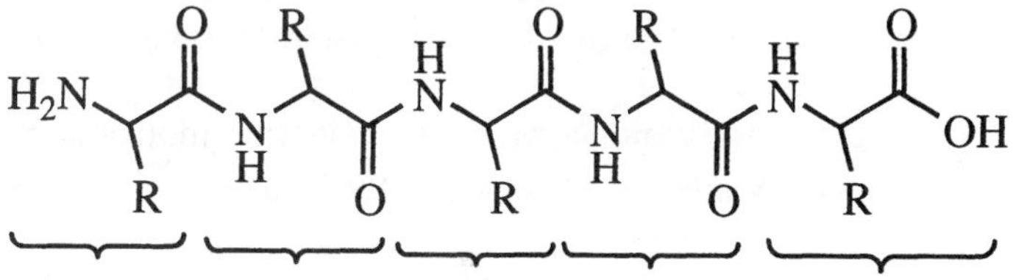

amino acid 1   amino acid 3   amino acid 5

amino acid 2   amino acid 4

You will notice that even the short piece of protein shown above, with only five amino acids (a "pentapeptide"), is 16 atoms long, with 15 bonds connecting them. The number of possible ways of orienting even such a short molecule as this is enormous, and if proteins had free rein to adopt any possible conformation it is unlikely that all copies of a given protein would be able to agree on a best one.

However, we already know that there is less than complete freedom here. We have already agreed that amides, the repeating unit in proteins, have resonance structures that are sufficiently important to constrain the geometry of these species:

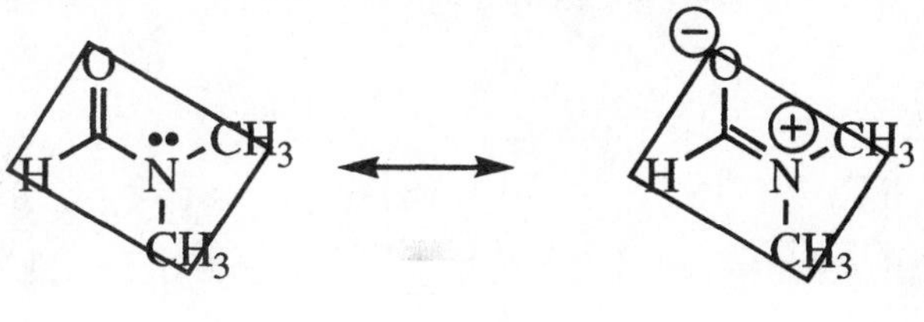

The six atoms constituting the amide system all try to remain in the same plane. Thus, the pentapeptide shown above does not in fact have limitless conformational possibilities open to it. The atoms in the boxes in the following picture are all constrained to specific planes. Each plane can rotate with respect to the next, but the end result is more like ravioli sewn together at the corners than it is like spaghetti. Many fewer conformational possibilities are available.

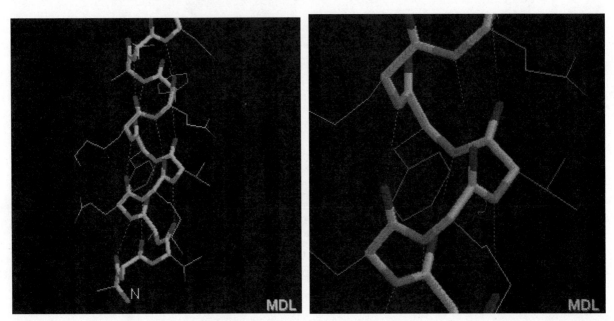

Within these constraints, the chain then goes on to the next stage and tries to make a bunch of hydrogen bonds.

Secondary Structure

Once the identity of a particular chain is established, the molecule wiggles around within the constraints of amide planarity to try to maximize attractive interactions between one part of the molecule and another. These generally take the form of hydrogen bonds.

There are two major approaches a peptide molecule can take to forming extensive arrays of hydrogen bonds: the α-helix and the β-sheet.

In an α-helix, the chain curls back on itself in a spiral shape in order to form hydrogen bonds along the axis of the spiral. A hydrogen bond forms between a carbonyl oxygen and an N–H bond that is 11 atoms away. Below are two pictures of this, but you really can't see things well from a flat picture. What you really ought to do is go to the web site listed and look at the moving models shown there. You will need the "Chime" plug-in, but you may already have that (I already recommended it in Chapter 5). The following two pictures are captured from this really nice web site at Carnegie Mellon University: http://info.bio.cmu.edu/Courses/BiochemMols/ProtG/ProtGMain.htm. See color plates for color versions of these and later pictures.

The second standard method of forming lots of hydrogen bonds at once is what is known as a β-pleated sheet. Here the chain is stretched out in a fairly linear fashion, then it doubles back, and a second portion of it lines up with the first in such a way as to form hydrogen bonds along the length of the chain and between the two pieces of the chain. The following picture is taken from another portion of the same web site.

Again, static pictures do not do this justice. Go to the web site and play.

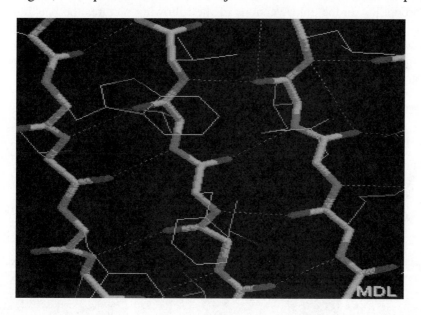

Tertiary Structure

Having maximized attractive hydrogen-bonding interactions, the molecule now wiggles around some more in an attempt to orient its side chains in a desirable way. It is easy to see from the picture of the α helix that the amino acid side chains (the R groups of the individual amino acids) stick out from the cylinder formed by the helix. It often happens that the R groups sticking out in one direction are polar, and interested in having water around, while the R groups sticking out in the other direction are non-polar and want to hide from water. Below are two pictures that illustrate this. In the color versions of these pictures, the red part is the backbone of the helix, green represents non-polar ("hydrophobic") side chains, and red represents polar ("hydrophilic") ones. The pictures show a side view and an end view of the same helix.

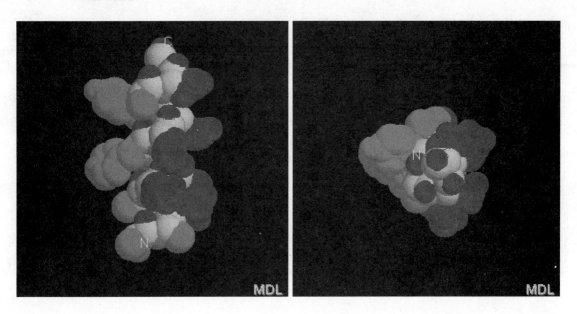

It is less obvious from the pictures I've shown, but the same principle applies to β sheets. You will be able to tell by manipulating the models on the web site that side chains stick out sort of perpendicular to the general plane of the sheet, going alternately up and down as you move along the chain. The sheet then develops surfaces that are either hydrophobic or hydrophilic. In the color versions of the following pictures the backbone of the sheet is in gray, the hydrophobic groups in green, and the hydrophilic groups in red. The left picture shows one side of the sheet, with almost exclusively hydrophobic groups. The middle picture shows the opposite side of the same sheet, with a large majority of hydrophilic groups. The right picture shows and edge-on view of both sides.

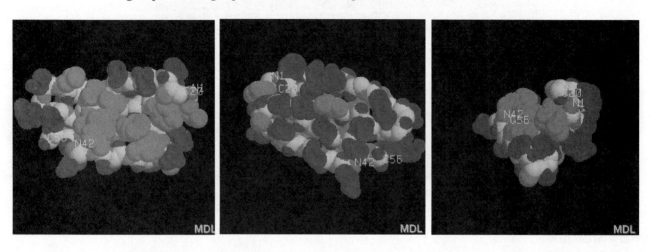

The protein now folds up in such a way as to bring the non-polar (hydrophobic) groups together on the inside of the glob, leaving the hydrophilic stuff hanging into the water surrounding it. The web site from which all these pictures are taken shows a protein called Protein G, which is composed of one helix and four strands of β sheet. There is a cartoon drawing to show these features—β sheet in yellow, α helix in red; the second picture fills in some of the side chains, in particular the non-polar ones in green. It is quite clear (especially in the moving pictures on the web site) how non-polar groups are sequestered in the middle of the molecule, hidden from the aqueous environment.

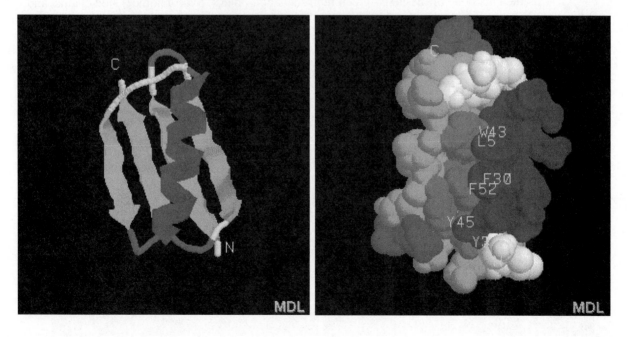

The way in which a protein folds its secondary structures into a full three-dimensional object is called tertiary structure. It is in this process that the "active site" of an enzyme is formed, a cleft in the surface that the substrate can fit into so that it can be operated on by other groups hanging around nearby. Below are two pictures of a substrate fitting into an enzyme pocket; one shows the enzyme and substrate as sticks, the other shows them as spheres. These come from another section of the Carnegie Mellon web site, http://stingray.bio.cmu.edu/~rule/bc1/lyso/lyso.html.

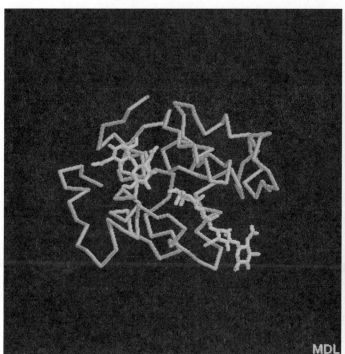 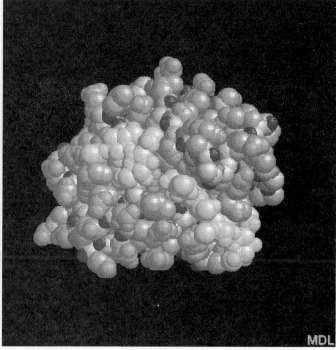

Some proteins go a step further, joining up with other protein molecules to form partnerships of two, three, eight, or some other number of proteins in a giant blob. The way in which proteins aggregate like this is called quaternary structure.

Here are some other web sites with very nice structures and descriptions of the concepts we have introduced.

http://tutor.lscf.ucsb.edu/mcdb108a/index.html

http://www.umass.edu/microbio/chime/hemoglob/2frmcont.htm

http://www.worthpublishers.com/lehninger3d/index.html

http://www.MoleculesInMotion.com/

Have fun exploring!

Problems

(Answers are provided at the end of the chapter for italicized problems.)

1. Show the organic products of the following reactions. Write NR if there is no reaction.

   a.

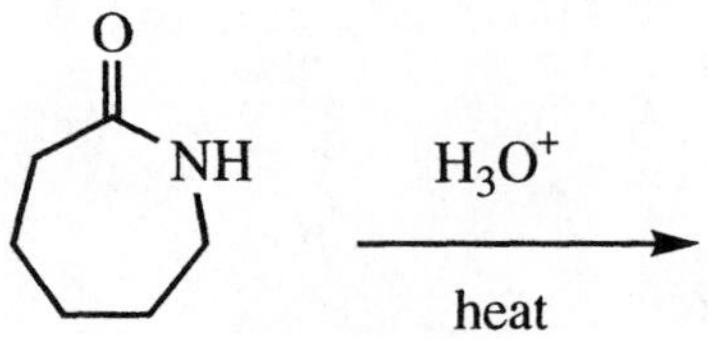

   b.

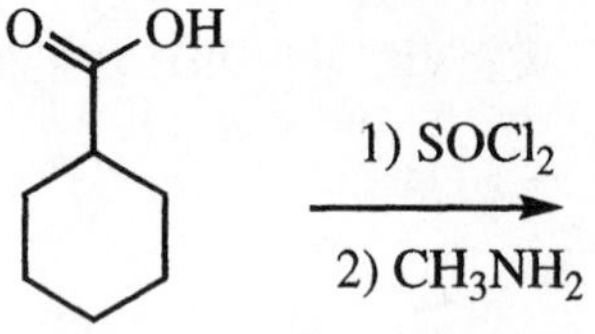

   c.

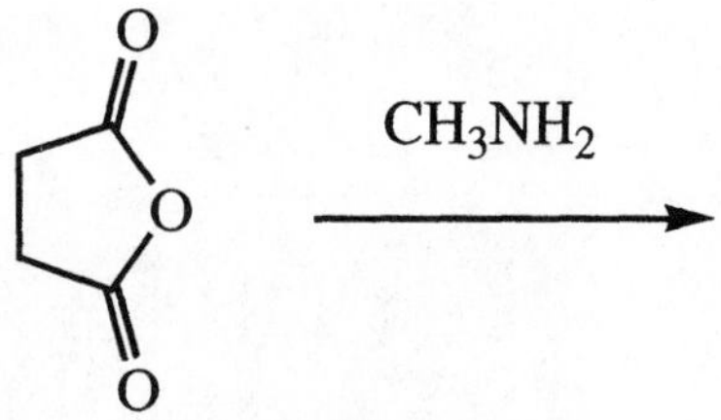

2. N,N-Diethyl-*m*-toluamide (DEET) is the active ingredient in many insect repellants. How might you synthesize this compound from *m*-bromotoluene (1-methyl-3-bromobenzene)?

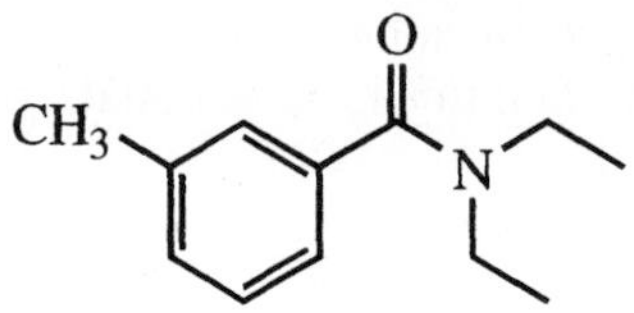

3. Explain why formation of an amide from an amine and an acid chloride requires two equivalents of amine (or added base). Would the same be true of formation of an amide from an ester? Why or why not?

4. *On treatment with strong base (e.g., LDA) followed by protonation, compounds **A** and **B** undergo cis-trans isomerization, but compound **C** does not. Show a mechanism for one of the isomerizations, and explain why **C** does not undergo the same reaction. Show all important resonance structures.*

A   B   C

5. Explain why amide hydrolysis is generally easier when done in aqueous acid rather than aqueous base.

6. *In spite of the statement in problem 4, some molecules cannot tolerate acid and have to be hydrolyzed in base. A clever method of efficient hydrolysis under basic conditions involves treatment of an amide with one equivalent of water and two equivalents of potassium* tert-*butoxide. Show the mechanism for this process and explain why this method (unlike the normal base hydrolysis you discussed in problem 4) leads to rapid and efficient basic hydrolysis of an amide to a carboxylic acid (anion).*

Selected Answers

Internal Problems

Test Yourself

NaCl in water

End of Chapter Problems

4.
In both **A** and **B** the most acidic proton is α to the carbonyl. These will be removed by strong base.

The anion, of course, is flat due to resonance, so a new proton could approach equally well from either side, leading to an equilibrium mixture of *cis* and *trans* isomers.

In **C** there are protons that are more acidic, and removal and replacement of these protons will not affect the stereochemistry:

6.

Instead of choosing between OH⁻ and NR_2^- as leaving groups, where OH⁻ will almost always be ejected, in this case we are choosing between NR_2^- and O^{-2}. NR_2^- is the better leaving group now.

Chapter 23
Aromatic Chemistry: Electrophilic Aromatic Substitution; Nucleic Acid Structure

Ever since we started discussing organic molecules we have been claiming that benzene and its relatives are special in some sense. We will spend most of this chapter exploring what the specialness is, and what consequences it has for the chemistry of benzene. In this chapter we will

- Define aromaticity and see when it applies

- Explore the reactions of benzene with electrophiles

- See what happens when a substituted benzene undergoes electrophilic aromatic substitution

- Discover biologically important aromatic compounds.

Normally, I start these chapters with nomenclature and properties of the functional groups under discussion. In this chapter, however, I want to start by taking a step back to examine the phenomenon we call "aromaticity."

AROMATICITY

We claimed long ago that benzene is special because of its resonance, which lets it spread out its electrons and lower its energy.

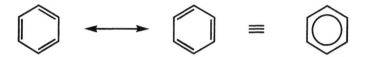

The questions I want to ask now are: first, what is our evidence that this is really the case, and second, how general is the phenomenon?

There are several lines of evidence concerning the specialness of benzene. Two relate to its structure, one to its spectroscopy, one to its energy, and one to its reactivity.

It was known long ago, before 1850, that benzene was unusual. Its formula was known to be C_6H_6, so it apparently had a lot of double bonds, but it did not act like other compounds that had double bonds (for example, the addition reactions we learned in Chapters 11 and 12 do not work with benzene). So what was its structure? No one could figure this out. The idea that the carbon chain might form a ring had not occurred to anyone yet. There is a famous story about the great chemist August Kekulé. He had been contemplating this riddle for some time, and one day, according to his own description,

> I was sitting at writing at my textbook, but the work did not progress; my thoughts were elsewhere. I turned my chair to the fire, and dozed. [Perhaps modern organic students can empathize with poor Kekulé.] Again the atoms were gamboling before my eyes. [Poor guy, he dreams chemistry! Has this happened to you yet?] This time the smaller groups kept modestly in the background. My mental eye, rendered more acute by repeated visions of this kind, could now distinguish larger structures of manifold conformations; long rows, sometimes more closely fitted together; all twisting and turning in snake-like motion. But look! What was that? One of the snakes had seized hold of its own tail, and the form whirled mockingly before my eyes. As if by a flash of lightning, I woke; ... I spent the rest of the night working out the consequences of the hypothesis. Let us learn to dream, ... and then, perhaps we shall learn the truth.

(In recent years this description has come under fire, both because there is a suspicion that Kekulé never had such a dream, and because there is evidence that others thought of the ring possibility before he did; there is even a debate about the accent in his name! But it is a neat story.)

Even postulating a ring did not solve the problem, however. Kekulé imagined benzene as a six-membered ring with alternating single and double bonds, a picture that we now call a "Kekulé structure":

Kekulé structure of benzene

If this were really the structure, there should be four different dibromobenzenes, as shown below:

The first and last structures should not be alike if the double bonds drawn in the ring are real double bonds, stuck where they are drawn, as Kekulé at first thought. In the first structure the two bromines are attached to the same double bond and in the last structure they are on different double bonds. The problem is, there *aren't* four dibromobenzenes: there are only three. Kekulé realized that his structure could not be right, and proposed that actually there are two such structures, going back and forth so rapidly that they cannot be separated. If this were the case, the first and last structures could not be distinguished.

Kekulé's solution to the isomer problem

We now believe that Kekulé was wrong about this. Instead of thinking about two structures jumping back and forth, we believe that benzene is a single structure halfway in between the two, a resonance hybrid:

Modern solution to the isomer problem

Kekulé was not so far off—for a guy in the 1860s before they even knew about shapes of molecules!

What makes us think that Kekulé was wrong? For one thing, if he had been right, there would be real double bonds in benzene, and it should react like other compounds with double bonds in it; but it doesn't, as we will see presently. For another, if Kekulé had been right, then at any given instant, the molecule should be distorted, with three long (single) bonds and three short (double) bonds. Carbon–carbon single bonds like this are usually about 1.48 Å long; carbon–carbon double bonds are typically 1.33 Å. *All* of the benzene bonds, by every measure, are 1.39 Å, almost exactly halfway in between, and as nearly as we can tell they do not fluctuate.

Resonance Energy

Finally, as you know, resonance results in a lowering in the energy of a species, and we can demonstrate that benzene is in fact lower in energy than it ought to be based on the structure Kekulé proposed. The proof presents a rather tricky problem, though, one we addressed in another context in Chapter 7 when we were discussing strain in small rings. At that time we measured the energy pent up in cyclopropane quite easily, simply by burning it, but then we had to compare the result with the energy we suspected the compound *ought* to have if it had the same bonds but no strain. Of course, no such compound exists, so we had to make a guess. It will be worth your time to go back and reread that section now to remind yourself of the problem we had then and how we solved it, because we are about to do something very similar.

Once again our job will be to measure something about a molecule, and compare the result to what we think the value *should* have been if the molecule were normal. Although we could start by burning benzene, as we did with cyclopropane, there is a simpler approach.

As you know, benzene does not undergo the normal reactions we expect of carbon–carbon double bonds. One of these reactions is catalytic hydrogenation, the addition of hydrogen across a double bond to make it a single bond. This reaction, though quite facile with normal double bonds, hardly happens at all with benzene, but if you *really* beat on it, you can make it occur. So although you should not expect benzene "double bonds" to hydrogenate easily, you *can* force the reaction:

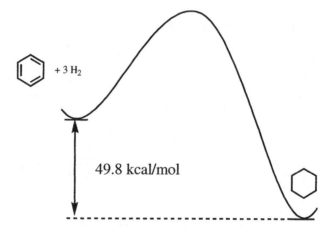

The reaction shown is exothermic, which means that energy is given off. (The reason it is difficult to accomplish is that the energy *barrier* is high, but this does not affect the exothermicity of the result.) Thus, the product is lower in energy than the starting material by the amount of energy released, which we can measure to be 49.8 kcal/mol.

This is the easy part. The tough question is figuring out how much energy this reaction *ought* to be worth if the "double bonds" in benzene were *real* double bonds. The most common way to estimate this is to recognize that the reaction shown involves hydrogenation of three "double bonds," each of which is in a six-membered ring. So if we could figure out the worth of one double bond in a six-membered ring, and multiply that by three, we ought to have a good comparison figure for the hydrogenation of benzene. This is easy: measure the heat of hydrogenation for cyclohexene. This reaction turns out to be exothermic by 28.6 kcal/mol. Multiply by three and you get 85.8 kcal/mol.

So what does all this mean? A six-membered ring containing three *real* double bonds ought to give off 85.8 kcal/mol upon hydrogenation. Benzene (*real* benzene) gives off only 49.8 kcal/mol. So benzene is lower in energy than it ought to be by 36 kcal/mol. This is illustrated in the following diagram.

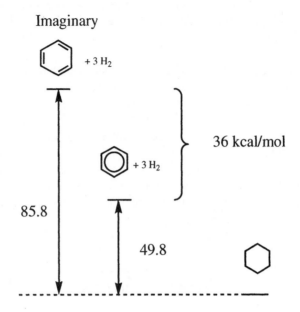

Imaginary

The 36 kcal/mol represents the extra stability benzene enjoys by being able to delocalize its electrons through resonance. This is often referred to as the "delocalization energy" or "resonance energy" of benzene. There are many problems with this number—more advanced treatments come up with different answers, and there are even debates in selected circles of organic chemists about whether delocalization of the π electrons is the real reason for the extra stability—but everyone does agree that benzene is more stable than it ought to be if it were composed of simply three double bonds in a ring.

Recall that I started this section with two questions: first, what is our evidence for benzene being special, and second, how general is the phenomenon? We have now looked at some of the evidence that benzene is special, and we will see more when we get to spectroscopy and chemical reactions; let's next consider the second question. Is this specialness restricted to the molecule benzene? No, it isn't. Actually, in the early 1800s, a number of molecules were discovered that had similar properties. Many also had pleasant odors, and that's why they were termed "aromatic" compounds. It was later discovered that the special properties and the pleasant odors had nothing to do with each other—many pleasant-smelling compounds don't share the special properties, and many molecules with these special properties don't smell good (indeed, many smell awful!). For reasons no one can explain, the term "aromatic compound" stuck with the properties, not the smells! So molecules that share the interesting properties of benzene are called aromatic compounds, and they are said to have "aromaticity." Aromaticity includes the unusual bond lengths and extra stability described above, and also certain spectral properties and special reactivities we will discuss later.

So what does it take for a molecule to be aromatic? For starters, benzene, or the benzene part of some larger molecule, is always aromatic. Here are a few examples of aromatic molecules.

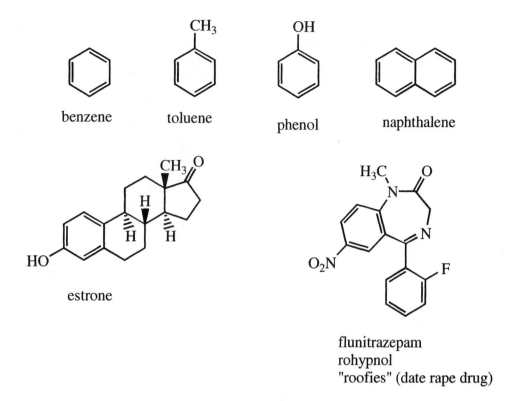

benzene   toluene   phenol   naphthalene

estrone

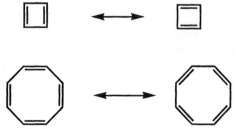

flunitrazepam
rohypnol
"roofies" (date rape drug)

Note that only the benzene ring portions of the lower two molecules would be considered aromatic. The rest of each of the molecules is normal. The official term for molecules (or parts of molecules) that are not aromatic is "aliphatic."

Hückel's Rule

But there are other aromatic systems also, and you should learn some more things about what makes a compound aromatic. Based on what I have said so far, you would probably guess that the key is the ability to delocalize electrons to make equivalent structures, as in benzene:

As we will see, this is not the whole story. After all, if this were all it took, then cyclobutadiene and cyclooctatetraene also would be aromatic:

And yet, by every measure, *neither* of these compounds shows the special stability and reactivity of benzene. Indeed, they are not only not aromatic, they seem to be actually *worse* off than normal aliphatic compounds—that is, instead of being *non*aromatic these molecules are sometimes referred to as *anti*aromatic. What is the matter with them? Why doesn't resonance stabilize these molecules the way it does benzene? The answer lies in quantum mechanics, and you can't understand it without doing a whole lot of math that most of you cannot do yet. But the bottom line is this: when resonance operates in a circle like this, the number of electrons involved is important. The number of electrons is almost always even, but if the number is a multiple of 4, this is bad; if not, it is good. Stated mathematically, if the number of electrons can be represented as $4n$, where n is any integer, this is bad for the system, and that system is not aromatic. If the number of electrons is even but *not* a multiple of 4, it could be written as $4n + 2$, where n is again some integer (including 0), and this kind of a number is good for the system; the compound will be aromatic. Cyclobutadiene has 4 electrons involved in the resonance shown above ($4n$ where $n = 1$), so even though we can *write* two nice resonance structures, the molecule does not think this is so nice, and is unstable. Benzene has six electrons ($4n + 2$ where $n = 1$), so the resonance shown makes the molecule stable. Cyclooctatetraene has 8 electrons ($4n$ where $n = 2$), so the resonance shown would make the molecule unstable. The same would be true of systems of 12, 16, 20, etc., electrons. On the other hand, systems of 2, 6 (benzene case), 10, 14, 18, etc., electrons are thrilled to have resonance in a circle and are aromatic.

The fact that the number of electrons is important was deduced by a German chemist named Hückel, and the conclusion is known as Hückel's rule: a planar, fully conjugated cyclic system is stabilized if the number of conjugated electrons is $4n + 2$, and destabilized if the number of conjugated electrons is $4n$.

This rule is deceptively simple, but many students have trouble counting electrons, so let's go through some examples. Let's start with pyridine: is this aromatic or not?

pyridine

This is too easy, you are thinking. There are six electrons, two in each double bond, so this is just like benzene. What's the big deal? Well suppose I remind you that there is also a lone pair on the nitrogen? Now how many electrons are there? Do you still think it is aromatic?

Hmm, now there are eight, and eight is bad. So this must not be aromatic. Wrong! It *is* aromatic! Why??? Because Hückel's rule involves the number of electrons that are *conjugated* with each other, not the total number of electrons that are present. Build a model of pyridine, or look at the side view shown below:

There is a p orbital on each atom in the ring—each atom is sp^2-hybridized—and these p orbitals are all conjugated with each other. There are six electrons involved in this set of orbitals, one from each atom. The two extra electrons on nitrogen are *not* part of the π system: they are in an sp^2 orbital sticking straight out from the ring and they do not even know about the six electrons happily whizzing around nearby. This is still a six-electron aromatic system.

OK, let's try another case. How about pyrrole?

pyrrole

Huh? Is this another trick question? You can't flip the bonds in this one and get back to more or less the same structure! What is he talking about? Again, go back to Hückel's rule: Hückel refers to a planar, fully conjugated cyclic system. A fully conjugated system is one with a p orbital on every atom, each lined up with the ones next to it. Pyrrole clearly has a p orbital on each of the carbons. What about the nitrogen? Based on Chapter 3 you'd probably label it as sp^3:

However, the nitrogen *could* morph into sp^2 hybridization, moving the lone pair into a p orbital, if that were advantageous. Normally such atoms choose to remain sp^3-hybridized in order to keep the electron pairs as far away from each other as possible. However, in this case, putting the lone pair into a p orbital puts those electrons in communication with the four electrons in the double bonds, making a cyclic array of six electrons, Nirvana to a molecule! So pyrrole actually does not look like the above picture, it looks like the one below, where the N–H bond is straight out from the ring and all six electrons are happily conjugated. It is aromatic.

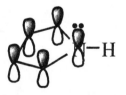

It is possible to show this conjugation by drawing resonance structures, but they are the kind of resonance structures that you should be loathe to write, since they require charge separation, so unless you pay attention to the number of electrons involved, it is easy to be led astray by such pictures. Nevertheless, I show them here.

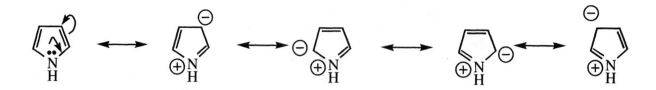

How about furan?

furan

Furan has two electrons in each double bond, for four, and *two* lone pairs, for another four. This adds to eight, a bad number. But be careful! Are they all conjugated with each other? Ignoring aromaticity, we would label the oxygen as sp^3 and draw it as on the left below; but recognizing that the molecule will choose aromatic conjugation if possible, we can allow *one* of the sp^3 orbitals to become a p orbital, as in pyrrole, while the oxygen atom changes from sp^3 to sp^2:

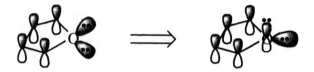

This allows six-electron aromatic conjugation around the ring, with the second lone pair perpendicular to the aromatic system and oblivious to it, as in pyridine.

As you can see, counting is not simply a matter of counting!

Test Yourself 1

Comment on the structure and properties of the following two ketones:

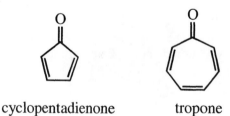

cyclopentadienone   tropone

Test Yourself 2

Comment on the properties you would expect for the following compound. Go ahead, draw some possible structures and think about it!

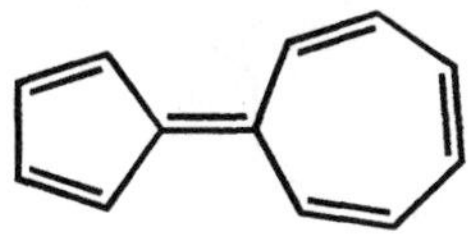

NOMENCLATURE

I am sorry to have to report that many aromatic compounds have special names. There is no need for you to remember them all, but you should certainly know what toluene, phenol, and naphthalene are.

benzene 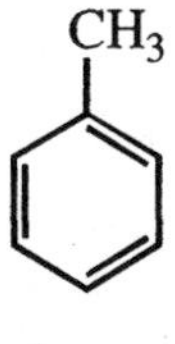   toluene 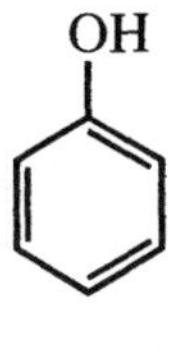   phenol 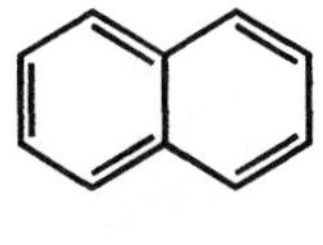   naphthalene

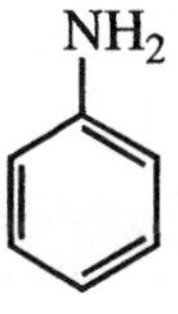

aniline 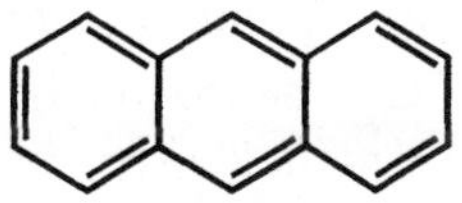   anthracene 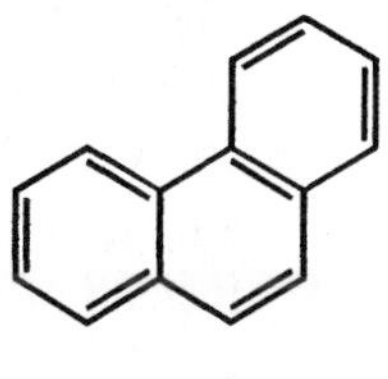   phenanthrene

You already know that a benzene ring, when considered as a substituent, is called "phenyl." This is often abbreviated in books as Ph, so that the following two structures represent the same thing, most easily named 3-phenylheptane:

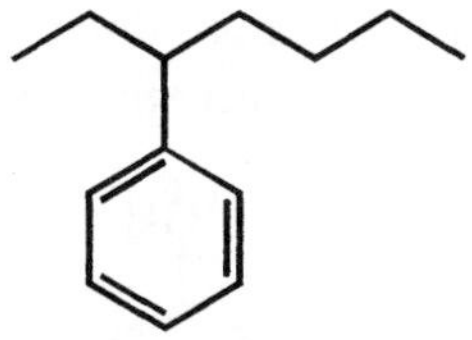

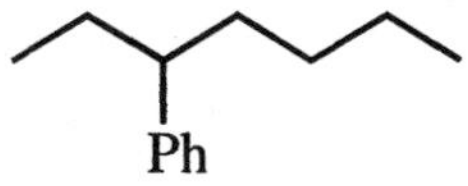

Ph

What I have not yet told you is that chemists writing by hand on paper or the blackboard often use the Greek letter "phi" to indicate a phenyl group, so that the structure below is also the same thing:

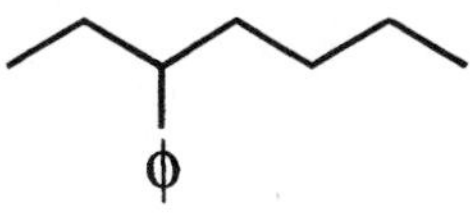

φ

You've also learned that the position next to a benzene ring is called the benzylic position, and the PhCH$_2$– group is the benzyl group. This position can also be referred to as an α position. Thus, all of the following names are acceptable for the structure shown, although the bold ones are used most often:

(bromomethyl)benzene
α-bromotoluene
benzyl bromide

There is another issue in naming that is different from anything we have seen before. This has some historical origins. Consider the three possible dibromobenzenes:

These are (or should be) 1,2- 1,3- and 1,4-dibromobenzene. As I told you before, it was known long ago that there were three such compounds, and all three had been carefully studied. One had a boiling point of 219 °C, one had a bp of 224 °C, and one was a solid with a melting point of 88 °C. The trouble is, no one knew which compound had which structure. Instead of referring to a particular compound as "the dibromobenzene with a boiling point of 219 °C," it was agreed that this one would be called "*meta*-dibromobenzene," the one with bp 224 °C would be called "*ortho*-dibromobenzene," and the solid one would be called "*para*-dibromobenzene." It was assumed that, sometime in the future, when they figured out which compound was which, these stupid, meaningless names would be dropped, and the correct, numbered names would be used. Unfortunately, the future happened, but the names did not change. So these three words, "*ortho*," "*meta*," and "*para*" have come to mean "next carbon over," "two carbons away," and "on the opposite side" in reference to benzene rings. Further, these words are almost always abbreviated by the single letter with which they start. So, for example, the following names are acceptable and often used:

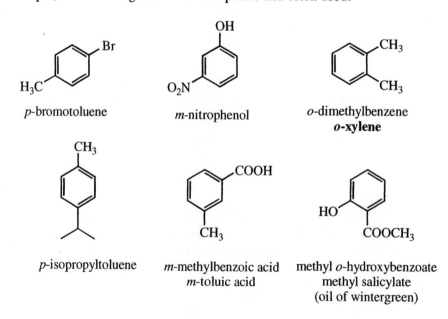

p-bromotoluene

m-nitrophenol

o-dimethylbenzene
***o*-xylene**

p-isopropyltoluene

m-methylbenzoic acid
m-toluic acid

methyl *o*-hydroxybenzoate
methyl salicylate
(oil of wintergreen)

STRUCTURE

Benzene is a six-membered ring. Each carbon is sp²-hybridized with a p orbital perpendicular to the plane of the ring. The p orbitals are all conjugated with each other.

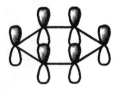

An sp² atom has a natural angle of 120°. A hexagon also has a natural angle of 120°. Thus, benzene, as a flat hexagon, could be called a perfect molecule. There is no strain whatsoever in its structure.

If you take benzene and start adding on additional benzene rings in random directions, you get other aromatic compounds. Each six-membered ring is a perfect hexagon, and the resulting huge molecule forms a flat sheet. If you keep going out in every direction like this, you get graphite. Bulk graphite is a bunch of such sheets layered on top of each other. It is extremely strong and stiff in the directions of the molecular plane, which makes it an excellent structural material for tennis rackets, skis, airplanes, etc., but each sheet can easily slide over the neighboring sheets, making graphite a good material for lubricants and pencil lead.

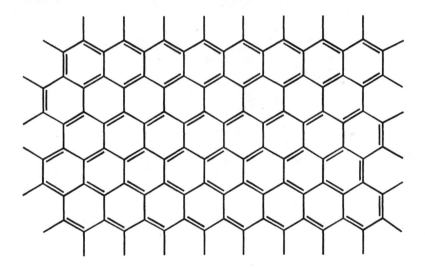

Segment of graphite

One of the important aspects of the structure of aromatic systems is that such compounds have a strong tendency to remain planar. What makes aromatic systems so low in energy is all those p orbitals talking to one another; in order for that to happen the orbitals have to be parallel, which requires the underlying framework to be flat. In general, aromatic systems *are* flat, and we will later see some consequences of that fact. However, recently a new kind of aromatic system has been found that incorporates a lot of benzene rings into spherical or tubular frameworks. Buckminsterfullerenes (fullerenes, or buckyballs, for short) are spheres of benzene rings interspersed with a few five-membered rings. They come in many sizes and shapes, but the most famous is the "soccer ball" molecule. Each six-membered ring in the molecule is a benzene ring, slightly distorted out of planarity. There are no hydrogens at all in this molecule: its formula is simply C_{60}.

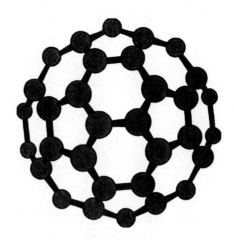

PROPERTIES

Acid–Base

Benzene is a very weak acid, because when you remove a proton, the electrons left behind are in an sp^2 orbital sticking out from the ring. These have no possibility of engaging in resonance. Be sure you understand this, because I cannot tell you how many times I have seen students draw multiple (and wrong) resonance structures for such anions (and similar cations). Recall that resonance structures are our poor way of representing communication among overlapping p orbitals. Look back in Chapter 8 for a review of this.

Benzene is slightly more acidic than a saturated hydrocarbon, because the electrons left behind are in an sp^2 orbital rather than an sp^3orbital, and as you know, an sp^2 orbital is slightly lower in energy than an sp^3 orbital (recall the discussion of acetylide anions), but the anion is still very unhappy. The pK_a for an aromatic proton is about 43. Benzene is also a very weak base, because in order to accept a proton it would have to use some of the electrons that are part of the aromatic "sextet."

Nevertheless, some very interesting acid–base chemistry is associated with certain aromatic molecules. The π system of an aromatic ring allows for great conjugation of anions and cations that are *adjacent* to the ring itself, i.e., in the benzylic position. Thus, the α protons of toluene are more acidic than other protons on sp^3 carbons, because the resulting negative charge can be delocalized into the ring. The pK_a of toluene is 41, compared to near 50 for most saturated centers.

If one benzene ring is good, two should be better, and three should be better yet:

$$pK_a = 34$$

$$pK_a = 32$$

By the way: it should not surprise you that this latter anion, called "trityl" anion, is colored. Look back at Chapter 14 to find out why. Also, before you go on, write out nine other resonance structures for the trityl anion, along with arrows that get you correctly from one to another.

There are other senses in which aromaticity affects acid–base properties. Consider pyrrole, which we discussed earlier:

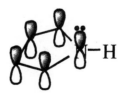

We determined that the nitrogen atom in pyrrole is sp^2-hybridized so that the lone pair will be in a p orbital to complete the aromatic sextet. As you know, lone pairs on nitrogen tend to be moderately basic, and they are easily protonated: ammonium ion has a pK_a of about 9, and the same is true for most amines. But this amine is different. Protonating the lone pair removes it from conjugation with the other four electrons. The nitrogen would become sp^3-hybridized, and there would no longer be any cyclic conjugation:

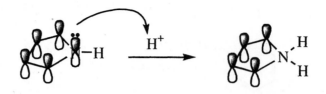

The molecule is very unhappy about losing its aromaticity and resists this process. Whereas a normal amine is about half protonated at pH 9, this one remains largely unprotonated even at pH 0. It is very hard to protonate pyrrole—the molecule is about the most non-basic amine there is.

Aromaticity sometimes works in the other direction as well. Consider cyclopentadiene:

What would happen if we removed a proton from this? With sp^2 hybridization, the new electron pair could interact with the four electrons already in the ring to make an aromatic system.

Now we have a situation where we have *created* an aromatic system in an acid–base reaction. Cyclopentadiene is the most acidic hydrocarbon you are ever likely to encounter. It has a pK_a of about 16, which means that this proton, attached to a carbon, is about as easy to remove as a proton from water! The resulting anion, cyclopentadienide anion, is a very common species in organometallic chemistry, which you may study in a later course.

COMMON MISTAKES: DON'T LET THIS HAPPEN TO YOU!

Many students attribute the acidity of cyclopentadiene to resonance. There are, in fact, five nice resonance structures you can write for the cyclopentadienide anion:

But if this were the whole story, cycloheptatriene would be even more acidic, since you can write *seven* resonance structures for its anion (do it!). But cycloheptatriene is even *less* acidic than normal hydrocarbons. See if you can explain why, based on the above discussion.

Spectroscopy

We learned long ago that the signals of aromatic compounds come between 7 and 8 ppm in the ^{1}H-NMR spectrum, where almost nothing else appears. This is well downfield of most other protons attached to carbon. Why do aromatic protons appear in such an unusual place?

The answer involves some physics, but not too much—you can handle it. Let's review the NMR experiment: you put the sample in a big magnet, and all parts of the sample are exposed to some magnetic field. But not all nuclei *experience* the same magnetic field, because each nucleus is buried inside its own little cloud of electrons, and these electrons shield the nucleus from the external magnetic field to a greater or lesser extent.

A physicist would describe this same situation in the following way: Electrons are related to magnetic fields in two ways. A magnetic field makes electrons move in a circle; furthermore, electrons moving in a circle *generate* a magnetic field (this is the principle behind electromagnets). So if we envision a molecule in a magnetic field and focus our attention on the electrons surrounding a particular nucleus, we can come to some conclusions, illustrated in the following picture: The external magnetic field will make the electrons surrounding the nucleus move, and those moving electrons will generate a new magnetic field. The rules of physics demand that the new magnetic field be opposed to the old one. This new magnetic field will therefore *subtract* from the old magnetic field, and the net field, the field that the nucleus experiences, will be less than what we put in. This is the same conclusion we came to above.

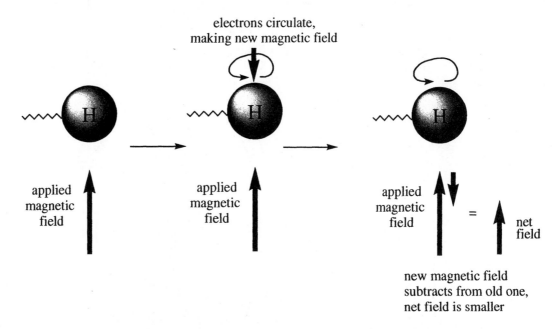

The nucleus will "resonate" when the *net* field arrow is a certain length. If there is a lot of electron density around the nucleus in question, the down arrow in the picture above will be long, and you will have to make the up arrow (applied field) extra long before the net arrow becomes the right length for resonance. That is, atoms with a lot of electron density around them are highly shielded, and appear upfield. On the other hand, electron-withdrawing groups remove electron density from the area of the observed nucleus, making the down arrow shorter and you do not need to apply so much field to get the net arrow to be the right length. That is, nuclei with low electron density around them are deshielded and appear downfield. All this describes what we previously knew about NMR.

So what's the deal with benzene? Benzene is special because it has a big, circular racetrack of p orbitals for electrons to run around. So when you apply an external magnetic field, not only do electrons around the individual nuclei begin to move, but all the electrons in the racetrack move also. Thus, a new opposing field is generated at each nucleus by the local electrons, but an even stronger opposing field is generated by the six π electrons of the ring.

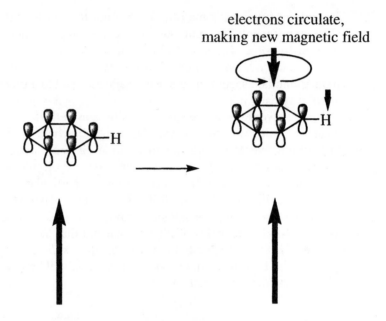

We need to focus on this latter field, often called a "ring current." Below is a blow-up of the situation.

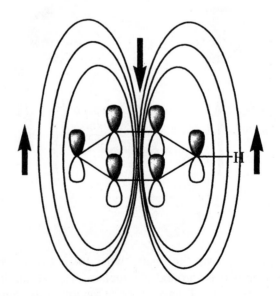

The important point is that magnetic field lines are always continuous. If the lines point *down* in the middle of the ring, opposing the applied field, then they must point *up* on the outside of the ring, reinforcing the applied field. Thus, in spite of the small field opposing the external field that is generated by the electron cloud surrounding the hydrogen shown, there is a larger effect from the aromatic ring that *adds* to the external field. The net result is that you need to apply *less* field than you might expect in order to get hydrogens on a benzene ring to feel the proper amount of magnetic field. This makes aromatic protons appear downfield in the spectrum, well to the left, between 7 and 8 ppm.

If this explanation is correct, you might expect protons *inside* an aromatic ring, or directly above or below it, to be shielded from the applied field by a double whammy: not only will local electrons shield the proton, but this time the aromatic field shown will operate in the down direction, and be very large. Indeed, such protons have been studied, and they are often so heavily shielded that they appear to the right of 0 ppm (i.e., they have *negative* chemical shift values!), off the charts of most NMRs. We will see an example of this near the end of this chapter.

There is another aspect of benzene NMR that I want to mention here, because I promised I would, way back in Chapter 14. We noticed at that time that monosubstituted benzenes often produced two groups of signals in the aromatic region, one group due to three protons and another group to two. For example, below is the aromatic portion of the spectrum of phenol. Clearly, there are three protons toward the right and two toward the left in the spectrum.

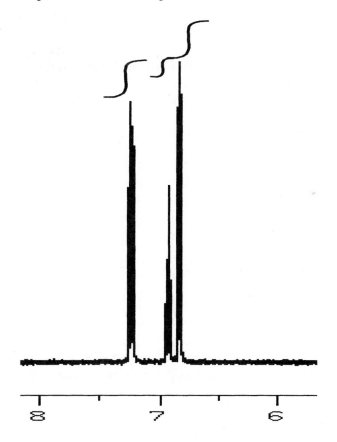

Most students naturally assume that the two protons that are different are the ones nearest to the OH group, and the other three protons are pretty similar:

similar?

This is a logical deduction, but it turns out to be wrong. To get the right answer, one must look at some resonance structures.

What these resonance structures show is that some of the carbons of the ring have a little extra electron density. Not much—these are minor resonance structures—but still some. More important, only three of the carbons have this extra electron density, and they are the carbons *ortho* and *para* to the OH group—*every other* carbon. Thus, these three carbons might be expected to have some properties in common, while the other two are different. As we will discover in the section on reactions, this is typical in benzene compounds.

The three carbons with extra electron density (and the protons attached to them) are slightly more shielded than the others, and therefore appear upfield, to the right. The group of three hydrogens you see in the NMR spectrum is not the three bottom hydrogens on the ring, but the *ortho* and *para* hydrogens grouped together, while the *meta* hydrogens are in a different location.

REACTIONS

Electrophilic Aromatic Substitution

The main point I have tried to establish so far in this chapter is that aromatic compounds are special in that they love having their cyclic conjugated system of (the right number of) π electrons. This one piece of information will serve to explain all the chemistry of aromatic compounds. If you ignore this, you will find a bewildering array of new reactions to learn. But in fact, all of aromatic chemistry is a straightforward extension of what we have already learned, provided we take into account the special stability of the aromatic π system.

The stability of the aromatic system leads to two consequences:

1. Aromatic molecules are very reluctant to use their π electrons for anything new; they want to keep their aromatic conjugation. Therefore, aromatic systems are much less willing to react with electrophiles than ordinary alkenes. This explains why all the alkene chemistry we talked about earlier does not apply to benzene and other aromatics.

2. If, somehow, the π electrons *do* get used for something new, they are exceptionally eager to get back to being part of an aromatic system.

Let's see how these generalizations play out in the chemistry of benzene.

One of the first reactions you learned was the reaction of alkenes with electrophiles. Among the electrophiles you studied were HX (where X is a halogen), H^+ in water or alcohol, and Br_2. In all these cases, the result was addition to the carbon–carbon double bond, with the electrophile attaching to one end to create the best possible carbocation, followed by arrival of an available nucleophile to take care of the resulting carbocation.

Given what you now know about aromatic systems, it should not surprise you that benzene and its relatives are not eager to donate their electrons to ordinary electrophiles like these. It takes a very special electrophile, one that is exceptionally desperate, to make benzene willing to share its electrons. This means that we have to go to special lengths to create such demanding electrophiles. There are two major approaches to this, and five major reactions that result.

Creating a Super Electrophile

Strong Acid Approach

Sulfuric acid has the structure shown below. Recall that the sulfur is able to accommodate more than eight electrons (and thus form more than four bonds) because of its d orbitals. Sulfuric acid is a very strong acid, because its conjugate base has several resonance structures (show them!), which means the negative charge is nicely delocalized.

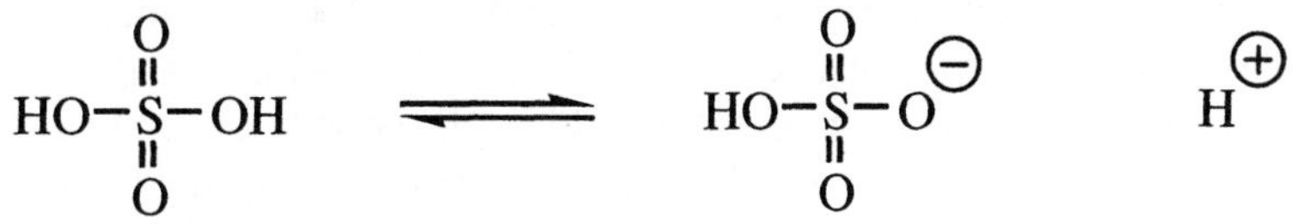

In solutions of extra strong sulfuric acid, neutral sulfuric acid molecules are occasionally protonated, leading to the HSO_3^+ cation:

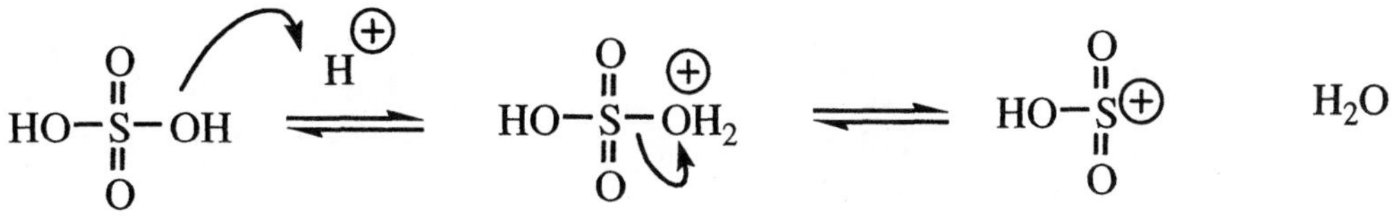

This cation is strong enough to attract electrons from a benzene ring. We will discuss later what happens next.

The same kind of thing occurs with nitric acid. To encourage the necessary protonation, nitric acid is usually mixed with sulfuric acid. An NO_2^+ cation is also strong enough to attract the electrons of a benzene ring.

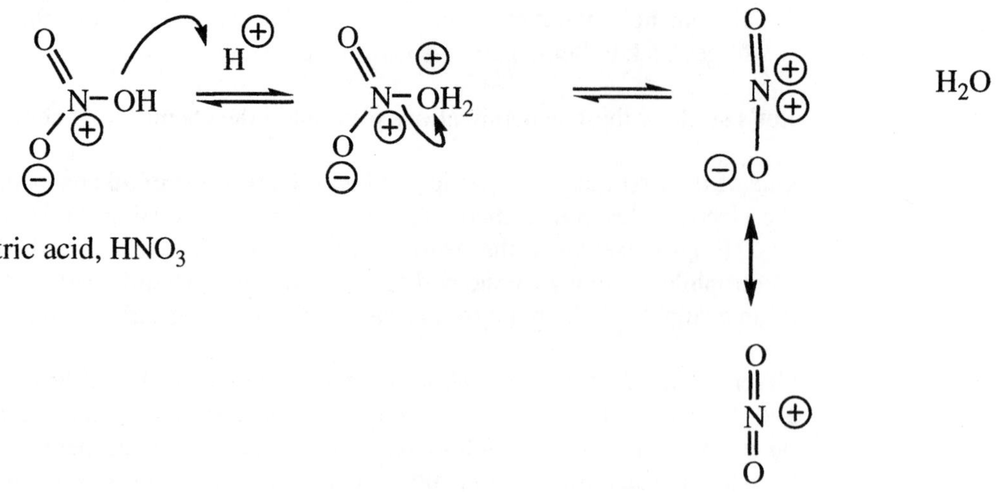

O=N-OH ... H⁺ ⇌ O=N-OH₂⁺ ⇌ (structure) + H_2O

nitric acid, HNO_3

Lewis Acid Approach

Bromine, Br_2, is an electrophile that is good enough to add to carbon–carbon double bonds, but it is not particularly desperate for electrons. We should expect that it would not react with benzene. However, if we could convince one of the bromine atoms that it was positively charged, that might do the job. This turns out to be true.

You may recall from Chapter 8 that I mentioned an important type of acid that is not a source of protons. According to the *Lewis definition* of acids, anything that accepts electrons (as protons do) is considered an acid. Carbocations, for example, are technically Lewis acids, although the term "electrophile" is more commonly used for them. The most common species referred to as "Lewis acids" are metal halides in which the metal has an incomplete octet. Two examples are BF_3 and $AlCl_3$, which you should easily recognize as having only six electrons in their outer shell, so the compounds should be able (and eager) to accept two more electrons. Another relevant species that will be less recognizable to you (because we have not studied transition elements) is $FeBr_3$. This behaves very much like the first two substances I mentioned.

When bromine is treated with $FeBr_3$, an interesting interaction takes place, shown below.

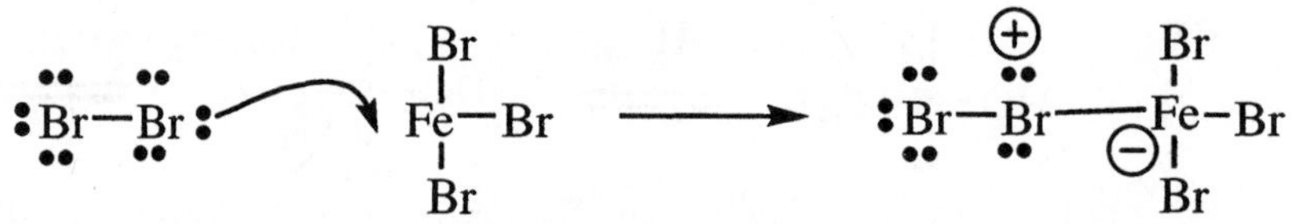

:Br—Br: + Fe(Br)₃ ⟶ :Br—Br⁺—Fe⁻(Br)₃

The bromine I have shown as positive now has a strong desire for electrons, so it pulls some away from the adjacent bromine atom, making *it* somewhat positive:

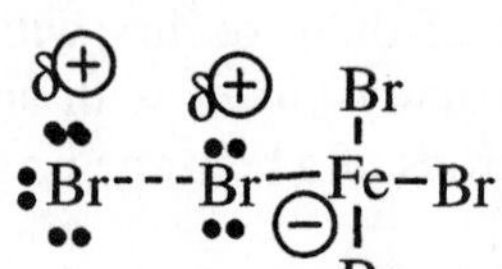

:Br---Br^(δ+)—Fe⁻(Br)₃

The bottom line is that this species, bromine-interacting-with-ferric-bromide, behaves very much as if it were Br^+, and we can treat it as such. So when we need the super-electrophile Br^+, we just use Br_2 and $FeBr_3$. Operationally, it is even easier than that, since Br_2 reacts with Fe metal to *make* $FeBr_3$. So if we take some Br_2 and toss in a little Fe, we will get a mixture of Br_2 and $FeBr_3$. Thus, a mixture of bromine and iron becomes functionally equivalent to Br^+. This is the mixture you will see written in reactions.

The other two ways we will look at for making super electrophiles are both very similar, and both are associated with the same name: Friedel–Crafts, after the two chemists who discovered the reactions. Both involve $AlCl_3$ as the Lewis acid, together with an organic halide, usually a chloride, as a partner. The following reaction occurs:

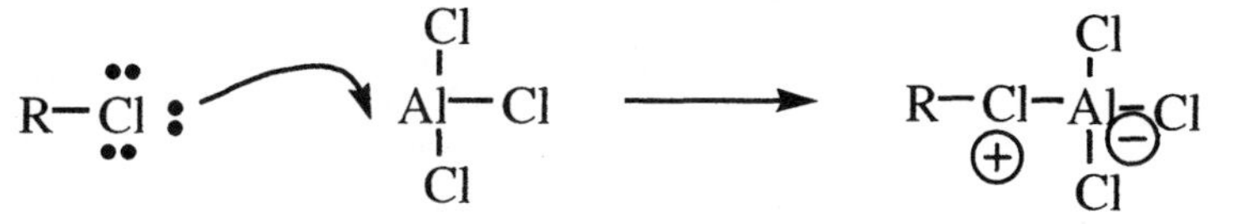

As was the case with bromine above, the positive chlorine begins to suck electrons away from its neighbor, with the result that the partner acquires a partial positive charge. The easiest way to think of these reactions is to pretend the charge that forms is real and full:

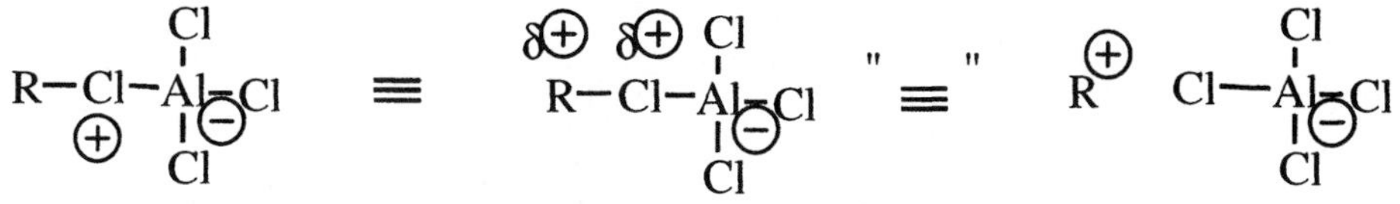

In other words, it's as if the Lewis acid sucks the chloride off your starting material, leaving a positive charge where the halide was. This is a neat way to get a positive charge almost anywhere you want it in a molecule.

The other Friedel–Crafts reaction begins with an acyl halide, and proceeds in the same way:

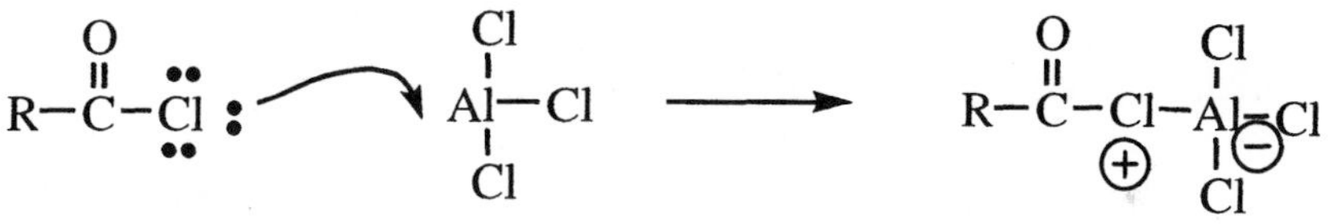

In this case, it is thought that the chlorine really does come all the way off, since the resulting carbocation can be resonance stabilized:

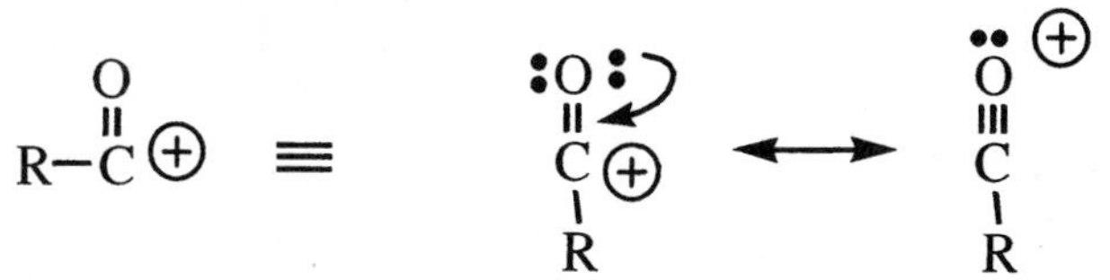

Again, the bottom line is that the Lewis acid pulls chlorine off the starting material, creating a carbocation.

In summary, there are five main ways of making electrophiles potent enough to react with benzene rings. Each of the following five reactions should lead to some new product. We now need to explore what that product is.

Reactions of Super Electrophiles with Benzene

Let's assume we chose one of the sets of conditions above, and therefore we have a benzene ring in the presence of a powerful electrophile. For convenience, let's use Br^+, although you should recognize that something completely analogous will occur with all the other electrophiles.

In the presence of a super-electrophile, benzene reluctantly agrees to lend its electrons, just like a double bond would. For bookkeeping purposes, it is easiest to show this by imagining that the double bonds in the benzene ring are real double bonds. That way, you already know how to move them:

Notice several things:

1. I wrote an H on one carbon. Of course, there are H's on all the others also. I just wrote this one to remind us that it is still there.

2. There is nothing special about the top right carbon. It was a convenient place to work in the picture I drew. The same reaction could occur on any carbon (after all, they are all alike!).

3. The resulting carbocation is resonance stabilized. It is pretty good as carbocations go, but it is still a carbocation, and high in energy, especially compared to what the system used to be, a benzene ring.

4. Conjugation in the ring is no longer cyclic. (The *ring* is still a ring, but the p orbitals suffer an interruption at the carbon atom carrying both a bromine and a hydrogen.)

5. The positive charge in this carbocation is spread among three (and *only* three) different positions.

6. This carbocation, like all others, could do any of three things: react with a nucleophile, eliminate a hydrogen, or rearrange. Rearrangement seems unlikely, since the carbocation is already resonance stabilized (it is unlikely to get any better while remaining a carbocation). Trapping a nucleophile would get rid of the positive charge, but the original benzene conjugation would have been destroyed. By far the best option for this cation is to drop a hydrogen in order to regain the wonderful situation of aromaticity. That is what occurs:

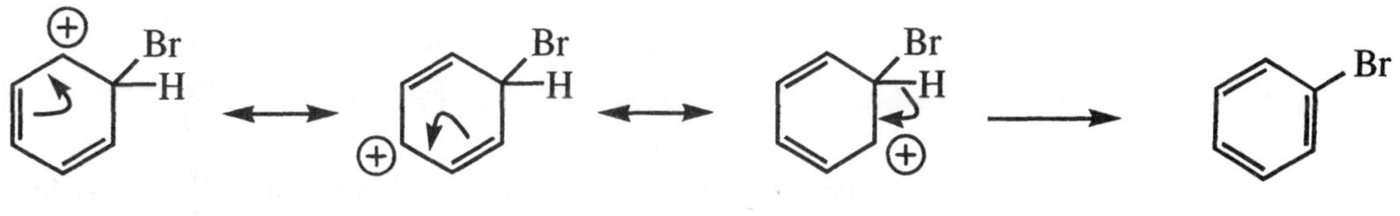

The net result of the process is that a bromine atom has taken the place of a hydrogen atom. This is therefore a substitution reaction, on an aromatic system, initiated by an electrophile. We call such reactions "electrophilic aromatic substitutions." Pretty clever, eh? One of the distinguishing properties of aromatic compounds is that, when presented with electrophiles, instead of undergoing addition (as alkenes would), they undergo substitution instead.

ORM: Under Aromatic Compounds, look at "electrophilic substitution of benzene."

You should now be able to show correct mechanisms for all of the following reactions. Do it now!

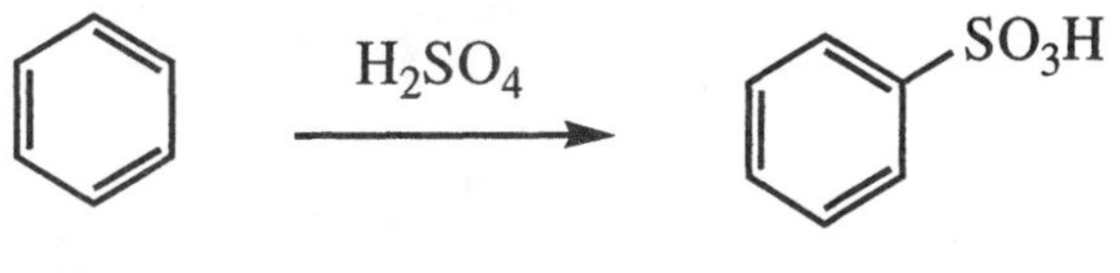

"sulfonation"

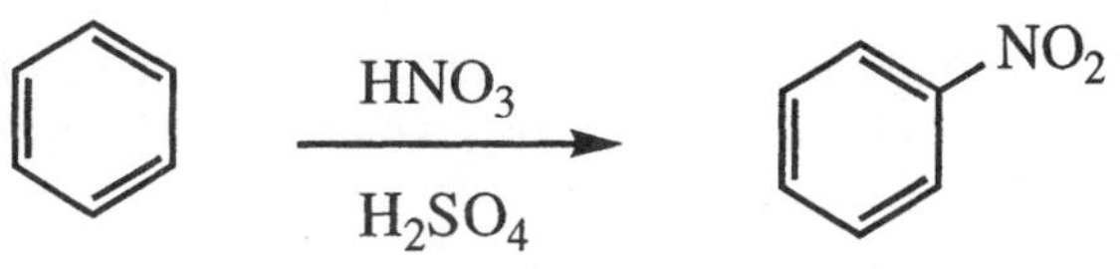

"nitration"

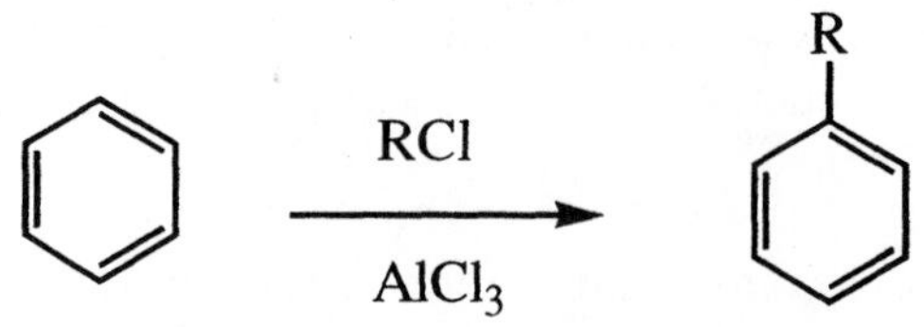

"Friedel–Crafts alkylation"

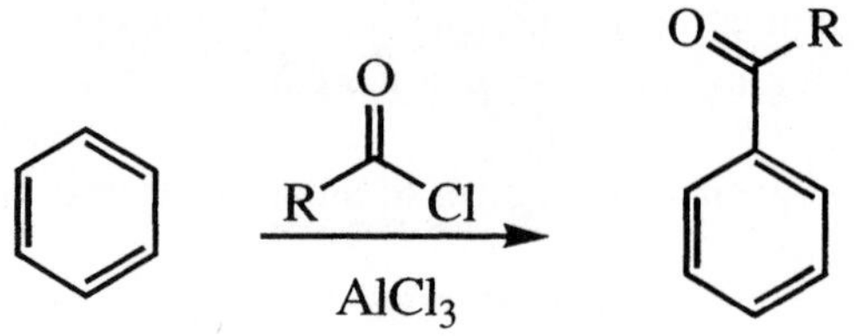

"Friedel–Crafts acylation"

I'll help you with the second one, because there is a slight wrinkle to it. The combination of HNO_3 and H_2SO_4 creates the electrophile NO_2^+, which looks like this:

What happens when benzene reacts with this? Many students will write the following:

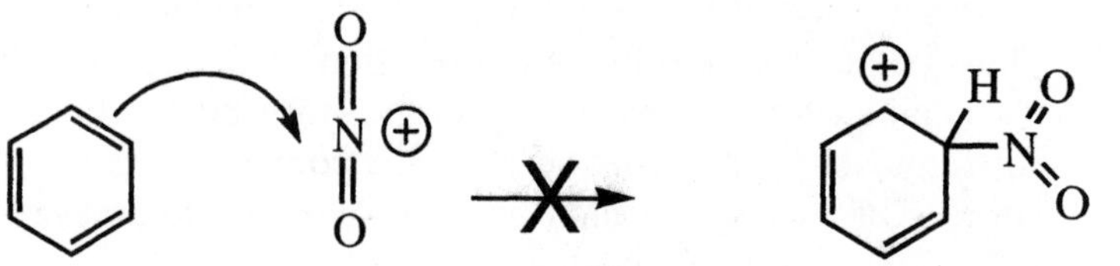

But this cannot possibly be right, because the nitrogen atom in the right-hand structure has five bonds, a major no-no. The nitrogen in the NO_2^+ already has four bonds, and can't accept any more. As the electrons from the benzene ring bond to the nitrogen, electrons from one of the nitrogen–oxygen bonds must retreat (just as in carbonyl chemistry):

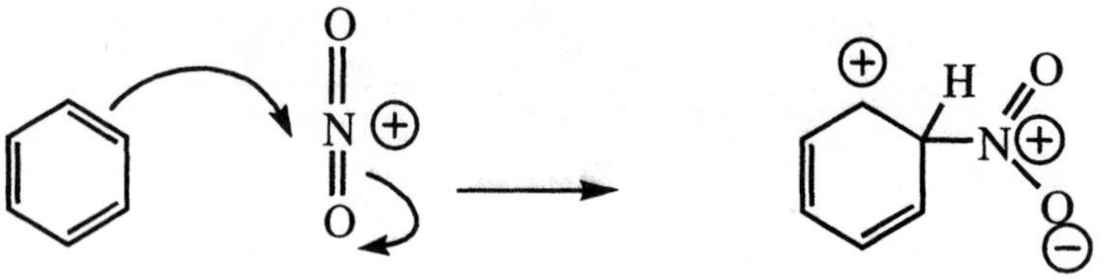

You can now show three resonance structures for the resulting carbocation, with subsequent loss of H^+ to form nitrobenzene, which has the following structure:

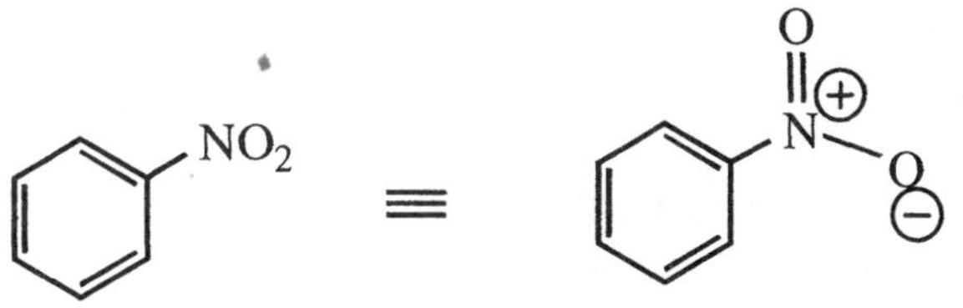

The point is that the nitro group, –NO$_2$, has an unusual structure with two charges, but it is OK with that. (Actually you should notice that a nitro group has its own resonance, independent of the benzene resonance, and that helps stabilize its structure. Show that resonance!)

There are two other points I need to make about these substitution reactions. Take a look at the Friedel–Crafts alkylation. The first thing that happens is that we get an alkyl carbocation. (Actually we do not quite get a carbocation, but it is close enough that we can consider it to be that.) As you know, carbocations can rearrange if they do not find something else to do first. In this case they should have plenty of time on their hands, since we are waiting for them to react with benzene, and that should be a slow process because benzene is so reluctant to donate its electrons. So carbocations that are prone to rearrange usually do so. This means that the following reaction is unlikely to occur as written:

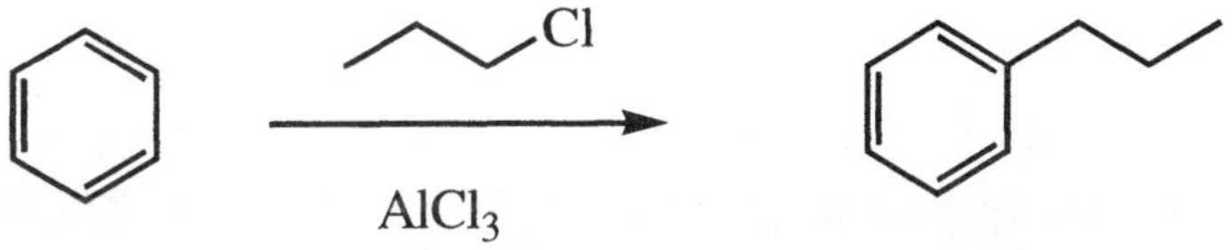

Why not? It looks like a perfectly ordinary Friedel–Crafts alkylation. But it requires forming the primary carbocation shown below, and that would be likely to rearrange to the corresponding (and better) secondary cation.

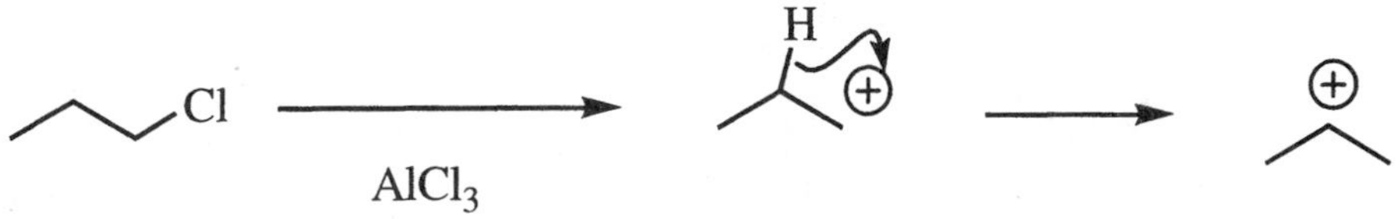

So in fact we would probably get the following mixture of products:

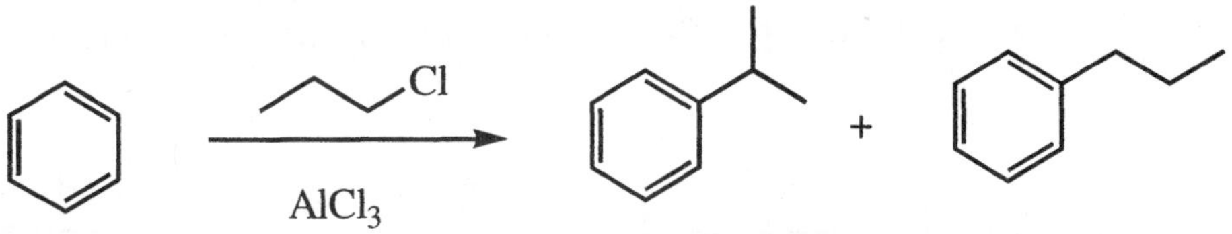

The conclusion is that it is very hard to make primary alkyl benzenes by Friedel–Crafts alkylation, with the exception of methyl and ethyl benzenes.

Test Yourself 3

Show a mechanism for the following reaction:

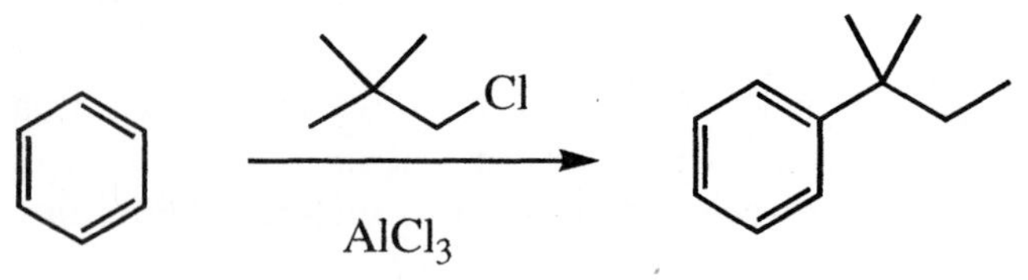

Test Yourself 4

Why is it possible to make methyl and ethyl benzene but not other primary alkyl benzenes?

The second point I want to make about these substitution reactions concerns the Friedel–Crafts *acy*lation, and it also relates to rearrangements. We determined that it is difficult to get an alkyl chain to attach to a benzene ring by its end, because rearrangements are likely. Is the same true of acyl groups? No, it isn't: the acyl cation is resonance stabilized, so rearrangement would not lead to something better.

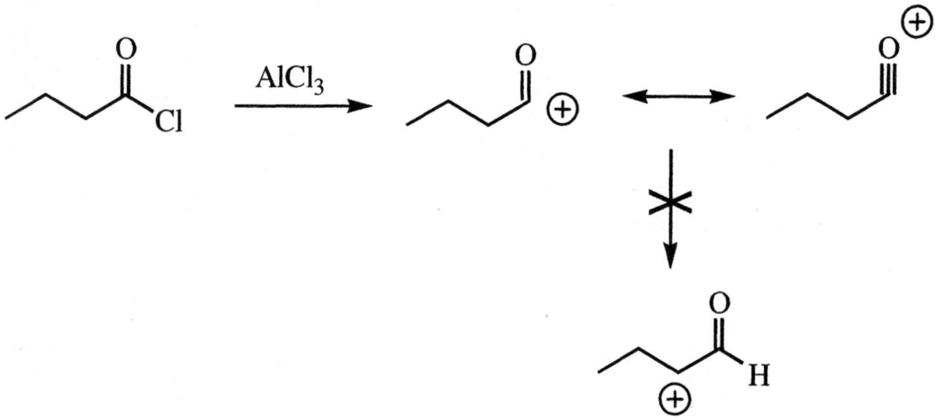

Thus, acyl groups can be reliably attached to a benzene ring by the terminal carbon. This can be used to our advantage: we *cannot* do the same with alkyl groups, but we can use acyl substitution to bypass this limitation. Check out the following sequence:

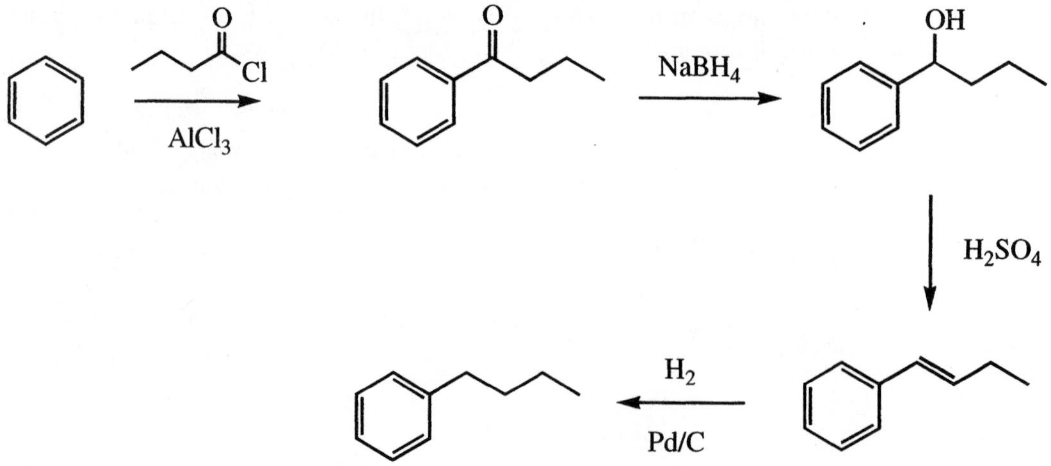

Of course, there are other ways to bypass the problem as well. For example, the following sequence accomplishes the same thing.

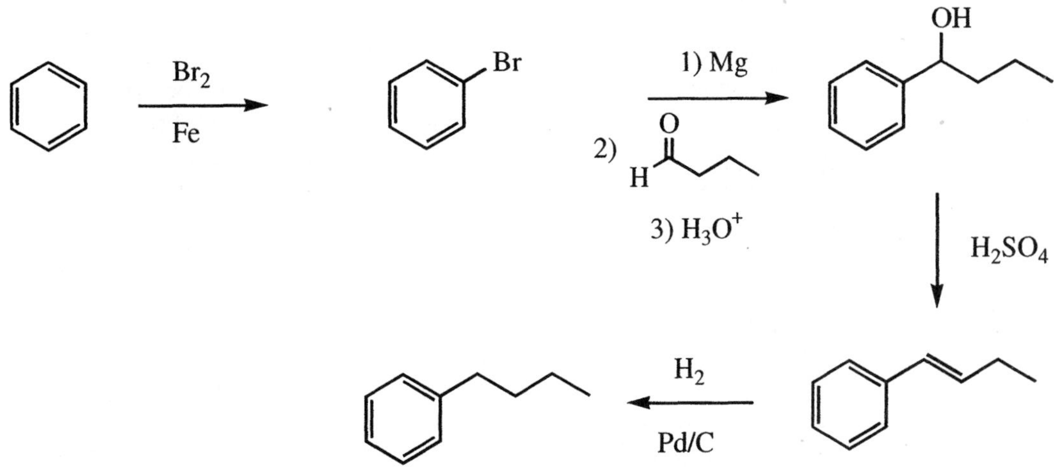

Reactions with Substituted Benzenes

You now know everything you need to know about electrophilic substitution reactions on benzene. Therefore, you now know ways to make benzenes that have various substituents on them. What happens when these *substituted* compounds undergo substitution reactions? There are two major effects to consider: an existing substituent influences *how fast* the next reaction will proceed, and also *where* the new group attaches.

Activating Effects

Electrophilic aromatic substitution depends on benzene donating its electrons to an electrophile. We already know that this is a difficult process, because benzene is jealous of those electrons and not interested in sharing them, so it takes an exceptional electrophile to force the reaction to occur. We can also anticipate that the ability of a particular benzene ring to donate its electrons should vary with the availability of those electrons—that is, a benzene ring that has an electron-donating substituent should be *more* willing to donate its electrons than benzene itself, and a ring with an electron-withdrawing substituent should be *less* eager to react than benzene itself.

This turns out to be true. Below is a list of a few substituted benzenes together with their rates of bromination compared to that of benzene itself.

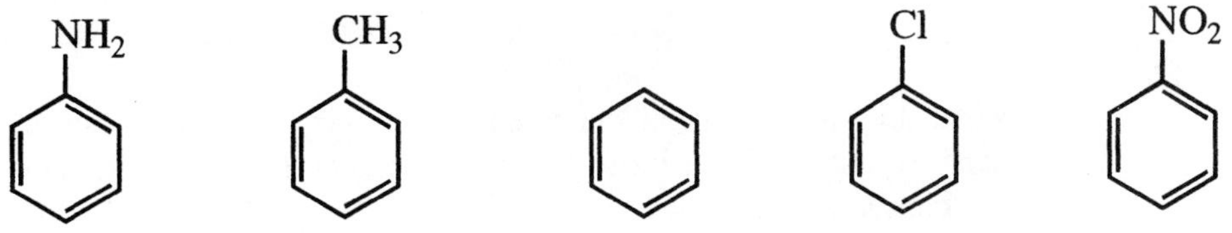

What you see is that placing a methyl group on a benzene ring makes the reaction go 25 times as fast. This should not surprise you: you already know that alkyl groups are electron donating—that is why tertiary carbocations are better than secondary. But you also know that this effect is what we call an "inductive" effect, a gentle push or pull of electrons through existing bonds. Far more effective is a resonance effect, where there is real sharing of electrons and actual resonance structures that can be written. As an example, see the previous section on NMR for the resonance structures of phenol. The ring in phenol should have genuine extra electron density compared to benzene itself. You can write the same type of resonance structures for aniline (aminobenzene), but they are even more important in this case, since nitrogen is less electronegative than oxygen and more willing to share its electrons. As you see, aniline is one million times as reactive as benzene.

In the other direction, we can first consider atoms that exert a gentle pull on electrons, like the halides, which are about 30 times less reactive than benzene, but there are other cases, like carbonyl groups and nitro groups, that can actually delocalize electrons from the ring out onto the group in question. Again we see that a resonance effect is much more important than an inductive effect: nitrobenzene is one million times *less* reactive than benzene.

Be sure you can show a complete, correct set of resonance structures for nitrobenzene. Here's one; you find the other two. (Ignore the resonance in the nitro group and in the original benzene ring—these are real but do not contribute to delocalization of the positive charge.)

The generalization is that in electrophilic aromatic substitution, electron-donating groups are "activating substituents," meaning that they make benzene more reactive; electron-withdrawing substituents are "deactivating." One can make a list of substituents and their relative activating abilities:

more activating

What does all this mean? It is harder to carry out a substitution on a nitrobenzene than on benzene itself; it's easier to substitute toluene (or other alkylbenzenes) than benzene. This has consequences.

Imagine, for example, doing a Friedel–Crafts alkylation of benzene, with isopropyl chloride:

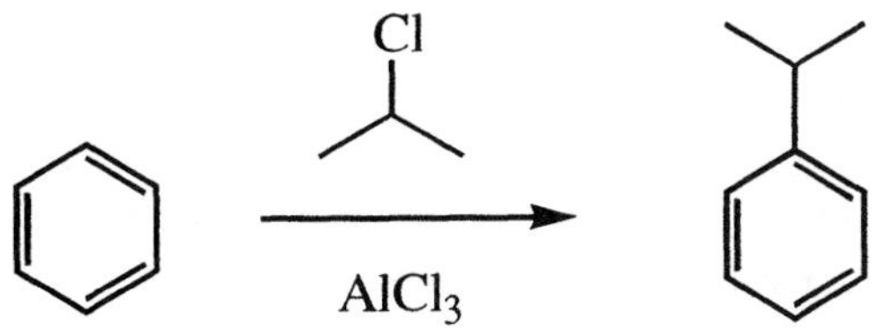

When the reaction is partway done, you will have both benzene and isopropylbenzene in your pot. What do you suppose will happen the next time you make an isopropyl cation? Isopropyl benzene is about 30 times as reactive as benzene! Given a choice, the cation would rather react with isopropylbenzene than with benzene, so you will get product with *two* isopropyl groups on it! The only way out of this mess is to use a large excess of benzene, more than 30 times extra, so that benzene will stand a good chance of competing for the cation. Typically this reaction is carried out in benzene solvent, so that could work, but you end up wasting a lot of benzene.

There is a second consequence to this activation/deactivation phenomenon. It turns out that Friedel–Crafts reactions, both alkylations and acylations, are right on the edge of working with benzene. A deactivated benzene will refuse to undergo such a reaction. You cannot alkylate or acylate a benzene ring that is deactivated past the halogens. That is, chlorobenzene will alkylate, but acetophenone will not:

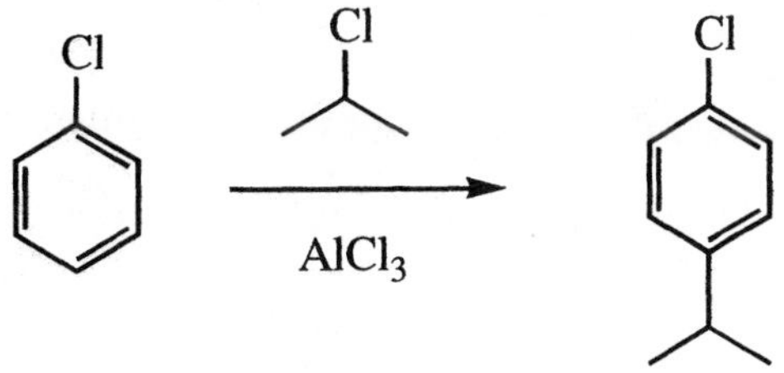

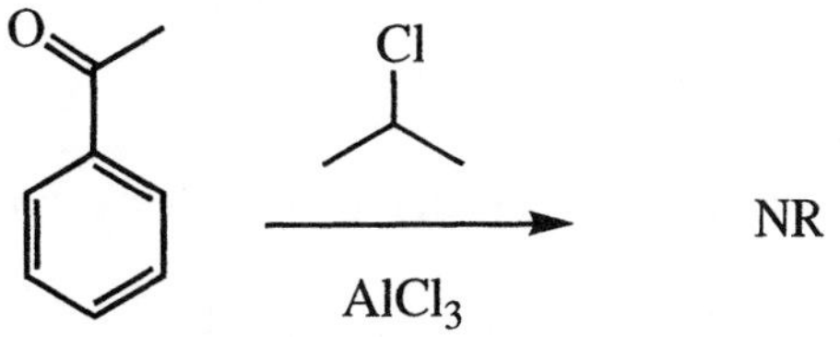

This can be a useful feature: we just got done determining that it is very difficult to put on only one substituent if you are doing a Friedel–Crafts alkylation, because the product is more reactive than the starting material. However, now we see that there is no such problem with acylation; the product is not only less reactive than the starting material, it is enough less reactive that the process cannot go any further. You do not even need to be careful with quantities: even if you use excess acetyl chloride, you can only get one acetyl group onto a benzene ring. And as we saw in the previous section, it is not difficult to convert an acyl group into an alkyl group, so you can use this roundabout method for making monoalkylbenzenes from benzene without resorting to use of a large excess of benzene.

Be careful with this! Although the reduced reactivity of deactivated benzenes can be very useful, it is frequently forgotten by beginning students. Remember: *No Friedel–Crafts on deactivated benzenes!!!*

Test Yourself 5

Propose a method for making isopropylbenzene starting from benzene and anything else you wish, but without using excess benzene.

Directing Effects

So far we have worried about the effect of a substituent on the *rate* of the next reaction. It turns out that a substituent on a benzene also influences *where* the next substitution will occur. It also turns out that what happens is very much within our capability to understand and even predict. Let's have a go at it.

Consider, for example, the bromination of toluene. There are three possible products: *ortho, meta,* and *para*-bromotoluene. Will we get all of them? Will any be preferred? To answer this we have to follow each reaction through its mechanism.

If the product is to be *ortho*, the bromine cation must react at the *ortho* position, like so:

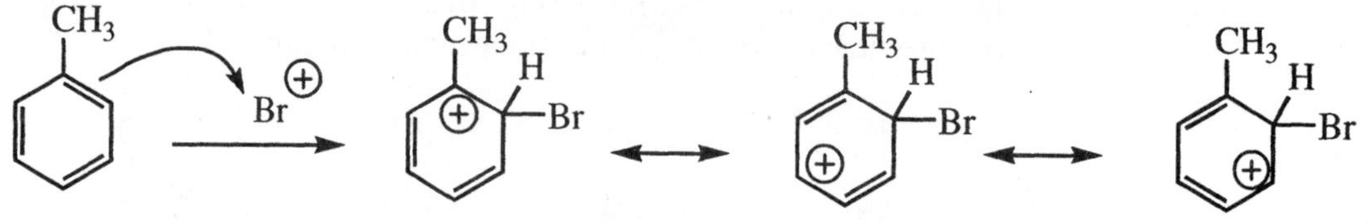

If the product is to be *meta*, the bromine cation must react at the *meta* position:

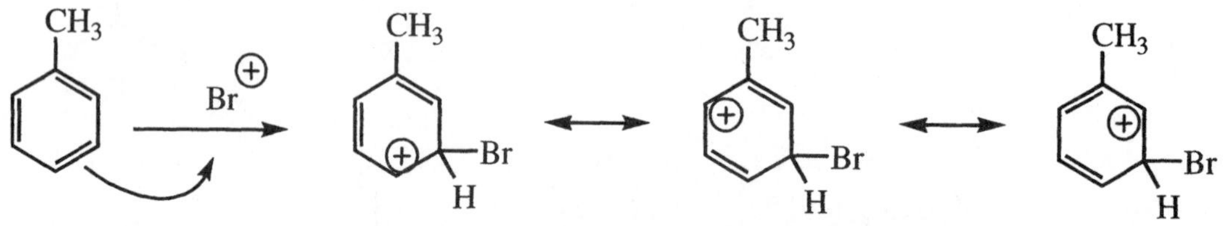

If the product is to be *para*, the bromine cation must react at the *para* position:

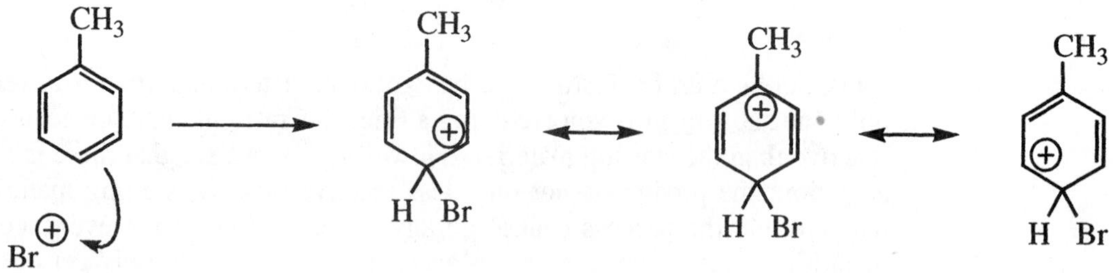

Examine these carbocations. Are any of them better or worse than others? Yes, one set is worse than the other two. The cation generated from *meta* attack shares the positive charge among three positions, all of which are secondary. But the other two cations have charge on one tertiary position and two secondary positions. Since tertiary cations are better than secondary ones, we conclude that the cations formed from *ortho* and *para* attack should both be about the same, and both better than the cation from the *meta* case. Therefore, we expect this reaction will produce mostly *ortho* and *para*-bromotoluene, and very little *meta*. This is true.

What about the distribution between *ortho* and *para*? We would expect these two to be equally favored from an electronic point of view. But there are two other factors we need to take into account. One is that there are two *ortho* positions and only one *para* position: thus, it is twice as likely that an electrophile will find an *ortho* position, from a statistical point of view. This would lead to more *ortho* than *para*, by a 2:1 margin. On the other hand, in order to form *ortho* product, the incoming group has to brush up against the existing substituent (methyl in this case). In other words, there ought to be a steric preference for *para*. This is also observed to be true. The two factors often come close to canceling each other out. Some reactions give a little more *ortho* product than *para*; some the other way around. To some extent it depends on the sizes of the two groups involved: when the existing (or incoming) group is particularly large, relatively little *ortho* product forms.

So, how general is this? Will we always get *ortho* and *para* products in preference to *meta*? You can answer this. Why did we get that distribution in this particular case? Because the cations produced by *ortho* and *para* attack were better than the one produced by *meta* attack. We should get a preference for *ortho* and *para* whenever this is true. When is that? Whenever the current substituent is one that carbocations like to be next to. Alkyl groups are good for carbocations; therefore, alkyl groups are "*ortho-para* directing." Atoms with lone pairs on them are also good for carbocations (for resonance reasons); therefore, these substituents, also, are *ortho-para* directing. This would include amine, amide, OH, OR, and halogen substituents. What we have so far come up with is almost identical to the list of *activating* substituents, and for good reason: activating substituents are activating, because they are electron donating, and electron-donating groups are good for carbocations. The only exception to the general rule is the halogens, which are *ortho-para* directing even though they are slightly deactivating. Otherwise, all activating groups are *ortho-para* directing, and *vice versa*.

Test Yourself 6

Show why an OH group is *ortho-para* directing. Your intermediate cation this time should have four, not three, resonance structures.

What about deactivating groups? Let's consider one, say, acetyl. Consider a bromination of acetophenone. Here are the three possibilities:

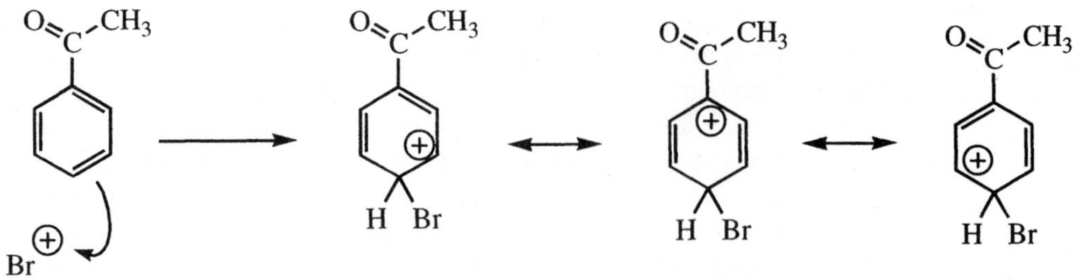

Once again, the *ortho* and *para* intermediates are different from the *meta* intermediate. In the *ortho* and *para* cases, one resonance structure has the carbocation on the same carbon as the original substituent. Is this good or bad? Let's take a look. Here is the unique structure. Do we like it?

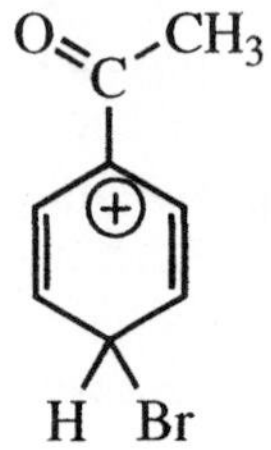

Well, it does appear to be tertiary, and tertiary is good for a carbocation. But we ought to be suspicious about this, because we know that *anions* like to be next to carbonyl groups. How likely is it that cations do too? Tertiary positions are good for carbocations, because alkyl groups are electron donating. Is this substituent electron donating? I think not: recall that there is a partial positive charge on the carbon of a carbonyl group. It is unlikely to be pushing electrons away from it. More likely, it will be pulling electrons *toward* it. Carbonyl groups are electron withdrawing, not donating. This should be bad for a carbocation.

What about resonance? Can't we draw an extra resonance structure for this cation, as shown below?

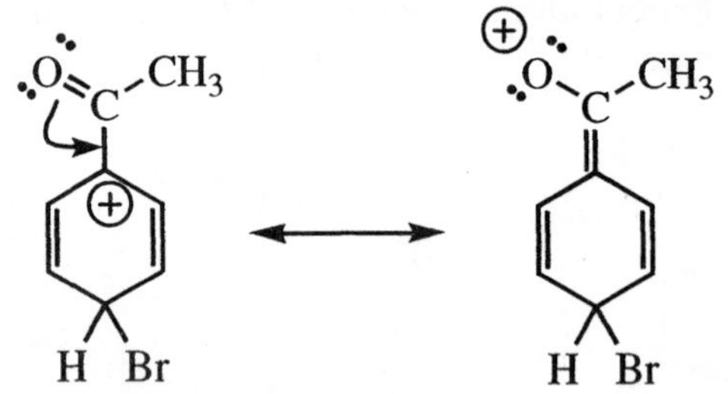

AARGH! This structure with a positive charge on oxygen is terrible! The oxygen has only six electrons, which is sheer torture for an electronegative atom. Sure, we have seen positive charge on oxygen plenty of times before, but always with an octet—the positive charge came from sharing too many electrons, not from having too few. This is a dismal resonance structure that we cannot even consider.

The bottom line is that the acetyl group, and all other electron-withdrawing groups (except the halogens) try to *avoid* having a positive charge immediately next to them. That makes the cation produced from *meta* attack *better* than the one produced by *ortho* or *para* attack. We predict for this reaction a large preponderance of *meta*-bromoacetophenone, with almost no *ortho* and *para*. This is true.

How general is the phenomenon? All deactivating substituents (except the halogens) are *meta* directors, because they are all electron withdrawing and therefore do not like positive charge next door.

Thus, we can divide substituents into two main camps: activators that are also *ortho-para* directors, and deactivators that are also *meta* directors. Only the halogens fall outside these categories.

ORM: Under Aromatic Compounds, look at "electrophilic bromination of toluene."
There is a lot there, so be sure you explore all the areas.

How can we use this information? Consider an attempt to synthesize *p*-nitrotoluene from benzene. I'm going to talk about this in text rather than pictures. If it helps you (and it probably will), feel free to draw pictures while I am talking. That is what professional chemists do! You should be able to follow the whole argument.

Starting with benzene, you need to get two groups on, a nitro group and a methyl group. We know how to do both things: a methyl group can be added using methyl iodide and aluminum chloride, in a Friedel–Crafts alkylation, and a nitro group will add with a combination of nitric and sulfuric acid. (Why did I choose methyl iodide rather than methyl chloride? On paper either is OK, but in fact methyl chloride is a gas, and so is methyl bromide. Methyl iodide is the only liquid methyl halide, so that is the most convenient one to use.) In what order should we carry out these steps? If we add the methyl group first, we will get toluene. Methyl is an *ortho-para* director, so the next group, nitro, will go on in either the *ortho* or the *para* position. We cannot determine which—we will be stuck with whatever mixture is obtained. This is one of the drawbacks of electrophilic aromatic substitution, and it is one we have to live with. On paper, the proper thing to do is to indicate that both are obtained, and that one will be separated and discarded.

What if you put the nitro group on first? This leads to two problems. Can you spot both? First, nitro is *meta* directing, so the next group would appear in the wrong place. Worse than that, nitro is deactivating, and you cannot carry out Friedel–Crafts reactions on deactivated rings, so in fact you will get no second reaction at all!

Test Yourself 7

Suggest a synthesis of *m*-bromobenzenesulfonic acid starting from benzene.

Competitions

It's easy enough to figure out where a second substituent will go on a benzene ring—we've just been through the analysis. But where will a third one go? The answer is, that depends.

Consider doing a substitution reaction on *p*-bromonitrobenzene. (How would you make that?) Draw a picture if you have to. For simplicity, let's number this with the bromo group at position 1. Now we ask, if we were to nitrate this, where would the new nitro group go?

Let's ask each of the current substituents. The nitro group that is already there says, go *meta* to me: in the 2 and 6 positions. The bromo group says, go *ortho* or *para* to me, and since the *para* position is already taken, better make it *ortho*. That's the 2 and 6 positions! Both groups are telling the new group to do the same thing. Guess where it will go? Of course, the product will be 2,4-dinitrobromobenzene.

Now consider the same reaction on *m*-nitrobromobenzene. (How would you make *that*??) Again, number it with the bromo group at position 1 and the nitro group at position 3. Where will a new nitro group come in now? The old nitro group says, go to position 5. But the bromo group says, no, go to 2, 4, or 6. Now we have a fight, and the only way to find out what happens is to run the reaction. What you get is mostly 1-bromo-3-4-dinitrobenzene and 1-bromo-3,6-dinitrobenzene. In other words, the bromine won. Its preference was stronger that the nitro group's preference.

By the way, what happened to position 2? Why didn't the new nitro group go there? Electronically, it could have, but there are steric problems. You would end up with three substituents in a row, a rather tight fit on a benzene ring.

In general, we can predict the outcome of many such fights: *ortho-para* directors almost always win against *meta* directors. So whenever you find a *meta* director urging one outcome and an *ortho-para* director urging another, go with the *ortho-para* director.

Not all such fights are as easy to predict. Consider *p*-bromotoluene. Now the bromo group says, go *ortho* to me (*para* is already taken), and the methyl group says no, go *ortho* to me! Who wins? There are rules for some such cases, but I will not burden you with them. For our purposes, we will assume that when our simple rule does not apply, all possible outcomes will occur. It should be clear that we will in general try to avoid such situations!

Test Yourself 8

Suggest a synthesis of 3-bromo-5-nitrobenzenesulfonic acid.

BIOLOGICAL AROMATIC COMPOUNDS

Nucleic Acids

We have already met several aromatic compounds that play important roles in biology, among them NAD^+ and pyridoxal. Probably the most ubiquitous, and important, biological aromatic compounds are the bases found in DNA and RNA: guanine, adenine, cytosine, thymine, and uracil. It is not immediately obvious that these are aromatic: you have to count the lone pairs of the saturated nitrogens but not those of the nitrogens that are already participating in double bonds, just as we did at the beginning of this chapter. Further, those molecules with carbonyl groups need to be looked at as their enols. Anyway, they *are* aromatic molecules, and therefore they are flat, rigid species. The first two are related to the molecule purine, and the other three to the molecule pyrimidine.

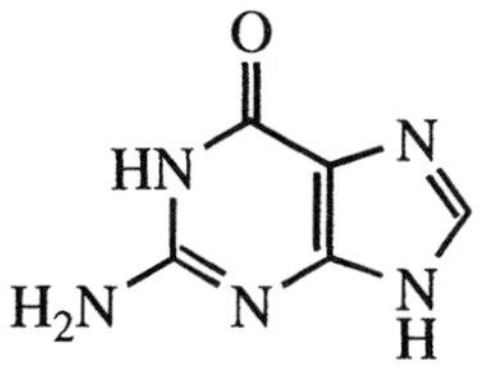

guanine 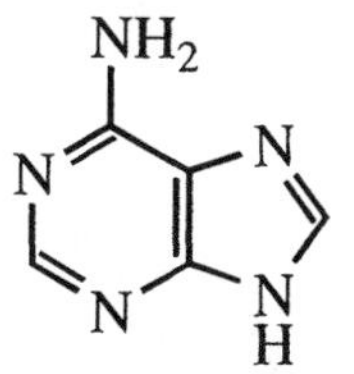   adenine   are related to   purine

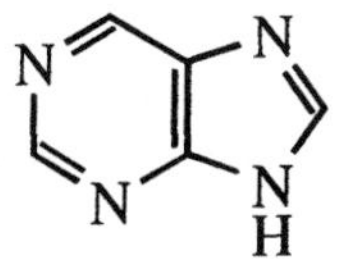

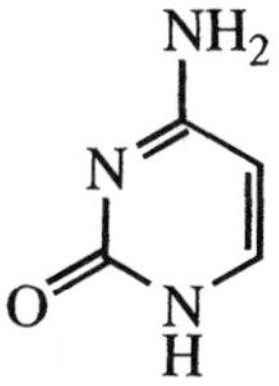

cytosine 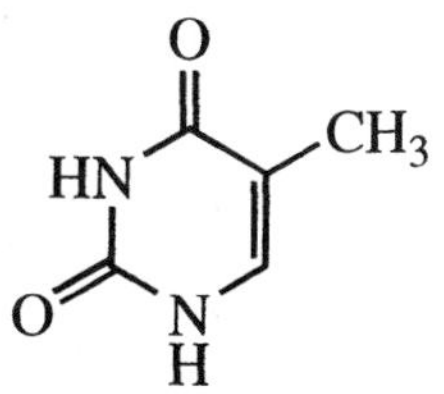   thymine 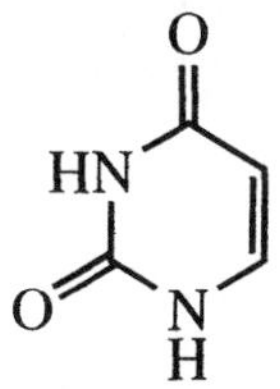   uracil   are related to   pyrimidine

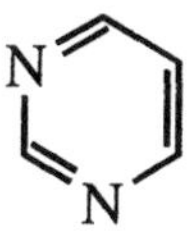

For simplicity, I am going to ignore uracil in the following discussion. I think you can see that everything I say about thymine will also be true of uracil. They are, in fact, interchangeable, in the sense that whatever thymine does in DNA, uracil does in RNA.

The most important thing about these molecules is that they form rather strong, pairwise attractions: cytosine loves guanine, and thymine loves adenine, because of the multiple hydrogen bonds that can be made between these particular pairs.

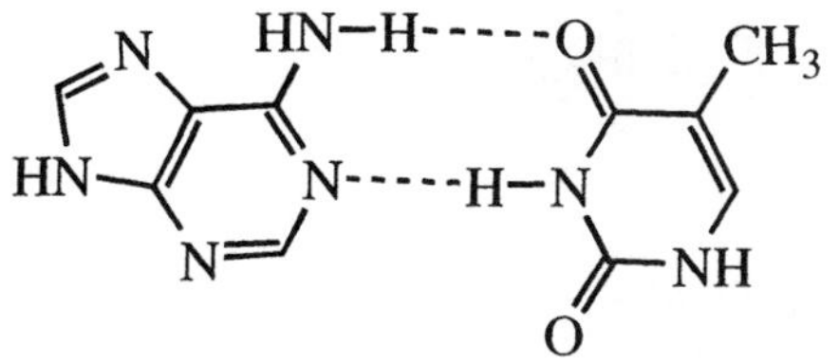

adenine   thymine 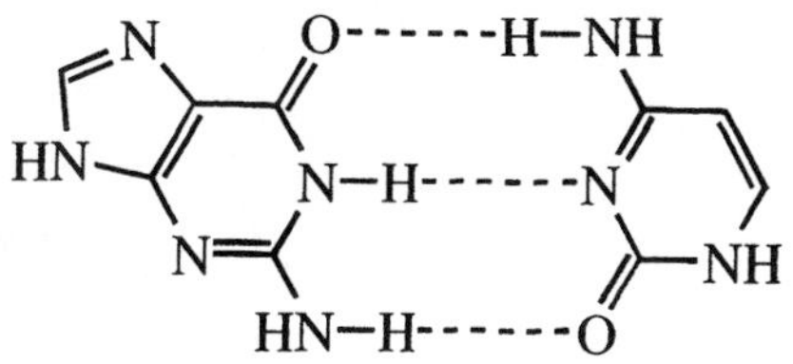   guanine   cytosine

The hydrogen bonds shown above are the entire basis of genetics.

These four molecules (usually abbreviated A, T, G, and C) attach themselves to sugar molecules, specifically deoxyribose in DNA (or ribose in RNA) to form what are known as "nucleosides."

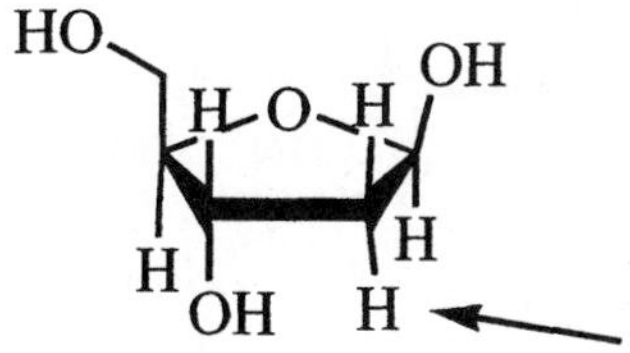

deoxyribose

Ribose itself would have an OH group here.

Nucleosides are combinations of the above bases with sugars. The following picture shows the four possible nucleosides formed from our bases and deoxyribose. These form the basis of DNA. If with these four structures you replace thymine with uracil (just remove the methyl group) and replace every deoxyribose with ribose (add an OH group to each), you have the four nucleosides that make up RNA.

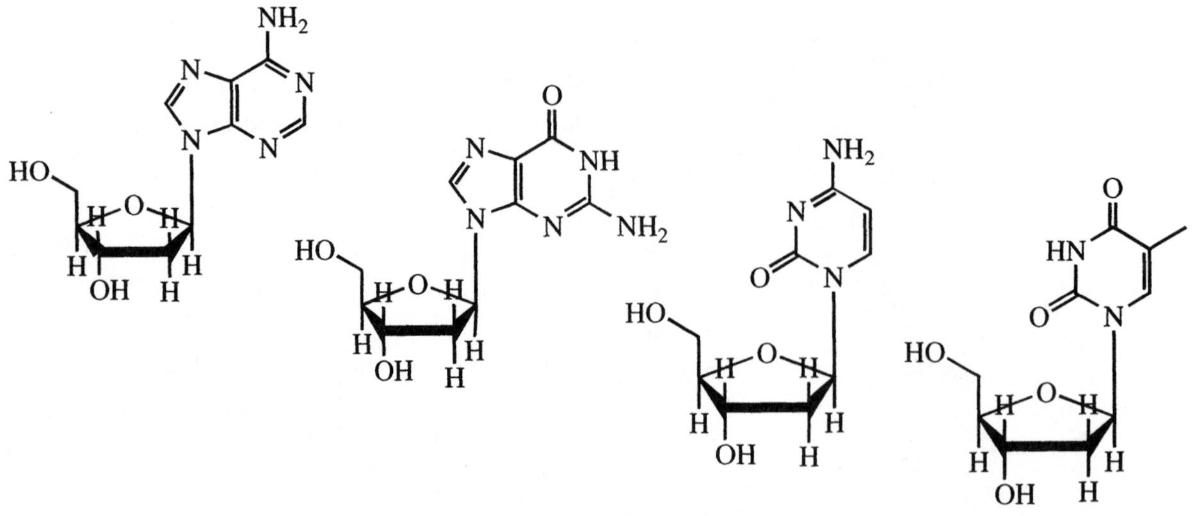

Nucleotides (t instead of s) are essentially the same things, but with a phosphate group (PO_3^{2-}) on one of the OH's. These nucleotides are then strung together by forming bonds between specific oxygens in the sugars and the phosphate groups. The result is what are called polynucleotides, a class to which the nucleic acids DNA and RNA belong. Below is a short strand of a polynucleotide, made from the pieces that make up DNA.

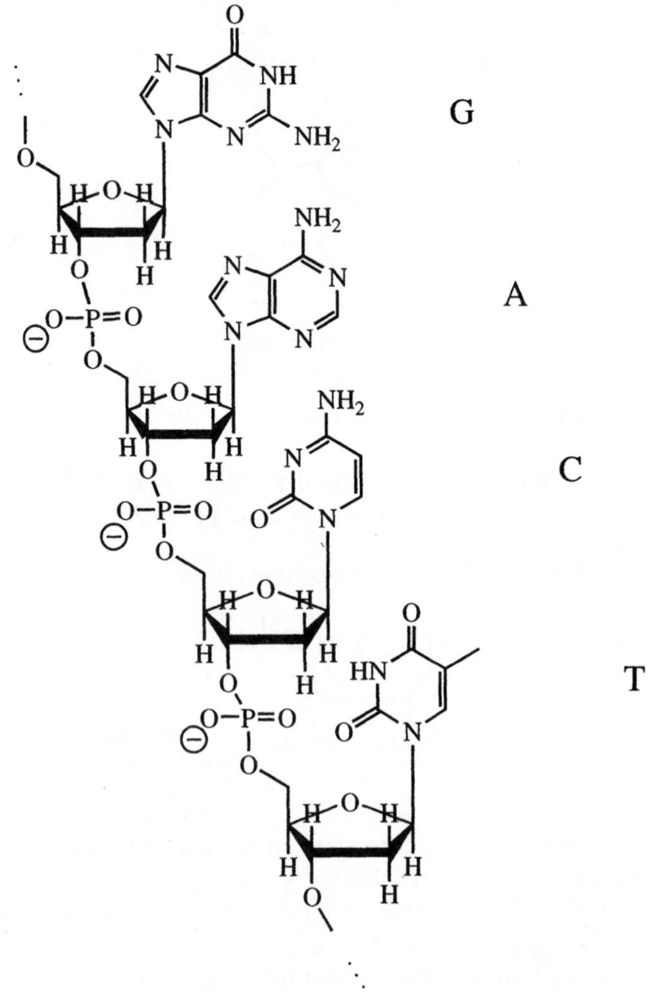

G

A

C

T

This structure would be given the shorthand notation ..GACT... A real DNA molecule would continue on for millions and millions of letters. You can see why we abbreviate with single letters instead of structures!

So let's review what we have here: a polymer of sugar molecules, each of which has a flat, aromatic structure hanging off it—like a piece of flagstone. Further, each of these pieces of flagstone has a special affinity for a partner piece that it wants to hold hands with. It is not hard to imagine sugar polymer "backbones" as the sides of a ladder, with bases forming the steps. This particular kind of ladder appears to have been sliced in half, and there is now glue holding the two pieces of each step together:

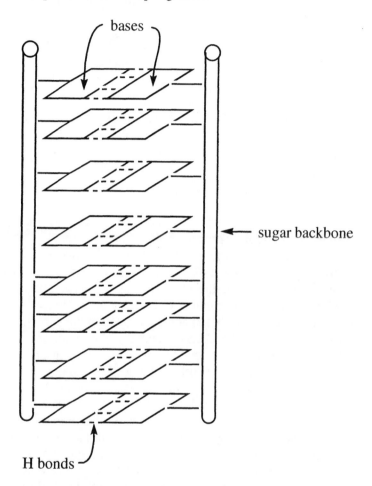

If you now take the ladder and twist it, you will have the familiar "double helix" of DNA.

How does a particular chain of DNA find a perfect partner chain to hold hands with? As a general rule, it doesn't actually find the partner chain—it makes it! Each base finds its perfect partner, and then the newly forming backbone stitches itself up to make a second strand. This is how DNA reproduces itself.

Intercalation

Take another look at the "ladder" picture above. There is a certain uniform distance between all the steps. These steps, you will recall, are the purine and pyrimidine aromatic bases that we started with in this section. They are flat molecules with clouds of electrons ("π-clouds") above and below their surfaces. At the distances between them along a straight backbone, they would not interfere with each other. But as you twist the ladder, the steps get closer and closer together. There is a limit, of course: at some point the π-clouds mash into each other so much that pieces begin to repel one another. This should not surprise you. What may be surprising, however, is that at some earlier point, before the bases start to repel each other, there is actually a slight attraction between the π-cloud of one step and the π-cloud of the next. The ladder twists itself just the right amount to maximize this "π-π attraction," also known as a "π-stacking interaction."

If there is an attraction from one flat, aromatic system to the next, it stands to reason that other flat, aromatic molecules might like to get in on this deal, and they do. You may remember the following picture of benzpyrene oxide from Chapter 17:

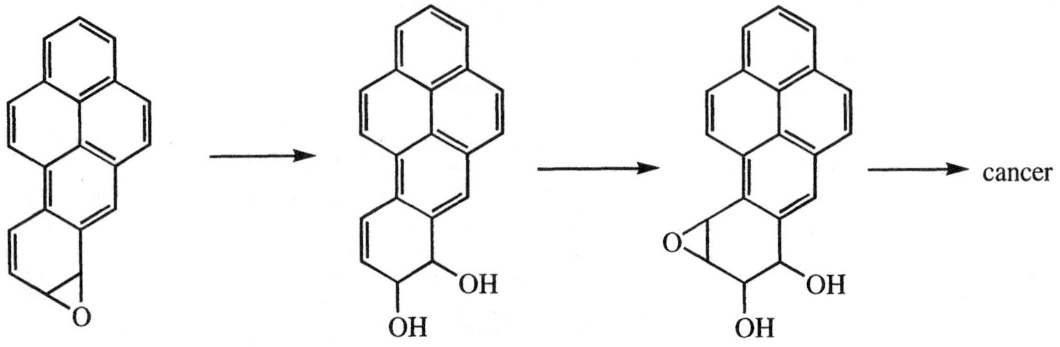

How does this work? The flat, aromatic part of this molecule sneaks in between some of the steps in DNA, in a process called "intercalation." Of course, this pushes the DNA bases apart, and unwinds the ladder slightly, but it makes a surprisingly good fit. Intercalation in itself is bad, because it disrupts the structure of DNA. But this particular molecule not only can insert itself into the DNA, but at the same time, it presents it with a highly reactive epoxide, which is begging to be sprung open. The DNA, in turn, is bristling with OH groups (from the sugar backbone) and amines (from the bases), and is only too happy to oblige. The result: a covalent bond forms between the DNA and this interloper, a bond that does not belong and that will prevent this particular section of DNA from being reproduced properly. Improper reproduction is often what leads to cancer.

Many anticancer molecules work exactly the same way, using a flat, aromatic piece to intercalate into DNA and a reactive piece that creates a bond to the DNA chain. Why can the same process cause cancer in some instances and cure it in others? It depends on your target. If you disrupt the process of a normal cell, you run the risk of making it abnormal. On the other hand, if you can somehow send your drug only to abnormal cells so you disrupt them, perhaps enough to kill them, you might have a useful drug. The key, of course, is being able to target just the cells you want. Most cancer drugs work on all cells, and since they disrupt reproduction, they are particularly effective against those cells that are actively growing. That is why most cancer drugs cause hair to fall out and make the patient throw up: hair cells and stomach lining, as well as cancer cells, are the fastest growing cells in the body, so they take the brunt of the effect of a cancer drug. If only we could figure out how to target just the cancer cells! Of course, you can bet there are a lot of smart people working on this problem, but it may not be solved until you make your contribution.

Heme

Another aromatic molecule important in biology is heme, which is a derivative of porphyrin.

heme   porphyrin

Why do I call these aromatic? Let's look at porphyrin. If you follow the bold trail you should be able to count a path of 18 electrons around the "periphery" of this molecule.

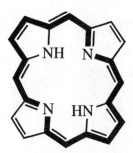

Eighteen electrons in a planar, cyclic array is, of course, one of the magic systems according to Hückel, so this should be an aromatic system. Is there any evidence that it is? Yes, indeed: you will notice that, unlike all the previous aromatic systems we looked at, this one has two hydrogens on the *inside* of the π system, both attached to nitrogen. If you recall our discussion of NMR properties of aromatic systems, you will remember that protons in this unusual environment would be doubly shielded from the applied magnetic field: once by their local electrons and again by the larger field created by circulation of the electrons in the aromatic system. Such highly shielded electrons appear in a very unusual place in the spectrum. In porphyrins, the two protons on the inside of the ring are found near –3 ppm, to the *right* of TMS.

So this is an aromatic ring. Like all others, it wants to remain flat. What we really have here is apparently a large disk with a hole in the middle, like a CD. The hole is just the right size for an iron atom to fit. Following is a front view and a side view of a heme molecule with iron already in place. See color plates for color versions of this and later pictures.

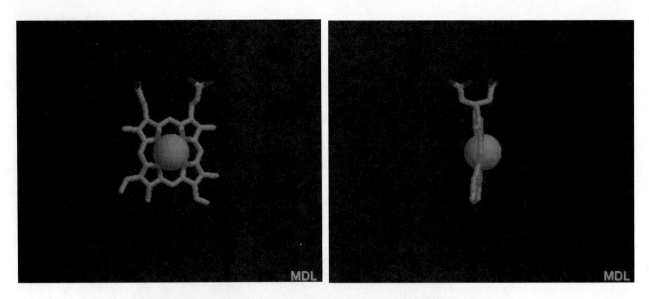

This disc fits perfectly in a cleft of a particular polypeptide.

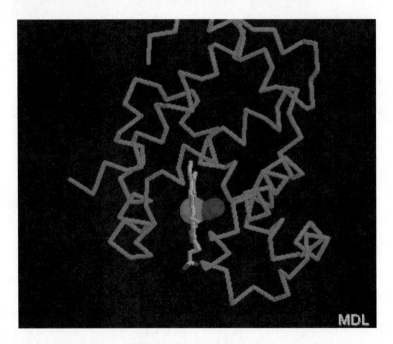

There it finds a nitrogen (blue in the color version) in just the right place to bind to the iron on one face (iron usually has six bonds, as you should learn next year), while the other face picks up an oxygen molecule (red dumbell), O_2.

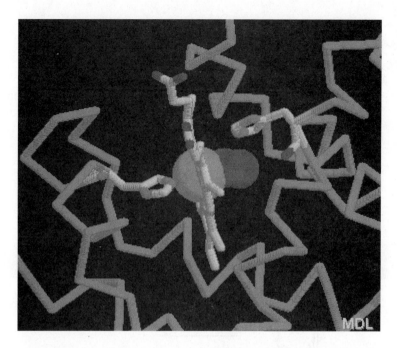

Four of these polypeptides group together to form hemoglobin, the protein that carries oxygen in your blood.

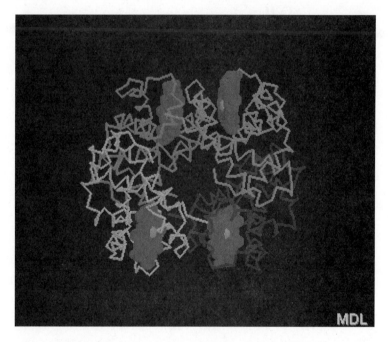

To see how this really fits together, you should explore the web site from which I took these pictures, http://www.umass.edu/microbio/chime/hemoglob/index.htm. Very similar heme molecules are involved in all other oxygen transport proteins.

Problems

(Answers are provided at the end of the chapter for italicized problems.)

1. Show the significant organic products of the following reactions. Write NR for no reaction. Do not worry about very minor products, but if there is more than one significant product, identify all of them.

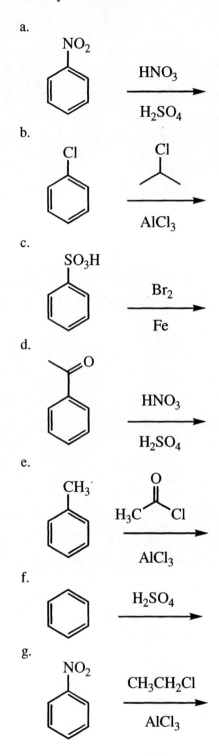

a.

NO₂

HNO₃

H₂SO₄

b.

Cl

Cl

AlCl₃

c.

SO₃H

Br₂

Fe

d.

O

HNO₃

H₂SO₄

e.

CH₃

H₃C   Cl
     O

AlCl₃

f.

H₂SO₄

g.

NO₂

CH₃CH₂Cl

AlCl₃

2. When 6.8 g of octahydronaphthalene is hydrogenated, 1.35 kcal of energy is released. When 3.20 g of naphthalene is hydrogenated, 1.85 kcal of energy is released. Based on these data, what is the "resonance energy" (extra stability) of naphthalene?

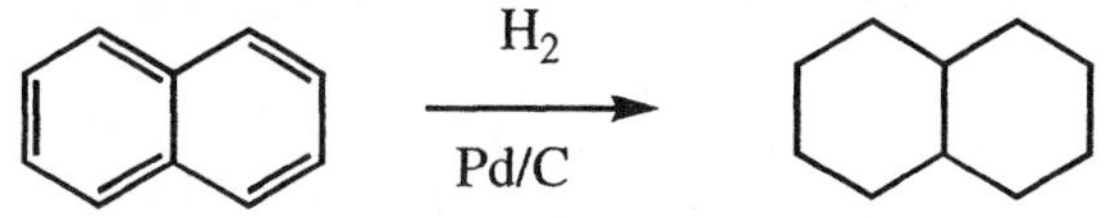

naphthalene

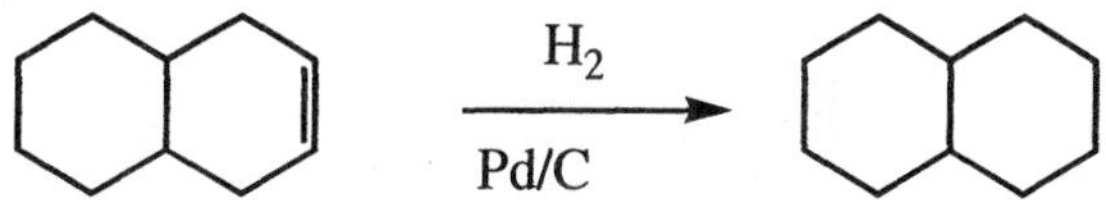

octahydronaphthalene

3. Show a reaction mechanism and use resonance structures to explain why a methoxy group (OMe) is *ortho-para* directing in electrophilic aromatic substitution.

4. Suggest a synthesis of the following molecule, starting with benzene and any other reagents you are familiar with.

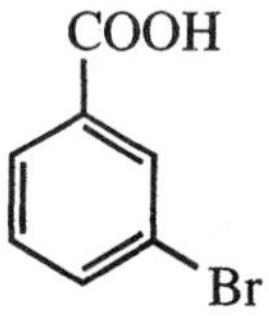

5. Suggest how the following might be synthesized starting with benzene and any other reagents with which you are familiar. Be sure to indicate any steps expected to lead to mixtures that would need to be separated.

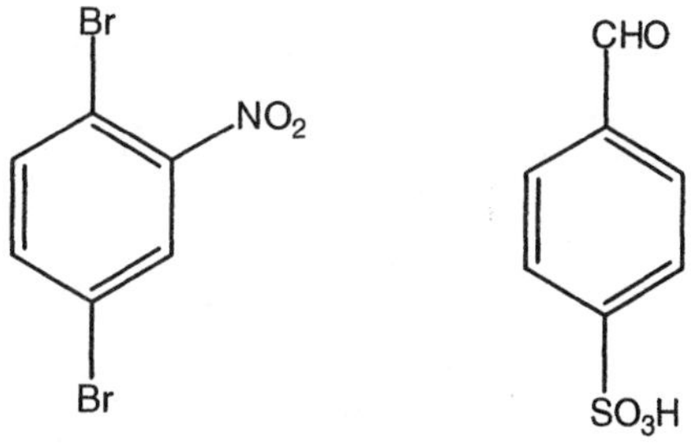

6. When aniline (aminobenzene, $NH_2C_6H_5$) is nitrated, the product obtained consists of roughly equal amounts of the ortho, meta, and para isomers. How might this result be explained? Write detailed mechanisms for the reactions involved. Hint: aniline, like ammonia and nearly all amines, is a base, and is subject to extensive protonation in the presence of an acid.

7. Student Chuckie Cheese wanted to make *n*-propyl benzene (1-phenylpropane) from benzene and one equivalent of 1-chloropropane and aluminum chloride. He was surprised to discover as his major product *para*-diisopropyl benzene. Explain to Chuckie:

   a. Why (i.e., how; show a mechanism) the attached group is isopropyl rather than *n*-propyl.

   b. Why there are two groups attached instead of one.

8. The sulfonation reaction, shown below, is reversible: treatment of a sulfonic acid with hot aqueous acid will regenerate benzene by a reversal of the sulfonation mechanism. Show a mechanism for the desulfonation of benzenesulfonic acid in acid.

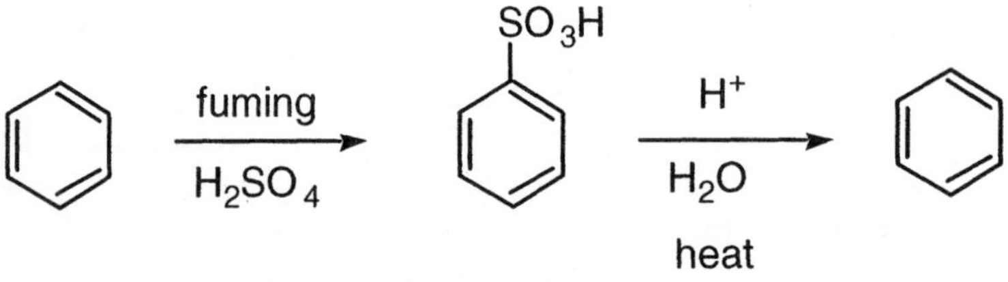

9. Show how to use the above process to convert toluene into *o*-bromotoluene in high yield (that is, without requiring a separation step).

10. *When p-di-tert-butylbenzene is nitrated, the following two products are observed. Show a mechanism for formation of the second product and explain why this pathway is not observed in the case of p-xylene.*

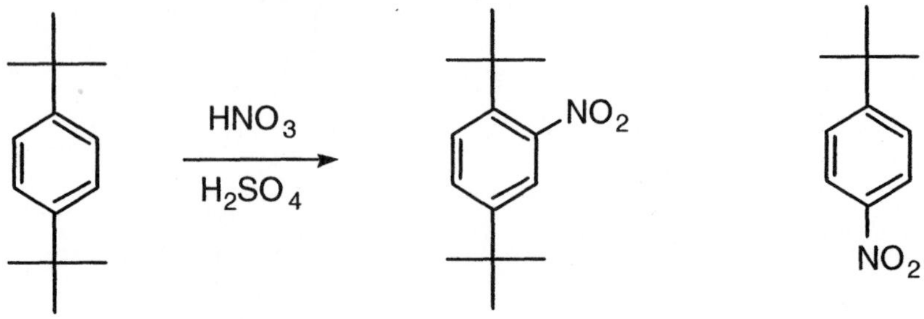

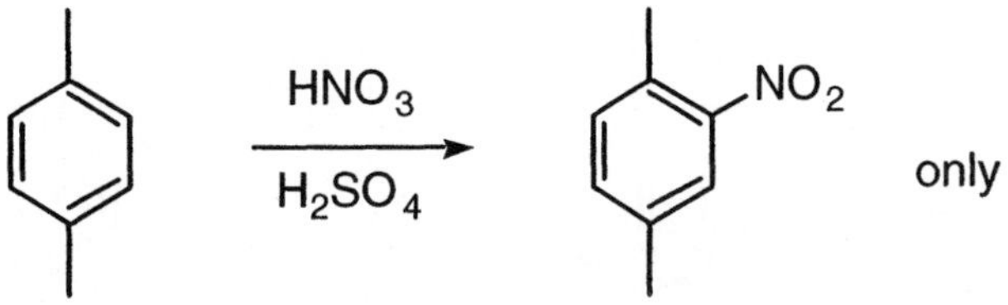

11. Suggest a synthesis of the following compound starting with benzene and anything else you like containing three or fewer carbons as your only sources of carbon. By this I mean that all the carbons in your product must originate in benzene or compounds containing three or fewer carbons. Other reagents that come and go may be larger.

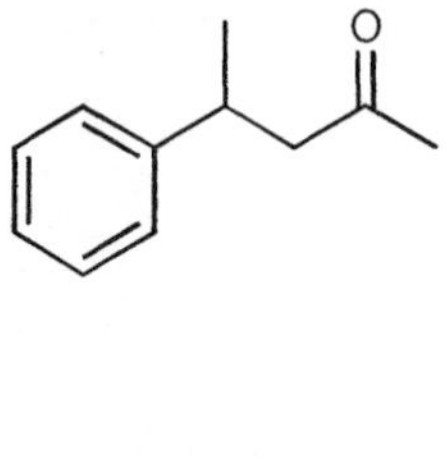

Selected Answers

Internal Problems

Test Yourself 1

The first thing to notice here is that these are ketones, and ketones, as you know, are polarized so that the carbon is positive and the oxygen is negative. We sometimes indicate this by drawing a minor resonance structure:

So how does this play in the above molecules? Let's look:

In the dipolar resonance structure, there is a positive charge in a p orbital on the carbon of the carbonyl group. This atom is in conjugation with the rest of the carbons of the ring. The number of electrons involved is four, two from each double bond and none from the carbocation. Four is a bad number, so we predict that the ring portion of this molecule is very unstable. The molecule is at war with itself: the oxygen of the carbonyl group wants to pull electrons away from the carbon of the carbonyl group, but that atom is very unhappy with a positive (or partial positive) charge. It should also be extremely difficult to protonate the oxygen of the carbonyl group, because that would lead to a full positive charge right where we do not want one. As it turns out, this molecule is extremely unstable. It reacts almost as soon as it forms, reacting with itself if there is nothing else around, in order to avoid perpetuating this unpleasant ring situation.

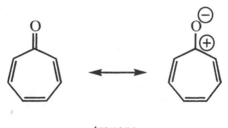

tropone

On the other hand, tropone with a seven-membered ring is extremely pleased with itself. The resonance structure that shows a positive charge in the ring corresponds to an aromatic system in the ring: there are seven conjugated atoms sharing six electrons, a good number. The two tendencies of the molecule reinforce each other: both the oxygen of the carbonyl group and the ring itself like the presence of positive charge on the ring carbon. Thus, the resonance structure with two charges should be much more important in this molecule than in a normal ketone. Tropone should be much more polar than a normal ketone (in fact, it is soluble in water); it should also be quite basic and protonate very easily (it does). Finally, it is very happy the way it is, so it should be less reactive than usual toward reagents that react with normal ketones.

Test Yourself 2

We have just determined that in order to be aromatic, a five-membered ring would like to have a negative charge and a seven-membered ring would like to have a positive charge. We should therefore predict that this molecule would have a significant resonance structure with a charge in both rings:

This would lead to a prediction of surprising polarity for this molecule; reaction with electrophiles in the smaller ring and with nucleophiles in the larger ring; and a bond length between the rings that is closer to a single bond than a double bond.

Test Yourself 3

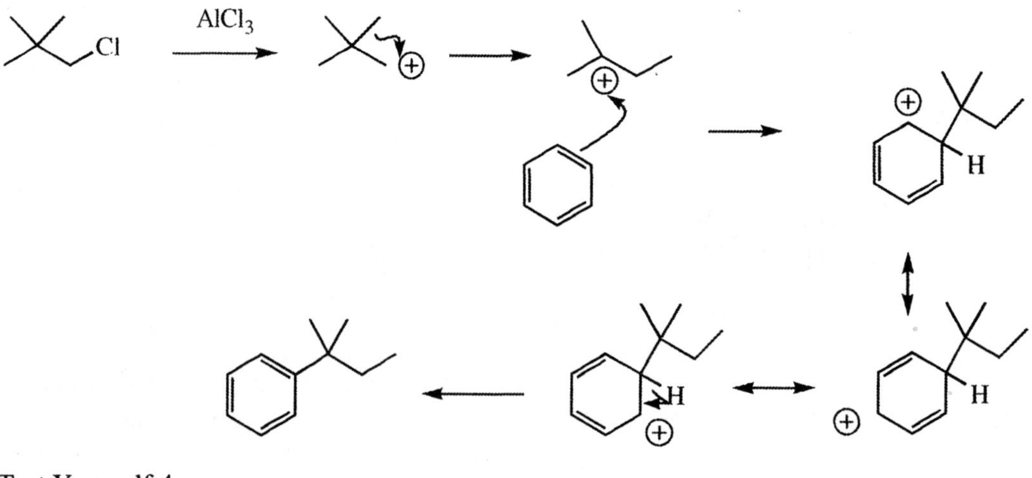

Test Yourself 4

Methyl cation cannot rearrange; ethyl cannot rearrange to anything other than itself.

Test Yourself 5

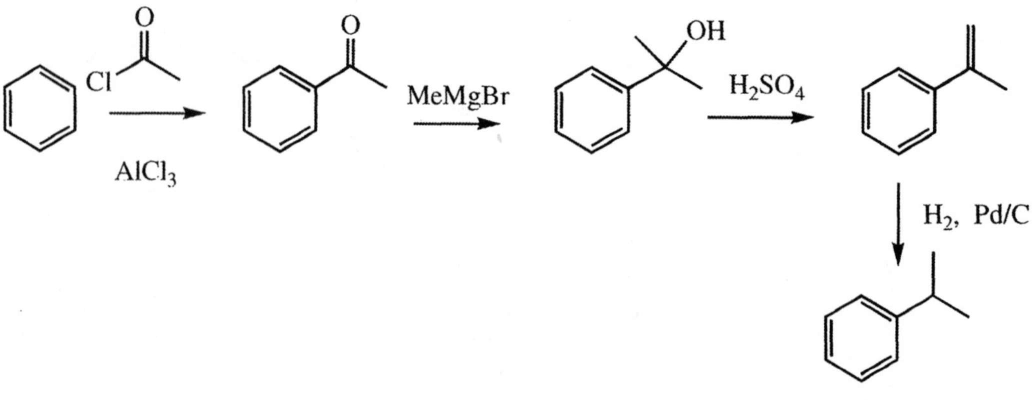

Test Yourself 6

Extra resonance structure makes this intermediate better. Works also for *para*, but *meta* intermediate does not have this extra structure.

Test Yourself 7

Test Yourself 8

End of Chapter Problems

2. 61 kcal/mol

6.

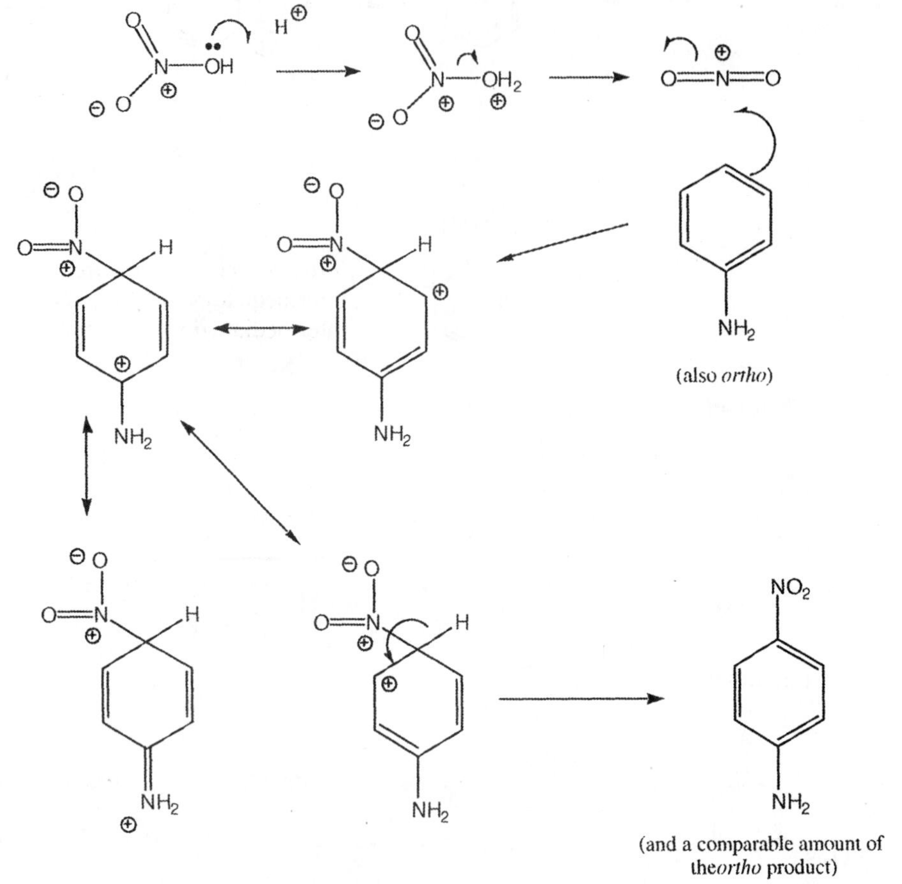

(also *ortho*)

(and a comparable amount of
the *ortho* product)

But in the presence of nitric and sulfuric acid most of the aniline is present as
the anilinium ion, the positive charge of which makes the substituent a *meta* director!

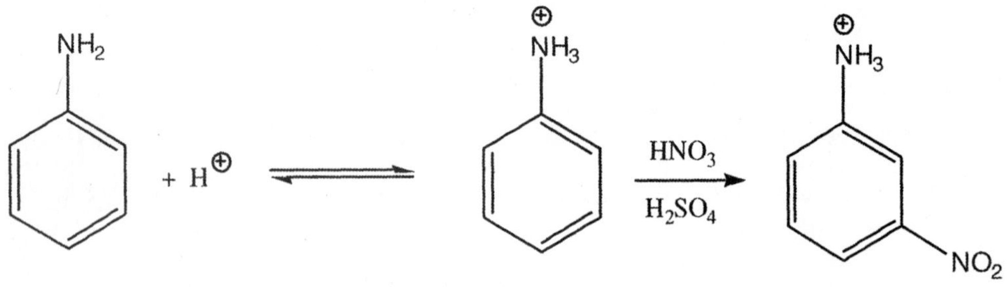

This reaction is *very* slow, however (due to deactivation), so nearly as much of
the substitution product actually obtained is *ortho/para*, derived from the small
amount of rapidly reacting free aniline still present.

10.

tert-butyl cation is good enough to be lost in second step. First step can occur in *p*-xylene also, but methyl cation will never be lost.

Afterword

How does one end a textbook on organic chemistry? Frankly, I have never gotten to the end of any organic text I have ever opened. When I look at the organic books on my shelf, they all end with the end of whatever material they were discussing in the last chapter. But it seems appropriate to me that there should be some kind of wrap-up, so here goes.

We have spent the year together discovering some of the chemistry of the compounds of carbon. In spite of what may seem to you to be an overwhelming amount of information in this book, this is in fact only a small amount of the organic chemistry that is known. I have deliberately left out of this book a great deal of organic chemistry that has traditionally been covered in standard undergraduate courses, because I felt it would be of little value to most of you (the rest will presumably see more in a later course); and even what has been traditional is still only the tip of the iceberg—so you have seen a piece of the tip! If you found this iceberg cold and forbidding, you can thank your lucky stars that this is as much of it as you will ever need to see; if you found it fascinating, full of challenge and adventure, then I hope you will take as much chemistry as possible and later join the thousands of men and women around the world who have made a career out of studying various aspects of this field.

What I have tried to do in this book is focus on those aspects of organic chemistry that are understandable at the beginning level and relevant to the processes of life. Of course, not everything in this book is directly applicable to the life sciences; some is background necessary to understand other stuff. But I have tried hard to weed out stuff that has no relevance at all. I have also tried to point out as often as possible *how* these concepts apply to living systems, so you would never lose sight of the fact that there is a reason why you are being asked to learn this stuff. I hope you will find it useful in the succeeding courses you take, both in chemistry and in other subjects.

I have enjoyed writing this book, and I hope most of you have enjoyed reading it. Nevertheless, I am certain that there are many places where it could be improved. I want to repeat now the invitation I made in the Foreword. My most important reader is you, the student. If you can think of any ways this book could be improved, please let me know by emailing me at Reingold@juniata.edu. Then I can use your suggestions to make this even better for the next group of students. Thanks for your help!

APPENDIX

MEASUREMENT

Like all scientists, chemists measure things, and they do so using the metric system. This is not in order to confuse you—it is simply the language scientists speak. It is important that you be able to speak it too. This means not only remembering a few conversion factors, but also developing a gut feeling for how large certain units are. For example, it is one thing to memorize that 1 g is one 454th of a pound, but does that really help you to understand what a gram is? It is also important to know that a gram is approximately the weight of a business card, or two paper clips, or one M&M candy. A milligram is one thousandth of a gram. How much is that? It is about the weight of the tiny shred of paper generated when you tear a page out of a spiral bound notebook. Why do you need to know these things? Because sometimes you do calculations wrong, and you can frequently tell just by thinking about the answer. If someone asks you to calculate the weight of a gallon of milk and your calculator tells you that it is 3.74 g, you might be inclined to write that down unless you know, in your gut, how much a gram is and recognize that this answer is ludicrous.

Below are some conversion factors that you should know, and also some feeling for what some of these quantities correspond to. You will notice that there are really only a few numbers you need to know, because whatever length you have in the English system, you can convert it into inches, then into centimeters, and then into whatever metric length unit is requested. *Don't waste your time memorizing other conversion factors!* With these numbers, and the ability to convert *within* each system, you can do any conversion required. We presume that you already know that there are 16 oz in a lb, 4 qt in a gal, and other English-to-English conversion factors. Within the metric system, all you need to know is that "milli" (m) means one thousandth, "micro" (μ) means one millionth, nano (n) means one billionth; "kilo" (k) means one thousand, "mega" (M) means one million, "giga" (G) means one billion. The only other prefix encountered with any frequency is "centi" (c), which means one hundredth, and that is seen almost exclusively in terms of length, as in centimeters, cm.

A word of warning: the English system has two *different* quantities that take the word "ounce." The weight ("avoirdupois") ounce is 1/16th of a pound. The volume ounce ("fluid ounce") is 1/16th of a pint. Because a pint is not a pound (in spite of the common saying—it is only approximate, and then only for water), the two kinds of ounce are not the same thing. Be sure to keep these straight: if you use only the conversion factors listed below, and keep in mind whether you are dealing with a weight or a volume, you cannot go wrong.

Purists will note that we have referred above to a gram as a weight. Technically, of course, a gram is a *mass*, the actual quantity of stuff, and not a weight, the push it exerts on the ground. The same mass of stuff would "weigh" less on the moon, because of the smaller gravitational pull of the smaller body. However, since all the chemistry we will discuss has been done on the earth, we will use the terms loosely and interchangeably.

	Conversion Factor	Gut Feeling
Weight	454 g = 1.00 lb	1 g ~ 2 paper clips, or 1 M&M candy
		1 kg ~ 1 qt milk (~2 lb)
		1 mg ~ piece of paper inside this O
Volume	0.946 L = 1.00 qt	1 L ~ 1 qt
		1 mL ~ a bit less than 1/4 teaspoon
Length	2.54 cm = 1 in (by definition)	1 cm ~ the thickness of your little finger at the fingernail
		1 mm ~ the thickness of a credit card, or blue jean cloth
		1 m ~ 1 yard
		1 km ~ 0.6 mi

Length --> Volume $1 \text{ cm}^3 = 1 \text{ mL}$

There is one other conversion factor you should know: for liquid water (but not generally for other substances), 1 mL weighs 1 g. Put technically, the density of water is 1 g/mL. (Actually this is true only at a certain temperature, but it is close enough for our purposes.)

Weight --> Volume   1 g = 1 mL *for water only!!*

Finally, you should be able to deal with temperature, which, everywhere else in the world, is measured on the Celsius (centigrade) scale. You probably know a formula that has 5/9 or 9/5 in it, and a 32, but, if you are like me, you keep forgetting which fraction to use and whether to add or subtract the 32, and whether to do it before or after the multiplication. Any time you memorize a formula, you risk remembering it wrong. You are much better off going with facts you already know. For example, you already know that the freezing point of water, 32 °F, is 0 °C. And you know that the boiling point of water, 212 °F, is 100 °C. That is all you need to know. Now I will give you two approaches to temperature conversion, the rational approach and the formula approach, neither of which requires memorization.

First, in the rational approach, you notice that the range of liquid water runs from 0–100°C (100 Celsius degrees) and 32–212 °F (180 Fahrenheit degrees). If you cut your cake into 180 pieces and I cut mine into 100 pieces, whose pieces are bigger? Clearly, a Celsius degree is bigger than a Fahrenheit degree, by the factor 180/100 (or 9/5). So wherever there is a temperature *difference* in Fahrenheit, there will be a difference in Celsius degrees corresponding to a smaller number of degrees (because each degree is bigger). To convert a temperature from one scale to another, you merely figure out how far away it is from a standard point and convert the *difference* into the other scale. The standard point is usually the freezing point of water.

To convert 364 °F to Celsius, you start by figuring out how far away it is from the freezing point of water, which is 32 °F. 364 – 32 is 332 °F above the freezing point of water. 332 Fahrenheit degrees = 5/9 x 332 = 184 Celsius degrees above the freezing point of water. Since the freezing point of water is 0 on the Celsius scale, this corresponds to an actual temperature of 184 °C. To convert –36 °C into Fahrenheit, you start by figuring out how far away it is from the freezing point of water, which is 0 °C. Obviously it is 36 °C away. 36 Celsius degrees = 9/5 x 36 = 65 Fahrenheit degrees below the freezing point of water. Since the freezing point of water is 32 on the Fahrenheit scale, and we are below it, this corresponds to an actual temperature of 32 – 65 = –33 °F.

This is illustrated below:

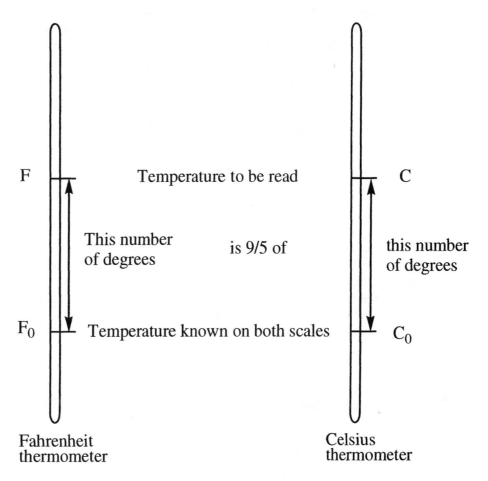

F — Temperature to be read — C

This number of degrees — is 9/5 of — this number of degrees

F_0 — Temperature known on both scales — C_0

Fahrenheit thermometer

Celsius thermometer

Mathematically, this picture an be expressed as follows:

$$F - F_0 = (C - C_0) \times 9/5$$

If we choose the freezing point of water as our basis, the above equation becomes

$$F - 32 = (C - 0) \times 9/5$$

which of course is the same as

$$F - 32 = (9/5)\,C \text{ or } F = (9/5)\,C + 32$$

The formula approach is to remember the 5/9 and the 32, and *figure out* where they must belong by converting temperatures you know about. You know that 0 °C corresponds to 32 °F, so to get from C to F you must multiply by something and *then* add 32: temp in F = fraction x (temp in C) + 32. Which fraction? You know that 100 °C corresponds to 212 °F, so obviously you need to make the number bigger, so you should multiply by 9/5. Thus,

Temperature: temp in F $= 9/5$ x (temp in C) $+ 32$ Comfortable room temp
                                                                       72 °F = 22 °C

After writing down the formula, check it on your known temperatures before using it to solve your problem.

Of course, you probably already know that the third temperature scale you'll need to work with, Kelvin, is the same as the Celsius scale but 273 degrees higher: $K = °C + 273$. Note that with Kelvins you do not write (or say) " ° " ("degrees").

CALCULATIONS

Most calculations in chemistry involve multiplying by 1. For example, since we know that 454 g = 1 lb, the number 454 g/1 lb, and the number 1 lb/454 g, are both ways of expressing "1." So we can take any weight, multiply it by one of these numbers, and still have the same equivalent weight:

$$3.7 \text{ lb} \times \frac{454 \text{ g}}{1 \text{ lb}} = 1680 \text{ g}$$

3.7 lb is the same amount of stuff as 1680 g—all we have done is multiply it by 1. A common problem for beginning students is how to know whether to multiply or divide by a conversion factor. The answer is, always multiply—but sometimes the conversion factor needs to be upside down (reciprocal). You arrange the conversion factor so that the *units* you want to change will cancel, and the units you want to end up with are in the numerator. Thus, to convert grams into pounds you will arrange your conversion factor to have g in the denominator and lb in the numerator:

$$358 \text{ g} \times \frac{1 \text{ lb}}{454 \text{ g}} = 0.789 \text{ lb}$$

Of course this is the same thing as dividing by the conversion factor, but it avoids the necessity of remembering whether to divide or multiply. This approach is often referred to as the "factor-label" or the "unit cancellation" method.

Conversion factors can be strung together, since each represents multiplying by 1, but care should be taken to be sure you know what each factor is doing for you. It should be possible to interrupt your string at any point and identify what you have at that point. For example, if you want to know how many km are in 0.68 miles, one could string together the following conversion factors:

$$0.68 \text{ mi} \times \underset{=1}{\frac{5280 \text{ ft}}{1 \text{ mi}}} \times \underset{=1}{\frac{12 \text{ in}}{1 \text{ ft}}} \times \underset{=1}{\frac{2.54 \text{ cm}}{1 \text{ in}}} \times \underset{=1}{\frac{1 \text{ m}}{100 \text{ cm}}} \times \underset{=1}{\frac{1 \text{ km}}{1000 \text{ m}}} = 1.1 \text{ km}$$

Note that each conversion factor equals 1. Also note that if you interrupted this string after, say, the 2.54 operation, you would still have an answer that means something, namely, how many cm are in 0.68 mi.

How to Report Your Answer

How Well Do You Know It?

In the previous section we determined that 358 g = 0.789 lb. If you ask your calculator how many lb are in 358 g, it does not really say 0.789. What it actually says is 0.7885462. We rounded it off to 0.789. Why did we do that, rather than round to 0.8, or 0.78855? The answer boils down to truth in advertising. With rare exceptions, the numbers we are dealing with in science are measurements: someone weighed something to be 358 g, and we want to know how many lb that corresponds to. In weighing the object, they could not determine that it was *exactly* 358 g, only that to the best of their ability to measure it, it was 358 g. But maybe the balance (or the reader's eye) was off, or the last digit flickered—it could have been 359, or 357 g. Measured quantities in science are usually assumed to be uncertain in the last digit reported. The number of digits that are certain, plus the last one that is a bit fuzzy, is called the number of significant figures. This is not a magic formula, and nothing needs to be memorized: it is nothing more than an admission that we are only certain of three digits (and not even quite certain about the last one) in the number 358 g. It has three significant figures.

How does this relate to calculations? Let's be honest about this. If the number 358 g is really anywhere from 357 to 359 g, then the number of pounds it corresponds to is really anywhere from 0.7863 lb to 0.7907 lb. To be completely accurate, we should report this as 0.7885 ± 0.0022. However, this gets to be cumbersome. The custom is to report the answer using the same number of significant figures as what you started with. That is why we reported the answer as 0.789 and not some other number. Clearly, this is not *completely* accurate, since there is still some possibility that the real answer is outside the range 0.788–0.790. But it is clearly a more truthful answer than 0.79 lb (implying that the real object might weigh anywhere between 0.78 and 0.80 lb) or 0.7885 (implying that the real object might weigh anywhere between 0.7884 and 0.7886 lb). The significant figure concept is an attempt to convey how accurately you know an answer you are reporting.

How does this work when there is a string of calculations to be done? Typically, you don't know your answer any more accurately than you know the measured quantity that was measured least accurately, that is, the piece of the calculation with the fewest significant figures. That is why the rule you learned in high school, to use the smallest number of significant figures that appears in your calculation, applies. But remember, this is not because some rule tells you to do it this way. It is because reason convinces you that any other approach would be lying.

There are two things you have to be careful of when applying this "rule." The first is that it applies only to *measured* quantities. There is no ambiguity in the statement that there are 12 inches in a foot. It is not true that there might be 11 or 13: there *are* 12 inches in a foot. Thus 12, as used here, does not contain only two significant figures, and using it does not force your whole calculation to be reported at two significant figures. The same applies to the 2.54 cm in an inch. This is a definition ("standard"), not a measurement. It contains, in a sense, an infinite number of significant figures. On the other hand, the 0.946 L in a qt *is* a measurement (0.9463525 would be more precise), and is good to only three significant figures. Likewise with the 454 g in a lb.

The second caution is that you should not round your answer to the proper number of significant figures until you get to the end of the calculation. Rounding off too early can lead to serious errors. For example, let us imagine converting 0.71 miles into inches and back to miles:

$$0.71 \text{ mi} \times \frac{5280 \text{ ft}}{1 \text{ mi}} \times \frac{12 \text{ in}}{1 \text{ ft}} \times \frac{2.54 \text{ cm}}{1 \text{ in}} \times \frac{1 \text{ in}}{2.54 \text{ cm}} \times \frac{1 \text{ ft}}{12 \text{ in}} \times \frac{1 \text{ mi}}{5280 \text{ ft}} = 0.71 \text{ mi}$$

Since each conversion factor appears once in the numerator and once in the denominator, all of them cancel, and you end up with the same number you started with. Ignoring math, this is of course a necessary (and obvious!) outcome: it is, after all, the same distance—it better not have changed! Nevertheless, if you carelessly round off the answer to two significant figures after each step, the answer comes out to be 0.68 mi. The only reason we know this is ludicrous is because we deliberately came back to where we started, and clearly 0.71 miles does not equal 0.68 miles. The point is that by rounding off a calculation in the middle, you can get wrong answers. *Keep all your digits, or at least several more than you will need, until the end of your calculation before rounding off to the correct number of significant figures!*

Those of you who remember the significant figure rules well will recall that there is a different rule for addition than for multiplication. Why is this? Again, it boils down to truth. If you have something that weighs 538 g (meaning 537–539 g) and add to it something that weighs 54 g (meaning 53–55 g), your total weight will be at least 537 + 53 = 590 and at most 539 + 55 = 594. Thus the true answer is 592 ± 2. The closest we can come using significant figures is 592. We still use three significant figures, even though one of our numbers only had two significant figures, because we still know our answer to within a gram or two. But let us start with the same 538 g object (meaning 537–539) and take 54 of them (meaning 53–55 of them). The math says 538 x 54 = 29,052 g. But the actual total weight could be anywhere from 537 x 53 = 28,461 g to 539 x 55 = 29,645 g. Clearly reporting our answer to three significant figures (29,100, meaning 29,000–29,200) suggests considerably more knowledge of the answer than we actually have. Similarly, reporting one significant figure (30,000 meaning 20,000–40,000) suggests more possibilities than actually exist. The practice is to use the number of significant figures in the number that had the fewest: 54 had two significant figures, so we report our answer to two significant figures: 29,000, meaning 28,000–30,000. Clearly this is still implying less accuracy than we really have, but it is the best we can do without going through this long process each time.

How Do You Tell People How Well You Know It?

The previous example brings up an ambiguity that we need to deal with. The number 29,000 was used to suggest 28,000–30,000, i.e., it had two significant figures. But it might have meant 28,900–29,100, meaning that it had three significant figures. The same number could also be read to have 4 or 5 significant figures. How do you, as someone reading this number, know what I want it to mean, without my writing down what range of values I intend? If I write it as 29,000, you can't tell. That is one reason scientists frequently convert their numbers into scientific notation: 3.74×10^8 Since the first part of every number written in scientific notation is between 1 and 10, you can indicate how accurately you know the number by specifying the right number of zeroes. Thus you would write 2.9×10^4 to indicate 28,000–30,000; 2.90×10^4 to indicate 28,900–29,100; 2.900×10^4 to indicate 28,990–29,010; and 2.9000×10^4 to indicate 28,999–29,001. Of course, the other reason for using scientific notation is that many of the numbers we use are so big or small that it makes no sense to write them any other way. This book assumes you know how to manipulate numbers in scientific notation.

Dealing with Percents

A percentage is not a quantity, it is a relationship between two quantities. You cannot manipulate percentages as if they were quantities; you must manipulate the underlying quantities. Let me try to illustrate this with an example.

On the first day of class, I give my students the following problem, which was actually sent in to the newspaper column of Marilyn vos Savant ("The World's Smartest Person," with an IQ around 200). "You pour one cup of 100% bran cereal in a bowl. Then you pour one cup of 40% bran cereal on top of it. What percent bran do you now have in the bowl? My wife says 140%, my brother-in-law-says 60%. I say it depends on whether you pour the 100% bran cereal or the 40% bran cereal in the bowl first—that is, whether you want to dilute or add. Who's right?"

Most of my students (about 75% of them) get the right answer to this: All the family members are wrong: there is 70% bran. However, many students get this by deciding that there is now 140% bran in 200% cereal, so 140% /200% = 70%. Now, remember I said in an earlier section that you should be able to stop a calculation at any point and still have a quantity that means something? Precisely what does the phrase "200% cereal" mean? Could you go to a restaurant and get 200% cereal? Could you bake a cake that called for 200% sugar? 200% is not an amount! Neither is 100%.

40% bran means that 40% *of whatever quantity you have* is bran. If you have one cup, 40% of it is bran. If you have 2 cups, 40% of that is bran. If you have one cup of 40% bran and add a second cup just like it, you do not now have 80%. It is true that the *quantity* of bran you have in two cups of 40% bran, namely, 0.8 cups of bran, is the same as the *quantity* there would be in one cup of 80% bran, but there is *no sense* in which there is 80% bran anywhere in 2 cups of 40% bran, and it is *wrong* to say that there is. The 0.8 cups of bran you have are contained in 2 cups of cereal, not 1, so you still have 40% bran in your bowl.

To solve the original problem, you have to deal with cups, not percents. In one cup of 100% bran, there is 1 cup of bran. In one cup of 40% bran there is 0.4 cups of bran. If you add these together (in any order) you now have 1.4 cups of bran (but *not* 140%), distributed in 2 cups of cereal (*not* 200%). 1.4 cups/2 cups = 0.7 = 70%. Notice that this approach will work regardless of how many cups of each kind of cereal are added, but adding percents will fail unless the amounts added together are the same.

Let's take another example, derived from this one. I told you that about 75% of my students got the "right" answer to this. So imagine I split the class up into two sections, composed of the students who got the question right and those who got it wrong. In section A, 100% of the students got it right. In section B, 0 % of the students got it right. So what percent of the students in the whole class got it right? Can we average 100% and 0% and decide that 50% of the students got it right? Of course not! We have to count how *many* were in section A and how *many* in section B. You have to count in measurable units (cups, people) and divide by the total *in the same units* to derive a percent.

Problems

Learn to show your work clearly and in an organized manner for full credit. Be sure to label all numbers with appropriate units.

1. You have 20 lb of peanuts and 80 lb of chocolate chips. Chocolate chips weigh 0.02 oz each and peanuts weigh 0.04 oz each. Both peanuts and chocolate chips have a volume of 0.5 cm^3 each. What total volume does your collection occupy?

2. You are making candy from the stuff in problem 1. The recipe calls for

   200 chocolate chips + 50 peanuts -----> 15 candies

Each candy weighs 0.4 oz. How much candy (in lb) can you make from your collection of ingredients? What will be left over?

3. Apples weigh 1 lb each and grapefruits weigh 2 lb each. You have a collection of apples and grapefruits and the average weight of a piece of fruit in your collection is 1.63 lb. What percent of your collection is apples and what percent is grapefruit?

4. You are living in a country where the tax rate is 73%, i.e., 73% of what you earn is taken away by the government. How much must you earn in order to take home $84?

5. An average tree makes 800 board feet of lumber. An average house requires 3000 board feet of lumber. How many trees must be cut down to make 5 houses?

6. You are a caterer. A call has come in for a bunch of those little triangle-shaped sandwiches served at fancy receptions. These are made by making a sandwich, cutting off the crust and cutting each sandwich into four triangles. Thus one could write

2 slices bread + 1 slice ham -----> 4 triangles + extra crust

Your customer has ordered 20 pounds of triangles. Looking in your storeroom, you find 11 pounds of ham and 10 pounds of bread. Bread comes 20 slices in a pound, ham is 8 slices in a pound, and a triangle weighs 0.05 pounds. Can you fill the customer's order without going to the store? If not, how many pounds of triangles CAN you make from what you have on hand? What would you need to buy at the store to complete the order?

7. Devise a conversion from Celsius to Fahrenheit using the boiling point of water as the starting point. Test it using other equivalencies you know already.

CREDITS

Pictures pp 21, 41, 64, 65, 66, 67 from Ebbing and Gammon, *General Chemistry, 6th Edition*, Houghton Mifflin. Used by permission.

Pictures p 20, 63 from Zumdahl, *Chemistry, 5th Edition*, Houghton Mifflin. Used by permission.

Pictures pp 22, 24, 25 from http://csep10.phys.utk.edu/astr162/lect/light/absorption.html, used by permission of Mike Guidry.

Picture p 90 from http://kristall.uni-mki.gwdg.de/homep1.htm , used by permission of Prof. Werner Kuhs.

Pictures p 91 from http://www.umass.edu/microbio/chime/dna, used by permission of Eric Martz.

Pictures p 93 from http://www.umass.edu/microbio/chime/hemoglob , used by permission of Eric Martz.

Pictures pp 342, 345, 346, 353-358, 361, 364, 365, 369, 372-375, 380-384, 386, 388, 391-398, 712 from http://www.aist.go.jp/RIODB/SDBS/, used by permission of K. Hayamizu.

Picture p 373 from Ege, *Organic Chemistry: Structure and Reactivity, 4th Edition*, Houghton Mifflin. Used by permission.

Picture p 715 from http://proxy.arts.uci.edu/~nideffer/Hawking/early_proto/game.html , used by permission of Robert Nideffer.

Pictures pp 720–723 from http://info.bio.cmu.edu/Courses/BiochemMols/ProtG/ProtGMain.htm and http://stingray.bio.cmu.edu/~rule/bc1/lyso/lyso.html, used by permission of William McClure.

Pictures pp 768-769 from http://www.umass.edu/microbio/chime/hemoglob/index.htm, used by permission of Eric Martz.